BIOLOGY

LARKMEAD
SCHOOL

ES198

PUPIL	Tutor Group	Teaching Group	Date of Issue	CONDITION
Tom Rawlings	12 St	/	20/10/97	Brand new

Book No.

97/5

This book is on loan and pupils are expected to take care of it. Loss or damage will be charged for.

Other titles in the Project

Physics   Robert Hutchings
Telecommunications   John Allen
Medical Physics   Martin Hollins
Energy   David Sang and Robert Hutchings
Nuclear Physics   David Sang

Applied Genetics   Geoff Hayward
Applied Ecology   Geoff Hayward
Micro-organisms and Biotechnology   Jane Taylor
Biochemistry and Molecular Biology   Moira Sheehan

Chemistry   Ken Gadd and Steve Gurr

UNIVERSITY OF BATH SCIENCE 16-19

Project Director: J. J. Thompson, CBE

BIOLOGY

MARTIN ROWLAND

Nelson

Thomas Nelson and Sons Ltd
Nelson House Mayfield Road
Walton-on-Thames Surrey
KT12 5PL UK

Thomas Nelson Australia
102 Dodds Street
South Melbourne
Victoria 3205 Australia

Nelson Canada
1120 Birchmount Road
Scarborough Ontario
M1K 5G4 Canada

First published by Thomas Nelson and Sons Ltd 1992

I(T)P   Thomas Nelson is an International
        Thomson Publishing Company

I(T)P   is used under licence

ISBN 0-17-438425-4
NPN 9 8 7 6

Printed in China

Contents

Theme 7: Ecology

The Project: an introduction

The University of Bath Science 16–19 Project, grew out of a reappraisal of how far sixth form science had travelled during a period of unprecedented curriculum reform and an attempt to evaluate future development. Changes were occurring both within the constitution of 16–19 syllabuses themselves and as a result of external pressures from 16+ and below: syllabus redefinition (starting with the common cores), the introduction of AS-level and its academic recognition, the originally optimistic outcome to the Higginson enquiry; new emphasis on skills and processes, and the balance of continuous and final assessment at GCSE level.

This activity offered fertile ground for the School of Education at the University of Bath and a major publisher to join forces with a team of science teachers, drawn from a wide spectrum of educational experience, to create a flexible curriculum model and then develop resources to fit it. This group addressed the task of satisfying these requirements:

- the new syllabus and examination demands of A- and AS-level courses;
- the provision of materials suitable for both the core and options parts of syllabuses;
- the striking of an appropriate balance of opportunities for students to acquire knowledge and understanding, develop skills and concepts, and to appreciate the applications and implications of science;
- the encouragement of a degree of independent learning through highly interactive texts;
- the satisfaction of the needs of a wide ability range of students at this level.

Some of these objectives were easier to achieve than others. Relationships to still evolving syllabuses demand the most rigorous analysis and a sense of vision – and optimism – regarding their eventual destination. Original assumptions about AS-level, for example, as a distinct though complementary sibling to A-level, needed to be revised.

The Project, though, always regarded itself as more than a provider of materials, important as this is, and concerned itself equally with the process of provision – how material can best be written and shaped to meet the requirements of the educational market-place. This aim found expression in two principal forms: the idea of secondment at the University and the extensive trialling of early material in schools and colleges.

Most authors enjoyed a period of secondment from teaching, which not only allowed them to reflect and write more strategically (and, particularly so, in a supportive academic environment) but , equally, to engage with each other in wrestling with the issues in question.

The Project saw in the trialling a crucial test for the acceptance of its ideas and their execution. Over one hundred institutions and one thousand students participated, and responses were invited from teachers and pupils alike. The reactions generally confirmed the soundness of the model and allowed for more scrupulous textual housekeeping, as details of confusion, ambiguity or plain misunderstanding were revised and reordered.

The test of all teaching must be in the quality of the learning, and the proof of these resources will be in the understanding and ease of accessibility which they generate. The Project, ultimately, is both a collection of materials and a message of faith in the science curriculum of the future.

J.J. Thompson
January 1992

How to use this book

To succeed in any biology course, you must be able to demonstrate a number of skills. When tested in an examination, these skills are called assessment objectives and include:

- recall of biological terms, facts and principles;
- understanding of these terms, facts and principles;
- application of biological knowledge and understanding to new problems;
- interpretation of biological information;
- conversion of biological information from one form to another;
- evaluation of biological information;
- selection and organisation of biological information;
- communication of your knowledge and understanding;
- a number of practical skills, such as accurate observation and recording of data.

A textbook can help you to acquire many of these skills.

Biology provides you with a concise summary of the biological terms, facts and principles which are included in the core of Advanced Level ('A' level) and Advanced Supplementary (AS) Biology syllabuses of the GCE Examination Boards in the United Kingdom. You should refer to it when consolidating your notes after each biology lesson, when preparing answers for homework exercises and when revising for examinations. If you are studying biology in an open-learning scheme, you should use this book to prepare for your tutorials.

The book is divided into seven themes containing closely related chapters. You could use it as a course book, working through the chapters of a particular theme in the order in which they are given, dip into selected chapters or use the in-text questions and summaries to help your revision.

Although you will probably be following a biology course in a school, sixth form college, tertiary college or college of further education, you are unlikely to realise your true potential unless you take an active responsibility for your own learning. Simply reading through a book is too passive to develop the important skills you need for your examination. This book has been designed to help you play an active part in your learning: the features designed to achieve this are listed below. You will get the best out of this book by using these features.

Learning Objectives
These are listed at the beginning of each chapter. You should read them to discover which skills you are expected to acquire in working through the chapter. As you cover the work in each chapter, you should refer back to the learning objectives to ensure that you are gaining these skills.

Summaries
Summaries of the main terms, facts and principles are given at the end of each chapter. Having worked through each chapter, you should read these summaries to ensure that you have grasped the main points of the chapter. You should also re-read these summaries regularly to help you retain your knowledge and as an aid to revision before examinations.

Questions
In-text questions regularly test your recall and understanding of what you have just read or prepare you for what you are about to read. These questions are of two types: quick questions (marked with a blue diamond) and longer questions (in blue boxes). Even if you choose not to produce written answers to these questions, you should check your learning by thinking about them before reading any new material. You should also return to these questions at regular intervals throughout your course to help you retain your knowledge of each chapter. Answers are given at the end of each chapter for the quick questions; answers to a selection of the longer questions are given at the end of the book.

Spotlight boxes

The information in these boxes supplements the text by highlighting recent applications of biology in a personal, economic or technological context.

Analysis boxes

The exercises within these boxes are designed to encourage you to practise important examination skills. Answers are provided at the back of the book for the exercises in Analysis boxes: they are given to help you check the development of important skills rather than as a short cut to obtaining correct answers.

Tutorial boxes

To encourage you to think for yourself, some of the material in this book is not given in a descriptive account. Instead, by working through a series of questions you develop an argument or line of enquiry and, by doing so, develop several skills at once.

Examination questions

Relevant questions from recent GCE Advanced Level and Advanced Supplementary examination papers in Biology and Human Biology are provided at the end of each theme. The first in each batch of questions has been answered to guide you in question analysis. You should attempt all of these examination questions and, if possible, have your answers checked by a tutor. In this way can you practise the skills which will be tested in your final examination.

Answers to in-text questions

Answers to in-text questions are given in the form that examination boards use for their mark scheme, i.e. a semi colon (;) separates answers which gain an individual mark and a solidus (/) separates alternative answers for the same mark. Use of this notation will help you to understand mark schemes which GCE 'A' level Examination Boards publish. (Publication of mark schemes began for the June 1991 examinations).

Diagrams, photographs, graphs and tables

Information is presented in a variety of ways both to maintain your interest and to encourage you to interpret and translate this information. Illustrations are usually referred to in the text, but you can test your ability to interpret data and convert data from one form to another by trying to interpret them yourself before you read the appropriate text.

Cross-references

These are used throughout the book to encourage you to make links between different parts of the subject matter. You should use the cross-references to help you to structure your knowledge in new ways.

Remember: you will be most successful if you develop all the skills to be tested in your examination. You will develop these skills more easily if you plan an active part in your own learning process; you need to constantly revise, practise and test what you know and can do.

Copies of syllabuses, past examination questions and mark schemes (only for papers from June 1991 onwards) can be obtained from the Publications Departments of the Examination Board whose syllabus you are following. In the examination question section at the end of each theme of this book, the examination boards are referred to by the initials of their names, i.e. AEB, JMB, NISEC, O&C, Scottish, UCLES, ULSEB (recently renamed ULEAC), UODLES and WJEC.

Acknowledgements

The author and publishers wish to thank the following for permission to use copyright material:

The Associated Examining Board, Joint Matriculation Board, Northern Ireland School Examinations and Assessment Council, Oxford and Cambridge Schools Examination Board, University of Cambridge Local Examinations Syndicate, University of London School Examinations Board and Welsh Joint Education Committee for questions from past examination papers; Cambridge University Press for material from *Basic Biology Course*, Book II, by Tribe and Evant; W.H. Freeman and Company for adapted material from *An Introduction to Genetic Analysis*, Fourth Ed. by David T. Suzuki, Anthony J.F. Griffiths, Jeffrey H. Miller, and Richard C. Lewontin. Copyright © 1989 by W.H. Freeman and Company; The Controller of Her Majesty's Stationery Office for Table 22 from *Manual of Nutrition*, 1973; The Open University for material from S102, S202 and S322 Science Foundation Courses; The Palaeontological Association for Simon Conway Morris's restoration of *Hallucigenia sparsai*.

The author and publishers wish to acknowledge, with thanks, the following photographic sources:
Pete Addis: *p* 121; AFRC, Rothampstead Experimental Station: *p* 678; Allsport: *pp* 259, 317; Ardea: *pp* 546, 555 Fig 24.6; Clive Barda: *p* 354; Biofotos: *pp* 164 (right), 202, 273, 274, 429 (right), 436 468, 476, 483 (upper), 499 (upper), 506, 542, 568 (upper left), 601, 602, 604 (lower), 609 (upper and lower), 611, 615 (left) and right), 627 (upper right), 635 (upper and lower right), 647, 653 (upper amd centre right), 676, 677, 682, 740; Biophoto Associates: *pp* 27, 51 (upper), 52 (left), 54 (upper right, lower left and right), 63 (left and centre), 85, 89, 96 (lower), 126, 187 (upper and lower), 218 (right), 221, 231, 408, 437, 461, 462, 491, 573, 645; Nick Birch: *p* 317; Anthony Blake Photo Library: *p* 560; J Allan Cash: *p* 160; Bruce Coleman: *pp* 159, 211, 218 (left), 240, 333, 363, 387 (lower left and right), 404, 430, 548, 561, 565, 568 (right and upper centre), 570 (upper and lower left), 627 (lower left), 649, 653 (lower left and right); J. Conaghan, Hammersmith Hospital: *p* 457; Connecticut Agricultural Experimental Station: *p* 442; Crown Copyright: *p* 643; Douglas Dickens: *p* 317; Environmental Picture Library: *p* 665 (right); Mary Evans Picture Library: *p* 134; FAO: *pp* 174, 666; Ford: *p* 68; Glasshouse Crops Research Institute: *p* 282; Geoscience Features: *p* 387 (upper); Glasgow Herald & Evening Times: *p* 435; Ronald Grant: *p* 336; Robert Harding Picture Library: *pp* 110, 307, 716; Philip Harris Biological: *pp* 414, 415; Hoechst: *p* 21; Holt Studios: *pp* 148, 157, 506, 600; Horticultural Research International: *pp* 269, 483; IBM: *p* 360; ICI: *p* 397; The Image Bank: *p* 691, Impact Photos: *p* 104; Lawes Agricultural Trust: *p* 115; Long Ashton Research Station: *p* 549; Microscopix: *p* 267 (upper); National Rivers Authority: *p* 720; The Natural History Museum: *p* 553 (upper); NHPA: *pp* 8, 199 (upper), 256 (left), 406, 433, 472, 473, 533, 553 (lower), 568 (lower left), 570 (upper and lower right), 536 (upper), 604 (upper), 625, 627 (upper left), 635 (upper and lower left), 653 (upper and centre left), 667, 700, 727, 736, 737; The Observer: *p* 718; Oxford Scientific Films: *pp* 128, 391, 502; Photo Researchers Inc: *pp* 23, 46, 71, 250; Planet Earth Pictures: *pp* 345, 445, 552; Nigel Press: *p* 734; Resound: *p* 365; Ann Ronan: *p* 49; Saint Bartholomew's Hospital, Department of Medical Illustration: *p* 230; Science Photo Library: *pp* 3, 29, 33, 51 (lower), 56, 63, 76, 87, 96 (upper), 112, 141 (right), 144, 164 (left), 182, 199, 200, 225, 241 (lower), 256 (right), 299, 321, 325, 338, 352, 375, 378, 410, 443, 449, 478, 501, 523, 536 (lower), 574, 57, 584; Tony Stone: *p* 368 (right); Spectrum Colour Library: *pp* 141 (left), 499, 728; Frank Spooner Pictures: *p* 301; Survival Anglia: *p* 665 (left); Tropix: *pp* 304, 529; University of Sheffield, Robert Hill Institute: *p* 147; C. J. Webb: *pp* 14, 37, 53 (right), 54 (upper left), 107, 136, 178, 185, 215, 267, 272, 278, 293, 309, 331, 434, 458, 607; ZEFA: *pp* 9, 209, 234, 241 (upper), 355, 368 (left), 393, 396, 400, 412, 524, 530.

Every effort has been made to trace all the copyright holders but if any have been inadvertently overlooked the publishers will be pleased to make the necessary arrangement at the first opportunity.

Theme 1

CELLS

Biology is the study of life. All living organisms are made of chemicals which they package into cells. In this theme we will look at the nature of these chemicals, the way they are packaged into structures that we can see and the way that some of their more important reactions occur.

In addition we will examine ways in which biologists find out how cells work.

- *Chemical consituents of cells* – the principal types of chemical in biology.
- *Movement of substances into and out of cells* – the uptake of raw materials and elimination of wastes.
- *Cell structure* – the way in which cells package themselves and separate different chemical processes.
- *Cell control and cell division* – the fascinating way in which cells control their activities and ensure that their blueprint is passed to their offspring cells.

(Left) This technician is examining cell ultrastructure using a transmission electron microscope.

(Right) A model showing part of a DNA molecule – the key to cell control.

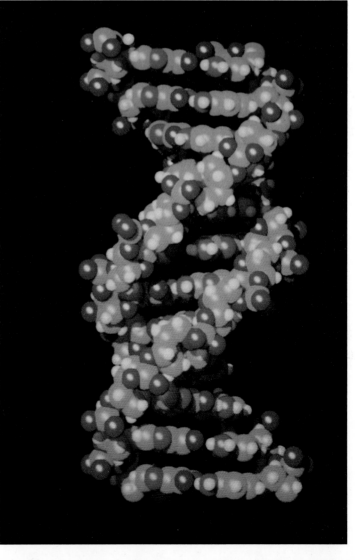

Chapter 1

CHEMICAL CONSTITUENTS OF CELLS

> **LEARNING OBJECTIVES**
>
> When you have studied this chapter you should be able to:
>
> 1. outline the structure of atoms, ions and molecules and use symbols to represent them;
>
> 2. explain ionic, covalent and hydrogen bonds;
>
> 3. account for the biological importance of water;
>
> 4. summarise the structure and biological importance of carbohydrates, lipids, proteins and nucleic acids.

Biology can be studied at many different levels. For example, biochemists are interested in the chemical and physical processes which take place within an organism, whereas ecologists are more concerned with whole groups of organisms (populations) and the way in which they interact with each other and the environment. In a textbook about biology it seems sensible to start with the smallest objects biologists investigate, i.e. cells, and work up to the largest, ecosystems. First though we need to look at some fundamental chemical concepts.

1.1 ATOMS, MOLECULES AND IONS

To survive, grow and reproduce, organisms need a constant supply of energy and materials. Energy will be considered in Theme 2. In this chapter we will look at matter and the building blocks from which it is made, atoms.

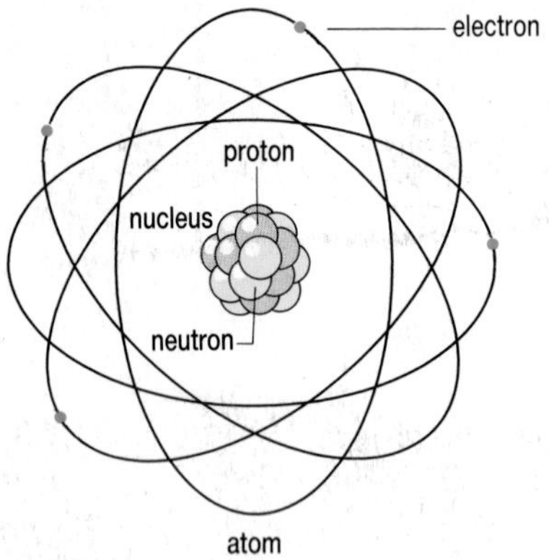

Atoms

All matter is made up of atoms which, in turn, are made up of three subatomic particles – **neutrons** and **protons** (which form the **nucleus**) and **electrons** which orbit around the nucleus (Fig 1.1). Neutrons have no charge, protons a positive charge and electrons a negative charge. Atoms contain equal numbers of protons and electrons, so they have no overall charge.

Ions

Since electrons have a negative charge, atoms or groups of atoms which gain electrons will develop a negative charge (**anions**) whilst those that lose them will develop a positive charge (**cations**).

- Ions are represented using the chemical symbols for their elements and a + sign or – sign to indicate their charge, e.g. Na^+ = **sodium** ion, Cl^- = **chloride** ion.

- Groups of atoms can also be charged, e.g. **nitrate** ion = NO_3^-, **phosphate** ion = PO_4^{3-}.

Fig 1.1 A simple model of an atom. Note **atomic number** = number of protons in the nucleus and **mass number** = number of protons + number of neutrons.

Element – A substance made up of atoms which all contain the same number of protons. For example, the element oxygen is made up of atoms which contain eight protons. Carbon atoms always contain six protons. Chemical elements are represented by **symbols** as a shorthand representation of their name. Table 1.1 summarises the symbols of some of the elements which are important in biology and their occurrence in the human body.

Table 1.1 Some chemical elements of biological importance and their occurrence in the human body

Element	Symbol	Percentage of body mass in humans
carbon	C	9.50
chlorine	Cl	0.03
hydrogen	H	63.00
iron	Fe	0.03
nitrogen	N	1.40
oxygen	O	25.00
phosphorous	P	0.20
potassium	K	0.06
sulphur	S	0.05

Molecule – Consists of two or more atoms chemically joined together. The atoms may be of the same element or of different elements combined in a whole-number ratio. If the atoms are of different elements, the molecule is a **compound**.

Molecular formula – Shows the number of atoms of each element in a molecule (Fig 1.2).

Structural formula – Shows the arrangement of atoms within a molecule (Fig 1.2).

molecule	hydrogen	oxygen	carbon dioxide
molecular formula	H_2	O_2	CO_2
structural formula	H — H	O $=$ O	O $=$ C $=$ O

Fig 1.2 Molecular and structural formulae. The molecular formula shows the type and ratio of the atoms present. The structural formula is a representation to show the arrangement of the atoms in space.

Ion – Formed when an atom or group of atoms gains or loses electrons. Because of this loss or gain, ions are positively or negatively charged.

 1 How many more electrons than protons does a phosphate ion have?

Isotopes

The atoms which form an element always contain the same number of protons. However, they can have different numbers of neutrons. For example, carbon atoms always have six protons in their nucleus and most carbon atoms also have six neutrons in their nucleus, so their **mass number** is twelve (six protons and six neutrons). Some carbon atoms have eight neutrons. Their mass number is not twelve but fourteen; they are, therefore, often referred to as carbon-14 (^{14}C). The two types of carbon atom show the same chemical properties but carbon-14 is radioactive. Atoms of one element which have different numbers of neutrons in their nucleus are called **isotopes**.

THE USES OF RADIOACTIVE IOSTOPES

Carbon-14 (^{14}C) is a radioactive isotope; carbon-12 (^{12}C) is a non-radioactive isotope. All naturally occurring carbon is a mixture of these isotopes. By finding out how much of a sample of carbon is carbon-12 and how much is carbon-14, the age of a carbon-containing structure, such as a fossil, can be estimated (a technique known as **carbon dating**). In 1988, samples of a famous religious relic, called the Turin shroud, were exposed to carbon dating. The result showed it to be much younger than any cloth in which the crucified Jesus would have been wrapped. Carbon-14 is also often used in biology to label substances so that their use or movement in the bodies of organisms can be followed (Fig 1.3).

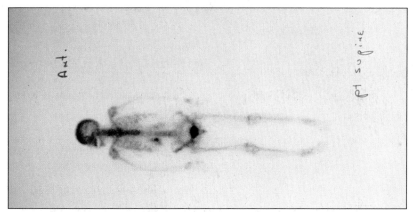

Fig 1.3 The use of a radioactive isotope in medicine. A gamma camera scan of a healthy person. This technique is used to detect cancerous cells. The patient is injected with the radio isotope Iodine 131, which is absorbed by cancerous thyroid cells to a greater degree than normal cells. Areas of uptake would be visible on the scan as darker areas. In this case no such areas are visible.

Acids, bases and buffers

Hydrogen atoms have one proton and one electron. A hydrogen ion (H^+) is, therefore, just one proton. The concentration of hydrogen ions in a solution determines the acidity, or **pH**, of that solution. Fig 1.4 shows how the pH value is derived from the concentration of hydrogen ions and also shows the pH of some common substances.

Living organisms only thrive within a narrow pH range. Very few can tolerate very acid or very alkaline conditions for long.

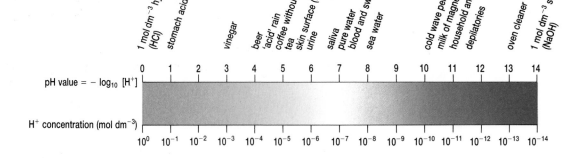

Fig 1.4 The pH scale. (The symbol 10^{-1} means 1 divided by 10^1, i.e. 1 divided by 10 = 0.1; the symbol 10^{-6} means 1 divided by 10^6, i.e. 1 divided by 1 000 000 = 0.000001, and so on. Notice that this means a solution with a pH of 1 is ten times more acidic than a solution with a pH of 2 and 1 million times more acidic than a solution with a pH of 7.)

- An **acid** is a substance which releases H^+ ions (protons) in solution.

For example, a common organic group, the carboxylic acid group, is an acid.

$$- COOH \rightarrow COO^- + H^+$$
carboxylic acid group   carboxylate ion + **proton**
   (acid)   **(base)**

- A **base** is a substance that accepts a proton.

For example, the amino group, $- NH_2$, is a base.

$$- NH_2 + H^+ \rightarrow NH_3^+$$
base   **acid**

Notice that when an acid loses a proton it becomes a base whereas a base which accepts a proton becomes an acid.

An alkali is a base which increases the concentration of **hydroxyl** ions (OH^-) in solution when it dissolves in water. This will raise the pH of the solution, i.e. make it alkaline. Living organisms maintain a narrow pH range within their cytoplasm because they possess substances which

- remove excess hydrogen ions by accepting protons, i.e. act as bases
- neutralise excess hydroxyl ions by donating a proton, i.e. act as acids.

Some substances are able to fulfill both roles; they have the capacity to act as bases and acids. Such substances are called **buffers**. Proteins, e.g. haemoglobin, can act as buffers.

QUESTIONS

1.1 (a) What is the difference between a sodium atom and a sodium ion?
   (b) What is the difference between an oxygen atom and an oxygen molecule?
   (c) If all atoms are made of neutrons, protons and electrons, how can there be so many different elements?

1.2 (a) How does carbon-12 differ from carbon-14?
   (b) What is the biological use of carbon-14?

1.3 Tomato juice has a pH of 4.3. Explain what this means.

1.2 CHEMICAL BONDS

Making and breaking chemical bonds requires energy.

When atoms combine they are held together by forces called chemical bonds. The human body is made of large molecules, some of which contain millions of atoms. If it were not for chemical bonds these molecules would fall apart. Making and breaking chemical bonds between atoms requires energy. There are three types of chemical bond with which biologists need to be familiar: ionic bonds, covalent bonds and hydrogen bonds.

Ionic bonds

Atoms are more stable when their outer orbitals are full. Stability is achieved either by filling an orbital, as in chlorine, or losing it, as in sodium (Fig 1.5).

An orbital is the part of an atom which contains electrons.

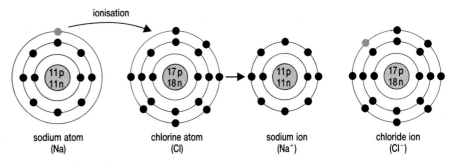

ionisation

| sodium atom | chlorine atom | sodium ion | chloride ion |
| (Na) | (Cl) | (Na⁺) | (Cl⁻) |

Fig 1.5 Ionic bond between sodium and chloride ions. Note how the electron from the sodium atom moves into the outer orbital of the chloride ion. (p=protons: n=neutrons).

Ionic bonds are formed when atoms either gain or lose electrons to form ions. For example, the electrons lost from a sodium atom as it ionises can be taken up by a chlorine atom as it ionises. Since sodium ions are positive and chloride ions are negative, they will attract each other. The **ionic bond** is the electrostatic force which attracts them and holds them near each other in a salt crystal.

 2 What happens to the ionic bonds which hold together a salt crystal when it dissolves in water?

Covalent bonds

The carbon-containing compounds found in living organisms are called organic because at one time they were thought to be made only by living organisms. In fact, they can be manufactured. The term organic compounds is still used, however.

These also involve the electrons in the outer orbitals of atoms. Look at Fig 1.6(a) which shows two hydrogen atoms. Each has a single electron in an orbital which is more stable if it has two electrons. Two hydrogen atoms can make themselves more stable if they share their electrons. The pair of shared electrons forms the single covalent bond which holds two hydrogen atoms together in a hydrogen molecule. This molecule is usually represented in a shorthand way by using H to represent the hydrogen atoms and H – H to represent the covalent bond.

Carbon plays a unique role in biology because of its ability to form strong covalent bonds with other carbon atoms.

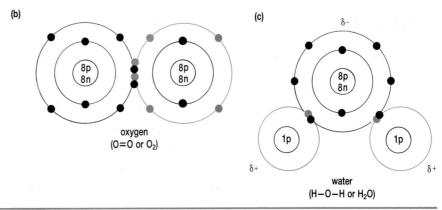

(a)

hydrogen
(H−H or H_2)

(b)

oxygen
(O=O or O_2)

(c)

$\delta-$

$\delta+$ water $\delta+$
(H−O−H or H_2O)

Fig 1.6 **(a)** Covalent bond between two hydrogen atoms.
(b) A double covalent bond between two oxygen atoms.
(c) Covalent bonds between an oxygen atom and two hydrogen atoms.

Molecules such as carbohydrates, proteins and nucleic acids are made of atoms of carbon, hydrogen, oxygen and nitrogen held together by covalent bonds. Some of these atoms may share two or even three pairs of electrons with another atom, forming double and triple covalent bonds. Two oxygen atoms in an oxygen molecule are held together by sharing two pairs of electrons (Fig 1.6(b)). This double bond is often represented by two lines between the atoms: $O = O$.

Hydrogen bonds

A molecule of water is made when two atoms of hydrogen each share their single electron with one atom of oxygen (Fig 1.6(c)). Each hydrogen atom forms a single covalent bond with the oxygen atom. Because the oxygen atom has more positively charged protons than the hydrogen atoms, it pulls on the shared electrons in the bond more strongly than the hydrogen atoms. The oxygen end of the molecule, therefore, has a slight negative charge relative to the hydrogen ends. This molecule is said to be **polar**, i.e. it has a positive pole (δ^+) and a negative pole (δ^-). When two water molecules get close to each other, the oppositely charged parts of the molecules attract each other (Fig 1.7(a)). This type of attraction is called a **hydrogen bond**. Whilst a hydrogen bond has only about 1/20th the strength of a covalent bond, its importance in biology cannot be overemphasised. For example, the role of hydrogen bonds in determining the properties of water will become apparent in the next section.

Hydrogen bonds can form between molecules other than water. Whenever hydrogen atoms form covalent bonds with other atoms that pull more strongly on the shared electrons, hydrogen bonding is possible (Fig 1.7(b)). The three-dimensional shape of nucleic acids and proteins is mainly held together by hydrogen bonds (Fig 1.7(c)). As these hydrogen bonds are weaker than covalent bonds the structure of nucleic acids and proteins is easily destroyed, for example by heating.

δ^+ means a small positive charge, δ^- a small negative one.

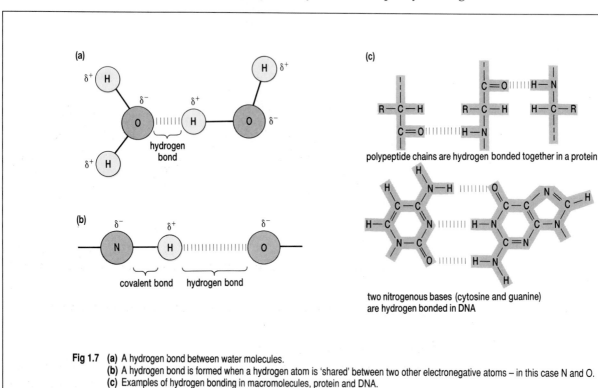

Fig 1.7 (a) A hydrogen bond between water molecules.
(b) A hydrogen bond is formed when a hydrogen atom is 'shared' between two other electronegative atoms – in this case N and O.
(c) Examples of hydrogen bonding in macromolecules, protein and DNA.

CHEMICAL CONSTITUENTS OF CELLS

1.4 In which type of chemical bond are
(a) electrons shared by atoms
(b) electrons lost from one atom to another?

1.5 Draw a diagram to represent the covalent bonds in
(a) NH_3
(b) CH_4
(c) NH_2CH_2COOH.

1.6 What is a hydrogen bond?

1.3 WATER

Most of the human body is water (Table 1.2). Substances with molecules of a similar mass to water, such as ammonia (NH_3) and methane (CH_4), are all gases at room temperature but water is a liquid. It is the polarity of water molecules and the fact that they can form hydrogen bonds which is the key to water's special properties.

Water has 'sticky molecules'

Water molecules form hydrogen bonds, one to another, which help to hold the molecules together. Look at Fig 1.8 which shows five water molecules joined together by hydrogen bonds forming a lattice. About 15% of the molecules in a glass of water at room temperature would be joined together at any one moment in this hydrogen-bonded lattice arrangement. Whilst each bond is fairly weak on its own, taken together all the hydrogen bonds in the lattice form a fairly strong structure. To break this structure would mean breaking all the hydrogen bonds and that would require a large amount of energy. This sticky nature of water molecules explains many of its properties and is called **cohesion**.

Table 1.2   Approximate chemical composition of a 70 kg human male

Substance	Body mass/%	Mass / kg
water	70.0	49.00
lipid	15.0	10.50
protein	12.0	8.40
carbohydrate	0.5	0.35
inorganic ions	2.5	1.76

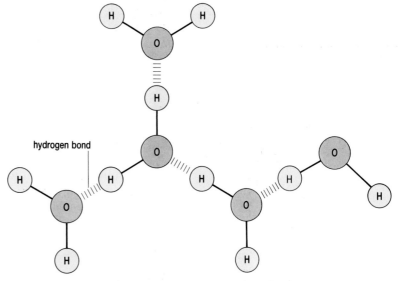

Fig 1.8 Water at room temperature forms a lattice held together by hydrogen bonds. The cohesive nature of water is responsible for many of its unusual properties, such as high surface tension, specific heat and heat of vaporisation.

 What do you think happens to the lattice when water boils?

Surface tension

At the surface of ponds and lakes, the cohesion between water molecules produces a **surface tension**. Things that are denser than water generally sink. The surface tension of water can make a solid-like surface and support relatively dense objects such as insects (Fig 1.9).

Fig 1.9 This pond skater (*Gervis lacustris*), is supported on the surface of the water by surface tension.

The cohesion of water helps its molecules to stick together when they are pulled. This is demonstrated when drinking a water-based drink through a straw by sucking: the pressure inside the mouth is lowered and the atmospheric pressure pushes the drink up the straw. If the water molecules did not stick together, the column of water would break under this pressure difference and the drink could not be sucked up. This is important for the larger land plants which pull water up through their tube-like water transport cells when water evaporates from their leaves (see Section 12.2).

Water as a solvent

Any substance which is polar attracts water molecules. For example, ions like Na⁺ and Cl⁻ in Fig 1.10(a), or organic chemical groups similar to the ones shown in Fig 1.10(b), will attract water. Polar substances, which can become part of water's hydrogen-bonded structure, are **hydrophilic** (water loving) and dissolve in water. Non-polar substances will not dissolve in water and are called **hydrophobic** (Fig 1.10(c)).

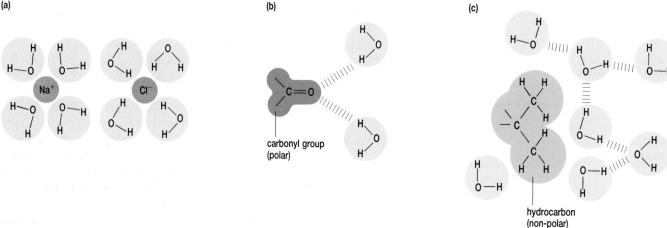

(a)

(b)

carbonyl group (polar)

(c)

hydrocarbon (non-polar)

Fig 1.10 Hydrophilic and hydrophobic molecules. Substances that take part in water's hydrogen-bonded structure such as **(a)** ions and **(b)** polar organic groups like the carbonyl group are **hydrophilic** and soluble in water. **(c) Non-polar molecules** do not become part of water's hydrogen-bonded structure. They are therefore **hydrophobic** and insoluble in water.

4 Use the terms hydrophilic, hydrophobic, polar and non-polar to describe the following: (a) salt (b) sugar (c) oil.

Water can dissolve a wide range of molecules, including salts, gases, sugars, amino acids, small nucleic acids and proteins. Whilst other solvents will dissolve some of these, none dissolves all of them. Of the important biological molecules, only lipids and the large polymers such as polysaccharides, small proteins and DNA do not dissolve in water. The chemical reactions which occur inside cells only occur quickly if the reacting molecules are in solution. Being in solution in water also helps them to be transported through the cell so that they are brought close together.

Water as a temperature buffer

Most cells tolerate only a narrow range of temperatures: this is partly the result of the properties of their enzymes (see Chapter 5). Three properties of water help to stabilise the temperature of cells.

1. **Relatively high heat capacity**. This means that water must absorb a relatively large amount of heat before its temperature will rise. The **specific heat capacity** of water is 4184 J kg⁻¹ K⁻¹ (compare this with 2520 J kg⁻¹ K⁻¹ for alcohol and 924 J kg⁻¹ K⁻¹ for sodium chloride).

Key units
kg⁻¹ means per kilogramme,
K⁻¹ means per degree Kelvin (SI unit of temperature),
J = joules (SI unit of energy).

CHEMICAL CONSTITUENTS OF CELLS

Fig 1.11 Water evaporating from the tongues of these dogs helps to cool them because water has such a high latent heat of vaporisation.

2. **Relatively large latent heat of vaporisation.** A large amount of heat energy (2.26 MJ kg^{-1} K^{-1}) must be absorbed by water to turn it into water vapour. Organisms can cool down when only a little of their water evaporates (Fig 1.11).

3. **Relatively large latent heat of fusion.** Water at $0\,^{\circ}C$ must lose a relatively large amount of energy before it forms ice crystals.

 Explain these properties of water in terms of its hydrogen-bonded structure.

Even when water does freeze it has an unusual property which is important to life. Ice expands, so that it becomes less dense than liquid water and thus floats. As ponds and lakes freeze, the ice floats on the surface and forms an insulating layer which helps the water underneath to avoid freezing. Most substances contract and become more dense as they solidify and so sink. If water did this, ponds and lakes would freeze from bottom to top, killing most of the life in them. In fact, water is densest at $4\,^{\circ}C$. There will be a layer of water at $4\,^{\circ}C$ at the bottom of lakes even when the lake surface is frozen over.

 Why would most cells burst if they froze?

Water as a metabolite

The chemical reactions which occur inside cells are called **metabolism** and the chemical reactants involved in these reactions are called **metabolites**. Water is either used or released in many metabolic reactions. It is used in photosynthesis and digestion; it is released when polysaccharides, lipids and proteins are made and during aerobic respiration.

QUESTIONS	
	1.7 The carbonyl group (see Fig 1.10(b)) is found in sugars. How do the properties of this group explain the solubility of sugar in water?
	1.8 Under what conditions might sweating cool humans? Explain your answer.

1.4 CARBOHYDRATES (SACCHARIDES)

Carbohydrates are used as a source of energy in all organisms and as structural materials in cell membranes, some cell walls and some skeletons, e.g. insects and plants. Carbohydrates contain only atoms of carbon, hydrogen and oxygen. There is often the same number of carbon and oxygen atoms but twice as many hydrogen atoms in simple carbohydrate molecules. The chemical shorthand of $(CH_2O)_n$ shows this ratio of atoms: one molecule of water (H_2O) combined with one carbon atom to form a subunit which can be repeated many times (n) to form the finished molecule. The (CH_2O) unit is repeated a number of times in some sugar molecules (Fig 1.12).

Monosaccharides (single sugars)

Monosaccharides are the building blocks of which larger carbohydrate molecules are made. They are also used by all living organisms as an energy source.

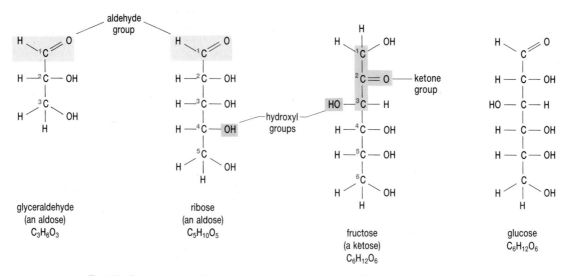

Fig 1.12 Four monosaccharides, each having a carbonyl group (C=O). The carbon atoms are numbered.

TUTORIAL

Monosaccharides (single sugars)

This tutorial will help you to understand the way monosaccharides are named and some important aspects of their chemistry.

Look at Fig 1.12; it shows some examples of monosaccharides. Do not worry at this stage what all the labels mean. You are going to use these diagrams to understand how monosaccharides are named and classified.

Monosaccharides can be classified on the basis of the number of carbon atoms they contain (n). **Trioses** have three carbon atoms ($n = 3$); **pentoses** have five ($n = 5$); **hexoses** have six ($n = 6$).
(a) Classify the four sugars on this basis.

Monosaccharides can also be classified on the basis of the chemical groups they contain. All the molecules in Fig 1.12 contain a carbonyl group (C = O). In glyceraldehyde and ribose, the carbonyl group is part of an **aldehyde** group so these are **aldose** sugars. Fructose has a **ketone** group so it is a **ketose** sugar.
(b) Is glucose an aldose or a ketose? Explain why.

Numbering the carbon atoms which form the backbone of each molecule is useful. These numbers are shown in blue in Fig 1.12. Notice that the carbon atom at the end of the molecule nearest to the aldehyde or ketone group is called carbon atom number one (^{1}C).
(c) Redraw the glucose molecule and number the carbon atoms appropriately.

Monosaccharides usually exist as **rings** when dissolved in water. You need to study Fig 1.13 carefully: it shows the convention used to depict these rings.
(d) Draw the ring forms of ribose and fructose. Number the carbon atoms on your drawing.

Molecules with the same molecular formula but different structural formulas are called **structural isomers**. There are two ways the structure of the ring shown in Fig 1.13 could be changed.

CHEMICAL CONSTITUENTS OF CELLS

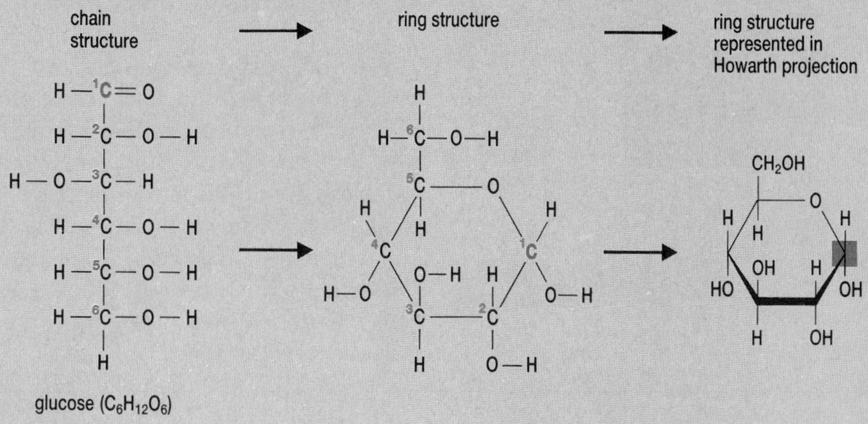

Fig 1.13 The relationship between straight chain and ring structure of glucose.

Look at Fig 1.14(a).
(e) What is the single difference between α-glucose and β-glucose?

The molecule shown in Fig 1.14(b) is another hexose sugar, **galactose**. It looks very similar to glucose but look at carbon atom number four (^4C): the OH group is the opposite way around. This small difference means that glucose and galactose behave quite differently in a cell – they are different molecules. Although the structures are so similar, they cannot be readily interconverted. The Analysis 'Using sugars for respiration' shows this.
(f) Does Fig 1.14(b) show the α or β form of galactose?

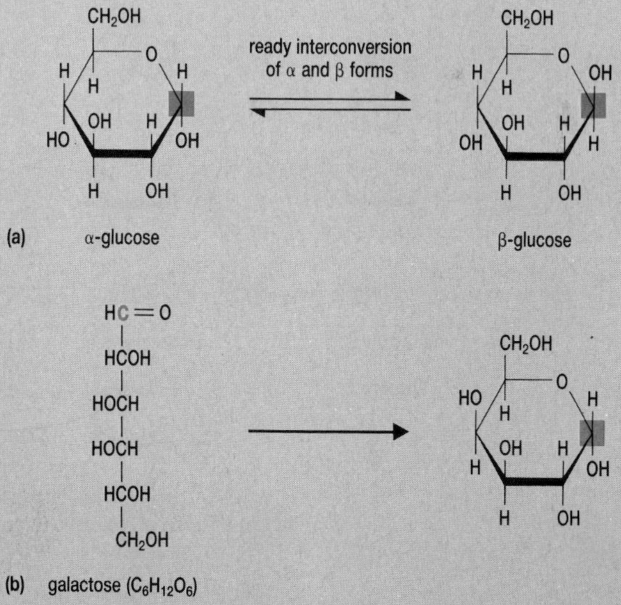

Fig 1.14 Structural isomerism.
 (a) The α and β forms of glucose. These are readily interconvertible. However, molecules made from α-glucose, e.g. amylose, have different properties to those built from β-glucose, e.g. cellulose.
 (b) Galactose.

test tube

CO_2 produced
by respiring
yeast cells

fermentation
tube

sugar solution
containing yeast

Fig 1.15

Using sugars for respiration

This exercise will help you to develop skills associated with experimental design. If you can, try and do the experiment.

The energy contained in a glucose molecule is released during the process of respiration. One of the products of respiration is carbon dioxide. If you measure the carbon dioxide released over a period of time you can measure how fast respiration is going. You can do this using the apparatus shown in Fig 1.15 and yeast. Design an experiment which investigates the efficiency with which yeast uses three different sugars, glucose, fructose and galactose, as respiratory substrates. Remember you only want to measure differences caused by the yeast cells using the different sugars as a respiratory substrate: everything else must be kept constant.

Disaccharides (double sugars)

Monosaccharides are used so quickly by living cells for their energy content that they may not always be suitable for transport and storage. They also affect a cell's osmotic balance (see Section 2.3). They are used less quickly and have a smaller effect on cells if linked together to form chains. The first step is to join just two monosaccharides together to form a disaccharide. This process is shown in Fig 1.16.

Note that

- water is eliminated during the reaction so this is a **condensation reaction**.
- the bond between the rings (shown in pink) is called a **glycosidic** bond.
- the glycosidic bond has formed between carbon atom number one (^1C) of one α-glucose ring and carbon atom number 4. (^4C) of the other; it is written as α 1, 4 glycosidic bond.

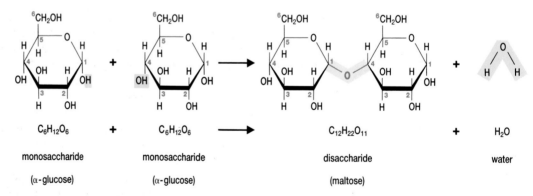

Fig 1.16   Two glucose units condense to form the disaccharide maltose.

7 Write a chemical equation for the reaction shown in Fig 1.16.

Three common disaccharides are **lactose** (milk sugar), **maltose** (malt sugar) and **sucrose** (cane sugar). They are all formed when one molecule of glucose combines with one molecule of another monosaccharide.

glucose	+	galactose	=	lactose	+	water
glucose	+	glucose	=	maltose	+	water
glucose	+	fructose	=	sucrose	+	water

Fig 1.17 A colorimeter is an instrument used to measure the intensity of the colour of a solution and is a useful biochemical tool. The test solution is put into the specimen chamber at the top.

REDUCING SUGARS

The blue colour of the Benedict's reagent, used in the test for reducing sugars, is caused by the presence of the copper ion Cu^{2+}. When a solution containing this ion is boiled with any molecule which contains a free carbonyl group, for example a monosaccharide or a disaccharide like maltose, the Cu^{2+} ion gains an electron from the carbonyl group, a process called **reduction**:

$$Cu^{2+} + e^- \rightarrow Cu^+$$

Benedict's reagent also contains sodium hydroxide which makes it very alkaline. Under these conditions the Cu^+ ions precipitate out as copper(I) oxide. This is the brick red precipitate, visible evidence of a positive result for reducing sugars. The colour changes with increasing concentrations of reducing sugar in the assay solution. At low concentrations it is green, going to yellow, orange and bright brick red as the concentration increases. This can be used to measure the concentration of sugar in fluids. Such a test is an essential part of quality control in the food industry.

Polysaccharides

These are formed when many hundreds of monosaccharide units condense to form chains (Fig 1.18). These chains of monosaccharides may be

- of a variable length, although usually they are very long
- branched or unbranched
- folded. (The molecule is then compact and ideal for storage, e.g. starch, glycogen.
- straight or coiled. (The molecule can be used to make meshes and is ideal for construction, e.g. cellulose Fig 1.18(c)).

Most of the important polysaccharides are made from hexose units, so are called **hexosans**. Many are made from glucose itself, so are called **glucosans**. Glucose units can form long chains in three ways (see Fig 1.18). **Amylose** and **cellulose** are completely straight. **Starch** is a mixture of both amylose and amylopectin. **Glycogen** is amylopectin with very short lengths between its branches.

 What is the difference in chemical bonding between (a) amylose and amylopectin (b) cellulose and amylose?

Unlike monosaccharides and disaccharides, polysaccharide molecules do not dissolve in water. They are, therefore, suitable as a storage substance, e.g. starch in plants and glycogen in animals and fungi, and as a building material for cell structures, e.g. cellulose in plant cell walls. Since they are used for storage, it is important that starch and glycogen can easily be broken down into their glucose subunits when required. Organisms possess chemicals called enzymes which enable them to speed up this breakdown (see Chapter 5).

The β **1, 4 glycosidic bonds** in cellulose give the molecule a different shape to other polysaccharides. This difference between amylose and cellulose is sufficient to prevent enzymes which can break down amylose, present in most animals, from attacking cellulose. Some bacteria can produce enzymes, called cellulases, which digest cellulose. In addition, cross-linking between parallel cellulose chains makes cellulose very strong, hence its structural role in plant cell walls. Both these factors make cellulose difficult to digest.

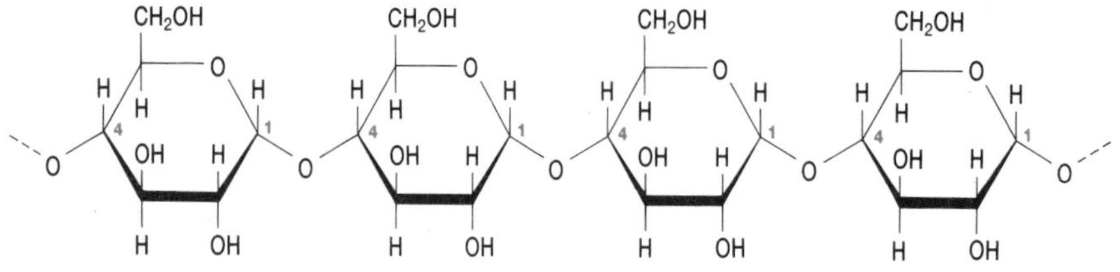

(a) amylose (α 1,4 linkages)

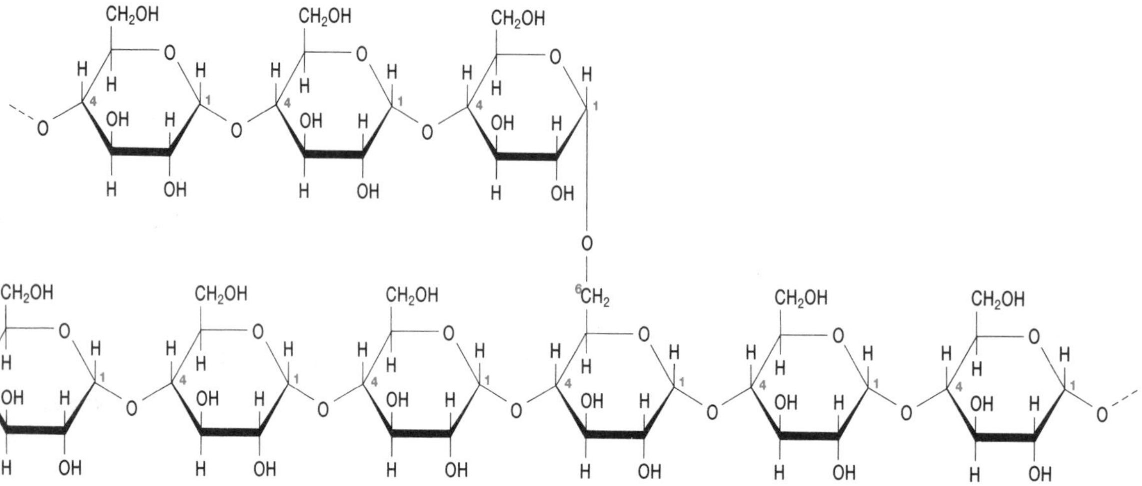

(b) amylopectin (α 1,4 and α 1,6 linkages)

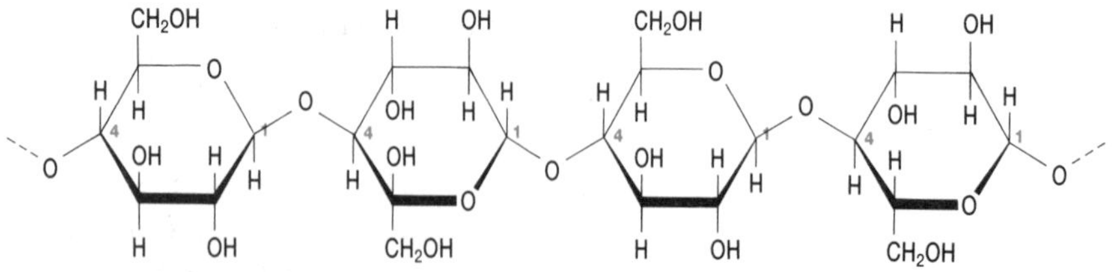

(c) cellulose (β 1,4 linkages)

Fig 1.18 Different types of polysaccharide. Notice the different sorts of glycosidic bonds and that cellulose is made of β-glucose sub-units not α-glucose subunits like the others.

THE COMMONEST ORGANIC MOLECULE ON EARTH

The structural strength of cellulose and its resistance to digestion make it a uniquely useful material (Fig 1.19). The words in this book are printed on cellulose. Cotton clothes are made from cellulose. The tough, synthetic fibre rayon, used for example in tyre manufacture, is made from cellulose. Cellophane, the clear film used in packaging, and celluloid, used to make photographic film, are also cellulose-based products.

Fig 1.19 All these products are made from, or contain materials made from, cellulose.

CHEMICAL CONSTITUENTS OF CELLS

Polysaccharides are also used to make other important molecules. When amino acids are included in a polysaccharide molecule, a mucopolysaccharide is formed. The exoskeletons of arthropods and the cell walls of fungi are made of such a mucopolysaccharide, chitin (Fig 1.20).

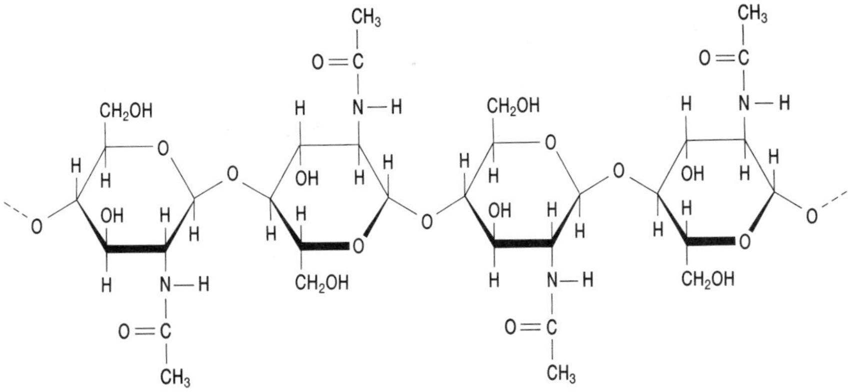

Fig 1.20   Chitin, a modified polysaccharide with side chains containing nitrogen, is used to make the tough and waterproof exoskeletons of insects.

9 How does chitin differ chemically from polysaccharides?

QUESTIONS

1.9 Give a biological role of
(a) glucose (b) starch (c) cellulose (d) chitin.

1.10 Represent the hydrolysis (breakdown) of sucrose into its monosaccharides.
(a) in words
(b) in molecular formulae
(c) in structural formulae.

1.11 Explain why cows can digest starch but depend on bacteria to digest cellulose in the plant food which they eat.

1.5 LIPIDS

The term lipid is used to describe a variable assortment of compounds. Lipids are all insoluble in water but can be extracted from cells using organic solvents, such as benzene, trichloromethane and ether.

 If water is described as a polar solvent, how would you describe a solvent like benzene?

Fats and oils

Fats and oils are made by joining two different types of molecule together: fatty acids and glycerol or other alcohols. Fatty acids have a carboxylic acid group at one end and a long hydrocarbon tail. A hydrocarbon is a substance which is made of only hydrogen and carbon.

 Predict what would happen if you mixed a fatty acid with water.

Triglycerides.

The hydroxyl (OH) groups from each of three fatty acids (Fig 1.21) react with those from the glycerol molecule to form a **triglyceride** (Fig 1.22), the commonest lipid in nature. Since there are hundreds of different kinds of fatty acid there are hundreds of different kinds of triglyceride. A major function of triglycerides is as an energy store, yielding much more energy than an equivalent mass of carbohydrate. Triglycerides are also stored under the skin of mammals, where they act as a heat insulator.

Fig 1.21 Formation of a triglyceride from the condensation of glycerol and three fatty acids. The bonds formed are called **ester bonds**.

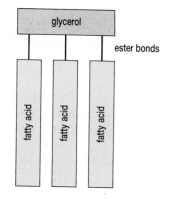

Fig 1.22 A simple diagram of a triglyceride.

12 Whales have a thick layer of blubber beneath their skin which keeps them warm. What do you think this is made of? What else could the whales use the blubber for?

Saturated and unsaturated fatty acids

If you look closely at the fatty acids in Fig 1.21 you will notice that one of them, oleic acid, has a double covalent bond between two of its carbon atoms; all the other covalent bonds between carbon atoms are single. Fatty acids with these double bonds are **unsaturated**; those without double bonds are **saturated**. Double bonds lower the melting point of fatty acids. For example, oleic acid, the principal component of olive oil, has a melting point of 13.4 °C whereas stearic acid and palmitic acid have melting points of 69.6 °C and 63.1 °C, respectively. So, triglycerides which contain unsaturated fatty acids are liquid at room temperature. They are called **oils** and are mainly produced by plants. Animal **fats** are solid at room temperature because their triglycerides contain saturated fatty acids.

13 The lid of a tub of sunflower margarine carries the message 'High in polyunsaturates'. What does this mean?

Waxes

These are similar to fats and oils except that the fatty acids are linked to long-chained alcohols instead of glycerol. They provide a waterproofing layer on the outer surface of most terrestrial animals and plants. The shiny appearance of many indoor houseplants and of most insects is a result of

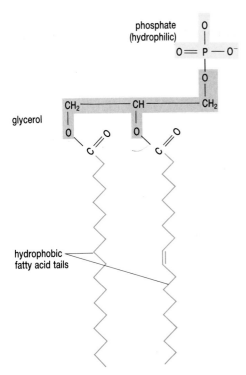

Fig 1.23 A phospholipid. These molecules are one of the major constituents of cell membranes. Notice they have a polar end and a non-polar end (shown in blue). The glycerol molecule is shown in grey.

this wax. As well as using wax to waterproof their bodies, honeybees use a stiffer wax to make the honeycombs in which they store food and house their young.

 14 Plants which live in water, hydrophytes, do not have wax-covered leaves. Why is this no disadvantage?

Phospholipids

Phosphoglycerides (Fig 1.23), the commonest group of phospholipids, are triglycerides in which one of the fatty acids is replaced by a phosphate group (PO_4^{3-}). The phosphate group is electrically charged (or **polar**) and so, unlike fatty acids, dissolves in water. A phospholipid molecule, therefore, has two contradictory ends: one dissolves in water (**hydrophilic**); the other will not mix with water (**hydrophobic**). Such a molecule is described as **amphipathic**. This property of phospholipids is important in determining the structure of cell membranes (Chapter 2).

Steroids

Except for being insoluble in water but soluble in organic solvents, steroids have very little in common with the lipids so far mentioned. They are lipids whose molecules contain four rings of carbon and hydrogen atoms with various side chains protruding from them, often only one of which, that from the carbon atom at position 17 in Fig 1.24, determines their physiological role. Many animal hormones, including ecdysone, which controls moulting (ecdysis) in arthropods, and the human sex hormones, are steroids. Human steroids are synthesised from cholesterol (shown in Fig 1.24) which is also an important constituent of cell surface membranes.

15 Should you have a diet which contains no cholesterol?

Fig 1.24 The structure of cholesterol and the steroid hormone testosterone which is made from it. Other steroid hormones include progesterone and β-oestradiol, one of the oestrogens. Their physiological importance is determined by the side chain from carbon number 17.

1.12 Name four biological functions of lipids.

1.13 List three similarities and three differences between carbohydrates and lipids.

1.14 Most lipids do not mix with water. How can a phospholipid mix with water?

1.15 Some fish such as mackerel and herring are described as oily. From what properties of oils are these fish benefiting?

1.6 PROTEINS

Proteins are important as

- enzymes (e.g. amylase)
- transport proteins (e.g. haemoglobin)
- contractile proteins (e.g. myosin in muscle)
- immunoproteins (e.g. immunoglobulins in the defence system)
- membrane proteins (in the structure and function of cell surface membranes)
- structural proteins (e.g. keratin in hair, fingernails and tendons)
- hormones (e.g. insulin).

Each different protein molecule is made under the direction of its own gene and performs its own precise function. Of particular importance is the shape of the protein molecule which is determined by its amino acid sequence.

Amino acids

The general formula of an amino acid is shown in Fig 1.25. **Amino acids** are the building blocks from which proteins are made. There are 20 common different amino acids found in naturally occurring proteins, each with a different side chain (**R**) attached to the α carbon atom. Different side chains are shown in Table 1.3.

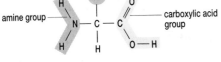

Fig 1.25 Generalised structure of an amino acid $NH_2CHRCOOH$. Examples of R groups are listed in Table 1.3.

Table 1.3 **R groups of some amino acids, arranged according to their chemical nature**

Amino acid name	R group
uncharged polar side chains	
glycine	–H
serine	–CH₂–OH
cysteine	–CH₂–SH
acidic side chains	
aspartic acid	–CH₂–COOH
basic side chains	
lysine	–CH₂-CH₂–CH₂–CH₂–NH₂
arginine	–CH₂–CH₂–CH₂–NH–C(=NH)(NH₂)
non-polar side chains	
alanine	–CH₃
phenylalanine	⬡–CH₂

Note the structure shown as ⬡ is a benzene ring and is composed of carbon and hydrogen only.

 Draw the structural formula of (a) glycine (b) alanine.

Amino acids ionise when dissolved in water (Fig 1.26). In a particular solution the positive and negative charges on this amino acid exactly balance, so the amino acid ion has no overall charge. An amino acid in this state, i.e. with balanced positive and negative groups, is called a **zwitterion**. A molecule which can both donate a proton (an acid) and accept a proton (a base) is described as being **amphoteric**.

$$NH_2CHRCOO^- \xleftarrow[\text{in alkaline solution}]{\text{loses H}^+} NH_3^+CHRCOO^- \xrightarrow[\text{in acid solution}]{\text{gains H}^+} NH_3^+CHRCOOH$$

| overall negative charge (anion); lost H⁺ makes solution more acidic | no overall charge **(zwitterion)** | overall positive charge (cation); gained H⁺ makes solution less acidic |

Fig 1.26 The ionization of amino acids in solutions of different pH. Note if R = H this molecule is glycine as referred to in the text.

 17 What is the overall charge on glycine in neutral conditions? Is the molecule polar or non-polar?

The common amino acids can be grouped together according to whether their side chains are **acidic, basic, uncharged polar** or **non-polar** (see Table 1.3). Acidic and basic side chains ionise when they dissolve in water.

 18 Which amino acid side chains would you expect to be (a) hydrophobic (b) hydrophilic?

The overall charge on an amino acid depends on the pH of its solution. At some characteristic pH an amino acid has no overall electric charge, i.e. it exists as a zwitterion. This pH is called the **isoelectric point** for the amino acid.

 19 What is the isoelectric point of glycine?

Amino acids are buffers. Look at Fig 1.26. If the pH falls below the isoelectric point (right of the diagram), i.e. the solution becomes more acidic, hydrogen ions are taken up by the carboxylate group. This reduces the concentration of free hydrogen ions in the solution, i.e. it becomes less acidic.

 20 What happens if the pH rises above the isoelectric point of the amino acid?

Protein structure

Proteins are long chains of **amino acids** held together by chemical linkages called **peptide bonds** (Fig 1.27). Such a chain of amino acids is called a polypeptide chain. Proteins may consist of one or more polypeptides.

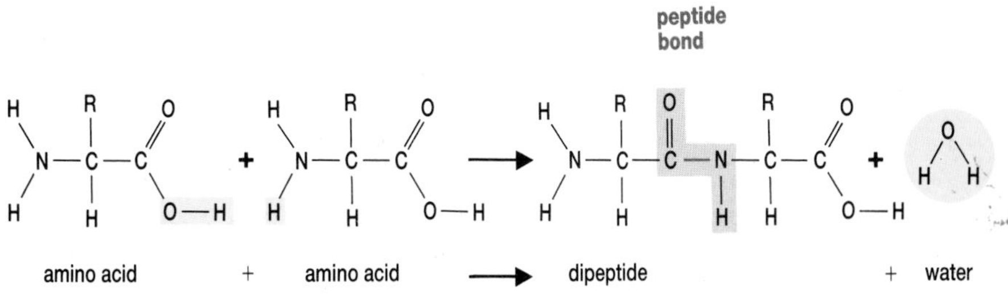

peptide bond

| amino acid | + | amino acid | → | dipeptide | + | water |

Fig 1.27 Formation of a dipeptide. Two amino acids are joined by a peptide bond.

 21 What does the formation of glycosidic, ester and peptide bonds have in common?

The sequence of amino acids, the **primary protein structure** (Fig 1.28(a), determines the eventual shape and biological function of a protein molecule. Even though there are only 20 different amino acids commonly found in proteins, these can be arranged in an enormous number of different ways. The number of different proteins is enormous.

The **primary structure** of proteins is specific for each protein and is determined by the DNA of the cell in which it is made. Once formed, the chain of amino acids folds spontaneously to make complex configurations Fig 1.28 (b) and (c). Such helices and pleated sheets are described as the **secondary structure** of the protein.

The secondary structure of structural proteins forms fibres. For example, those of the fibrous protein **keratin**, found in hair, horns, nails and feathers, twist to form a two-stranded helix, held in shape by hydrogen bonds. The protein **collagen**, found in mammalian connective tissues including tendons and bones, forms a three-stranded helix held together by hydrogen bonds. Yet others, like the proteins in silk, produce pleated sheets.

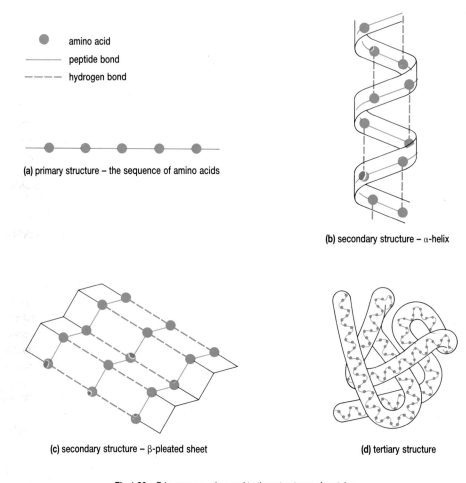

● amino acid

——— peptide bond

- - - - hydrogen bond

(a) primary structure – the sequence of amino acids

(b) secondary structure – α-helix

(c) secondary structure – β-pleated sheet

(d) tertiary structure

Fig 1.28   Primary, secondary and tertiary structures of proteins.

 What type of bonding holds the secondary structure together?

Polypeptide chains may also fold into a globular shape. This is described as a **tertiary protein structure** (Fig 1.28(d)). Such globular proteins, which include enzymes and immunoglobulins, may also contain regions of α-helix and β-pleated sheet. The exact three-dimensional shape of these molecules is vital in allowing those proteins which are enzymes, or which act as receptor sites on the cell surface, to recognise the specific molecules to which they attach.

CHEMICAL CONSTITUENTS OF CELLS

23 What would happen to this three-dimensional shape if the sequence of amino acids was changed? How would this affect the biological function of these proteins?

Tertiary protein structure is stabilised by hydrogen bonds, ionic bonds and covalent bonds between sulphur atoms in the side chains of the amino acid cystine (**disulphide bridges**). In addition, hydrophobic side chains on the non-polar amino acids will tend to be pushed towards the inside of the molecule whereas hydrophilic side chains will tend to congregate on the outside.

 24 Why will this arrangement of side chains allow globular proteins to mix with water?

In unborn babies the haemoglobin is made of two α-polypeptide chains and two γ-polypeptide chains. The switch to production of adult haemoglobin takes place at birth.

Some proteins, which consist of more than one polypeptide chain, exhibit **quaternary structure**. For example, human haemoglobin is a protein consisting of two α-polypeptide chains and two β-polypeptide chains arranged around an iron-containing **haem** group.

Separating proteins

The charge on a protein molecule depends on its amino acid composition and on the pH of its solution. Proteins can be separated according to their charge and shape by placing them on moist filter paper or gel in an electric field, the technique of **electrophoresis**. Depending on the conditions applied, each protein will move a characteristic distance. Look at Fig 1.29 which shows how a number of proteins have been separated using this technique on a polyacrylamide gel.

Fig 1.29 A polyacrylamide gel stained to show proteins separated on the basis of their electrical charge – the technique of electrophoresis.

1.16 The genetic code is a code controlling the structure of proteins. Explain why a genetic code is not needed for carbohydrates and lipids.

1.17 (a) Draw a diagram to represent the general structure of an amino acid.
(b) How do two of these molecules join together?
(c) What is the name given to the chemical bond which joins two amino acids together?

1.18 Distinguish between the primary, secondary, tertiary and quaternary structure of proteins.

1.19 What happens to an amino acid
(a) in acidic solution
(b) in alkaline solution
(c) in a solution at the isoelectric point of the amino acid?

1.7 NUCLEOTIDES AND NUCLEIC ACIDS

A **nucleotide** consists of a nitrogen-containing base, a 5-carbon sugar (a pentose) and one or more phosphate groups. Nucleotides have many functions within cells. For example, they

- carry chemical energy around the cell, e.g. adenosine triphosphate (ATP)
- combine with other groups to form coenzymes, e.g. coenzyme A (CoA)
- are used as signalling molecules inside cells, e.g. cyclic AMP (cAMP).

Probably the best-known role of nucleotides is as the building blocks of DNA and RNA (Fig 1.30), both of which are nucleic acids.

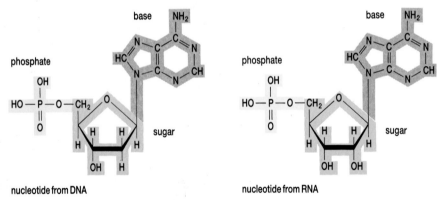

Fig 1.30 The deoxyribose and ribose nucleotides of DNA and RNA.

DNA AND THE DOUBLE HELIX

During the early 1950s, a number of scientists were studying crystals of DNA by bombarding them with X-rays and examining the photographic image made by the rays after being diffracted by the crystal. One of these scientists, Rosalind Franklin, found a cross-shaped pattern like the one in Fig 1.31. This pattern indicates that DNA has a helical molecule. Unfortunately for Franklin, two other scientists, Francis Crick and James Watson, used her X-ray diffraction plates and, using modelling techniques together with inspired guesswork, won the race to discover the structure of DNA. Nine years later, Watson, Crick and Maurice Williams won the Nobel Prize for this discovery. Rosalind Franklin's premature death from cancer prevented her sharing the prize because it cannot be awarded posthumously.

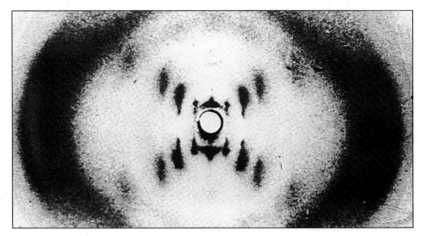

Fig 1.31   This pattern, produced on photographic film by X-rays diffracted through a DNA helix, shows that DNA has a helical structure.

Deoxyribonucleic acid (DNA)

This molecule, which carries genetic information, is a polymer made up of the four subunits shown in Fig 1.32(a).

25   In what ways are these four deoxyribonucleotides similar and different?

(a)

phosphate
deoxyribose sugar
nitrogenous base — A

G

C

T

(b)

G

A

T

C

C

(c)

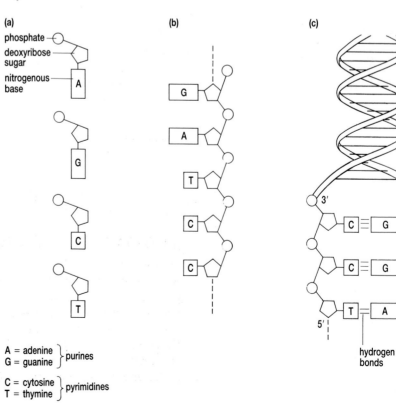

3'   5'

C ≡ G

C ≡ G

T = A

5'   3'

hydrogen bonds

Fig 1.32   **(a)** The four deoxyribonucleotides found in DNA.
(b) Polymerization of deoxyribonucleotides to form a single-stranded DNA molecule. Notice that the two ends of the sugar–phosphate backbone are different.
(c) A small part of a double-stranded DNA molecule. Notice the complementary base pairing stabilised by hydrogen bonds (see Fig 1.7 (c)). Can you see that the two strands are turning in opposite directions? They are said to be anti-parallel.

A = adenine
G = guanine } purines

C = cytosine
T = thymine } pyrimidines

COMPLEMENTARY BASE PAIRING

This is the key to
- how DNA copies itself
- how the genetic information is read
- all the processes involved in recombinant DNA technology (genetic engineering).

The polymer is held together by links between the sugar molecules and the phosphate groups as shown in Fig 1.32(b). A single DNA molecule is made of two strands of nucleotides, each like the one represented in Fig 1.32(b), millions of nucleotides long. These two strands are held together by hydrogen bonds between the nitrogen-containing bases (Fig 1.32(c)); a process called **complementary base pairing**.

 Which base always pairs with which in DNA?

The process of complementary base pairing means that the order of bases on one strand of the double helix determines the order on the opposite strand. The presence of guanine on the right-hand strand in Fig 1.32(c) determines that cytosine must be opposite it on the left-hand strand, and so on. Finally, the double strand is wound into a helix: the famous double helix. The activities of DNA are detailed in Chapter 4.

Ribonucleic acid (RNA)

This differs from DNA in the following ways.

- It contains the sugar ribose rather than deoxyribose.
- The pyrimidine base thymine is replaced by the pyrimidine base uracil.
- It is a much shorter molecule and is single stranded.
- There are three different sorts of RNA in a cell, each of which has a different function (see Chapter 4).

QUESTIONS

1.20 What is a nucleotide?

1.21 Name three ways in which the structure of DNA differs from that of RNA.

1.22 DNA is often represented as a ladder. What part of the DNA molecule is represented by
(a) the rungs of the ladder
(b) the uprights of the ladder?

1.23 If adenine, cytosine, guanine and thymine are represented by their initial letters, which **one** of the following arithmetic statements is true about a DNA molecule?

(a) $\dfrac{A + T}{C + G} = 1$ (b) $\dfrac{A}{T} = \dfrac{C}{G}$ (c) $AT = CG$

SUMMARY

Living organisms are made of large molecules composed mainly of the elements carbon, hydrogen, oxygen, nitrogen, sulphur and phosphorus. Atoms of these elements are joined by a variety of bonds, including ionic, covalent and hydrogen bonds, to form four major classes of biological molecule: carbohydrates, lipids, proteins and nucleic acids.

Water is an important compound for life. It is able to dissolve most biological materials, is involved in many biochemical reactions and helps maintain fairly constant temperatures.

Carbohydrates are used as an energy source, as a structural component of cell walls and in the construction of other molecules, including mucopolysaccharides and nucleic acids. Sugars, starch and cellulose are examples of carbohydrates.

Lipids are a diverse group of compounds which are insoluble in water.

They include fats, oils, waxes and steroids. The uses of lipids include long-term energy stores, waterproofing, insulation, construction of membranes and formation of hormones.

Proteins form a large part of the dry mass of cells and most enzymes are proteins. The structure of proteins is determined by the genetic code in the chromosome(s) of each cell.

Nucleic acids form the genetic code for proteins. Nucleic acids are made of chains of smaller units called nucleotides; the nucleotides also form many coenzymes, cellular messengers and energy-carrier molecules.

Answers to Quick Questions: Chapter 1

1. 3
2. They must break.
3. It is destroyed using heat energy.
4. Salt – hydrophilic, polar
   Sugar – hydrophilic, polar
   Oil – hydrophobic, non-polar.
5. Large amounts of energy are needed to break up or change the hydrogen-bonded lattice holding the water molecules together.
6. Increase of cell volume as water inside freezes and causes internal pressure to increase on cell surface membrane until it breaks; pointed ice crystals may also rupture cell surface membrane.
7. $C_6H_{12}O_6 + C_6H_{12}O_6 \rightarrow C_{12}H_{22}O_{11} + H_2O$
8. Amylose has a straight chain structure whereas amylopectin is branched with α 1, 4 and α 1, 6 glycosidic bonds. Cellulose has β 1,4 glycosidic bonds whilst amylose has α 1,4 glycosidic bonds.
9. Chitin contains nitrogen.
10. Non-polar solvent.
11. Fatty acid and water would form separate layers.
12. Triglycerides. Energy store.
13. Contains lipids with unsaturated fatty acid chains.
14. Do not need to conserve water.
15. No. Cholesterol is essential to make cell surface membranes and some hormones.
16. Replace R in Fig 1.25 with the structural formula for the side groups of glycine and alanine shown in Table 1.3.
17. There is no net charge but the molecule is polar.
18. Alanine and phenylalanine will be hydrophobic because they have non-polar side chains; all the others are polar and therefore hydrophilic.
19. pH 7.0
20. Hydrogen ions are released from the amino group, so increasing the concentration of hydrogen ions in the solution, i.e. the solution becomes less alkaline.
21. All formed by condensation reactions (loss of water).
22. Hydrogen bonding
23. It might change, impairing the biological function of the molecule by stopping it attaching to other molecules.
24. The outer hydrophilic side chains will be able to interact with the hydrogen-bonded water lattice.
25. Each type of nucleotide is similar in having three components – phosphate, deoxyribose and a nitrogenous base. Each differs in respect of the identity of its nitrogenous base – adenine, guanine, cytosine or thymine.
26. Adenine with thymine, guanine with cytosine.

Chapter 2

MOVEMENT OF SUBSTANCES INTO AND OUT OF CELLS

LEARNING OBJECTIVES

When you have studied this chapter you should be able to:

1. describe a modern model of the structure of biological membranes;

2. distinguish between diffusion, facilitated diffusion, active transport and osmosis and relate these to the structure of the cell surface membrane;

3. explain the term water potential and perform calculations, using given data, to find the water potential of a cell;

4. appreciate that proteins can act as receptor sites on a cell's surface.

Eukaryotic cells are full of membranes which play a central role in the cell's metabolism. To understand the role of cell membranes, their properties are investigated first. The easiest membrane to study is the one on the outside of the cell – the **cell surface membrane**. This chapter concentrates on the cell surface membrane, but the basic structure and properties are equally applicable to all the membranes within a eukaryotic cell.

2.1 CELL MEMBRANES

The cell of a freshwater alga called *Nitella* is represented in Fig 2.1. This cell has such a large vacuole that it is relatively easy to extract the fluid (**cell sap**) from within it. The diagram shows the relative concentration of sodium ions in the cell sap, the cytoplasm and the freshwater around the *Nitella*: the sodium ion concentration is different in the three locations. This could be explained if movement of sodium ions is prevented because any of the following are impermeable to sodium ions:

- the cell wall
- the cell surface membrane
- the membrane which separates the vacuole from the cytoplasm (the **tonoplast**).

In fact the plant's cell wall allows the passage of all molecules and ions through it, i.e. it is **freely permeable**. The data in Fig 2.1 could still be explained if either of the membranes restricts the passage of sodium ions. Sodium ions *can* pass across membranes, however. This is shown in Table 2.1, which also shows that the *rate* at which sodium ions cross cell membranes is different from that of some other ions. Another explanation for the data in Fig 2.1 is, therefore, needed. To find this explanation we need to know something about the structure of the cell's membranes.

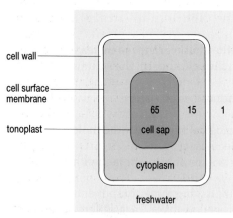

numbers represent the relative
concentrations of sodium ions in solution

Fig 2.1 The concentration of sodium ions in the cell sap and cytoplasm of *Nitella* and in the freshwater in which this alga lives.

Table 2.1 The rate at which a number of particles cross cell membranes

Particle name	Particle diameter /nm	Relative rate of movement across cell membranes
water molecule	0.300	5.0×10^7
urea molecule	0.360	4.0×10^7
hydrated chloride ion	0.386	3.6×10^7
hydrated potassium ion	0.396	100.0
hydrated sodium ion	0.512	1.0
glucose molecule	0.860	0.4

 Large non-polar molecules such as steroids (see Section 1.5) diffuse rapidly through cell surface membranes. What does this suggest about the chemical composition of the membrane?

The structure of cell membranes

The electronmicrograph (Fig 2.2) shows the appearance of a highly magnified, thin section of cell surface membrane. It appears as three lines. The two outer layers are darker than the inner layer because they have reacted differently to treatment carried out in preparation for examination under the electron microscope. This means that the inner layer must be chemically different from the outer layer, but tells us little else. Any further knowledge about the structure of cell membranes must come from other sources.

 Calculate the relative thicknesses of the parts of the membrane shown in the electron micrograph.

Chemical analysis of cell membranes

Human red blood cells (**erythrocytes**) have cytoplasm which is full of the red, oxygen-carrying pigment **haemoglobin** (see Chapter 10) but little else. If erythrocytes are broken up and centrifuged, most of the cell debris which collects at the bottom of the centrifuge tube consists of cell membrane. For this reason, erythrocytes have often been used in studies of cell membrane composition. Table 2.2 shows the chemical composition of membranes from erythrocytes. The most common compounds are lipids and proteins.

 Would it be fair to conclude that all cell membranes would have the same composition as erythrocytes? Explain why.

Do the features of erythrocytes which make them suitable for chemical analysis also make them unique? The only way to test this idea is to analyse the composition of membranes from other cells. This has been done and, in fact, their composition does show some differences. Differences in the chemical composition of membrane from different parts of the same cell have also been found. However, all membranes are made mainly of lipid and protein, and most animal and plant membranes have a composition in which lipids make up 30–50% of the dry mass, carbohydrates make up about 5% of the dry mass and the rest is chiefly protein.

The lipids found in cell membranes belong to a class known as **trigly-cerides**, so called because they have one molecule of glycerol chemically linked to three molecules of fatty acid (see Section 1.5). The majority belong to one subgroup of triglycerides known as **phospholipids**.

How do phospholipids differ from triglycerides? Firstly, the phosphate group has the effect of making the phospholipid molecule **amphipathic**. This means that it has a part which will mix with water (**hydrophilic**) and a part which will not mix with water (**hydrophobic**). The fatty acids form the hydrophobic part whereas the glycerol and phosphate group form the

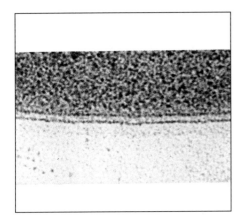

Fig 2.2 An electronmicrograph showing the tramline appearance of the cell surface membrane at high magnification.

Table 2.2 The chemical composition of membranes from human erythrocytes

Compound	Composition of dry mass /%
carbohydrate	8–10
lipid	35–45
protein	50–60

KEY CONCEPT

Cell membranes are primarily made up of lipids and proteins. However, cell membranes differ in their relative proportions of lipids and proteins.

hydrophilic part of the molecule. The hydrophilic and hydrophobic groups are shown in Fig 2.3.

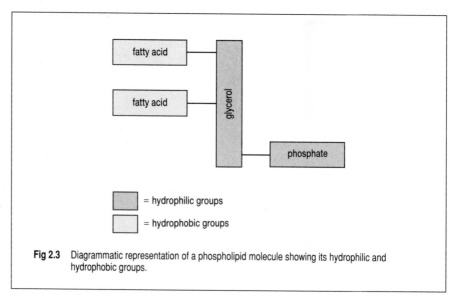

Fig 2.3   Diagrammatic representation of a phospholipid molecule showing its hydrophilic and hydrophobic groups.

4   Draw a diagram to show what would happen to phospholipid molecules if you poured a single layer of them onto the surface of a bowl of water.

ANALYSIS

The experiments of Gorter and Grendel

In this exercise you will use the skills of analysing data, drawing conclusions and proposing hypotheses.

The Dutch scientists, Gorter and Grendel, extracted the phospholipids from the surface membranes of red blood cells and spread them out on the surface of a trough of water so they formed a layer that was one molecule thick (Fig 2.4).

(a) Why do the molecules adopt the orientation shown in Fig 2.4?

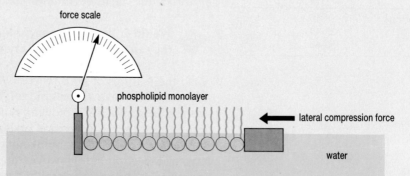

Fig 2.4   Gorter and Grendel's experiment. Lateral compression of the phospholipid film packs the molecules tightly together: their surface area can then be estimated.

They next measured the surface area occupied by the molecules on the water using the apparatus shown in Fig 2.4.

(b) If the phospholipids were arranged in a single layer in the red blood cell membrane, what relationship would there be between the surface area of a red blood cell and the surface area of the lipids measured by Gorter and Grendel?

In fact, Gorter and Grendel found that the area of the lipid monolayer in their trough was about twice the surface area of a red blood cell.

(c) Suggest a hypothesis to explain the arrangement of phospholipids in cell membranes consistent with these observations.

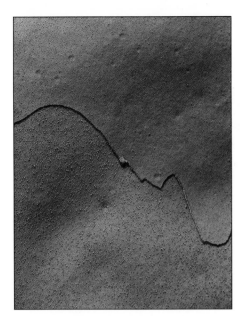

Fig 2.5 Electronmicrograph of freeze-fractured cell membrane. The little bumps on the photograph represent the proteins in the cell surface membrane.

Analysing membranes

In one method of preparing cells for examination under an electron microscope, they are plunged into liquid nitrogen at –196 °C. If these frozen cells are pushed against a sharp razor blade in a very precise way, they tend to fracture along lines of weakness (**freeze fracture**). Very often this line of weakness is through the middle of a membrane.

 How does freeze fracture support the idea of a bilayer?

When a membrane has been split into its two layers, some of its protein can be removed by fairly gentle chemical methods; other protein cannot be extracted without totally destroying the membrane. This observation has led to the interpretation that some proteins are embedded in only one of the membrane layers (**extrinsic proteins**) whilst others lie in both membrane layers (**intrinsic proteins**). Freeze fracture photographs show these proteins (Fig 2.5).

Carbohydrates form a relatively minor part of the membrane. They are attached to some of the lipids, forming **glycolipids**, or to some of the proteins, forming **glycoproteins**.

The fluid mosaic model of membrane structure

Since cell biologists cannot see the chemical structure of membranes, they need a model which best fits all the evidence from studies such as those already described. Such a model is the **fluid mosaic model**.

The inner and outer layers of the fluid mosaic model (Fig 2.6) are made of phospholipid molecules arranged so that their hydrophobic tails are in the middle of the membrane. The fluidity of the membrane is affected by the nature of these lipids; phospholipids with short fatty acids or with unsaturated fatty acids are more fluid than others. Using chemical labelling techniques it has been shown that lipids may move about in the membrane. Usually, they move horizontally staying in their own layer but, very occasionally, they move from one layer to the other in a process called **flip-flop**.

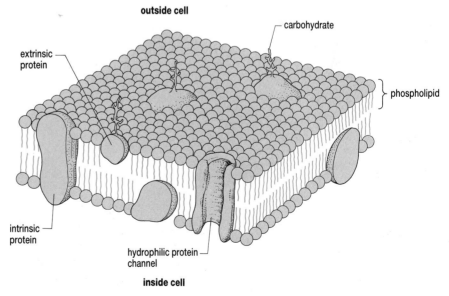

Fig 2.6 The fluid mosaic model of membrane structure.

Cholesterol is important in regulating membrane fluidity. The more cholesterol a membrane contains the more fluid it is.

Extrinsic proteins are arranged in the outside or inside layers while intrinsic proteins run through both layers of the membrane. Like the phospholipids, these proteins may also have hydrophobic and hydrophilic parts. Some of these intrinsic proteins form hydrophilic channels through which some ions and molecules can move (see Fig 2.6).

Notice the carbohydrates are all on the outside of the cell surface membrane where they form a **glycocalyx**. The glycocalyx is thought to be important in cell recognition and interactions between cells. It is essential to the immune response, for example.

QUESTIONS

2.1 What evidence is there that membranes
  (a) are made of phospholipids
  (b) are arranged in a bilayer
  (c) have globular proteins inserted into the lipid phase of the membrane?

2.2 (a) What do the terms hydrophobic and hydrophilic mean?
  (b) What is a hydrophilic channel in the cell membrane?

2.3 Why is the model of cell surface membrane structure referred to as
  (a) fluid
  (b) mosaic?

2.4 The human immunodeficiency virus (HIV) causes AIDS in humans. It attacks certain cells in the bloodstream called T-cells by attaching to a specific site on the surface membrane of these cells. What is the probable chemical nature of these receptor sites on the surface membrane? Explain your answer.

2.2 PASSIVE TRANSPORT

Fig 2.7 A model to show how molecules cause a pressure when they strike an object.

To understand transport mechanisms across cell surface membranes, you first need an understanding about molecular motion.

Motion and pressure

Solid materials, like this book, keep their shape. The particles from which they are made do move, but their movement is restricted to vibration around a fixed position. Liquids and gases do not keep their shape. The particles in liquids and gases, such as atoms, ions and molecules, are not held in a fixed position and constantly move about. They are spaced further apart and move more rapidly in gases than they do in liquids. At 25 °C hydrogen molecules move about at a speed of 1930 m s^{-1} (equivalent to 4300 miles per hour!).

When the moving molecules in a liquid bump against a solid object, they push against it. This was first seen by a Scottish biologist who used an optical microscope to watch pollen grains on a water surface. They were buffeted around as water molecules bumped into them (**Brownian motion**). Small particles of smoke can be seen to be buffeted around by collisions with air molecules too. This bumping of molecules causes a pressure. A spring weighing scale with its pan turned upside down is shown in Fig 2.7. The marbles which are being poured over it represent molecules. As they hit the pan they force the needle to move. This force is the pressure exerted by the marbles on the pan and will increase the greater the number, mass and speed of the marbles. Exactly the same happens when atoms, ions and molecules strike objects except, being much smaller than marbles, the forces they exert are much smaller.

Although atoms, ions and molecules move randomly (**random thermal movement**), there is a net movement from regions where they are abundant and exert a high pressure to regions where they are less abundant and create a low pressure. If this net movement occurs across a cell membrane, the cell does not need to use energy to make it happen. Such movement is, therefore, called **passive transport**. If cells do use energy to make particles move across their membranes, the process is called **active transport**.

Diffusion

You may have watched what happens in a room where some people smoke tobacco. The smoke from their cigarettes does not stay close to them; it moves throughout the room until everywhere is equally smokey. This illustrates **diffusion**. The particles of smoke, like ions and molecules, show random thermal movement. The overall result of their movement is that they move from where people are smoking (where the smoke is concentrated) into the rest of the room (where, initially, there was no smoke) until the smoke is evenly spread out. (The air breathed in by non-smokers inevitably contains tobacco smoke!) Other particles, including ions and molecules, do exactly the same thing as smoke particles, i.e. they move down a **concentration gradient**. The greater the difference in concentration, the steeper the concentration gradient and the faster the rate of diffusion.

ANALYSIS

The factors affecting diffusion

This exercise requires skills of data handling and interpretation of results.

A group of students set out to investigate the factors which affect the rate of diffusion of two molecules, ammonia and hydrogen chloride. Both of these are gases at room temperature. Their apparatus is shown in Fig 2.8, and the data they obtained from their experiments are presented in Table 2.3. They repeated each experiment several times.

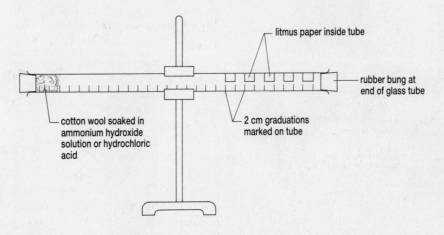

Fig 2.8   Apparatus used for investigating the rate of diffusion of ammonia and hydrogen chloride. The cotton wool is soaked in the appropriate solution and placed in the end of the tube. The movement of each gas can be monitored by changes in the colour of the litmus paper.

Table 2.3   Experimental result. Mol dm^{-3} R is a measure of concentration. Thus 2 mol dm^{-3} is twice as concentrated as 1 mol dm^{-3}.

Distance travelled cm	Time taken to 'reach' distance/s							
	Hydrogen Chloride				Ammonia			
	20 °C		30 °C		20 °C		30 °C	
	1 mol dm^{-3}	2 mol dm^{-3}	1 mol dm^{-3}	2 mol dm^{-3}	1 mol dm^{-3}	2 mol dm^{-3}	1 mol dm^{-3}	2 mol dm^{-3}
10	133	65	87	42	62	30	40	19
12	248	127	168	78	115	58	78	38
14	333	159	221	117	155	75	103	55
16	573	291	370	184	267	136	174	85
18	790	384	527	276	367	180	244	128

(a) Which factors affecting the rate of diffusion were investigated in this experiment?

(b) Plot the data and fit **lines of best fit** through the points. You should now have eight lines.

(c) What conclusion can you reach from your graph about the effect of the factors you have identified on the rate of diffusion?

(d) Why are your curves not straight lines?

DIFFUSION

- Diffusion is the process by which substances spread from regions of high concentration to regions of low concentration.
- It involves the net movement of ions and molecules from a region of high concentration of the ions or molecules to a region of their low concentration until the concentrations are equal.
- Even when diffusion stops, ions and molecules still move randomly about but, on average, they move equally in all directions.

Diffusion through membranes

Non-polar molecules which can dissolve in lipid diffuse very quickly through cell membranes. They do so by dissolving in the lipid part of the membrane and moving from a high concentration on one side of the membrane to a low concentration on the other side of the membrane. Polar molecules, such as amino acids and glucose, as well as ions, pass more slowly through membranes. Their charge stops them dissolving in the lipid but they can pass through some of the hydrophilic channels formed by the intrinsic proteins.

ARTIFICIAL BILAYERS

Much of the present understanding about how molecules and ions pass across membranes is based on experiments with artificial lipid bilayers. These are formed when a small amount of lipid dissolved in hexane is placed in the 1 mm diameter opening between the two chambers shown in Fig 2.9(a). The chambers can then be filled with a test solution, say, water on one side and salt solution on the other (Fig 2.9(b)).

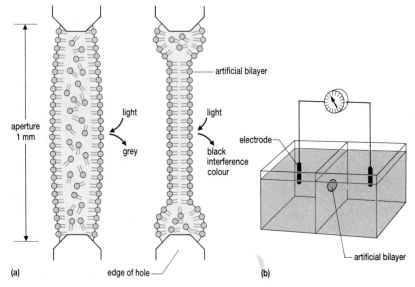

Fig 2.9 (a) An artificial bilayer of lipid molecules forming in a 1 mm opening between two chambers. When the artificial membrane reaches the correct thickness the interference pattern on the surface changes from grey to black.
(b) The test chamber is then filled with test medium and the movement of substances, e.g. ions, can be measured electrically.

Selected substances can be added to test their effects on membrane permeability, for example antibiotic ionophores – molecules that assist the diffusion of ions across membranes by forming hydrophilic channels through them. Some antibiotics are antibiotic ionophores and they kill bacterial cells by allowing ions essential to the bacterium's metabolism to diffuse rapidly out of the bacterial cell into the surrounding medium. The bacterium literally leaks to death. This method allows comparison of the effectiveness of different antibiotics.

ANALYSIS

Fig 2.10 Regular haemodialysis keeps patients suffering from kidney failure alive until either their kidneys recover or they receive a transplant.

Haemodialysis

This exercise requires you to apply your knowledge to a new situation.

People who have suffered kidney failure need help to remove excess water and solutes, such as urea, from their blood. Such help is given by a dialysis machine (Fig 2.10).

The main features of one type of dialysis machine are shown in Fig 2.11. A dialysis membrane is partially permeable, i.e. it contains small holes which let small molecules through but not bigger molecules. The dialysis fluid is kept at about 40 °C and consists of water with carefully measured concentrations of salts and glucose. Urea and excess solutes from the blood are removed in the effluent fluid.

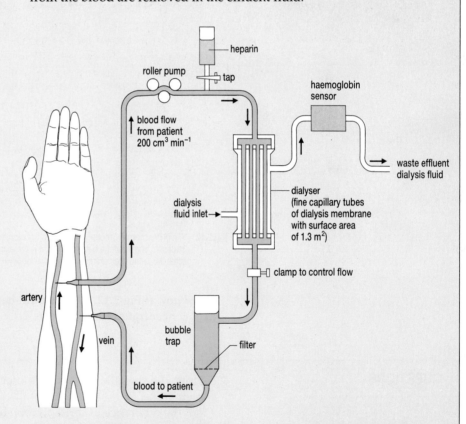

Fig 2.11 A schematic diagram of one type of artificial kidney machine used for haemodialysis.

Use the diagram and the knowledge you have gained in this section to answer the following questions.
(a) Calculate the volume of blood which is processed by the dialyser in three hours.
(b) During dialysis, what process removes the solutes and urea from the patient's blood?
(c) Why is the dialysis fluid continually replaced with fresh solution?
(d) Why is the dialysis fluid kept at 40 °C?

Facilitated diffusion

Many of the intrinsic proteins in a cell's membrane have **receptor sites** which recognise and attach to molecules which do not dissolve in lipid. Such protein molecules change shape (a conformational change) when attached to a lipid-insoluble molecule resulting in that molecule being deposited on the other side of the membrane (Fig 2.12). This process is completely reversible so that lipid-insoluble molecules can pass in either direction depending on their concentration on each side of the membrane. The cell does not have to provide energy for this process to proceed.

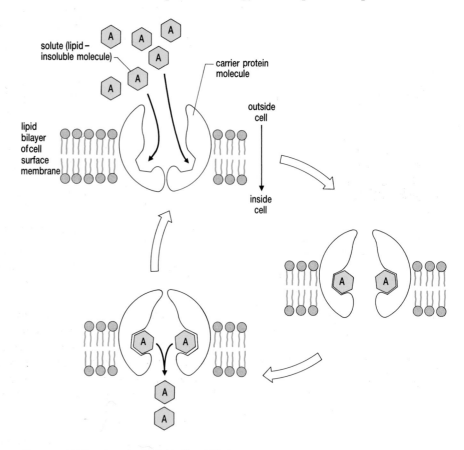

Fig 2.12 A highly schematic model of facilitated diffusion for a solute A. The carrier protein can exist in two states: in one state the binding sites for A are exposed on the outside of the membrane; in the other state the same sites are exposed on the inside of the membrane.

 6 Look at Fig 2.12. **What would happen if molecule A were more concentrated inside the cell?**

QUESTIONS

2.5 How will the following factors affect the rate of diffusion across a membrane
   (a) the difference in concentration on either side of the membrane
   (b) the size of the molecules
   (c) the temperature
   (d) the polarity of the molecules?

2.6 How does the diffusion of a charged ion through a membrane differ from that of a lipid?

2.7 How do proteins in the membrane help diffusion of substances through the membrane?

2.3 OSMOSIS: BASIC PRINCIPLES

Look at Fig 2.13. On the left-hand side of the membrane is a solution; a mixture of **solute** molecules (shown in grey) and **solvent** molecules (shown in blue). The solute molecules are dissolved in the solvent. In all biological systems the solvent is water whose molecules are smaller than the solute molecules. On the right-hand side of the membrane in Fig 2.13 is pure water.

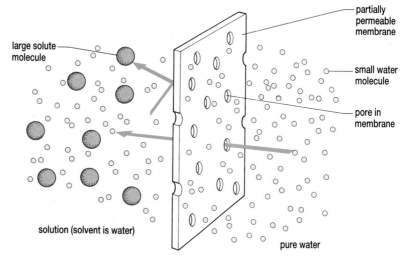

Fig 2.13 Diagrammatic representation of osmosis through a partially permeable membrane.

> **7** Imagine you stirred some sugar into a glass of hot water. Which molecule is the solute and which the solvent?

Look again at the membrane shown in Fig 2.13. The membrane has pores which let the smaller water molecules through but not the larger solute molecules. Such a membrane is described as **partially permeable**.

Since there are more water molecules on the pure water side of the membrane than there are on the solution side of the membrane (the solute has diluted the water), there is a water concentration gradient across this membrane.

> **8** Predict which way the water molecules will diffuse. Explain why an equilibrium will never be reached.

This diffusion of water molecules down their concentration gradient across a partially permeable membrane is called **osmosis**.

Osmosis and cell volume

Look at Fig 2.14 which shows what happens to erythrocytes when placed in solutions of different concentration. The cytoplasm of these cells is a colloidal suspension of haemoglobin, a molecule which is too large to pass through the cell surface membrane. When erythrocytes are placed in a hypertonic solution, water leaves the cytoplasm by osmosis and the cells shrink and become **crenated**. When erythrocytes are placed in a hypotonic solution, they gain water by osmosis; the volume of their cytoplasm expands and pushes against the cell surface membranes. Almost instantly, the cell surface membranes rupture and the cells' contents leak out. Erythrocytes which have been disrupted in this way are called erythrocyte ghosts and have been used to study the structure of the cell membrane.

Animals and animal-like protoctists which live in freshwater constantly suffer entry of water by osmosis. If they are not to burst, they must regulate this entry of water. Many animal-like protoctists (see Chapter 25) do this by

KEY CONCEPT

Osmosis is the net movement of water molecules from a region of their higher concentration to a region of their lower concentration through a partially permeable membrane. This movement usually takes place through a membrane which is permeable to water but not to solutes (a partially permeable membrane). Like diffusion, osmosis is a passive process.

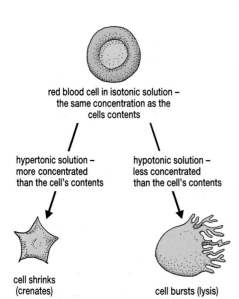

Fig 2.14   Human erythrocytes placed in strong (hypertonic) salt solution lose water by osmosis. Human erythrocytes placed in pure water (hypotonic) gain water by osmosis until they lyse (lyse).

means of special vacuoles, called **contractile vacuoles**, which use energy to pump water from the cytoplasm back into the environment. This is an active process which is described in Section 2.5.

 9 **Explain why a red blood cell becomes crenated when placed in a hypertonic solution.**

ANALYSIS

The problems of living in sea water

This exercise develops data analysis skills.

In an experiment, the marine ciliate *Cothurnia* was placed in a series of dilutions of sea water and the output of its contractile vacuole was measured. In another experiment, the change in volume of the organism in different dilutions of sea water was recorded.

Table 2.4

Added fresh water /%	Contractile vacuole output /dm^3 s^{-1}	Relative body volume
0	0.65	1.000
10	0.56	1.075
20	1.10	1.175
30	1.00	1.280
40	1.51	1.451
50	2.40	1.600
60	6.31	1.785
70	18.25	2.010
80	35.10	2.092
90	9.55	2.035

The results are given in Table 2.4.
(a) Plot these data using a single set of axes.
(b) With reference to the contractile vacuole output curve, explain the effects of dilution on the activity of the contractile vacuole.
(c) What do the changes in relative body volume indicate about the effect of contractile vacuole activity?
(d) Some species of marine protozoa form contractile vacuoles only when the animal begins to feed. Suggest an explanation for this observation.

Osmosis in plant cells

The erythrocyte in Fig 2.14 burst because a pressure developed inside the cell as water entered by osmosis. The cells of prokaryotes, fungi and plants have rigid walls which prevent them from bursting. The walls exert a pressure on the expanding cell which eventually stops osmosis.

In hypertonic solutions, cells with cell walls behave in a similar way to cells without walls; the presence of a cell wall cannot protect them from water loss by osmosis. As their volume decreases and they shrink, the cells lose contact with their cell wall. In plant cells, this is known as **plasmolysis**

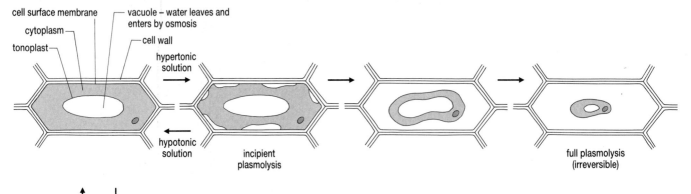

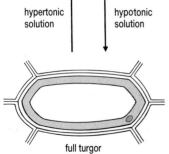

Fig 2.15 Turgor and plasmolysis in plant cells when surrounded by hypotonic and hypertonic solutions.

MOVEMENT OF SUBSTANCES INTO AND OUT OF CELLS

and the point at which the cytoplasm just begins to lose contact with the cell wall is called **incipient plasmolysis**. The different effects on plant cells of placing them into hypotonic and hypertonic solutions are shown in Fig 2.15.

QUESTION

2.8 Explain why foods can be preserved by storing them in strong solutions of salt or sugar (Fig 2.16).

Fig 2.16 Which of these foods depends on osmosis for its preservation?

2.4 OSMOSIS AND WATER POTENTIAL

An understanding of the physical processes involved in osmosis is essential to understand, for example, how the human kidney and circulatory system work and how plants absorb water from the soil. To understand osmosis, you will need to be confident in working with negative numbers; remember that −1 is a less negative number than −10.

Water potential

Look at Fig 2.17. It shows a cell surrounded by a partially permeable membrane. As water molecules hit the membrane they generate a pressure (see Fig 2.7). This pressure is called the **water potential**. The more water molecules that hit the membrane per unit time the higher the pressure, i.e. the higher the water potential.

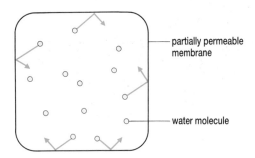

partially permeable membrane

water molecule

Fig 2.17 Water molecules generate pressure as they collide with a partially permeable membrane.

Water potentials and osmosis

Osmosis is the diffusion of water molecules. It occurs when there is a net movement of water molecules from a region of their higher concentration to a region of their lower concentration through a partially permeable membrane.

10 Regions with a high concentration of water molecules are said to have a high water potential. Rewrite the definition of osmosis using the term 'water potential' instead of the term 'concentration'.

Consider two cells separated by a partially permeable membrane as shown in Fig 2.18. Since there are more water molecules in cell A, more water molecules will hit the partially permeable membrane on the A side than on the B side.

 11 Which solution exerts the greatest water pressure on the partially permeable membrane? Which solution has the highest (least negative) water potential? Which way will osmosis occur?

Look again at Fig 2.18. Cell A contains pure water whilst cell B contains a solution: cell A has a water potential of zero whilst cell B has a negative water potential. As the cells are separated by a partially permeable membrane, osmosis will occur. It occurs from cell A to cell B, from a cell with a water potential of zero to a cell with a negative water potential.

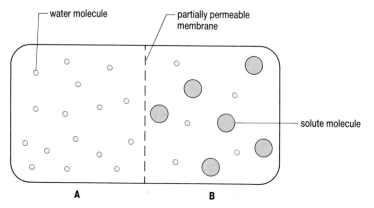

Fig 2.18 Two cells, A and B, separated by a partially permeable membrane. Cell A contains pure water; cell B contains a solution.

QUESTIONS

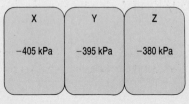

Fig 2.19

2.9 Which is the greater water potential, $\Psi = 0$ kPa or $\Psi = -1$ kPa?

2.10 (a) Look at the three cells X, Y and Z in Fig 2.19. They have water potentials of value −405 kPa, −395 kPa and −380 kPa , respectively. In which direction will osmosis occur between these cells?
(b) Between which pair of cells will the net rate of water movement be greatest? Justify your answer.

Solute potential

Look back at Fig 2.18. As well as the water molecules, solute molecules move about hitting the membrane around the cell and creating a pressure. This pressure is called the **solute potential**. It has the symbol Ψ_s and always has a negative value.

 12 Why does solute potential (Ψ_s) always have a negative value?

The osmotic potential of a solution can be expressed in a simple equation:

water potential of solution = water potential of pure water + solute potential

$$\Psi_{sol} \quad = \quad \Psi_w \quad + \quad \Psi_s$$

 13 **Remembering that $\Psi_w = 0$, what is the solute potential of the three cells X, Y and Z in Fig 2.19?**

The effect of external pressure on water potential

Both the cells in Fig 2.20(a) have partially permeable membranes. They each contain pure water, i.e. they both have water potentials of 0 kPa.

 14 **Why will osmosis not occur between these cells? Will any movement of water occur between them?**

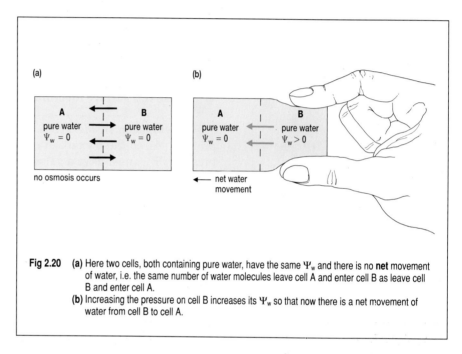

(a)

A
pure water
$\Psi_w = 0$

B
pure water
$\Psi_w = 0$

no osmosis occurs

(b)

A
pure water
$\Psi_w = 0$

B
pure water
$\Psi_w > 0$

← net water movement

Fig 2.20 (a) Here two cells, both containing pure water, have the same Ψ_w and there is no **net** movement of water, i.e. the same number of water molecules leave cell A and enter cell B as leave cell B and enter cell A.

(b) Increasing the pressure on cell B increases its Ψ_w so that now there is a net movement of water from cell B to cell A.

Look at Fig 2.20(b) to see what would happen if one of the cells is squeezed.

 15 **What would happen to the pressure inside cell B? What would therefore happen to the water potential of cell B? Will osmosis now occur from B to A?**

The contribution of external pressure on water potential (like the squeeze in Fig 2.20) is called the **pressure potential**, Ψ_p. At atmospheric pressure, $\Psi_p = 0$. If this term is added to the equation, we have:

water potential = water potential + solute potential + pressure potential
of solution   of pure water

 16 **Write this equation using appropriate symbols.**

Since the water potential of pure water has a value of zero, the equation can be simplified:

$$\Psi_{sol} \quad = \quad \Psi_s \quad + \quad \Psi_p$$

You should practise using these symbols in future.

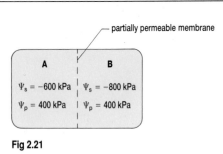

partially permeable membrane

A	B
$\Psi_s = -600$ kPa	$\Psi_s = -800$ kPa
$\Psi_p = 400$ kPa	$\Psi_p = 400$ kPa

Fig 2.21

A worked example

In the system shown in Fig 2.21, two cells are separated by a partially permeable membrane.

Ψ_{sol} of cell A has a less negative value than Ψ_{sol} of cell B. Therefore, water moves by osmosis from A to B.

Cell A

$$\begin{aligned} \Psi_{sol} &= \Psi_s + \Psi_p \\ &= -600 \text{ kPa} + 400 \text{ kPa} \\ &= -200 \text{ kPa} \end{aligned}$$

Cell B

$$\begin{aligned} \Psi_{sol} &= \Psi_s + \Psi_p \\ &= -800 \text{ kPa} + 400 \text{ kPa} \\ &= -400 \text{ kPa} \end{aligned}$$

 Predict the way osmosis will occur in the examples shown in Fig 2.22.

Cell water potential, Ψ_{cell}

So far water potential and osmosis in theoretical cells have been considered. This section considers real cells.

 Are cell surface membranes partially permeable?

The cytoplasm inside a cell is a solution of salts, sugars and colloidal proteins. All of these exert an osmotic potential. Therefore, at atmospheric pressure, Ψ_{cell} must have a value less than zero.

(a)
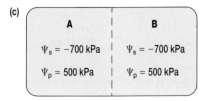

A	B
$\Psi_s = -600$ kPa	$\Psi_s = -700$ kPa
$\Psi_p = 300$ kPa	$\Psi_p = 200$ kPa

(b)

A	B
$\Psi_s = -400$ kPa	$\Psi_s = -600$ kPa
$\Psi_p = 100$ kPa	$\Psi_p = 600$ kPa

(c)

A	B
$\Psi_s = -700$ kPa	$\Psi_s = -700$ kPa
$\Psi_p = 500$ kPa	$\Psi_p = 500$ kPa

Fig 2.22 Examples of osmosis. Predict the directions of water movement.

Cells without walls

An animal cell in pure water is shown in Fig 2.23. The water is hypotonic to the cell solution, i.e. Ψ_w is less negative than Ψ_{cell}. As a result, water enters the cell increasing its volume and hence its pressure. Eventually, the cell bursts. This is cell lysis which has already been described for red blood cells.

 Use the terms Ψ_{cell} and Ψ_{sol} to explain what happens when an animal cell is placed in a hypertonic solution.

Cells with walls: pressure potential

Imagine blowing into a balloon; eventually, it will burst. Similarly, when water continues to enter an animal cell by osmosis, it bursts. Now imagine a balloon inside a metal box (Fig 2.24). Blowing into this balloon will cause it to expand and push against the sides of the box. Eventually it will not be possible to blow any more air into the balloon. The reason for this is shown in the diagram: the box pushes back on the expanding balloon as strongly as the balloon pushes against the walls of the box. There is a box pressure.

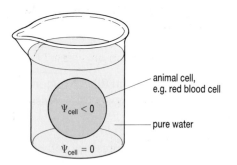

$\Psi_{cell} < 0$

animal cell, e.g. red blood cell

pure water

$\Psi_{cell} = 0$

Fig 2.23 An animal cell, e.g. an erythrocyte, suspended in a beaker of pure water. What will happen?

 What will happen when the balloon pressure and box pressure are equal?

Cells with a cell wall behave like a balloon inside a box. When water enters such a cell, the cell expands and its cell surface membrane starts to push against the inside of the cell wall. In turn, the wall pushes back on the expanding cell (Fig 2.25). This is another example of a pressure potential, Ψ_p. Its effect on the water potential of a cell can be written:

water potential of cell = solute potential of cytoplasm + pressure potential

$$\Psi_{cell} = \Psi_s + \Psi_p$$

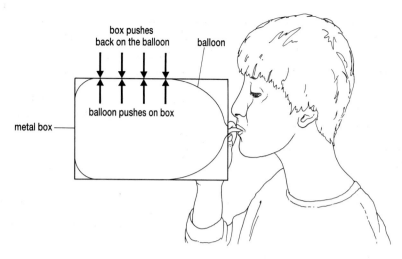

Fig 2.24 A simple model to explain how a plant cell behaves.

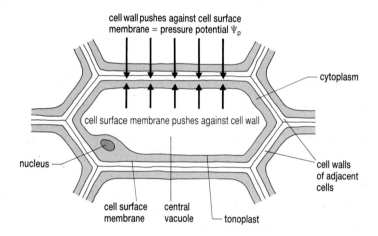

Fig 2.25 Pressure potential in cells with a cell wall.

Turgidity in plant cells

Plant cells have cell walls. Eventually, there comes a point at which there is no further net movement of water into a plant cell. This happens when $\Psi_s \equiv \Psi_p$ and a cell in this condition is called **turgid**.

 21 **What is the advantage of a cell wall to a freshwater organism?**

ANALYSIS

Water potential, turgor and plasmolysis

In this exercise you have to interpret graphical information.

Water potential, the solute potential of the cell contents and the pressure potential of a plant cell with extensible walls each change as the plant cell takes up or loses water (Fig 2.26). The cell also changes in volume.

(a) Why is ψ_p always positive?

(b) Make a copy of Fig 2.26. Indicate on it
    (i)   the point of incipient plasmolysis
    (ii) the point at which the cell is fully turgid.

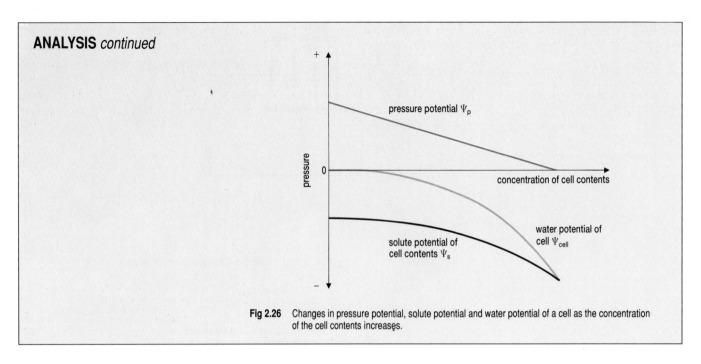

Fig 2.26 Changes in pressure potential, solute potential and water potential of a cell as the concentration of the cell contents increases.

QUESTIONS

2.11 **(a)** Explain in words what is meant by the term 'water potential'.
(b) How does water potential govern the movement of water molecules by osmosis?
(c) Name two factors which affect water potential and write an equation to show their relationship.

2.12 The two systems shown in Fig 2.27 are separated by a partially permeable membrane. In which direction will osmosis occur? Explain your answer.

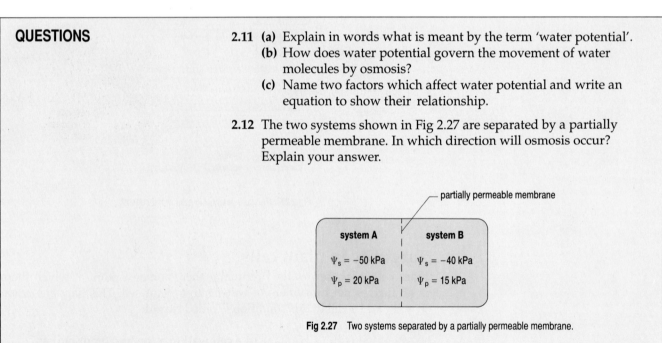

Fig 2.27 Two systems separated by a partially permeable membrane.

2.5 ACTIVE TRANSPORT

At some stage in their life cycle, most cells take substances from their environment which are already more concentrated inside the cell than they are outside. They may also get rid of substances from their cytoplasm which are already more concentrated outside the cell. This movement of substances against a concentration gradient can be achieved by **active transport**, a process which uses energy. This energy is usually released when molecules of ATP are broken down, a process described in Chapter 5.

Protein pumps

The proposed mechanism for active transport is shown in Fig 2.28. Like facilitated diffusion, active transport depends on intrinsic proteins in the cell surface membrane which have specific receptor sites for substances to

KEY CONCEPT

Active transport is the movement of molecules and ions against a concentration gradient and requires an input of metabolic energy from the breakdown of ATP.

be transported. Unlike facilitated diffusion, however, these proteins can bind with their molecules or ions on one side of the membrane only. Having done so, the protein either moves within the membrane or changes shape (a conformational change), so that the molecule or ion is carried across the membrane and released. Unlike facilitated diffusion, this conformational change requires the energy released by the breakdown of one molecule of ATP.

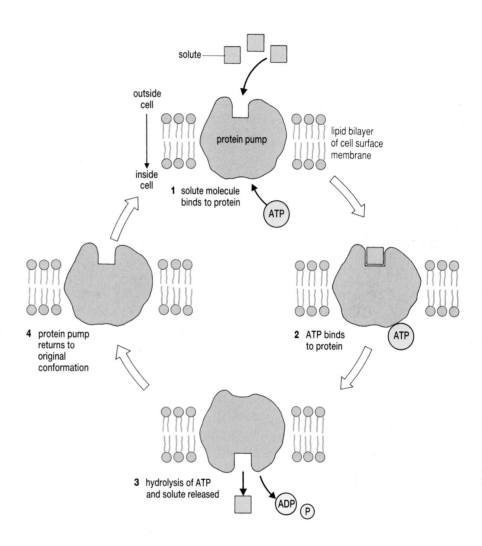

Fig 2.28 Proposed mechanism for active transport. Binding of the solute (1) and attachment of ATP to the inside of the membrane protein (2) causes the protein to change shape transferring the molecule across the membrane and releasing it on the inside (3). This is accompanied by the hydrolysis of ATP (ATP→ ADP + P$_i$) which provides energy. The membrane protein pump then flips back to its original shape (4).

Since they are pumping molecules from one side of the membrane to the other, these membrane proteins are called protein pumps. There are two major protein pumps.

- **Sodium–potassium (Na$^+$–K$^+$) pumps** (Fig 2.29) are important, for example, in passing impulses along nerve cells. They pump sodium ions (Na$^+$) out of, and potassium ions (K$^+$) into, a cell against concentration gradients. In reality, by the hydrolysis of each molecule of ATP three Na$^+$ are pumped out and two K$^+$ pumped in.
- **Proton pumps** are found on the inner membranes of mitochondria and chloroplasts. These pump hydrogen ions (H$^+$) from one side of the membrane to the other and play a vital role in ATP synthesis.

MOVEMENT OF SUBSTANCES INTO AND OUT OF CELLS

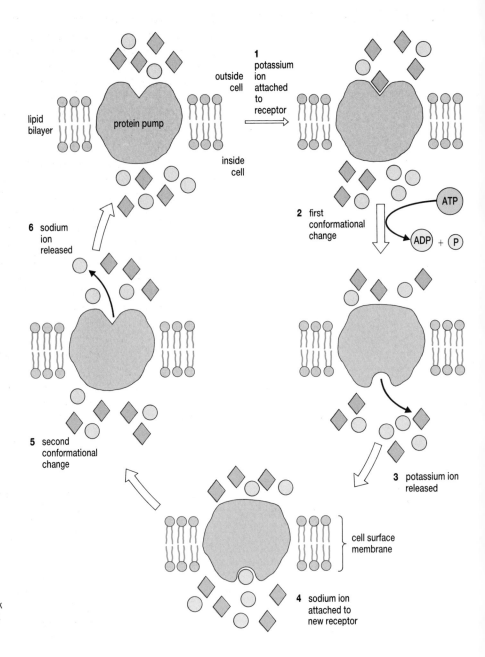

Fig 2.29 The Na⁺ – K⁺ pumps in cell surface membranes cause the movement of Na⁺ shown as pink circles and K⁺ shown as pink diamonds against a concentration gradient. This active transport uses energy.

Labels within the figure:
- lipid bilayer
- protein pump
- outside cell
- inside cell
- **1** potassium ion attached to receptor
- **2** first conformational change
- ATP → ADP + P
- **3** potassium ion released
- cell surface membrane
- **4** sodium ion attached to new receptor
- **5** second conformational change
- **6** sodium ion released

QUESTIONS

2.13 How does active transport
(a) resemble
(b) differ from facilitated diffusion?

2.14 The data in the Table 2.5 show the relative rate of uptake of glucose and xylose (a pentose) from living intestine and from intestine which had been poisoned with cyanide. Cyanide greatly reduces the availability of ATP. Discuss these data.

Table 2.5 The relative rate of uptake of two sugars by cyanide-poisoned and unpoisoned intestines

| | Relative rate of uptake by intestine | |
Sugar	Without cyanide	With cyanide
glucose	100	28
xylose	18	18

2.6 ENDOCYTOSIS AND EXOCYTOSIS

Endocytosis is the transport into the cytoplasm of substances which are too large to be transported by protein carriers. In humans, cholesterol is one of the most common of these substances. If the substances are solid, for example particles of debris or entire cells, the process is termed **phagocytosis** (cell eating). If the material is liquid or colloidal, the process is referred to as **pinocytosis** (cell drinking). **Exocytosis** is the reverse of endocytosis, i.e. the transport of substances, such as waste products of digestion or material to be secreted, from the cytoplasm to the outside environment.

Endocytosis and exocytosis may occur in the cyclic series of events shown in Fig 2.30. The endocytic cycle involves specific receptor proteins at the cell surface which bind to nutrients outside the cell. The bound nutrients are called **ligands**. Receptors of various kinds, each with their bound ligands, move sideways through the cell surface membrane until they are in groups of one thousand or so. The cell surface membrane where these groups of ligands collect starts to indent to form a pit. Since the cytoplasm side of these pits becomes coated with a protein called clathrin, they are called **coated pits**. The electronmicrograph (Fig 2.31) shows one of these coated pits forming. The coated pits bud inwards and pinch off to form **coated vesicles** which move into the cytoplasm. Once in the cytoplasm, the clathrin peels away from the coated vesicle and, in a process which is not well understood, the ligands are released. The vesicle returns to the cell surface membrane and fuses with it in the process of exocytosis. In certain human cells, this cycle takes about one minute. Since about 2% of their cell surface membrane is in the form of coated pits, these cells must take up an area equivalent to their own surface every 50 minutes.

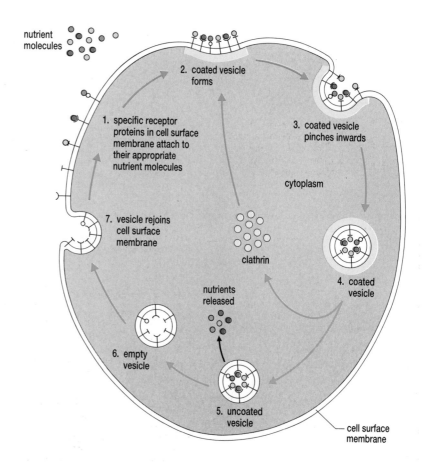

Fig 2.30   The endocytic-exocytic cycle.

MOVEMENT OF SUBSTANCES INTO AND OUT OF CELLS

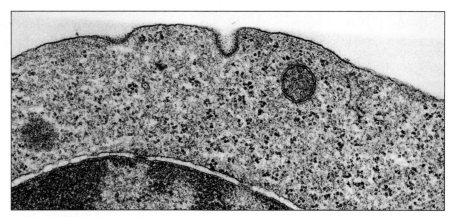

Fig 2.31 Electronmicrograph of a clathrin-coated pit forming during endocytosis on a cell surface membrane.

2.15 Copy and complete Figs 2.32(a) and (b) to remind yourself of the movement of substances. Write the missing labels for the structure of the cell membrane (A to E) and transport processes across membranes (F to I). Label the inside and outside of the membrane.

2.16 Red blood cells have $Na^+ - K^+$ pumps in their cell surface membrane. These continually pump Na^+ out of the cell so that the concentration of Na^+ in the plasma surrounding the cell is very high whilst the concentration of Na^+ inside the cell is very low. Some red blood cells, kept in plasma, were treated with cyanide, a substance which inhibits the production of ATP. Within a few minutes all the red blood cells had burst (lysed).

(a) What happened to the $Na^+ - K^+$ pumps once the cyanide had been added?

(b) What would then have happened to the concentration of Na^+ inside the cell?

(c) Why does this help you to explain why the cell burst?

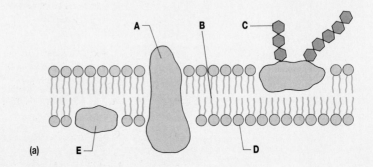

(a)

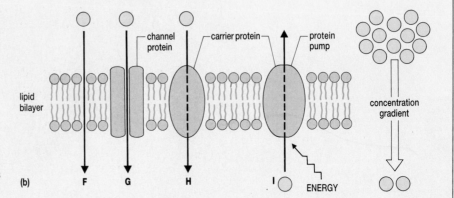

(b)

Fig 2.32 (a) A simplified picture of the fluid mosaic model.
(b) A summary of transport mechanisms across cell membranes.

MOVEMENT OF SUBSTANCES INTO AND OUT OF CELLS

SUMMARY

The cell surface membrane around each cell is made mainly of phospho-lipids and proteins. It regulates the movement of molecules and ions into and out of the cell.

Phospholipids are arranged in a double layer within the cell surface membrane; protein molecules are embedded within this phospholipid bilayer. This model of cell membrane structure is the fluid mosaic model.

Lipid-soluble molecules can cross the cell surface membrane by dissolving in its phospholipids. The embedded protein molecules form channels through which water-soluble molecules and ions can pass.

Molecules and ions in solution constantly move about. They tend to move from a region where their concentration is higher to a region where their concentration is lower until the concentrations are equal. This process is called diffusion.

Cell surface membranes are partially permeable because their embedded proteins form small pores which allow the passage of small particles but prevent the passage of larger particles. Osmosis is the net movement of water molecules through a partially permeable membrane from a solution with a lower concentration of dissolved substances (less negative water potential) to a solution with a higher concentration of dissolved substances (more negative water potential) until the concentrations are equal. The water potential of a plant cell (ψ_{cell}) is found by adding the pressure potential of its wall (ψ_p) to the solute potential of its cytoplasm (ψ_s).

A hypotonic solution contains fewer dissolved particles than the contents of a cell. When cells are placed in hypotonic solutions they gain water by osmosis; their volume increases causing a rise in pressure on their cell surface membranes. Cells without cell walls may burst as a result of this pressure increase. A hypertonic solution contains more dissolved particles than the cytoplasm of a cell. When cells are placed in hypertonic solutions they lose water by osmosis and shrink.

Cells do not expend energy when molecules and ions cross their cell surface membranes by osmosis and diffusion. Such movement is called passive transport. Cells may expend energy to cause molecules and ions to cross their cell surface membranes, e.g. in moving molecules against a concentration gradient. This is active transport and is thought to involve carrier molecules embedded in the cell surface membrane.

Relatively large particles may enter cells by endocytosis and leave cells by exocytosis. In both cases, small vesicles of membrane contain the substances during transport and movement of particles across the cell surface membrane occurs when these vesicles are formed from (endocytosis) or when they fuse with (exocytosis) the cell surface membrane.

Proteins which are embedded in the cell surface membrane often act as receptor sites. They recognise specific molecules or ions and aid their passage through the membrane.

Answers to Quick Questions: Chapter 2

1 Made of non-polar molecules, e.g. lipid.
2 Inner layer is 2–3 times the thickness of the outer layers.
3 No. Different cell membranes will have different functions and therefore different compositions. However, all cell membranes will have the same basic structure.
4

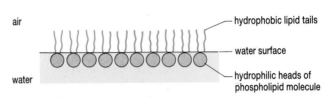

5 You might expect a bilayer to split down the middle.
6 A would diffuse out of the cell by reversing the process seen in Fig 2.12.
7 Sugar = solute, water = solvent.
8 There will be a net movement of water molecules from right to left. Water molecules will always be more concentrated on the right-hand side of the membrane.
9 Water is lost from the erythrocyte to surrounding solution by osmosis. Volume of cell contents gets less and cell surface membrane folds.
10 Osmosis is the net movement of water molecules from a higher water potential to a lower potential through a partially permeable membrane *or, better still,*
Osmosis is the net movement of water molecules from a less negative water potential to a more negative water potential through a partially permeable membrane.
11 Cell A; Cell A; from Cell A to Cell B.
12 It reduces the water potential of its solution compared with pure water which has $\psi_s = 0$.
13 –405 kPa, –395 kPa, –380 kPa ($\Psi_{sol} = 0 - \Psi_s$).
14 They have the same water potential. Water molecules will move equally in both directions.
15 The pressure will increase so Ψ_w will increase and water molecules will move from B to A.
16 $\psi_{sol} = \psi_w + \psi_s + \psi_p$.
17 (a) From cell A to cell B.
    (b) From cell B to cell A.
    (c) Water will diffuse equally in both directions; no net movement of water.
18 All cell surface membranes are partially permeable. Their protein channels allow small molecules (including water) to pass through.
19 In hypertonic solution Ψ_{cell} is less negative than Ψ_{sol} so the cytoplasm loses water by osmosis and the cell shrinks.
20 It will not be possible to blow up the balloon any further.
21 Such cells would gain water by osmosis and burst but the cell wall stops this cell lysis.

Chapter **3**

CELL STRUCTURE

LEARNING OBJECTIVES

When you have studied this chapter you should be able to:

1. describe the structure of a cell seen using an optical microscope;

2. describe the ultrastructure of a cell seen using an electron microscope;

3. begin to explain the function of the cell structures revealed by electron microscopy;

4. explain the relative merits of optical and electron microscopy;

5. describe some of the techniques employed in the study of cell structure and function.

Cork is made of dead bark, usually from the cork tree. You can see from Fig 3.1(b) that it has a honeycombed appearance. The small, dark spaces are empty holes. They are surrounded by cork substance, which is a complex mixture of different chemicals. In 1665, Robert Hooke thought that the solid cork substance was the living material from which cork was made. He did not know what the small, dark holes were, but it is obvious what they reminded him of; he called them **cells**. Following his discovery, Hooke and other scientists found that all the plant material they looked at under their microscopes was made of cells. In 1839, a Belgian botanist, Mathias Schleiden, proposed a theory that all higher plants were made of cells. At

Fig 3.1 **(a)** *(left)* Robert Hooke's microscope. Note the complicated mechanism for shining light onto the specimen.
(b) *(right)* Cork cells were the first cells seen and recorded by Robert Hooke. This is a copy of Robert Hooke's drawing of cork. He called the empty spaces cells because they looked like monks' cells. After Hooke, other people saw the same units in living tissue and the same name was used – the cell.

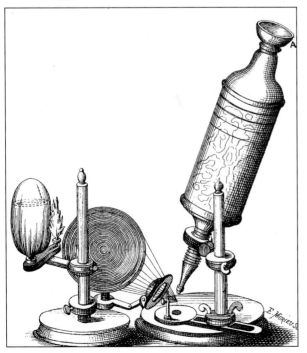

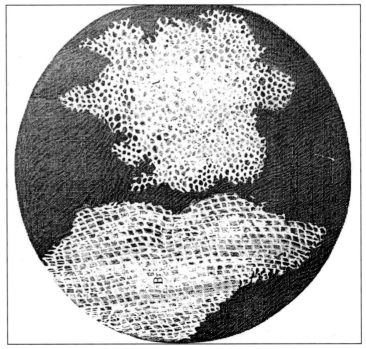

the same time a German zoologist, Theodor Schwann, proposed that the same was true for higher animals. This **cell theory** was an important milestone in biology because it suggested that there were some underlying similarities between the enormous range of animals and plants which people could see. In 1858, Rudolph Virchow proposed that the cell was the basic metabolic, as well as structural, unit of living organisms, thus expanding the cell theory. He also emphasised that all cells arise from pre-existing cells.

3.1 SEEING CELLS

Two different types of microscope are used in the study of cells, the optical or light microscope, and the electron microscope. During your course of study you will use a light microscope to look at cells. In addition, photographs (electronmicrographs) may be studied taken with an electron microscope.

Optical microscopes

An optical microscope consists essentially of a series of lenses through which ordinary white light can be focused. The exact arrangement of the lenses will depend upon the make, but the path of light rays through a student microscope will look something like Fig 3.2(a). Such a microscope enables more of the fine structure of an object to be seen than could be seen with the naked eye. One thing that helps is the magnification of the microscope. However, even more important is the **resolution** of the microscope. No matter how well made or how high the magnification, optical microscopes cannot resolve objects which are less than a few hundred nanometres apart because of the wavelength of light. This places real limitations on what can be seen.

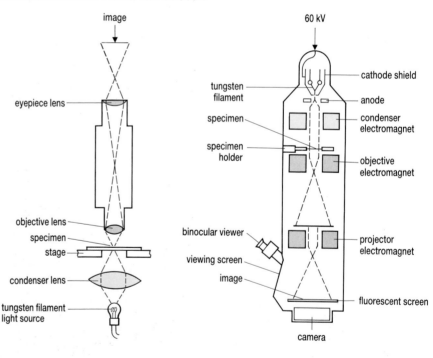

Fig 3.2 The pathway of light through an optical microscope compared with that of electrons through a transmission electron microscope. Note the similarities of overall design. An electron microscope requires the specimen to be placed in a vacuum.

(a) compound optical microscope

(b) transmission electron microscope

The resolving power of microscopes

The limit of resolution is defined as the smallest distance between two points that can be seen using a microscope. So a microscope with a high resolving power will enable two small objects close together to be seen as two distinct objects, whereas with a microscope of low resolving power the two distinct objects will merge into one. Optical microscopes have improved in both their magnification and resolution since 1665 but there are both practical and theoretical limits to resolving power. In particular the resolving power is inversely proportional to the wavelength of the light being used.

To improve resolution beyond that of a light microscope requires radiation with a much shorter wavelength, e.g. the electron beam of an electron microscope.

Electron microscopes

Electrons can behave as waves. The wavelengths of electrons (about 0.01 nm) are many thousands of times smaller than wavelengths of light (about 500 nm), so the limit of resolution of an electron microscope is thousands of times smaller. In other words, electrons allow much smaller structures to be 'seen' than does light.

Of course, electrons cannot be seen, but their effects can be seen in the pictures produced on a screen. Electrons are 'fired' from an electron gun at the back of a television tube and hit the fluorescent screen at its front, making the screen glow. An electron gun (hot cathode) is used inside an electron microscope and fires electrons at the specimen being viewed and onto a fluorescent screen or a photographic plate (Fig 3.2(b)). In **transmission electron microscopes**, the electrons pass through the specimen. The specimen must, therefore, be extremely thin (usually between 10 nm and 100 nm) so the electrons can pass through. In **scanning electron microscopes**, the electrons bounce off the surface of the specimen. They produce images with a better three-dimensional appearance than transmission microscopes.

 Suggest why electron microscopes contain a vacuum.

Fig 3.3 A modern transmission electron microscope.

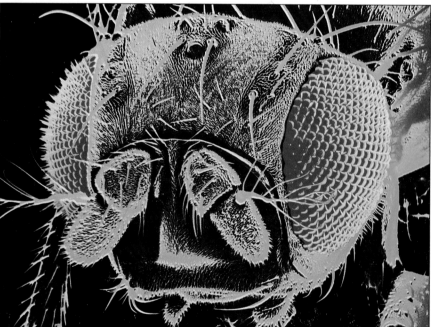

Fig 3.4 The head of a fruit fly (*Drosophila*) seen in a scanning electron microscope. Even though the magnification is not enormous the detail is very sharp because of the high resolution of the electron microscope.

Preparing specimens for the optical microscope

For an optical microscope, thin slices of animal or plant material can be placed on a glass slide and flooded with water. A thin glass square, called a **coverslip**, is used to cover this **wet mount** before it is examined using a microscope. Wet mounts like this are quickly prepared; they only last for 30 minutes or so. Since many cells are transparent, dyes (called **stains**) are often used to make their structure easier to see. These stains combine with certain chemicals inside the cell and their colour shows where they are. Structures inside cells which stain differently do so because they have different chemical compositions.

Table 3.1 Stages in the processing of living material for examination under the optical microscope

Stage	Description of process	Reason for process
fixation	biological structures left to soak in a solution such as formalin	kills the material in a life-like position, stops it decomposing and helps the stains to work
dehydration	structures left to soak in alcohol solutions of increasing strengths	removes water from the cells (which would not mix with wax)
embedding	structures left to soak in hot wax for several hours	when cool, wax supports the structure, making it rigid for sectioning
sectioning	embedded structure cut into thin slices (**sections**) using a **microtome**	produces sections which are thin enough to let light pass through
mounting	sections placed on a glass slide	sections can now be carried about in the laboratory
clearing	sections are left in organic solvents	dissolves the wax which has served its purpose
rehydration	sections immersed into decreasing strengths of alcohol and finally into water	removes anything which will cloud the section when water-based stains are used
staining	sections are soaked in stain(s)	colours different parts of the cells
dehydration	structures left to soak in alcohol solutions of increasing strengths	removes water from cells (which would not mix with mountant)
clearing	sections left to soak in organic solvent like xylene	removes alcohol from the section and makes it transparent
mounting	mountant, such as DPX, added to section and coverslip lowered into place	holds section and coverslip permanently in place

To make **permanent mounts**, the plant and animal parts need to be processed in a special way. Table 3.1 summarises the stages in the commonest processing method. This processing kills the cells. If biologists wish to examine structures in unstained, living cells, they must use a different technique (Fig 3.5).

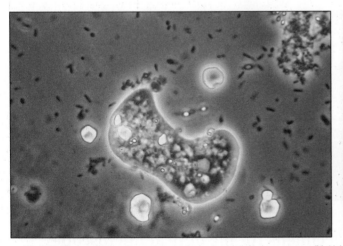 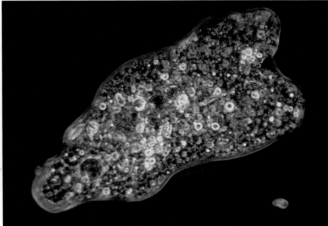

Fig 3.5 These cells have been photographed using **(a)** *(left)* phase-contrast **(b)** *(right)* dark-field light microscopy allowing greater detail to be seen without killing the cell.

Specimens for electron microscopy are processed using much the same stages as those for optical microscopy given in Table 3.1. Because the sections must be much thinner than for optical microscopy, the specimen

Uranyl salts are salts of the metal uranium.

from which they are cut is embedded in a resin, rather than in wax, which tends to crumble. They are cut using an **ultra-microtome** which cuts very thin sections – almost one cell thick. Glass slides would absorb electrons, so the sections are mounted on small copper grids, where the electrons can pass through the holes in the grid. Solutions of lead salts and uranyl salts, which are taken up by different chemicals in the cell, are used instead of coloured stains. These salts do not allow electrons to pass through them, as they are **electron dense**, so they provide contrast between different parts of the cell.

 What would part of the cell which had taken up the salt look like on a black and white television screen compared to a part which had not?

Artefacts

An important point to realise is that the treatments to which specimens have been subjected, such as fixing, sectioning, staining and so on, are all extremely harsh. Such treatments may alter the cell, producing artefacts, so you need to be very careful about interpreting what you see. This problem is particularly acute in electron microscopy because living cells are never observed.

QUESTIONS

3.1 Why was the cell theory important to biologists in the nineteenth century?

3.2 What are the advantages and disadvantages of using an electron microscope?

3.3 Explain to an intelligent non-scientist why electron microscopes show more detail of cell structure than highly magnified images using an optical microscope ever could.

3.4 Suggest why the images revealed by electron microscopy might not accurately represent structures in living cells.

3.2 CELL STRUCTURE

This section deals with what can actually be seen and what can be inferred to be present inside a cell using a light microscope. How the invention of the electron microscope caused a revolution in biological knowledge will then be better appreciated.

ANALYSIS

The structure of cells as seen in the light microscope
This exercise involves drawing and interpreting light micrographs.

Look at Fig 3.6. It shows the appearance of an animal cell, an *Amoeba*, a plant cell and a fungal cell as seen through a compound optical microscope. Each cell has a dark-staining **nucleus**, surrounded by **cytoplasm** which has not stained as darkly.
(a) Why has the nucleus stained more darkly than the cytoplasm?

Since the cytoplasm of the animal cell comes to an abrupt end, we must assume that there is a **cell surface membrane** around it which we cannot see. We cannot see whether there is a cell surface membrane around the cytoplasm of the plant and fungal cells. There is obviously a **cell wall** around them both.
(b) Why can you not see the cell surface membrane?

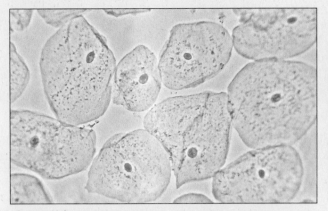
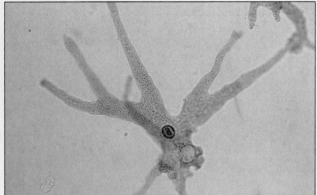
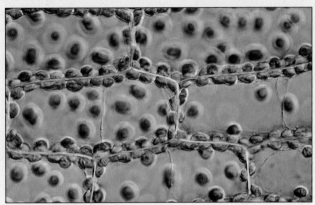
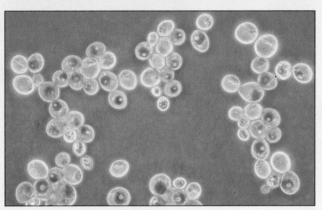

Fig 3.6 Light micrographs of
- **(a)** *(top left)* animal cells
- **(b)** *(top right)* an *Amoeba*
- **(c)** *(bottom left)* plant cells
- **(d)** *(bottom right)* fungal cells.

The plant and fungal cells also have a large, fluid-filled **vacuole** in their cytoplasm. Table 3.2 summarises the function of the structures you can see or infer exist.

(c) Make line drawings of each of the cells in Fig 3.6. Do not shade or colour your drawings, but label the parts of the cell which you have drawn with their name and function (an annotation).

(d) What structures are present in plant and fungal cells but absent from animal cells and the *Amoeba*?

Table 3.2   The structures which can be seen (or inferred) from light microscopy

Structure	Position	Function
nucleus	within cytoplasm	contains genetic code which controls the activities of the cell
cytoplasm	around nucleus	does the work of the cell, e.g. produces proteins, releases energy
cell surface membrane	around cytoplasm	controls exchange of substances between cytoplasm and its surroundings
cell wall	around cell membrane	gives cell rigidity, stops it bursting if put into water
cell vacuole	within cytoplasm	affects concentration of cytoplasm; is a store of inorganic ions
tonoplast	around cell vacuole	controls exchange of substances in plant cells between vacuole and cytoplasm
large granules	within cytoplasm	usually stores of food, e.g. starch, oils

Prokaryotic and eukaryotic cells

Cells which have a nucleus, like all those you have seen so far, are called **eukaryotic cells**. All protoctists (e.g. *Amoeba)*, animals, plants and fungi have eukaryotic cells. Red blood cells of mammals and the mature phloem cells of flowering plants are exceptional eukaryotic cells because they have lost their nuclei during development. The single **prokaryotic cells** of bacteria do not have a nucleus. The rest of this chapter is concerned with eukaryotic cells only. Prokaryotic cells are described in Chapter 25. Table 25.1 compares these two fundamentally different types of cell.

Cell specialisation

Most organisms have bodies which are made up of many cells. The cells within their bodies are not all the same, however. During development, some cells specialise (or **differentiate**) to carry out one particular function. These specialised cells interact together (Fig 3.7) to form tissues and organs. Fig 3.8 shows examples of some specialised cells.

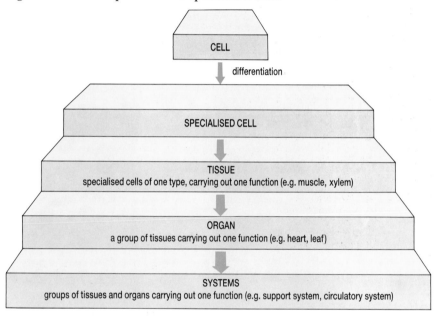

Fig 3.7 The organisation of cells within the body of multicellular organisms.

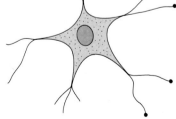
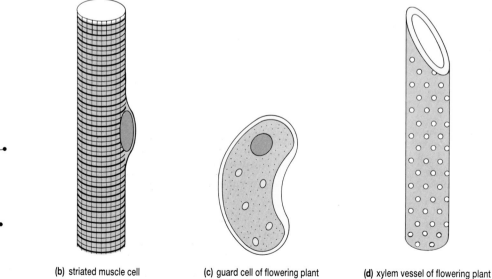

(a) mammalian nerve cell   (b) striated muscle cell   (c) guard cell of flowering plant   (d) xylem vessel of flowering plant

Fig 3.8 Some specialised animal and plant cells.

CELL STRUCTURE

3.5 Why is a red blood cell eukaryotic even though it contains no nucleus?

3.6 A leaf is a plant **organ** which contains photosynthetic **tissue**. Explain the meaning of the emboldened words.

3.3 CELL ULTRA-STRUCTURE

The appearance of a cell seen through an electron microscope is its **ultrastructure**. An animal cell photographed through a transmission electron microscope is shown in Fig 3.9(a) and Fig 3.9(b) is a diagrammatic representation of this cell. Just as with the optical microscope, the nucleus can be seen but it is now clear that there are membranes around it. The cell surface membrane, invisible using an optical light microscope, can now be seen. Perhaps the most striking difference between this cell and the ones viewed in Fig 3.6 is that the cytoplasm is full of structures, called **organelles**. Some of these, for example mitochondria, are surrounded by cell membranes but others, for example ribosomes, are not.

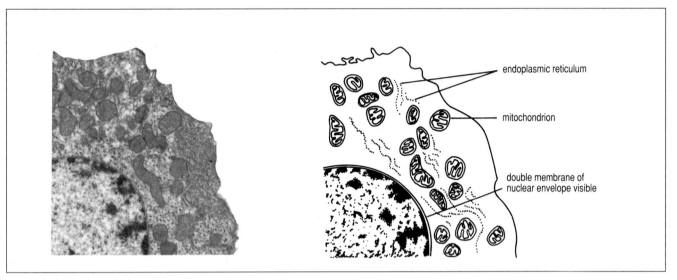

endoplasmic reticulum

mitochondrion

double membrane of nuclear envelope visible

Fig 3.9 **(a)** An animal cell photographed through a transmission electron microscope.
(b) A diagrammatic representation of the same cell.

KEY CONCEPT

Cell membranes: A profusion of internal membranes is a basic feature of all eukaryotic cells. One function of the membranes is to divide the cell into compartments within which specialised chemical reactions can occur. In addition the cell membranes provide a large surface area within the cell on which chemical reactions can occur.

 Why can you see all these structures with an electron microscope but not with an optical microscope?

Cell wall

The cells of plants, plant-like protoctists and fungi have outer cell walls (Fig 3.10). These walls have tough fibres of cellulose surrounded by a matrix of other polysaccharides, rather like the structure of reinforced concrete or fibre-glass.

Cell surface membrane (or plasma membrane)

This is the outer boundary of the cell (Fig 3.11). It is a continuous sheet of lipid molecules about 4–8 nm thick within which proteins are embedded. Some of these proteins are pumps and channels which transport specific molecules into and out of cells. Chapter 2 investigates the cell surface membrane further.

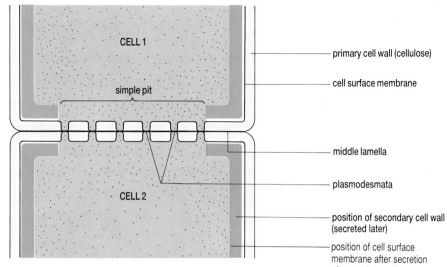

Fig 3.10 Plant cell wall. Notice the pits in the cell wall between adjacent cells which are connected by **plasmodesmata.**

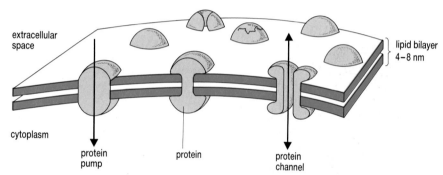

Fig 3.11 Cell surface membrane – the lipoprotein outer boundary.

The nucleus

This is large enough to be easily seen with an optical microscope. The nucleus is separated from the cytoplasm by a **nuclear envelope**, a double membrane, the outer of which is part of the endoplasmic reticulum, perforated with holes, the **nuclear pores** (Fig 3.12). These are partly plugged by complexes of ribonucleic acid and proteins which regulate the passage of molecules between nucleus and cytoplasm. The nucleus

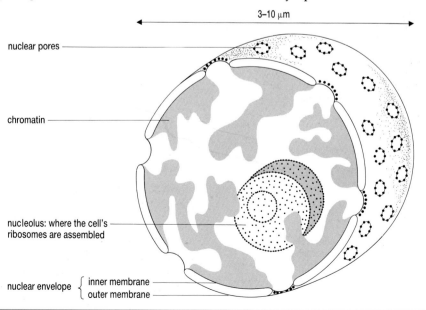

Fig 3.12 The nucleus contains chromosomal DNA and has a double membrane (the nuclear envelope) through which various molecules can pass between cytoplasm and nucleus.

contains all the chromosomal DNA of the cell. The chromosomes are surrounded by nuclear sap, or nucleoplasm. Most eukaryotic cells have one or more nucleoli in this nucleoplasm. They may be attached to the inside of the nuclear envelope and make ribosomal ribonucleic acid (rRNA).

Undulipodia

These are eukaryotic flagella and cilia and are basically similar in structure though eukaryotic flagella tend to be longer and cells have fewer of them. Both of these structures are involved in moving cells and in moving materials over the surface of cells. Note the characteristic 9 (peripheral) and 2 (central) arrangement of microtubules of undulipodia (Fig 3.13).

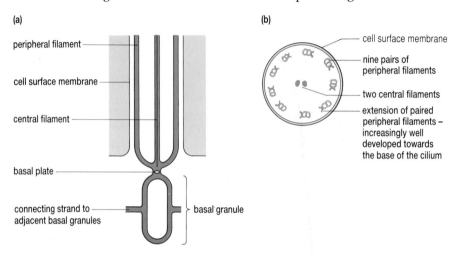

Fig 3.13   **(a)** Longitudinal section of a cilium, a type of undulipodium present in large numbers.
(b) Cross-section of a cilium.

Cytoplasmic organelles

Endoplasmic reticulum (ER)

A network (reticulum) of hollow sacs and tubes which is surrounded by cell membrane. The membrane of the endoplasmic reticulum is continuous with the nuclear envelope. Its functions are to synthesise and transport proteins and lipids. The flattened **rough endoplasmic reticulum** (Fig 3.14(a)) is studded on its cytoplasmic side with small spherical **ribosomes**. The more tubular **smooth endoplasmic reticulum** (Fig 3.14(b)) lacks ribosomes and is involved with lipid metabolism.

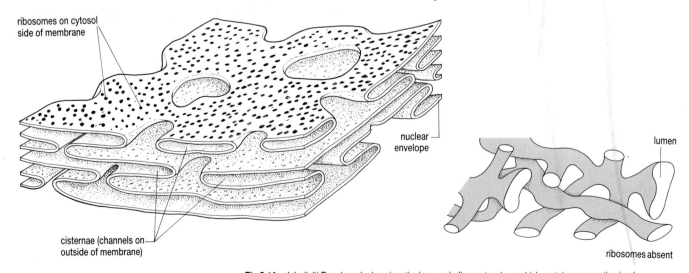

Fig 3.14   **(a)** *(left)* Rough endoplasmic reticulum – a hollow network on which proteins are synthesised.
(b) *(right)* Smooth endoplasmic reticulum lacks ribosomes.

CELL STRUCTURE

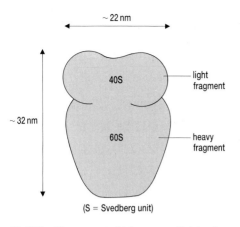

Fig 3.15 Ribosomes are rich in enzymes; their function is the synthesis of proteins.

Ribosomes

These consist of two subunits (Fig 3.15). Chemical analysis shows that half of their mass is ribonucleic acid (**ribosomal RNA or rRNA**) which is made inside the nucleus. The other half is protein. The ribosomes are assembled in the cytoplasm, where they make proteins. Some of the RNA they contain may act as enzymes (see Chapter 5), speeding up the formation of these proteins. Depending on the type of protein they are making, ribosomes might attach to the rough endoplasmic reticulum or form clusters in the cytoplasm called **polysomes**.

Golgi apparatus

The membrane-bound sacs (**cisternae**) of the Golgi apparatus are involved in modifying, sorting and packaging macromolecules either for secretion across the cell surface membrane or for delivery elsewhere in the cell. The small membrane-bound **vesicles** (Fig 3.16) are thought to transport material between the Golgi apparatus and other membrane-bound compartments in the cell.

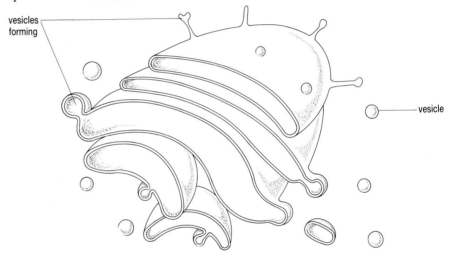

Fig 3.16 The Golgi apparatus is involved in temporary storage and then transport of many different types of macromolecules within the cell.

Mitochondria

Glucose never enters mitochondria. It is first broken down into smaller molecules such as pyruvate.

These are found in almost all eukaryotic cells. Mitochondria have a double membrane with the inner one being elaborately folded (Fig 3.17). They are involved in the oxidation of molecules such as glucose to make ATP (respiration). Subsequent breakdown of this ATP provides energy, needed, for instance, to make peptide bonds.

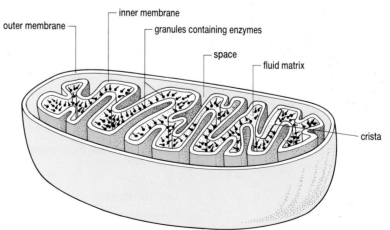

Fig 3.17 Mitochondria are rich in enzymes for the oxidation of respiratory substrates and production of ATP.

Summarise how the nucleus, endoplasmic reticulum, ribosomes, Golgi apparatus and mitochondria interact in cells which secrete insulin (a protein).

Chloroplast

These chlorophyll-containing organelles are surrounded by a double membrane (Fig 3.18). Found in all plants and plant-like protoctists, they have an elaborate internal membrane system where some of the reactions of photosynthesis are catalysed (see Chapter 7).

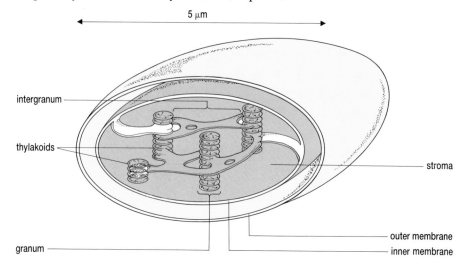

Fig 3.18 Chloroplasts contain the photosynthetic pigments of plants and protoctists. This diagram shows a cross-section of a single chloroplast to show the internal structures.

Vacuole

Plant cells have a central vacuole (Fig 3.19) which is so big that it can be seen using an optical microscope. It may occupy up to 90% of the space inside a plant cell. The membrane around it is called the **tonoplast**. These vacuoles contain a solution of ions and sugars. They also contain enzymes which are capable of digesting proteins. A major function is controlling cell shape and volume.

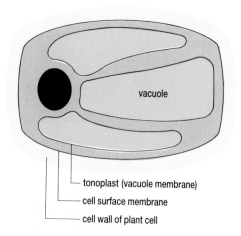

Fig 3.19 Cell vacuoles are of variable size and solute concentration but are largely responsible for cell shape and volume.

Lysosomes

These are membrane-bound vesicles (Fig 3.20) which contain enzymes involved in **intracellular digestion**. The enzymes, produced by the rough endoplasmic reticulum and transported via the Golgi apparatus, break down unwanted cell debris and foreign matter. On the death of the cell, the enzymes (**lysozymes**) are released from the vesicles and digest the cytoplasm in a process called **autolysis**.

Peroxisomes

These vesicles (Fig 3.21) also contain enzymes which are produced by the rough endoplasmic reticulum. The most common enzyme in these peroxisomes, **catalase**, speeds up the breakdown of hydrogen peroxide into water and oxygen. Hydrogen peroxide is a poisonous waste product of certain cell reactions, so its elimination is vital to all cells.

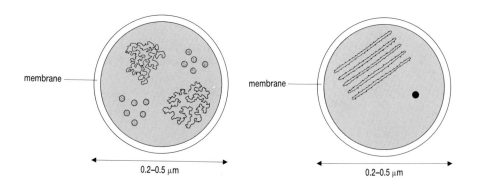

Fig 3.20 Lysosomes are membrane-bound vacuoles containing enzymes for intracellular digestion.

Fig 3.21 Peroxisomes are membrane-bound vesicles known to contain the enzyme catalase.

SPOTLIGHT

ARE MITOCHONDRIA AND CHLOROPLASTS ENDOSYMBIONTS?

Chloroplasts and mitochondria are both about the same size as prokaryotic cells. Their fluid interior contains ribosomes which are similar to those of prokaryotic cells but smaller than those of eukaryotic cells. Like prokaryotic cells, they have circular DNA lying freely in their interior (though the diameter of this DNA is much smaller than in prokaryotic DNA). The DNA and ribosomes inside the mitochondria and chloroplasts produce proteins which are quite different from those produced by the rest of the cell. This protein production is inhibited by the antibiotic chloramphenicol, which also inhibits protein production in prokaryotic cells but not in eukaryotic cells.

These observations have led biologists to wonder about the origins of chloroplasts and mitochondria. One theory, which is consistent with these observations, is that, in the evolution of life on Earth, prokaryotic cells might have entered the cytoplasm of eukaryotic cells and lived together for their mutual benefit. These prokaryotic cells are proposed to have evolved into the mitochondria and chloroplasts we see today. This is called the **endosymbiont theory**.

The cytoskeleton

The **cytosol** is the semi-fluid matrix which surrounds the cell organelles. It consists of a watery solution of ions, sugars and amino acids with suspended fatty acids, nucleotides and proteins.

The **cytosol** is not a formless disorganised blob of jelly. It is supported by a complex system of **protein filaments** which make up the **cytoskeleton**. This network of protein filaments gives the cell its shape and provides a basis for its movement. In protoctistan and animal cells, the cytoskeleton is thought to radiate from an area near the centrioles. Two sorts of cytoskeletal filaments are shown in Fig 3.22.

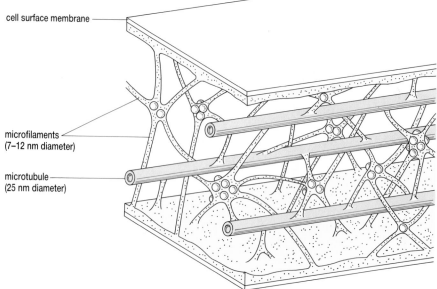

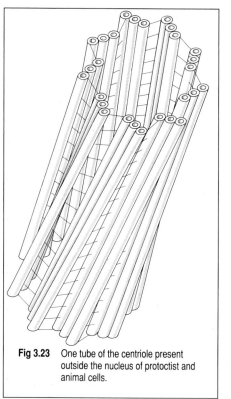

Fig 3.23 One tube of the centriole present outside the nucleus of protoctist and animal cells.

Fig 3.22 Cytoskeleton showing two types of fibres: microfilaments linked to elongated microtubules.

A **centriole** (Fig 3.23) is found in a particular region of the cytoplasm (the **centrosome**) just outside the nucleus of protoctistan and animal cells. It consists of two hollow tubes about 0.5 μm long and 0.2 μm diameter which lie at right angles to each other. The wall of each tube is made of nine triplets of microtubules. The centriole is involved in producing a fibrous network which pulls chromosomes apart during cell division.

QUESTION

3.7 (a) Make a large copy of Table 3.3. Complete it to show the function of each cell structure.
   (b) Which of these cell structures are (i) only found in plant cells (ii) seldom found in plant cells?

Table 3.3 Cell structures and functions

Region	Structure	Function
cell surface	cell wall	
	cell surface membrane	
cytoplasm	rough endoplasmic reticulum	
	smooth endoplasmic reticulum	
	ribosome	
	Golgi apparatus	
	mitochondrion	
	chloroplast	
	centriole	
	peroxisome	
	cytoskeleton	
nucleus	nuclear envelope	
	nuclear pore	
	nucleolus	

CELL STRUCTURE

Identifying and measuring organelles

This exercise is to give practice in identifying stuctures in photographs and in measuring and scaling microscopic structures.

Measure the diameter of the organelle shown in Fig 3.24(a).

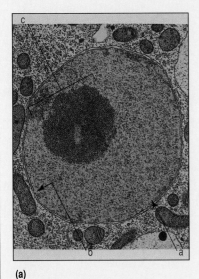

(a)

Fig 3.24

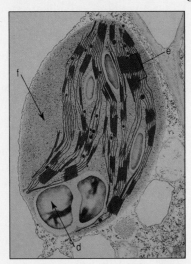

(b)

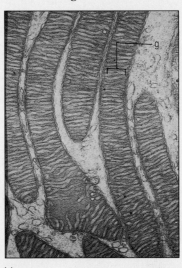

(c)

Depending on which diameter you measure you should find that it is between 41 and 44 mm (i.e. 41×10^{-3} and 44×10^{-3} m).

To find the real diameter, the measured length must be divided by the magnification of the photograph, in this case $\times 11\ 000$. The diameter of this organelle is about:

$$\frac{44 \times 10^{-3}\ m}{11\ 000} = 4 \times 10^{-6}\ m = 4\ \mu m$$

(a) Identify the organelle you have been measuring.
(b) Identify each of the objects labelled **a–g** in the photographs of Fig 3.24.
(c) Is the cell in Fig 3.24(a) eukaryotic or prokaryotic? Give reasons for your answer.
(d) Is the cell in photograph Fig 3.24(b) a plant or animal? Give reasons for your answer.
(e) Approximately how wide is the object labelled **h** in Fig 3.24(b)? The organelle in Fig 3.24(c) is magnified 30 000 times.

The organelles in Fig 3.24(c) are magnified 30 000 times
(f) Name the organelles in Fig 3.24(c).
(g) What is the mean real width of these organelles?

3.4 STUDYING THE FUNCTION OF CELL ORGANELLES

In reading Section 3.3, you may have wondered how the function of structures as small as organelles could ever have been determined. Whilst the use of most of the techniques are beyond the skills expected at the level associated with this textbook, an outline of some of them is given below so that you can appreciate the likely source of artefacts.

Cell fractionation

In order to separate organelles, cells have to be broken up by grinding them with a pestle and mortar or a Waring blender. The use of a more sophisticated tissue homogeniser, consisting of a ground glass pestle which precisely fits into a tube, helps to regulate the extent to which the cell organelles are broken up. The use of a saline or sugar solution whose water potential is the same as that of the cells also restricts damage to cell organelles. Cells can also be broken up using ultrasonic waves.

5 **During fractionation, cells and the fractionation solutions are usually kept very cold. Suggest why.**

The **homogenate** which is produced after the cells have been broken up contains a mixture of nuclei, fragments of membrane and cell organelles suspended in saline or sugar solution. These are too small to be easily separated by filtration, but can be separated by **centrifugation**. Tubes containing homogenate are placed inside a centrifuge and are spun at high speeds. As a result, the solid fragments form a pellet at the bottom of the tubes with a fluid **supernatant** lying above (Fig 3.25), which can be removed using a fine pipette. Table 3.4 shows how some of the major cell components can be separated. The type of bench centrifuge which is likely to be in a school or college laboratory can accelerate to about 1000 g (g denotes the acceleration due to gravity and has a value of 9.8 m s^{-2}). Once isolated, pure pellets can be resuspended in a saline or sugar solution and the single type of organelle inside it investigated using chemical analyses.

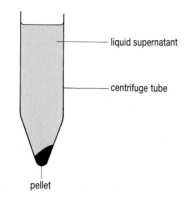

Fig 3.25 A mixture after centrifugation.

liquid supernatant

centrifuge tube

pellet

Table 3.4 Separation of cell components by centrifuging a cell homogenate

Centrifugation specification		Organelles in pellet
500–1 000 g	for 10 minutes	nuclei and chloroplasts
10 000–20 000 g	for 20 minutes	mitochondria and lysomes
100 000 g	for 60 minutes	rough endoplasmic reticulum and ribosomes

6 **How could the contents of a pellet be determined?**

Chromatography

Chromatography is a common technique used to separate mixtures of compounds. The particular type of chromatography most commonly used in biological research employs either absorbent paper or a dried, thin layer of powder on a glass or plastic base. Spots of a mixture are put at one end of the paper or thin layer, which is then dipped into a suitable solvent (Fig 3.26). The solvent moves through the paper or thin layer by capillarity until

Gas liquid chromatography (GLC) and high performance liquid chromatography (HPLC) are other chromatographic techniques important in biology.

Fig 3.26 Paper chromatography. Spots of mixture are placed on a pencil line (the origin) near the bottom of the chromatogram. This is then dipped in solvent inside a container whose atmosphere is saturated with solvent vapour and left undisturbed until the solvent has nearly reached the top.

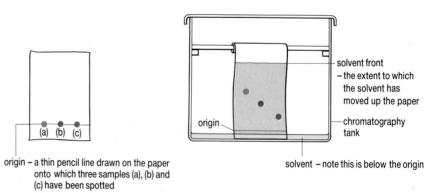

origin – a thin pencil line drawn on the paper onto which three samples (a), (b) and (c) have been spotted

origin

solvent front – the extent to which the solvent has moved up the paper

chromatography tank

solvent – note this is below the origin

(a) (b) (c)

it reaches the spot(s) of mixture. The compounds in the mixture dissolve in the solvent and are carried with the solvent as it soaks further through the chromatogram. Depending on its properties, each compound in the mixture will move at a different characteristic speed. When the chromatogram is removed and the solvent front is marked, the distance moved by each compound divided by the distance moved by the solvent is unique. Its value is called the R$_f$ value:

$$R_f \text{ value} = \frac{\text{distance moved by compound}}{\text{distance moved by solvent}}$$

 7 **Calculate the R_f values of the substances shown in Fig 3.26.**

Radioactive labels

The sequence of many biochemical pathways has been discovered by labelling certain of the chemicals involved and following the path of the radioactivity. Carbon is a key element in biology. Normal carbon atoms have a **mass number** (their total number of protons and neutrons, see Section 1.1) of twelve and are denoted ^{12}C. A different form of carbon, ^{14}C, is radioactive. By introducing organic compounds containing some ^{14}C atoms, the uses of these compounds and how they are changed by cell organelles can be studied.

Some other isotopes used in biology include ^{15}N (not radioactive) ^{32}P (radioactive) and ^{35}S (radioactive).

 8 **For each of these isotopes, suggest one biological molecule you could label with it.**

Autoradiography

Radioactive substances give off sub-atomic particles which affect photographic film. If a mixture of radioactive and non-radioactive substances is left against a photographic plate in the dark for two days, the position of the radioactive substances will be shown by bright spots on the developed film. This technique, known as autoradiography, can be used to find the whereabouts of radioactively labelled substances in dried, pressed organisms, or in sections through organs. Used in conjunction with chromatography, it can be used to identify chemicals in the middle of a series of reactions which contain a radioactive atom.

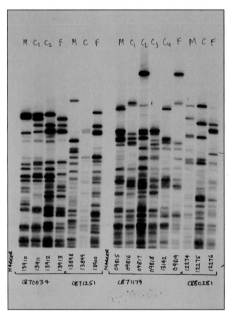

Fig 3.27 **(a)** *(left)* An autoradiograph of samples of DNA which was used for establishing family relationships. M indicates mother, C = child, F = father. Some bands will be identical in closely related people.

(b) *(right)* Technicians analysing autoradiographs of DNA. They are each comparing the banding patterns in 2 autoradiographs to see how similar they are. If the patterns are identical then both samples come from the same individual, for the pattern of banding is unique to an individual.

Following a biochemical pathway

This exercise involves the interpretation of experimental data.

Imagine you are studying the breakdown (catabolism) of glucose in the cytosol. From the work of others, you think that the metabolic pathway is:

glucose → X → Y → pyruvate (a 3-carbon compound)

(a) Explain how, using a radioactive isotope and autoradiography, you could identify X and Y.

After further research, you have developed a hypothesis that the catabolism of glucose occurs in two stages.

- glucose → pyruvate (occurs only in the cytosol)
- pyruvate + oxygen → carbon dioxide and water (occurs in mitochondria)

(b) Which of the following results support this hypothesis? Explain your answer.
  (i) Pyruvate labelled with ^{14}C yielded $^{14}CO_2$ when added to an oxygenated homogenate of liver cells.
  (ii) Glucose labelled with ^{14}C yielded $^{14}CO_2$ when added to an oxygenated homogenate of liver cells.
  (iii) Pyruvate labelled with ^{14}C when added to an oxygenated cytosol-free suspension of mitochondria yielded $^{14}CO_2$.
  (iv) Glucose labelled with ^{14}C yielded ^{14}pyruvate when added to a liver cell homogenate from which oxygen had been excluded.

3.8 An animal cell homogenate is centrifuged at 10 000–15 000 g for 20 minutes and the pellet which is formed discarded.
  (a) What cell components would be in the discarded pellet?
  (b) What cell component must the scientist be interested in studying?

3.9 The sequence below shows the location of radioactively labelled molecules of amino acid, following their introduction into a cell. Use the information in Sections 3.3 and 3.4 to explain this sequence.
  Sequence: cytosol → ribosomes → sacs of rough ER

The modern cell theory can be summarised: cells are the structural and functional units of all living organisms; all cells arise from pre-existing cells. Cells are of two major types: prokaryotic cells occur exclusively in the kingdom Prokaryotae; eukaryotic cells occur in all other kingdoms. Only eukaryotic cells are described in this chapter.

Cells can be investigated using optical microscopes and electron microscopes. Because electrons have a shorter wavelength than light, the electron microscope allows better resolution of structures within cells. In either case, tissues must be prepared for examination. Such preparation involves the fixation, dehydration, embedding, sectioning and staining of tissues. Each of these processes is likely to distort the original tissue, producing artefacts. The use of electron microscopes has the added disadvantages that tissues must be placed in a vacuum and bombarded by electrons, which may cause further artefacts.

The appearance of cells using an optical microscope is termed cell structure. All eukaryotic cells have a cell surface membrane, nucleus and cytoplasm. Cells of plants, fungi and some protoctists also have a rigid cell wall which is secreted by the surface membrane. Plant cells also contain central, fluid-filled vacuoles, as do many other cells. During differentiation the cells of multicellular organisms become specialised for one function or group of functions. Groups of similar specialised cells are arranged into tissues, tissues into organs and organs into systems.

The appearance of cells using an electron microscope is termed cell ultrastructure. Electron microscopy reveals that the nucleus of a eukaryotic cell is enclosed within a double membrane system, called the nuclear envelope, which separates it from the cytoplasm. Pores in the nuclear envelope regulate the passage of molecules between nucleus and cytoplasm. Inside the nuclear envelope is semi-fluid nucleoplasm which surrounds the chromosomes. One or more stores of ribonucleic acid, called nucleoli, may also be present.

The cytoplasm contains semi-fluid cytosol, which is supported by a network of proteinaceous microtubules and microfilaments, called the cytoskeleton. Within the cytoplasm are a number of organelles. Many of these are themselves membranous structures: mitochondria synthesise ATP, using energy released during cell respiration; endoplasmic reticulum synthesises and transports materials through the cytoplasm; Golgi apparatus stores, chemically modifies and secretes a number of molecules; chloroplasts in the cells of plants and some protoctists carry out photosynthesis; centrioles in the cells of animals and some protoctists produce a proteinaceous spindle which separates chromosomes during cell division; lysosomes are small vacuoles containing proteolytic enzymes. Ribosomes are not membranous organelles. They may be found attached to endoplasmic reticulum or free in the cytosol and manufacture proteins from amino acids.

Fungi, plants and some protoctists secrete a cell wall which consists of fibres surrounded by a matrix.

Cells can be studied by breaking them up (cell fractionation) and separating their constituent parts using a centrifuge. Chromatography and autoradiography are techniques which help to reveal the function of cell organelles.

Answers to Quick Questions: Chapter 3

1 Gas molecules in the air would deflect/absorb electrons.
2 The part that takes up the salt appears black.
3 Because of the increased resolution of the electron microscope.
4 The nucleus provides the information needed to assemble insulin, the ribosomes make insulin using energy supplied by the mitochondria; insulin is then transported through the ER to the Golgi apparatus where it is chemically modified before being secreted.
5 To slow down rates of chemical reactions, including decomposition.
6 Look at a sample of it under the electron microscope.
7 0.55, 0.38, 0.22 (Measure from the centre of the dot.)
8 ^{15}N label proteins and nucleic acids; ^{32}P label nucleic acids and phospholipids; ^{35}S label proteins.

Chapter 4

CELL CONTROL AND CELL DIVISION

LEARNING OBJECTIVES

When you have studied this chapter you should be able to:

1. describe the process of protein synthesis;

2. understand how genes are thought to be controlled;

3. describe DNA replication;

4. summarise the nuclear events which occur during the cell cycle;

5. interpret photographs of mitosis and meiosis;

6. summarise the major differences between mitosis and meiosis.

Cells need to work to stay alive. A cell's work involves chemical reactions, collectively called metabolism. Enzymes have a central role in metabolism: without them chemical reactions would not occur quickly enough to keep cells alive. The vast majority of enzymes are proteins (see Chapter 5). Like all other proteins, enzymes are made from amino acids joined together in exactly the right order. If the order is wrong an enzyme may not work efficiently, if at all. Producing proteins with the correct amino acid sequence is therefore essential to the success of any cell. This chapter examines how cells do this.

Fig 4.1 To build cars, these robots need information. To build proteins, cells need information contained in the DNA in the nucleus.

DNA carries the genetic information needed to make proteins.

Before examining how cells produce proteins, think of a car factory. In this factory, the products (cars) are assembled correctly (by humans and by robots – see Fig 4.1) according to information from the design engineers. This factory model can be used to describe cells. A cell's products (proteins) are assembled (by ribosomes) in the right order using plans in its **genetic information**. This genetic information is supplied by DNA. Understanding the whole process in cells involves knowing

- how the genetic information is transferred from the DNA to the ribosomes;
- how the flow of genetic information to ribosomes is controlled so that cells only make the proteins they need and not all possible proteins (for example, cells lining your gut produce digestive enzymes but your brain cells do not, even though both types of cell contain exactly the same genetic information);
- how the genetic information is passed from parent to daughter cells during the process of cell division;
- how the genetic information is passed from parents to offspring during the process of reproduction.

ANALYSIS

How do we know DNA carries the genetic information?

This exercise requires you to interpret experimental results.

For many years it was thought that such an apparently simple molecule as DNA could not possibly encode the information needed to make all the different proteins found in a cell. However, an experiment, performed by Frank Hershey and Martha Chase using a bacterial virus, bacteriophage T2 (phage for short), demonstrated that DNA does indeed carry the genetic information.

A phage is a very simple structure consisting of DNA contained within a protein head (Fig 4.2). During their life cycle (Fig 4.3) the phages inject their DNA into the host bacterial cell, in this case the gut bacterium *Escherichia coli*, but the protein heads (called **phage ghosts**) remain on the outside of the bacterium.

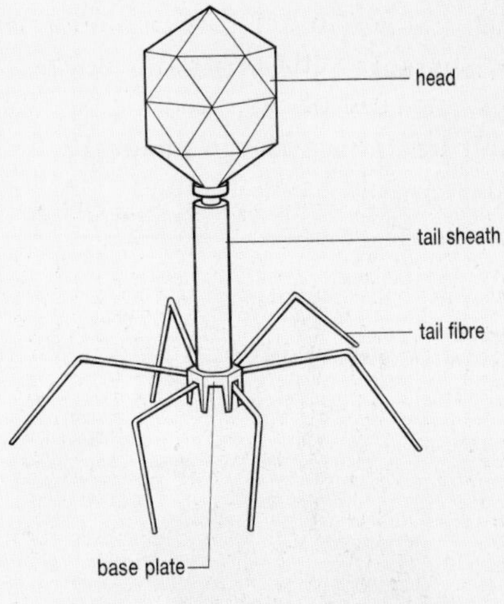

Hershey and Chase's experiment rests on the fact that DNA contains phosphorus whereas protein does not, and that protein contains sulphur whereas DNA does not. They developed two different sorts of phage. Type A phage had DNA which included the radioactive isotope of phosphorus, ^{32}P: type B phage had proteins which included the radioactive isotope of sulphur, ^{35}S.

Fig 4.2 The anatomy of T2 phage – a bacterial virus. All the structures visible are made of protein. The phage DNA is inside the head. The tail fibres are used to 'lock on' to the cell surface membrane of the bacterial host, in this case *Escherichia coli*. The tail sheath then contracts, injecting the phage DNA into the cell.

Separate cultures of *E. coli* were infected with either type A or type B phage. After sufficient time for DNA injection had passed, the empty phage heads (called phage ghosts) were separated from the bacterial cells by agitating the culture in a blender. The bacterial cells were then isolated from the phage ghosts by centrifugation and the two fractions were tested for radioactivity. In a different experiment, the phages were then allowed to complete their life cycle and the offspring (progeny) phages were tested for radioactivity. The results of both experiments are summarised below.

Phage type (label)	Bacterial fraction	Phage ghost fraction	Phage progeny
A (^{32}P)	radioactive	non-radioactive	radioactive
B (^{35}S)	non-radioactive	radioactive	non-radioactive

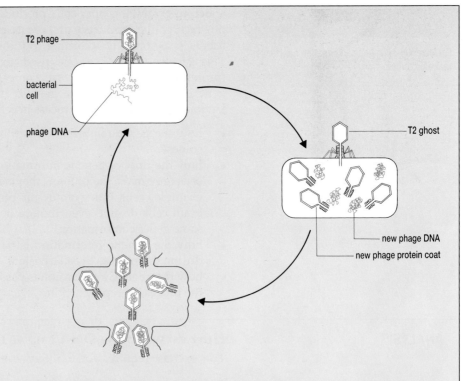

Fig 4.3 The life cycle of T2 phage. Notice how the protein coat (ghost) remains on the outside of the bacterial cell. Once the phage DNA has been injected, it takes over the protein-synthesising machinery of the bacterial cell, 'forcing' it to produce new virus particles. The complete life cycle takes about 40 minutes to complete under ideal conditions.

(a) What do these results suggest had happened to the DNA in type A phages?

(b) Why were the progeny of type A phages radioactive whilst the progeny of type B phages were not?

(c) Why do these results support the view that DNA carries the genetic information?

(d) If protein and not DNA were the information carrier in phage T2, predict the results Hershey and Chase would have obtained.

4.1 PROTEIN SYNTHESIS AND THE GENETIC CODE

Look at Fig 4.4 which summarises the process of protein synthesis in a eukaryotic cell.

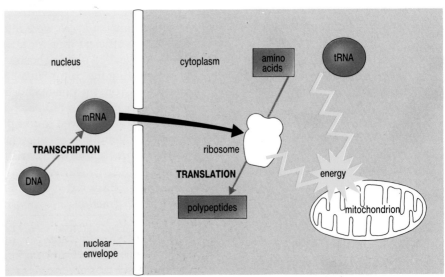

Fig 4.4 Protein synthesis. Notice how energy, from ATP hydrolysis, is needed to join tRNA to amino acids in the cytoplasm and to form peptide bonds during the process of translation.

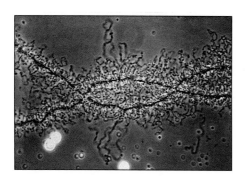

Fig 4.5 Transcription is occurring on the loops projecting from this lampbrush chromosome.

Note the following points.

- DNA molecules are too large to fit through the pores in the nuclear envelope.
- Part of the information carried by the DNA is copied (the process of **transcription**) into smaller messenger RNA (mRNA) molecules which can pass through the nuclear pores to the ribosomes.
- Ribosomes interact with mRNA and amino-acid-carrying transfer RNA (tRNA) molecules to translate the information in the mRNA molecule into a polypeptide with the correct amino acid sequence, the process of **translation**.

The details of transcription and translation are dealt with in Section 4.2. First we shall consider the nature of the code which carries the genetic information.

The genetic code

The genetic code is a code for the amino-acid sequence of proteins. The code is stored in the molecular structure of DNA and copied into the molecular structure of mRNA during the process of transcription.

The structure of DNA and RNA is described in Section 1.7. Both are polymers (long chains) of **nucleotides**. Each nucleotide has a **pentose** sugar, inorganic **phosphate** group and organic **base**. A molecule of RNA has two types of purine base, **adenine** and **guanine**, and two types of pyrimidine base, **cytosine** and **uracil**. The sequence of purine and pyrimidine bases along a mRNA molecule determines the sequence of amino acids in the protein molecules which ribosomes make from that RNA. Table 4.1 shows just how this is done. A combination of three bases (a **base triplet**) encodes the inclusion of one specific amino acid in a protein. This base triplet is called a **codon**.

Table 4.1   The mRNA genetic codes for amino acids

First base	Second base				Third base
	G	**A**	**C**	**U**	
G	GGG glycine GGA glycine GGC glycine GGU glycine	GAG glutamic acid GAA glutamic acid GAC aspartic acid GAU aspartic acid	GCG alanine GCA alanine GCC alanine GCU alanine	GUG valine GUA valine GUC valine GUU valine	G A C U
A	AGG arginine AGA arginine AGC serine AGU serine	AAG lysine AAA lysine AAC asparagine AAU asparagine	ACG threonine ACA threonine ACC threonine ACU threonine	AUG **start** AUA isoleucine AUC isoleucine AUU isoleucine	G A C U
C	CGG arginine CGA arginine CGC arginine CGU arginine	CAG glutamine CAA glutamine CAC histidine CAU histidine	CCG proline CCA proline CCC proline CCU proline	CUG leucine CUA leucine CUC leucine CUU leucine	G A C U
U	UGG tryptophan UGA **stop** UGC cysteine UGU cysteine	UAG **stop** UAA **stop** UAC tyrosine UAU tyrosine	UCG serine UCA serine UCC serine UCU serine	UUG leucine UUA leucine UUC phenylalanine UUU phenylalanine	G A C U

The left-hand column of Table 4.1 represents the first letter of the code, e.g. all the codes in the top rows begin with the base guanine. The four central columns of the table represent the second base in the code, e.g. all the codes in the top left begin with guanine and have guanine as their second base; those in the next column have guanine as their first base and cytosine as their second. The third base in each codon is shown by the extreme right-hand column. The base codon cytosine-cytosine-cytosine is found by looking in the third row down on the left-hand side, third column along, third line within that box. The name of the amino acid for which that combination of bases encodes is proline.

Proline is also encoded by the codon cytosine-cytosine-uracil. A code like this, with more than one sequence of three bases coding for a particular amino acid, is said to be **degenerate.** Look through Table 4.1; many other amino acids are encoded by more than one sequence of three bases.

To make any sense, a code has to be read from a particular starting point. The letters:

NOWDIDYOUFIXTHEOLDCARENDNOWHOWFARCANTHEBOYRUNEND

only make sense if they are split into three-letter words. With start instructions (now) and stop instructions (end), they make sense:
(now)DID YOU FIX THE OLD CAR(end).
(now)HOW FAR CAN THE BOY RUN (end).

Four of the codons in Table 4.1, AUG, UGA, UAG and UAA, do not carry a code for an amino acid. Instead, they code a reading instruction for ribosomes.

SPOTLIGHT

WHY A TRIPLET CODE?

Consider the following problem. There are twenty amino acids commonly found in polypeptides but only four bases in DNA to encode the sequence of those amino acids in a polypeptide. What, then, is the minimum number of bases which could encode each amino acid?

If only one base were used per amino acid the DNA could only specify the position of four amino acids. Clearly, this will not do. If there were two-base (doublet) codes, how many different amino acids would this encode for? There are 16 (4^2) ways of arranging four bases in a doublet code but 20 amino acids have to be encoded. Again, this would be no use. With a triplet code there are $4^3 = 64$ different combinations, more than enough to encode the 20 amino acids found in proteins. In fact the code now has too many combinations, i.e. it is degenerate or redundant, so that some amino acids are encoded by more than one triplet.

QUESTIONS

4.1 (a) Why is it essential to specify so accurately the primary structure of a protein?
(b) What is a codon?
(c) What chain of amino acids would be encoded by a portion of mRNA with the base sequence AUGAAUCGAUUCGGAUCUGGC?

4.2 If you remove the nucleus from a single-celled marine organism called *Acetabularia* the resulting anucleate cell continues to synthesise proteins in the cytoplasm for at least two weeks. Why does this experiment suggest that DNA is only indirectly involved in protein synthesis?

4.2 PROTEIN SYNTHESIS

The information contained in the sequence of bases in a gene determines the amino acid sequence of the protein. However, the information contained in the DNA molecule is not used directly by the cell's protein synthesising mechanism. Rather, it is expressed indirectly via other molecules. The genetic information in the DNA molecule within the

CELL CONTROL AND CELL DIVISION

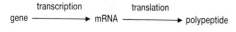

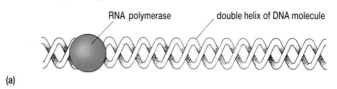

Fig 4.6 The flow of information during protein synthesis.

nucleus is first copied into a molecule of messenger RNA (mRNA) in the process of **transcription** (Fig 4.6). The mRNA molecule, which usually carries the information encoded by one gene, then moves out of the nucleus to the cytoplasm. Here the information it carries is used to direct the synthesis of a polypeptide by a ribosome in the process of **translation** (see Fig 4.6).

Transcription

This process, controlled by the enzyme **RNA polymerase**, can be divided into three stages (Fig 4.7).

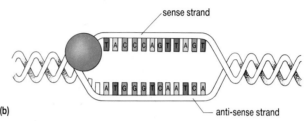

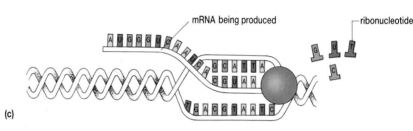

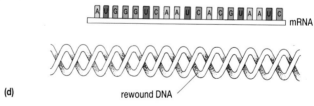

Fig 4.7 The process of transcription. Genetic information in DNA is transferred to mRNA aided by RNA polymerase.

KEY TERM

A gene is the portion of nucleotides in a DNA molecule which carries the genetic code for the production of functional protein.

Table 4.2 The specific base pairs which exist throughout a molecule of DNA. The dashed lines represent hydrogen bonds between bases: it is these bonds which hold the two strands of a DNA molecule together.

Base name	Structure of base pair	Base name
adenine		thymine
	1.11 nm	
cytosine		guanine
	1.08 nm	

Stage 1

RNA polymerase binds to a region of the DNA, the promoter, near the beginning of the gene which is going to be transcribed (Fig 4.7(a)). This causes the DNA double helix to unwind (Fig 4.7(b)).

Stage 2

One of the exposed strands of DNA, called the **sense strand**, is used as a template for mRNA production (Fig 4.7(c)). As RNA polymerase moves along the sense strand, ribonucleotides are taken from the nucleoplasm and matched up in the precise way shown in Table 4.2.

Stage 3

When the RNA polymerase reaches the end of the gene, it releases the fully formed mRNA and the DNA rewinds (Fig 4.7(d)).

Essentially, transcription involves transferring information stored in one sequence of nucleotide bases in DNA to another sequence of nucleotide bases in mRNA.

 What will be the sequence of bases encoded by a mRNA molecule transcribed from the following DNA sequence – GTAACGATC?

Ribosomes and transfer RNA (tRNA)

The mRNA molecules leave the nucleus via the nuclear pores and enter the cytoplasm. During translation, ribosomes become loosely attached to the mRNA and pass along the length of the molecule forming a polypeptide. The sequence of amino acids in the polypeptide will be determined by the triplet code contained in the mRNA molecule. Ribosomes consist of two subunits, each made of ribosomal RNA (rRNA) and protein (Fig 4.8). The smaller of the two subunits has an information processing region which is just big enough to hold two codons of mRNA.

 How many bases will there be in two codons? How many amino acids will this number of bases code for?

Transfer RNA (tRNA) molecules (Fig 4.9) form the final part of the protein synthesising mechanism. Each tRNA molecule is a single-stranded polymer of ribonucleotides held in shape by hydrogen bonds between base pairs. Three bases form the **anticodon** (more about this later). At the 3' end of the molecule there is a region with the three bases CCA which bind to a specific amino acid.

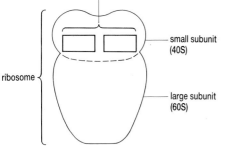

Fig 4.8 The structure of a ribosome. Two codons of mRNA fit into the information processing region of the ribosome.

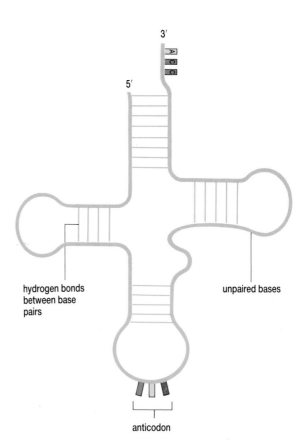

Fig 4.9 Structure of a transfer RNA molecule.

 Given that there are 20 commonly occurring amino acids found in polypeptides, what is the minimum number of different tRNA molecules which must be present in the cytoplasm? Why, in fact, will there be more than this minimum number?

Translation

For convenience, this process can also be divided into a number of stages (Fig 4.10).

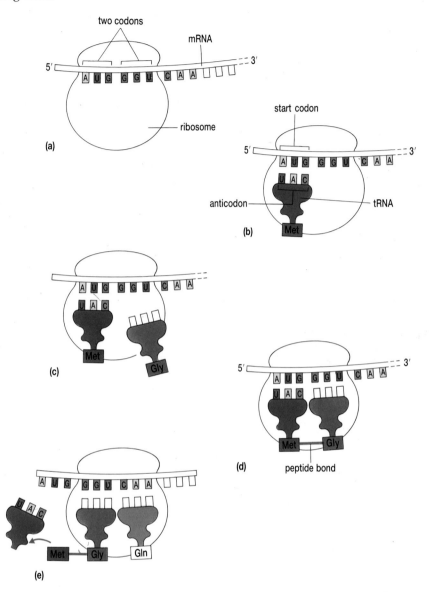

Fig 4.10 The process of translation occurs in five stages. A chain of a protein molecule forms as amino acids are bonded together in a ribosome.

Stage 1

Translation begins when the 5′ end of the mRNA molecule with the start codon (AUG) binds to the information processing region of the ribosome (Fig 4.10(a)).

Stage 2

The tRNA with the complementary anticodon UAC binds to the start codon (AUG) held in the information processing region. The tRNA is held in place by the larger subunit of the ribosome. Note that this particular tRNA molecule always carries the amino acid methionine (Fig 4.10(b)).

Stage 3

A second tRNA molecule now binds to the second mRNA codon in the information processing region (Fig 4.10(c)).

What will the sequence of bases in the anticodon of this tRNA molecule be? Which amino acid will it be carrying?

Stage 4

A peptide bond is now formed between the two amino acids (Fig 4.10(d)). This condensation reaction involves a catalytic site on the larger subunit of the ribosome and ATP hydrolysis. Catalysts and ATP are discussed in the next chapter.

Stage 5

The ribosome now moves one codon along the mRNA molecule. The first tRNA molecule, the one which was carrying methionine, is released back into the cytoplasm. The information processing region now contains a tRNA molecule attached to a dipeptide in its first site and a new mRNA codon in the second site (Fig 4.10(e)).

 What will happen to the first tRNA released from the ribosome? Predict what will happen next to form a tripeptide. Which amino acid will be added?

This process of translation is repeated again and again. At each step, a further amino acid is added to the polypeptide chain. Eventually, the information processing region will encounter a stop or terminate codon. This stops polypeptide synthesis and the complete polypeptide is released from the ribosome.

Both the ribosome and the mRNA molecule are re-usable: each mRNA molecule will usually direct the synthesis of many molecules of a particular polypeptide. However, it is not immortal and eventually the mRNA will be degraded by enzymes.

 Predict what happens to the ribonucleotides produced when the mRNA is broken down?

The same mRNA molecule is usually read by many ribosomes at the same time. Such a group of working ribosomes attached to a single mRNA molecule is called a **polysome** (Fig 4.11).

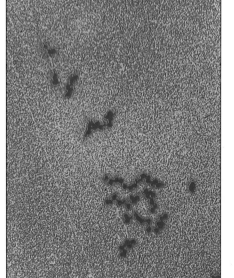

Fig 4.11 Electronmicrograph showing DNA, mRNA and polysomes in *Escherichia coli.* The DNA fibres are the yellow lines running from the top left with numerous ribosomes (in red) attached to each mRNA chain.

SPOTLIGHT

HOW FAST ARE TRANSLATION AND TRANSCRIPTION

The rate of translation is staggering. In one second, a single bacterial ribosome adds 20 amino acids to a growing polypeptide chain. Each second, your body synthesises 5×10^{14} haemoglobin molecules. High rates of transcription and translation ensure that large numbers of protein molecules can be made using the information contained in one gene. For example, consider the cells in the silk glands of the larvae of the silk moth. The protein fibroin is the major component of silk. In each silk gland cell, a single fibroin gene makes 10^4 copies of its mRNA during the four days of silk synthesis. Each mRNA molecule directs the synthesis of 10^5 molecules of fibroin – producing a total of 10^9 molecules of fibroin per cell in a period of four days.

CELL CONTROL AND CELL DIVISION

4.3 Read the following statements. Decide which are true or false, giving brief reasons for your answer.
   (a) In eukaryotic cells, transcription takes place in the nucleus.
   (b) mRNA, tRNA and ribosomes are all essential in the translation of genetic information.
   (c) During translation, amino acids attach directly to their respective codons on the mRNA.
   (d) mRNA molecules can be used many times to direct the synthesis of a polypeptide.
   (e) During transcription, ribosomes attach themselves to the DNA sequence being transcribed.
   (f) Translation within eukaryotic cells occurs in the nucleus.

4.4 Imagine a polypeptide containing 20 amino acids being synthesised by a ribosome using the information contained in a mRNA molecule.
   (a) How many mRNA bases code for the polypeptide?
   (b) How many bases in the coding strand of the DNA molecule would be needed to produce the mRNA molecule?
   (c) Hydrolysis of the polypeptide reveals that it contains only 12 of the possible 20 amino acids commonly found in polypeptides. What is the minimum number of different tRNA molecules needed to synthesise the polypeptide?

4.5 Copy Table 4.3 and complete each blank box using the information contained in Section 4.2.

Table 4.3

DNA double helix	non-coding strand					
	coding strand			TGT		
transcribed mRNA			AGC			
tRNA anticodon			GUA			UCC
amino acid incorporated into the polypeptide		Tyr				Met

4.3 SPLIT GENES

In eukaryotes, genes contain numerous sections of DNA which carry a 'meaningless' code, called **introns**. These introns may be up to 100 000 nucleotides long and they separate the sections of DNA carrying the 'sense' code for amino acid sequence, called **exons**. For example, the gene which encodes the egg white protein, ovalbumen, has eight exons and seven intervening introns.

When a single gene is transcribed, the whole of the DNA sequence is used, including the introns. This wholesale transcription produces a molecule of RNA which is full of 'genetic nonsense'. This RNA molecule is called messenger RNA precursor, or **pre-mRNA**.

 Why could this pre-mRNA not be used for protein synthesis?

The nucleus contains a number of complexes of RNA and a variety of proteins, called **snRNPs** (pronounced snurps and standing for **small nuclear ribonucleoproteins**). Some of the snRNPs edit newly formed pre-mRNA. They splice the pre-mRNA, removing the introns and reconnecting the exons in the same order that they occurred in the DNA template. The edited molecule is now mRNA proper and is the molecule which migrates through the nuclear pores to the cytoplasm.

Further modification may also be needed after translation to produce the functional protein, as shown in Fig 4.12 in the synthesis of insulin.

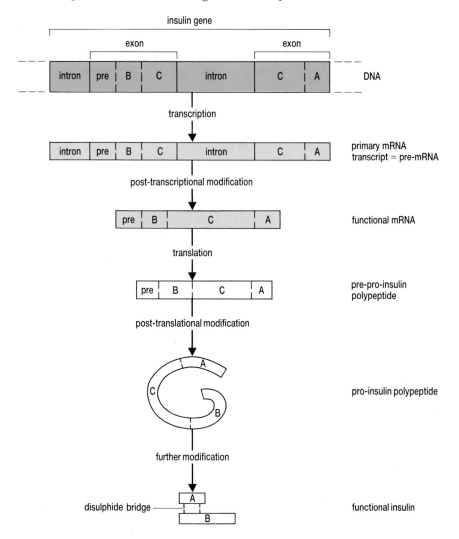

Fig 4.12 Steps in the synthesis of insulin.

4.4 GENE REGULATION

Knowing how the information contained in genes is translated into the primary structure of a protein does not fully tell us how protein synthesis is controlled. Returning to the specialised cells in Fig 3.8, the plant cells all have the same DNA in their chromosomes but do not all produce the same proteins. The cells in your stomach have the same DNA in their nuclei as those in your thyroid gland, yet some cells of the stomach produce hydrochloric acid and others a protein-digesting enzyme whereas the cells of the thyroid produce the hormone thyroxine. To take another example, a moth has the same DNA in its cells whether it is a larva, a pupa or an adult but the three stages in its life cycle are completely different.

We must conclude that a nucleus can switch on and off the transcription and translation of its genes.

Gene regulation in prokaryotes

Prokaryotic organisms have one circular chromosome with no nuclear envelope separating it from the cytoplasm. Their chromosome is made of DNA but, unlike eukaryotic cells, contains no histones. In these circular chromosomes the genes controlling related functions are located together to form **operons**.

The intestinal bacterium *Escherichia coli* is able to absorb the disaccharide lactose and break it down into its component monosaccharides, glucose and galactose. To do this the bacterium needs to make two enzymes: one, called **lactose permease**, speeds up the absorption of lactose from the growth medium; the second, called β-**galactosidase**, speeds up the breakdown of lactose. Normally, *E. coli* does not make either of these enzymes: they are made only if the bacterium is growing on a food containing lactose. The production of the enzymes lactose permease and β-galactosidase is controlled by two genes. Fig 4.13 shows a model to suggest how transcription of these genes might be controlled. It is known as the Jacob and Monod model, after the two Frenchmen who proposed it.

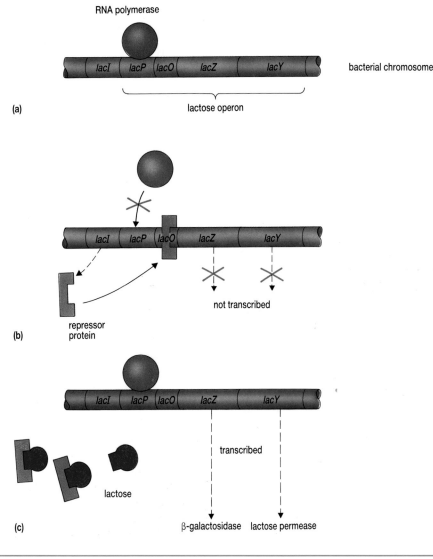

Fig 4.13 Operon model of gene regulation in bacteria.

CELL CONTROL AND CELL DIVISION

8 Why is it advantageous for a cell of *E. coli* to produce lactose permease and β-galactosidase only when it meets a food source containing lactose?

Look at Fig 4.13(a) which shows how the gene controlling production of lactose permease (called *lacY*) and the gene controlling production of β-galactosidase (called *lacZ*) are thought to lie next to each other on the chromosome of *E. coli*. Next to these structural genes are two overlapping regulatory regions of the chromosome, the **promoter** (*lacP*) and the **operator** (*lacO*). These four regions, *lacY, lacZ, lacP* and *lacO* form the **lactose operon**. A third regulatory gene, the **regulator** (*lacI*), is closely linked to the lactose operon but is not part of it.

Before the lactose permease and β-galactosidase genes can be transcribed, the enzyme RNA polymerase must bind to the promoter region of the lactose operon (Fig 4.13(a)). Jacob and Monod proposed that this transcription does not normally happen because RNA polymerase is prevented from binding with the promoter. It is prevented from binding by the product of the regulator, a protein called the **repressor**, which binds with the adjacent operator region (Fig 4.13(b)). If lactose is present, it combines with the repressor (Fig 4.13(c)). As a result, RNA polymerase can bind with promoter so that the lactose permease and β-galactosidase genes are transcribed.

Gene regulation in eukaryotes

In eukaryotic organisms, genes for related functions are often found on different chromosomes, so the simple operon model proposed for bacteria cannot be applied. Some of the features of gene control are thought to be similar, though. Although eukaryotic genes do not have an operator region near them, they do have a promoter region which is recognised by RNA polymerase as the place to start transcription of DNA. Near to this promoter is a region of chromosome called the **transcriptional enhancer** which controls access of RNA polymerase to the promoter region. In some way which is still being investigated, changes in the transcriptional enhancer allow or prevent DNA transcription.

In most bird species, females produce eggs only in a breeding season. The eggs contain stores of food for the growth of the embryo chick which are made in cells lining the females' oviducts. Among these food stores is the protein **ovalbumen**. During the breeding season, oestrogen is released by the ovaries and travels in the bloodstream to the oviducts. In the cytoplasm of the cells lining the oviducts this oestrogen combines with receptor proteins to form an oestrogen-protein complex. This complex travels into the nucleus and combines with the transcriptional enhancer of the ovalbumen gene (Fig 4.14). As a result, RNA polymerase has access to the promoter region of the ovalbumen gene, the gene is transcribed and ovalbumen is produced.

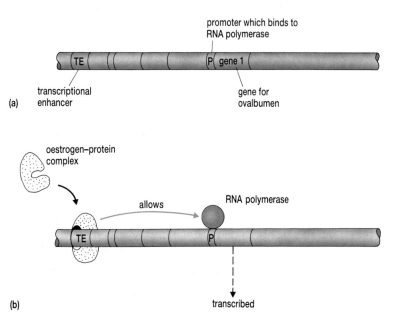

promoter which binds to
RNA polymerase

(a)

transcriptional
enhancer

gene for
ovalbumen

oestrogen–protein
complex

allows

RNA polymerase

(b)

transcribed

Fig 4.14 Model of regulation of the ovalbumen gene in birds (see text for further details).

QUESTIONS

4.7 Imagine that, on a planet in a distant galaxy, evolution has resulted in organisms which share the same basic biochemistry as humans. However, the organisms have proteins built out of 30 rather than 20 amino acids and they have eight rather than four different bases in their DNA. (Otherwise transcription and translation proceed in an identical fashion to that described in Section 4.1).

In theory, for these extra-galactic organisms, what would be
(a) the minimum number of bases per DNA codon
(b) the minimum number of kinds of tRNA
(c) the maximum number of kinds of tRNA (assuming that two codons are stop codons and that codons consist of the minimum number of bases).

4.8 How is the DNA of a bacterial cell different from that of an animal or plant cell?

4.9 List the genes which are present in the β-galactosidase operon of *E. coli* and summarise the suggested role of each.

4.10 Compare and contrast the models of gene regulation in bacterial and eukaryotic cells.

4.5 DNA REPLICATION

In common with most organisms, humans start off life as a single cell – a fertilised egg. During development, that single cell divides to produce billions of cells. Such a process raises many interesting biological questions.

- How does a cell know when to divide?
- How does a cell know which type of cell to become?
- How is the genetic information contained in the original fertilised egg transmitted to each of the daughter cells?
- How was the genetic information contained in the parents' cells passed to the offspring?

These last two aspects of cell division will be discussed in the rest of this chapter.

Semi-conservative replication

We have already considered two vital features about DNA structure.

1. DNA is a double helix in which each strand is composed of many nucleotides joined together by covalent bonds.
2. Adenine (A) in one strand always pairs with thymine (T) in the other strand and guanine (G) always pairs with cytosine (C). This complementary base pairing holds the key to accurate DNA copying.

 What type of bonds join the bases in one strand with the bases in the other strand? Are such bonds easy to break? Are they weaker or stronger than the bonds which hold the nucleotides in each strand together?

This structure suggests a model, called semi-conservative replication, by which DNA can be copied. The model is summarised in Fig 4.15. Note the following points.

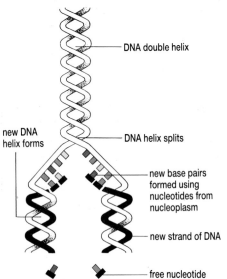

DNA double helix

new DNA helix forms

DNA helix splits

new base pairs formed using nucleotides from nucleoplasm

new strand of DNA

free nucleotide

- Free nucleotides are manufactured in the cytoplasm and are present in the nucleoplasm prior to replication starting.
- The DNA helix unwinds and the hydrogen bonds holding the two DNA strands together break. This leaves unpaired bases exposed on each strand.
- The sequence of unpaired bases serves as a template on which to arrange the free nucleotides from the nucleoplasm.
- The enzyme DNA polymerase moves along the unwound parts of the DNA, pairing complementary nucleotides from the nucleoplasm with each exposed base.
- The same enzyme connects the nucleotides together to form a new strand of DNA hydrogen bonded to the old strand.

ANALYSIS

The evidence for semi-conservative replication

This exercise involves using your understanding of DNA replication to interpret experimental results.

The model of replication shown in Fig 4.15 seems reasonable but how do we know what is actually taking place? After all, the process cannot actually be seen. Indeed two other models of DNA replication (Fig 4.16) could also be suggested. A classic experiment, by Matthew Meselsohn and Frank Stahl, will allow you to differentiate between the three models.

The key to this experiment is the different properties of the two isotopes of nitrogen, ^{14}N and ^{15}N. We can distinguish the isotopes on the basis of their mass: ^{14}N is lighter than ^{15}N; neither is radioactive. The experiment is described below.

Bacteria are grown in a medium where their nitrogen source, used to synthesise nucleotides and eventually DNA, is the ammonium ion (NH_4^+) containing heavy ^{15}N. After several generations, all the DNA in these bacteria is heavy – shown in black.
(a) Where are the N atoms in the DNA?

Some of the parental bacteria are now placed in a fresh medium containing light (^{14}N) ammonium ions and are allowed to divide, i.e. copy their DNA, just once, so producing generation 1.

CELL CONTROL AND CELL DIVISION

(b) If DNA replication is semi-conservative, all the generation 1 bacteria are expected to have hybrid DNA with one heavy strand and one light strand. Explain why.

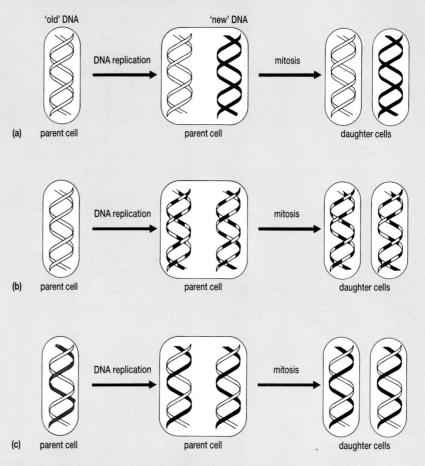

Fig 4.16 **(a)** Conservative model of DNA replication. The two new DNA strands occur together in one daughter cell, the two old in the other.
(b) Dispersive model of DNA replication. New and old DNA occur together in both daughter cells. The old DNA is distributed amongst all strands in the daughter cell.
(c) Semi-conservative model of DNA replication. New and old DNA occur together in both daughter cells but old DNA and new DNA each form their own strand.

Some generation 1 bacteria are now allowed to divide just once in ^{14}N medium, so producing generation 2 bacteria.

(c) If DNA replication is semi-conservative, two types of bacterial DNA are now expected to be present in generation 2, i.e. light and hybrid. Explain why.

To confirm whether semi-conservative replication is occurring, we now need to isolate and separate the DNA from generation 1 and generation 2 to see if the pattern of light and hybrid DNA is as predicted. This can be done using the technique of density gradient centrifugation. Essentially, this technique allows separation of molecules on the basis of their density. The process is shown in Fig 4.17.

After centrifugation, the tubes are irradiated with ultraviolet light of wavelength 260 nm. DNA absorbs this wavelength very strongly so the position of the DNA in the centrifuge tube shows up as a black band.

1. DNA isolated from the cells is placed on top of a density gradient of CsCl solution
   (i.e. on top of layers of CsCl solutions of increasing concentrations)

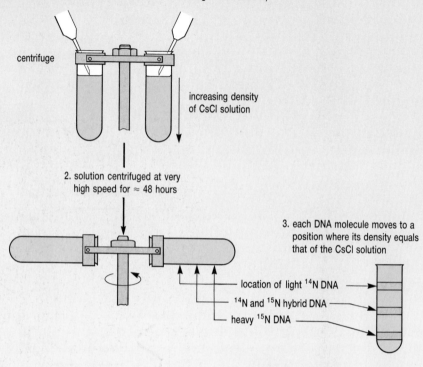

centrifuge

increasing density
of CsCl solution

2. solution centrifuged at very
   high speed for ≈ 48 hours

3. each DNA molecule moves to a
   position where its density equals
   that of the CsCl solution

location of light ^{14}N DNA

^{14}N and ^{15}N hybrid DNA

heavy ^{15}N DNA

Fig 4.17 Density gradient centrifugation, a technique for separating substances with different densities.

(d) Meselsohn and Stahl's results are shown in Fig 4.18. Why do these support the hypothesis that DNA replication is semi-conservative?

(e) If replication had been (i) conservative (ii) dispersive, what pattern of DNA bands would be seen in the centrifuge tubes for generations 1 and 2?

(f) If the bacteria had been left to divide for another two generations in the light ammonium medium, in what ratio would the hybrid and light DNA occur in the cells of generation 4?

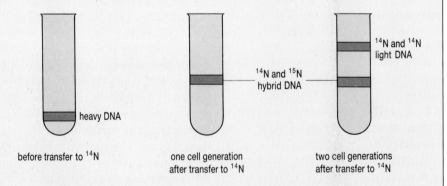

^{14}N and ^{14}N
light DNA

^{14}N and ^{15}N
hybrid DNA

heavy DNA

before transfer to ^{14}N

one cell generation
after transfer to ^{14}N

two cell generations
after transfer to ^{14}N

Fig 4.18 Meselsohn and Stahl's results support the semi-conservative replication of DNA.

DNA and chromosomes

A typical human cell, for example an epithelial or brain cell, contains about 1 m of DNA. However, the nucleus is only 5 μm in diameter. The way this problem has been overcome in the process of evolution is to package the DNA into chromosomes. In the nucleus of one epithelial cell of the human body there are 46 chromosomes, and although they differ in size (Fig 4.19), estimates suggest that each human chromosome consists of a single DNA molecule about 5 cm long containing about 1.3×10^8 nucleotides.

 What does this suggest must happen to the DNA if it is going to fit inside the nucleus?

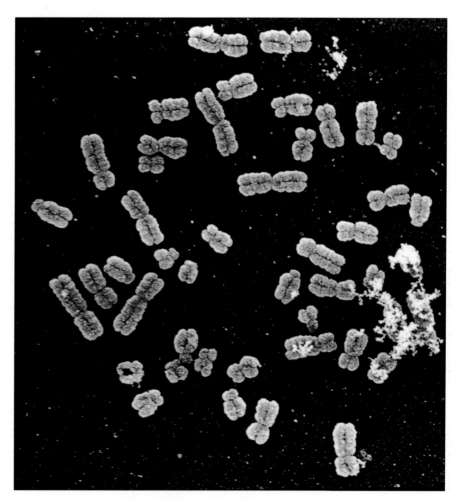

Fig 4.19 Scanning electronmicrograph of human chromosomes.

A group of proteins, called histones, play a major role in packaging the DNA, as shown in Fig 4.20. Here you can see that the DNA helix is first wound around clusters of histone molecules giving a beads-on-a-string appearance. This structure is then further condensed by packing the histone beads close together (Fig 4.20(c)). The chromatin is then further looped, twisted and coiled (Figs 4.20(d) and (e)).

Clearly, the chromosomes are enormously intricate and complex structures. Their physical complexity raises two interesting biological problems which we cannot yet answer.

1. How do these structures unravel to allow the DNA to be replicated?
2. How does the metabolic machinery involved in transcription find the gene to be transcribed amongst all this chromatin?

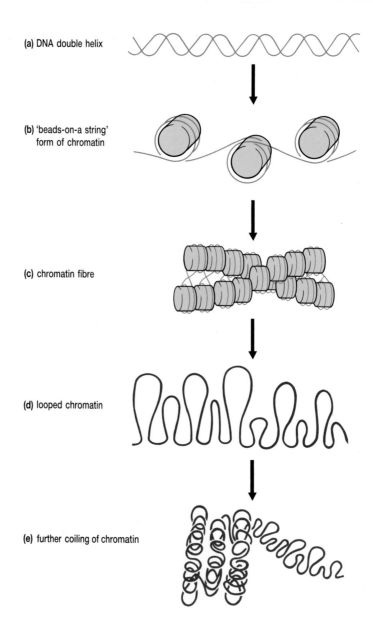

(a) DNA double helix

(b) 'beads-on-a string' form of chromatin

(c) chromatin fibre

(d) looped chromatin

(e) further coiling of chromatin

Fig 4.20   How eukaryotic cells package DNA.

QUESTIONS

4.11 The following string of bases is found on the coding strand of a DNA molecule: CGACGGCTACCA. What are the complementary sequences of
(a) the non-coding strand of the DNA
(b) a mRNA molecule transcribed from the coding strand?

4.12 If you look at Fig 4.19, the chromosomes appear to be double structures. Suggest why.

4.13 What is the importance of base pairing during the replication of DNA?

4.14 Explain why DNA replication is said to be semi-conservative.

4.6 MITOSIS

This form of cell division produces daughter cells which are identical to the parent cells. Daughter and parent cells contain exactly the same number of chromosomes and the same genetic information. Cells only undergo mitosis when they reach some optimum size. This optimum might be

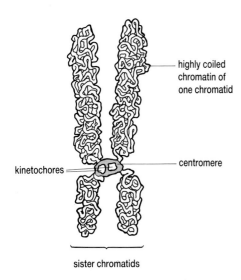

highly coiled
chromatin of
one chromatid

kinetochores

centromere

sister chromatids

Fig 4.21 The structure of a chromosome.

related to the volume of cytoplasm which a nucleus can control or to the ratio of cell surface area of membrane (through which substances enter and leave the cell) to volume (most of the cell's work is done in the cytoplasm). There must also be a genetic component since cells of different species divide when they are at different sizes and some organisms, e.g. filamentous fungi, rarely separate their cytoplasm into cells.

Before a cell divides it must make

- new cell organelles, so as there are enough in the cytoplasm of each daughter cell
- a copy of all its DNA, so that each daughter cell can control protein synthesis in exactly the same way as the parent cell did.

At the end of replication there are two identical DNA helices, usually held together at a point along their length called the **centromere** (Fig 4.21). As long as they are held together the DNA molecules are called **chromatids**; as soon as they are pulled apart during cell division they are referred to as individual chromosomes again.

At an early stage of cell division in eukaryotic cells, chromosomes coil up (**condense**) so that they become easily visible with an optical microscope. The two thick strands are the chromatids, products of DNA replication; the 'waist' in the middle is the centromere holding them together. At the edges of the chromatids are fine strands where the DNA molecule in each chromatid is not tightly coiled (Fig 4.22).

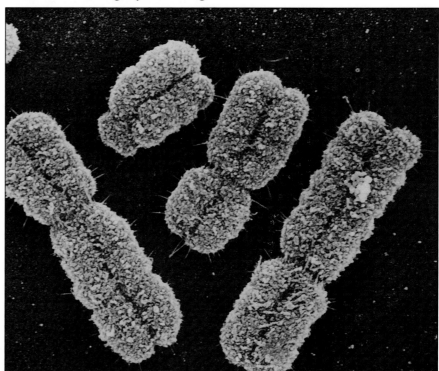

Fig 4.22 Scanning electronmicrograph of condensed chromosomes. Notice that each consists of a pair of chromatids held together by a centromere. During mitosis, these chromatids are separated to form the chromosomes in the daughter cells.

The stages of mitosis

Cell division is most easily studied in cells which have been fixed and stained. Studying such cells, it is easy to get the impression that cell division occurs as a series of discrete phases but this is not true. The study of living cells, using time-lapse techniques with a phase-contrast microscope, shows the continuity of cell division. As a convenient shorthand way of summarising a number of events, however, it is still often described in discrete phases.

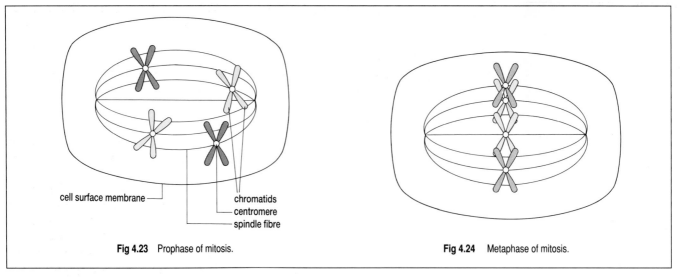

cell surface membrane

chromatids
centromere
spindle fibre

Fig 4.23 Prophase of mitosis.

Fig 4.24 Metaphase of mitosis.

Prophase

Chromatids shorten and thicken (condense) and can be clearly seen. Each of the chromatids in a pair has a region called a **kinetochore**, located at the centromere which holds them together (see Fig 4.21). A system of microtubules, the **spindle**, forms. In animal cells, this is controlled by movement of the centrioles. Some of the spindle fibres become attached to the kinetochores of the chromatids (Fig 4.23). In each pair of sister chromatids, fibres run from one end (**pole**) of the spindle to the kinetochore of one chromatid and other fibres run from the opposite pole of the spindle to the kinetochore of the second chromatid.

Metaphase

The chromosomes are moved by the spindle fibres so that their centromeres line up on the middle (**equator**) of the spindle (Fig 4.24). The kinetochores of each pair of chromatids face opposite poles of the spindle.

Anaphase

The centromeres holding the chromatids together split in two and the paired chromatids separate, becoming two individual chromosomes. The newly independent chromosomes of each pair are pulled by the spindle fibres to opposite ends of the cell (Fig 4.25).

Telophase

The two groups of chromosomes reach opposite poles of the cell and a new nuclear envelope forms around each group (Fig 4.26). The spindle fibres

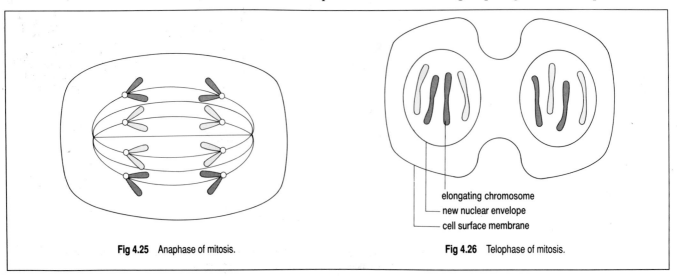

elongating chromosome
new nuclear envelope
cell surface membrane

Fig 4.25 Anaphase of mitosis.

Fig 4.26 Telophase of mitosis.

CELL CONTROL AND CELL DIVISION

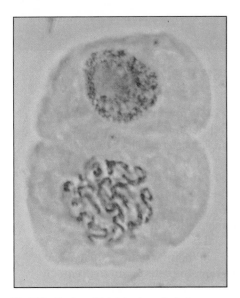

Fig 4.27 The top cell is in interphase, the bottom one has just moved into prophase and its chromosomes are condensing.

disintegrate and nucleoli reform. The cytoplasm may divide (**cyto-kinesis**).

Interphase

This is the normal state of the cell between divisions, during which time the chromosomes are not visible (Fig 4.27).

Cytokinesis

The division of the cytoplasm, cytokinesis, often begins during telophase of mitosis but may be delayed indefinitely. Where it does occur, the division is not always into two equal-sized portions of cytoplasm.

Cytokinesis in animal cells occurs when a ring of actin microfilaments develops around the cell, usually in the region of the equator of the disintegrating spindle. These microfilaments are attached to the cell surface membrane. When they contract they pull the surface membrane inwards to make a 'waist' or **division furrow** (see Fig 4.26), in much the same way as you might pull the drawstrings around the waist of a pair of jogging pants. Eventually, the division furrow contracts to nothing; the cell surface membranes on each side join up and the two cells separate.

Cytokinesis in plants is rather different (Fig 4.28). The Golgi apparatus is moved to the equator of the disintegrating spindle, producing carbo-hydrate-filled vesicles. These fuse together to make a hollow **cell plate** in the middle of the cell. As more vesicles are added, this cell plate grows until it stretches right the way across the cell and its membranes fuse with the cell surface membrane. The membrane on one side of the cell plate becomes the cell surface membrane of one cell; the membrane on the other side of the cell plate becomes the cell surface membrane of the other cell; the contents of the cell plate become the middle lamella and primary walls between the two new cells.

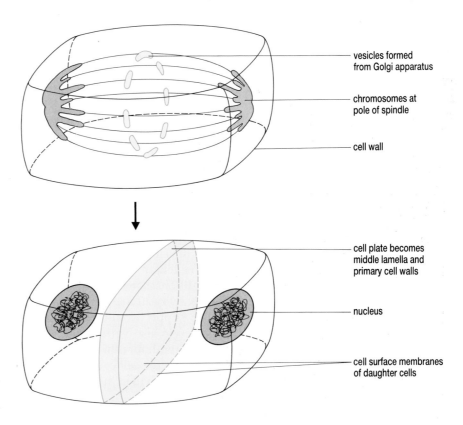

Fig 4.28 Cytokinesis in a plant cell. The cell plate grows from the centre towards the edges of the cell to form the middle lamella and cell wall between two new cells.

CAN DIVISION OF CANCER CELLS BE STOPPED?

Biologists working at the University of Oxford and at Scripps Clinic in La Jolla, California may have found the answer to this question through studies of the yeast *Schizosaccharomyces pombe*.

Cells of *S. pombe* have been found that are unable to divide by mitosis. It is through investigating these cells that the normal control of cell division has been discovered. The abnormal cells have a mutation in a gene known as *cdc25* that codes for a protein called p80^{cdc25}. This protein is found in normal cells and, as cells of *S. pombe* grow, the protein p80^{cdc25} accumulates in their cytoplasm. Just as the protein reaches its highest concentration, the cells begin to divide by mitosis. Their inability to produce protein p80^{cdc25} explains why the abnormal cells cannot divide.

The biologists propose that mitosis is initiated when a critical level of p80^{cdc25} has accumulated in a cell. It seems that p80^{cdc25} activates a protein–kinase enzyme. Since enzymes similar to the protein kinase in *S. pombe* have been found in many animals, including humans, this method of control of mitosis might be common to all eukaryotic organisms. If so, this discovery might have exciting consequences since it may lead to a way of stopping the uncontrolled division of unwanted cells in humans, i.e. a cure for cancer.

The Cell cycle

Replication of chromosomes occurs before mitosis, in a period called **interphase**. At mitosis the replicated copies of each chromosome are separated into two new nuclei after which cytokinesis often occurs. Interphase, mitosis and cytokinesis are the three phases of the **cell cycle**, i.e. the sequence of events which occurs throughout the life of a cell, from its formation as a new cell by division of a mother cell, through its growth and development to its own division into two daughter cells.

The length of the cell cycle is affected by external conditions: under ideal conditions, some bacteria complete their cell cycle and divide into two new cells every 20 minutes; too low a temperature, absence of water or suitable food supply would increase the time between divisions. The length of the cell cycle also varies from species to species: some bacteria divide about every ten hours. In spite of these differences, the relative time interval of stages in the cell cycle is fairly constant (Fig 4.29). Like mitosis, interphase has been divided into a number of stages, referred to as G$_1$, S, and G$_2$ in Fig 4.29.

Table 4.4 summarises the events which occur during these stages of the cell cycle. Just as with the stages of mitosis, however, all these events really occur in a continuous process and the division into stages is artificial.

Table 4.4   Events occurring during the cell cycle

Stage of cell cycle	Phase	Major events occurring in cell
Interphase	G_1	Period of high metabolic activity resulting from cell growth.
		New proteins are synthesised; rRNA, mRNA and tRNA are produced in the nucleolus; all cell organelles are produced.
	S	DNA replication occurs.
		Each new DNA double helix is surrounded by proteins, called histones, to form two chromatids.
	G_2	Centrioles (if present) replicate and mitotic spindle of microtubules is formed.
		Mitochondria and chloroplasts divide.
Mitosis	prophase metaphase anaphase telophase	see Figs 4.23 to 4.26.
Cytokinesis (C)		Division of cytoplasm and cell organelles into two approximately equal halves.

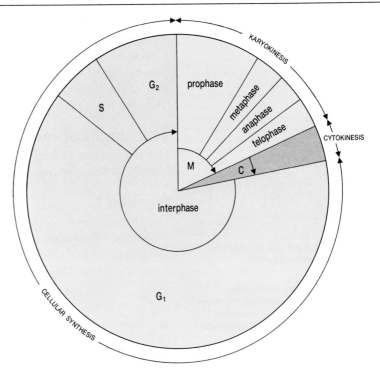

Fig 4.29   The cell cycle. Cellular synthesis (interphase) is a much longer phase than nuclear and cytoplasmic divisions.

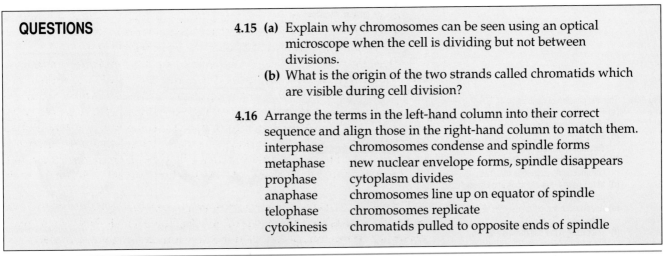

QUESTIONS

4.15 (a) Explain why chromosomes can be seen using an optical microscope when the cell is dividing but not between divisions.

(b) What is the origin of the two strands called chromatids which are visible during cell division?

4.16 Arrange the terms in the left-hand column into their correct sequence and align those in the right-hand column to match them.

interphase   chromosomes condense and spindle forms
metaphase   new nuclear envelope forms, spindle disappears
prophase   cytoplasm divides
anaphase   chromosomes line up on equator of spindle
telophase   chromosomes replicate
cytokinesis   chromatids pulled to opposite ends of spindle

4.7 MEIOSIS

KEY CONCEPT

Haploid cells have one copy of each of their chromosomes.

Diploid cells have two copies of each of their chromosomes, forming homologous pairs.

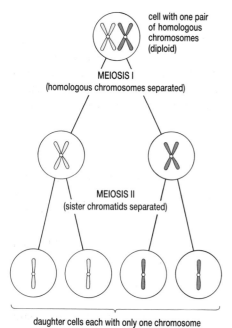

cell with one pair of homologous chromosomes (diploid)

MEIOSIS I
(homologous chromosomes separated)

MEIOSIS II
(sister chromatids separated)

daughter cells each with only one chromosome from the original pair (haploid)

Fig 4.30 An overview of meiosis. Homologous chromosomes are separated during meiosis I. The chromatids which make up each chromosome are separated during meiosis II.

Any cell in the human body which is not a gamete, i.e. either a sperm or an egg cell, is called a somatic cell. Skin cells, nerve cells and muscle cells are all somatic cells. The number of chromosomes found in somatic cells is always twice the number found in the nucleus of gametes in humans, i.e. 46 chromosomes in somatic cells, 23 in gametes.

Since each human somatic cell contains two copies of every chromosome, **homologous pairs**, somatic cells are diploid. The human diploid number is 46. Since gamete cells contains only one copy of every chromosome, i.e. one member of every homologous pair, they are haploid.

11 **What is your haploid number?**

When a human somatic cell undergoes mitosis, there will be exactly the same number of chromosomes in the daughter cells as in the parent cell. But, to produce a human gamete, the number of chromosomes has to be halved. Clearly, another type of cell division is going on in order to produce human gametes; this is the process of **meiosis**.

Meiosis is a form of cell division in which the number of chromosomes is halved to give the haploid number of chromosomes (Fig 4.30). The importance of this process cannot be over-emphasised since it holds the key to the pattern of inheritance of the different forms of a particular character and to genetic variation.

The stages of meiosis

Meiosis consists of two nuclear divisions. The second, **meiosis II**, usually follows on immediately from the first, **meiosis I**.

Prophase I

The chromosomes condense and appear as two chromatids. In Fig 4.31(a) the white chromosomes have come from the female parent, the black chromosomes from the male parent.

Fig 4.31 The first division of meiosis.
   (a) Early prophase I: the homologous chromosomes appear.
   (b) Mid prophase I: the homologous chromosomes form pairs.
   (c) Late prophase I: the homologous chromosomes exchange material.
   (d) Metaphase I: the chromosomes in each pair line up along the equator.
   (e) Anaphase I: the chromosomes are pulled apart.
   (f) Telophase I: the cell divides into two.

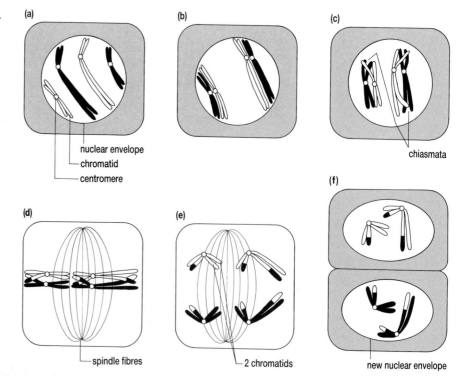

nuclear envelope
chromatid
centromere
chiasmata
spindle fibres
2 chromatids
new nuclear envelope

What is the diploid number of the cell in Fig 4.31? How many pairs of homologous chromosomes does it contain?

By an unknown mechanism, the chromosomes in each homologous pair lie side by side to form **bivalents** (Fig 4.31(b)). Note the two pairs: chromosomes 1 have formed one pair and chromosomes 2 have formed the other. The centromere of each chromosome pair forms kinetochores and kinetochore fibres which attach the centromeres to the spindle.

 How does this differ from the events occurring in prophase of mitosis?

As the chromosomes in each bivalent condense, their chromatids become entwined (Fig 4.31(c)) and tension develops at these points (**chiasmata**). Under the control of enzymes, the DNA may be cut at these points and spliced back onto the wrong chromatid. If so, we say that **genetic crossing over** occurs. These chiasmata hold the chromosomes together as they drift apart during late prophase I.

Metaphase I

The bivalents are pulled to the equator of the spindle (Fig 4.31 (d)). Whether the centromere of one chromosome is attached to spindle fibres from the 'north' pole or 'south' pole of the spindle is completely at random.

 How does this arrangement of chromosomes differ from that seen during metaphase of mitosis?

Anaphase I

The chromosomes of each bivalent are pulled by spindle fibres to opposite poles of the spindle (Fig 4.31 (e)). **Note:** each chromosome is still made of two chromatids.

 How does this differ from anaphase of mitosis?

Telophase I

The spindle disappears (Fig 4.31 (f)). Sometimes a new nuclear envelope forms around the new haploid nuclei, sometimes not. The chromosomes usually stay in their condensed form and meiosis II follows.

 What has happened to the number of chromosomes in each daughter cell as a result of meiosis I?

Prophase II

New spindles are formed at right angles to the old spindle (Fig 4.32 (b)).

Metaphase II

The chromosomes are pulled to the equators of the new spindles (Fig 4.32 (b)). The spindle fibres from each pole are now attached to the kinetochores of the centromere of sister chromatids in each chromosome.

Anaphase II

The spindle fibres pull the kinetochores of the sister chromatids to opposite ends of the spindle (Fig 4.32 (c)). The separated chromatids, now referred to as chromosomes again, are pulled to the poles of the new spindle.

Telophase II

The chromosomes elongate, the spindles disappear and the nuclear envelopes reform. Cytokinesis occurs, producing four haploid cells, two of which are shown in Fig 4.32(d).

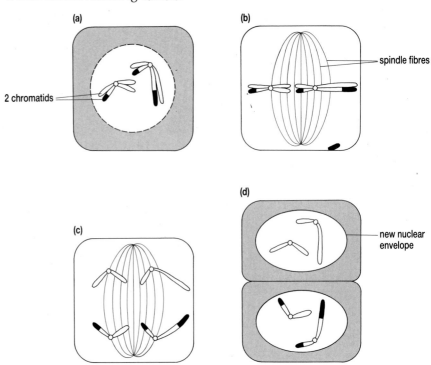

Fig 4.32 The second division of meiosis in one of the products of Fig 4.31.
(a) Prophase II: two new spindles are formed.
(b) Metaphase II: the chromosomes line up along the equator.
(c) Anaphase II: the chromosomes are pulled apart.
(d) Telophase II: each haploid cell divides into two.

The timing of mitosis and meiosis in life cycles

Meiosis results in a precise halving of the chromosome number. This is needed during sexual reproduction when, unless unchecked, chromosome number would double when gametes fused during fertilisation. Some organisms halve their chromosome number when they produce gametes whereas others do so after the gametes have fused at fertilisation (Fig 4.33).

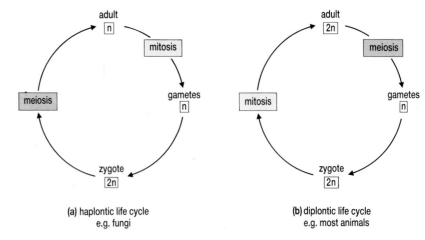

(a) haplontic life cycle
e.g. fungi

(b) diplontic life cycle
e.g. most animals

Fig 4.33 The different timing of mitosis and meiosis in the life cycle of (a) haplontic (b) diplontic organisms.

Plants have a very complex life cycle in which there are actually two completely different types of adult, a haploid **gametophyte** and a diploid

sporophyte (Fig 4.34). The gametophyte reproduces sexually to produce the sporophyte which reproduces asexually, forming spores which grow into the gametophyte. Mitosis and meiosis both occur during this **alternation of generations** (see Section 26.2). Notice how, in this life cycle, gamete formation during sexual reproduction involves mitosis whereas asexual reproduction involves meiosis.

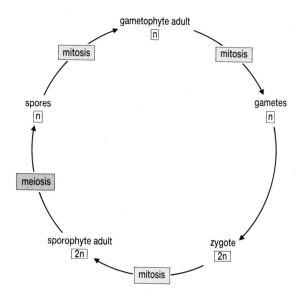

Fig 4.34 The alternation of generations in the life cycle of plants.

QUESTIONS

4.17 (a) During genetic crossing over, is material exchanged
   (i)   between homologous chromosomes or non-homologous chromosomes
   (ii) between sister chromatids or chromatids from a different chromosome?
(b) Why does genetic crossing over not occur during prophase II?

4.18 Copy Table 4.5 and complete it to show events during mitosis and meiosis. Line 1 has been done for you.

4.19 Look carefully at the four photographs of dividing cells in Fig 4.35. For each photograph say **(a)** which type of division is occurring **(b)** the stage which is shown.

4.20 The salivary glands of some insect larvae have chromosomes which are very large and can easily be seen with an optical microscope. Fig 4.36 shows one of these giant chromosomes. Using your knowledge of chromosome structure and replication and of cell division, suggest how these giant chromosomes might have originated.

Table 4.5

Feature	Mitosis	Meiosis I	Meiosis II
chromosomes shorten and thicken	✔	✔	✕
chromosomes present as two sister chromatids			
homologous chromosomes pair together			
chiasmata occur			
homologous chromosomes separated			
sister chromatids separated			

QUESTIONS *continued*

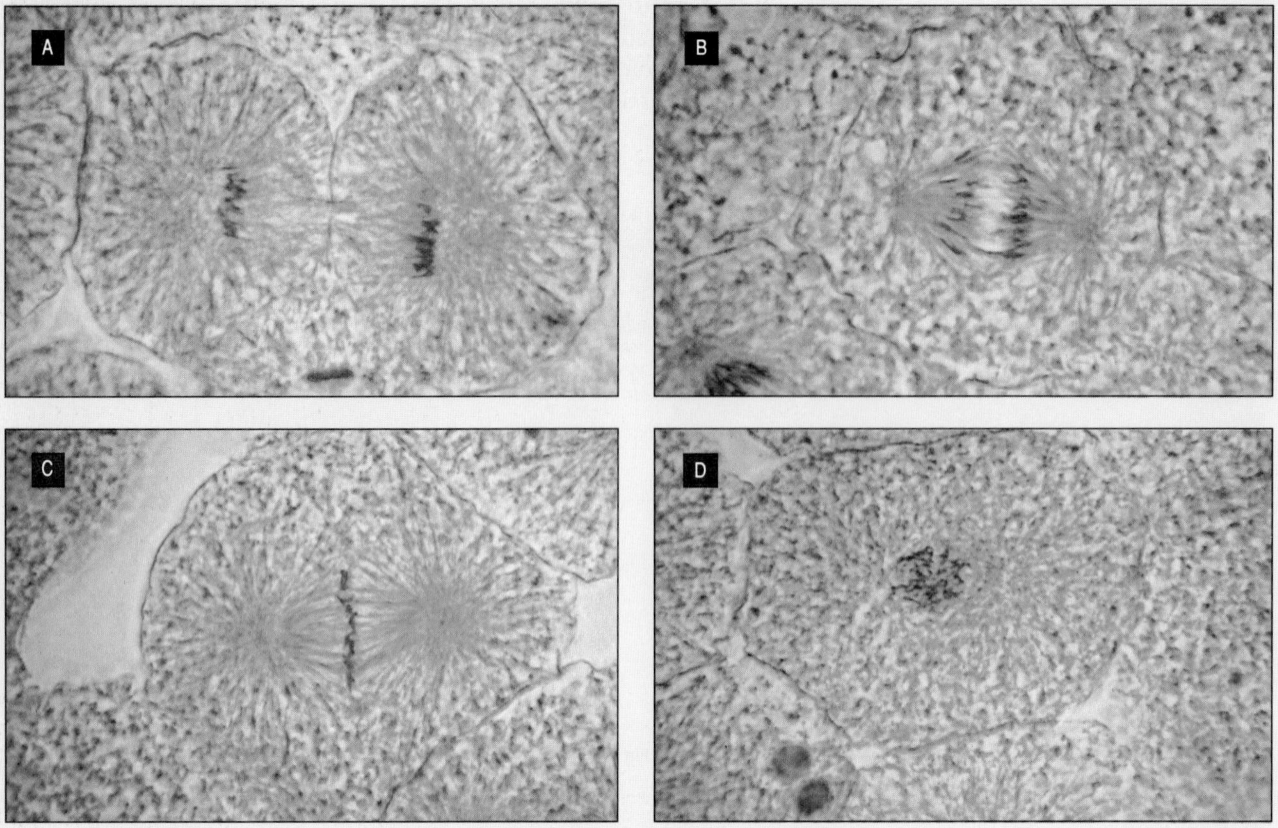

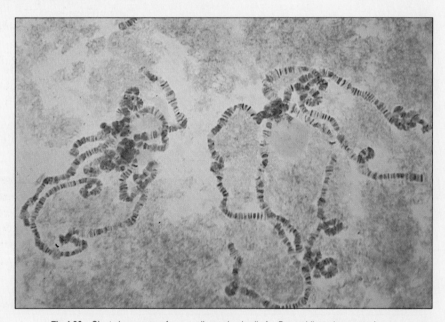

Fig 4.35 Dividing cells: identify the type of division and the stage.

Fig 4.36 Giant chromosome from a salivary gland cell of a *Drosophila melanogaster* larva.

ASSIGNMENT

Use modelling clay to make a homologous pair of chromosomes. Use your models to represent the processes involved in:

(a) chromosome replication
(b) mitosis
(c) meiosis with genetic crossing over.

SUMMARY

The genetic code determines the sequence of amino acids in the molecule of a protein or polypeptide. It is carried in the base sequence of DNA which forms the chromosomes in the nucleus of each cell. The base sequence which carries the genetic code for the production of a functional protein is called a gene.

DNA molecules are too large to move out of the nucleus. Their code for a single protein is copied in a smaller molecule of messenger RNA (mRNA) which leaves the nucleus via pores in the nuclear envelope. The process by which mRNA is produced from DNA is called transcription.

Once in the cytoplasm, the base sequence of mRNA is read by ribosomes. Each combination of three bases in the mRNA, called a codon, codes for one amino acid. Molecules of transfer RNA (tRNA) carry specific amino acids to the ribosomes which assemble them into polypeptide chains in the process of translation.

The transcription of genes can be controlled by the cell. In prokaryotes, promoter, operator and regulator genes interact to regulate the transcription of genes: environmental chemicals may also be involved in this regulation. Similar control mechanisms have been postulated for eukaryotes.

The genetic code is passed from cell to cell during cell division. Prior to cell division, DNA in the nucleus is exactly copied in a process called replication. During this process the two strands of a DNA double helix unwind, exposing their bases. Base pairing between these DNA strands and free nucleotides in the nucleus ensures that each DNA strand builds up an exact copy of its old partner. Because each new DNA molecule contains one complete DNA strand from the old molecule, this process of replication is said to be semi-conservative.

The two copies of each chromosome, called chromatids, are separated during cell division. Two types of cell division occur. During mitosis, one chromatid from each pair is pulled into two new nuclei, so that each nucleus is identical. After mitosis the cytoplasm usually divides into two halves. During meiosis, cells are formed which are genetically different from each other and have half the number of chromosomes of the parent cell.

Mitosis and meiosis occur at different times during three types of life cycle. In a haplontic life cycle, mitosis occurs in gamete formation and meiosis after fertilisation, whereas in a diplontic life cycle, meiosis occurs during gamete formation and mitosis after fertilisation. Plants have a complex life cycle involving alternation of generations. In this cycle, meiosis is involved in the production of spores and mitosis in gamete production during sexual reproduction.

Answers to Quick Questions: Chapter 4

1 CAUUGCUAG.
2 6; 2
3 20, because the genetic code is redundant.
4 CCA; glycine.
5 It will be recycled to pick up another methionine. Another tRNA molecule, anticodon GUU, will bind to the mRNA. Glutamine.
6 They will go back to the nucleus to be converted into new mRNA.
7 It still contains introns which do not carry code from which a functional protein can be made.
8 Reduces energy consumption – proteins are energetically expensive molecules to make.
9 Hydrogen bonds. Yes. Weaker.
10 Must be tightly coiled.
11 23
12 4; 2
13 In mitosis, homologous chromosomes do not pair up.
14 The chromosomes are not paired up on the mitotic spindle and so no chiasmata are visible.
15 At this stage of mitosis, the chromatids in each chromosome have separated.
16 It has been halved.

Theme 1
EXAMINATION QUESTIONS

1.1 **(a)** Describe a test that you could perform to identify the presence of a protein in solution. (2)

(b) Figure **A** shows the primary structure of a protein in its inactive form. Figure **B** shows the same protein in its active state with some features of its tertiary structure shown. The numbers refer to the amino acid sequence. The structures of the R groups of two amino acids are shown:
Alanine (ala) –CH₃ Glycine (gly) –H
Show the chemical structure of the dipeptide formed from amino acid numbers 1 and 2. What is the name given to the chemical bond formed between these amino acids? (4)

(c) With reference to figures **A** and **B**:
   (i) explain the terms **primary** and **tertiary** structure of a protein molecule. (4)
   (ii) suggest **two** things that happen to this protein molecule during the process of activation. (2)

(d) Name a protein that requires to be activated. (1)

(e) (i) The amino acids phenylalanine (phe) and tyrosine (tyr) are examples of aromatic amino acids. Aspartic acid (asp) and glutamic acid (glu) are dicarboxylic amino acids.

During digestion pepsin hydrolyses polypeptides between an adjacent aromatic amino acid and a dicarboxylic acid. Later, chymotrypsin hydrolyses polypeptides on the carboxyl side of aromatic amino acids. Between which amino acids will pepsin and chymotrypsin hydrolyse the active form of the protein? (3)
   (ii) In the small intestine the peptides are further hydrolysed by the digestive enzyme aminopeptidase. Describe the role of this enzyme. (3)

(f) Describe **two** possible fates, other than protein synthesis, for the products of protein digestion after they have been absorbed into the bloodstream. (2)

(g) Figure **C** shows the translation phase of part of the synthesis of the inactive protein molecule.
Describe what happens in each of stages 1, 2 and 3. (3)

(AEB 1989)

Figure A

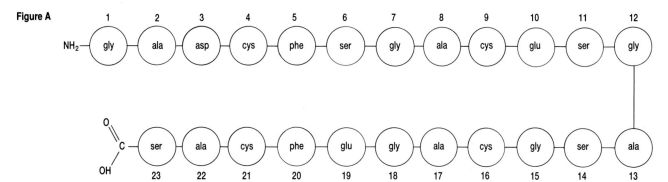

Figure B

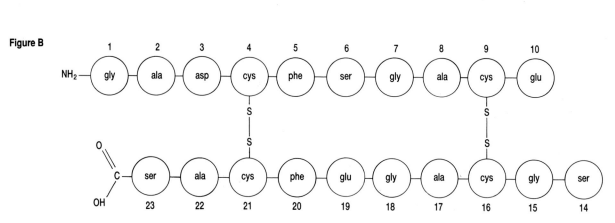

Figure C

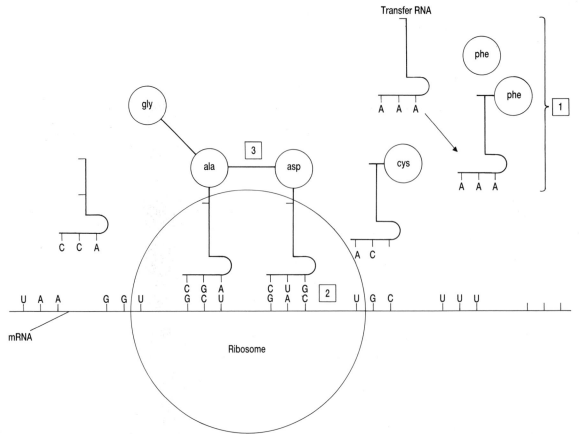

Figure C

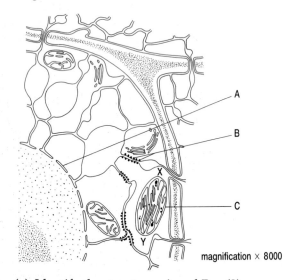

1.2 The diagram is drawn from an electron-micrograph and shows the structure of part of a cell and the barrier between it and two of its neighbours.

Figure C

(a) Identify the structures **A** and **B**. (2)

(b) Give **two** features of the cell which show that it is **not** a prokaryotic cell. (2)

(c) Is it a plant cell or an animal cell? Give **two** reasons for your identification. (2)

(d) Measure the length **XY** of the organelle **C** and calculate its actual length in μm. Show your working. (2)

(AEB 1988)

1.3 (a) Describe the biological occurrence and function(s) of each of the compounds listed below. Explain briefly how each function is dependent on the molecular structure.
  (i) deoxyribonucleic acid (DNA) (5)
  (ii) glycogen (4)
  (iii) myoglobin (4)
  (iv) phospholipid. (3)

(b) Starch and cellulose are plant polysaccharides of economic importance. Account for the importance of each, in terms of its natural occurrence, properties and use. (4)

(NISEC 1989)

1.4 *Either* **(a)** Give an account of the structure and functions of cell membranes. (20)
or **(b)** Discuss the way in which structure is related to function in eukaryotic cells. (20)

(ULSEB 1988)

1.5 Certain cells in the pancreas produce the protein insulin from amino acids.

(a) Describe the roles of different parts of these cells in obtaining amino acids and in the synthesis and secretion of insulin. (14)

(b) Given a fresh extract of pancreatic tissue, explain how an investigator might
  (i) determine that it contained insulin,
  (ii) separate the insulin from other proteins present in the extract. (6)

(JMB 1990)

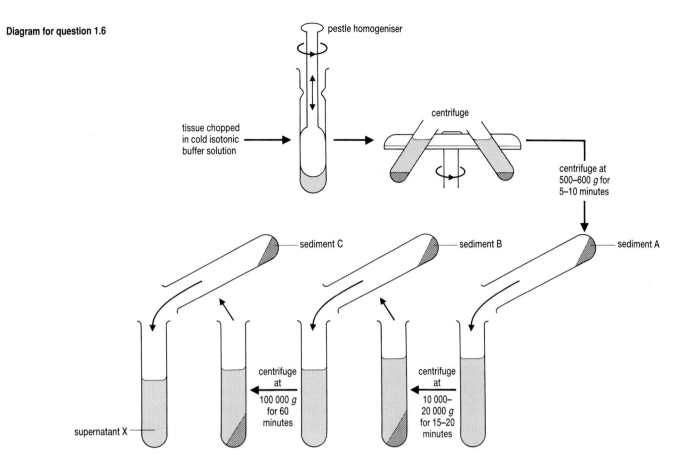

1.6 In order to separate ribosomes, nuclei and mitochondria, liver tissue was chopped, homogenised and centrifuged at different speeds for different lengths of time as indicated above. "g" is a measure of gravitational force.

(a) (i) In which sediment, A, B or C would you expect to find the following cell organelles?
Ribosomes, Nuclei, Mitochondria.

(ii) Give a reason for your choice. (4)

(b) In the preparation of the tissue for homogenisation, a buffer solution was used which was isotonic with the liver tissue. What would happen to the organelles if this solution were
(i) hypotonic
(ii) hypertonic? (2)

(c) This technique was used for the isolation of mitochondria in order to study their enzymic activity. However, liver tissue contains many lysosomes. Suggest why this feature makes the study of mitochondrial activity difficult. (2)

(d) The biochemical activity of isolated organelles is rapidly lost. Suggest **two** features of this technique which will serve to reduce this loss. (2)

(e) Supernatant X was found to contain several amino acids. Name a technique that would be used to separate and identify them. (1)

(WJEC 1989)

1.7 The graph below shows the relationship between the volume, water potential (Ψ_{cell}) and solute potential (osmotic potential) (Ψ_s) of a plant cell immersed in a series of sucrose solutions of increasing concentration. In each solution, the cell was allowed to reach equilibrium with the bathing solution, so that water was being neither lost nor gained, before the measurements were made.

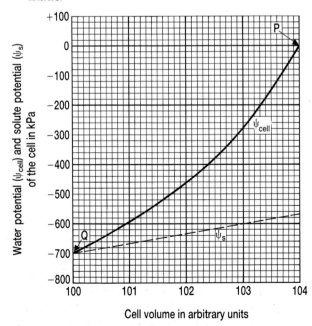

(a) What terms are used to describe the condition of the cell at (i) P and (ii) Q? (2)

(b) What is the volume of the cell (in arbitrary units) when in equilibrium with a sucrose solution of solute potential –400 kPa? (1)

(c) Calculate the pressure potential (turgor pressure) (Ψ_p) of the cell when its volume is 103 units. Show your working. (2)

(d) Suggest *two* ways in which reversible changes in cell volume may be important in flowering plants. (2)

(ULSEB 1989)

1.8 (a) Outline the functions of a cell nucleus. (4)

(b) Describe how a nucleus divides during the process of meiosis. (10)

(c) Explain the significance of mitosis and meiosis in the life of a human being. (6)

(UCLES 1988)

1.9 (a) Describe the structure of a meristematic cell from a root tip as you would expect to observe it under high power of a 'light' microscope. (5)

(b) (i) Define the term mitosis.

(ii) What is the significance of mitosis? (4)

(c) Outline, with the aid of representative labelled diagrams, the events that occur during mitosis. (16)

(d) List the practical steps you would take to prepare and observe microscopically, stages of mitosis in some suitable **named** material. (5)

(UODLE 1988)

1.10 Restriction enzymes are widely used in genetic engineering. There are many different restriction enzymes, each one cutting DNA at a particular base sequence. One restriction enzyme, Eco R1, acts on DNA by cutting it at the sequence shown:

(a) Complete the diagram by writing in the base sequence on the complementary strand of the DNA molecule. (1)

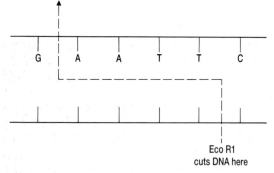

Eco R1
cuts DNA here

(b) Explain why an enzyme like Eco R1 can only cut DNA at a specific base sequence. (2)

(c) Explain why Eco R1 cannot be used for cutting out **all** genes. (2)

(AEB 1990)

1.11 (a) (i) How have our ideas on the structure of plasma membranes been influenced by the introduction of the electron microscope?

(ii) Describe the *fluid mosaic model* of a plasma membrane. (7, 8)

(b) How do substances move across cell membranes? (15)

(UODLE 1987)

1.12 (a) Give an illustrated account of meiosis. (10)

(b) What does meiosis have in common with mitosis and in what way does it differ? (4)

(c) Discuss the significance of the differences. (6)

(Oxf and Camb SEB 1986)

Answer guidelines for Question 1.1

1.1 This question covers information contained in several Themes of this book. Although it looks complicated at first sight, it is split into small sub-questions each carrying only a few marks. Many of the answers are contained within the information given. You should always check each part of questions like this to make sure that you pick out those questions you can answer.

(a) Mix equal volumes of the protein solution and (5%) potassium hydroxide solution; add drops of (1%) copper sulphate solution; purple colour shows protein is present;

(b) $NH_2CH_2CONHCHCH_3COOH$ (1 mark for correct structure of each amino acid and 1 mark for structure of peptide bond); peptide;

(c) (i) primary structure is sequence of amino acids; e.g. amino acids 1 to 23 in Figure A; tertiary structure is the folding of the amino acids chains into a three-dimensional shape; e.g. disulphide bridges holding chains 1 to 10 and 14 to 23 together in Figure B;

(ii) disulphide bridges formed between amino acids 4 and 21/9 and 16; portion of polypeptide chain (amino acids 11 and 12) removed;

(d) Pepsinogen/trypsinogen / chymotrypsinogen/fibrinogen;

(e) (i) pepsin hydrolyses between amino acids 19 and 20; chymotrypsin hydrolyses between amino acids 5 and 6; and between amino acids 20 and 21;

(ii) breaks down polypeptides; produces individual amino acids; removes terminal amino acid from amino end of polypeptide;

(f) Broken down in ornithine cycle to form urea; used as an energy source;

(g) 1 – tRNA combines with appropriate amino acid;

2 – anticodon of tRNA links to codon on mRNA molecule;

3 – ATP is used to make a peptide bond between amino acids held by tRNA molecules;

Theme 2

ENERGY

Within the second law of thermodynamics is a concept that systems move towards states of maximum molecular disorder or chaos.

A living organism can be thought of as a system. Throughout an organism's life it maintains itself in a state of molecular order, not chaos. It expands this molecular order as it grows and replicates it when it reproduces. In other words, living organisms seem to defy the second law of thermodynamics. They can do this only by using vast amounts of energy. Once they are dead, the molecules in their bodies move towards disorder again.

In the following chapters you will see how organisms link the energy changes of chemical reactions to their own advantage. So fundamental to life is this ability that the same chemical processes are almost universal to all forms of life. Even so, the number and details of these chemical reactions are complex, so you should note only the overview presented in these chapters.

In learning about energy relationships within organisms, you will also discover how structures are adapted to facilitate the processes that occur within them.

- *Energy and enzymes* – metabolism involves a reduction in the energy needed for essential chemical reactions to occur and the regulation of these chemical reactions.
- *Respiration* – the linkage of specific energy-releasing reactions to those involved in forming ATP.
- *Autotrophic nutrition* – the way in which some organisms are able to trap energy from sunlight or inorganic reactions and use it to make their own organic molecules from inorganic ones.
- *Heterotrophic nutrition* – the dependence of some organisms on ready-made organic molecules and the energy they contain.

(Top left) A plant provides the nutrients for the growth of this animal.

(Bottom left) Sunlight and leaves: the starting point for all nutrition.

(Right) This small, active animal obtains energy from its food.

Chapter **5**

ENERGY AND ENZYMES

Fig 5.1 Industrial chemical reactions often require high temperatures whilst the chemical reactions in your body occur at low temperatures.

> **LEARNING OBJECTIVES**
>
> When you have studied this chapter you should be able to:
>
> 1. explain the energy changes which occur during anabolism and catabolism;
>
> 2. define activation energy and explain how enzymes affect it;
>
> 3. critically describe the lock and key and induced-fit models of enzyme activity;
>
> 4. describe and explain how enzyme activity is affected by pH, temperature, concentration of enzyme and substrate, inhibitors and promoters;
>
> 5. describe how reactions in cells may be linked in a series, or coupled.

If you were to visit a chemical factory, you would immediately notice how hot it was. High temperatures are needed to make the commercial chemical reactions occur quickly. However, thousands of chemical reactions are occurring in the cells of your body at 37 °C. The ability of your cells to support rapid low-temperature chemistry depends on a group of catalysts called enzymes.

The result of enzymatic catalysis can be quite spectacular. A good enzyme can accelerate a reaction by 10^{10} times. Without such acceleration, a process that takes five seconds, such as reading this sentence, would take 1500 years. During that time, other unwanted chemical reactions would occur instead, so life would not just be slow, it would be impossible.

5.1 CHEMICAL REACTIONS IN CELLS: METABOLISM

To stay alive and grow, cells must break down large molecules into smaller ones, rearrange the atoms within existing molecules and build large molecules from smaller ones. Hundreds of these chemical changes, called chemical reactions, occur at any one time in a single eukaryotic cell. These chemical reactions are collectively called **metabolism**, a term which may refer to the activities of an entire multicellular organism as well as those of a single cell. Metabolic processes which involve large molecules breaking down into smaller ones are called **catabolism**, those which involve small molecules combining to form larger ones are called **anabolism**. The reactants, i.e. those chemical particles which are involved in the chemical reactions of metabolism, are called **metabolites**.

Here are examples of anabolic and catabolic reactions occurring in cells.

1. **anabolic reaction**

   amino acid + amino acid $\longrightarrow$ dipeptide + water

   $NH_2CHRCOOH \quad NH_2CHRCOOH \quad NH_2CHRCONHCHRCOOH \quad H_2O$

2. catabolic reaction

hydrogen peroxide ⟶ water + oxygen

$$2H_2O_2 \qquad\qquad 2H_2O \qquad O_2$$

Which of the following are catabolic and which anabolic reactions: the synthesis of a protein; the digestion (hydrolysis) of a protein; the synthesis of carbohydrates from CO_2 and H_2O?

Chemical reactions

Particles in gases and solutions continually move about. Inevitably, these particles will bump into each other as they move. According to chemical theory, these collisions are needed before chemical reactions can occur. However, chemical reactions will not always occur when reactant particles collide. The particles must be correctly aligned: the reaction only occurs when they collide in a very precise way. In addition, a certain amount of energy is needed to break existing chemical bonds and form new ones.

Energy changes which occur during a chemical reaction are shown in Fig 5.2. Two reactants collide, combine to form an intermediate **activated complex** and then form products. In Fig 5.2(a) the **potential energy** (hidden energy) within the reactants is more than that within the products. The change from reactants to products, therefore, releases energy. A reaction which releases energy is referred to as an **exergonic** reaction. Some of this energy might be useful for doing work in the cell, some will be lost as heat energy.

The energy changes occurring in a reaction in which the potential energy of the products is greater than that of the reactants (Fig 5.2(b)) requires a supply of external energy: it is said to be **endergonic**. In both types of reaction a certain amount of energy is needed for the reactants to form the activated complex. This energy is known as the **activation energy (E_a)** and without it chemical reactions cannot proceed.

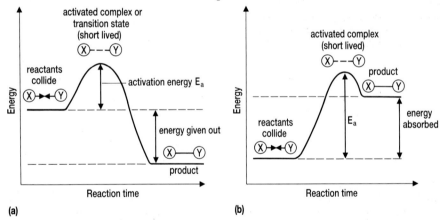

(a)

(b)

Fig 5.2 **(a)** An **exergonic reaction** is one in which the products have less potential energy than the reactants.
(b) An **endergonic reaction** requires energy input since the products contain more potential energy than the reactants.
The activation energy, E_a, is the energy required to bring the reactants into the correct position to interact.

Is fuel burning in the engine of a motor car an example of an exergonic or endergonic reaction?

Catalysis

In test tubes, the energy needed to make a reaction proceed can be supplied by heating the reactants or, in the case of gases, increasing the pressure. The reaction will then proceed at a faster rate. For reasons that will be explained in Section 5.3, cells can only survive within a narrow range of relatively

KEY CONCEPT

CATALYSIS

Catalysts are substances which increase the rate of a chemical reaction without being used up in the chemical reaction. However,

- they do not make reactions happen which would otherwise not occur
- they do not change the amount of product which would be formed.

low temperatures: heating will often kill them rather than make their metabolism happen faster. The rate of reactions can also be increased using substances called **catalysts**.

Hydrogen peroxide is commonly used to bleach hair. It decomposes to form water and oxygen, a reaction which takes place very slowly unless energy, in the form of heat or light, is supplied. However, if a small amount of manganese(IV) oxide is added to a test tube of hydrogen peroxide the reaction occurs so rapidly that a froth of oxygen bubbles is produced. At the end of the reaction the test tube of water contains just as much manganese(IV) oxide as was originally put in with the hydrogen peroxide, i.e. none of it has been used up. Such substances which speed up the rate of a chemical reaction without being used up themselves are catalysts.

 3 **The breakdown of hydrogen peroxide is an exergonic reaction. Why does it not occur spontaneously in the absence of an external supply of energy?**

If a catalyst is not used up in the reaction it catalyses, how does it work? Normally, when molecules react to form new molecules, they pass through an unstable **transition state** (an activated complex) during which some of their chemical bonds break and others form spontaneously. The transition state has a higher energy level than either the reacting molecules or the product molecules. The higher the energy level of the transition state, the slower the rate of reaction. Although the catalyst is recoverable at the end of the reaction it catalyses, it does become involved in the reaction. In fact, the catalyst lowers the energy level of the transition state of the reacting molecules by binding with it, i.e. the activation energy is reduced (see Fig 5.3). As a result, the reaction occurs more rapidly.

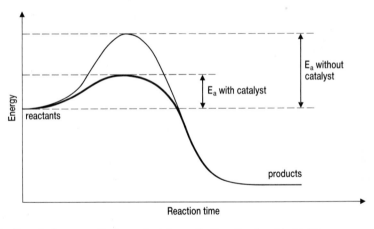

Fig 5.3 The activation energy, E_a, of a reaction is lowered by the action of a catalyst, in this case an enzyme. Note that the overall energy change is the same with or without the catalyst.

KEY CONCEPT

LOWER ACTIVATION ENERGY

Catalysts, including enzymes, work by providing an alternative pathway for a chemical reaction. The pathway has an activation energy lower than the pathway followed in the absence of the catalyst.

The combination of catalyst and reactants is represented in the equations below where a catalyst (**C**) combines with the reactant, known as the **substrate** (**S**), to form a catalyst-substrate complex (**C-S**). This catalyst-substrate complex breaks down forming the product (**P**) and free catalyst.

Stage 1	$C + S \rightarrow C\text{-}S$
Stage 2	$C\text{-}S \rightarrow P + C$
Overall reaction	$S \longrightarrow P$

Notice how the catalyst is used up in Stage 1 and regenerated in Stage 2 of this reaction.

Like manganese(IV) oxide, iodide ions (I^-) can also speed up the rate of decomposition of hydrogen peroxide into water and gaseous oxygen. The two-stage reaction can be represented as shown on the following page.

ENERGY AND ENZYMES

$$\text{Stage 1} \quad H_2O_2 + I^- \longrightarrow H_2O + IO^-$$
$$\text{Stage 2} \quad H_2O_2 + IO^- \longrightarrow H_2O + O_2 + I^-$$
$$\text{Overall} \quad 2H_2O_2 \longrightarrow 2H_2O + O_2$$

The catalyst, therefore, provides an alternative pathway for the reaction. The activation energy for this alternative is lower than that without the catalyst (Fig 5.3). As a result, the reaction with the catalyst occurs more readily and the rate of reaction is faster.

The chemical reactions making up a cell's metabolism occur rapidly and in an organised way because of special biological catalysts called **enzymes**.

QUESTIONS

5.1 What is meant by activation energy of a chemical reaction?

5.2 Explain why many chemical reactions occur faster if the reactants are heated in a test tube.

5.3 **(a)** Define the term catalyst.
(b) List five properties shown by all catalysts.
(c) Explain how catalysts are thought to help reactions occur faster.

5.2 ENZYMES

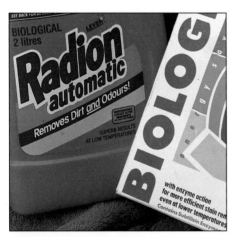

Fig 5.4 Biological washing powders not only get some clothes cleaner, they do so at lower temperatures than conventional washing powders.

Enzymes are catalysts made within cells according to instructions encoded by the cell's DNA. Whilst some are not associated with organelles, many are bound to membranes within the endoplasmic reticulum, mitochondria, chloroplasts and Golgi apparatus. A small number may be secreted and work outside the cell. Enzymes speed up the rate of metabolic reactions; without them these reactions would occur so slowly that life would not be possible. As with all catalysts, enzymes are not used up and can be recovered at the end of the reactions they catalyse. Unless secreted, a small amount of enzyme will, therefore, last for a long time: a cell need not waste energy producing vast amounts of each enzyme. All enzymes have a complex globular shape and are relatively large molecules, e.g. the bacterial enzyme used in biological washing powders, called subtilisin, has a relative molecular mass of 27 600.

 Why is only a small amount of catalyst required to increase the rate of a reaction?

Enzymes reduce the activation energy

Like other catalysts, enzymes (**E**) react with their substrates (**S**) forming an intermediate enzyme-substrate complex (**E-S**):

$$\text{Stage 1} \quad E + S \longrightarrow E\text{-}S$$
$$\text{Stage 2} \quad E\text{-}S \longrightarrow E + P$$
$$\text{Overall} \quad S \longrightarrow P$$

The activation energy for this pathway is less than the alternative without the enzyme. A single reaction, therefore, happens more readily; more reactions happen in a specified time interval, i.e. the reaction is faster. Like manganese(IV) oxide and iodide ions, the enzyme **catalase** speeds up the decomposition of hydrogen peroxide into water and gaseous oxygen. Catalase lowers the activation energy of this reaction from around 75 kJ mol^{-1} without a catalyst to around 21 kJ mol^{-1}. At 0 °C, a temperature at which enzyme-controlled reactions are usually thought to be slow, one molecule of catalase catalyses the decomposition of about 50 000 molecules of hydrogen peroxide each second.

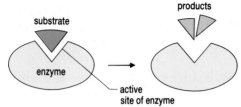

Fig 5.5 The lock and key model of enzyme action. A substrate molecule of the correct shape combines with the active site of an enzyme to form an enzyme-substrate complex.

Lock and key model of enzyme activity

All enzymes have a complex globular shape (see Section 1.2 for the tertiary structure of proteins). The substrate(s) combine with only one small part of this globular enzyme molecule, a cleft or cavity in the enzyme, called the **active site** (Fig 5.5). An enzyme's active site has regions which form chemical bonds to hold the substrate(s) in place. Separate from the binding groups, other groups of atoms within the active site speed up the chemical reaction of substrate(s) to product(s). Most enzyme reactions involve two or three such catalytic groups. Notice in Fig 5.5 how the shape of the enzyme's active site exactly mirrors the part of the substrate molecule which combines with it. This configuration has led to the name of this **lock and key model** of enzyme action.

Like all models, the lock and key model is misleadingly simple. For, although the lock and key model explains the specificity of enzymes, it does not provide a totally satisfactory explanation of how enzymes work. In particular, it does not explain how enzymes reduce the activation energy of a chemical reaction.

The **induced fit model** of enzyme action proposes that the structure of the active site, the lock, is not rigid. Once a suitable substrate combines with its binding groups, a conformational change (change in shape) occurs so that the enzyme closes up to enfold the substrate. The lock is, therefore, fluid wrapping itself around a key of the correct type to fit it more closely. This change in the shape of the enzyme promotes the formation of the transition state (Fig 5.2), so reducing the activation energy of the reaction.

QUESTIONS

5.4 Represent the stages in the catalase-catalysed decomposition of hydrogen peroxide
(a) in a word equation
(b) in labelled diagrammatic form using your own invented molecular shapes.

5.5 At 0 °C, catalase is said to have a **turn-over number** of 50 000. Suggest what this means.

5.3 THE PROPERTIES OF ENZYMES

Enzymes share the properties of other catalysts described in Section 5.1. They

- remain chemically unaltered by the reactions they catalyse
- are not used up in the reaction they catalyse
- do not make reactions occur which would otherwise not happen
- do not alter the amount of product formed.

They also show other, unique properties which result from their complex globular shape and the need for their active site to mirror precisely that of the substrate molecules.

Specificity

Whilst catalysts may be specific in the types of reaction they catalyse, enzymes always are. It is easy to understand this if you look back at the lock and key model of enzyme action in Fig 5.5. The active site of the enzyme molecule must have an exact chemical configuration if it is to form chemical bonds with its substrate. Sometimes only part of the substrate is involved in the bonding of the enzyme-substrate complex; a number of similar molecules with the same arrangement of atoms will fit into the active site and their reaction may be catalysed. In other cases, no other molecule has exactly the same arrangement of atoms as the substrate and so will not fit into the active site of the enzyme. The hydrolysis of most lipids is catalysed by an enzyme called lipase. An amylase, on the other hand, only catalyses the hydrolysis of the α1,4 glycosidic bonds between glucose units in starch to produce maltose and cannot catalyse the hydrolysis of the β1,4 glycosidic bonds in another chain of glucose units, cellulose (these bonds are represented in Fig 1.18). Hydrolysis of β1,4 glycosidic bonds is catalysed by cellulase.

 5 **Why can you not survive on a diet of grass whilst a cow can?**

NAMING ENZYMES

The names of most enzymes refer to the specific type of reaction which they catalyse. These names are formed by adding the suffix 'ase' to the name of the substrate. For example, amylase catalyses the hydrolysis of amylose, proteases catalyse the hydrolysis of proteins and β-galactosidase catalyses the hydrolysis of the terminal β-D-galactose units in β-galactosides such as lactose. Enzymes can also be grouped together according to the types of reaction they catalyse. The names of these groups, the major six of which are shown in Table 5.1, are also formed by adding the suffix 'ase' to the type of reaction. Enzymes are not always named so informatively, however. Although the suffix 'ase' is present, the name catalase does not refer to the action of the enzyme. Many of the mammalian digestive enzymes described in Chapter 8 do not conform to the 'ase' system at all, e.g. pepsin.

Table 5.1 Enzymes are classified according to the reactions they catalyse. This table shows the major six groups of enzymes.

Group of enzymes	Type of reaction catalysed
hydrolase	hydrolysis (splitting of a molecule into smaller parts using water)
isomerase	rearrangements of atoms within a molecule
ligase	formation of a larger molecule from two smaller molecules using energy released by the breakdown of ATP
lyase	breakage of a double bond without addition of water and addition of a new group to the 'freed' bonds
oxidoreductase	oxidation–reduction (redox) reactions
transferase	transfer of a group of atoms from one molecule to another

Susceptibility to temperature changes

All chemical reactions can be speeded up by heating the reactants. Heat energy is transferred to the reactant molecules, ensuring that more colliding molecules have sufficient energy for the reaction to occur.

Over a certain temperature range, an increase in temperature increases the rate of an enzyme-catalysed reaction (Fig 5.6). The substrate and enzyme molecules collide with more energy so more of those collisions result in the formation of product. However, heat also causes a change in the shape of protein molecules, called **denaturation**. Denaturation occurs when the weak bonds, which stabilise a protein's secondary and tertiary structure, break. You have seen the results of this change if you have

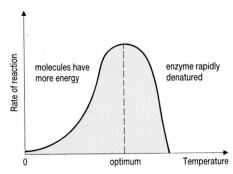

Fig 5.6 The overall effect of temperature on the rate of an enzyme-catalysed reaction is the result of two underlying and opposing processes: the effect of temperature on the thermal motion and hence collision rate of molecules and thermal denaturation of enzymes.

KEY CONCEPT

TEMPERATURE COEFFICIENT Q$_{10}$

The effect of a 10 °C rise in temperature on the rate of a chemical reaction is termed its **temperature coefficient (Q$_{10}$)**. It can be calculated from the equation

$$Q_{10} = \frac{\text{rate of reaction at 10 °C above certain temperature}}{\text{rate of reaction at certain temperature}}$$

As a rule of thumb, the rate of a chemical reaction doubles for each 10 °C rise in temperature (Q$_{10}$ = 2). This is true for most enzyme-catalysed reactions between about 4 °C and the optimum temperature, whereafter the denaturation of the enzyme reduces the reaction rate.

cooked an egg. The white of a fresh egg, which is mainly a protein called albumen, is colourless and fluid; as it cooks it becomes white and solidified as the albumen is denatured. When a cooked egg cools, the egg-white stays white and solid: denaturation is irreversible.

 6 **What effect will denaturation have on the shape of an enzyme's active site? Why will this prevent catalysis occurring?**

The temperature at which an enzyme works best is a compromise between the increase in the number of successful collisions between reactants and the rate of destruction of enzyme molecules – both caused by an increase in temperature. The **optimum temperature** for an enzyme is this compromise temperature.

The **optimum temperature** (see Fig 5.6) for most enzymes is in the range of 30–40 °C and is one factor which limits the environments in which organisms can survive. Note that many enzymes work well outside this range of temperatures. The ability of catalase to catalyse the decomposition of 50 000 molecules of hydrogen peroxide per second at 0 °C has already been mentioned; enzymes of the sulphur bacterium *Desulfotomaculum* sp. have an optimum temperature of about 70 °C.

7 **How could you use the temperature dependence of biochemical reactions to determine if a particular reaction was catalysed by an enzyme?**

Susceptibility to pH changes

Proteins are denatured by changes in pH as well as by changes in temperature. Section 1.6 describes how changes in pH affect the ionisation of amino acids. As the ionisation of its constituent amino acids changes, the ionic bonds which help to stabilise protein shape are broken. As a result, the protein changes shape. For example, casein, the protein present in milk, is denatured when the pH of milk falls to about 4.5. It is the denatured casein which precipitates, forming the solid part (curds) of sour milk. Adding lactic acid bacteria to milk to precipitate the casein is a starting point in the manufacture of cheeses. Changes of pH may also affect the charge of key amino acids at an enzyme's active site, preventing the active site from combining with its substrate.

Fig 5.7   Cheese manufacture depends on a bacterium which produces lactic acid and so lowers the pH in the cheese-making vat.

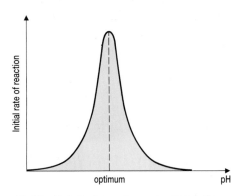

Fig 5.8 Typical bell-shaped curve showing how the initial rate of an enzyme-catalysed reaction varies with pH.

 If the pH of your blood goes outside a narrow range of pH, about 7.0–7.8, you will die. Explain this observation. What type of substance must your blood contain to maintain its pH within this narrow range?

The effect of changes in pH on the rate of an enzyme-catalysed reaction is shown in Fig 5.8. The enzyme molecule changes its shape over the range of pH values and only over a narrow pH range is the active site correctly formed.

 The enzyme pepsin works well at the low pH values found in your stomach. What does this suggest about pepsin?

Enzyme concentration

The rate of an enzyme-controlled reaction is dependent on the number of successful collisions between molecules of enzyme and substrate(s). A typical concentration of a particular enzyme in a cell is about 1 μmol dm^{-3} and of its substrate about 10 mmol dm^{-3}, i.e. about 10 000 times greater. If enzyme molecules are added, more successful enzyme-substrate collisions will occur and the rate of reaction will increase. Thus, the rate of an enzyme-catalysed reaction is directly proportional to the concentration of enzyme, as long as there is excess substrate (Fig 5.9).

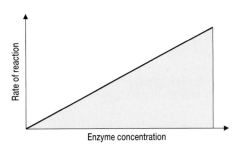

Fig 5.9 The straight line relationship between enzyme concentration and the rate of an enzyme-catalysed reaction. Such a relationship is described as directly proportional.

Substrate concentration

For a fixed enzyme concentration, the rate of reaction is affected by increases in substrate concentration as shown in Fig 5.10. An increase in the number of substrate molecules (region A) increases the number of successful enzyme-substrate collisions, so the rate of reaction is faster. At higher substrate concentrations (region B) the active site of every enzyme is occupied at any given moment. Any added substrate molecules have to 'wait' until an existing enzyme-substrate complex dissociates to release product(s) and free enzyme. Consequently, adding more substrate has no effect on the rate of the enzyme-catalysed reaction. If further enzyme is added, the reaction rate can increase again.

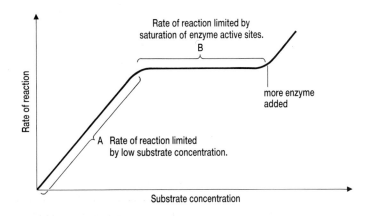

Fig 5.10 The relationship between substrate concentration and the rate of an enzyme-catalysed reaction.

ENZYMES AND DIAGNOSING DISEASE

The linear relationship between rate of reaction and enzyme concentration, shown in Fig 5.9, provides a valuable diagnostic tool for doctors. By measuring the rate of a particular enzyme-catalysed reaction under standard conditions, i.e. same temperature and substrate concentration, it is possible to compare the concentration of an enzyme in two different solutions. This procedure, called enzyme assay, can be done on different body fluids, especially plasma, and so provide valuable diagnostic information for doctors. For example, an increase in the concentration of the enzyme alcohol dehydrogenase in the plasma would indicate that something was wrong with the patient's liver since this enzyme is only found within liver cells.

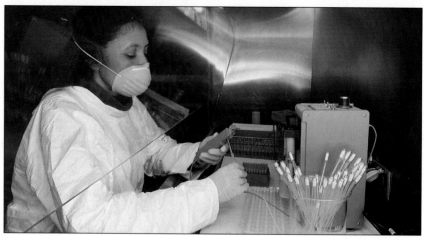

Fig 5.11   Enzyme assays are used routinely as a diagnostic tool. A hospital technician prepares a sample of human blood serum for enzyme assay tests.

ANALYSIS

An enzyme investigation

This exercise develops an understanding of experimental procedures for investigating an environmental factor and the ability to handle data.

The following procedure was used to discover the optimum pH for a bacterial amylase, an enzyme which digests starch.

Separate solutions of amylase and starch were prepared in two glass beakers <u>using distilled water</u> and kept at <u>37 °C in a water bath.</u> Into each of five test tubes, also kept in the water bath, 5 cm³ of different buffer solution, pH 4, 5, 6, 7 and 8, were pipetted. All these solutions <u>were left for 30 minutes.</u>

At the end of this time, 1 cm³ of enzyme solution was placed into each of the five buffer solutions. <u>After 10 minutes,</u> 0.1 cm³ of starch solution was added to each of the tubes containing the enzyme and buffer.

At one minute intervals, a drop of the mixture in each tube was tested for the presence of starch. The time taken for all the starch to be hydrolysed, the achromatic time (the time when the colour disappears) was noted for each tube. The results are shown in Table 5. 2.

(a) Explain the reason for each of the underlined procedures.
(b) Why were buffer solutions used?
(c) How could the investigator have tested for the presence of starch?
(d) Plot the data and determine the optimum pH range for this enzyme. How could you define the optimum pH more precisely?

As a **control**, the investigator added starch to five more tubes of buffer solution at pH 4, 5, 6, 7 and 8.

(e) What do you understand by the term control?
(f) What would the investigator have found was occurring in each of the control tubes?
(g) Why do these tubes serve as controls for this investigation?
(h) Write a detailed experimental procedure which would enable you to determine the optimum temperature for this enzyme, now you know its optimum pH. Do not forget to include controls.

Table 5.2

pH	Achromatic time / s
4	720
5	540
6	240
7	180
8	240

5.6 5 cm³ of substrate and 1 cm³ of enzyme were both pipetted into two tubes. One tube was put in a water bath at 20 °C and the second in a water bath at 30 °C. At 20 °C the reaction took eight minutes to complete, at 30 °C the reaction was completed in four minutes.
(a) What was the rate of reaction at the two temperatures?
(b) Calculate the Q_{10} for this reaction.

5.7 Examine Fig 5.6.
Explain the relationship between temperature and rate of reaction
(a) up to the optimum temperature
(b) above the optimum temperature.

5.8 Explain why enzymes work within a narrow pH range to catalyse only one type of reaction.

5.9 As the manager of a detergent company which manufactures Whammo 'the new wonder, washes whiter than all other leading brands' biological washing powder, you have received the following complaint:
    '... my tea towels, after having had a good hot wash with Whammo, came out as dirty as they had gone in.'
Explain to this customer, who is *not* a biologist, why a low temperature wash produces the desired results rather than a 'good hot wash'. Simple diagrams may help your explanation.

5.4 ENZYME INHIBITION

Inhibitors are substances which, when added to a mixture of enzyme and substrate solutions, reduce the rate of reaction. They do this by combining with the enzyme molecules to form enzyme-inhibitor complexes, which cannot combine with substrate molecules. If an enzyme-inhibitor complex can dissociate,

i.e. $E + I \rightleftharpoons EI$, the inhibition is **reversible**.

If an enzyme-inhibitor complex cannot dissociate,

i.e. $E + I \longrightarrow EI$, the inhibition is **irreversible.**

Competitive and non-competitive inhibition

Competitive inhibitors combine with the active site of an enzyme molecule (Fig 5.12), thereby stopping substrate molecules from attaching.

 What property must the competitive inhibitor have in common with the substrate molecule?

The effect of a competitive inhibitor can usually be reduced by increasing the concentration of substrate, since this increases the likelihood of an enzyme-substrate collision over an enzyme-inhibitor collision.
A **non-competitive inhibitor** (see Fig 5.13) does not combine with the active site of the enzyme but with another part of the enzyme molecule. In doing so it changes the shape of the enzyme molecule (**conformational change**), so that the substrate can no longer combine with the active site.

Allosteric inhibition and activation

Some enzyme molecules have another specific site, positioned well away from their active site, which can combine with substances other than the substrate. These sites are called **allosteric sites** and enzymes possessing them are called **allosteric enzymes**. By combining with an enzyme's

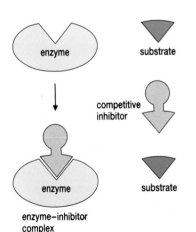

Fig 5.12 A **competitive inhibitor** binds to the active site of an enzyme, thereby preventing the binding of the substrate. It can be displaced, however, by a substrate molecule, so the substrate and inhibitor **compete** for the active site.

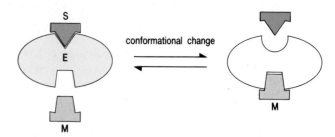

Fig 5.13 A **non-competitive inhibitor (M)** binds to a site away from the active site, altering the configuration of the enzyme's active site and so rendering the enzyme (E) inactive. Notice that the substrate (S) and the inhibitor are not competing for this binding site hence the term non-competitive inhibitor.

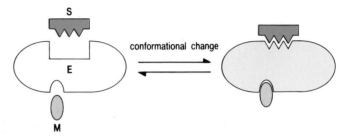

Fig 5.14 An enzyme's active site may be activated by the binding of an allosteric modulator. S = substrate; E = enzyme; M = allosteric modulator.

allosteric site, a substance causes a reversible change in the structure of the enzyme's active site. **Allosteric inhibitors** reversibly combine with an enzyme's allosteric site and slow down the rate of the enzyme-controlled reaction by changing the enzyme's active site and reducing its ability to combine with the substrate. Thus, allosteric inhibition is one particular type of non-competitive inhibition (Fig 5.13). Alternatively, an enzyme may be activated by an **allosteric modulator** (Fig 5.14). An example of allosteric inhibition is included in the series of reactions depicted in Fig 5.15.

EXPLOITING ENZYME INHIBITION

If one of their vital enzymes is inhibited, organisms are killed. Enzyme inhibition has, thus, been of great use in controlling disease-causing organisms, disease-carrying vectors and agricultural pests.

Before the invention of poisons such as warfarin, rats were often killed using the poison cyanide. This was effective because it is an irreversible inhibitor of the enzyme cytochrome oxidase, a vital enzyme in aerobic respiration. Poisoned rats died because they were unable to produce ATP, the source of energy in their cells. Unfortunately, cyanide has an identical effect on humans, farm animals and pets, restricting its general usefulness. Warfarin itself exploits a biological reaction, antagonising the action of vitamin K in promoting blood clotting thus inhibiting blood clotting and causing rats to bleed to death when wounded.

Animals coordinate many of their activities using nerve impulses and an enzyme, cholinesterase, essential for the passage of nerve impulses. Aphids (greenfly and blackfly), which are serious agricultural pests, are controlled using an aphicide that is a

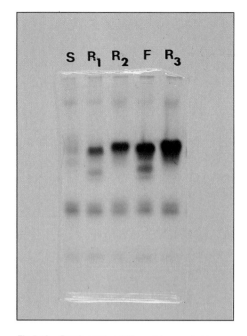

Fig 5.16 Peach-potato aphids contain an esterase enzyme that degrades insecticides by hydrolysis before they reach their target in the insect nervous system. This photograph shows the electrophoretic separation of esterases from susceptible (S) and insecticide-resistant aphids. The increasing amount of one of these esterases from the aphid cultures R₁ through to R₃ is responsible for a progressive increase in resistance to insecticides.

Acetylcholinesterase is sometimes referred to simply as cholinesterase.

SPOTLIGHT *continued*

competitive inhibitor of their acetylcholinesterase enzyme. Whilst it performs exactly the same function, the acetylcholinesterase of other insects, birds and mammals is sufficiently different from that of aphids for the pesticide to have no effect.

Folic acid is an important coenzyme. Many bacteria which cause diseases in humans need para-aminobenzoate in their 'diet' in order to produce this coenzyme. Sulphanomide drugs, which were used extensively during World War II, are similar in structure to para-aminobenzoate. They, therefore, act as competitive inhibitors of one of the bacterial enzymes involved in making folic acid. Human cells, which do not possess the enzymes needed to make folic acid from para-aminobenzoate, are unaffected by sulphonamide drugs, making them safe to use.

End-product inhibition

Many metabolic processes involve a series of enzyme-catalysed reactions in which the product from one reaction is the substrate for the next. The reactions involved in aerobic respiration and in photosynthesis are examples with which you will become familiar. The enzymes which catalyse such chain reactions often form a linear series bound to membranes within the cell, forming a **multi-enzyme complex**. Such close proximity of enzymes is efficient since collisions of enzyme and its substrate are made more likely.

A common feature of multi-enzyme complexes is that the final product is an allosteric inhibitor of one of the earlier enzymes in the pathway. An accumulation of final product will, thus, slow or stop its further production. Such inhibition of an earlier stage in a process is called **negative feedback**. Look again at Fig 5.15: it shows how a multi-enzyme complex might operate and how it might be regulated by end-product inhibition.

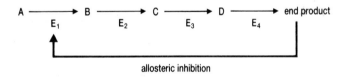

Fig 5.15 Allosteric inhibition of a rate-limiting enzyme by the end product of a reaction sequence. Enzyme E_1 is, in effect, a molecular switch which is turned on or off depending on the concentration of the end product, i.e. low concentration = on, high concentration = off.

Enzyme cofactors

Many enzymes only work efficiently in the presence of a second substance, called a **cofactor**. These cofactors may be individual ions or may be complex organic molecules (though never protein). Three types are recognised.

1. **prosthetic groups**

    These are organic cofactors which are permanently combined with the enzymes they assist. Catalase, the enzyme which speeds the decomposition of hydrogen peroxide, has an iron-containing **haem** prosthetic group.

2. **coenzymes**

These are organic cofactors which are not bonded to the enzyme molecules they assist. A number of the vitamins needed in our diet are used to make coenzymes. Nicotinamide adenine dinucleotide (**NAD**), an important coenzyme in respiration (see Section 6.1), is made from nicotinamide, one of the B-group vitamins.

3. **enzyme activators**

These are inorganic ions which act as cofactors. By combining with either the enzyme or the substrate, they are thought to make the formation of enzyme-substrate complex occur more easily, so increasing the rate of enzyme-catalysed reaction. For example, the activity of salivary amylase is increased by chloride ions.

QUESTIONS

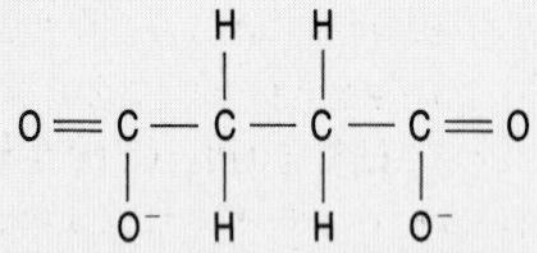

succinate (substrate)

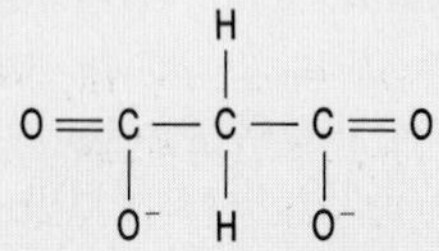

malonate (inhibitor)

Fig 5.17 The structural formulae of succinate and malonate.

5.10 The enzyme succinic dehydrogenase, which catalyses the conversion of succinate to fumarate during cell respiration, is inhibited by malonate. The structural formulae of succinate and malonate is given in Fig 5.17.
(a) Explain the likely method by which succinic dehyrogenase is inhibited.
(b) How could you test your hypothesis?

5.11 Food decays when digestive enzymes inside the food or those released by bacteria and fungi living in the food catalyse the breakdown of its molecules.
(a) Before canned food is sealed into its can it is heated to high temperatures. Explain why (i) heating and (ii) sealing the can reduce food decay.
(b) The complaints department of a large canned-food manufacturer received a dead insect which had reportedly been found in a tin of their beans. Their technician squashed the insect in water and tested this extract for the presence of amylase. How would the results of this test show whether the insect had fallen in the can during processing or after the can had been opened?

5.5 COUPLED REACTIONS

Look back at Fig 5.2 which shows the energy changes during two types of chemical reaction which occur in cells. Although the exergonic reaction in Fig 5.2(a) releases energy, a small amount of energy is needed to start it. The endergonic reaction in Fig 5.2(b) requires a greater input of energy since the potential energy of its products is greater than that of the reactants. Many of the chemical reactions occurring in cells are endergonic, for example the synthesis of enzymes, the transport of sodium ions across the cell surface membrane and the movement of endocytic vacuoles through the cytoplasm. To ensure these endergonic reactions occur, cells need a source of energy in a form that can be released instantaneously in amounts which are small enough for the cell to use. The most important source of energy in all cells is **adenosine triphosphate**, **ATP** (Fig 5.18).

11 ▸ **Why would the release of too much energy harm the cell?**

When a phosphate group is removed from ATP, **adenosine diphosphate** (**ADP**) is formed and 30.6 kJ mol^{-1} of energy is released. The removal of a second phosphate group forms adenosine monophosphate (AMP) and 30.6 kJ mol^{-1} of energy is again released. Removal of the final phosphate group to form adenosine results in only 13.8 kJ mol^{-1} of energy being released, leading to the notion that the second and third phosphate groups

---

ENERGY AND ENZYMES

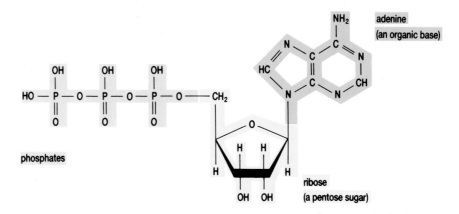

Fig 5.18 A molecule of adenosine triphosphate (ATP). Note ATP is a nucleotide consisting of the 5-carbon sugar **ribose**, the nitrogen-containing base **adenine** and three phosphate groups. Phosphate groups can be removed from ATP by hydrolysis, releasing energy.

are linked to the adenosine by 'high energy bonds'. In fact the release of energy involves changes within the whole molecule other than just the breakage of one bond and so the notion of 'high energy bonds' is not accurate. These reactions can be summarised in the equations below.

$$ATP + H_2O \rightarrow ADP + H_3PO_4 + 30.6\,kJ\,mol^{-1}$$
$$ADP + H_2O \rightarrow AMP + H_3PO_4 + 30.6\,kJ\,mol^{-1}$$
$$AMP + H_2O \rightarrow Adenosine + H_3PO_4 + 13.8\,kJ\,mol^{-1}$$

ATP breakdown releases energy which can be used by other energy-requiring reactions occurring in cells. Such linking of the release of energy from the breakdown of ATP with an energy-requiring reaction is called **coupling**. The amount of energy released from ATP breakdown is more than is required by the reaction to which it is coupled: some energy is always lost as heat. One example of coupled reactions is shown in Fig 5.19. Further examples are shown in Fig 6.1 and Fig 6.5.

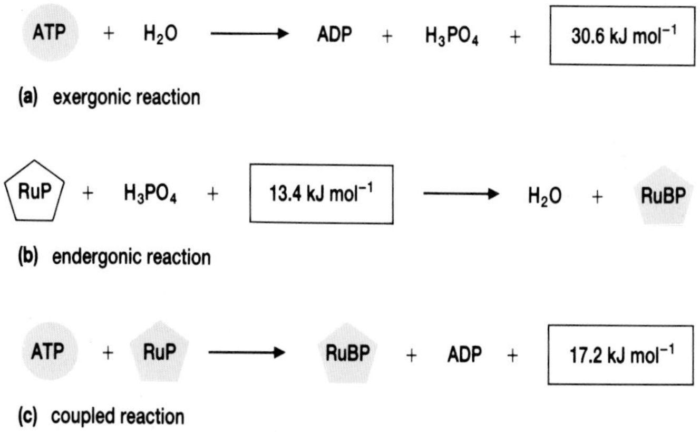

$$ATP + H_2O \longrightarrow ADP + H_3PO_4 + \boxed{30.6\ kJ\ mol^{-1}}$$

(a) exergonic reaction

$$RuP + H_3PO_4 + \boxed{13.4\ kJ\ mol^{-1}} \longrightarrow H_2O + RuBP$$

(b) endergonic reaction

$$ATP + RuP \longrightarrow RuBP + ADP + \boxed{17.2\ kJ\ mol^{-1}}$$

(c) coupled reaction

Fig 5.19 Coupled reactions occur when an exergonic reaction is used to 'power' an endergonic reaction. In this case about 17.2 kJ mol^{-1} of energy might be released as heat when ribulose bisphosphate (RuBP) is formed from ribulose phosphate (RuP).

SUMMARY

Hundreds of chemical reactions, collectively called metabolism, occur in the cytoplasm of cells and keep the cells alive. The molecules and ions involved in metabolism are called metabolites. In catabolic reactions large molecules are broken down into simpler ones whilst in anabolic reactions, small molecules are joined to form larger ones. Catabolic reactions release energy, i.e. are exergonic, whilst anabolic reactions require energy, i.e. are endergonic.

An input of energy, called the activation energy, is needed before any reaction can proceed. Catalysts reduce the activation energy of the reactions they catalyse, making the reaction proceed faster. They do not alter the nature or balance of products, neither are they consumed during the reaction they catalyse.

All metabolic reactions are catalysed by enzymes, the vast majority of which are proteins. Enzymes have a complex three-dimensional shape. When an enzyme and its specific substrate molecule(s) collide in a precise way, the active site of the enzyme molecule combines with the substrate. As a result, an enzyme-substrate complex is formed which rapidly breaks down to release free enzyme and product molecule(s). The complementary structure of substrate and active site are likened to a lock and key.

The catalytic properties of enzymes are affected by any agent which changes the chemical nature of their active site. Excessively high temperatures, changes in pH and certain chemical inhibitors have this effect. Some types of inhibitor do not change the chemical nature of the active site but combine reversibly with it. The effect of these competitive inhibitors can be reduced by increasing the concentration of substrate.

The rate of enzyme-catalysed reactions can be increased by increasing the concentration of enzyme. Increases in the concentration of substrate may increase the rate of reaction only if the enzyme concentration is not a limiting factor. Because a rise in temperature increases the random thermal movement of molecules, it increases the rate of enzyme-catalysed reactions. At a certain value, further increases in temperature cause progressive denaturation of enzyme molecules, with a consequent fall in the rate of reaction.

Many enzymes work only in the presence of cofactors, which may be inorganic ions, called enzyme activators, or organic molecules. Some organic cofactors, called prosthetic groups, are permanently bound with the enzyme they help; others, called coenzymes, are not bound to their enzyme. Some of the vitamins needed in our diet are used to make coenzymes.

Some enzyme-catalysed reactions within the cytoplasm are linked in a series, where the product of one reaction is the substrate of the next. Others are coupled, where the energy released by an exergonic reaction is used to drive an endergonic reaction.

Chapter 5: Answers to Quick Questions

1　Anabolic; catabolic; anabolic.
2　Exergonic
3　The reaction has a high activation energy.
4　The catalyst is rapidly regenerated.
5　You lack the enzyme cellulase needed to break the β1,4-glycosidic bonds in cellulose, but a cow's symbiotic gut flora produce this enzyme.
6　Destroys its shape. This will prevent the enzyme from locking onto its substrate, so preventing the formation of an enzyme-substrate complex.
7　Boiling the reaction mixture will inhibit an enzyme-catalysed reaction.
8　The majority of enzymes in your body work best in the pH range 7.0–7.8. If the pH goes outside this range the enzymes are denatured and do not work efficiently so reducing the efficiency of your metabolism. Buffers in the blood, e.g. haemoglobin, maintain the pH.

9 The amino acid sequence in the molecule is such that it is able to maintain its tertiary structure only at low pH.
10 Must have a similar shape.
11 Lead to an increase in the cell's temperature, so denaturing the cell's enzymes.

Chapter **6**

ENERGY RELEASE: RESPIRATION

LEARNING OBJECTIVES

When you have studied this chapter you should be able to:

1. outline the processes by which ATP is made during the aerobic and anaerobic respiration of glucose;

2. compare the efficiency of the aerobic and anaerobic respiration of glucose;

3. explain how amino acids, fatty acids and glycerol can also be used to make ATP;

4. calculate and interpret data relating to respiratory quotients (RQ);

5. appreciate how complex metabolic pathways can be controlled.

Fig 6.1 Autotrophs make sugars by photosynthesis using water, carbon dioxide and light energy. These sugars form the starting point for the biosynthesis of other compounds, e.g. cellulose, proteins and lipids. Heterotrophs digest and assimilate the organic compounds made by autotrophs. In both types of organism, the endergonic reactions of biosynthesis are driven by ATP hydrolysis. This molecule, the universal energy currency of cells, is made during the process of respiration.

Two biological processes which are explored in this and the next chapter are summarised in Fig 6.1. Autotrophic organisms, such as trees, use sunlight to synthesise sugar molecules from carbon dioxide and water: the process of **photosynthesis**. In turn these sugar molecules form the starting point for the synthesis of other organic molecules. The organic compounds may then pass to heterotrophic organisms, like elephants. The oxidation of organic molecules in the cells of autotrophs and heterotrophs releases energy which is used to synthesise ATP: the process of **respiration**. Since respiration is common to the metabolism of all organisms it is discussed first. The next chapter deals with photosynthesis.

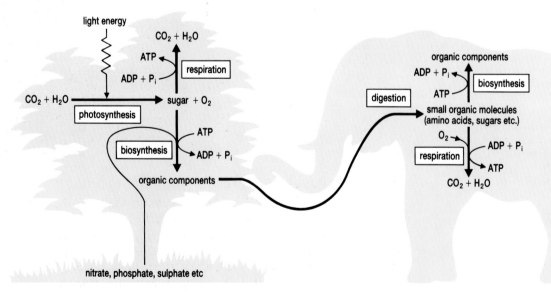

What types of compounds will the tree in Fig 6.1 synthesise using sulphates, phosphates and nitrates?

6.1 ENERGY FROM FOOD

To make ATP during respiration, organic molecules are needed to act as fuel. The first step in releasing their energy is to break these complex molecules into simple ones by digesting them, e.g. starch is broken down into glucose, protein to amino acids and so on. These small molecules can then serve as fuel once they have been absorbed into a cell.

For the moment we will concentrate on carbohydrates, in particular glucose, as a fuel source. This sugar is a major fuel molecule in all cells. Indeed, in the mammalian brain it is the only molecule which can be used for this purpose. Section 6.4 describes how cells can also catabolise fatty acids, glycerol and amino acids to make ATP.

Fig 6.2 This chocolate represents a store of chemical energy which is inaccessible to your body until it has been transformed.

The role of ATP in cells

The idea was introduced in Chapter 5 that endergonic metabolic reactions can be made to occur by coupling them with the hydrolysis of ATP to ADP and P_i (P_i is the shorthand way of writing H_3PO_4). To keep endergonic reactions going, cells must continually resynthesise ATP from ADP and P_i. This is itself an endergonic reaction and so requires a source of energy to drive it. One way of providing this energy is to oxidise glucose.

ENERGY RELEASE: RESPIRATION

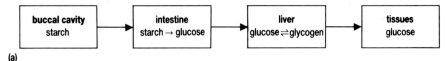

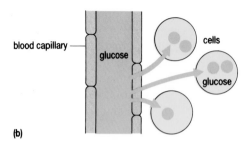

(a)

(b)

Fig 6.3   Once formed by digestion, glucose enters the bloodstream and is transported to the liver. Here it may be temporarily stored or sent, in the blood, to the tissues where it is absorbed by cells.

You might wonder why the breakdown of glucose is not directly coupled to the cell's endergonic reactions, eliminating the need for ATP. Practically all cells use ATP as their energy source for metabolism for two reasons.

- Energy release from ATP hydrolysis is instantaneous: catabolism of glucose takes some time.
- Catabolism of one molecule of ATP releases only a small amount of energy: catabolism of one molecule of glucose releases much more.

 Why do these features make ATP a better immediate energy source than glucose during cell metabolism?

In addition, linking all endergonic reactions to ATP hydrolysis means that the cell can economise on enzymes. Only one enzyme is needed to hydrolyse ATP whereas many are needed to release the energy contained in a glucose molecule. These three reasons – instantaneous access to energy that is released in small, controllable amounts using only one enzyme to release it – probably explain why ATP has become the universal energy currency in cells.

QUESTIONS

6.1 Why must cells always have a supply of ATP molecules in their cytoplasm if they are to stay alive?

6.2 Energy for ATP synthesis is released from the breakdown of glucose. Explain why the breakdown of glucose cannot be used directly to power the cell's work.

TUTORIAL

Oxidation and reduction

- Oxidation is often represented as the addition of oxygen to an element or compound:

$$A + O_2 = AO_2$$

- Thus, during the oxidation of glucose to carbon dioxide and water, oxygen is added.

$$C_6H_{12}O_6 + 6O_2 \rightarrow 6CO_2 + 6H_2O$$

- More accurately, the terms oxidation and reduction apply to any reaction in which electrons (e^-) are transferred from one reactant to another. Oxidation is the removal of electrons, reduction the addition of electrons.

- Thus, in the reaction

$$Fe^{3+} + e^- \rightarrow Fe^{2+}$$
$$\text{Iron(III)} \qquad\qquad \text{Iron(II)}$$

the iron(III) ion has been reduced to form the iron(II) ion.

(a) Write an equation to show the oxidation of an iron(II) ion to an iron(III) ion.

- Electron transfer also occurs when hydrogen atoms move from one substance to another

$$AH_2 + B \rightarrow A + BH_2$$

(Remember a hydrogen atom consists of a proton, H^+, and an electron, e^-)

- The above equation also demonstrates another important point: whenever one substance is reduced (i.e. $B \rightarrow BH_2$) another is oxidised ($AH_2 \rightarrow A$). Such reactions are therefore called **redox reactions**.

- Two important redox reactions in respiration involve coenzymes: **nicotinamide adenine dinucleotide (NAD)** and **flavine adenine dinucleotide (FAD)**. Both these molecules act as electron transporters by carrying hydrogen atoms, and the electrons they contain, from glucose, or other fuel molecules, to other molecules in the respiratory pathway.

(b) In the reactions shown below, which is the oxidised and which the reduced form of NAD and FAD?

$$NAD^+ + 2H \rightarrow NADH + H^+$$
$$FAD + 2H \rightarrow FADH_2$$

(c) Identify which of the following reactants have been oxidised and which reduced. Explain your answers.

$$Cu^{2+} + e^- \rightarrow Cu^+$$
$$NADH + H^+ + FAD \rightarrow NAD^+ + FADH_2$$

6.2 GLUCOSE CATABOLISM

Before starting Section 6.2, you should work through **Tutorial: Oxidation and reduction** since these processes lie at the heart of ATP synthesis.

Before examining the details of the pathway by which glucose is broken down, you need to know why cells can obtain energy by oxidising glucose. The carbon and hydrogen atoms in cells, for example in glucose molecules, are not in their most stable form. The most energetically stable form of carbon is carbon dioxide and the most energetically stable form of hydrogen is water. A cell can therefore obtain energy from sugar molecules or amino acids or fatty acids by allowing the carbon and hydrogen atoms in these molecules to combine with oxygen, producing carbon dioxide and water respectively.

However, cells do not oxidise these molecules in one step (Fig 6.4(a)), as would occur if we were burning glucose on a spoon. Instead, the oxidation occurs through a series of reactions, called **aerobic respiration**. Molecular oxygen is involved only at the very end of this metabolic pathway (Fig 6.4(b)). Notice, however, that the total amount of energy released in both cases, 3000 kJ mol⁻¹, is the same.

 Into what form(s) of energy is the chemical energy contained in glucose converted when you burn glucose on a spoon or catabolise it inside your cells?

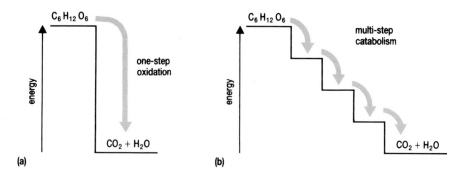

Fig 6.4 **(a)** Oxidising glucose to carbon dioxide and water, in this case by burning it, releases a relatively large amount of energy in one step.
(b) In cells, the energy contained in glucose is released in a stepwise manner through a large number of catabolic reactions. It is these reactions which provide the energy to make ATP.

ANALYSIS

1 mole (usually abbreviated to mol) of a substance contains 6.022×10^{23} particles (atoms or molecules) of that substance. 1 mol of glucose weighs 180 g.

The energy yield from glucose

This exercise involves simple calculations.

To make 1 mol of ATP from ADP and P_i requires about 30 kJ of energy. Complete oxidation of 1 mol of glucose to carbon dioxide and water yields 38 mol of ATP. Given that the complete oxidation of 1 mol of glucose yields 3000 kJ of energy calculate:
(a) the energy used to produce ATP during the catabolism of 1 mol of glucose
(b) the percentage of the chemical energy available from glucose oxidation that is converted into chemical energy stored in ATP
(c) the amount of the chemical energy available from glucose oxidation which is converted into heat.

Glucose catabolism can be divided into the four stages shown in Fig 6.5. In the following pages, you will discover how the four stages shown in Fig 6.5 can produce up to 38 molecules of ATP for each molecule of glucose which is completely oxidised to carbon dioxide and water.

Glycolysis (the Embden-Meyerhof pathway)

The first stage in the catabolism of a glucose molecule occurs in the cytosol and is called **glycolysis** – literally the lysis (splitting) of glucose. It does not involve oxygen and is common to both aerobic and anaerobic respiration. Its individual steps are shown in Fig 6.6. You should not try to memorise this sequence of reactions. Instead you should try to understand what happens during this first stage of respiration, identifying its important features.

- One molecule of glucose (with six carbon atoms) is broken down into two molecules of pyruvate (with three carbon atoms).
- Two molecules of ATP are used in the very first steps but later four molecules of ATP are synthesised by **substrate level phosphorylation** (a net gain of two molecules of ATP for each molecule of glucose broken down to pyruvate).

ENERGY RELEASE: RESPIRATION

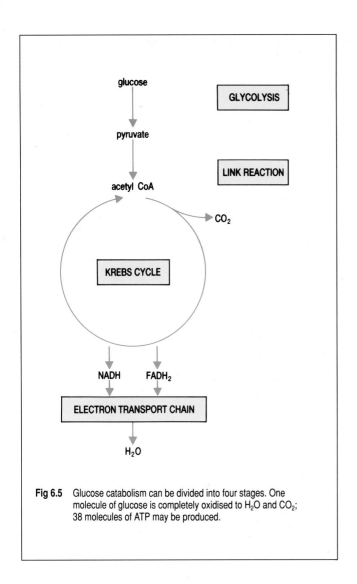

Fig 6.5 Glucose catabolism can be divided into four stages. One molecule of glucose is completely oxidised to H_2O and CO_2; 38 molecules of ATP may be produced.

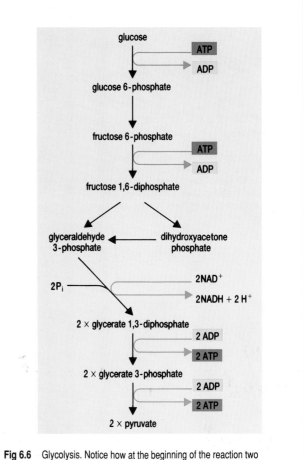

Fig 6.6 Glycolysis. Notice how at the beginning of the reaction two molecules of ATP are used. Later in the pathway four molecules of ATP are made by substrate level phosphorylation. Notice two molecules of NAD^+ are reduced to NADH. You will meet these again later.

- Hydrogen atoms are removed from glucose molecules when two molecules of NAD^+ are reduced to NADH (remember moving hydrogen atoms from one molecule to another also transfers electrons).

The energy changes which occur during glycolysis are shown in Fig 6.7. For simplicity only those stages of glycolysis which involve energy changes have been shown.

Fig 6.7 Energy changes occurring during glycolysis. The glucose molecule is first activated by the addition of two phosphate groups, eventually forming fructose 1,6-diphosphate. The energy contained in this molecule is then released as it is catabolised to pyruvate.

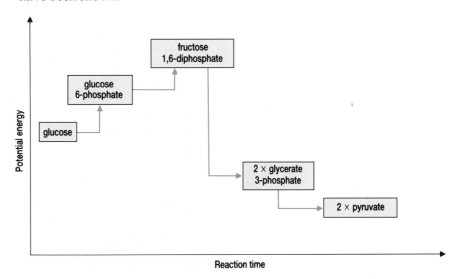

ENERGY RELEASE: RESPIRATION

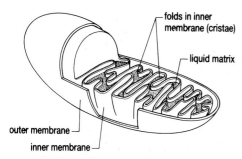

folds in inner
membrane (cristae)

liquid matrix

outer membrane

inner membrane

Fig 6.8 **(a)** Diagram of a mitochondrion. The inner membrane, containing the compounds which form the electron transport chain, is highly folded to produce shelf-like cristae. Compare this diagram with the electronmicrograph of a mitochondrion.

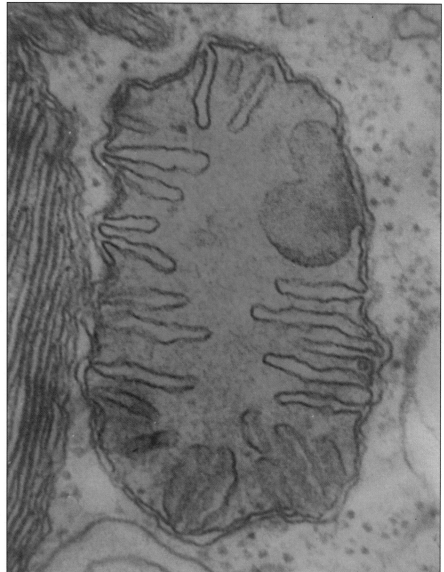

Fig 6.8 **(b)** An electronmicrograph of a mitochondrion. Notice the highly folded inner membrane.

'Link reaction' in aerobic respiration

This occurs in the fluid-filled centre of mitochondria (the matrix).

The pyruvates produced by glycolysis enter the mitochondrial matrix from the cytosol. Here they are converted to acetate radicals:

$$2CH_3 COCOO^- + 2NAD^+ + 2H_2O \rightarrow 2CH_3COO^- + 2NADH + 2H^+ + 2CO_2$$

pyruvate   acetate

In reality, the product of the above reaction is a compound formed by acetate and another coenzyme, called coenzyme A. The name of this compound, acetylcoenzyme A, is usually abbreviated to acetyl CoA.

 Write a word equation for the 'link reaction' to show the formation of acetyl CoA.

The Krebs cycle

This series of reactions (Fig 6.9) also occurs in the mitochondrial matrix. Each molecule of acetyl CoA combines with a four-carbon compound (oxaloacetate) to form a six-carbon compound (citrate). This citrate

ENERGY RELEASE: RESPIRATION

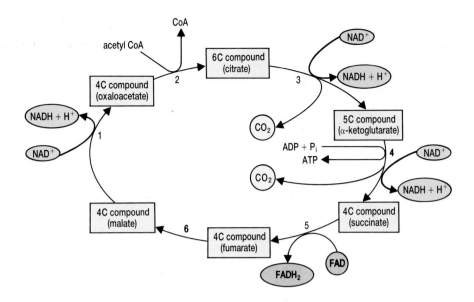

Fig 6.9 The Krebs (tricarboxylic acid) cycle. For each circuit, one acetate group is fed in at Step 2. As it goes round the cycle carbon, oxygen and hydrogen atoms are removed as CO_2 and $H^+ + e^-$, which are used to reduce either NAD^+ or FAD^+. One molecule of ATP is synthesised by substrate level phosphorylation.

compound then undergoes four redox reactions (dehydrogenations in this case – the removal of hydrogen) and two decarboxylations.

As a result of these reactions:

- oxaloacetate is regenerated and can start the cycle off again
- three molecules of NAD^+ are reduced to NADH
- one molecule of FAD is reduced to $FADH_2$
- two molecules of carbon dioxide are produced
- one molecule of ATP is made by substrate level phosphorylation.

 How many times will the cycle 'turn' during the catabolism of one molecule of glucose?

The overall reaction accomplished by two 'turns' of the Krebs cycle is:

$$2 \text{ acetyl CoA} + 6NAD^+ + 2FAD + 2ADP + 2P_i \rightarrow$$
$$4CO_2 + 6NADH + 6H^+ + 2FADH_2 + 2ATP$$

The electron transport chain (cytochrome system)

If you look back at the equations representing the catabolism of a single glucose molecule, you will notice that a number of reduced coenzyme molecules, either NADH or $FADH_2$, have been produced as the glucose molecule has been stripped of its electrons.

 How many of each type of reduced coenzyme have been produced so far?

In the electron transport chain these reduced coenzymes are reoxidised. The enzymes which catalyse this reoxidation process are located on the folded inner mitochondrial membrane where they are arranged in a series.

 What is the likely advantage of this high degree of folding of the inner membrane of the mitochondria to form cristae?

ENERGY RELEASE: RESPIRATION

The last step in the catabolism of a single glucose molecule directly involves molecular oxygen dissolved in the mitochondrial matrix. The overall reaction of the electron transport chain for each glucose molecule is:

$$10NADH + 2FADH_2 + 6O_2 + 10H^+ \rightarrow 10NAD^+ + 2FAD + 12H_2O$$

During this oxidative reaction, ATP is made by the process of **oxidative phosphorylation.**

A simplified version of the electron transport chain including the points at which ATP is made is shown in Fig 6.10.

 How many ATP molecules are made for each molecule of NADH and FADH₂ reoxidised by the electron transport chain?

As you can see in Fig 6.10, NADH passes the pair of electrons and the H⁺ ion it is carrying to a flavoprotein carrier (FP). The FP picks up another proton from the matrix, producing reduced flavoprotein (FPH₂). The H⁺ ion and electrons are then passed to coenzyme Q (CoQ) which becomes reduced (CoQH₂).

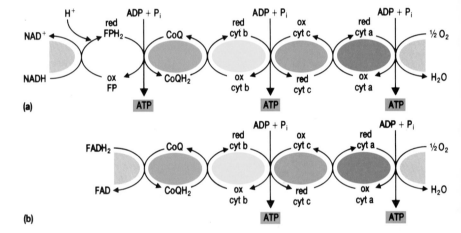

Fig 6.10 A simplified version of the electron transport chain showing how electrons are transferred from **(a)** NADH **(b)** FADH₂ to oxygen via a series of redox reactions. During this process ATP is synthesised by oxidative phosphorylation. The molecules of the electron transport chain are bound to the inner mitochondrial membrane.

 What happens to the flavoprotein as a result of this reaction? What is the significance of this?

Fig 6.11 This plant achieves some protection against insect damage by having cyanide in its tissues.

The next group of components in the electron transport chain are the iron-containing **cytochromes (cyt)**. The iron atom within each cytochrome is alternatively reduced to Fe^{2+} and then oxidised to Fe^{3+} as it gains and loses an electron. The final component of the electron transport chain is cytochrome a.

 10 **How is the reduced form of cytochrome a reoxidised?**

The ATP tally during aerobic respiration

This exercise will help you to summarise and consolidate the work done so far in this section.

Under ideal conditions within a cell, 38 molecules of ATP are produced when one molecule of glucose is completely oxidised. In reality cells rarely reach this ideal of efficiency!

(a) Copy and complete the following table using the information you have gained so far in this chapter.

Stages in aerobic respiration	Number of molecules of			
	ATP produced by substrate level phosphorylation	ATP used	NADH produced	FADH$_2$ produced
Glycolysis				
Link reaction				
Krebs cycle				
Total				

You should find this still leaves us 34 molecules of ATP short of the maximum possible total of 38. These missing ATP molecules are produced by oxidative phosphorylation.

(b) How many molecules of ATP are produced following the oxidation of 10NADH molecules?

(c) How many molecules of ATP are produced following the oxidation of 2FADH$_2$ molecules?

(d) Therefore, how many molecules of ATP are produced by oxidative phosphorylation?

(e) How many molecules of ATP are produced by substrate level phosphorylation?

(f) How many molecules of ATP are used during glycolysis?

(g) What is the total number of molecules of ATP produced as a result of the complete oxidation of one molecule of glucose?

(h) Suggest why cells may not actually produce this number of ATP molecules from the oxidation of a single glucose molecule.

(i) Glucose catabolism is spread out over many reactions. Suggest an advantage of this.

6.3 MITCHELL'S CHEMIOSMOTIC HYPOTHESIS

The description of the electron transport chain given in the previous section is very simplified. In particular it does not show how ATP is actually made in the series of redox reactions. The most widely accepted theory seeking to explain how ATP is synthesised by the electron transport chain, the chemiosmotic hypothesis, was proposed by Peter Mitchell.

Mitchell's hypothesis depends on the fact that most of the reactions in the electron transport chain do not involve the transport of hydrogen atoms. Rather, each hydrogen atom is split into a H^+ ion and an electron. The H^+ ion is then ejected and the electron transported, hence the name electron transport chain.

However, according to the chemiosmotic hypothesis it is the H^+ ions which are the key to ATP synthesis. Look at Fig 6.12. Here you can see how the H^+ ions released by the electron transport chain are discharged to the intermembrane space outside the inner mitochondrial membrane. This means that the concentration of H^+ ions in the solution on the outside of the inner membrane becomes greater than the concentration of the H^+ ions in the matrix on the inside of the inner membrane. In other words the electron transport chain establishes a H^+ ion concentration gradient across the inner mitochondrial membrane.

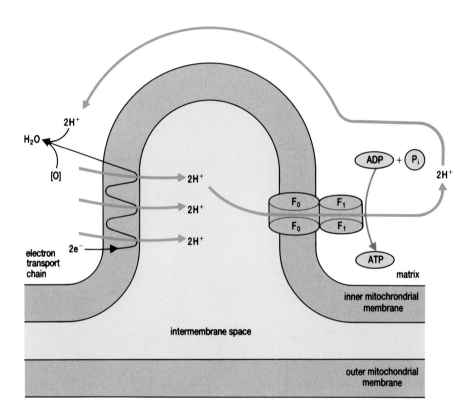

Fig 6.12 Mitchell's chemiosmotic hypothesis. The flow of electrons ($2e^-$) along the electron transport chain leads to an increased H^+ ion concentration on the outside of the inner membrane. The H^+ ions flow back into the matrix down the concentration gradient through an ATP-making complex (F_0–F_1 protein complex) in the inner mitochondrial membrane.

 Which way will the H^+ ions tend to move along this concentration gradient? Name two processes driving this movement of H^+ ions.

The idea was introduced in Chapter 2 that pumping ions against a concentration gradient, active transport, requires the expenditure of energy. This is supplied by hydrolysing ATP to ADP and P_i. What happens on the inner mitochondrial membrane is the reverse of active transport.

The H^+ ions diffuse back from the outside to the inside of the membrane

ENERGY RELEASE: RESPIRATION

through hydrophilic protein channels, called F_0 **proteins**. Associated with each F_0 protein is an enzyme, ATP synthetase (called the F_1 **protein**) which catalyses the synthesis of ATP from ADP and P_i. For each pair of H^+ ions that diffuses into the mitochondrial matrix through the $F_0 - F_1$ protein complex, sufficient energy is released to synthesise one molecule of ATP.

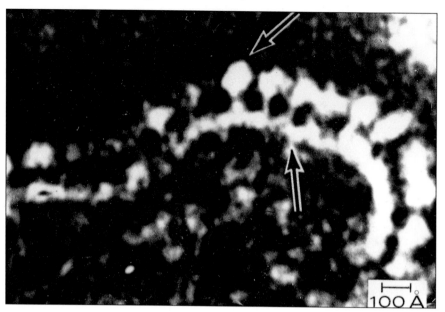

Fig 6.13 The lollipop-like structures you can see here are the ATP synthetase molecules on the inner mitochondrial membrane.

QUESTION

6.3 How would the pH of the mitochondrial matrix and intermembranal space differ?

6.4 THE CATABOLISM OF OTHER FUELS

In addition to carbohydrates, lipids and proteins can be used as energy sources for ATP synthesis. After hydrolysis, these large molecules yield fatty acids plus glycerol and amino acids respectively. These three types of molecule can also be catabolised to yield energy (Table 6.1) and so produce ATP. The central pathway used for glucose catabolism is also used for these compounds.

Glycerol

In the cytosol of cells, glycerol is converted into dihydroxyacetone phosphate which is one of the molecules in the glycolytic (glycolysis) pathway (Fig 6.14).

Fatty acids

These molecules are long chains of carbon and hydrogen atoms. These chains can be broken into two-carbon acetyl fragments which are then converted to acetyl CoA in the mitochondrion. This process is called β-**oxidation** (Fig 6.14). The acetyl CoA so formed can then be fed into the Krebs cycle.

Amino acids

Excess amino acids cannot be stored but are catabolised immediately. Since there are 20 different common amino acids there are 20 different beginnings to the catabolic pathway for amino acids. However, the general principles are the same.

Table 6.1 Energy yields on complete combustion

Substrate	Maximum energy yield for ATP synthesis/kJ g^{-1}
carbohydrate	17.22
lipid	39.06
protein	22.68

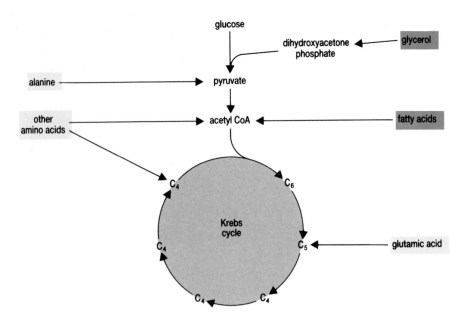

Fig 6.14 Catabolism of other substances (glycerol, fatty acids and amino acids) also occurs via the central pathway already described for glucose.

The steps are as follows

- removal of the nitrogen-containing amino group(s) – the process of **deamination** which, in mammals, occurs in the liver
- the remaining carbon-containing fragment is then fed into the central pathway at different points depending upon the amino acid (Fig 6.14)
- the amino group is excreted in a form which depends upon the organism (e.g. urea in humans, ammonia in freshwater fish, uric acid in birds).

The processes involved in the aerobic respiration of different fuel molecules to yield ATP are summarised in Fig 6.14.

Table 6.2 Aerobic RQ values of different respiratory substrates

Respiratory substrate	RQ
glucose	1.0
triglycerides	0.7
protein	0.9

The respiratory quotient (RQ)

The ratio of carbon dioxide produced and oxygen consumed per unit time by an organism is known as the respiratory quotient (RQ). As it is a ratio, the RQ has no units.

$$RQ = \frac{\text{volume of carbon dioxide produced}}{\text{volume of oxygen used}} \text{ per unit of time}$$

Look at the simplified equation for the aerobic respiration of glucose:

$$C_6H_{12}O_6 \ + \ 6O_2 \ \rightarrow \ 6H_2O \ + \ 6CO_2$$

Six carbon dioxide molecules are produced and six oxygen molecules are consumed. The RQ for this reaction is therefore $6/6 = 1$.

 12 Which tissue in your body would you expect *always* to have an RQ of 1?

The RQ value can be used to tell us something about the substrate which an organism is using for respiration (Table 6.2).

Measuring respiratory quotients

This exercise involves interpreting the way in which an item of experimental apparatus works.

The RQ of small organisms, such as woodlice or germinating seeds, can be determined using the apparatus in Fig 6.15. This apparatus actually allows us to measure changes in pressure inside the boiling tubes. To understand how it works you need to know that potassium hydroxide absorbs carbon dioxide.

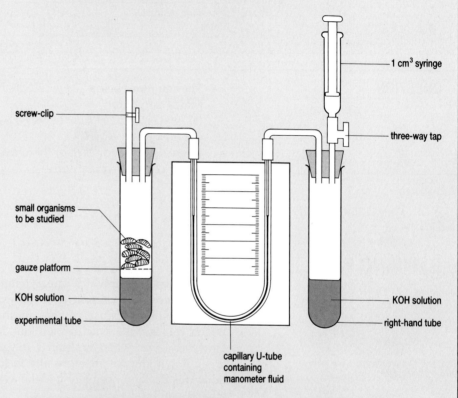

Fig 6.15 A respirometer: this apparatus is used to measure oxygen consumption, from which the respiratory quotient (RQ) is calculated.

(a) What will happen to the number of oxygen molecules in the experimental tube as the organisms respire?

(b) What will happen to the carbon dioxide molecules produced by respiration?

(c) As a result, what will happen to the total number of gas molecules in the left-hand tube as the organisms respire?

(d) How will this affect the gas pressure in the left-hand tube? (Look back at Fig 2.7 if you are not sure about this.)

(e) As a result of this pressure change, which way will the fluid in the manometer be pushed?

(f) If you knew the internal diameter of the manometer tube, how would you calculate the volume of oxygen consumed by the organisms in the respirometer?

(g) Why are the tubes (containing KOH solution) kept in the waterbath throughout the investigation?

(h) What is the purpose of the right-hand tube?

(i) In a second experiment, the potassium hydroxide was replaced with water. How would this enable you to determine the amount of carbon dioxide produced by the respiration of the organisms in the respirometer?

(j) The results of an experiment using this apparatus are given in Table 6.3. Use this information to calculate the RQ of the respiring organisms. What does this value suggest about the nature of their respiratory substrates?

(k) Suggest possible sources of error in this experiment.

Table 6.3   Results of experiment using respirometer

Solution in tube	Direction of movement of fluid in left-hand arm of manometer	Volume change /mm³ h⁻¹
potassium hydroxide	upwards	20
water	upwards	2

QUESTION

6.4 The equations below show the catabolism of (a) glycerol (b) oleic acid, a fatty acid.

(a) glycerol: $2C_3H_8O_3 + 7O_2 \rightarrow 6CO_2 + 8H_2O$

(b) oleic acid: $2C_{18}H_{34}O_2 + 51O_2 \rightarrow 36CO_2 + 34H_2O$

What is the RQ for each of these reactions?

6.5 LIVING WITHOUT OXYGEN

Respiration without oxygen is called **anaerobic respiration**. *Clostridium tetani*, the bacterium which causes tetanus (lock-jaw), can thrive only in the total absence of oxygen: it is an **obligate anaerobe**. Without oxygen you would soon die because you are an **obligate aerobe**. Yeast is an organism which can survive with or without oxygen: it is a **facultative aerobe**. However, even when molecular oxygen is present yeast cells normally break down glucose anaerobically to form ethanol. Only when they have done this do they normally use the available molecular oxygen to respire ethanol aerobically.

 Suggest how *C. tetani* makes ATP.

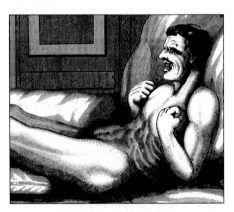

Fig 6.16 A contemporary drawing of a soldier in the Napoleonic wars suffering from tetanus. The disease results from toxins produced by the anaerobic bacterium *Clostridium tetani*. Tissue necrosis (death) at the site of a wound provides a locally anaerobic environment in which the bacterium can grow. The tetanus toxin causes muscles to stay contracted.

Anaerobic respiration: the regeneration of NAD⁺ without oxygen

If you look back at Fig 6.6 you can see that glycolysis does not involve molecular oxygen. However, for glycolysis to continue a constant supply of oxidised NAD⁺ is required. Oxidised NAD⁺ is generated from reduced NADH by the electron transport chain (Fig 6.10(a)). Since molecular oxygen is needed for the electron transport chain to function, in the absence of oxygen, regeneration of oxidised NAD⁺ will stop.

 Explain why regeneration of NAD⁺ will stop.

Since each cell contains only a few micrograms of NAD⁺ and FAD, you might expect that, in the absence of oxygen, all the NAD⁺ and FAD would become reduced to NADH and FADH₂. With the electron transport chain not working, neither of the molecules could be reoxidised.

 Why would this cause pyruvate production to stop?

Under these circumstances ATP production would stop and death would rapidly follow, yet this clearly does not happen to anaerobes. They survive because they have an alternative metabolic pathway which can regenerate NAD^+ and so keep glycolysis going. This alternative pathway (Fig 6.17) involves the oxidation of NADH by pyruvate itself, producing ethanol and carbon dioxide (plants and fungi) or lactate (animals and many bacteria). This alternative pathway releases a little energy but no further ATP is produced.

$$CH_3COCOO^- + NADH + 2H^+ \longrightarrow CH_3CH_2OH + CO_2 + NAD^+$$
$$\text{(pyruvate)} \qquad\qquad\qquad \text{(ethanol)}$$

$$CH_3COCOO^- + NADH + H^+ \longrightarrow CH_3CHOHCOO^- + NAD^+$$
$$\text{(pyruvate)} \qquad\qquad\qquad \text{(lactate)}$$

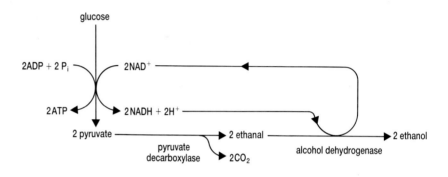

Fig 6.17 Fermentation – the anaerobic catabolism of glucose in yeast. Note the effect of this reaction is to regenerate NAD^+ and keep glycolysis going.

 16 **What is the yield of ATP from fermentation?**

Fermentation

Louis Pasteur, the famous Frenchman after whom the process of pasteurisation is named, defined fermentation as life without air. Obligate anaerobes are totally dependent on fermentation to obtain their energy. Yeast cells use fermentation whether oxygen is present or not, thus explaining their typical diphasic growth curve shown in Fig 6.20. Yeast fermentation is exploited in every bakery, brewery and wine vat.

 17 **What would be the RQ value of yeast if it were to respire glucose (a) aerobically; (b) anaerobically?**

EXPLOITING MICROBIAL FERMENTATION

Since prehistoric times, microbial fermentation has been exploited by humans to make and preserve food. Fermentation is also essential to many modern industrial processes, including those in the booming field of biotechnology. For example, in the clothing industry, the freeing of plant fibres in making linen and jute (known as retting) is the result of the anaerobic activity of butyric acid bacteria. In the chemical industry, acetone (propanone), butanol and the dextrans, used to make molecular sieves, can all be made by microbial activity.

The agricultural production of silage depends on the fermentation of vegetation by bacteria. An increasing number of antibiotics are also made by micro-organisms. You may read more fully about such uses of micro-organisms in Jane Taylor's book in this series, *Micro-organisms and Biotechnology.*

Fig 6.18   All these products are made by fermentation.

Lactate production and the oxygen debt

For a short period of time, cells such as those in your muscles can respire anaerobically. However, rather than producing ethanol or butyric acid, your muscle cells regenerate NAD^+ by converting pyruvate to lactate. Such anaerobic respiration only yields two molecules of ATP from each molecule of glucose catabolised. Even so, this allows muscles to keep contracting when they have exhausted their oxygen supplies. Although muscle cells have relatively high tolerance, as lactate accumulates in muscles it causes fatigue.

This limited ability of muscle cells to respire anaerobically is vital to sprinters (or to you when you run for a bus). After a sprint race is over, the lactate in the athlete's muscles is carried in the blood to the liver. Here about 20% of the lactate is oxidised to carbon dioxide and water, releasing energy that enables further ATP production. This ATP is then used to convert the remaining 80% of the lactate to glycogen. The oxygen required to do this is called the **oxygen debt**.

 Intestinal tapeworms excrete lactate directly into the gut of their host. Suggest the advantage to the tapeworm of this behaviour.

The regulation of glucose catabolism

This exercise is designed for you to apply previous understanding and to make predictions.

When you run for a bus your muscles use a large amount of ATP. However, when you sit down the demand of your muscles for ATP is reduced. Your cells avoid wasting valuable fuel by increasing or decreasing the rate of ATP synthesis as required.

Chapter 5 discussed how complex biochemical pathways can be controlled using allosteric inhibition and promotion. This idea is summarised in Fig 6.19.

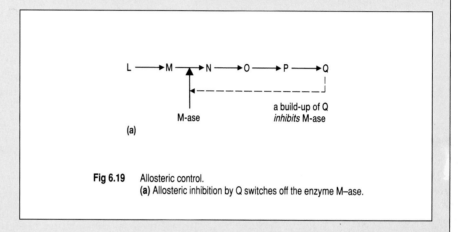

Fig 6.19   Allosteric control.
(a) Allosteric inhibition by Q switches off the enzyme M–ase.

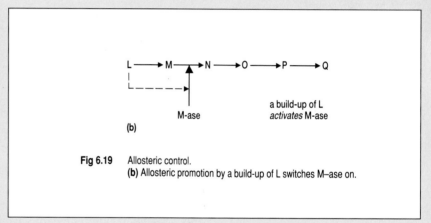

Fig 6.19   Allosteric control.
(b) Allosteric promotion by a build-up of L switches M–ase on.

The enzyme isocitrate dehydrogenase (ICDH) occurs in the Krebs cycle where it catalyses the dehydrogenation of citrate. Its activity is controlled by the concentration of ATP within mitochondria.

(a) Imagine that ATP concentrations in the mitochondrial matrices are high. To reduce ATP levels, should the activity of the Krebs cycle increase or decrease?

(b) Will ATP activate or inactivate ICDH?

(c) Predict the effects of high ADP concentrations on ICDH and consequently on the activity of the Krebs cycle.

(d) ICDH activity is also affected by NADH. Will NADH promote or inhibit ICDH activity?

6.5 (a) Calculate the efficiency of anaerobic respiration in yeast.
(b) Compare this with the efficiency of aerobic respiration.
(c) Explain what the data in Fig 6.20 show about respiration in yeast.

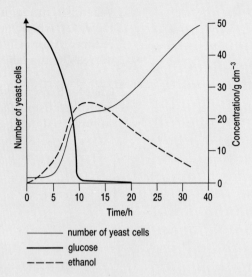

number of yeast cells
glucose
----- ethanol

Fig 6.20

6.6 The muscles of a sprinter respire anaerobically throughout a race.
(a) Explain how the sprinter can complete a 100 m race without breathing.
(b) Why does the sprinter need to pant to get extra oxygen at the end of the race?
(c) Where is this extra oxygen being used?
(d) Why do you think it is advantageous to transport the lactate produced during the race from the sprinter's muscles to the liver cells?

6.7 *Clostridium botulinum* is an obligate anaerobe which causes botulism, a potentially lethal form of food poisoning. Suggest why this bacterium might be a particular problem in canned foods.

SUMMARY

During metabolism, the hydrolysis of ATP is commonly coupled to endergonic reactions, such as the synthesis of macromolecules. Consequently, cells need a constant supply of ATP. Respiration is the process by which cells make ATP from ADP and P_i.

Aerobic respiration of glucose involves four stages. In glycolysis, glucose is broken down in the cytosol to produce two molecules of pyruvate and reduced NAD (NADH). The 'link reaction' converts pyruvate to acetyl CoA, producing carbon dioxide and more NADH. The acetyl CoA is taken up by the Krebs cycle which generates carbon dioxide, NADH and reduced FAD ($FADH_2$). Using molecular oxygen, the electron transport chain reoxidises NADH and $FADH_2$, so regenerating NAD^+ and FAD.

During glycolysis, substrate level phosphorylation yields sufficient energy to resynthesise two ATP molecules per glucose molecule: a further two ATP molecules per molecule of glucose are synthesised using energy released from the Krebs cycle. Oxidative phosphorylation in the electron transport chain can yield sufficient energy to resynthesise 34 ATP molecules per molecule of glucose: a theoretical total of 38 ATP molecules per molecule of glucose when catabolised to CO_2 and H_2O under ideal conditions.

Anaerobic respiration occurs without the use of molecular oxygen. Here there is no electron transport chain activity to regenerate NAD^+. However, glycolysis can continue because its end product, pyruvate, is used to regenerate NAD^+ from NADH, producing ethanol or lactate. Only two ATP molecules, produced entirely by substrate phosphorylation, are produced per molecule of glucose catabolised.

Chapter 6: Answers to Quick Questions

1   Proteins and nucleic acids.
2   Reactions require energy instantaneously and in small packets otherwise the organism would die.
3   Heat and light; heat and ATP.
4   2 pyruvate + 2NAD$^+$ + 2CoA + 2H$_2$O
     → 2 acetyl CoA + 2NADH + 2H$^+$ + 2CO$_2$.
5   Two
6   18NADH; 2FADH$_2$
7   Increase surface area for location of enzymes of electron transport system.
8   3 and 2 respectively.
9   It is reoxidised ready to pick up more electrons and hydrogen from NADH.
10 By donating its electrons to oxygen.
11 Back into the matrix. The two processes are diffusion along a concentration gradient and movement along an electrochemical gradient.
12 The brain (catabolises only glucose, remember).
13 Anaerobic respiration involving only glycolysis.
14 Because cytochrome a cannot be reoxidised the electron transport chain ceases to function.
15 Because glycolysis depends on a supply of NAD$^+$.
16 2 ATP
17 The RQ will be 1 during aerobic respiration but greater than 1 during anaerobic respiration (since the amount of oxygen used is the divisor in our formula for calculating RQ).
18 Since there is little oxygen in the host's gut, anaerobic respiration produces lactate. Excretion of the lactate avoids the need to waste energy in repaying an oxygen debt.

Chapter 7

ENERGY CAPTURE: AUTOTROPHIC NUTRITION

> **LEARNING OBJECTIVES**
>
> When you have studied this chapter you should be able to:
>
> 1. describe how plants absorb energy from sunlight and use it to make ATP;
>
> 2. describe how electrons from water are used to reduce $NADP^+$;
>
> 3. explain how carbon fixation is carried out in C_3 and C_4 plants;
>
> 4. explain how a number of environmental factors affect the rate of photosynthesis;
>
> 5. explain how plant leaves are adapted for efficient photosynthesis;
>
> 6. outline the sources of energy used by chemoautotrophic bacteria.

All cells need energy. The Law of Conservation of Energy states that energy can neither be created nor destroyed, but energy in one form can be changed into another form. **Autotrophs** are organisms which can change energy from one form into the chemical energy of organic molecules, which they make using an inorganic source of carbon. **Photoautotrophs** make organic molecules using light energy. **Chemoautotrophs** make organic molecules using energy released by exergonic chemical reactions mostly involving inorganic compounds. **Heterotrophs** cannot manufacture the energy-storing molecules they need using inorganic sources of carbon and must rely on consuming ready-made organic molecules. This chapter describes how autotrophs trap energy and make organic molecules. The activities of heterotrophs are described in Chapter 8.

7.1 PHOTOSYNTHESIS: AN OVERVIEW

Photosynthesis can be represented by the equation

$$CO_2 \; + \; H_2O \xrightarrow{\text{light}} (CH_2O) \; + \; O_2$$
$$\text{carbohydrate}$$

which shows that photoautotrophs synthesise carbohydrate using carbon dioxide, water and light energy.

 Is photosynthesis an endergonic or exergonic reaction and is it a reduction or an oxidation?

This simple summary hides the fact that photosynthesis is a series of reactions controlled by specific enzymes.

(a)

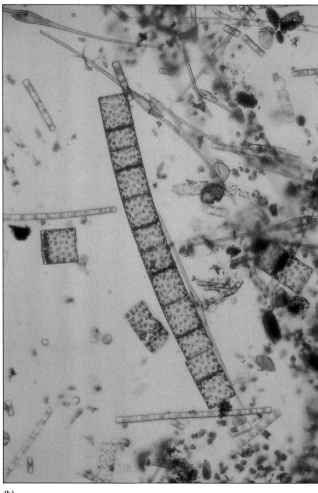

(b)

Fig 7.1 Giant trees, redwoods **(a)** and minute phytoplankton **(b)** both produce food by the same mechanism, photosynthesis.

Light-dependent and light-independent reactions

The reactions of photosynthesis can be divided into two distinct stages

- the light-dependent reactions in which ATP and a reduced coenzyme, NADPH, are made
- the light-independent reactions in which the products of the light-dependent stage are used to reduce carbon dioxide to carbohydrate.

The relationship between the two stages is summarised in Fig 7.2. In the light-dependent reactions, the energy from light is trapped by chlorophyll and used to make ATP molecules (**photophosphorylation**). At the same time, in a reaction which is poorly understood, water molecules are split (**photolysis**) into molecular oxygen, electrons and hydrogen ions. These electrons react with a carrier molecule called **nicotinamide adenine dinucleotide phosphate** (**NADP**) and change it from its oxidised state (represented **NADP$^+$**) to its reduced state (represented **NADPH**).

$$NADP^+ + 2e^- + 2H^+ \rightarrow NADPH + H^+$$

The reduction of NADP$^+$ requires a large amount of energy. As you will shortly see, this energy is supplied by light energy trapped by chlorophyll.

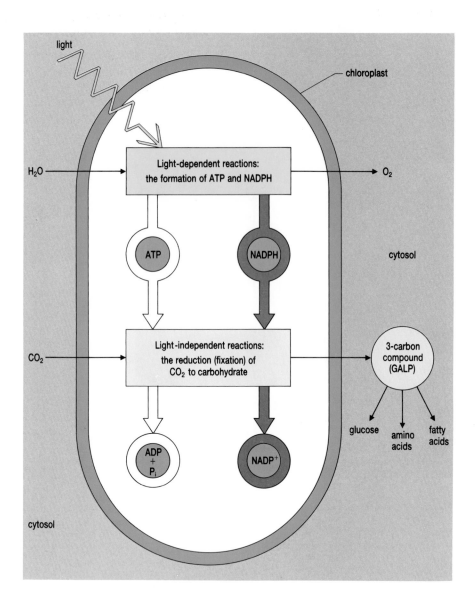

Fig 7.2 Photosynthesis occurs in two stages. The light-dependent reactions produce ATP and NADPH which are used in the light-independent reactions to reduce CO_2 to the 3-carbon sugar phosphate, GALP. This then forms the raw material for the plant cell's biosynthetic machinery.

In the light-independent reactions, the ATP and reduced NADPH produced by the light-dependent reactions, are used to reduce carbon dioxide. The initial product of this reduction is glyceraldehyde 3-phosphate (**GALP**), a three-carbon compound.

Chloroplasts: the site of photosynthesis

The two stages of photosynthesis occur separately in chloroplasts (Fig 7.3). The light-dependent reactions occur on the **thylakoid membranes** whilst the light-independent reactions occur in the solution, **stroma**, that fills the chloroplast between the membranes. Notice how the thylakoids are stacked into **grana** (singular granum).

 What would a thylakoid look like in three dimensions?

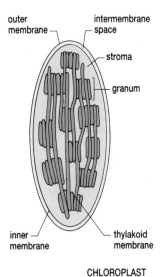

CHLOROPLAST

Fig 7.3 A section through a chloroplast. The light-dependent reactions occur on the thylakoid membrane. The light-independent reactions occur in the stroma.

ENERGY CAPTURE: AUTOTROPHIC NUTRITION

7.1 Suggest how a photosynthetic heterotroph (photoheterotroph) might make energy-rich organic molecules.

7.2 Copy the table and complete it to show the raw materials used in, and the products of, the light-dependent and light-independent reactions in photosynthesis.

	light-dependent reactions	light-independent reactions
raw materials		
products		

7.3 The equation below represents a process which occurs in the bacterium *Chlorobium* sp.

$$12H_2S + 6CO_2 \xrightarrow[\text{bacteriochlorophyll}]{\text{light}} C_6H_{12}O_6 + 12S + 6H_2O$$

(a) How is this process similar to photosynthesis in plants?
(b) How does this process differ from photosynthesis in plants?

7.4 Oxygen has two isotopes, the normal oxygen-16 and the heavier oxygen-18. In an experiment, a plant was provided with carbon dioxide containing only oxygen-16 and water containing only oxygen-18. After photosynthesis had occurred, which type of oxygen atom would you expect to find in **(a)** the carbohydrate produced **(b)** the oxygen released? Explain your answer.

7.2 THE LIGHT-DEPENDENT REACTIONS

To understand this series of reactions you need to know

- how light energy is trapped
- how this trapped energy is converted into chemical energy in the form of ATP and NADPH.

The nature of light

Look at Fig 7.4: it shows the continuum of electromagnetic radiation. Only a part of this electromagnetic spectrum stimulates the human eye. This is called **visible light** or, more simply, light. Light can be considered both as

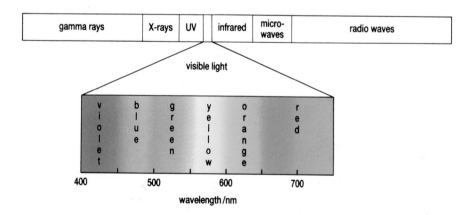

Fig 7.4 The electromagnetic spectrum. Visible light forms just a small part of the spectrum which ranges from long wavelength radio waves to very short wavelength gamma rays.

waves and as particles, called **photons**. The **intensity** of light is dependent on the number of photons falling on a given area each second. Its **energy level** is determined by the distance between successive peaks in its wave form (**wavelength**). The energy associated with a photon of light is called a **quantum** of energy. Light of different wavelength not only has different energy levels, but also different colours. Blue light has a relatively short wavelength (about 450 nm) whilst red light has a relatively long wavelength (about 700 nm). White light is a mixture of all colours of light seen together.

UV LIGHT AND CANCER

You will be aware that there is currently great concern over the disappearance of the ozone layer over Antarctica. Ozone is a gas which absorbs ultraviolet light. The short wavelength of UV light means that it has a high energy level and, since DNA strongly absorbs UV light, this energy can cause serious damage to the DNA. Such mutation can lead to the formation of cancerous cells, particularly in the skin. The incidence of skin cancer among people with white skin in Australia already gives cause for concern. A reduction in the UV shielding around the earth by further depletion of the ozone layer will only further exacerbate the situation. Too much sun really is not good for you.

Fig 7.5 Scientists sending up instruments attached to balloons like this were the first to realise ozone depletion was occurring over the Antarctic. The cause of this depletion is the result of complex chemical reactions in the upper atmosphere between ozone and chlorine derived from chlorofluorocarbons (CFCs).

ANALYSIS

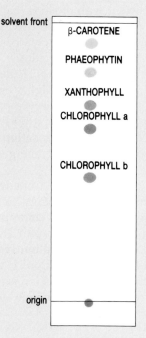

Fig 7.6 Separation of plant pigments on a TLC plate.

solvent front

β-CAROTENE

PHAEOPHYTIN

XANTHOPHYLL

CHLOROPHYLL a

CHLOROPHYLL b

origin

Plant pigments and light absorption

This exercise involves the interpretation of, and drawing conclusions from, experimental results.

Chlorophyll is found in the thylakoid membranes of plant chloroplasts. It can be extracted by grinding a leaf with a solvent such as propanone. If spots of such a leaf extract are separated using thin layer chromatography (TLC) the pattern shown in Fig 7.6 emerges. Clearly the extract contains several pigments.

(a) Calculate the Rf value (see Section 3.4) for each pigment shown in Fig 7.6.

The blue line in Fig 7.7 shows the wavelengths of light absorbed by the total plant extract, i.e. its **absorption spectrum**.

(b) Which wavelengths and colours of light are absorbed? Why, then, do plants appear green?

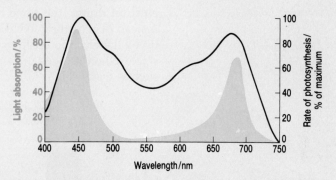

Fig 7.7 The relationship between absorption (blue) and action spectra (black).

The black line in Fig 7.7 shows the efficiency of photosynthesis using light of different wavelengths, i.e. the **action spectrum** for photosynthesis.

(c) What is the relation between the absorption spectrum and the action spectrum?

Look at Fig 7.8 which shows the absorption spectrum for a pure solution of chlorophyll a and of xanthophyll.

(d) Suggest the role of xanthophyll in trapping light energy.

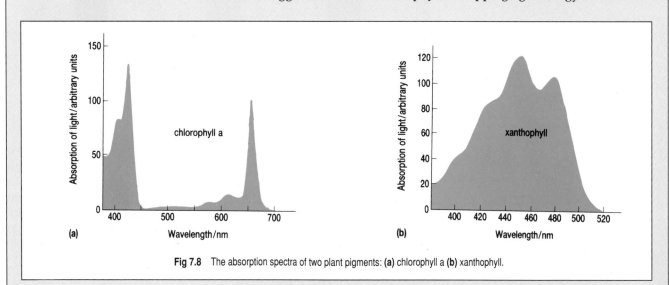

Fig 7.8 The absorption spectra of two plant pigments: **(a)** chlorophyll a **(b)** xanthophyll.

The effects of light on chlorophyll

Photosynthesis occurs more quickly at those wavelengths of light that are absorbed by chlorophyll (Fig 7.7). This is consistent with the idea that light energy is converted into chemical energy and that this drives the endergonic reactions of photosynthesis. To understand how light does this, you need to understand the effect of light on chlorophyll molecules.

When light energy is absorbed by a single chlorophyll molecule, the energy is absorbed by a pair of electrons which move to a higher energy level (**photoexcitation**). When they absorb sufficient light energy, a pair of energised electrons actually leave their chlorophyll molecule: a positively charged chlorophyll ion is left behind. This is called **photoionisation**. It is this process that holds the key for the conversion of light energy into chemical energy.

In a solution of pure chlorophyll, the pair of energised electrons would rapidly return to the unstable chlorophyll ion, releasing energy as light (**bioluminescence**) and heat (Fig 7.9(a)). However, in whole chloroplasts, each chlorophyll molecule is associated with an **electron acceptor** and an **electron donor**; the three molecules forming the heart of a **photosystem**. Instead of returning to chlorophyll, the pair of energised electrons is taken up by the electron acceptor. The positively charged chlorophyll ion is made stable when it takes up a pair of electrons from a nearby electron donor, such as water (Fig 7.9(b)).

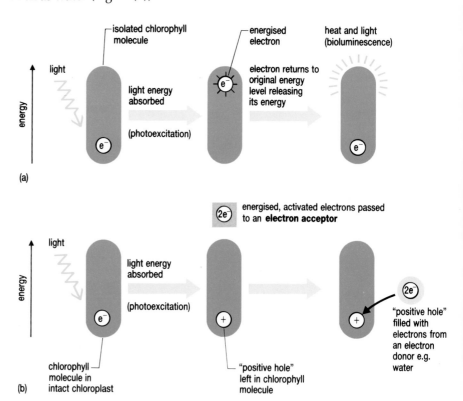

Fig 7.9 The effects of light on chlorophyll **(a)** isolated chlorophyll molecule **(b)** chlorophyll molecule in intact chloroplast.

Having been captured by the electron acceptor, each pair of energised electrons is carried by an electron transfer system back and forth through the thylakoid membrane. The light energy used to drive this process is absorbed at different steps during the electron transport by two different photosystems: **photosystem I (PSI)** and **photosystem II (PSII)**. A 'graph' showing the energy changes which occur during electron transfer (Fig 7.11) has a Z-shaped curve: for this reason the electron transfer process is often referred to as the Z scheme. The important point about the Z scheme is that sufficient energy is released during electron transfer to enable the synthesis of ATP from ADP and P_i.

Light harvesting

The light-absorbing pigments, a mixture of chlorophylls and accessory pigments such as xanthophyll, are grouped into clusters of several hundred molecules, called an **antenna complex**, by special proteins which anchor them to the thylakoid membrane. The special proteins allow the light energy absorbed by any pigment molecule in the antenna complex to be funnelled to a special chlorophyll molecule in the complex, the **reaction centre** chlorophyll. Thus each antenna complex acts like a funnel (Fig 7.12), collecting light energy and directing it towards the single reaction centre. It is the reaction centre chlorophyll which then produces the excited electrons which are passed to the electron transport chains of the Z scheme.

Fig 7.10 The photoluminescence of chlorophyll.

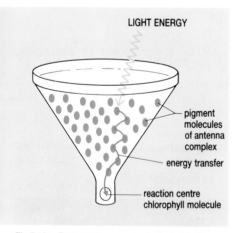

Fig 7.12 Each antenna complex in the thylakoid membrane consists of several hundred chlorophyll and accessory pigment molecules which funnel the energy they absorb from sunlight to a single reaction centre chlorophyll.

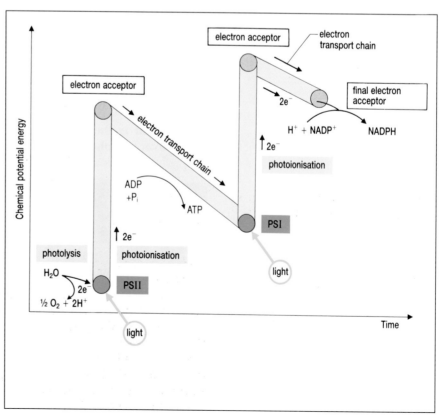

The Z scheme: non-cyclic photophosphorylation

In this mechanism of photosynthesis (see Fig 7.11), which produces both NADPH (reduced coenzyme) and ATP (by **non-cyclic photophosphorylation** in the electron transport chain), two photosystems in series are used to energise each pair of electrons.

 What does the term 'photophosphorylation' suggest is the source of energy for the manufacture of ATP?

In the first photosystem, called photosystem II **(PSII)** for historical reasons, excited electrons are passed from the chlorophyll in the reaction centre to an electron acceptor. Each electron requires a quantum of light energy to excite it. The positive holes in the reaction centre left by photoionisation are filled by electrons removed from water molecules. This reaction (**photolysis** of water) also produces molecular oxygen and hydrogen ions.

$$2H_2O \rightarrow 4H^+ + 4e^- + O_2$$

 How many quanta of light energy will the chlorophyll in the reaction centre of PSII have to absorb for this photolytic reaction to occur?

 How many energised electrons will be produced during this reaction?

The electron acceptor now passes its electrons down an electron transport chain which closely resembles the electron transport chain found in the inner mitochondrial membrane. The final electron acceptor of this electron transport chain is the second photosystem of the Z scheme, photosystem I **(PSI)**.

 What must have happened to PSI before it can accept the electrons from the electron transport chain?

Photosystem I gives the electrons another energy boost by absorbing light energy. Each pair of electrons has now reached such a high energy level as a result of absorbing two quanta of light, (one each from PSII and PSI) that two of these electrons (plus two H^+ ions from the stroma) can drive the reduction of $NADP^+$ to NADPH.

What is the likely evolutionary advantage of two photosystems over one photosystem?
Hint: think in terms of the energy of the electrons.

Fig 7.13 Many herbicides work by blocking the flow of electrons through the Z scheme, producing the results you can see here. In this soya bean crop the weeds on the left have been treated, those on the right have not.

ENERGY CAPTURE: AUTOTROPHIC NUTRITION

HERBICIDES STARVE PLANTS TO DEATH

The flow of electrons during the light-dependent stage of photosynthesis is the key to the conversion of light energy into chemical energy. Clearly anything which prevents this flow of electrons will stop photosynthesis and so prevent a plant from growing. Such an effect is achieved by two weed killers DCMU (dichlorophenyl methyl urea) and CMU (p-chlorophenyl dimethyl urea).

These substances block the electrons from the electron transport chains and so prevent the production of ATP and NADPH. The plant dies because it has no energy for maintenance or growth.

Chemiosmosis and ATP synthesis

The components of the Z scheme, i.e. the chlorophyll molecules, electron acceptors and transport chains, (Fig 7.11) are located in the thylakoid membranes of the chloroplast (Fig 7.14).

The flow of electrons down the electron transport chain of PSII and PSI provides the energy needed to pump H^+ ions from the stroma, across the thylakoid membrane into the thylakoid compartment (**X** on Fig 7.14). This creates an electrochemical gradient: H^+ ions are more concentrated in the thylakoid compartment than in the stroma. Diffusion of hydrogen ions (protons) down this electrochemical gradient drives the synthesis of ATP by an ATP synthetase complex (**Y** in Fig 7.14) similar to the F_0–F_1 complex 'in the inner mitochondrial membrane (Section 6.3).

8 ▷ The photolysis of water occurs in the thylakoid compartment. What effect will this have on the H^+ electrochemical gradient across the thylakoid membrane?

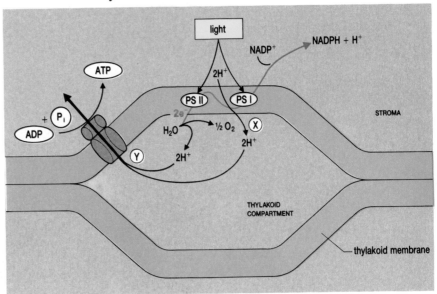

Fig 7.14 ATP production in chloroplasts.

Cyclic photophosphorylation

The overall effect of the Z scheme is to pass electrons from water (low energy) to NADP (high energy). For every pair of electrons that flows through the scheme, sufficient energy is released to produce one molecule

LIGHT-DEPENDENT REACTION

Photolysis of water (a low energy compound) releases electrons which are passed by the Z scheme (via PSII and PSI) to form ATP and NADPH (energy-rich compounds).

of ATP. However, much more ATP than this is needed to drive the light-independent reactions.

Chloroplasts can produce this extra ATP by **cyclic photophosphorylation**. This mechanism, which only involves PSI produces ATP without producing NADPH. Excited electrons produced by PSI are transferred to the electron transport chain between PSII and PSI, rather than to NADP$^+$. The electrons then pass down the electron transport chain back to PSI, completing the cycle.

 Will oxygen be produced as a result of cyclic photophosphorylation? Explain your answer.

Summarising the light-dependent reactions of photosynthesis

- non-cyclic photophosphorylation involves the reduction of NADP$^+$ using electrons derived from water which have been boosted to a high energy level by light absorbed by PSII and PSI
- the products of non-cyclic photophosphorylation are NADPH, ATP and oxygen
- cyclic photophosphorylation involves only PSI and produces only ATP.

ATP provides the energy to drive the light-independent reactions whilst the NADPH provides the 'reducing power' needed to convert carbon dioxide into carbohydrate.

ANALYSIS

Herbicides and photosynthesis

This is an exercise involving data analysis skills.

The data below were obtained from suspensions of isolated chloroplast kept in the light.

Experiment	ATP produced per unit time / μmol dm^{-3}
1. Control	18
2. CMV added	2
3. CMV and ascorbate added	10

CMV (a herbicide) inhibits photolysis.
Which of the following statements is true (T); could be true (X); or is false (F)?

(a) ATP production per unit time is a convenient measurement of photosynthetic rate.

(b) Oxygen is evolved in all three experiments.

(c) Ascorbate donates electrons to cytochrome f. (Note cytochrome f is a component in the electron transport chain.)

(d) Low ATP production in experiment 2 is due to inhibition of photosystem I.

(e) ATP production in experiment 3 must be due to cyclic photophosphorylation.

(f) ATP production can only occur when both photosystems are activated.

QUESTIONS

7.5 An experiment conducted with motile oxygen-sensitive bacteria is shown in Fig 7.15. Explain its results.

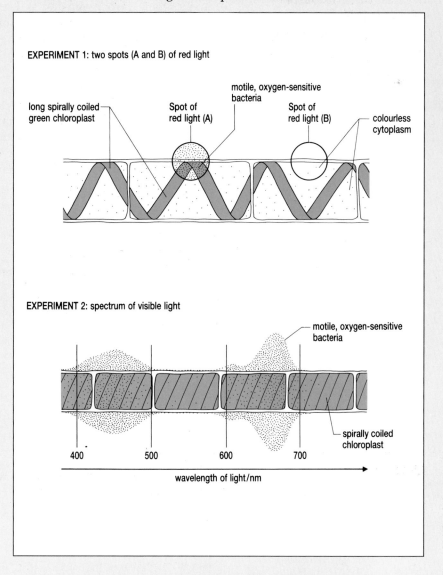

EXPERIMENT 1: two spots (A and B) of red light

long spirally coiled green chloroplast

Spot of red light (A)

motile, oxygen-sensitive bacteria

Spot of red light (B)

colourless cytoplasm

EXPERIMENT 2: spectrum of visible light

motile, oxygen-sensitive bacteria

spirally coiled chloroplast

400   500   600   700

wavelength of light/nm

7.6 Which of the following are true for the absorption of light by chlorophyll molecules in intact chloroplasts?
(a) All the visible wavelengths of light can provide energy for photosynthesis.
(b) Absorption of light transforms electrons in specific chlorophyll molecules from a ground energy state to an activated energy state.
(c) Absorption of light by a chlorophyll molecule takes place at characteristic wavelengths, called the absorption spectrum of the molecule.
(d) Each chlorophyll molecule donates electrons to its own electron transport chain.
(e) During photophosphorylation electrons are pumped by an electron transport chain from the stroma into the thylakoid space.

7.7 Which of the following are products of the light-dependent stage of photosynthesis?
(a) NADH   **(b)** O_2 **(c)** CO_2 **(d)** NADPH   **(e)** ATP

7.3 THE LIGHT-INDEPENDENT REACTIONS

In the stroma, ATP is used as the source of energy to drive the synthesis of organic compounds. This process is called **carbon fixation** since the carbon in inorganic carbon dioxide is used to make (fix) organic compounds. Reduced NADPH is the source of electrons needed to reduce carbon dioxide.

C₃ pathway

Look at Fig 7.16: it summarises a cyclic pathway, named the **Calvin cycle** after its discoverer, in which individual molecules of carbon dioxide are captured and used to form organic compounds. Since the principal components of the pathway are three-carbon compounds, the cycle is also known as the **three-carbon cycle** or **C₃ cycle**. Only the major steps of the C₃ cycle are shown in Fig 7.16, the actual cycle is much more complex with many more enzyme-controlled reactions.

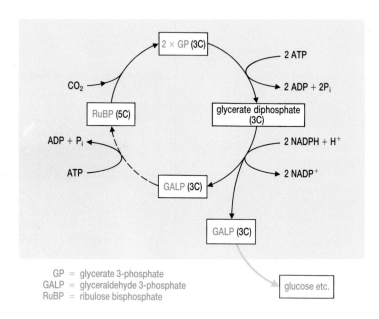

GP = glycerate 3-phosphate
GALP = glyceraldehyde 3-phosphate
RuBP = ribulose bisphosphate

Fig 7.16 The Calvin cycle. This series of reactions, which lies at the heart of the light-independent stage of photosynthesis, occurs in the stroma.

- Start to consider the cycle with the entry of carbon dioxide which combines with a five-carbon sugar, ribulose 1,5-bisphosphate (RuBP). This reaction is catalysed by the enzyme RuBP carboxylase, possibly the most abundant enzyme on earth.
- An unstable six-carbon compound is formed which spontaneously breaks down to form two three-carbon molecules of glycerate 3-phosphate (GP).
- ATP is used to phosphorylate both these molecules into glycerate diphosphate molecules which are then reduced, using the NADPH from the light-dependent reactions, to form two molecules of glyceraldehyde 3-phosphate (GALP).
- One of these GALP molecules is converted by a series of reactions into RuBP. The second molecule of GALP is the initial end product of photosynthesis: it is rapidly converted into glucose and other carbohydrates, amino acids or lipids.

 How could you, in principle, make glucose from GALP?

Determining the steps in the Calvin cycle

This exercise involves the interpretation of experimental data.

Calvin worked with the single-celled organism *Chlamydomonas*. This organism can be grown in large fermenters under carefully controlled conditions. The photosynthesising cells were fed with carbon dioxide containing ^{14}C, a radioactive isotope of carbon. At frequent intervals throughout the course of the experiment, cells were removed through the tap of the fermenter and dropped immediately into boiling methanol.

(a) What was the purpose of dropping the cells into boiling methanol?

Extracts from the cells were then subjected to two-dimensional paper chromatography (Fig 7.17). After the chromatogram had dried, a piece of X-ray film was laid over it and left for several days in a dark-room.

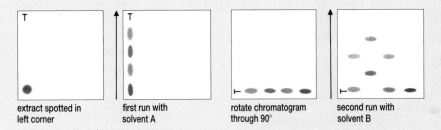

| extract spotted in left corner | first run with solvent A | rotate chromatogram through 90° | second run with solvent B |

Fig 7.17 Two-dimensional paper chromatography. By using two different solvents in sequence, this technique gives better separation of the compounds contained in the original extract compared with one-dimensional paper chromatography.

(b) When developed, why will this X-ray film show the position of compounds which contain ^{14}C?
(c) Where must the ^{14}C in these compounds have come from?
(d) How have the ^{14}C-containing compounds on the chromatogram been produced inside the cell?

This technique, called **autoradiography**, enabled Calvin to discover the position of radioactive compounds on the chromatogram which were invisible to the naked eye. The radioactive spots could now be cut out of the chromatogram and the molecules they contained could be analysed and identified. Some of the results are shown in Fig 7.18.

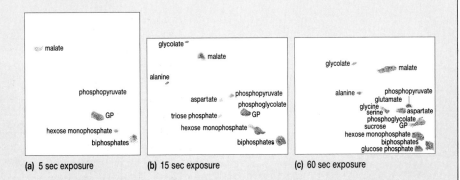

(a) 5 sec exposure   (b) 15 sec exposure   (c) 60 sec exposure

Fig 7.18 Autoradiographs made from two-dimensional chromatograms. In this case a single-celled alga, *Scenedesmus*, was the experimental organism. Note the darker the spot the more ^{14}C it contains.

(e) Using these data, suggest the order in which organic molecules are synthesised from the initial products of photosynthesis.

7.8 Explain why each of the following is essential for the light-independent stage of photosynthesis **(a)** ATP **(b)** NADPH **(c)** CO_2.

7.9 Prepare an essay plan to answer the following question. 'How is the structure of chloroplasts related to their function in photosynthesis?' Using your essay plan, try to complete your essay in about 40–45 minutes.

7.4 THE C₄ (HATCH-SLACK) PATHWAY

KEY CONCEPT

C₃ AND C₄ PATHWAYS

C_3: CO_2 is accepted by RuBP then forming 2 molecules of GALP, a 3-carbon compound, before glucose is made.
C_4: CO_2 is accepted by PEP then forming a 4-carbon compound; C_4 plants are mostly tropical.

The initial step in the C_3 cycle, the combination of RuBP and carbon dioxide, is very inefficient. The reaction is very slow at low carbon dioxide concentrations (the partial pressure of carbon dioxide in atmospheric air is only about 0.04 kPa). Also, the active site of the enzyme controlling the reaction can combine with oxygen instead of with carbon dioxide. Combination with oxygen leads to the production of glycolate which the plant breaks down with the release of carbon dioxide. This process, called **photorespiration** (although no ATP is produced), is the reverse of carbon dioxide fixation. Some plants trap carbon dioxide in a way which overcomes the limitations of the C_3 pathway. In this new pathway, carbon dioxide is used to make a four-carbon compound, oxaloacetate, so it is called the C_4 pathway. Plants using the C_4 pathway include maize and many tropical grasses.

The main stages of the C_4 pathway are summarised in Fig 7.19. Cells in which photosynthesis normally occurs combine carbon dioxide with phosphoenolpyruvate (PEP) forming oxaloacetate. Other four-carbon compounds are formed from this oxaloacetate. These four-carbon compounds are transported to cells called **bundle sheath cells** which lie around the vascular bundles of the leaf. Here the carbon dioxide is released from the four-carbon compounds and the C_3 pathway occurs. The reaction between PEP and carbon dioxide is fairly fast at low carbon dioxide concentrations and at tropical temperatures.

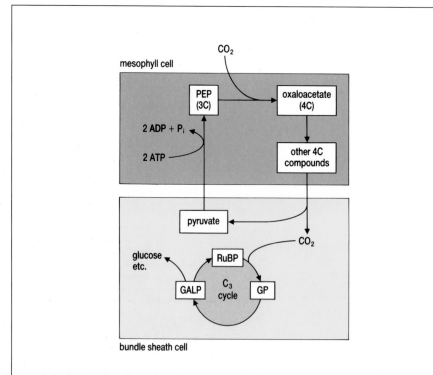

Fig 7.19 The C_4 or Hatch-Slack Pathway. By maintaining high CO_2 concentrations in the bundle sheath cells, where the light-independent reactions occur, the competitive inhibition of RuBP carboxylase by oxygen (photorespiration) is minimised.

7.10 Suggest, with reasons, which carbon fixation pathway is more efficient
  (a) if the carbon dioxide concentration is low
  (b) in hot climates where the air spaces (stomata) on the leaves are closed to reduce water loss
  (c) if the oxygen concentration is high
  (d) in bright light.

7.11 Suggest why bundle sheath cells have chloroplasts with poorly developed grana.

7.5 ENVIRONMENTAL FACTORS AND THE RATE OF PHOTO-SYNTHESIS

You should now have a basic understanding of the biochemistry of photosynthesis and be in a position to ask how environmental factors affect the rate of photosynthesis. Such investigations are not merely idle curiosity: they are essential if biologists are to help farmers grow crops more efficiently. The main factors limiting photosynthesis are light intensity and quality, carbon dioxide concentration, water availability, temperature and the availability of inorganic ions (e.g. nitrate). A method for measuring the rate of photosynthesis is an essential prerequisite for investigations of photosynthesis itself.

Photosynthesis: $6CO_2 + 6H_2O \rightarrow C_6H_{12}O_6 + 6O_2$

Measuring the rate of photosynthesis

The simple equation for photosynthesis suggests that photosynthetic rate may be measured in a number of ways, e.g. as the rate of carbon dioxide uptake or the rate of carbohydrate production. One of the simplest ways is to collect the oxygen produced over a period of time and measure its volume.

 How would this enable you to measure the rate of photosynthesis?

Having established a method, you now need to think about experimental design. Photosynthetic rates will be affected by all the factors listed above, light intensity and so on. Imagine performing an experiment in which the light intensity and temperature were increased simultaneously.

 Why is this not a good experimental design?

Clearly, what should be done is to vary either light intensity and keep temperature constant or to keep light intensity constant and vary the temperature, but not to vary both factors together.

Finally, you need to think about the sort of question you are trying to answer. What we are interested in is, 'Given factors, such as light intensity, temperature, carbon dioxide concentration and so on, which one limits the rate of photosynthesis, i.e. is the **limiting factor**, under these particular conditions?'. You can practise answering such questions by working your way through the following tutorial.

TUTORIAL

The law of limiting factors

This 'law' states that when a process is influenced by several factors the rate at which the process occurs is determined by the factor in shortest supply.
(a) What are the potential limiting factors for photosynthesis?
(b) Which of these will be limiting in (i) the middle of the night
  (ii) the middle of a warm, sunny day?

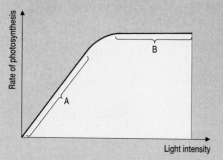

Fig 7.20 The relationship between photosynthesis and light intensity.

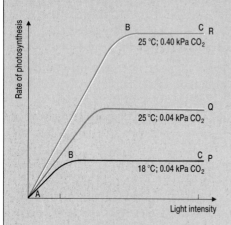

Fig 7.21 The results of three experiments which investigate the effects of changing light intensity on the rate of photosynthesis under different conditions of temperature and CO_2 concentration.

Look at Fig 7.20 which shows the relationship between rate of photosynthesis and light intensity. Over region A of the graph, an increase in light intensity results in an increase in the rate of photosynthesis. So, over this part of the curve a reduction in light intensity would cause a reduction in the rate of photosynthesis. Light is therefore a limiting factor over this part of the curve.

(c) Is light a limiting factor over region B of the curve? Explain your answer.

(d) Suggest what factor(s) might be limiting the rate of photosynthesis over region B of the curve.

(e) Suggest an experimental approach which would allow you to test your hypothesis, bearing in mind that you should vary only one factor at a time. Try to predict the result of your experiment if your hypothesis is correct.

Now look at Fig 7.21: this is a more complex graph which shows the results of three different experiments, P, Q and R.

(f) Which factor was varied in all the experiments?

Look first at the curves P and Q.

(g) What is the difference(s) between the curves?

(h) Which factor has been changed to produce the difference(s) you can see between curves P and Q?

(i) What does this tell you about the factor(s) limiting the rate of photosynthesis in the regions A–B and B–C of curve P?

Now compare curves Q and R.

(j) Which factor has been altered to produce the differences you can see between the two curves?

(k) What does this experiment tell you about the factor(s) limiting the rate of photosynthesis in the regions A–B and B–C of curve R?

(l) Why do the three curves level off at different values of light intensity?

(m) In a few sentences, summarise what the results of these experiments tell you about the factors which limit photosynthesis under different environmental conditions.

Market gardeners often use articificial lights and heaters in their glasshouses and enrich the air with carbon dioxide.

(n) Suggest one way in which the gardener could heat the glasshouse and increase the carbon dioxide level simultaneously.

(o) Suggest why the expenditure on only one of these methods might be an uneconomic way of attempting to increase the growth of glasshouse crops.

(p) Would it be feasible to use any of these methods to increase the growth rate of crops grown in open fields?

Compensation point

Plants do not just photosynthesise they also respire. During the night, the plant will not be photosynthesising so

> rate of respiration > rate of photosynthesis.

However, if we consider the same plant in the middle of a warm, sunny day, then

> rate of respiration < rate of photosynthesis.

13 **Why will this change occur?**

There is an intermediate light intensity where

rate of respiration = rate of photosynthesis.

The light intensity at which this occurs is called the **compensation point**.

Light

In the light-dependent reaction, light energy is trapped by chloroplasts and converted into ATP. Several properties of light determine its effect on photosynthesis.

Light intensity

Light intensity increases as more photons of light fall on a unit area. The more photons of light falling on the leaf the greater the number of chlorophyll molecules which will be ionised and the more ATP generated. However, light of very high intensity damages chlorophyll and actually reduces the rate of photosynthesis.

Plant species do not all react in the same way to light of a particular intensity. **Shade-tolerant plants** are able to photosynthesise in light of fairly low intensity and so are able to grow in shaded places, such as the floor of woodlands. **Shade-intolerant plants** cannot photosynthesise sufficiently rapidly at low light intensities to support their growth. These species are found in bright environments.

Fig 7.22 The plants growing on the floor of this wood are adapted, physically and biochemically, to tolerate low light intensities.

14 **How will the compensation point of a shade-tolerant plant differ from that of a shade intolerant species?**

Wavelength of light

Since chlorophyll is only affected by photons from light of certain colours (Fig 7.7), the wavelength of incident light affects the rate of photosynthesis. Photons of red light have less energy than photons of blue light, so that wavelength also affects light intensity. In most environments, natural light is white. The leaves of woodland trees absorb light from the red and blue ends of the spectrum, so that light within a dense woodland has a greeny hue. This green light is of less value to plants than white light and reduces the ability of many plant species to survive on the floor of woodlands.

Light duration (photoperiod)

The onset of flowering and seed germination in plants is affected by the photoperiod: this is described in Chapter 17. Except that photosynthesis can only occur in the light, it is unaffected by photoperiod.

Carbon dioxide concentration

Carbon dioxide is fixed in the light-independent reactions of photosynthesis to produce organic compounds. The partial pressure of carbon dioxide in atmospheric air is only about 0.04 kPa and carbon dioxide concentration is the most common limiting factor in field conditions. At normal low carbon dioxide concentrations, C_4 plants (see Section 7.4) fix carbon dioxide more efficiently than C_3 plants. At partial pressures greater than about 1.0 kPa, carbon dioxide can damage plants.

Water availability

Although water is an essential donor of electrons during photophosphorylation, the effects of lack of water on the rate of photosynthesis are difficult to demonstrate experimentally. Even slight wilting as a result of water deficiency causes great physiological stress on plants, affecting most of their metabolism.

Temperature

The individual reactions occurring during photosynthesis are speeded up by enzymes. Chapter 5 describes how, as temperature increases, the rate of an enzyme-controlled reaction is a balance between the increased energy of reactants and the progressive denaturation of enzyme. The optimum temperature for photosynthesis in C_3 plants is about 25 °C and in C_4 plants about 35 °C, showing that different enzymes are used in these pathways.

Availability of inorganic ions

A low concentration of chlorophyll within the leaves of a plant will reduce its rate of photosynthesis. Plants need a number of inorganic ions in order to make photosynthetic pigments. For example, both nitrogen and magnesium are constituents of chlorophyll molecules. Although not a constituent part of its molecule, chlorophyll cannot be synthesised in the absence of iron. Consequently, plants grown in soils with a deficiency of iron, magnesium or nitrate ions produce little chlorophyll so that their leaves are yellow rather than green, a condition called **chlorosis**.

Inhibitors

The action of competitive inhibitors was described in Section 5.4. Oxygen is a competitive inhibitor of the enzyme RuBP carboxylase which catalyses the reaction between carbon dioxide and ribulose bisphosphate (RuBP) in the Calvin cycle (see Fig 7.16). At partial pressures above the normal atmospheric value of about 21 kPa, oxygen reduces the rate of photosynthesis in C_3 plants.

Pollutants, e.g sulphur dioxide, inhibit photosynthesis. Some herbicides (chemicals which are used to kill weeds) inhibit enzymes controlling photosynthesis.

ENERGY CAPTURE: AUTOTROPHIC NUTRITION

7.12 A shade-intolerant plant has a low rate of photosynthesis on the floor of woodlands. Suggest how you could determine whether this low rate of photosynthesis results from the low intensity of light or from its wavelength (green).

7.13 When potting plant cuttings, horticulturists often put the pot and plant in a plastic bag and blow into the bag before sealing its top. Suggest how the growth of the plant cutting is helped by **(a)** blowing into the bag **(b)** sealing the bag.

7.6 THE ADAPTATIONS OF LEAVES FOR PHOTOSYNTHESIS

In spermatophytes, photosynthesis occurs mainly in the leaves. Leaves generally have a large surface area to volume ratio, allowing efficient absorption of sunlight. They are often placed at right angles to the sunlight by the growth of shoots towards a light source (**phototropism**, described in Chapter 16). In some plants, the leaves grow from the stem in such a way that they do not shade each other, a so-called leaf mosaic. In spite of these features, most of the sunlight falling on a plant's leaves is lost through reflection or as heat energy used to warm the plant and evaporate water from its surface.

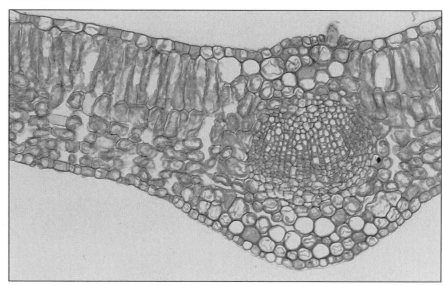

Fig 7.23 Vertical section of privet leaf *(Ligustrum* sp.). The area stained pink is the xylem.

Note the following adaptations shown in a vertical section of a privet leaf (*Ligustrum* sp) (Fig 7.23). Because the leaf is thin, photosynthesising cells are not far from the leaf surface where gas exchange and light absorption occur. Transport tissues are common, allowing water to be brought to the photosynthesising cells (in the xylem) and the products of photosynthesis to be transported away from the photosynthesising cells (in the phloem).

Cells in the middle of the leaf (**mesophyll cells**) contain chloroplasts in their cytoplasm. The mesophyll in the lower half of the leaf (**spongy mesophyll**) contains large air spaces. Gas exchange between these air spaces and the atmosphere can occur via the numerous pores (**stomata**) in the lower epidermis. The two **guard cells** around each stoma (singular of stomata) can regulate the opening and closing of the pore (Chapter 12). The upper epidermis contains few stomata in most plant species.

The cuticle which reduces water loss through the epidermis is transparent and so absorbs little light. Cells in the upper mesophyll, the **palisade layer**, have a box-like shape (Fig 7.24) enabling them to be closely packed together, an efficient arrangement to trap light. Palisade cells

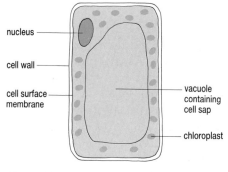

nucleus

cell wall

cell surface membrane

vacuole containing cell sap

chloroplast

Fig 7.24 A palisade cell. Note the large number of chloroplasts.

contain numerous chloroplasts, which are able to move within the cytoplasm allowing maximum absorption of light under dim conditions but protection from the bleaching effects of intense sunlight.

QUESTIONS

7.14 Draw a palisade cell to show the distribution of chloroplasts under **(a)** bright **(b)** dim conditions. Label the top and bottom of your cell clearly.

7.15 Soot particles from chimneys may form films on the surface of leaves and block the stomata. Explain why this would reduce the rate of photosynthesis.

7.7 CHEMOSYNTHESIS

Chemoautotrophs are bacteria which are able to manufacture energy-storing molecules from inorganic carbon sources. Instead of using the energy in sunlight to manufacture ATP, they use energy released during the oxidation of certain inorganic compounds. Many of the bacteria involved in the **nitrogen cycle** are chemoautotrophs

e.g. *Nitrosomonas* $NH_4^+ + 2O_2 + 2e^- \rightarrow NO_2^- + 2H_2O$ + energy

e.g. *Nitrobacter* $2NO_2^- + O_2 \rightarrow 2NO_3^-$ + energy

Fig 7.25 Chemoautotrophic bacteria are increasingly used in extracting copper from ores. The lorry gives an idea of the size of this copper mine in Australia and the quantity of ore produced.

Such bacteria can also be used in extracting metals, e.g. copper, from ores. The bacterium *Thiobacillus ferooxidans* oxidises iron pyrites to obtain energy, producing iron(III) sulphate and sulphuric acid in the process. These chemicals then react with the copper ore, chalcolite, converting insoluble copper(I) sulphide to soluble copper(II) sulphate. Water is continually added to the ore heap producing a solute of copper(II) sulphate which can then be drained off and the copper extracted.

SUMMARY

During photosynthesis, light energy is trapped by plants and then used to manufacture organic compounds from carbon dioxide and water. This process involves two series of enzyme-catalysed reactions: the former called the light-dependent reactions and the latter called the light-independent reactions (or Calvin cycle).

During the light-dependent reactions two types of electron flow may occur.
1. A non-cyclic flow which involves two photosystems linked in series that transfer electrons from water to $NADP^+$, producing NADPH and ATP. Oxygen, produced by the photolysis of water, is evolved.
2. A cyclic flow, involving only one photosystem, during which electrons flow in a circuit producing only ATP.

Both of these electron transport processes occur in the thylakoid membranes of chloroplasts and they cause H^+ ions to be pumped into the thylakoid space. The resulting electrochemical gradient between the thylakoid space and the stroma is used to synthesise ATP.

The products of the light-dependent reaction are then used to reduce CO_2 to carbohydrate in the light-independent reactions or Calvin cycle. The carbohydrate produced by these reactions, and exported to the cell cytosol, is a three-carbon sugar phosphate (GALP) which forms the starting point for many of the plant cell's synthetic pathways.

The rate of photosynthesis is affected by a number of environmental factors: light is needed as a source of energy; carbon dioxide and water are needed as substrates; temperature affects the rate of the enzyme-catalysed reactions; inorganic ions are needed for the manufacture of chlorophyll. If any one of these is in insufficient supply to allow the maximum rate of photosynthesis it is said to be a limiting factor. In addition, photosynthesis is affected by a number of chemical inhibitors.

Chemoautotrophs, a small number of genera of bacteria, have the ability to generate ATP from chemical reactions with inorganic substrates.

Chapter 7: Answers to Quick Questions

1 Endergonic; CO_2 is reduced, water is oxidised.
2 A flattened sack – see diagram.

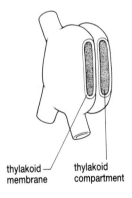

thylakoid membrane

thylakoid compartment

3 Light
4 4
5 4
6 It too must have lost electrons (by photoionisation).
7 To boost the electrons to a sufficiently high energy level to reduce $NADP^+$ to NADPH.
8 Make it steeper.
9 No. The same electrons are being recycled and are not replaced by electrons derived from the photolysis of water, the source of oxygen.
10 By joining two molecules of GALP together.

11 Rate of photosynthesis is proportional to the rate of evolution of oxygen.

12 Two variables are changing at the same time so you cannot say whether the change in light intensity or the change in temperature or both affects the rate of photosynthesis.

13 As light intensity increases so the rate of photosynthesis will increase until it exceeds the rate of respiration (which does not increase).

14 It will be lower.

Chapter **8**

HETEROTROPHIC NUTRITION

LEARNING OBJECTIVES

When you have studied this chapter you should be able to:

1. explain why food must be digested before it can be absorbed and distinguish between intracellular and extracellular digestion;

2. describe, with examples, the feeding strategies of organisms with holozoic, saprophytic, symbiotic and parasitic nutrition;

3. understand the concept of a balanced diet;

4. interpret the histology and gross structure of the human gut in terms of its function;

5. understand how the dentition and gut of carnivorous and herbivorous mammals are adapted to their diet.

Autotrophs use the energy in sunlight or from inorganic chemicals to convert an inorganic carbon compound, carbon dioxide, into organic molecules. Heterotrophs ('other feeders') are organisms which cannot manufacture all the organic molecules they need for their metabolism. They must take at least some of these molecules ready-made into their bodies, so are often called **consumers**. The source of these organic molecules is their **food**, which is always the bodies or remains of another organism. Chapter 28 describes some of the ecological relationships between autotrophic and heterotrophic organisms.

8.1 DIGESTIVE STRATEGIES

Most of the organic molecules from which organisms are made are large and complex (Chapter 1). Such molecules are unable to pass through cell surface membranes. All heterotrophs release enzymes which speed up the breakdown of organic molecules into their smaller components which can pass through cell surface membranes. Chapter 4 describes how the production of digestive enzymes may be controlled by the presence of the appropriate substrate in the food. Consumers which are unable to produce a certain digestive enzyme cannot utilise that substance as a food source. Many humans cannot produce sufficient quantities of lactase and so are unable to utilise milk sugar (lactose) as a food source.

Intracellular digestion

Protozoa are single-celled heterotrophs. They lack a rigid cell wall outside their cell surface membrane so are able to change shape. An amoeba changes its shape during locomotion, by forming temporary extensions of its cytoplasm, called pseudopodia (Chapter 18.2).

An amoeba uses its pseudopodia to engulf its prey (Fig 8.1), a process called phagocytosis ('cell eating'). Notice how the engulfed prey are totally

---

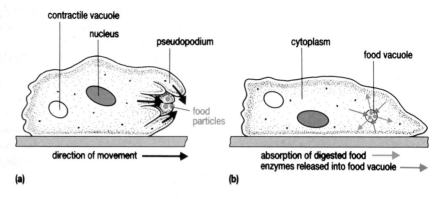

Fig 8.1 **(a)** Phagocytosis and **(b)** intracellular digestion in amoeba. (The amoeba is shown in an imaginary horizontal section).

enclosed within a food vacuole in the cytoplasm. Smaller vacuoles, containing digestive enzymes, fuse with the membrane of the food vacuole and release their enzymes into the food vacuole. The enzymes digest the food and the small molecules from the digested food are eventually absorbed across the membrane surrounding the food vacuole into the cytoplasm. Since the food is actually digested inside the vacuole of the cell, this process is called intracellular digestion.

Amylase, an enzyme produced in large amounts by the fungus *Aspergillus oryzae* growing in liquid culture, is extracted and used in starch and paper industries.

> **1** Why is the food initially digested inside a food vacuole rather than in the cytoplasm itself?

Extracellular digestion

Figure 8.2(a) is a photograph of a pin mould. This fungus has cells surrounded by a rigid cell wall so it cannot change its shape like the amoeba. Fungal cells release (secrete) digestive enzymes through their wall and cell surface membrane onto the food (Fig 8.3). The food is digested by these enzymes and the soluble products of the digested food are then absorbed by the fungus. Since digestion occurs outside the fungus this process is called extracellular digestion.

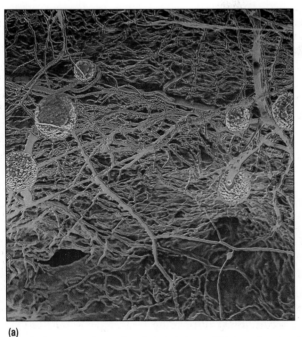

(a)

(b)

Fig 8.2 Both **(a)** the pin mould decomposing bread and **(b)** the predatory sea anemone are examples of heterotrophs. They both obtain the chemicals they require for metabolism by breaking complex organic molecules into smaller ones and absorbing these products of digestion into their bodies.

The sea anemone (Fig 8.2(b)) employs both digestive strategies. It secretes enzymes into its gut, or enteron, (extracellular digestion) whilst pseudopodia from the endodermal cells lining the gut ingest small fragments of food, forming food vacuoles in the cytoplasm. Digestive enzymes are released into these food vacuoles where digestion is completed intracellularly.

Digestion in your small intestine involves both extracellular digestion and intracellular digestive strategies (Section 8.4).

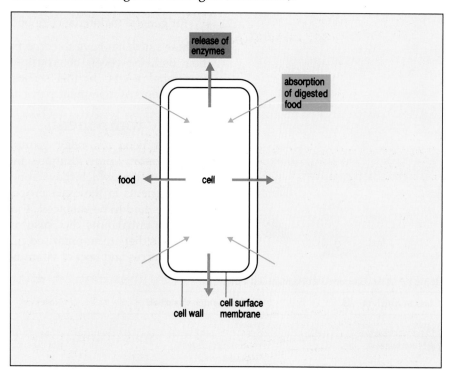

QUESTIONS

8.1 Explain
   (a) why heterotrophs usually need to digest their food
   (b) why intestinal parasites, e.g. tapeworms, can survive without a digestive system.

8.2 Table 8.1 shows how heterotrophs may be classified according to the source of their food. Use this table to classify the following heterotrophs
   (a) cows in a field of grass
   (b) a tapeworm living in the gut of a cow
   (c) fungi growing on the surface of the faeces deposited in the field by the cows.

Table 8.1   Heterotrophs can be classified according to their source of food

Class of heterotroph	Source of organic molecules (food source)	Biological groups
holozoan	ingested organisms or remains of organisms	herbivores, carnivores, detritivores
herbivore	living or recently dead plants	animals, some protoctists
carnivore	living or recently dead animals	animals, some protoctists, a few plants
detritivore	particles of organic debris	animals
saprophyte	dead organisms, waste products	bacteria, fungi
symbiont	organism lives on or in the body of its food source (host)	most
parasite	symbionts which cause harm to their host	most

8.3 Most digestion in animals is extracellular. Suggest how this might be advantageous to animals.

8.2 THE MAMMALIAN DIET

Nutrients are substances that are needed

- as sources of metabolic energy
- as raw materials for growth or repair of tissues
- for general maintenance of body functions.

These nutrients have to come from the food consumed by heterotrophs; their diet. However, heterotrophs vary widely in their species-specific nutritional needs. In this section we will concentrate on the nutrient requirements of mammals, particularly humans.

Dietary components

Mammals need six major groups of nutrients in their diet – proteins, carbohydrates, lipids, vitamins, inorganic ions and water – plus a variable amount of fibre which they cannot themselves digest. If the diet contains these nutrients in the exact proportions needed to meet the body's needs the diet is said to be balanced. Too much or too little of any single nutrient causes an imbalanced diet, resulting in malnutrition. The components of a mammal's diet are summarised in Table 8.2: refer also to Tables 8.4, 8.5 and 8.6 for sources and uses of vitamins and essential inorganic ions.

Table 8.2 The components of a mammal's diet

Class of nutrient	Form in which absorbed	Notes
water	water	
carbohydrate	monosaccharides disaccharides	not essential if other energy-containing substances eaten
lipid	lipid droplets some fatty acids glycerol	some fatty acids are essential, e.g. linoleic acid in humans
protein	amino acids dipeptides	some amino acids are essential, e.g. leucine in humans
vitamin	vitamins	vitamin C can be made by all mammals except primates, including humans, and guinea pigs
inorganic ion	hydrated ions	absorption of some is aided by vitamins

Minerals are more accurately referred to as inorganic ions.

A balanced diet must provide an animal with

- sufficient energy to power all body processes
- enough amino acids to maintain a positive nitrogen balance and so avoid net losses to body proteins
- enough water and inorganic ions to compensate for losses through, for example, urine and sweat, or incorporation, for example into bones
- those nutrients, e.g. essential amino acids, essential fatty acids and vitamins which cannot be synthesised within the body.

 Why will a small animal require more food for energy per kilogram of body mass than a large animal? Why will a mammal require more food for energy than a reptile of similar size?

A perfectly balanced diet is difficult to establish. Even within one species of mammal, requirements of each nutrient vary according to the individual's genetic make-up, age, size, sex, reproductive status, state of health and work rate. Recommended diets for a mammalian species are those which cause no adverse effects; neither deficiency symptoms resulting from a lack of any nutrient nor harmful effects from an excess of any nutrient. Usually a range of values for each nutrient is recommended. Table 8.3 shows the recommendations of the United Kingdom's Department of Health and Social Security regarding human diet. Most information regarding the composition of a balanced diet for humans has been obtained by observing the effects of natural dietary imbalances. Such data are not always easy to interpret. For example, recent data show that the incidence of heart disease is higher in groups with a high dietary intake of saturated fatty acids (see Section 1.5). However, heart disease is also more likely in people who take little exercise, smoke tobacco, are under stress or had a parent who suffered heart disease: diet is only one contributing factor.

Table 8.3 **Recommended daily amounts of nutrients for population groups**
(adapted from Department of Health and Social Security, 1979.
The department was renamed the Department of Health in 1988).

Age range in years		Energy/MJ	Protein/g	Calcium/mg
Male				
Under 1		3.25	19	600
1		5.0	30	600
2		5.75	35	600
3 – 4		6.5	39	600
5 – 6		7.25	43	600
7 – 8		8.25	49	600
9 – 11		9.5	56	700
12 – 14		11.0	66	700
15 – 17		12.0	72	600
	(Sedentary)	10.5	62	500
18 – 34	(Moderately active)	12.0	72	500
	(Very active)	14.0	84	500
	(Sedentary)	10.0	60	500
35 – 64	(Moderately active)	11.5	69	500
	(Very active)	14.0	84	500
65 – 74		10.0	60	500
75 and over		9.0	54	500
Female				
Under 1		3.0	18	600
1		4.5	27	600
2		5.5	32	600
3 – 4		6.25	37	600
5 – 6		7.0	42	600
7 – 8		8.0	48	600
9 – 11		8.5	51	700
12 – 14		9.0	53	700
15 – 17		9.0	53	600
18 – 54	(Most occupations)	9.0	54	500
	(Very active)	10.5	62	500
55 – 74		8.0	47	500
75 and over		7.0	42	500
Pregnant		10.0	60	1200
Lactating		11.5	69	1200

ANALYSIS

A balanced diet

This exercise involves using data handling and interpretation skills.

Use Table 8.3 and your general knowledge to answer the following questions.

ANALYSIS *continued*

(a) How do (i) the energy (ii) the protein (iii) the calcium requirements of males and females change up to 17 years of age?
(Hint: you will find it easier to make such comparisons if you draw graphs of the changes rather than just relying on the numbers in the table. Examiners expect you to make your answer as quantitative as possible at this level, i.e. do not just say that energy requirements increase between age x and y, say they increase z times.)

(b) Explain the changes you have described. Suggest why the pattern of change is different for males and females.

(c) How do pregnancy and lactation alter the nutritional requirements of women? Suggest explanations for these changes.

(d) How does activity affect the nutritional requirements of adults?

(e) Discuss the effects of ageing on nutrition.

Essential nutrients

Given the appropriate raw materials, mammals can link small molecules together to form polymers such as polysaccharides, proteins and lipids. If mammals absorb monosaccharides from their food they can rearrange the atoms in the molecule to make other monosaccharides, converting fructose into glucose for example. Mammals can also rearrange the atoms in absorbed amino acids and fatty acids to make other amino acids and fatty acids. However, the ability to synthesise amino acids and fatty acids differs among species. Those which cannot be synthesised must be taken in with the diet, i.e. they are **essential**. Thus amino acids which cannot be synthesised by an animal, but are required for the synthesis of its proteins, are the **essential amino acids** for that animal. Essential amino acids for humans include lysine, leucine and methionine: linoleic acid is an example of an essential fatty acid for humans. In addition inorganic ions and many vitamins are essential nutrients which must be supplied by the diet if a mammal cannot synthesise them.

 Vitamin C is essential for primates. Explain what this means.

 Explain why vegetarians should eat a wide range of vegetables.

Vitamins

The members of this diverse and chemically unrelated group of organic substances are generally needed only in small quantities. Some vitamins important in human nutrition, along with their diverse functions, are listed in Tables 8.4 and 8.5. Notice how many are coenzymes.

Lack of any vitamin leads to specific deficiency symptoms (Tables 8.4 and 8.5). The ability to synthesise different vitamins differs among species, and those essential vitamins which a mammal cannot synthesise for itself must be obtained from other sources, primarily plants but also from animal flesh or products and intestinal microbes. Thus humans require vitamin C (ascorbic acid) in their diet and depend upon intestinal bacteria to make vitamins K and B_{12}.

Vitamins can be divided into two groups: water-soluble and fat-soluble. Fat-soluble vitamins, e.g. A, D, E and K, can be stored in body fat but water-soluble vitamins, e.g. ascorbic acid, are not stored and are lost from the body in urine. Consequently water-soluble vitamins must be ingested or produced continually if adequate levels are going to be maintained.

Vitamins are so called because they were once thought to be amines vital for health.

Table 8.4   Vitamins – fat soluble

Vitamin	Function	Comment and source	Deficiency symptoms and disorders
A	maintains general health and vigour of epithelial cells formation of rhodopsin (light-sensitive chemical in rods of retina)	stored in liver. sources – fish-liver oils, milk, butter	atrophy and keratinisation of epithelium, leading to dry skin and hair increased incidence of ear, sinus respiratory, urinary and digestive infections inability to gain weight drying of cornea with ulceration (**xerophthalmia**) nervous disorders and skin sores **night blindness** or decreased ability for dark adaptation
D	absorption and utilisation of calcium and phosphorus from gastrointestinal tract may act with parathyroid hormone in controlling calcium metabolism	in presence of sunlight provitamin D_3 (derivative of cholesterol) converted to vitamin D in the skin, liver and kidneys dietary vitamin D requires moderate amounts of bile salts and fat for absorption stored in tissues to slight extent, most excreted via bile sources – fish-liver oils, egg yolk, milk	defective utilisation of calcium by bones leads to **rickets** in children and **osteomalacia** in adults possible loss of muscle tone
E	believed to inhibit catabolism of certain fatty acids that help form cell structures, especially membranes involved in formation of DNA, RNA and red blood cells	stored in liver, fatty tissue, and muscles sources – fresh nuts and wheat germ, seed oils, green leafy vegetables	may cause the oxidation of unsaturated fats, abnormal structure and function of mitochondria, lysosomes and plasma membranes haemolytic anaemia muscular dystrophy in monkeys and sterility in rats
K	coenzyme believed to be essential for synthesis of prothrombin and several blood clotting factors by liver	produced by intestinal bacteria stored in liver and spleen sources – spinach, cauliflower, cabbage, liver	delayed blood clotting time results in excessive bleeding

Table 8.5   Vitamins – water soluble

Vitamin	Function	Comment and Source	Deficiency symptons and disorders
B$_1$ (thiamine)	coenzyme for many different enzymes involved in metabolism of pyruvate to CO_2 and H_2O synthesis of acetylcholine	not stored in body sources – whole-grain products, eggs, pork, nuts, liver, yeast	abnormal carbohydrate metabolism, build up of pyruvate and lactate, insufficient energy for muscle and nerve cells partial paralysis of smooth muscle of gut causing digestive disturbances, skeletal muscle paralysis, atrophy of limbs (**beriberi**)
B$_2$ (riboflavin)	component of certain coenzymes (e.g. FAD) concerned with carbohydrate and protein metabolism	not stored in large amounts in tissues small amounts supplied by gut bacteria sources – yeast, liver, beef, veal, lamb, eggs, whole-grain products, asparagus, peas, beets, peanuts	may lead to abnormal utilisation of oxygen blurred vision, cataracts, corneal ulcerations, dermatitis and cracking of skin lesions of intestinal mucosa, development of anaemia
Niacin (nicotinamide)	essential component of coenzyme (NAD) concerned with energy-releasing reactions inhibits production of cholesterol and assists in fat breakdown in lipid metabolism	derived from amino acid tryptophan sources – yeast, meat, liver, fish, whole-grain products, peas, beans, nuts	**pellagra** – dermatitis, diarrhoea, psychological disturbances
B$_{12}$ (cyanocobalamin)	coenzyme necessary for red blood cell formation, formation of amino acid methionine, entrance of some amino acids into Krebs cycle	only B vitamin not found in vegetables only vitamin containing cobalt absorption from gut dependent on HCl and intrinsic factor secreted by gastric mucosa sources – liver, kidney, milk, eggs, cheese, meat	**pernicious anaemia** malfunction of nervous system due to degeneration of axons of spinal cord
C (ascorbic acid)	essential for iron absorption promotes many metabolic reactions, particularly protein metabolism, including laying down of collagen in formation of connective tissue acts with antibodies promotes wound healing	rapidly destroyed by enzymes on exposure to air some stored in glandular tissue and plasma sources – citrus fruits, tomatoes, green vegetables	**scurvy**, including such symptoms as simple anaemia, many symptoms related to poor connective tissue growth, tender swollen gums, loosening of teeth, poor wound healing, bleeding, vessel walls fragile because of connective tissue degeneration retardation of growth

Inorganic ions

A high proportion of a mammal's mass consists of inorganic ions in the body. Like vitamins they are needed for specific purposes (Tables 8.6 and 8.7).

Thus chloride, sulphate, phosphate and carbonate are important constituents of intra- and extracellular fluids. Calcium phosphate lends hardness and rigidity to bone. Iron, copper and other metals are needed for redox reactions and for oxygen transport. Many enzymes need metal atoms to function. Some ions are needed in moderate quantities (Ca^{2+}, PO_4^{3-}, K^+, Na^+, Mg^{2+}, Cl^-) whilst others are needed only trace amounts (Mn^{2+}, Fe^{2+}, SO_4^{2-}, I^-, Co^{2+}, Cu^{2+}, Zn^{2+}).

Table 8.6 **Major essential inorganic ions**

Ions	Importance	Comment and sources
Calcium (Ca^{2+})	formation of bones and teeth, blood clotting, normal muscle contraction and nerve activity	most abundant cation in body; about 99% is stored in bone and teeth; remainder stored in muscle, other soft tissues, blood plasma
		absorption occurs only in the presence of vitamin D
		sources – milk, egg yolk, shellfish, green leafy vegetables
Chlorine (chloride Cl^-)	important role in acid-base balance of blood and water balance of body	found in extracellular and intracellular fluids
	formation of HCl in stomach	principal anion of extracellular fluid
Iron (Fe^{2+})	component of haemoglobin, carries O_2 to body cells	about 66% found in haemoglobin; remainder distributed in skeletal muscles, liver, spleen, enzymes
	component of cytochromes	normal loses of iron occur by shedding of hair, epithelial cells, mucosal cells in sweat, urine, faeces and bile
		sources – meat, liver, shellfish, egg yolk, beans, legumes, dried fruits, nuts, cereals
Magnesium (Mg^{2+})	normal functioning of muscle and nervous tissue	component of soft tissues and bone
	bone formation	widespread in various foods
	constituent of many coenzymes	
Phosphorus (phosphate PO_4^{3-})	formation of bones and teeth	about 80% found in bones and teeth; remainder distributed in muscle, brain cells, blood
	a major buffer system of blood	
	important role in muscle contraction and nerve activity	more functions than any other mineral
	component of many enzymes	sources – dairy products, meat, fish, poultry, nuts
	involved in transfer and storage of energy in ATP	
	component of DNA and RNA	
Potassium (K^+)	functions in transmission of nerve impulses and muscle contraction	principal cation in intracellular fluid
		normal food intake supplies required amounts
Sodium (Na^+)	most abundant cation in extracellular fluid, strongly affects distribution of water through osmosis. Part of hydrogencarbonate buffer system	most found in extracellular fluids, some in bones
	functions in nerve impulse conduction	normal food intake supplies required amounts
Sulphur (sulphate SO_4^{2-})	as component of hormones and vitamins, regulates various body activities	constituent of many proteins (e.g. insulin) and some vitamins (thiamine and biotin)
		sources include beef, liver, lamb, fish, poultry, eggs, cheese, beans

Table 8.7   Essential inorganic ions needed in trace amounts

Ions	Importance	Comment and source
Cobalt (Co^{2+})	as part of vitamin B_{12}, required for maturation of erythrocytes	
Copper (Cu^{2+})	required with iron for synthesis of haemoglobin	some stored in liver, spleen
	component of enzyme necessary for melanin formation	sources – eggs, whole-wheat flour, beans, beets, liver, fish, spinach, asparagus
Fluorine (fluoride F^-)	appears to improve tooth structure and inhibit formation of dental caries	component of bones, teeth, other tissues
		major source is drinking-water
Iodine (iodide I^-)	required by thyroid gland to synthesise thyroid hormones that regulate metabolic rate	essential component of thyroid hormones
		sources – include seafood, fish-liver oil, vegetables grown in iodine-rich soils and iodised salt
Zinc (Zn^{2+})	component of carbonic anhydrase, important in carbon dioxide metabolism	important component of certain enzymes
	normal growth, proper functioning of prostate gland, normal taste sensations and appetite, normal sperm counts in males	widespread in many foods, especially meats
	as a component of peptidases, involved in protein digestion	

Energy balance

To ensure that the body stays in energy balance, the energy intake over a period of time must equal the amount of energy used in tissue maintenance, repair, work and, in mammals and birds, the production of body heat. Insufficient intake of energy can be temporarily offset by breaking down tissue fat, carbohydrates or proteins, with a resulting loss in body mass. Excessive energy intake will result in increased storage of body fat (obesity).

 5   Obesity is a major problem in the developed world. Why is this considered to be a form of malnutrition?

Energy content of diet (carbohydrates and fats)

Table 8.3 shows how energy requirements of humans vary throughout life. These requirements must balance the energy expenditure which may be calculated in a number of ways. **Indirect calorimetry** involves measurement of the oxygen consumption or carbon dioxide excretion of an individual over a period of time. By making assumptions about the person's diet, the energy released by each cubic decimetre of oxygen consumed (kJ dm^{-3} O_2) can be calculated. A second method, **direct calorimetry,** is more accurate but less convenient. Heat lost by a person warms water which circulates in pipes around an insulated cubicle. The energy released by this person over many hours is calculated from the rise in temperature of the water.

The World Health Organisation recommends that about one-third of our energy intake should be lipids and two-thirds should be carbohydrates.

ANALYSIS

Measuring the energy content of food

This exercise involves the skills of calculation and critical appraisal of experimental procedure.

If you have ever been on a weight-reducing diet then one of the things you would have been very conscious of is the need to control your intake of energy-rich foods.

HETEROTROPHIC NUTRITION

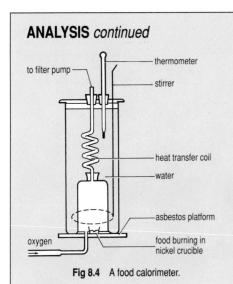

Fig 8.4 A food calorimeter.

A calorimeter is used to find the energy content of a food (Fig 8.4). When food is burned in pure oxygen, the energy it contains is released as heat which is transferred to the water in the outer jacket. The temperature rise of the water is then measured. Since you know the mass of water in the jacket, here 500 g, and you know 4.18 J of energy are needed to raise the temperature of 1 g of water by 1 °C, you can calculate the amount of energy transferred to the water when the food was burned.

When 1 g of dried bread was burned in such a calorimeter the temperature of the water was raised by 5 °C.

(a) Calculate the energy content (kJ) of the bread.
(b) Suggest why the bread was dried before being used in the experiment.
(c) Why is the bread burned in pure oxygen and not in air?
(d) What inaccuracies might arise in using this piece of apparatus?

Dietary fibre

Cellulose, which forms the cell wall around plant cells, makes up the bulk of the fibrous material eaten by mammals. A large proportion of the diet of herbivorous mammals is fibre and this is reflected in adaptations of their teeth and gut described in Section 8.6. Carnivores may eat very little fibre. The amount of dietary fibre eaten by humans has received a great deal of attention during the 1980s. Lack of fibre has been associated with a number of dietary disorders, including cancer of the large intestine. Rich sources of fibre are wholemeal cereals, unpolished (brown) rice, fresh vegetables and fruit.

SPOTLIGHT

OUR HUNGRY WORLD

The human population of the world is fast approaching 6 million million and many of these people, particularly children, do not receive enough energy in their diet.

In the short term, this energy shortfall can be met using reserves of carbohydrate and fat but eventually the body has to start catabolising muscle and tissue protein to meet its energy needs. This results in wasting and death.

Even more people are malnourished because their diet lacks other essential nutrients, e.g. vitamins. Protein represents a particular problem since your body cannot store excess amino acids; a continual supply of protein in the diet is therefore needed. Children who develop the deficiency disease **marasmus** are short of both energy and protein. Such children, often weaned too early or given a poor substitute for their mothers' milk, are emaciated with wrinkled skin and large, sunken eyes. Children suffering from the protein-deficiency disease, **kwashiorkor**, are often children who have been weaned too early and placed on a diet of maize which lacks the essential amino acids tryptophan and lysine. Consequently, a child cannot synthesise essential proteins and develops a bloated belly due to oedema, grows slowly and has retarded brain development.

Fig 8.5 Sometimes hunger turns into famine, with an appalling loss of life and long-term impact on social structures.

8.4 Explain what is meant by a balanced diet.

8.5 In a number of experiments to determine the nutritional requirements of a flour beetle, *Tribolium*, the larval stages were reared on diets lacking one of a number of amino acids. Matched groups of larvae were weighed after 15 days on each diet. The average mass of the animals on each of the diets was:

complete diet	1.11 g	no hydroxyproline	1.09 g
no proline	1.00 g	no threonine	0.01 g
no leucine	0.003 g	no glycine	1.31 g
no phenylalanine	0.01 g	no serine	0.84 g

Which of the amino acids listed are essential?

8.3 THE HUMAN DIGESTIVE SYSTEM

Mammalian digestive systems have five basic functions

- ingestion – taking food into the body
- peristalsis – the movement of food along the alimentary canal
- digestion – the breakdown of large, insoluble food molecules into smaller, soluble molecules
- absorption – the uptake of digested food from the gut into the blood stream
- egestion (or defaecation) – the elimination of undigested food, bacteria and dead cells from the body.

Digestion, absorption and egestion are dealt with in Sections 8.4 and 8.5.

The organs involved in human digestion can be divided into two main groups.

1. The gut or **alimentary canal**, is a continuous muscular tube running from the mouth to the anus (Fig 8.6). In adults it is about 10 m long and is subdivided into organs whose functions are summarised in Tables 8.9, 8.10 and 8.11.

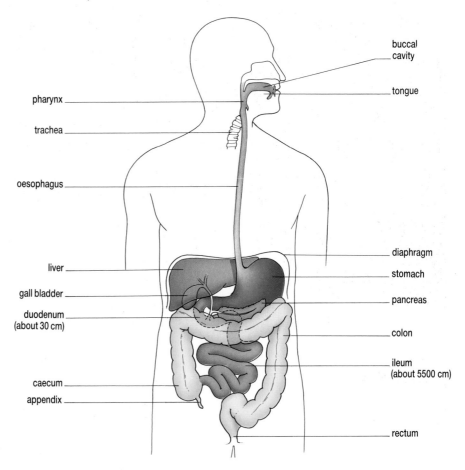

Fig 8.6 The human digestive system (alimentary canal or gut and accessory structures).

2. **Accessory structures** include salivary glands, liver, gall bladder and pancreas. These accessory structures lie outside the alimentary canal and either produce or store secretions which aid the digestion of food. These secretions are released into the alimentary canal through ducts.

Histology of the gut

The four different layers of tissue shown in Fig 8.7 are present throughout the entire length of the gut. The innermost layer, the **mucosa**, has an epithelium which produces mucus. This mucus reduces damage to the wall of the intestine by lubricating the passage of food, thus reducing friction, and by placing a barrier between damaging chemicals and the intestine wall. The epithelium also secretes hormones and digestive enzymes in different regions of the gut as well as acting as an absorptive layer. The mucosa has a layer of involuntary muscle (the **muscularis mucosae**) which allows movement of the finger-like villi in the small intestine. The **submucosa** contains a rich supply of blood and lymph vessels which carry absorbed nutrients away from the intestine.

The **muscle layer** has two layers of involuntary muscle. The inner circular layer has its muscle fibres arranged around the gut: contraction of these cells causes the gut to become longer and thinner. The outer longitudinal layer has its muscle fibres arranged along the gut: contraction

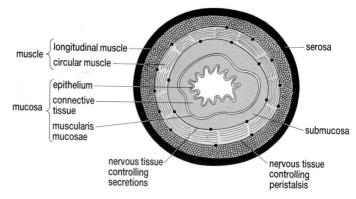

Fig 8.7 A generalised section across the alimentary canal showing the layout of tissues. The exact details vary between different organs but the general layout of the tissues is the same.

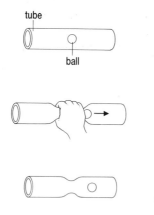

Fig 8.8 Peristalsis. You can imagine this process as a squeezing action forcing the bolus of food along. The squeezing action is caused by the action of circular and longitudinal muscles of the gut wall.

of these cells causes the gut to widen and shorten. Wave-like contraction of the circular muscle behind a ball of food (a **bolus**) and relaxation of circular muscle in front of a ball of food push food through the gut (Fig 8.8). As circular muscle contracts, the longitudinal muscle that is level with it relaxes, and vice versa when circular muscle relaxes. These contractions are called **peristalsis**. At points along the gut the circular muscle thickens to produce **sphincter muscles** which can be contracted to temporarily prevent further passage of food.

The outermost **serosa** layer has connective tissue which protects the rest of the gut from friction with other organs in the abdomen. It is surrounded by an epithelium which lines the inside of the abdominal cavity and the outside of every abdominal organ, the **mesentery**.

> **6** Suggest why sphincters are essential to the normal functioning of the human gut.

Teeth and dentition

Teeth break down food during mastication. Although food is compressed back into a loose ball during swallowing, mastication makes its physical breakdown in the stomach easier, allowing a larger surface area for enzyme action.

Humans have two sets of teeth during their lifetime, a condition known as **diphyodonty**. The milk teeth (**deciduous teeth**) become progressively inadequate as the head grows and are subsequently replaced by **permanent teeth**. Human teeth are **heterodont** (Fig 8.9), which means that they have different shapes and sizes.

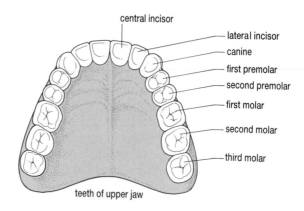

Fig 8.9 Permanent human dentition of the upper jaw. Notice the different shapes and sizes of teeth. The lower jaw has similar teeth.

7 **What are the advantages of a heterodont dentition?**

The teeth of other mammals are rather more specialised than human teeth allowing specialised diets (Section 8.6). A convenient way to represent the number of each type of tooth is the **dental formula**. This represents the number of each type of tooth in one half of the upper and in one half of the lower jaw. In total there are four incisors (i), two canines (c), four premolars (p) and six molars (m) in both the upper and lower jaw of humans. Drawing an imaginary line in Fig 8.9 to separate the incisors on the upper jaw and then counting one side only, there are two incisors on the upper jaw, represented i2. Dentition in the lower jaw is identical. Again from the imaginary line, the lower jaw has two incisors: the dental formula is represented i2/2. Continuing through the rest of the teeth, the full dental formula is represented

$$ i\frac{2}{2} \quad c\frac{1}{1} \quad p\frac{2}{2} \quad m\frac{3}{3} $$

The formula for milk dentition is

$$ i\frac{2}{2} \quad c\frac{1}{1} \quad p\frac{2}{2} $$

8 **How many teeth does a child have?**

The bulk of a tooth, the **root**, is embedded in the jaw (Fig 8.10). It is made from a bone-like substance called **dentine** and held in place by **cement** and by **periodontal fibres** of protein. This anchorage prevents the teeth falling out but allows movement of the teeth in their sockets, lessening the likelihood that they will shear during chewing. Holes at the base of the root allow the passage of blood vessels and nerves to and from the **pulp cavity** in the centre of the tooth. The part of the tooth which is normally visible, the **crown**, is covered by **enamel**, an extremely hard substance.

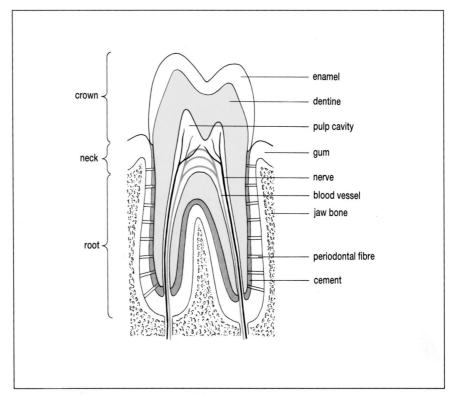

Fig 8.10 The parts of a typical tooth as seen in a vertical section through a molar.

Fig 8.11 Dental caries is still a major health issue despite modern preventative measures such as fissure sealing of teeth and fluoridation of water.

QUESTIONS

8.6 (a) Describe how peristalsis occurs and suggest why dietary fibre may aid this action.
   (b) What is the role of the muscularis mucosae in the gut?

8.7 The main cause of dental disease in children is dental caries but in adults is periodontal disease. Explain the meaning of these terms and suggest why such differences might occur.

8.4 HUMAN DIGESTION

Digestion involves the breakdown of large, insoluble food molecules into smaller, soluble compounds which can be absorbed. For example, starch is broken down into smaller monosaccharides and disaccharides; proteins are broken down into polypeptides, then into tripeptides, dipeptides and finally amino acids.

 What type of chemical bonds are being broken when a protein and a polysaccharide are digested?

Physical digestion

This is brought about by the action of teeth (chewing or mastication), the churning movements produced by muscular contractions of the alimentary canal, particularly the stomach, and by bile salts. It helps mix the food with the digestive juices and breaks the food into smaller pieces, so increasing the surface area available for enzyme attack.

 Why will increasing the surface area increase the rate of chemical digestion?

Chemical digestion

This involves the type of chemical reaction called **hydrolysis**: the chemical bonds which hold the large food molecules together are broken by adding water to them. This is the opposite of the condensation reactions used to make large molecules such as proteins and starch from their smaller subunits (see Chapter 1). Enzymes which catalyse hydrolysis reactions are called **hydrolases**: all digestive enzymes are therefore hydrolases. Some of these enzymes are produced by accessory glands, e.g. the pancreas, and released into the alimentary canal through ducts; others are contained within the cell surface membrane or cytoplasm of cells of the gut mucosa (see Table 8.8). The general processes involved in chemical digestion in the gut are summarised in Fig 8.12.

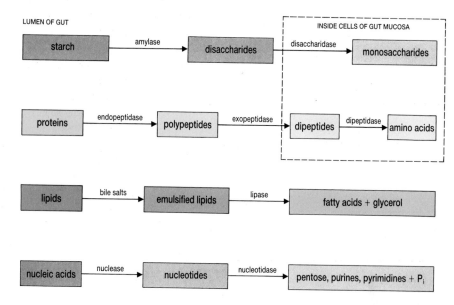

Fig 8.12 A summary of digestion in humans. Most of the processes occur in the lumen of the gut but a few occur in the cells of the gut mucosa. Individual enzyme names are given in Table 8.8.

 11 **What are the two different types of simpler molecules produced by hydrolysis of a lipid?**

Table 8.8 Summary of digestive enzyme activity throughout the human gut

Region of gut	Secretion	pH	Enzyme	Substrate	Product(s)
buccal cavity	saliva	6.5 – 7.5	amylase	starch	maltose
stomach	gastric juice	2.0	pepsin	protein	polypeptides
				pepsinogen	pepsin
duodenum	pancreatic juice	7.0	amylase	starch	maltose
			trypsin	protein	polypeptides
			chymotrypsin	protein	polypeptides
			carboxy-peptidase	polypeptides	short peptides
			lipase	lipids	fatty acids and glycerol
			nuclease	nucleic acids	nucleotides
	bile (from liver)	7.0	none	lipids	emulsified lipids
	other enzymes*		maltase	maltose	glucose
			sucrase	sucrose	glucose and fructose
			lactase	lactose	glucose and galactose
			tripeptidases	tripeptides	amino acids
			dipeptidases	dipeptides	amino acids

* these are all located in the cell surface membrane or cytoplasm of the epithelial cells of the mucosa

Digestion in the buccal cavity

Chewing food breaks it into small particles and mixes it with saliva, secreted from the three pairs of **salivary glands** – sublingual (under the tongue); submandibular (below the lower molars) and parotid (above the upper molars). Saliva, with a pH of 6.5–7, consists primarily of water and mucus which soften and lubricate the food. In most people, saliva also contains an amylase enzyme which starts the digestion of carbohydrate (Fig 8.12). Chloride ions in the saliva activate the amylase.

After chewing, the food is formed into a bolus by the action of the tongue and then swallowed. Peristalsis forces the bolus down the oesophagus to the stomach. This takes about 5 to 7 seconds for solid food.

Table 8.9   Functions of the human digestive system – mouth to stomach

Organ	Description	Activity
buccal (mouth) cavity	teeth	cut, grind and pulverise food, reducing it to smaller particles so increasing surface area for the action of enzymes
	tongue	manoeuvres food for chewing; rolls food into bolus for swallowing
	openings of salivary glands	saliva, pH 6.5, moistens food; mucus sticks bolus together and lubricates it for swallowing; salivary amylase, if present, starts digestion of starch
oesophagus	25 cm long muscular tube connecting buccal cavity and stomach	bolus of food pushed along the oesophagus by peristaltic waves of muscle contraction
stomach	mucosal bag which can hold up to 2 dm³ of food	muscular contractions churn and mix the food with gastric juices producing semi-liquid chyme
	mucosal cells secrete pepsinogen, HCl and mucus under hormonal and nervous control	HCl converts pepsinogen into pepsin, a protease; low pH kills bacteria; mucus prevents digestion of the stomach wall
	cardiac and pyloric	storage of food by sphincters eliminates need for frequent small meals

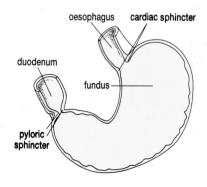

Fig 8.13   A vertical section through the stomach.

Digestion in the stomach

The stomach is essentially a large, distensible bag which can hold up to $2\ dm^3$ of food (Fig 8.13).

 Suggest an advantage in carnivores, such as lions, of having stomachs which are relatively more distensible than herbivores, such as horses.

The mucosa of the stomach is indented to form **gastric pits** (Fig 8.14). The **oxyntic cells** in these gastric pits secrete hydrochloric acid; the **chief cells** secrete **pepsinogen**. Inactive pepsinogen is converted to the active protein-digesting enzyme **pepsin** by the action of hydrochloric acid and by pepsin itself.

 Suggest an advantage to explain why the chief cells make and store an inactive form of pepsin rather than the active form?

In addition to activating pepsinogen, hydrochloric acid creates optimum conditions of low pH necessary for the maximum activity of pepsin. This low pH also kills most of the bacteria and inactivates the salivary amylase in swallowed food. The stomach wall is protected from this corrosive mixture of hydrochloric acid and pepsin by a thick layer of mucus produced by **goblet cells**. Factors such as irregular eating, nervous tension

HETEROTROPHIC NUTRITION

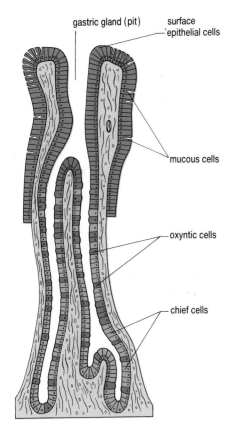

gastric gland (pit)　surface epithelial cells

mucous cells

oxyntic cells

chief cells

Fig 8.14　A gastric gland. The epithelium lining the gland contains chief cells (secrete pepsinogen) and oxyntic cells (secrete HCl and intrinsic factor, essential for the absorption of vitamin B_{12}).

and smoking can increase the rate of acid secretion, increasing the risk of peptic ulcer formation. If the cardiac sphincter at the top of the stomach is not working properly then the acid contents of the stomach can leak upwards into the oesophagus giving rise to heartburn.

　14　**Explain the bactericidal action of gastric juice.**

The cardiac and pyloric sphincters (Fig 8.13) ensure that food is retained in the stomach for about two hours; the more fat the food contains the longer it stays in the stomach. During this time muscular contractions of the stomach wall churn the food, mixing it with the gastric juices and turning it into semi-liquid **chyme**. It is this chyme which eventually passes, by means of controlled squirts, through the pyloric sphincter into the first part of the small intestine, the duodenum.

Digestion in the small intestine

Once in the duodenum, chyme is mixed with secretions from the liver and pancreas. Whilst **bile** from the liver contains no digestive enzymes, its **bile salts** cause emulsification of lipids to form water-miscible droplets: just like washing-up liquid causes grease on plates to form an emulsion. Like bile, the **pancreatic juice** contains hydrogencarbonate ions which neutralise the acidic chyme and create the optimum pH for enzymes which are active in the small intestine. Table 8.8 shows the effect of enzymes which are active in the small intestine.

There are deep folds in the wall of the duodenum, called **crypts of Lieberkuhn** (Fig 8.15). Lying beneath these are the **Brunner's glands** (Fig 8.16) which secrete a viscous alkaline fluid, the succus entericus, containing water, hydrogencarbonate ions and mucoprotein. It contains no enzymes but probably serves to protect the wall of the duodenum from the ulcerating effects of hydrochloric acid and pepsin in the chyme.

Both extracellular and intracellular digestion occur in the small intestine of mammals. The mammalian small intestine is lined by a single layer of cells (a simple epithelium). Although about 20% of these cells release a secretion, they do not secrete digestive enzymes. Extracellular digestion is brought about by enzymes released into the small intestine by the pancreas. However, most of the epithelial cells of the small intestine have

Table 8.10　Functions of the human digestive system – small intestine

Organ	Description	Activity
small intestine	6 m long muscular tube about 2.5 cm in diameter, lined by glands, inner surface folded to form villi, epithelial cell surface membrane folded to form microvilli	churning motions keep chyme in motion
		folding provides a greatly increased surface area for secretion, digestion and absorption
		further digestion occurs, mainly in the duodenum, whilst jejunum and ileum are the main sites of absorption
	three regions – duodenum, jejunum, ileum	
pancreas	large gland which opens into the duodenum via common duct	pancreatic juice secretion – rich in sodium hydrogencarbonate; neutralises acid chyme from stomach, raises pH ~ 7.5; release is stimulated by a hormone, secretin, released by the duodenum
	produces pancreatic juice	– rich in enzymes (see Table 8.8), release controlled by a hormone, CCK-PZ, from the duodenum
liver	produces bile, stored in the gall bladder; bile duct connects gall bladder to common duct	release of bile from gall bladder controlled by CCK-PZ
		bile contains no digestive enzymes but bile salts emulsify fats (breaks them into small droplets called micelles) increasing surface area for action of lipase

digestive enzymes embedded within their cell surface membranes and in their cytoplasm. Thus, digestion occurs outside these cells, at their cell surface membranes and in their cytoplasm (see Fig 8.17).

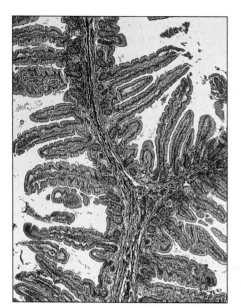

Fig 8.15 Micrograph of the wall of the duodenum showing the deep folding and the villi. The mucous membrane cells are stained pink.

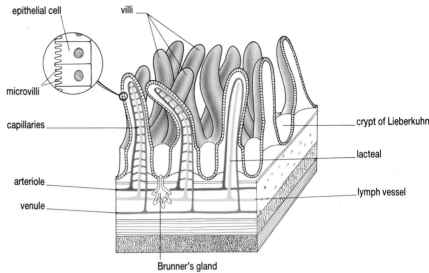

Fig 8.16 A stereogram of the wall of the duodenum.

Resulting from digestion in the lumen of the small intestine, carbohydrates have been hydrolysed to disaccharides and proteins have been hydrolysed to di- and tripeptides. Their further digestion takes place in the cell surface membrane of the microvilli of the intestinal mucosa. Disaccharidases, dipeptidases and tripeptidases are intrinsic proteins in these cell surface membranes. Disaccharides are taken up by the microvilli and broken down into their constituent monosaccharides. Most of these monosaccharides are then released back into the gut lumen. A similar process occurs with di- and tripeptides, which are taken up by the microvilli and broken down into their respective amino acids. However, some peptidases have been found in the cytoplasm of the cells of the villi, as well as in their surface membranes. It is thought that some di- and tripeptides are absorbed directly into these cells where their digestion takes place.

Mucosal cells are lost from the upper part of the villi at a rapid rate and are found in the succus entericus. It is thought that the action of their membrane-bound enzymes is no longer significant, so they are unlikely to give the succus entericus any enzyme activity.

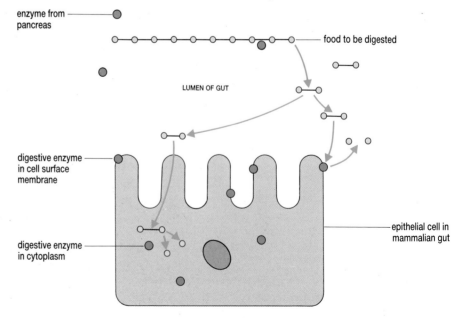

Fig 8.17 Digestion in the small intestine is both extra- and intracellular. Enzymes such as trypsin and lipase act in the lumen of the gut. Disaccharidases act in the cell surface membrane of epithelial cells; active dipeptidases are found in the cell surface membrane and cytoplasm of the epithelium.

HETEROTROPHIC NUTRITION

8.8 Explain what is meant by **(a)** physical digestion **(b)** chemical digestion. Why is each necessary?

8.9 Explain why the buccal cavity, stomach and intestine have a different pH. Describe how these pH values are brought about.

8.10 Draw two summary diagrams to represent stages in **(a)** the digestion of starch **(b)** the digestion of proteins. In each diagram, include **(i)** the named enzymes involved in digestion **(ii)** the part of the gut in which each stage of digestion occurs.

8.5 ABSORPTION AND ASSIMILATION OF DIGESTED FOOD

Section 8.4 has shown a progressive digestion of food to produce a solution of glucose, fructose, amino acids, nucleotides, fatty acids and glycerol. However many of these small molecules are still in the lumen of the intestine, i.e. outside the body. To be of any use they must be moved from the lumen into the bloodstream or lymph – the process of **absorption**. Nutrient molecules are then transported to the tissues where they are **assimilated** and used either for energy, growth, repair or maintenance.

Absorption

Some relatively small molecules can be absorbed from the food in the buccal cavity and the stomach. For example, the effects of drinking alcohol are experienced quickly because alcohol absorption occurs in the stomach. However, most absorption occurs in the small intestine which, in adult humans, is about 6 metres long. Numerous folds in the wall of the small intestine increase its surface area. The folds themselves have tiny, finger-like projections, called **villi**, the mucosal cells of which have **microvilli**.

 What is the effect of all this folding?

The single layer of cells of the intestinal epithelium provides a short distance for the uptake of digested food. Absorption of water-soluble nutrients, such as disaccharides, monosaccharides, tripeptides, dipeptides, amino acids, vitamins and inorganic ions, occurs by diffusion and active transport (Section 2.2). The successful oral administration of vaccines shows that absorption of intact proteins must also occur, although natural protein absorption occurs mainly in the newborn. Uptake of glucose and amino acids is linked to the active transport of sodium ions (see Fig 8.18); uptake of calcium ions is enhanced by the presence of vitamin D; uptake of iron ions is enhanced by the presence of vitamin C. Digested fats and lipid-soluble vitamins do not dissolve in water. They form globular **micelles** about 5 nm in diameter which dissolve in the surface membrane of the cells lining the villus and diffuse into their cytoplasm. Here, triglycerides and phospholipids are reformed and packaged with protein and cholesterol into **chylomicrons** of about 100 nm diameter.

Fate of absorbed food materials (assimilation)

The water-soluble products of digestion are passed from epithelial cells to the blood capillaries within each villus. Transfer of monosaccharides is aided by sodium ions; Fig 8.18 summarises current views to explain how this transfer occurs. Short-chain fatty acids may also pass directly into the blood capillaries. From the villus, these nutrients are carried to the liver by the hepatic portal vein. Immediately after a meal the concentration of these nutrients may be very high, at other times it may be low. The activities of the liver (Chapter 11) ensure that blood passed to the rest of the body does not suffer these fluctuations in concentration. Regulation of the composition of the blood is part of **homeostasis** and is described in Chapter 11.

The chylomicrons, containing lipids and lipid-soluble vitamins, are passed into the lymph capillaries. These are called **lacteals** because their fatty contents make them appear white. The lymph system eventually empties into the veins near the heart. In the blood system, these lipid-soluble compounds eventually reach the liver.

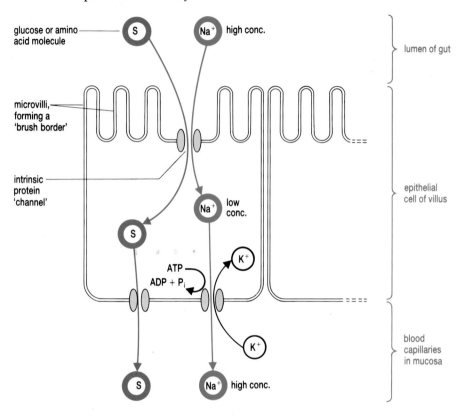

Fig 8.18 Transport of amino acids and sugars from the lumen of the gut to the plasma of the blood capillaries. A coupled sodium–substrate transport carries the substrate, S, into the absorptive epithelial cell. The substrate leaves the cell by facilitated diffusion (bottom left) and enters the plasma. The Na^+–K^+ pump (bottom right) maintains the sodium gradient essential to the coupled Na^+-substrate transport into the cell.

Table 8.11 Functions of the human digestive system – the large intestine

Organ	Description	Activity
caecum and appendix	sack-like structures at the junction of small and large intestines	appendix has no function (vestigal) in humans
colon and rectum	muscular tube about 6 cm diameter, 1.5 m long	peristalisis moves contents along colon and compacts faeces, which are stored in rectum
	mucosa secretes mucus	mucus lubricates and protects mucosa from action of digestive enzymes
	contains large numbers of bacteria	absorbs water and other soluble compounds, so solidifying faeces
		vitamins and ions absorbed
		bacteria break down undigested food (either absorbed or expelled in faeces), certain B vitamins and vitamin K synthesised
		about 250 g of compacted faecal material consisting of undigested food, bile pigments (colour faeces), bacteria (10-50% faeces) and dead cells sloughed off small intestine, egested daily

HETEROTROPHIC NUTRITION

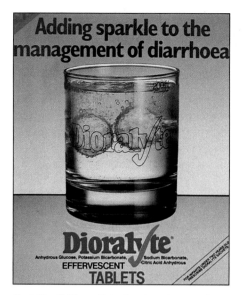

Fig 8.19 Children suffering from loss of water and minerals during diarrhoea, a major killer in the developing world, can be cured easily and cheaply by administering a mixture of glucose, salt and water until the diarrhoea ceases.

Absorption in the colon

About 10 dm³ of fluid may be secreted into the human gut each day. Most of this volume is water. Whilst water and unused bile salts are reabsorbed in the ileum, much water is absorbed by the colon. Similar to the lower part of the ileum, the colon contains a large population of bacteria (notably *Escherichia coli*), and protozoa which feed on them (such as *Entamoeba*). To some extent, these bacteria are able to utilise undigested remains of human food, including fibre. In doing so, some synthesise vitamins, such as K, B_{12} and riboflavin, which are absorbed by the colon. Gases, such as carbon dioxide, methane and hydrogen sulphide, are the result of bacterial activity. Bacteria in the colon also excrete toxins which may cause irritation and even cancer of the lining of the colon. Occasionally, micro-organisms may cause disease. A most common symptom of such disease is the diarrhoea which results when the ileum no longer absorbs sufficient water. The largest single cause of child deaths throughout the world is the dehydration and loss of salts associated with diarrhoea.

SPOTLIGHT

ANOREXIA NERVOSA AND BULIMIA

Anorexia nervosa is a psychological disorder, characterised by loss of appetite and strange patterns of eating. The physical consequence of the disorder is severe and progressive starvation. Amenorrhoea (absence of menstruation) and a lowered metabolic rate reflect the depressant effects of starvation. Individuals may become emaciated and may ultimately die of starvation. Treatment consists of psychotherapy and dietary regulation.

Bulimia, or binge-purge syndrome, is characterised by uncontrollable overeating followed by forced vomiting or overdoses of laxatives. The binge-purge cycle seems to occur in response to fears of being overweight, stress or depression. Since it involves continual vomiting, bulimia can upset the body's electrolyte balance and may increase tooth decay as the acid in the vomit attacks the enamel of the teeth. Treatment consists of nutrition counselling, psychotherapy and medical treatment to rectify electrolyte imbalance.

For reasons unknown, both these eating disorders are more common among single, young, white females than in any other group and their incidence has increased in recent years.

Egestion

The rectum stores the dried, undigested remains of food. Together with dead cells which have sloughed off the gut, dead bacteria and bile pigments, these form the **faeces**. Relaxation of the sphincter muscle around the anus allows these faeces to be expelled during **defaecation**. Since the bulk of faeces is material which has never been part of the body (the gut lumen is outside the body, remember), defaecation is correctly termed egestion, not excretion.

Control of secretions in the gut

Stimulation of cells occurs by hormones and by nerve cells (Fig 8.20). Both types of coordination are involved in controlling the secretion of digestive juices into the gut (Chapters 14 and 15). Stimulation of the salivary glands

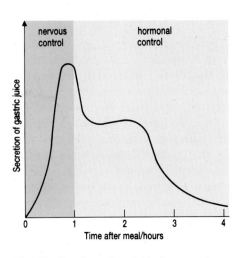

Fig 8.20 The release of gastric juice in response to nervous and hormonal stimulation.

is entirely by the nervous system in response to the stimulus of real or imagined food. For the first hour or so after a meal, the release of gastric juice is stimulated by nerves in response to the presence of food in the buccal cavity. This produces the first peak of the curve in Fig 8.20. The presence of food in the stomach stimulates the stomach lining to release the hormone **gastrin**. Like most hormones, gastrin is passed directly into the bloodstream. On its return to the stomach it stimulates the continued release of gastric juice for several hours, producing the second peak in Fig 8.20. A second hormone, **enterogasterone**, is released by the stomach mucosa in response to lipid in the stomach. This hormone decreases the flow of gastric juice and slows churning of the stomach, delaying the exit of fatty food from the stomach.

Chyme entering the small intestine stimulates its lining to release a number of hormones. **Secretin** stimulates the release of a pancreatic secretion rich in sodium hydrogencarbonate.

 What is the function of this sodium hydrogencarbonate?

Cholecystokinin-pancreozymin (CCK-PZ) stimulates the gall bladder to contract. The gall bladder stores bile, which is continuously secreted by the liver. Hence, CCK-PZ causes a sudden release of bile into the small intestine. In addition CCK-PZ stimulates the release of digestive enzymes from the pancreas.

QUESTIONS

8.11 (a) List the end-products of the digestion of carbohydrates, lipids and proteins which are absorbed in the small intestine.
   (b) How is the small intestine adapted to allow efficient absorption of these end-products of digestion?

8.12 (a) Make a table to summarise the control of secretions in the gut.
   (b) Why is it important that such secretions are controlled?

8.6 ADAPTATIONS OF MAMMALS TO THEIR DIET

Humans are **omnivores**, eating both plant and animal food. The guts of mammals which feed exclusively on meat or on grass are adapted to these diets in a number of ways.

Carnivores

Meat is a richer source of nutrients than plant food, enabling carnivores to spend less time feeding than herbivores. The stomach of most carnivores is able to tolerate sudden stretching to accommodate infrequent large meals but otherwise their gut is little different from that of humans. The major adaptations for eating meat are in their skull and teeth.

Look at Fig 8.21: it shows the skull and teeth of a dog. Its incisors, not used for slicing food from its prey, are small and insignificant. Instead, the meat is skewered by the long, pointed and curved canine teeth as the head is pulled back to tear meat from the prey. The canines are also useful to kill prey. Premolars and molars have sharp cusps and flat sides which overlap like scissors. They are used to slice meat before it is swallowed. The last upper premolar and first lower molar (the **carnassial teeth**) are particularly well developed for this purpose. The slicing action of the teeth requires only an up and down movement of the jaw. The lower jaw articulates with a deep surface on the skull, restricting all but the vertical movements needed for slicing meat and killing prey. The temporalis muscle, whose contraction raises the lower jaw, is large. The crest on the dog's cranium, which is visible in Fig 8.21, provides a large surface for the attachment of the temporalis muscle. The masseter muscle, whose contraction would move the jaw from side to side, is small.

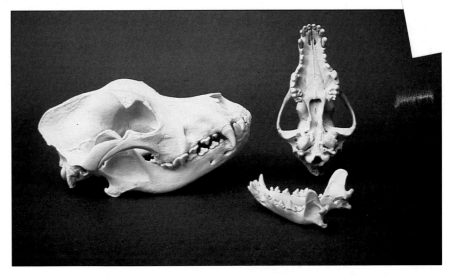

Fig 8.21 The characteristics of teeth (shape, size, dental formula) and skull (size, muscle attachments and jaw movements) typify the adaptations of a dog (carnivore) for eating meat.

Herbivores

Plant food is a fairly tough material and must be ground into fine particles to aid its digestion. This places different demands on the teeth and jaws of herbivores in comparison with carnivores. Plant cells are surrounded by cellulose, yet no herbivorous mammal produces a cellulase enzyme. They rely on micro-organisms in their gut, mainly bacteria, to digest cellulose. Their gut is adapted to harbour these bacteria.

 17 **Given that bacteria digest their food extracellularly how does their presence help the herbivores to gain access to nutrients locked up in cellulose?**

Dentition

The skull of a sheep is shown in Fig 8.22. Sheep cut grass using a sideways movement of the lower incisors and canines against a horny pad on the upper jaw. The incisors and canines are the same height and chisel-shaped. Their roots are open, allowing continuous growth throughout the sheep's life; otherwise the teeth would eventually be worn away by the constant

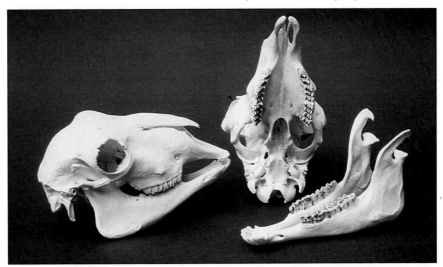

Fig 8.22 Sheep are herbivores. Chisel-shaped lower incisors cut grass against a horny pad in the upper jaw. Upper and lower premolars and molars grind the grass by sideways movement.

friction against grass. The premolars and molars are the same height and have flat surfaces with zig-zag shaped ridges. The sideways movement of the lower jaw, used for cutting grass, rubs these ridged surfaces against each other, grinding the grass in much the same way as rubbing it between two files would do. The masseter muscle, used to move the jaw sideways, is very large, whilst the temporalis is relatively small. The lower jaw articulates with a shallow surface on the skull so that sideways movement of the clenched jaw is easy. (It is hard for humans to do this.) Separation of food which is being chewed from food which is being cut is allowed by movements of the tongue in the large, toothless gap (**diastema**) between the cheek teeth and cutting teeth.

Adaptations of gut to harbour cellulose-digesting bacteria

Ruminants are animals with a 'stomach' which consists of the four chambers shown in Fig 8.23. Swallowed food is fermented by bacteria in the first two of these chambers, the **rumen** and **reticulum**. The products of cellulose fermentation (ethanoic, proprionic and butyric acids) may be used by these bacteria but most are absorbed by the host's rumen and reticulum and used as energy sources in respiration. The inner surface of the rumen and reticulum (sold as tripe) have villi and a honeycomb-like surface, respectively, providing a large surface area for this absorption. Carbon dioxide and methane gases are also produced from the fermentation and are expelled via the mouth. Since both bacteria and host may be considered to benefit from the relationship, it is described as **mutualistic** (Table 28.1).

Partially digested food can be regurgitated and chewed (**rumination**). On the second swallowing, food passes rapidly through to the third chamber, the **omasum** and thence to the true stomach, the **abomasum**. From here, digestion occurs in the same way as has been described for humans.

Bloat is a condition of ruminant animals when the gut becomes dangerously extended with gases from the digestive fermentation process. The rumen may press on the lungs to restrict breathing. A cow produces 800 dm³ CO_2 and 500 dm³ CH_4 per day. About 75% of the gases are expelled via the mouth and 25% are absorbed into the blood and escape through the lungs.

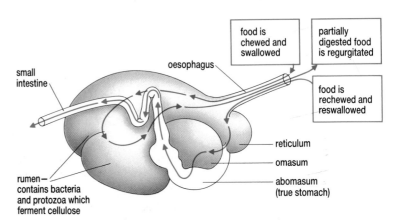

Fig 8.23 The activities of a rumen.

Herbivores which do not have a rumen may harbour their cellulose-digesting bacteria elsewhere in the gut. For example, they are found in the caecum and appendix of rabbits and horses. On first passage through its stomach and small intestine, the cellulose in a rabbit's food remains undigested until the caecum and appendix. Bacterial fermentation then digests cellulose. However, few of the products of this fermentation can be absorbed in the rabbit's colon. The partly digested food is egested as moist, white faeces which the rabbit eats. On its second passage through the rabbit's intestine, this food is digested in a similar way to human food; any undigested remains are egested as dry, brown faeces.

HETEROTROPHIC NUTRITION

INCREASING ANIMAL YIELDS

The yield from agricultural animals, meat, milk, wool and so on, depends on their diet. Thus, adding essential amino acids, like lysine, to the diet of chickens has dramatically increased poultry yields.

Fermentation of proteins in a sheep's rumen results in the breakdown of most of the ingested amino acids. Since the growth of wool is limited by the absorption of sulphur-containing amino acids, this fermentation reduces wool growth. Albumin from pea plants resists fermentation in the rumen. Genetic engineers in Australia are trialling alfalfa plants, used for sheep fodder, into which the gene for pea albumin has been transferred. This may result in an estimated A$300 million per year in increased wool production.

You may read about genetic engineering in Geoff Hayward's book in this series, *Applied Genetics*.

ANALYSIS

Molecular adaptations to diet

This exercise involves the skills of data analysis and translating information from one form to another.

In addition to morphological adaptations, e.g. teeth and jaws, heterotrophs also show molecular adaptations to their diet.

The mid gut of insects is normally a region of production and release of digestive enzymes.
(a) Suggest how you could test for the presence of (i) amylases (ii) proteases (iii) lipases in this region of an insect's gut.

The results of such tests on the mid-gut secretions of a number of insects are summarised in Table 8.12. Use information in the table to answer the following questions.
(b) Blood contains significant amounts of sugar, mainly glucose, and a number of different proteins. Which of these insects feeds mainly on blood?
(c) Leafhoppers are small insects which feed by inserting their piercing mouthparts into leaf cells containing the full range of plant chemicals, or into phloem cells containing sucrose. To which of these two categories do leafhoppers A and B belong?
(d) The adult and larva of the blowfly have quite different diets; one feeds on rotting meat and the other on sugars. Which feeds on meat, the adult or the larva? Explain your answer.
(e) It would also seem advantageous if an animal only synthesised an enzyme if stimulated by the presence of a specific substrate. State the term used to describe the stimulation of enzyme synthesis by the presence of its substrate (see Section 5.4).

Table 8.12 Presence of enzymes in mid-gut secretions of some insects

Insects	Protease	Lipase	Amylase	Sucrase	Maltase
1 leafhopper (A)	+	+	+	+	+
leafhopper (B)	–	–	–	+	–
2 *Glossina*	+	–	trace	–	–
3 caterpillar (plant feeding)	+	+	+	+	+
moth (non-feeding)	–	–	–	–	–
4 blowfly larva	+	+	–	–	–
blowfly adult	trace	–	+	+	+

+ = present;   – = absent

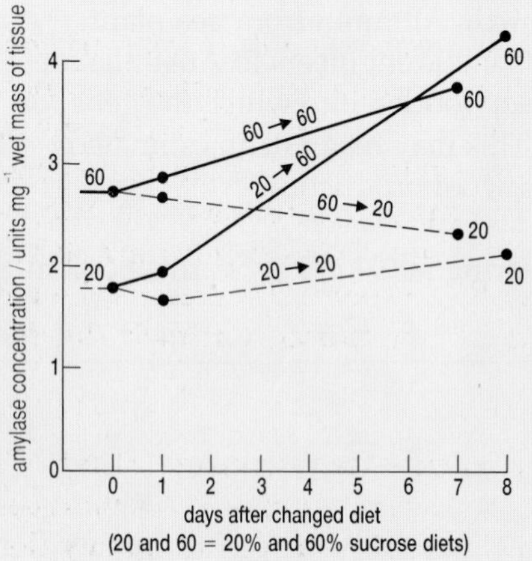

Fig 8.24 Changes in amylase concentration in fresh samples of rat pancreas following changes in diet.

In a separate experiment, rats were fed a diet containing either 20% sucrose or 60% sucrose for one week. Then half of each group of rats were put on the other diet. The activity of pancreatic amylase was determined for some rats on day 0 (the day of diet change), day 1, day 7 or day 8 (eight days after diet change). The results of the experiment are shown in Fig 8.24.

(f) What was the effect of changing the diet from 20% to 60% sucrose?

(g) Did the rats fed on a high sucrose diet (60%) have a higher pancreatic amylase concentration than those fed on a low sucrose diet (20%)?

(h) What conclusions can you draw from the results of these experiments?

QUESTIONS

8.13 Compare the role of gut bacteria in humans and in herbivorous mammals.

8.14 Look at Table 8.13 which shows the mean relative salivary amylase activity of four groups of humans. Fully explain these data.

Table 8.13 Relative mean salivary amylase activity in four groups of humans

Group of humans	Mean relative amylase activity
Tswans of Botswana	248
Europeans	101
Kalahari bushmen (in normal desert environment)	22
Kalahari bushmen (following 3 months imprisonment)	95

SUMMARY

Organisms which cannot manufacture their own food from inorganic sources are called heterotrophs. Before they can absorb their food, it must be made soluble. This is achieved during the process of digestion, which includes the enzyme-catalysed hydrolysis of large molecules into smaller ones. Digestive enzymes may be secreted onto food which is outside the cytoplasm of the heterotroph, resulting in extracellular digestion, or may operate within the cytoplasm of the heterotroph, resulting in intracellular digestion. Intracellular digestion occurs in protozoans, which engulf food into food vacuoles. Bacteria and fungi are unable to engulf food since they have a rigid cell wall outside their cell surface membranes; these organisms perform entirely extracellular digestion. The initial stages of digestion in animals is extracellular but the final stages may be intracellular.

Heterotrophs may be classed according to their food source. Holozoans are animals and some protoctists which ingest other organisms. Herbivores are holozoans which feed on plants whilst carnivores feed on animals. Detritivores ingest small particles of organic debris. Saprophytes cannot ingest their food but digest the dead organic remains of organisms, so causing decay. Symbiotic organisms live in or on the body of their food source (called a host). Parasites are symbionts which cause harm to their host.

To stay healthy, mammals need a supply of carbohydrates, lipids, proteins, vitamins, inorganic ions and water. These substances are called nutrients and form the diet. For a mammal to have a balanced diet it must consume each of its essential nutrients in the same amount as the nutrient is being used. The composition of a balanced diet varies depending on such factors as the age and sex of a mammal, the amount of work it performs, its state of health and pregnancy. As well as all the nutrients listed above, humans need indigestible dietary fibre to exercise the muscles of their digestive system.

The digestive system consists of the gut and a few associated glands. The gut has four major layers of tissue, the innermost mucosa, submucosa, muscle layer and outermost serosa. It is specialised at various points along its length to form the buccal cavity, oesophagus, stomach, small intestine and large intestine. Muscles in the gut move food along during peristalsis and form sphincters, which are used to stop progression of food along the gut and allow its digestion or storage.

The teeth break food into smaller particles which can be digested more easily. Mammals have two sets of teeth during their lifetime: the deciduous and permanent dentition. An adult mammal has four types of teeth, incisors, canines, premolars and molars, each adapted to perform a particular type of mechanical action. The number of each type of tooth in the dentition of a mammal can be expressed as a dental formula. The root of each tooth is held in the jaw bone by cement and by periodontal fibres of protein. The body of the tooth is composed of dentine but the crown, which protrudes above the gum, is covered by hard-wearing enamel. Cells within each tooth are supplied by blood vessels within the inner pulp cavity.

Digestive enzymes are released into the gut from secretory tissues in the salivary glands, stomach and pancreas. Bile, which emulsifies lipids, is secreted into the gut from the liver (following storage in the gall bladder in humans). The pH varies along the gut providing the optimum conditions for the activity of enzymes. Secretion from the salivary glands is controlled by nervous stimulation, from the pancreas and liver by hormonal stimulation and from the stomach by both types of stimulation. Enzymes released into the gut catalyse the hydrolysis of polysaccharides to disaccharides and proteins to tri- and dipeptides. Enzymes embedded in the surface membrane of cells lining the small intestine catalyse the hydrolysis of disaccharides to monosaccharides, whilst others embedded in the membrane or within the cytoplasm of these cells catalyse the hydrolysis of short peptides to amino acids. Some digestion of lipids occurs.

Short-chain fatty acids and the water-soluble products of digestion are absorbed in the small intestine and carried to the liver by the hepatic portal vein. Here they may be metabolised, distributed or stored. Following their absorption in the small intestine, digested lipids and lipid-soluble vitamins form chylomicrons which are transported from the intestine in capillaries of the lymphatic system, called lacteals. The absorptive surface of the small intestine is enlarged by folds, villi and microvilli. Any undigested food is stored in the rectum before being egested as faeces.

Carnivorous and herbivorous mammals show adaptations related to their diet. Carnivores have small incisors, large peg-like canines and sharp scissor-like premolars and molars. Their jaw and skull are articulated in such a way that little sideways movement is possible and have large surface areas for the attachment of the temporalis muscles which raise and lower the jaw. In contrast, the incisors and canines of herbivores have sharp cutting edges and their premolars and molars are ridged to form flat grinding surfaces. Their jaw and skull articulation allows sideways movement of a clenched jaw and their temporalis muscles are much smaller than their masseter muscles, which move the jaw from side to side. Herbivores rely on the activity of cellulose-fermenting bacteria in their guts to enable them to digest their food. These bacteria are found in swellings at the base of the oesophagus of ruminants and in the caecum and appendix of other herbivores.

Chapter 8: Answers to Quick Questions

1   The digestive enzymes would digest the cytoplasm.
2   Small animals have a larger surface area to volume ratio than larger animals and so lose heat more rapidly per kilogram of body mass. Therefore they will need proportionally more food. Endotherms need more food than ectotherms to sustain their higher metabolic rate.
3   Cannot be synthesised from other dietary components.
4   To obtain all the essential amino acids.
5   Result of incorrect balance of nutrients in the diet.
6   Regulate flow of food through the gut in spite of irregular ingestion patterns.
7   Allows the teeth to perform a wide range of functions, e.g. biting, chewing, and so process food efficiently.
8   20
9   Peptide and glycosidic bonds, respectively.
10 Present a larger surface area for the digestive enzymes to work on.
11 Glycerol, fatty acid.
12 Carnivores take large infrequent meals whilst herbivores can graze practically all the time.
13 Pepsin is a protease and would digest the cells producing it if it were stored in an active form.
14 Acid in the stomach will denature bacterial proteins so killing bacteria.
15 Increases the surface area (to an estimated 300 m^2) and so increases the area over which the products of digestion can be absorbed.
16 It neutralises the acidic chyme.
17 Once the cellulose cell walls have been broken down by the bacteria, the breakdown products of cellulose will be available in the gut lumen for absorption by the herbivore.

Theme 2
EXAMINATION QUESTIONS

2.1 **(a)** (i) Give an illustrated account of the structure of a mitochondrion.
(ii) Describe how ATP is produced inside a mitochondrion. (14)
(b) Outline the ways in which ATP is utilised in living organisms. (6)

(JMB 1990)

2.2 The results of an investigation into the effect of increasing substrate concentration on the rate of an enzyme reaction are shown in the graph.

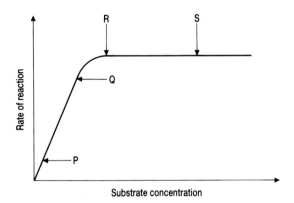

(a) (i) Name the factor which determines the rate of reaction between points P and Q.
(ii) Name **two** factors which could account for the shape of the curve between R and S. (3)
(b) (i) State **two** conditions which should be kept constant in this investigation.
(ii) What should be measured in order to determine the rate of an enzyme reaction? (4)
(c) The investigation was repeated with the addition of a competitive inhibitor. The same amount of inhibitor was added to the substrate at **each** concentration.
(i) Draw the expected curve on the graph below.

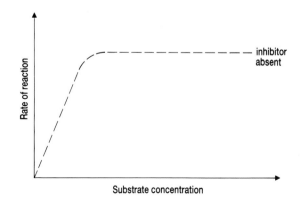

(ii) Briefly explain how a competitive inhibitor would bring about this effect. (5)
(d) Name **three** compounds which might have been used in the investigation. (3)
(i) Enzyme; (ii) Substrate; (iii) Competitive inhibitor.

(WJEC 1988)

2.3 **(a)** What are the main features of
(i) *autotrophic* nutrition; (2)
(ii) *heterotrophic* nutrition? (1)
The diagram shows part of a transverse section through a mammalian ileum.
(b) Name the parts labelled **A** to **D** on the diagram. (4)

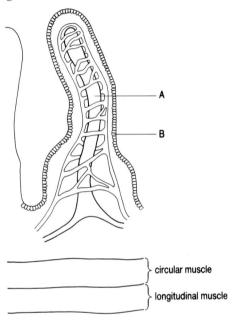

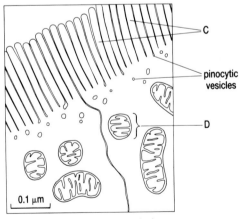

Drawing of electronmicrograph of part of structure B.

(c) Briefly describe how **three** features, shown in the diagram, enable the ileum to carry out its function of absorption. (6)
(d) (i) Of what type of muscle do the layers of circular and longitudinal muscle consist? (1)

(ii) What is the function of this muscle in the ileum? How is this function achieved? (2)

(e) The duodenum and ileum receive secretions from the pancreas and liver.
   (i) List the components of the secretion derived from the pancreas. (2)
   (ii) List the components of the secretion derived from the liver. (2)
   (iii) Describe how the flow of the pancreatic secretion is controlled. (4)

(AEB 1989)

2.4 (a) Give an illustrated account of the structure of a chloroplast. (6)
(b) Describe the physical and chemical mechanisms by which solar energy is converted into the chemical energy of ATP during the light stage of photosynthesis. (9)
(c) Name the other products of the light stage and indicate their general importance in biological systems. (5)

(WJEC 1988)

2.5 Apparatus used in an investigation to establish the sequence of biochemical changes in photosynthesis is illustrated below.

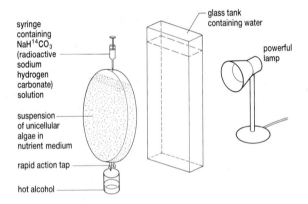

syringe containing NaH^{14}CO$_3$ (radioactive sodium hydrogen carbonate) solution

glass tank containing water

powerful lamp

suspension of unicellular algae in nutrient medium

rapid action tap

hot alcohol

(a) Suggest **one** reason for each of the following.
   (i) the flattened shape of the glass vessel containing the algae
   (ii) the water tank
   (iii) the hot alcohol
   (iv) the NaH^{14}CO$_3$ solution (4)
(b) Using this apparatus, explain how you would obtain a sample of algae that could be used to investigate the first products of photosynthesis. (2)
(c) State what techniques would be used for
   (i) the separation and identification of the photosynthetic products.
   (ii) the estimation of the relative ^{14}C content of these products. (4)

The results of this investigation are shown in the table below, which gives the ^{14}C content (in μ moles cm^{-3}) of four organic compounds

(A to D) after five different periods of photosynthesis.

Compound	Time, in seconds, allowed for photosynthesis				
	5	15	60	180	600
A	0.3	2.5	6.2	10.3	7.9
B	1.0	2.0	3.1	3.2	3.2
C	0.05	0.11	0.16	1.0	1.0
D	0.01	0.02	0.08	0.17	1.7

(d) (i) Use the data in the table to place the compounds **A** to **D** in the order in which they would be formed.
   (ii) Using your knowledge of photosynthesis, suggest a reason why the level of compound **B** remained steady in later samples. (3)

(JMB 1989)

2.6 (a) Distinguish between autotrophic and heterotrophic nutrition. (5)
(b) Describe the processes of digestion and absorption of lipids in a mammal. (10)
(c) Explain why lipids are suitable storage compounds in living organisms. (5)

(ULSEB 1988)

2.7 (a) Describe the composition of a balanced diet. (4)
(b) Outline the mechanical and the chemical digestion of protein in mammals. (10)
(c) Outline a method for determining the energy requirements of a mammal. (4)

(UCLES 1988)

2.8 Give an account of the ways in which the structural features and modes of life of mammals are related to their diets. (20)

(ULSEB 1989)

2.9 (a) In textbooks, the equation for aerobic respiration is often shown as
$C_6H_{12}O_6 + 6O_2 = 6 CO_2 + 6 H_2O.$
Give **four** different reasons why this equation does not fully represent the respiratory process. (4)
(b) (i) The respiration rate in wheat leaves has been quoted as 310 mg CO$_2$ kg^{-1} fresh mass h^{-1} at 20 °C. Write out in full what the units *kg^{-1} fresh mass h^{-1}* mean. (2)
   (ii) Do you consider that the respiration rate of leaves would be better expressed in terms of leaf area rather than on a mass basis? Give a reason for your answer. (2)
   (iii) State **two** other ways (apart from carbon dioxide output) by which the respiration rate of plant tissues may be measured. (2)
(c) The respiratory quotient (R.Q.) is defined as the ratio of the volume of carbon dioxide

produced divided by the volume of oxygen uptake over the same period of time.

The table below shows the changes in R.Q. as a soaked seed germinates.

Treatment		R.Q.
1	After 4h soaking in water	6.0
2	4h soaking + 4h in air	1.8
3	4h soaking + 24h in air	1.0

(i) What would be the value of the respiratory quotient for the alcoholic fermentation of glucose by yeast cells. (1)

(ii) Suggest a reason for the high R.Q. value obtained in treatment 1; (2)
the fall in the R.Q. value in treatment 2; (2)
the R.Q. value obtained in treatment 3. (2)

(O and C SEB 1989)

2.10 Photosynthesis is said to consist of light and light-independent (dark) reactions.

(a) Outline **one** piece of evidence to support this contention. (3)

(b) (i) What are the products of the light reaction, apart from oxygen? (2)

(ii) Indicate briefly how these products are subsequently used in the light-independent reactions. (4)

(c) An inhibitor of photosynthesis has its greatest inhibitory effect when the light intensity is low and the carbon dioxide concentration is high. State, giving a reason, whether such an inhibitor is affecting the light or the light-independent reactions. (2)

(d) Indicate briefly how **one** *named* isotope has been used to investigate the process of photosynthesis. (4)

(e) The rate of photosynthesis was determined for a maize plant during a warm sunny day. The results, in terms of carbon dioxide uptake $(cm^3/m^2/hour)$, are given in the table. Comment on these results in relation to the factors which may affect the rate of photosynthesis during the course of the day. (5)

Time	07.00–09.00h	09.00–11.00h	11.00–13.00h	13.00–15.00h	15.00–17.00h
Uptake	1.2	3.8	2.4	1.9	1.2

(O and C SEB 1986)

2.11 An experiment was performed in which the action of an enzyme on its substrate was investigated. At each of **five** temperatures 1 cm³ of enzyme was mixed with 10 cm³ of substrate. The concentration of substrate was measured at

the start of the experiment, and after one minute. A suitable control was also set up at each temperature. The results are shown in the table.

(a) Complete the table to show the rate of substrate conversion in milligrams per minute at the temperatures shown. (2)

Temperature/C°		Concentration of substrate/mg cm⁻³		Rate of substrate conversion in mg per minute
		At start	After 1 min.	
Experimental tubes	15	10	4.4	
	20	10	3.2	
	25	10	2.1	
	30	10	1.4	
	35	10	0.8	
Control tubes	15	10	10	
	20	10	10	
	25	10	10	
	30	10	10	
	35	10	10	

(b) (i) What is meant by the temperature coefficient (Q_{10}) of a reaction? (1)

(ii) Calculate the Q_{10} for this reaction between 20 °C and 30 °C. (2)

(c) In this reaction, the Q_{10} between 40 °C and 50 °C is lower than the Q_{10} between 20 °C and 30 °C. Explain why this is so. (1)

(d) Describe the control experiment that should have been set up at 15 °C. (1)

(AEB 1990)

2.12 Investigations were carried out using two strains of the same species of unicellular alga, one of which was a mutant that could not survive long periods of intense illumination. Light of known wavelength was passed through a tube containing the alga and measurements were taken both of the oxygen produced and of the light transmitted. The experimental arrangement is represented in the following diagram:

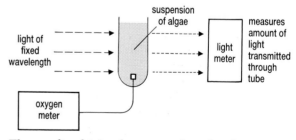

The results obtained were used to plot the absorption and actions spectra for each strain of alga. These are shown on the following page.

(a) (i) What is meant by the term *action spectrum*?

(ii) What information from the experiment would have been used to plot the action spectra? (2)

(b) (i) The amount of light **transmitted** through tubes without algae in them was 100%.

Suggest how the figures plotted for the *absorption spectra* were derived from the results obtained with the light meter.

(ii) Apart from temperature and pH (which have little effect), state **two** factors which should be standardised when using the apparatus shown above to measure the absorption spectra. (3)

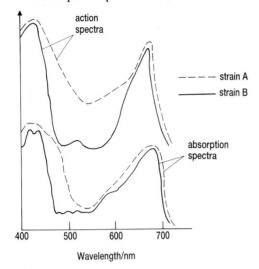

Information about the photosynthetic pigments found in these unicellular algae is given in the table below.

| Pigment | Absorption maxima / nm | Rf values | |
		solvent I	solvent II
P	620	0.20	0.89
Q	545 and 575	0.60	0.29
R	420 and 660	0.65	0.11
S	490	0.91	0.19
T	430 and 645	0.82	0.92

The pigments from each strain were extracted and separated using two-dimensional paper chromatography, to give the chromatograms shown.

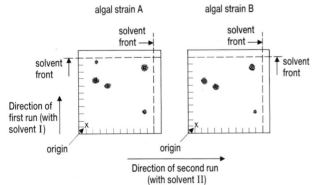

(c) Explain clearly the advantage of using two-dimensional rather than one-dimensional chromatography in separating these pigments. (1)

(d) One of the strains of alga lacks one of the pigments.

(i) Draw a box around this pigment on the appropriate chromatogram.
(ii) Give the letter of the pigment concerned.
(iii) Explain what other evidence is available to confirm the absence of this pigment from one of the strains. (5)

(e) From the information provided in the question, suggest the physiological role of the pigment in this species of alga. (2)

(JMB 1988)

2.13 The table below refers to the alimentary canals of a sheep and a dog.

If the statement is correct for that animal place a tick (✔) in the appropriate box and if the statement is incorrect place a cross (✗). (5)

Statement	Sheep	Dog
Crowns of teeth entirely covered with enamel		
Diastema present		
Stomach region of adult expanded into pouches containing symbiotic micro-organisms		
Rennin present in gastric juice in young		
Alimentary canal connected by mesentery to body wall		

(ULSEB 1989)

2.14 (a) Distinguish between the structure of a mononucleotide and a dinucleotide. (5)

(b) (i) How is ATP formed in respiration and in photosynthesis?
(ii) Describe the functions of ATP in these processes. (10)

(c) Comment on the distribution of chloroplasts in plants. (5)

(ULSEB 1990)

Answer guideline for Question 2.1

In free-response questions, individual marks are allocated for each relevant fact, concept or relationship which is clearly and unambiguously given in the answer. Some examination boards (e.g. AEB) publish guidelines to show you how marks are allocated for the separate skills tested in their free response questions. This free-response question has been structured in such a way that you can deduce that six relevant statements are needed for full marks in part (b); you cannot tell how the fourteen marks for part (a) are to be allocated though.

(a) (i) Mitochondrion is rod-shaped (you could give dimensions); mitochondrion is surrounded by two membranes; outer membrane is smooth; inner membrane folded to form cristae; space exists between membranes; inner membrane has (stalked)

particles; inner fluid (matrix) contains enzymes/proteins/DNA/ribosomes.
(*maximum: 4 marks*)

(ii) Acetyl coA enters the mitochondrion; then combines with oxaloacetate to form citrate; undergoes oxidative decarboxylation; to form succinate; oxidised to form oxaloacetate; NAD accepts electrons; FAD accepts electrons during conversion of succinate to malate; electrons passed from NAD to FAD; yields 1 ATP; electrons then passed from FAD to cytochromes; electron transport down cytochromes yields 2 ATP; eventually electrons pass to oxygen to form water; substrate phosphorylation at one stage; energy released during electron transfer drives protons (H^+) across inner membrane into intermembranal space: pH of fluid in intermembranal space is lowered; protons diffuse back via ATP complex in inner membrane; diffusion of protons generates energy for ATP formation from ADP and P_i.
(*maximum: 10 marks*)

(b) energy to drive endergonic reactions; any named example; substrate phosphorylation; e.g. glucose to form glycogen; movement of muscles/cilia/flagella; ATP needed for movement of proteins; active transport/ion pumps; named example.
(*maximum 6 marks*).

Theme 3

MOVEMENT AND REGULATION OF BODY FLUIDS

To stay alive, cells must take in nutrients and excrete waste products. These processes occur at permeable exchange surfaces.

The exchange surface of a single-celled organism is its cell surface membrane: nutrients and waste products pass throughout the cytoplasm by diffusion and osmosis. Even in small multicellular organisms, all the cells may be close enough to each other and to the exchange surfaces for diffusion and osmosis to supply them efficiently with nutrients and to remove their waste products.

Most cells in large animals and plants are too far from exchange surfaces for diffusion and osmosis to be an efficient method for transporting nutrients and waste products. Multicellular animals and plants have transport systems which link their cells and exchange surfaces.

We begin this theme by considering a specific exchange surface and then consider how nutrients and waste products are transported in animals and the part played by the transport system in regulating the animal's internal environment and preventing disease. Transport of nutrients in plants is also discussed.

(Left) The blood vessels can clearly be seen in the gills of this Axolotl.

(Below) A scanning electronmicrograph of erythrocytes densely packed into a capillary. Blood flowing through capillaries carries gasses, nutrients, waste products, heat and hormones.

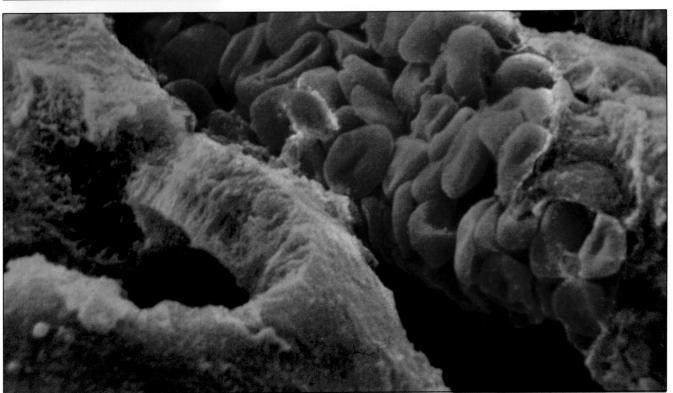

9

GAS EXCHANGE

LEARNING OBJECTIVES

When you have studied this chapter you should be able to:

1. describe the essential features of an efficient gas exchange surface;

2. explain the nature of specialised gas exchange systems in relation to the size, metabolic rate and environment of organisms;

3. compare and contrast the gas exchange systems of mammals, fish and insects;

4. describe the sites of gas exchange in plants;

5. use spirometer traces to calculate lung volume and rates of oxygen consumption for humans.

During aerobic respiration, mitochondria (Fig 9.1) use oxygen and release carbon dioxide. Although oxygen is not used in anaerobic respiration, carbon dioxide is still produced. Both these gases diffuse across a suitable body surface in respiring organisms during the process of **gas exchange**.

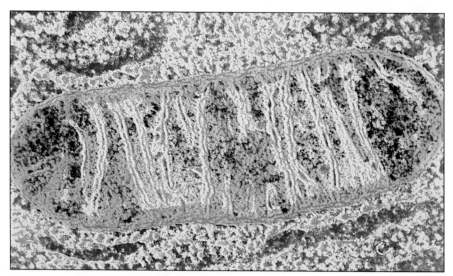

Fig 9.1 Mitochondrion – the cell organelle in which aerobic respiration takes place. Oxygen gas is used and carbon dioxide gas is a waste product.

The rate of gas exchange is affected by

- the area available for diffusion
- the distance over which diffusion occurs
- the concentration gradient across the gas exchange surface
- the speed with which molecules diffuse through membranes.

Efficient gas exchange systems must

- have a large surface area to volume ratio
- be thin
- have mechanisms for maintaining steep concentration gradients across themselves
- be permeable to gases.

1 **Suggest why gas exchange surfaces are inevitably moist. Is their moistness advantageous for gas exchange?**

9.1 BASICS OF GAS EXCHANGE SYSTEMS

Different organisms have different gas exchange surfaces, such as the cell surface membrane, skin, gills or lungs. The nature of these surfaces is affected by several factors.

The environment in which the organism lives. Being denser than air, water supports delicate thin structures. However, water contains much less oxygen per unit volume than does air.

The thickness of the organism's outer surface. Large organisms usually have thicker surfaces than small organisms. Terrestrial organisms have thicker surfaces than aquatic ones to protect them against desiccation.

The organism's rate of respiration. Multicellular organisms usually have a faster respiration rate than single-celled organisms of the same size: homoiothermic animals have a higher metabolic rate than poikilothermic animals of the same size.

The size of the organism. Generally, large organisms use more oxygen and produce more carbon dioxide than small organisms of the same species. Since cytoplasm consists mainly of water, volume is often used as an approximate measure of mass (1 dm^3 of water has a mass of 1 kg). However, gas exchange occurs across a surface. Understanding the relationship between volume and surface area is crucial to the study of gas exchange systems.

ANALYSIS

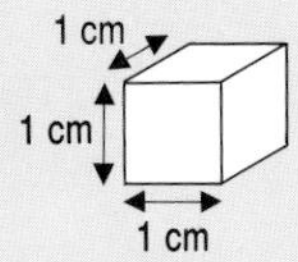

Fig 9.2 An imaginary cuboidal organism. Each of its sides has the same length.

Surface area to volume relationships

This exercise will help you to understand the relationship between the surface area and volume of organisms. It will also give you the opportunity to practise simple mathematical skills, tabulate results and plot graphs.

Since real organisms have complex shapes, imaginary cuboidal organisms have been used in this exercise. However, the principles you will discover hold true for more complex shapes.

(a) Calculate
  (i) the surface area
  (ii) volume
  (iii) ratio of surface area to volume for cubes with sides of 1, 2, 3, 4, 5 and 10 cm in length.

$$\text{surface area : volume} = \frac{\text{surface area/cm}^2}{\text{volume/cm}^3}$$

Summarise your results in a table of your own design.

(b) Using the x-axis to represent length of cube (why?), plot surface area, volume and ratio of surface area to volume against length of cube. Describe what your curves show about the relationship

---

between the size of a cube and these three variables.
(If you have time, you might like to repeat this exercise for spheres of diameter 1, 2, 3, 4, 5 and 10 cm. Do the relationships you discovered using cubes still hold true?)

(c) Imagine one cell at the centre of a large cuboidal organism and another cell at the centre of a small cuboidal organism. How will diffusion to and from these cells be affected by their position?

An increase in size seems to be associated with both a decrease in the ratio of surface area to volume and an increase in the distance over which diffusion occurs.

(d) Explain why diffusion across its surface membrane is likely to meet the gas exchange needs of an organism smaller than 0.5 mm in diameter.

(e) Would diffusion across the small organism's cell surface membrane still be sufficient if its metabolic rate were doubled? Explain your answer.

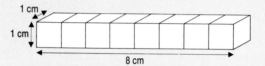

Fig 9.3   An imaginary elongated organism.

Now look at the imaginary organism in Fig 9.3.

(f) Calculate its surface area, volume and surface area : volume ratio. Account for any differences between these values and those you calculated for a cuboidal organism of similar volume in **(a)** above.

(g) Suggest why the large, multicellular animal shown in Fig 9.4 is successful even though it lacks a specialised gas exchange surface.

Fig 9.4   The nemertine worm *Lineus longissinius:* these can be several metres in length.

The oxygen needed by the cells of large organisms is supplied more efficiently when there are large surfaces for gas exchange. These can be achieved if the organism

- is flat (thalloid) or long and thin
- has a specialised gas exchange surface that is large and thin.

The gas exchange surface in your own lungs consists of a single layer of cells only 0.5–15 μm thick with a surface area of 50–100 m^2 (the size of a tennis court). By contrast, the rest of your body has a surface area of about 2 m^2. Large gas exchange surfaces solve the surface area : volume ratio problem faced by large organisms, but they do not solve a second problem of having a large body volume.

 What other gas exchange problem is faced by large organisms?

To answer this question, think about a respiring muscle cell in your foot, consuming oxygen and producing carbon dioxide. This cell is about 1.5 m from your lungs. Diffusion cannot operate efficiently over such a large distance. Transport of gases within the blood overcomes this problem of diffusion distance. For the sake of convenience, animal circulatory systems are dealt with in the next chapter, but remember the link.

So far, you have seen the importance of a relatively large surface area for diffusion and of a circulatory system to carry the gases to and from respiring cells (summarised in Fig 9.5). However, there is one further gas exchange problem for large animals. Efficient diffusion depends on the presence of a concentration gradient across the gas exchange surface (Fig 9.6).

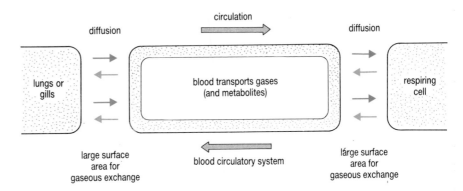

Fig 9.5 The gaseous exchange mechanism. Gases move into and out of the circulatory system by diffusion down concentration gradients. Pink represents oxygen, blue arrows indicate diffusion of carbon dioxide that also occurs.

 What would happen to the oxygen concentration gradient across the gas exchange surface separating your lungs from your blood if the air in your lungs were stagnant?

The final component of many animal gas exchange systems is the movement of air or water over the gas exchange surface by a **ventilation mechanism**. Ventilation is brought about by undulipodia (e.g. in mussels) or by muscle contraction (e.g. your own breathing movements) and requires energy. Since air has a low density it can be forced in and out through the same pathway, e.g. your trachea, with relatively little muscular effort.

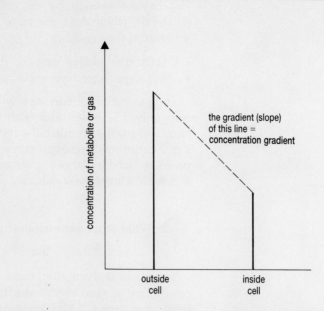

Fig 9.6 The relationship between concentration gradient and diffusion. The steeper the gradient in this diagram, the faster will be the rate of diffusion from outside the cell to inside the cell.

 4 Suggest a mechanical reason to explain why most aquatic organisms move water over their gills and out of their body in one direction only.

To summarise, gas exchange in large animals usually involves the following four stages

- renewal of fresh air or water over the gas exchange surface by ventilation movements
- diffusion of gases across the gas exchange surface into or out of the blood
- bulk transport of gases in the blood
- diffusion of gases between the blood and the mitochondria of respiring cells.

The first two of these stages are dealt with in this chapter, the last two are described in Chapter 10.

QUESTIONS

9.1 **(a)** In terms of maintaining concentration gradients across exchange surfaces, explain the advantage of keeping blood moving in a circulatory system.
(b) How will the rate of blood flow affect the steepness of concentration gradients across exchange surfaces?

9.2 List the properties you would expect to find associated with the exchange surface involved in transferring gases between the blood and respiring tissues of a mammal.

9.3 Interpret the process of gas exchange in the aquatic organism shown in Fig 9.7. In your answer, take account of such features as shape and size, environment, metabolic activity and maintenance of diffusion gradients.

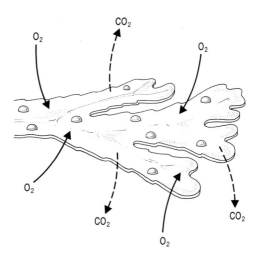

Fig 9.7 A seaweed *(Fucus)* which has a thin body. Such seaweeds are common on rocks in the inter-tidal zone on the sea shore.

9.2 GAS EXCHANGE IN MAMMALS

Cigarette smoke impairs the function of the cilia so that mucus tends to accumulate in the lungs. The rasping cough of a heavy smoker first thing in the morning is the body's way of removing the mucus which has built up in the lungs overnight.

The gas exchange surface of mammals is the wall of the air sacs (**alveoli**) in the lungs. The position of the lungs in the human thorax is shown in Fig 9.8. The internal structure of one lung is also shown. Air can enter and leave the pharynx via the nostrils, which are adapted to filter and warm the air, or via the mouth and buccal cavity. The **trachea** runs from the pharynx and branches into two **bronchi** which enter the lungs. At the top of the trachea is the **larynx**, or voice box. The trachea and bronchi have involuntary muscle which can be used to change their internal diameter. Their collapse is prevented by C-shaped rings of cartilage in their walls. The lining of the trachea secretes mucus to which particles in the air stick. The mucus is moved to the pharynx by cilia which line the trachea.

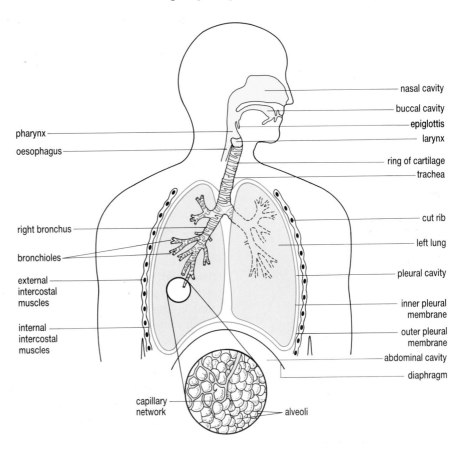

Fig 9.8 The position of human lungs. Inspiration takes place when lung volume is increased by contraction of external intercostal muscles and diaphragm muscles, and relaxation of internal intercostal muscles.

Within the lungs, each bronchus (singular of bronchi) branches into smaller **bronchioles** which divide throughout the whole lung tissue before ending at a group of alveoli. The fluid-filled **pleural cavity** separates the lung surface from the inner surface of the thoracic cavity. Normally, surface tension between the moist **pleural membranes** on each side of this cavity keeps the membranes together, and so prevents the lung from collapsing.

 5 As a result of an injury, air may get into the pleural cavity. What effect would this have on the lung?

Breathing

By changing the volume of the lungs, the air pressure inside the lungs is varied. Air moves from a region of high pressure to a region of low pressure. When the pressure in the lungs is lower than atmospheric pressure, air moves into the respiratory system (**inspiration**); when the pressure in the lungs is higher than atmospheric pressure, air moves out of the respiratory system (**expiration**). Because the lungs do not normally collapse, there is always some air in the lungs (the **residual volume**, which is about 1.5 dm^3 in humans).

Enlargement of the lungs during inspiration is brought about when the **diaphragm muscles** and the **external intercostal muscles** contract and the **internal intercostal muscles** relax. The resulting raising of the ribs and flattening of the diaphragm makes the thoracic cavity bigger and, because the pleural membranes are held together by surface tension, the lungs expand. Air in an expanded lung exerts a lesser pressure, so air moves from the higher pressure in the atmosphere to the lungs.

Expiration is brought about by relaxing the diaphragm and external intercostal muscles. The ribs fall back to their resting position and the elastic recoil of the intestines pushes the diaphragm upwards. Aided by its own elasticity, the volume of the lungs is decreased, squeezing the air inside them. During hard exercise, this squeezing is aided by the contraction of the internal intercostal muscles, allowing much greater volumes of air to be expelled than normal.

Gas exchange

A regular volume of air is breathed into and out of the lungs. This is called the **tidal volume** and is about 500 cm^3 in humans when at rest. Of this volume only about 350 cm^3 reaches the lungs, the rest, the **dead space air**, stays in the trachea, bronchi and bronchioles where gas exchange does not occur. Inspired air has an oxygen partial pressure of 21 kPa. After it is mixed with the residual air in the alveoli the partial pressure of oxygen is reduced to about 13 kPa. However, this concentration is still higher than that in the blood vessels around the alveoli, so that diffusion of oxygen occurs into the blood. At the same time, carbon dioxide diffuses from the blood into the alveolar air. The alveoli and capillaries are well adapted for this exchange (Fig 9.9); each is bounded by a single layer (**simple epithelium**) of flat (**squamous**) cells which are in intimate contact with each other, so providing the shortest possible distance over which diffusion has to occur. The mechanism for carrying oxygen from the lungs to the respiring tissues in the blood system allows mammals to grow much bigger than the distance for efficient diffusion alone would allow.

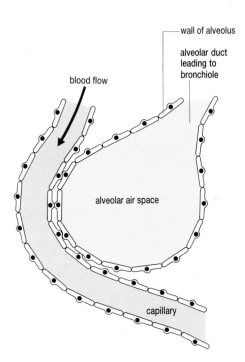

Fig 9.9 An alveolus. The gas exchange surfaces in the lungs are the walls of the blood capillary and of the alveolus. Both are thin, moist and extensive.

Measurement of human lung volumes

This exercise gives you practice in analysing data.

The volume of air moved into and out of a human lung can be measured using a spirometer. A simple laboratory spirometer is essentially a clear box, or bell, sealed at the bottom by water. The bell moves up as you breathe out and down as you breathe in. These movements are recorded on a chart recorder and the trace can then be used to calculate the volumes of air entering and leaving the lungs. Note the container for soda lime, a substance which absorbs carbon dioxide. A one-way valve ensures that the air you breathe out passes through this container on its way to the bell, whilst the air you breath in comes directly from the bell (Fig 9.10).

Now look at the results of two experiments.

Experiment 1

In the first experiment the soda lime container is left empty. The bell is filled with oxygen and a subject is asked to breathe normally through the mouthpiece. The amount of air moved in and out of the lungs during this normal breathing is called the **tidal volume**.

(a) Use the trace shown in Fig 9.11(a) to calculate the tidal volume (TV) of this subject in dm^3.

The subject is now asked to breathe in as hard as possible, then to return to normal breathing. The extra volume of air, over and above that taken in during normal breathing, is called the **inspiratory reserve volume** (IRV). The subject is now asked to breathe out as strongly as possible. The extra air pushed out by the forced expiration is the **expiratory reserve volume** (ERV).

Vital capacity of lungs = TV + IRV + ERV.

Fig 9.10 A spirometer. The apparatus is used to measure the volumes of gases moved into and out of the human lungs. You breathe in and out of the blue mouthpiece. The container for soda lime is on the right-hand side.

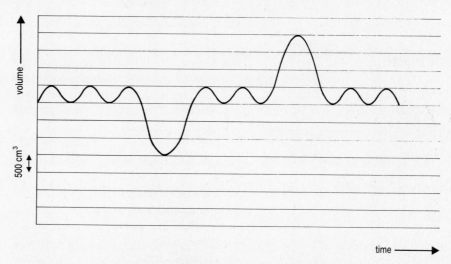

Fig 9.11(a) A trace from a spirometer without soda lime.

(b) Calculate, in dm^3, IRV, ERV and vital capacity for this subject.

(c) Make a copy of the trace shown in Fig 9.11(a) and annotate it to show the various lung volumes.

(d) Given that the residual volume in this subject is 1.5 dm^3, calculate the total lung volume of the subject.

(e) During pulmonary infections, such as pneumonia, fluid can collect in the air spaces of the lung (pulmonary oedema). How will this affect vital capacity? How could you use changes in vital capacity to diagnose pulmonary oedema?

(f) Emphysema is a respiratory disease in which the alveolar walls lose their elasticity so that during expiration they fail to recoil. How could you use spirometry to diagnose the early stages of this illness?

Experiment 2

In a second experiment the soda lime container is filled and the subject again breathes normally into the spirometer to produce the trace shown in Fig 9.11(b).

(g) Why does the trace fall with time?

(h) Using Fig 9.11(b) calculate
  (i) the breathing rate in breaths min^{-1}
  (ii) the minute volume of respiration, i.e. the total amount of air pumped into and out of the lungs in one minute.

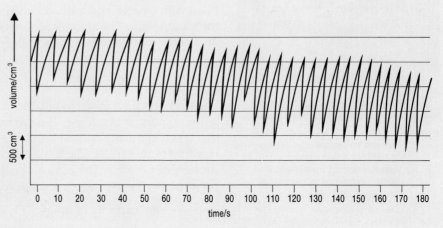

Fig 9.11(b) A trace from a spirometer with soda lime.

9.4 Crocodiles do not have a diaphragm. Rather, they are able to move their liver, stomach and intestines backwards and forwards into and out of their thoracic cavity. Explain how this mechanism enables a crocodile to ventilate its lungs.

9.3 GAS EXCHANGE IN BONY FISH

The skin of bony fish is covered by impermeable scales and gas exchange occurs through gills. Four pairs of gills (Fig 9.12) lie in the **pharynx**, the area between the buccal cavity and the oesophagus. Each gill is supported by a bony **gill arch** and has two stacks of thin **lamellae** which lie on top of each other like the pages in a book. Unless kept apart by water, these lamellae stick together (like wet pages of a book) so that their surface area is too small for efficient gas exchange. The gills lie in a cavity which is surrounded by a movable **operculum**.

The upper and lower surface of each lamella has projections called **gill plates** which further increase the surface area of the gill. Blood flows from an afferent vessel in each gill arch down the inner side of each lamella, across the gill plate and back to an efferent vessel (see Fig 9.13). Water passes over the gill plates in the opposite direction to the flow of blood so that, across the whole gill plate, blood meets water with a higher concentration of oxygen than its own. This **counter current mechanism**

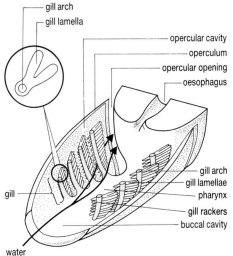

gill arch
gill lamella
opercular cavity
operculum
opercular opening
oesophagus

gill arch
gill lamellae
pharynx
gill rackers
buccal cavity

gill

water

Fig 9.12 The gas exchange system of a bony fish. Water moves in a one-way system entering the mouth then flowing between the four pairs of gills to exit through the operculum opening.

increases the rate of diffusion from water to blood and improves the efficiency with which the fish's gill extracts oxygen from water.

Bony fish ventilate their gills. Muscular contractions cause water to move into the buccal cavity through the mouth, through the pharynx, over the gills and out through the valves of the opercula. Some fish, such as mackerel, achieve this by swimming with their mouths and opercular valves open. Their forward movement through stationary water results in water moving across their gills (Fig 9.14). Most fish use muscles to change the volume of their buccal cavity, pharynx, gill cavity and opercular cavity. As the volume of a chamber becomes less its pressure increases, squeezing water to where the pressure is less.

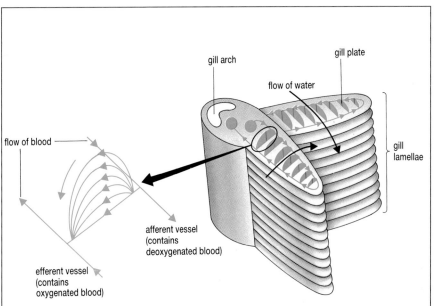

gill arch
gill plate
flow of water
flow of blood
gill lamellae

afferent vessel (contains deoxygenated blood)

efferent vessel (contains oxygenated blood)

Fig 9.13 The structure of a gill. Notice how blood flows from the back of the gill plate to the front. This ensures that the blood is always in contact with water which contains a higher concentration of oxygen than the blood itself does – an example of a counter current mechanism.

Fig 9.14 A shark is not a bony fish, but belongs to the Chondrichthyes (cartilagenous fish). Like a mackerel, it normally breathes in water by keeping its mouth open as it swims.

9.5 Why would fish gills not be an effective gaseous exchange system on land?

9.6 Look at Fig 9.15; it shows the pressures measured inside the buccal and opercular cavities of a bony fish during normal breathing. Identify, with reasons, the part of the graph in which

(a) water enters the mouth
(b) water is pushed from the mouth to the pharynx
(c) the mouth closes
(d) water is pushed over the gills from the pharynx
(e) the opercular valve is open.

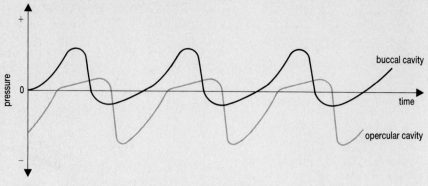

Fig 9.15 Pressure in buccal cavity (black line) and opercular cavity (grey line) during breathing of a bony fish.

9.4 GAS EXCHANGE IN INSECTS

Insects (see Chapter 27) have a segmented body with a rigid exoskeleton. On the outside of the exoskeleton is a layer of wax which is impermeable to water and gases. The second and third thoracic segments, and the first eight or so abdominal segments, of the insect's body have a small hole on each side, the **spiracle**, through which gases can diffuse. The spiracles lead into a system of large tubes, the **tracheae**, which branch in a tree-like network into smaller **tracheoles**. The tracheoles grow between, and even into, the insect's body cells. Whilst the outer tracheae are impermeable, the tracheoles are permeable. Carbon dioxide can diffuse from the cells into the tracheoles and oxygen can diffuse from the tracheoles into the cells. This is shown in Fig 9.16.

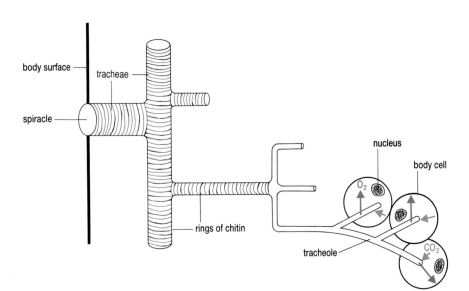

Fig 9.16 The insect tracheal system. It is a very short distance from the spiracle to the body cells. Note: this diagram is not drawn to scale.

6 Suggest a function for the chitin rings around the tracheae.

Movement of air in the insect tracheal system depends mainly on the diffusion of oxygen from, and carbon dioxide to, the outside air. This occurs 10 000 times more rapidly in air than in blood. Diffusion is only an efficient method of gas exchange if the distance involved is fairly short. The distance between the spiracle and the finest tracheoles must be short or the insect will die. This is probably one reason why insects are relatively small animals. Insects can ventilate their tracheoles, but only to a limited extent. By compressing their bodies, they squeeze air from their tracheal tubes; fresh air moving into the tubes when the body returns to its normal size. During times of activity, lactic acid accumulates in muscle cells so that water moves from the lining of the tracheoles into the muscle cells by osmosis. This movement of water increases the free volume within the tracheoles which, provided the spiracles are open, results in fresh air moving deeper into the tracheoles.

SPOTLIGHT

ARTHROPOD AQUALUNGS

Many aquatic insects carry their own oxygen supply with them as they dive below the surface. For example, *Notonecta* carries air bubbles clinging to hair-like structures, setae, on its ventral surface. During a dive, oxygen within the bubbles is transferred via the tracheal system to the respiring cells. As the insect uses up the oxygen in the bubble, there will come a point when the oxygen concentration in the bubble is less than that in the surrounding water. Oxygen will now diffuse from the water into the bubble. It has been calculated that up to seven times the initial amount of oxygen contained in the bubble diffuses into the bubble from the water, so this means the insect can stay underwater for longer.

The water spider, *Diomedes*, actually constructs a diving bell which is replenished with bubbles of air brought from the surface. The spider can then hide in the bell, ready to leap out to capture small fish and invertebrates.

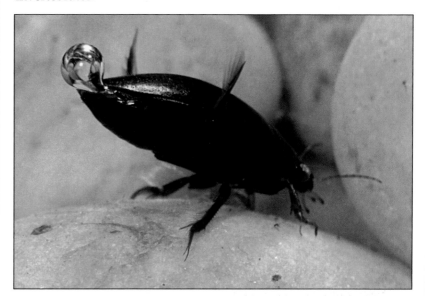

Fig 9.17 Diving beetle showing its air bubble.

9.7 (a) Insects come in various shapes and sizes, but none is more than about 1 cm wide. Explain how their gas exchange system tends to limit their size.

(b) The spiracles of insects have a hinge-like valve and are surrounded by bristle-like structures called setae. Suggest what the function of these structures might be.

9.8 (a) What are the essential features of an efficient gas exchange surface?

(b) Describe how the gas exchange systems of an insect, a bony fish and a mammal are adapted as efficient gas exchange surfaces.

9.5 GAS EXCHANGE IN TERRESTRIAL PLANTS

Terrestrial plants have a body which is divided into three organ systems: the roots, stem and leaves. The leaf is an active organ which respires at all times and photosynthesises when adequately illuminated. In respiration, the leaf uses oxygen and produces carbon dioxide; in photosynthesis it uses carbon dioxide and produces oxygen. The ways in which leaves are adapted to allow efficient gas exchange is described alongside their role in photosynthesis in Section 7.6.

A plant's roots use energy in processes such as growth and the active uptake of water and inorganic ions from the soil. Only a part near the tip of each root, referred to as the **piliferous region**, is permeable to water, ions and gases. Epidermal cells in this region have finger-like projections, the **root hairs**, which increase their surface area to volume ratio. In soil which is not waterlogged, roots hairs are surrounded by air spaces which lie between aggregates of soil particles, called **crumbs**. Diffusion of gases occurs through the cell surface membrane and cell wall of these root hairs (Fig 9.18).

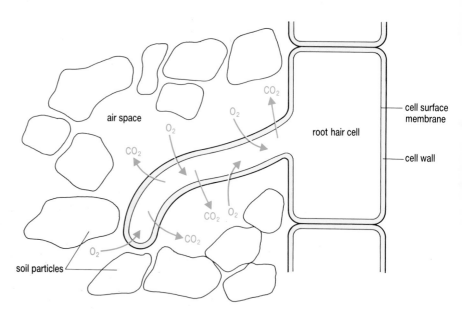

Fig 9.18 Gaseous exchange in a root hair. The extensive, thin and moist surface of the cell walls and cell surface membranes in the piliferous layer of the root permit gas exchange by diffusion in roots growing in the soil.

Living cells are most abundant around the periphery of a plant stem, especially in woody stems. The stem's outer covering is usually water-proofed to reduce the loss of water to the air and so is not permeable to gases. Small pores, called **lenticels**, are found on stems. Diffusion of gases occurs through a lenticel to and from the loosely packed cells behind the pore (Fig 9.19).

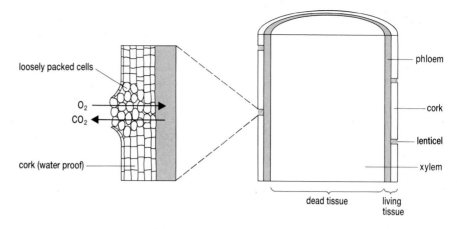

Fig 9.19 VS of a stem showing pathway of gases. The loosely packed cells of the lenticel permit gas exchange. Such a process cannot take place through the waterproof layer of cork found on the outside of woody stems of plants.

<table>
<tr><td>

QUESTION

</td><td>

9.9 Plants do not have a circulatory system for moving gases yet they can be extremely large. Explain why. (Hint: look carefully at Fig 9.19.)

</td></tr>
</table>

SUMMARY

Gas exchange occurs as a result of respiration, when carbon dioxide is excreted and oxygen taken up, and photosynthesis, when oxygen is excreted and carbon dioxide is taken up.

Efficient gas exchange surfaces share common features: they are permeable to gases; they have a large surface to volume ratio; they are thin, presenting only a short distance over which diffusion must occur. Since they are permeable, they allow loss of water in terrestrial organisms.

Single-celled organisms are aquatic and their cell surface membrane has a sufficiently large surface area to volume ratio to act as an efficient gas exchange surface.

In larger organisms, permeable, thin, flat structures have all the properties of efficient gas exchange surfaces but need water to prevent their dehydration and give them mechanical support. The thalloid bodies of aquatic algae and flatworms and the gills of fish, molluscs and many crustacea are examples of such gas exchange structures. Since the solubility of oxygen in water is low, organisms that obtain their oxygen from water can maintain only a low metabolic rate.

Gills are not suitable for terrestrial animals as they would collapse in air. Gas exchange surfaces of these animals are formed from permeable in-pushings of the external surface: the alveoli of mammals and the fine tracheae of insects are examples of internal gas exchange surfaces.

In small and thin organisms, the distance from gas exchange surface to the inside of the organism is short enough for diffusion of gases to be efficient. Diffusion gradients are maintained because gases are continually used up or produced. In larger organisms, simple diffusion is not an efficient way of transporting gases between cells in the body and the gas exchange surface. In many animals a blood circulatory system carries gases to and from the gas exchange surface. The gas-carrying capacity of the blood is increased by respiratory pigments, such as haemoglobin. Insects have an inefficient blood circulatory system: their tracheae branch directly into their cells, acting as an air transport system.

Animals with an internal gas exchange surface ventilate it by passing fresh air or water through their respiratory system. Air usually flows in and out through the same pathway; being light this requires very little muscular effort. Denser water is passed in a one-directional pathway over gills.

In terrestrial plants diffusion of gases through pores is sufficient to service the few living cells in its stem cortex and its thalloid leaves. In the roots, gas exchange is restricted to a small permeable area.

Answers to Quick Questions: Chapter 9

1   Surfaces which are freely permeable to CO_2 and O_2 will also be permeable to H_2O; consequently water will always be diffusing across the exchange surface from body to air, so keeping it moist. There is some doubt as to whether gases diffuse across gas exchange surfaces faster in solution than as gases.
2   Gases have to be efficiently transported to and from the gas exchange surface and the cells consuming and producing the gases, i.e. a circulatory system is needed.
3   It would become less steep.
4   Water is fairly dense. To stop its flow in one direction and then push it back along the same pathway would require more effort (and hence more energy) than pushing it along a one-way path. Air is less dense than water, requiring much less effort and energy to move it about.
5   Cause them to collapse.
6   Prevent the tracheae from collapsing during gas exchange.

Chapter 10

ANIMAL BODY FLUIDS AND THEIR TRANSPORT

LEARNING OBJECTIVES

When you have studied this chapter you should be able to:

1. explain the advantage which an animal gains by having a blood circulatory system;

2. distinguish between an open and a closed circulatory system and between a single and a double circulatory system;

3. describe the composition of mammalian blood and explain the role of each component;

4. recognise the internal structure of a mammalian heart, artery, vein and capillary and relate the structure of each to its function;

5. outline the arrangement of the major blood vessels in a mammal and explain how blood flow through them is brought about;

6. explain the formation of tissue fluid and its reabsorption as lymph.

All cells require a supply of raw materials for metabolism, e.g. aerobic respiration in a mitochondrion uses both oxygen and glucose. Metabolism produces toxic waste products which must be removed from the cells, e.g. aerobic respiration produces carbon dioxide and water. Movement of metabolites and waste products between a cell's organelles and its cell surface membrane occurs by diffusion. However, diffusion is only efficient over very small distances. Single-celled organisms, such as *Amoeba*, are small enough for diffusion to meet their internal transport needs. However, the distance over which metabolites must be moved in large multicellular animals is too great for diffusion to be adequate. This chapter is concerned with the way the circulatory system necessary for larger organisms, and of mammals in particular, works.

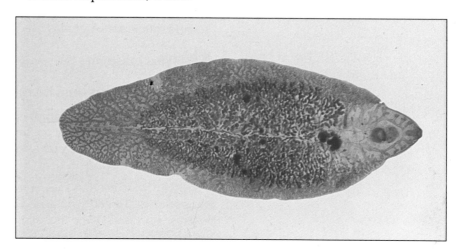

Fig 10.1 A flatworm, *Fasciola hepatica*. The branched, pink structures are the testes (in the centre) and the outer, paler pink branches show the yolk glands. The gut is also branched, but is not visible here because it has not been stained. The mouth is the dark pink blob on the right.

Explain how, in the absence of a blood circulatory system, efficient transport of respiratory gases and digested food occurs in the multicellular organism in Fig 10.1.

10.1 TYPES OF CIRCULATORY SYSTEMS

The main function of a circulatory system is to link exchange surfaces. Look at Fig 10.2; this illustrates one exchange mechanism.

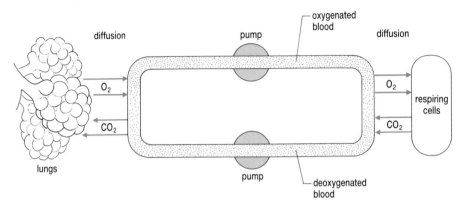

Fig 10.2 Circulatory systems link exchange surfaces. In this example the respiratory system is linked to respiring body cells by the circulatory system.

 List some other exchange surfaces in your body.

Chemicals move across exchange surfaces by diffusion. However, the efficiency of diffusion depends upon the steepness of the concentration gradients across the exchange surface. The steeper the concentration gradients the faster the rate of diffusion and the more efficient the exchange of materials.

 What would happen to the concentration gradient across the exchange surfaces shown in Fig 10.2 if the blood in the circulatory system stopped flowing?

A steep concentration gradient is maintained if the blood flows quickly. To make a fluid flow we have to subject it to pressure. For example, to make washing-up liquid come out of the 'squeezy' bottle we compress the bottle. This raises the pressure of the fluid inside the bottle so it is now greater than the air pressure outside the bottle and the washing-up liquid squirts out. The harder we press, the greater the pressure we create inside the bottle and the faster the liquid comes out. In fact, what is important in determining the rate of flow of a fluid in a pipe, e.g. blood flowing in a blood vessel, is not the absolute pressure at one end of the pipe but the difference in pressure between the two ends of the pipe. To create such a pressure difference a pump is needed.

 What constitutes the pump for your circulatory system?

Blood circulatory systems have three components

- a transport fluid, for example blood
- a series of tubes to carry the fluid
- a pump to create a pressure differential in the transport fluid, for example a heart.

The way in which these key components are arranged in different animal groups varies. To help you assess the efficiency of the different circulatory systems you should keep the following three key concepts in mind.

- Efficient diffusion across exchange surfaces depends upon the maintenance of steep concentration gradients across those surfaces.
- Fast blood flow helps to maintain steep concentration gradients across exchange surfaces.
- The greater the difference in the pressure between two points, the faster a fluid will flow between those two points.

Open and closed systems

Circulatory systems can be classified as

- open systems, e.g. insects
- closed systems, e.g. fish and mammals.

Animals with a closed circulatory system have their blood enclosed within blood vessels. Blood is moved from a muscular pump to the tissues of the body and back again entirely through blood vessels. This is summarised in Fig 10.3(a). High pressure in the capillaries forces water and other small molecules through the thin capillary walls to form a **tissue fluid** which bathes the cells.

 Suggest how glucose dissolved in the tissue fluid would reach the mitochondria of the cells it is bathing.

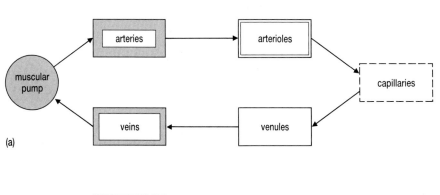

(a)

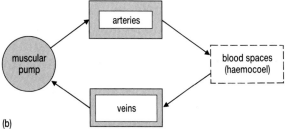

(b)

Fig 10.3 **(a)** Closed circulatory system – present in fish and mammals. Blood is contained in vessels and moved by a muscular pump, the heart.
(b) Open circulatory system of insects in which blood is pumped out to all parts of the body in arteries and then collects in open spaces to be returned to the pump (the heart) in a few veins.

In a few animal groups, notably arthropods (crabs, insects and spiders), the artery which leaves the heart branches into short arteries which themselves open into large, blood-filled spaces, collectively called the **haemocoel**. Blood from these spaces gradually returns to the heart through a few open-ended veins. This pattern of blood flow is called an **open circulatory system** and is summarised in Fig 10.3(b).

 Which of these two systems will generate the highest blood pressure and so maintain the fastest rates of blood flow?

(a)

(b)

Fig 10.4 Two highly successful arthropods. **(a)** Termites which often live in huge colonies and **(b)** a coral crab from the Australian coral reef.

SUCCESS OF INSECTS

Blood travels faster in a closed system, therefore the concentration gradients across the exchange surfaces in an animal with a closed circulatory system should be steeper than in one with an open system. This might suggest that closed circulatory systems are more efficient and so better than open circulatory systems. However, this is the wrong way to look at these two types of circulatory system: they are just different solutions to the same problem.

You cannot think of one being 'better' than the other. Indeed, arthropods are the most successful group of animals on Earth; there are more types of insect than any other type of animal. By this criterion open systems appear better than closed ones. This apparent paradox, a highly successful animal group with an apparently inefficient circulatory system, is resolved when you consider that insects do not rely on blood to transport respiratory gases. Instead they have a unique gas exchange system consisting of tiny tubes which run throughout their bodies (see Section 9.4). It is this tracheal system, not the blood, which transports oxygen to the insect's cells and carbon dioxide away, largely by the process of diffusion.

Single and double circulatory systems

All chordates have closed circulatory systems. There are two types

- single circulatory systems, e.g. fish
- double circulatory systems, e.g. mammals.

The circulatory system of a fish is summarised in Fig 10.5(a). Blood flows from the heart to a capillary network in the gills, then to a second capillary network in the tissues of the rest of the body and finally back to the heart. Such a system is described as a **single circulatory system** because the blood is moved by a single pump.

In chordates other than fish, the blood flow between the heart and the lungs is, to different degrees, separated from the blood flow between the heart and the body tissues. Such a system is called a **double circulatory system**.

The features of the circulatory system of a healthy adult mammal, in which separation of blood to the lungs (**pulmonary circulation**) and blood to the body tissues (**systemic circulation**) is complete, is shown in Fig 10.5(b). (In fetal mammals, an additional blood vessel, the **ductus arteriosus**, allows blood to flow from the pulmonary circulation to the systemic circulation.)

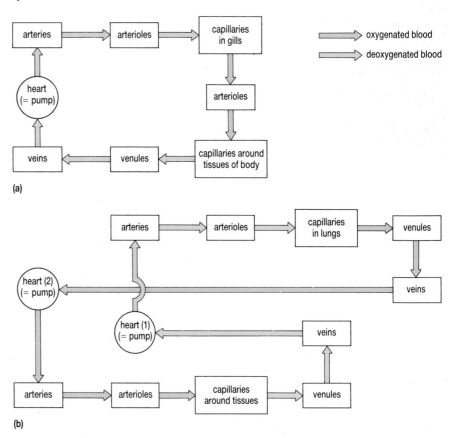

Fig 10.5 **(a)** Single circulatory system of a fish. Blood is moved by a single pumping action of the heart in a simple route from the heart to the gills to the tissues of the body and back to the heart.
(b) Double circulatory system of a mammal. The heart has a double pumping action and two (almost) separate circulation pathways for the blood. In one route, blood is pumped and circulated between the heart and the lungs. In the second, blood is pumped from the heart to all the other tissues of the body in a separate circulation route, and returns to the heart.

ANALYSIS

Single and double circulations

In this exercise you will consider the mechanical differences between single and double circulatory systems. Producing the logical train of thought required to do this is an essential skill that you need to practise.

First imagine blood flowing from an artery into the narrow blood vessels, capillaries, in a fish gill.

(a) What will happen to the blood pressure as the blood is squeezed through the narrow capillaries?

(b) What effects will this have on
- the pressure of the blood going to the fish's body tissues?
- the speed at which the blood flows through the capillaries in the fish's body tissues?
- the steepness of the concentration gradients in the gills and the body tissues?
- the efficiency with which metabolites and waste products enter and leave the blood in the gills and the body tissues?

(c) Fish manage well despite an apparently inefficient system. Can you suggest why?

Now consider the mammalian double circulation.
(d) Explain how this system will work in terms of blood pressure, rates of blood flow and the steepness of concentration gradients across exchange surfaces.

QUESTIONS

10.1 (a) Why are possession of large body size and high metabolic rate related to the possession of an efficient circulatory system?

(b) Why, despite the wilder fantasies of science fiction film makers, is it unlikely that an ant could grow as large as an elephant?

10.2 Look at Fig 10.6 which represents the circulatory system of an octopus, a fast swimming, predatory mollusc. How does the octopus overcome the problem of low pressure blood from the gills going to its body?

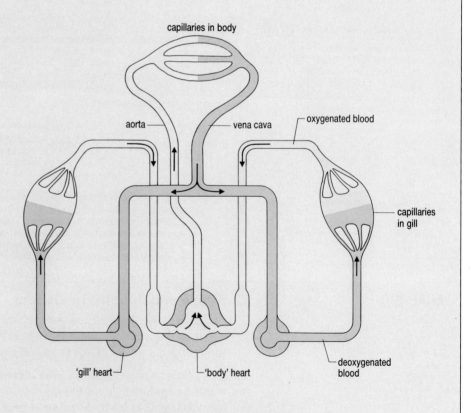

Fig 10.6 Octopus circulatory system.

10.3 Prepare an essay plan to answer the following:
 (a) Using examples, describe what is meant by:
  (i)   open circulation   (iii) single circulation
  (ii) closed circulation   (iv) double circulation.
 (b) What are the advantages of a double circulation compared
   with a single circulation?

Use your essay plan to complete your essay in about 45–50
minutes. Include simple, fully labelled diagrams wherever
appropriate.

10.2 MAMMALIAN BLOOD

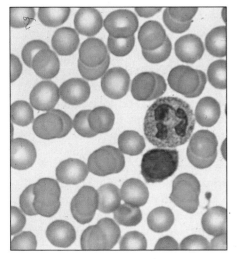

Fig 10.7 This smear of human blood shows the many erythrocytes. They look pale in the centre because they are thinner there. The larger, pinker cell with red blobs in it is a leucocyte.

Blood appears to be a sticky red fluid. In fact, only about 55% of its volume is made of a fluid, called **plasma**. The remaining 45% of its volume consists of **erythrocytes** (red cells), a variety of **leucocytes** (white cells) and cell fragments, called **thrombocytes** (Fig 10.7).

Plasma

Plasma is mainly water. The materials which are dissolved and suspended in this water comprise about 10% of the plasma and give it a pale yellow appearance and sticky texture. Apart from the absorbed products of digestion and waste products of cell metabolism, the plasma contains a number of proteins and inorganic ions. The major components of plasma and their functions are shown in Table 10.1.

Table 10.1   Components of plasma and their functions

Component	Function
1. Water	solvent
	metabolite
	maintains blood pressure
2. Proteins (produced in liver)	
albumin	pH buffer; maintains colloidal osmotic potential and viscosity of blood; transport of steroid hormones
α-globulin	transport of hormones (e.g. insulin) and lipid-soluble vitamins
β-globulin	as α-globulin
γ-globulin	antibodies binding to specific antigens on cell surface membranes
prothrombin	converted to enzyme thrombin, active in blood clotting
fibrinogen	converted to insoluble fibrin forming network of blood clot
hormones	coordination
complement proteins	coat outside of bacterial and yeast cells, aiding their phagocytosis
3. Inorganic ions	
$Na^+, K^+, Cl^-, PO_4^{3-}$	maintain osmotic potential; metabolites
Ca^{2+}	blood clotting; muscle contraction
HCO_3^-	carriage of carbon dioxide; pH buffer

Erythrocytes (red blood cells)

The cytoplasm of these cells contains the oxygen transporting pigment **haemoglobin**, which gives the blood its red colour. The role of erythrocytes in the transport of respiratory gases is described in Section 10.3.

Mature erythrocytes do not contain a nucleus. As a result, the cell volume available for haemoglobin is increased. Their biconcave, rather than spherical, shape increases the surface area over which gas exchange can occur.

One cubic decimetre of blood from a healthy adult human contains about

5×10^{12} erythrocytes. After about 120 days in the circulation, they are destroyed in the spleen and liver. Between 2 and 10 million erythrocytes are destroyed each second in an adult human. The four polypeptide chains of the haemoglobin molecule are broken down into the amino acids from which they are made; these can be reused to make new proteins. The haem group is broken down into the iron component and the **bile pigments**, biliverdin and bilirubin. The iron may be stored in the liver and used for making new erythrocytes; the bile pigments are excreted through the bile duct into the gut.

The rate of synthesis of new erythrocytes normally balances their rate of destruction. The synthesis rate is stimulated by a hormone called **erythropoietin**. This hormone is secreted by the kidneys in response to a drop in the oxygen concentration in the blood such as might occur during rapid growth or excessive loss of blood.

SPOTLIGHT

ANAEMIA

Erythrocytes are produced by **haemopoietic tissue** which is found in the marrow of all the bones of an infant but only in the flat bones of the axial skeleton of an adult. Because haemoglobin colours this marrow, it is called **red bone marrow**. Vitamin B_{12} and folic acid are needed for the production of erythrocytes. However, the absorption of vitamin B_{12} from the small intestine requires the presence of another substance, **intrinsic factor**, produced by cells lining the stomach. People who do not produce intrinsic factor have blood which contains fewer and larger erythrocytes than normal blood, symptoms of **pernicious anaemia**. This should not be confused with **simple anaemia**, caused by a lack of iron, in which erythrocytes of normal size and number are produced but each has less haemoglobin than normal.

ANALYSIS

Why erythrocytes?

As a biologist you always need to ask the question, 'What is the advantage of ...?' This exercise is designed to help you to ask such questions.

Many animals have respiratory pigment free in the blood plasma. In mammals the respiratory pigment is inside blood cells.

At first sight it would seem simpler to have haemoglobin free in the plasma rather than packaged in cells. This would mean, for example, that during the loading and unloading of oxygen there would be one less cell surface membrane to diffuse through. However, the question you might have asked is, 'What is the advantage of packaging the haemoglobin into red cells?'

Blood viscosity is lowered with the haemoglobin packed in cells. (Viscosity is a measure of how easily fluids flow.)
(a) Why is it advantageous for the blood to have a low viscosity?

A large number of enzymes and chemicals synthesised within erythrocytes are in close proximity to haemoglobin if they are within the cell.

(b) Why is it chemically more efficient to put enzymes, chemicals and the haemoglobin inside a membrane-bound bag?

The water potential of the blood is made less negative by packing haemoglobin inside blood cells.

(c) What effect would a more negative blood water potential have on body cells?

Leucocytes (white blood cells)

Leucocytes are less common than erythrocytes; in one cubic decimetre of normal blood there may be between 4×10^9 and 11×10^9 leucocytes. Abnormally high numbers of leucocytes is a symptom of **leukaemia**.

Leucocytes are larger than erythrocytes, lack pigment and possess a

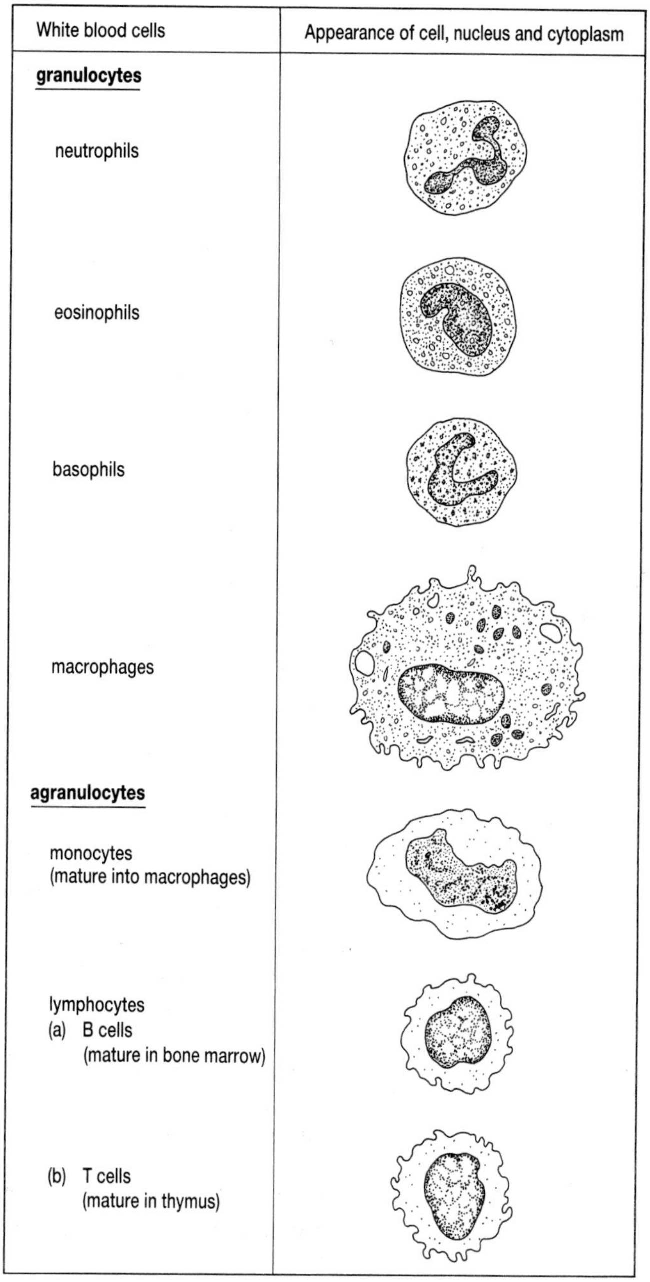

White blood cells	Appearance of cell, nucleus and cytoplasm
granulocytes	
neutrophils	
eosinophils	
basophils	
macrophages	
agranulocytes	
monocytes (mature into macrophages)	
lymphocytes (a) B cells (mature in bone marrow)	
(b) T cells (mature in thymus)	

Fig 10.8 White blood cell types. These all have a nucleus. The granulocytes have cytoplasm with a granular appearance and the nucleus of each type has a characteristic shape. The macrophages are the largest type.

nucleus. Despite the diversity shown in Fig 10.8, all leucocytes are formed from **pluripotent stem cells** in the bone marrow. During their development, one type of stem cell develops along a pathway which gives rise to the **granulocytes** and **monocytes**: a second type of stem cell develops along a different pathway, giving rise to **lymphocytes**. All these cells are concerned with immunity, which is described in more detail in Chapter 13. The most common leucocyte is the **neutrophil**, produced at a rate of about 80 million per minute by an adult human. These cells, together with the monocytes, spend about two days in the circulatory system before squeezing between the cells lining the capillaries **(diapedesis)** and migrating to the tissues. Here they engulf micro-organisms and other foreign bodies by phagocytosis. Unlike monocytes, neutrophils are very short-lived, surviving no more than a few days. The other leucocytes secrete chemicals into the blood.

Leukaemia or cancer of the blood is an uncontrolled, greatly accelerated production of white blood cells. Many of these cells fail to mature. As with most cancers, the symptoms result from the interference of the cancer cells with normal body processes. Thus, the anaemia and bleeding problems commonly associated with leukaemia result from the crowding out of normal bone marrow cells, which results in a reduction in production of red blood cells and thrombocytes (see below).

Thrombocytes (platelets)

These fragments of cytoplasm are formed by the disintegration of large cells in the bone marrow, called **megakaryocytes**. There are about 150×10^9 to 400×10^9 platelets dm^{-3} of healthy blood.

Thrombocytes stick together **(agglutinate)** when blood is exposed to air or when blood flows through a damaged blood vessel. Agglutinated thrombocytes disintegrate to release **thromboplastin**. In the presence of a number of other blood-borne factors, thromboplastin aids the formation of a blood clot which prevents entry of foreign organisms and excessive blood loss. The main stages in the formation of a clot are summarised in Fig 10.9. Thromboplastin, together with a number of plasma enzymes, catalyses the conversion of inactive prothrombin to active thrombin. Thrombin catalyses the conversion of soluble fibrinogen to the insoluble fibrin which forms a network over the area of the wound.

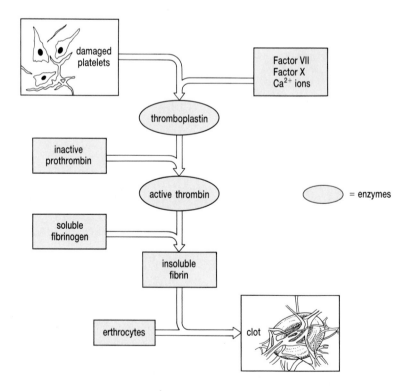

Fig 10.9   The process of clotting. A blood clot is formed after platelets in the blood are broken. The enzymes thromboplastin and active thrombin, together with several other essential substances present in blood, react in an orderly sequence to form fibrin. This insoluble meshwork of fibres encloses red blood cells at the site of the wound, preventing further escape of blood and entry of pathogens.

ANIMAL BODY FLUIDS AND THEIR TRANSPORT

Absence of any of the clotting factors results in a prolonged blood-clotting time, a condition called **haemophilia**. Haemophiliacs are usually male because the condition is inherited through a gene on the X chromosome (Section 22.5). They may need blood transfusions for fairly minor injuries and may need to inject themselves with the blood-clotting factor they lack. Contaminated blood samples used to supplement the most common missing factor, Factor VIII, have been implicated in the spread of the virus which causes AIDS to some haemophiliacs.

In many western European countries, blood clots which form inside blood vessels account for more deaths than any other cause. Blood is normally prevented from clotting inside vessels by a number of **anticoagulants** released within the body, notably **heparin** from the liver and **prostacyclin** from the lining of healthy blood vessels. Blood vessels which are damaged, or whose lining is roughened by cholesterol deposits, do not release prostacyclin. As a result thrombocytes agglutinate and thromboplastin is released. The resulting clot (**thrombus**) may block that vessel (**thrombosis**) leading to the death (**infarction**) of the tissues which it serves. A heart attack (**myocardial infarction**) results if a thrombus forms in a coronary artery supplying the heart muscle with blood. A thrombus may sometimes be dislodged and carried as an **embolus** by the flow of blood. Eventually, an embolus will block a small artery (**embolism**) causing the same effects as a thrombosis.

Warfarin is a poison which is used to kill rats. It works by inhibiting blood clotting, so that rats die of internal bleeding. At suitable dilutions, it can be used to treat humans whose blood is in danger of clotting in their blood vessels. Aspirin is also widely used as a blood-thinning agent.

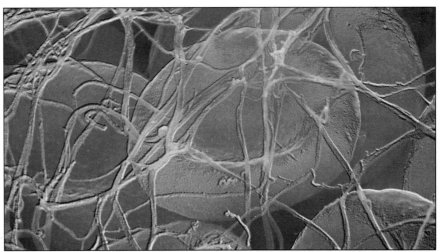

Fig 10.10 A blood clot showing the strands of fibrin around erythrocytes. Fibrin is not really green, it is so here because this photograph has been taken by a technique called false-colour photography.

QUESTIONS

10.4 If fibrinogen is removed from a sample of blood plasma, the remaining fluid is called serum.
(a) What important blood function is stopped by removing fibrinogen?
(b) List the major chemical components of serum.
(c) Explain why blood proteins act as buffers.
(d) What will the effect be of 'dissolved' proteins on the osmotic potential (colloid solute potential) of the blood?

10.5 Monocytes and lymphocytes are types of agranulocyte cell.
(a) Explain how these cells are (i) similar and (ii) different.
(b) Which type of granulocyte has a similar function to the monocyte?

10.6 (a) Why is it important that a person's blood should clot? Explain how this clotting occurs.
(b) In a very rare condition, some people produce too many thrombocytes. Explain how this could cause a heart attack.

In most animal groups a major function of blood is to transport respiratory gases, oxygen and carbon dioxide, to and from respiring cells. However, since oxygen only dissolves poorly in water most animals have some type of respiratory pigment which will increase the oxygen carrying capacity of their blood. The commonest respiratory pigment is the red, iron-containing haemoglobin but others do exist. For example, haemocyanin is a blue copper-containing respiratory pigment common amongst molluscs. This section deals with only one pigment, haemoglobin.

Oxygen dissociation curves

Haemoglobin is a complex protein consisting of four polypeptide chains and an iron-containing haem prosthetic group. One molecule of haemoglobin readily reacts with four molecules of oxygen to form **oxyhaemoglobin**, the form in which most oxygen is carried from the lungs to the tissues. To a lesser extent, haemoglobin can react with carbon dioxide to form **carboxyhaemoglobin**, in which form some carbon dioxide is carried from respiring tissues to the lungs.

The reaction between haemoglobin (abbreviated to Hb) and oxygen can be represented:

$$\underset{\text{haemoglobin}}{Hb} \quad + \quad \underset{\text{oxygen}}{4O_2} \quad \rightleftharpoons \quad \underset{\text{oxyhaemoglobin}}{HbO_8}$$

This equation shows that one molecule of oxyhaemoglobin carries four molecules of oxygen. It also shows that the reaction is reversible: if the concentration of oxygen is high, oxyhaemoglobin is produced; if the concentration of oxygen is low, oxyhaemoglobin breaks down (**dissociates**) to release oxygen. When air of different oxygen concentrations (oxygen tensions) is bubbled into human haemoglobin an S-shaped (or **sigmoid**) curve is formed, called the **oxygen dissociation curve** of haemoglobin (Fig 10.11).

The significance of the sigmoid shape is best understood by referring to the black curve first.

- At relatively high oxygen tensions, such as would be found in the lungs, almost all the haemoglobin in the blood is in the form of oxyhaemoglobin.
- As the oxygen tension falls, as it would in the veins carrying blood from the lungs to the heart and in the arteries carrying blood from the heart to the tissues of the body, most of the haemoglobin is still in the form of oxyhaemoglobin.
- However, over the lower range of oxygen tensions that would be found in the tissues of the body, oxyhaemoglobin dissociates, releasing oxygen. Thus, oxyhaemoglobin carries oxygen in the bloodstream from the lungs with very little loss until it reaches the tissues.

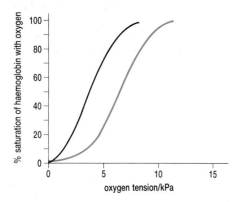

Fig 10.11 Oxygen dissociation curve. The curves show the dissociation of oxyhaemoglobin at different carbon dioxide concentrations. The blue line illustrates the Bohr effect. It means that actively respiring cells with high carbon dioxide concentration cause oxyhaemoglobin to dissociate more readily.

The Bohr effect

The blue curve in Fig 10.11 shows the oxygen dissociation curve for human haemoglobin which is held at a higher carbon dioxide concentration than the previous curve. The shape is still sigmoid but it is shifted to the right of the y-axis. At any particular oxygen tension, the proportion of haemoglobin in the form of oxyhaemoglobin is less in the blue curve than in the black curve. This is called the **Bohr effect**. Its significance is that groups of cells which are actively respiring cause a high concentration of carbon dioxide around themselves which, in turn, causes oxyhaemoglobin to dissociate more readily. In this way, cells which need more oxygen obtain it by promoting the dissociation of oxyhaemoglobin.

To explain the Bohr effect we need to know what happens to carbon dioxide that diffuses into the blood plasma and then into erythrocytes. A small proportion of this carbon dioxide reacts with haemoglobin to produce **carboxyhaemoglobin**. Most reacts with water in the erythrocytes to produce carbonic acid. This reaction is catalysed by an enzyme present in the erythrocytes, called **carbonic anhydrase**.

$$H_2O + CO_2 \xrightleftharpoons{\text{carbonic anhydrase}} H_2CO_3 \quad \text{carbonic acid}$$

As carbonic acid accumulates, it dissociates to form hydrogen and hydrogencarbonate ions:

$$H_2CO_3 \rightleftharpoons H^+ + HCO_3^-$$

Eventually, the concentration of hydrogencarbonate ions inside the erythrocytes will exceed that in the plasma.

 7 **What will now begin to happen to the accumulating hydrogencarbonate ions?**

The carbon dioxide from cell respiration is, thus, in the form of hydrogencarbonate ions in the erythrocytes and in the plasma. Diffusion of negatively charged hydrogencarbonate ions from the cytoplasm of the erythrocyte to the plasma, is balanced by diffusion of negatively charged chloride ions in the opposite direction (the **chloride shift**).

The accumulation of hydrogen ions from the dissociation of carbonic acid would reduce the pH of the erythrocytes and of the blood, reducing the efficiency of enzymes throughout the body. Hydrogen ions react with oxyhaemoglobin forming reduced haemoglobin (abbreviated to HHb) and oxygen:

$$H^+ + HbO_8 \rightleftharpoons HHb + 4O_2$$

It is this reaction with hydrogen ions which explains the Bohr effect. Look at Fig 10.12 which summarises the processes (numbers 1–6) that take place in erythrocytes in the blood around the tissues. In the high oxygen tensions of the capillaries of the lungs, all the reactions are reversed; oxyhaemoglobin is formed and carbon dioxide is released for excretion.

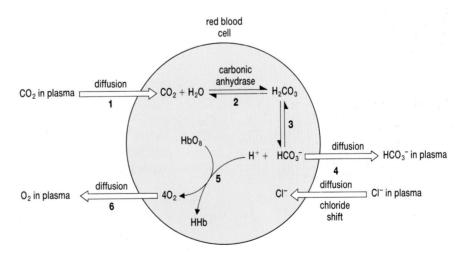

Fig 10.12 Biochemical reactions (1–6) in the erythrocyte occur at low oxygen tension in the capillaries. Carbon dioxide is taken in to form hydrogencarbonate ions. Chloride ions enter by diffusion to balance hydrogencarbonate ions lost to the plasma. Hydrogen ions react with oxyhaemoglobin forming reduced haemoglobin and oxygen, which is released into the plasma. These reactions are reversed at high oxygen tension.

Control of respiratory gases and blood pH

Carbon dioxide is an acidic gas, produced by cell respiration. If allowed to accumulate, the resultant fall in pH would affect the rate of enzyme-controlled reactions throughout the body. A very small increase in the concentration of carbon dioxide in the blood is detected by **chemoreceptors** in the wall of the aorta, in the walls of **carotid bodies** in the carotid arteries and in the medulla oblongata of the brain (see Fig 15.17). When the carbon dioxide concentration of the blood rises, or when the blood pH falls, these chemoreceptors are stimulated and send nerve impulses to a respiratory centre in the medulla oblongata. This centre in turn sends out nerve impulses which increase the rate and depth of breathing, so that the carbon dioxide is exhaled and the blood pH rises again. The carotid bodies also have cells which are sensitive to oxygen concentrations. When the oxygen concentration of the blood falls, the carotid bodies stimulate the medulla oblongata, and the rate and depth of breathing increase. These effects are summarised in Fig 10.13.

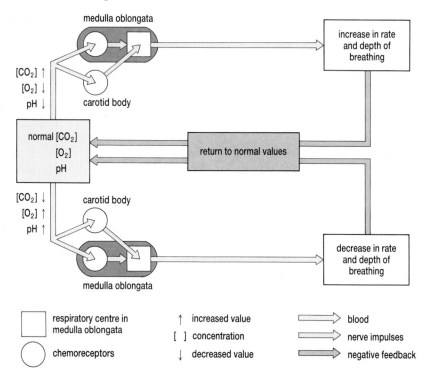

Fig 10.13   The control of respiratory gases and blood pH. A negative-feedback mechanism operates via the medulla oblongata.

QUESTIONS

10.7 (a) Explain, in words, what is meant by an oxygen dissociation curve.

(b) Myoglobin is a protein found in muscle cells. It acts in a similar way to haemoglobin but has a higher affinity for oxygen. Use this information to sketch a graph to show an oxygen dissociation curve for haemoglobin and for myoglobin. Justify your answer.

(c) Where would the oxygen dissociation curve for fetal haemoglobin lie in relation to that for maternal haemoglobin? Explain your answer.

(d) Aquatic insect larvae of the family Chironomidae (blood worms) live at the bottoms of lakes and rivers, often in conditions of very low oxygen tension. Explain the value of haemoglobin to these animals.

ANIMAL BODY FLUIDS AND THEIR TRANSPORT

10.8 **(a)** Explain how the pH of the blood affects (i) oxygen dissociation curves (ii) the rate of breathing.

(b) Pearl divers in the Far East used to hunt for pearls by diving without aqualungs. Prior to a dive the diver would hyperventilate, taking short rapid breaths which reduced the level of carbon dioxide in the blood. Why would this practice enable the diver to stay down longer without the urge to breathe?

10.4 THE MAMMALIAN CIRCULATORY SYSTEM

The general pattern of blood flow in a mammal is shown in Fig 10.14. Blood flows away from the heart in arteries. These arteries eventually supply the tissues with blood via capillaries. Blood from the capillaries returns to the heart via veins.

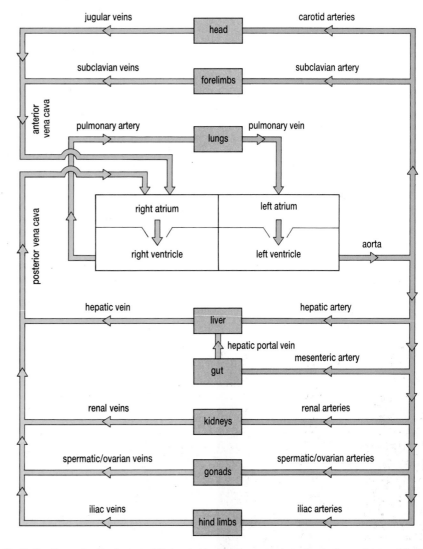

Fig 10.14 Mammalian blood system. This is a double circulatory system but notice the liver and gut are linked by the hepatic portal vein. Blue=deoxygenated blood. Pink=oxygenated blood.

tunica adventitia

lumen

tunica media

tunica intima

(a) artery

(b) vein

(c) capillary

Fig 10.15 Comparison of an artery, a vein and a capillary. Note: drawings are not to scale.

Blood vessels

Arteries and veins have a similar structure, shown in Fig 10.15. The outer layer (**tunica adventitia**) protects the blood vessel from wear as it rubs against other organs which surround it. The inner layer (**tunica intima**) has an epithelium whose smoothness reduces the friction caused by blood

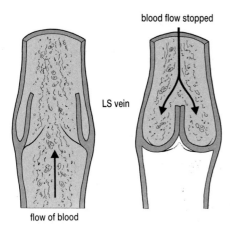

blood flow stopped

LS vein

flow of blood

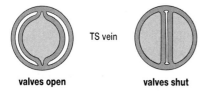

TS vein

valves open　　　**valves shut**

Fig 10.16 The action of semi-lunar valves. Any blood attempting to flow back is caught in the 'pockets', so shutting the valves.

rubbing against it as it flows along the vessel. The middle layer (**tunica media**) consists of involuntary muscle and connective tissue.

Some of the differences in structure between arteries, veins and capillaries are shown in Table 10.2. Arteries always carry blood away from the heart. Blood which has recently left the heart is at high pressure and the thick middle layer of an artery resists this pressure. The middle layer of an artery also contains elastic tissue. When blood is forced into an artery, the elastic tissue allows it to bulge outwards. After stretching, this elastic tissue recoils, helping to push blood along the artery. Friction against the walls of blood vessels reduces blood pressure. Consequently the blood pressure in veins is low.

Table 10.2　Comparison of structure of artery, vein and capillary

Feature	Artery	Vein	Capillary
tunica adventitia	present, with collagen fibres	present with collagen fibres	absent
tunica media	present, with smooth muscle and elastic fibres	present, with smooth muscle and a few elastic fibres	absent
tunica intima	present	present	present

 Suggest advantages of the differences in the structure of veins and arteries shown in Fig 10.15.

Veins are usually surrounded by skeletal muscle whose contraction squeezes the vein and pushes blood along it. **Semi-lunar** valves in the veins ensure that when the veins are squeezed the blood within them usually flows in one direction. A semi-lunar valve is shown in Fig 10.16. It works in a similar way to the rear pockets in a pair of denim jeans. If you run your hand upwards against your jeans you push the pocket flat, just as blood flowing against a valve pushes it against the wall of a vein. When you run your hand downwards against your jeans you open the entrance to the pocket and your hand slips inside. Blood flowing backwards opens the flaps of tissue making each valve and they fill with blood. As they do this, they fill the lumen of the vein, stopping any blood going backwards past the valve.

Capillary walls are so thin that they are leaky, allowing small soluble molecules to pass between blood and the fluid which surrounds the cells of the body (**tissue fluid**, described in Section 10.5). Neutrophils and monocytes can also leave the capillaries, squeezing between adjacent cells, at sites of infection.

The heart

The internal structure of the heart of an adult mammal is shown in Fig 10.18. The heart has two sides, separated by a muscular **septum**. The right side passes deoxygenated blood to the lungs; the left side passes oxygenated blood to the rest of the body. Both sides of the heart have two chambers. The upper **atria** (singular atrium) collect blood from veins; the lower **ventricles** force blood into arteries.

Atrioventricular valves separate the atria and ventricles on each side of the heart. These work in the same way as the semi-lunar valves in the veins, closing the passage between atria and ventricles when blood fills their 'pockets'. Strands of strong, inelastic tissue (**tendinous chords**) prevent the flaps of the atrioventricular valves being turned inside out by the high pressure generated when the ventricles contract.

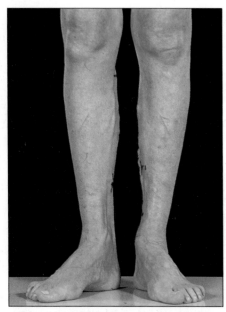

Fig 10.17 This person suffers from varicose veins. The veins swell like this when the valves malfunction and so reduce the blood flow.

 A 'hole in the heart' is a small opening in the ventricular or atrial septum. Predict the consequences of such a hole.

ANIMAL BODY FLUIDS AND THEIR TRANSPORT

Your heart muscle is capable of contracting 200 times per minute during vigorous exercise. When you are exercising the flow of blood from the heart, the cardiac output, may be 25 dm³ min⁻¹, a rate 40% faster than petrol coming out of a petrol pump on a garage forecourt. This 300 g pump may continue to work for 100 years – a truly remarkable organ!

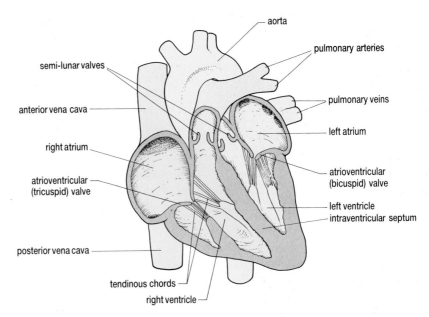

Fig 10.18 Structure of the mammalian heart. The muscular wall of the left ventricle is able to develop the high blood pressure needed for the blood flow out through the aorta.

TUTORIAL

Fig 10.19 Photomicrograph of cardiac muscle cells.

Purkyne fibres used to be called Purkinje fibres.

The heart beat

The heart consists mainly of **cardiac muscle** (Fig 10.19) which can contract without becoming fatigued for the whole of a mammal's life. Cardiac muscle cells form a network linked by the ends of their cylindrical cytoplasm and short lateral arms. This network is able to conduct electrical impulses from cell to cell, ensuring that the cardiac muscle contracts rapidly and smoothly. The striations in the muscle cells result from the arrangement of two proteins, actin and myosin, which are involved in making the cells contract.

Heart muscle does not need to be stimulated by hormone or nerve impulses to make it contract. The heart beat is **myogenic**, meaning that the contraction originates in the heart itself. Contraction of the heart muscle is known as **systole** and relaxation as **diastole**.

The heart beat starts at the **sinoatrial node** (Fig 10.20(a)). This small patch of tissue, sometimes called the pacemaker, has its own natural rhythm which can be altered by nervous impulses and some hormones, e.g. adrenalin. Impulses from the sinoatrial node spread rapidly over the atria causing them to contract together. This is called **atrial systole** (Fig 10.20(b)).

(a) What effect will atrial systole have on the pressure of the blood in the atria?

(b) The base of the veins entering the atria contract during atrial systole. What effect will this have on blood flow?

Impulses from the sinoatrial node can pass across the atrioventricular septum through one patch of tissue, called the **atrioventricular node**. This 'gate' temporarily slows down the impulse. Once through the gate the impulse is rapidly conducted by specially adapted muscle cells, the **Purkyne fibres**, down the interventricular septum to the walls of the ventricles. The ventricles now contract – **ventricular systole** (Fig 10.20(c)).

(c) What is the advantage of the impulse being slowed down at the atrioventricular node?

(d) What will happen to (i) the atrioventricular valves (ii) the semi-lunar valves as blood pressure starts to rise in the ventricle? Explain the significance of these valve movements.

(e) At the end of systole the walls of the heart relax, **diastole** (Fig 10.20(a)). Explain in terms of pressure differences
- why the atria now start to fill with blood
- the effect of this atrial filling on the atrioventricular valves
- why the semi-lunar valves now close.

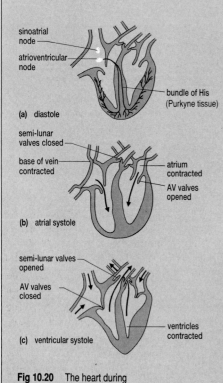

(a) diastole

(b) atrial systole

(c) ventricular systole

Fig 10.20 The heart during
(a) diastole (relaxation)
(b) atrial systole (contraction)
(c) ventricular systole (contraction).

Contraction of the heart is accompanied by a characteristic set of sounds produced as the heart valves close. Recording such sounds produces a **phonocardiogram** (Fig 10.21). The sounds can be used to diagnose heart malfunctions, particularly those associated with valves failing to close properly.

The electrical changes occurring as the heart contracts can be measured, by electrodes placed at various points on the body, to produce an **electrocardiogram (ECG)** (Fig 10.22). Curve P represents the spread of impulses from the sinoatrial node across the atria. Curve QRS represents the spread of impulses through the Purkyne tissue and over the ventricles. Curve T represents diastole. Electrocardiography is a major diagnostic tool.

(f) Using the ECG calculate
- the heart rate of this individual
- the delay of the impulse at the atrioventricular node.

Disruption of the heart's conduction system will interfere with normal heart function. For example, blockage of the impulse at the atrioventricular node means that the atria and ventricles will start to beat out of sequence. This problem can be overcome by fitting an artificial pacemaker, a small device which sends out electrical signals which stimulate the heart and so regulate the heart beat.

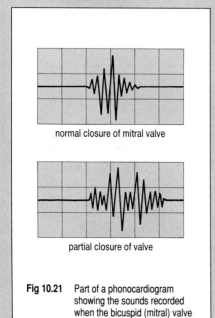

normal closure of mitral valve

partial closure of valve

Fig 10.21 Part of a phonocardiogram showing the sounds recorded when the bicuspid (mitral) valve closes. A malfunction or partial closure of a valve produces a distinctive trace compared with the normal closure.

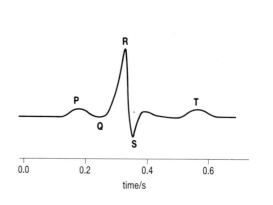

Fig 10.22 An electrocardiogram – a record of the electrical changes occurring as the heart contracts. P = impulses from sinoatrial node; QRS = impulses over ventricles; T = diastole.

ANIMAL BODY FLUIDS AND THEIR TRANSPORT

SPOTLIGHT

DYNAMICS OF HEART CONTRACTION

Scientists in America have recently studied the function of the heart by implanting up to fourteen coils of tantalum (an unreactive metal) at specific points in the hearts of patients undergoing open-heart surgery. Each coil is only about 2 mm long and less than 1 mm in diameter. Once surgery is complete, each patient's chest is X-rayed 60 times per minute over three-and-a-half beats. Fed into a computer, these data have yielded new and accurate information about the dynamics of heart contraction, showing for example that the heart twists as it beats and that the septum does not move as the heart beats. They have also allowed much more accurate estimates of ventricular volumes than had previously been possible.

Blood pressure and flow

Changes in the pressure of the blood of a reclining adult human as it flows from the left ventricle through the circulatory system and back to the heart via the vena cava are shown in Fig 10.23. Blood leaves the ventricle in spurts, corresponding to contractions of the ventricles. As the aorta fills with blood, it stretches. Moments later, as the ventricle relaxes, the stretched aorta recoils, increasing the pressure on the blood within it. The actions of the ventricle and aorta result in the **pulsatile** flow of blood. This pulsatile flow continues throughout the arteries and arterioles and, where an artery can be pressed onto a bone, may be felt as your pulse.

As soon as blood leaves the left ventricle, there is friction between blood and the wall of the aorta. This friction causes a drop in blood pressure. We say that there is a **flow resistance**, defined as the ratio of the drop in pressure to the flow rate.

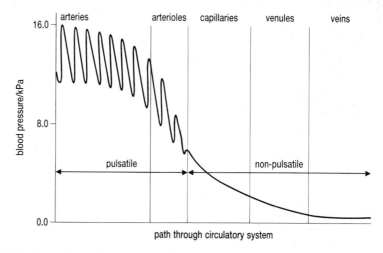

Fig 10.23   Changes in blood pressure as blood flows from the heart through the arteries, arterioles, capillaries and venules. The lowest pressure is recorded in the veins. Exercise of skeletal muscle pushes the muscles against the veins; this helps blood circulation.

10 **Where does the greatest drop in pressure occur? Explain why.**

The high flow-resistance of arterioles has two consequences.

- Arterial pressure depends on whether the arterioles are dilated or constricted.
- The pressure within capillaries is largely independent of the pressure in the large arterial trunks.

ANIMAL BODY FLUIDS AND THEIR TRANSPORT

A further large drop in pressure occurs in the capillaries, since the millions of capillaries in the body form a large total internal surface area which results in much friction. A relatively large drop in pressure also results from the leakage of plasma from the capillaries during the formation of tissue fluid (see Section 10.5),

As a result of the drop in pressure across capillary beds, the pressure of blood in veins is very low. Continued friction between blood and the walls of the veins reduces this pressure still further. The pressure in the large veins is so low (the mean pressure in the large veins is only 1.07 kPa compared to 15.79 kPa in the arteries during systole) that it is insufficient to return the blood to the heart. An increase in pressure, caused by exercising skeletal muscles pushing against the veins, helps to push blood back to the heart. Semi-lunar valves in the veins prevent backflow, as shown in Fig 10.16.

 11 **How will the low pressure which develops in the chest cavity during inhalation help the flow of blood back to the heart?**

ANALYSIS

Blood pressure
This exercise allows you to practise the skills required for the interpretation of data.

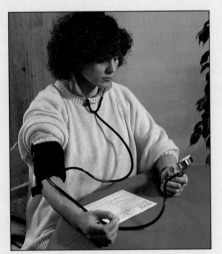

Fig 10.24 Using a sphygmomanometer to measure blood pressure.

Human blood pressure can be measured using a **sphygmomanometer**. This instrument has a cloth cuff, containing a rubber bag, which is wrapped around the arm just above the elbow. Inflation of the rubber bag, using a hand pump, pushes against the brachial artery and stops the flow of blood through it. By releasing the pressure and listening to the brachial pulse inside the elbow, the person in Fig 10.24 can hear characteristic tapping sounds as the blood begins to move under the bag into the lower arm. Since the arm is at the same level and very close to the left ventricle, the pressure in the bag at which these sounds are heard corresponds to the systolic pressure. As the pressure in the bag is released still further, the sounds in the artery disappear. The pressure at which this happens corresponds to diastole. Thus, everyone's blood pressure has two values: the systolic written on top of the diastolic. The mean value for resting adult human males is 15.79/10.53 kPa (more usually expressed in outdated units: 120/80 mm of mercury).

An adult human has a resting blood pressure of 15.79/10.53 kPa.

(a) Explain the meaning of this statement.
(b) Suggest, with reasons, how the blood pressure of this adult might be different if measured (i) immediately after a short run (ii) around the leg instead of around the arm.

The blood pressure measured in the left ventricle and in large arteries in the head and feet of a human is shown in Table 10.3.

(c) Explain why the blood pressures in the head and feet of a reclining person were different from that measured in the heart.
(d) Explain why the blood pressures in the head and feet of a reclining person were different from those of a standing person.
(e) Use these data to suggest why rabbits die if supported in an upright position for too long.

Table 10.3 Some measurements of blood pressure of a human

Part of body	Blood pressure / kPa	
	standing person	reclining person
head	9.3	13.2
heart	13.3	13.3
feet	26.8	13.1

ANIMAL BODY FLUIDS AND THEIR TRANSPORT

10.5 TISSUE FLUID AND LYMPH

If the mammalian circulatory system were completely closed, metabolites would neither leave nor enter the blood. As a consequence, cells outside the circulatory system would die. This does not occur because there is regular exchange between the blood in the capillaries and a fluid which surrounds the cells they serve, called the **tissue fluid**.

Tissue fluid formation

Capillaries have walls consisting of a single layer of flat, irregularly shaped cells (**squamous cells**). Tiny gaps exist between these cells which enable the passage of small molecules. The two pressures within capillaries are shown in Fig 10.25: the **hydrostatic pressure** (also known as the pressure potential), caused by the contraction of the ventricles of the heart, and the **colloid solute potential** of the plasma proteins. The hydrostatic pressure tends to force water, ions and small molecules out of the capillary, rather as the hydrostatic pressure forces water out of a leaking hose. The colloid solute potential of the plasma proteins tends to pull water into the capillary. At the arteriole end of the capillary, the outward pressure is greater than the inward pressure; ions and small molecules are forced from the capillary into the tissue fluid. As a result, the hydrostatic pressure falls so that at the venular end of the capillary the inward pressures are greater than the outward pressures and so water diffuses from the tissue fluid back into the capillary. Newly formed tissue fluid has the same composition as plasma except that the larger plasma proteins are absent.

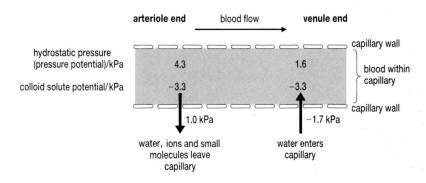

Fig 10.25   Tissue fluid formation. Water, ions and small molecules are moved between the capillaries and surrounding tissue by the hydrostatic pressure and by the colloid solute potential in the capillary.

 12 Why will ions and other small molecules follow the water back into the capillary at its venular end?

Lymph and the lymph system

In an average human, about 3 dm³ more fluid leaves the capillaries as tissue fluid each day than is reabsorbed at the venular end of the capillaries. If this fluid accumulated, the tissues would swell, a condition known as **oedema**. Surplus tissue fluid is normally collected by the blind-ended capillaries of the lymph system. These lymph capillaries have thin walls with small gaps between their cells which allow tissue fluid to enter. Lymph capillaries join to form the larger lymph vessels shown in Fig 10.26, which finally empty into veins in the neck. Like veins, the larger lymph vessels have valves that ensure a one-way flow. Lymph vessels also have swellings, the **lymph nodes**, along their length. These are most common around the small intestine (the so-called **Peyer's patches**) and where the head and limbs join the trunk of the body. Within each lymph node are numerous channels lined by macrophages which engulf foreign material. The lymph nodes also contain large numbers of antibody-producing lymphocyte cells.

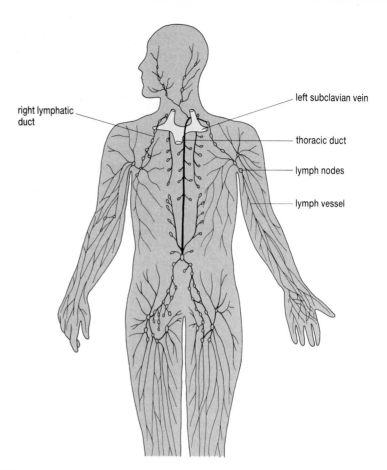

Fig 10.26 Lymph system. At intervals throughout the body there are lymph nodes. The lymph returns to the subclavian vein of the blood system via the thoracic duct.

QUESTIONS

10.9 Oedema (surplus tissue fluid) may result from the following clinical disorders
(i) excessively high blood pressure
(ii) a protein-deficient diet
(iii) blockage of lymph vessels of the legs by a parasitic worm (elephantiasis).
In each case, explain why oedema occurs.

10.10 (a) Suggest how the composition of tissue fluid might be changed by the activities of cells in the body tissues.
(b) Use your suggestions to deduce the likely composition of lymph.

SUMMARY

Blood circulatory systems link body cells with exchange surfaces, thus overcoming the limitations of diffusion over great distances in large and active animals. The efficiency of a circulatory system is increased if blood is moved through vessels (a closed system) by a muscular pump. Insects do not have a closed circulatory system, instead their blood occupies a large cavity within the body. Mammals and fish have closed circulatory systems. Their blood is pumped in sequence through arteries, arterioles, capillaries, venules, veins and back to the heart. Each vessel is adapted for its specific role.

Fish have a single circulatory system in which blood flows from heart to gills to body tissues and back to the heart. The resulting loss of pressure results in slow circulation of blood. Such a system can support only a low

ANIMAL BODY FLUIDS AND THEIR TRANSPORT

metabolic rate. Mammals have a double circulation in which blood circulation between heart and lungs is separate from that between the heart and the rest of the body. By returning oxygenated blood from the lungs to the heart, the pressure of blood is maintained and a higher metabolic rate can be supported.

The mammalian heart has four chambers: two upper atria separated from the two lower ventricles by atrioventricular valves. A muscular septum separates the chambers on each side of the heart. The heart beat is myogenic and coordinated by two specialised groups of muscle cells. The sinoatrial node of the right atrium serves as a pacemaker, initiating a regular heart beat. Impulses from the sinoatrial node activate the atrioventricular node, from which they pass to the walls of the ventricles, causing them to contract. The heart rate can be modified by nervous and hormonal stimulation of the sinoatrial node.

Mammalian blood consists of plasma and cells. Plasma is mainly water with dissolved proteins, gases, nutrients and waste products. The most common cells are erythrocytes; they contain haemoglobin which is involved in the transport of respiratory gases: oxygen is transported as oxyhaemoglobin and carbon dioxide as hydrogencarbonate ions. Leucocytes are less common but essential in their action against infections. Thrombocytes are fragments of cells which release enzymes initiating the process of blood clotting.

Water and dissolved solutes are forced out of the blood in the thin-walled capillaries, forming tissue fluid. Diffusion of metabolites occurs between this fluid and the body cells. Much of the tissue fluid is reabsorbed by the capillaries but some is taken up by capillaries of the lymphatic system where it becomes lymph. As lymph is returned to the blood circulatory system, it is filtered in lymph nodes.

Answers to Quick Questions: Chapter 10

1. Flat, all cells close to an exchange surface, their metabolic needs met by diffusion. Branched gut distributes food throughout flatworm's body.
2. Anywhere materials are transferred to and from the blood, e.g. gut, kidney.
3. Becomes less steep.
4. Heart aided by elastic recoil of arteries.
5. By diffusion.
6. Closed circulatory system.
7. Diffuse out of erythrocytes into plasma.
8. Arteries carry blood at greater pressure than veins; arteries thicker, more muscular walls. Elastic material in arterial walls provides elastic recoil needed to maintain blood pressure.
9. Deoxygenated blood in right ventricle (or atrium) passes directly into left ventricle (or atrium) and so reduces oxygen content of blood being pumped to the body. This leads to cyanosis or blueing of the skin.
10. Entering arterioles; high resistance to flow.
11. Pressure of blood in vena cava in chest will fall below pressure of blood in rest of venous system; blood will flow down this pressure gradient.
12. The concentration of ions in the tissue fluid will be greater than in the blood. Therefore, they will diffuse down this concentration gradient.

Chapter 11

REGULATION OF BODY FLUIDS IN ANIMALS: HOMEOSTASIS

LEARNING OBJECTIVES

When you have studied this chapter you should be able to:

1. explain the terms homeostasis and negative feedback;

2. discuss the regulation of body temperature in poikilotherms and homoiotherms;

3. explain how mammals control the temperature, water potential, pH and concentrations of glucose, carbon dioxide and oxygen of their blood;

4. outline the way in which a variety of animals control their water potential and relate this to their method of nitrogenous excretion;

5. relate the structure of skin, kidneys and liver of a mammal to their respective functions.

Cells in an animal's body can only function normally within a narrow range of conditions. The enzymes which control cell metabolism are sensitive to changes in temperature and pH (Section 5.3). Cells need raw materials for growth and energy release; waste products must be eliminated. Since the immediate environment of each mammalian cell is the tissue fluid that surrounds it, the composition of tissue fluid must be kept within narrow limits. Section 10.5 describes how tissue fluid is formed from, and eventually drains back into, the bloodstream. By regulating blood composition, the composition of tissue fluid is regulated. Control of blood composition is an example of **homeostasis**. The term derives from two Greek words: *homo* means same and *stasis* means state; hence homeostasis is often referred to as the maintenance of a constant state.

Control systems must have a **sensor**, which monitors the component being controlled, and an **effector**, which can change the properties of that component. How a sensor and effector interact in a simple system such as the electrically controlled water bath you might use in a laboratory is shown in Fig 11.1. A fall in temperature of the water bath is detected by a thermostat which sends electrical signals to switch on the heating element. As a result the temperature of the water bath rises. Once the temperature has risen beyond the desired temperature, signals from the thermostat to the heating element stop. The temperature of the water bath then falls as heat is lost to the environment. The action of the heating element is controlled by information about the temperature of the water, called **feedback**. Since the effect of the heater causes a change in temperature of the water bath which is opposite to that already occurring, the thermostat control system is said to operate by **negative feedback**.

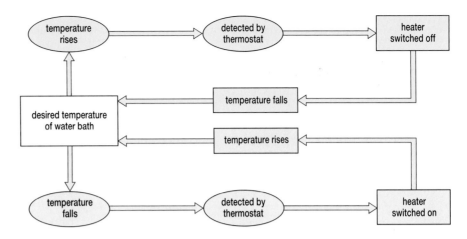

Fig 11.1 The temperature of a water bath is regulated by a negative feedback control system.

 1 Suggest what positive feedback might be.

 2 Assume the pH of the blood of an imaginary animal begins to fall from its normal value of pH 7.0 to pH 6.5. If this animal were able to control its blood pH, what would happen to blood pH in (a) a negative feedback system and (b) a positive feedback system?

11.1 REGULATION OF BODY TEMPERATURE

Organisms can remain active only if they maintain a high enough temperature to enable chemical reactions to occur at a reasonable rate. At the same time too high a temperature may denature the organism's enzymes, so reducing the rate of metabolism. Effectively, biological systems maintain a balance between being too cold and too hot.

Losing and gaining heat

An organism's temperature changes as it gains or loses heat (Fig 11.2). If the heat gain is greater than the heat loss, the temperature of the organism will rise and vice versa. To maintain a steady temperature, an organism must balance its heat gains and losses.

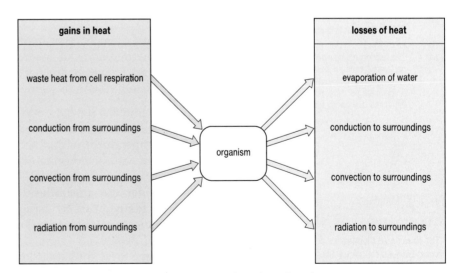

Fig 11.2 How an organism gains and loses heat.

Chemical inefficiency inevitably means that heat is released as a by-product of metabolism. Organisms can also gain heat from, and lose heat to, the environment by three mechanisms.

Fig 11.3 The Arctic fox can survive at temperatures of –40°C.

- **Radiation** is energy that travels from one place to another as electromagnetic waves. The spectrum of electromagnetic radiation is shown in Fig 7.4. Infrared radiation accounts for most of the heat energy radiated and absorbed by animals and, at normal temperatures, animals both radiate and absorb infrared waves.
- **Conduction** is the transport of heat by the collision of molecules. Molecules in a hot region move faster than those in a colder region. When molecules collide, energy is transferred from the faster to the slower molecule. Air does not conduct heat as well as liquids and solids. Therefore, conduction is greatest when the external or internal surface of an animal is in contact with a liquid or solid of different temperature. Since fatty adipose tissue does not conduct heat very well, its presence beneath the skin of a mammal greatly reduces heat loss to, or heat gain from, the environment.
- **Convection** is the transfer of heat by currents of air or water. Hot air and water rises and cooler air and water sinks, setting up convection currents around relatively hot objects. Prevention of such currents, e.g., by trapping air in a fibrous layer outside the skin, can greatly reduce heat gains and losses. Fur in mammals and feathers in birds perform this function.

In addition, heat is lost when liquid water is evaporated, e.g. on skin surfaces or lung surfaces.

 Consider an Arctic fox lying on snow during a spring day. List all the ways this animal will be losing and gaining heat.

Thermal classification of animals

Animals can be classified on the basis of the stability of their body temperature (Fig 11.4). When exposed to changing air or water temperatures

- **homoiotherms**, e.g. mammals and birds, regulate their body temperature close to a value of 37–38 °C in mammals and 40 °C in birds by controlling heat gain and loss
- **poikilotherms**, like most fish and reptiles, allow their temperatures to fluctuate more or less with that of the air or water temperature around them.

At one time, animals other than birds and mammals were considered to be poikilotherms and all mammals and birds were thought to be homoiotherms. However, further studies of the biology of these animals revealed difficulties with this simplistic view. It is known that

- many mammals, e.g. bats, and birds have unstable body temperatures
- many so-called poikilotherms, e.g. lizards, are able to regulate their body temperatures within narrow limits by controlling heat exchange with their natural environment.

 Consider the crocodiles in Fig. 11.5. They maintain a constant body temperature by using their environment. Suggest how, using the clues available from the photograph.

To overcome these difficulties, physiologists introduced another thermal classification of animals on the basis of their source of body heat.

- **Endothermic** animals generate their own body heat.
- **Ectothermic** animals depend almost entirely on the environment for their body heat.

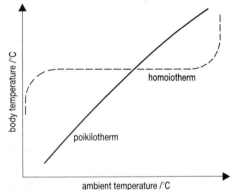

Fig 11.4 The relationship between body temperature and ambient temperature. The body temperature of poikilotherms is similar to the ambient temperature; homoiotherms have an almost constant body temperature.

KEY CONCEPT

The terms cold-blooded for poikilotherms and warm-blooded for homoiotherms are wholly inappropriate since many poikilotherms can become quite warm. For example a 'cold-blooded' locust flying in the desert may have a blood temperature greater than that of a 'warm-blooded' mammal.

REGULATION OF BODY FLUIDS IN ANIMALS: HOMEOSTASIS

Fig 11.5 Crocodiles are poikilotherms which live in or near water.

You should appreciate that the concepts of endothermy and ectothermy are idealized extremes. Most organisms are intermediate.

 5 **Think of a cat basking on a sunny window sill. Does it act as an endotherm, an ectotherm or both?**

By generating their own body heat, endotherms increase their body temperature above the ambient environmental temperature. They usually have good insulation which enables them to conserve their body heat despite a low environmental temperature. In addition to mammals and birds, a few large fish and flying insects can also be considered endotherms. Whilst endothermy has enabled these organisms to colonise habitats which are too cold for most ectotherms, they pay a penalty: the metabolic rate of an endotherm at rest is at least five times that of an ectotherm of equal size and body temperature.

 6 **What does this high metabolic rate mean in terms of the food requirements of an endotherm compared with an ectotherm of a similar size?**

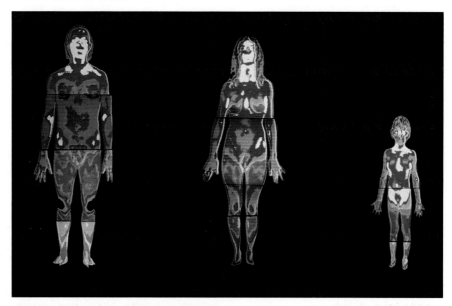

Fig 11.6 This thermograph shows quite clearly where heat loss of the human body is greatest. White is the hottest part i.e. losing most heat, through red, orange, green and blue to purple for the coolest parts.

Heat exchange with the environment is much more important to ectotherms than endotherms. Ectotherms regulate their body temperature by behavioural means. A snake, for example, might bask with its body at right angles to the sun's rays early in the morning, so warming up, but during the heat of the day it might crawl under a rock, so cooling down.

 7 **Why is it an advantage for an ectotherm to be poorly insulated?**

QUESTIONS

11.1 Explain the survival value of the fur of

    (a) a polar bear living in the Arctic
    (b) a camel living in the Sahara desert.

11.2 **(a)** Fish and reptiles lack fur or feathers. How will this affect their heat gain and loss?
      (b) Explain why beached whales die from overheating.

SPOTLIGHT

HYPOTHERMIA

In a cold winter thousands of old people in Britain die because their bodies become too cold. Whilst the needless death of so many is a national scandal, hypothermia is an essential part of life-saving open-heart operations. Cooling a patient's blood enables his or her metabolism to be reduced so reducing the oxygen demands of tissues. Thus during heart-lung bypass operations, the heart and, more importantly, the brain can withstand short periods of interrupted or reduced blood flow.

The reduced tissue demand for oxygen as a result of hypothermia also explains the survival of those people, usually young children, who fall through ice into the chilly waters below. Even though such individuals are submerged so long under the ice they should drown, a lucky few survive.

11.2 TEMPERATURE REGULATION IN MAMMALS

Mammals gain and lose heat through any surface which comes into contact with their environment. Heat exchange through the surfaces of their lungs and gut cannot be controlled; only heat exchange through their skin can be controlled.

Mammalian skin

Whilst differences occur between species and between the parts of a single mammal's body, mammalian skin has the general structure shown in Fig 11.7. The complexity of its structure reflects the various functions which skin performs. These functions and the way that the skin is adapted to perform them are summarised in Table 11.1.

Capillaries

The amount of heat lost from blood in the skin capillaries is directly proportional to the volume of blood flowing into them. Capillaries have no muscle and so can neither change their position in the skin nor change their volume. A capillary network in the skin and the blood vessels which supply it with blood are shown in Fig 11.8. Notice the position of the **arterio-venous shunt vessel** that carries blood into the capillary network

Table 11.1 Functions of mammalian skin

Function	Part of skin involved	Notes
reduces damage to underlying tissues by friction	epidermis	many layers of expendable cells are lost before damage is done to tissues below
reduces damage to underlying tissues by mechanical shock	sub-cutaneous fat in adipose cells	fat is a poor conductor of mechanical shock
reduces damage to underlying tissues by radiation	Malpighian layer in epidermis	pigment (melanin) absorbs radiation
reduces entry of pathogens	epidermis	compact layers of dead cells act as a barrier
		sebum (from sebaceous glands) keeps skin supple and prevents cracks in surface
		sebum contains fatty acids which are toxic to some bacteria
reduces water loss	epidermis	compact layer of dead cells reduces evaporation
		oily sebum reduces water loss
		hair traps a humid layer of air near skin which reduces diffusion of water vapour
sensitivity	receptors in dermis	receptors specific to touch, pain, heat, cold and pressure
excretion	sweat glands in dermis	sweat contains water, inorganic salts and urea
production of vitamin D	region just below Malpighian layer	vitamin D produced under action of UV light
controls heat loss	subcutaneous fat	fat is a poor conductor of heat, reduces conduction of heat
	hair in epidermis	hair traps a layer of still air near skin, reduces convection of heat
	capillaries in dermis	carry warm blood near surface
	sweat glands in dermis	absorb water and dissolved substances from their capillary network; evaporation of this sweat on skin surface removes heat from skin.

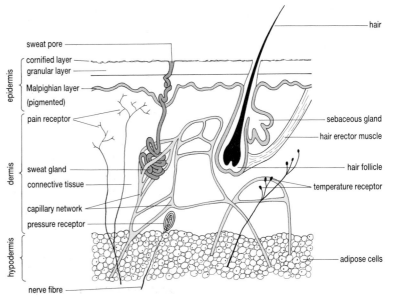

Fig 11.7   Vertical section through mammalian skin. Notice the sensory receptors in the dermis in addition to the capillary network and the nerve fibre. The hair and sweat pore reach the air through the outermost cornified layer.

(a)   (b)

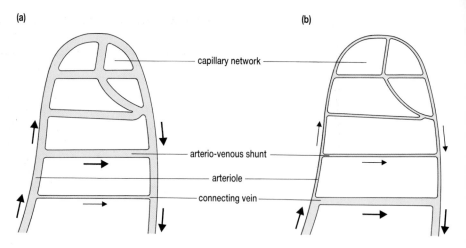

Fig 11.8   The capillary network receives more blood in **(a)** than in **(b)** due to the dilation of the arterio-venous
shunt vessel and arteriole, and the contraction of the connecting vein.

and of the **connecting vein** that allows blood to bypass the capillary
network. Fig 11.8(a) shows that when the arteriole and arterio-venous
shunt vessels are dilated, more blood passes through both the shunt and
skin capillaries in a given time: as a result more heat will be lost through
the skin. The muscles within the blood vessels are controlled by a
temperature control centre in the hypothalamus, a region of the brain.

 8 What happens to the blood flow when the arteriole and arterio-
venous shunt vessels contract (Fig 11.8(b))? How will this affect
heat loss?

Sweat glands

Each sweat gland has a capillary supply around it. The coiled gland
absorbs tissue fluid from this capillary network and secretes it into the
sweat duct and thence onto the surface of the skin via the sweat pore.
Liquid sweat on the surface of the skin does not cool the body. Only when
the sweat evaporates is heat lost from the skin surface. Release of sweat by
the sweat glands is also under control of the temperature control centre in
the brain.

Hair

Most mammals have a lot of hair covering much of their body. The angle
between each hair and the skin surface can be changed by the action of its
erector muscle. When the hairs stand erect, they trap a thicker layer of still
air which reduces heat gain or loss by convection. Control of the erector
hairs is by the temperature control centre of the brain.

 9 Polar bears have hairs which are hollow and contain air. Suggest
how this helps them to keep warm in an Arctic environment.

Control of blood temperature in mammals

The position of the **hypothalamus**, a small body at the base of the brain, is
shown in Fig 15.17. One function of the hypothalamus is to monitor and
control the temperature of the blood. When the temperature of the blood
passing through the hypothalamus is lower than 'normal', nerve impulses
pass from its **heat gain centre** to the skin causing a decrease in the rate of
heat loss by reducing blood flow through surface capillaries and reducing
the formation of sweat. At the same time, impulses from the **heat gain
centre** increase the rate at which heat is released by metabolism and by
involuntary muscle contractions (**shivering**). A second part of the

hypothalamus, the **heat loss centre**, reduces the metabolic rate, stops shivering and increases heat loss through the skin when the temperature of the blood passing through the hypothalamus is above 'normal'. These effects are summarised in Fig 11.9.

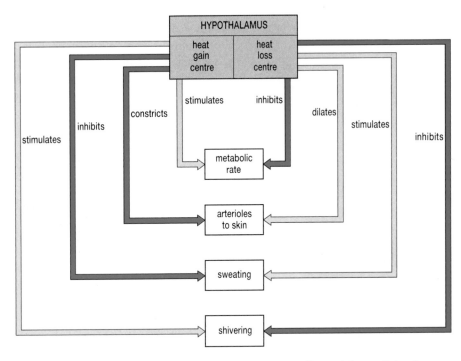

Fig 11.9 A summary of temperature control mechanisms in homoiotherms. The metabolic rate, dilation of arterioles in the skin, sweating and shivering are under the control of the hypothalamus. Darker blue shows stimulation. Paler blue shows inhibition.

QUESTIONS

11.3 Suggest an explanation for each of the following observations.

(a) Animals of the same genus tend to be small in warm regions and large in cold regions (Bergman's rule).

(b) Tails, ears, beaks and other projecting parts of the body tend to be shorter in animals that live in a cold climate than in animals that live in a warm climate (Allen's rule).

(c) A sudden drop in the core body temperature to below 35 °C (hypothermia) is more common in elderly people and young babies than in humans of other age groups.

(d) In mammals with dense body hair, sweat glands are more efficient in cooling the body if they are present on areas of skin where hair is absent than on areas of skin where hair is abundant.

(e) The artery supplying blood to the leg of a duck lies deep in the centre of the leg and is surrounded by veins returning from the leg to the body.

11.4 The Galapagos marine iguana is an ectothermic species of lizard. On land, the iguana absorbs heat from the sun. Whilst it basks, its heart beat is rapid and arterioles in its skin dilate. However, when it dives into the cold sea, its heart beat slows and the skin arterioles constrict. How will this behaviour help the lizard to maintain a constant body temperature?

11.3 OSMOREGULATION IN MAMMALS

Fig 11.10 This desert rodent, the gerbil, has a very efficient mechanism for conserving water. It can survive with almost no liquid intake at all, yet it must excrete waste nitrogenous substances in liquid form through its kidneys.

The water potential of blood results from the relative amounts of water and dissolved solutes within it. Both a loss of water from the blood or a gain of dissolved solutes within it will make the water potential of the blood more negative. The water potential of blood will become less negative following a gain in water or a loss of solutes. The way in which changes in the relative amounts of water and solutes may arise is summarised in Fig 11.11. Since such changes would interfere with the normal metabolism of cells animals need to regulate the water potential of their blood, a process called **osmoregulation**. This involves regulating both the water content and solute composition of the blood. In particular, mammals must control the balance of such ions as Na^+, Cl^- and K^+ whilst excreting poisonous products of protein metabolism like urea.

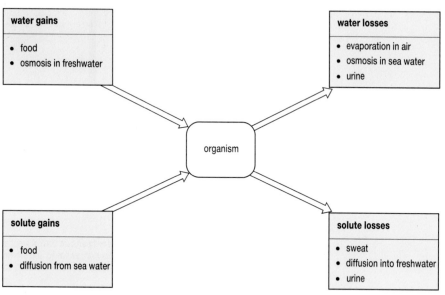

Fig 11.11 How changes in the blood's water potential may occur. Water is gained and lost; solutes, e.g. salt are gained and lost.

The kidneys

A mammal's kidneys control the water potential of the blood that passes through them. Substances which cause an imbalance in the water potential of the blood are removed from it and excreted in urine. For example, nitrogenous wastes from the breakdown of amino acids and nucleic acids are always removed from the blood and excreted. Because it is a solvent, water is inevitably present in the urine.

The kidneys, surrounded by protective fat, lie at the back of the abdominal cavity. Urine leaves each kidney via the **ureters** and passes to the **urinary bladder**, where it is stored. Relaxation of the sphincter muscle at the exit from the bladder allows urine to be released via the **urethra** during **micturition**. The position of these structures in humans and the position of the blood vessels serving each kidney is shown in Fig 11.12. The gross structure of a kidney is shown in Fig 11.13. Each renal artery supplies capillary networks which occur mainly in the outer **cortex** of the kidney. These rich capillary networks make the cortex much darker than the inner **medulla**. **Collecting ducts** carry urine through the medulla to a central chamber at the top of the ureter, called the **pelvis**.

Urine formation

There are three processes that affect the ultimate composition of mammalian urine

REGULATION OF BODY FLUIDS IN ANIMALS: HOMEOSTASIS

- **ultrafiltration** of blood in Bowman's capsule
- **tubular reabsorption** of approximately 99% of the water, most of the ions and all the glucose present in the fluid formed by ultrafiltration (the **ultrafiltrate**)
- **tubular secretion**, a selective process in which K^+, H^+, HCO_3^- ions and foreign substances such as drugs are eliminated from the blood.

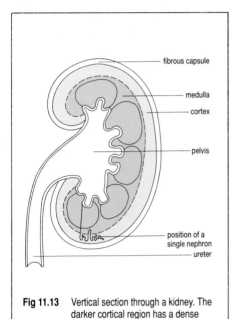

Fig 11.13 Vertical section through a kidney. The darker cortical region has a dense capillary network supplied by the renal artery. Each nephron extends from the cortex to the collecting duct.

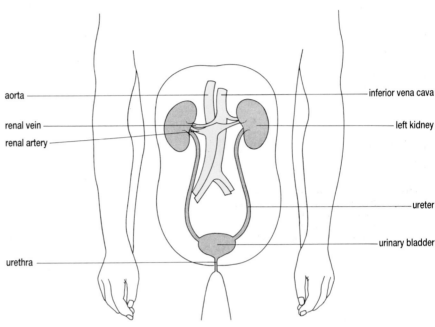

Fig 11.12 Human urinary system. The blood system links all parts of the body to the kidneys. Urine leaves the body via the urethra.

The functional unit of the kidney is the **nephron**. Each human kidney contains 1–1.5 million nephrons. The position of a single nephron within the kidney is shown in Fig 11.13; its structure is shown more clearly in Fig 11.14. Each nephron is about 12–14 mm long and up to about 5 μm in diameter. The first part of each nephron forms a cup-shaped **Bowman's capsule**. From this capsule the rest of the nephron forms a tube with three

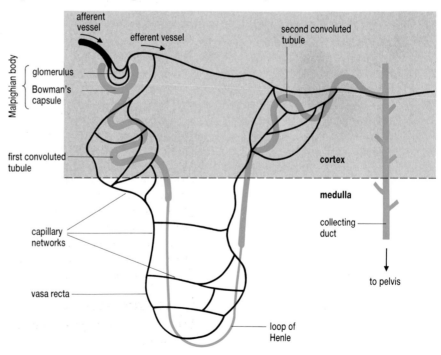

Fig 11.14 A single nephron. Follow the pathway of a single nephron from the Bowman's capsule, first convoluted tubule, loop of Henle, second convoluted tubule to the collecting duct.

distinct regions; the **first convoluted tubule**, the **loop of Henle** and the **second convoluted tubule**. Arterial blood enters each capsule through an **afferent vessel** which branches to form a capillary network, called the **glomerulus**. Blood leaves the glomerulus via an **efferent vessel** which then branches again to form a network of capillaries, the **vasa recta**, surrounding the loop of Henle.

 10 **In which regions of the kidney do the Bowman's capsule, first and second convoluted tubules, loop of Henle and collecting ducts lie?**

Ultrafiltration

Ultrafiltration occurs in the Bowman's capsule and glomerulus. To understand how blood is filtered from the glomerulus into the lumen of the Bowman's capsule you need to know something of their structure.

The outer wall of the capsule consists of a single layer of squamous (flat) epithelial cells (Fig 11.15(a)). Between the outer and inner wall is the capsular space, which is continuous with the lumen of the first convoluted tubule. The inner wall of the capsule consists of special cells called **podocytes** which surround the capillaries of the glomerulus (Fig 11.15(b)). Notice that the capillary wall (the **endothelium**) consists of a single layer of squamous cells.

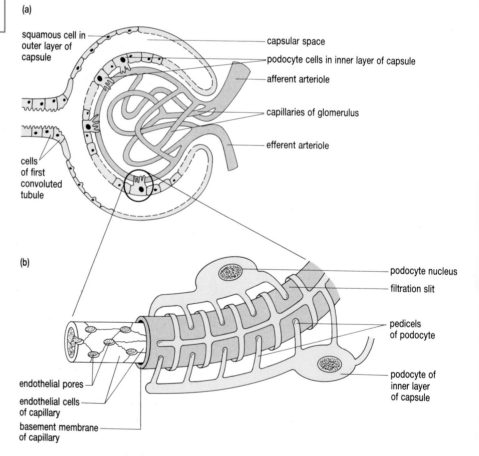

(a)

squamous cell in outer layer of capsule

capsular space

podocyte cells in inner layer of capsule

afferent arteriole

capillaries of glomerulus

efferent arteriole

cells of first convoluted tubule

(b)

podocyte nucleus

filtration slit

pedicels of podocyte

podocyte of inner layer of capsule

endothelial pores

endothelial cells of capillary

basement membrane of capillary

Fig 11.15 **(a)** Bowman's capsule
(b) endothelial capsular membrane.

 11 **Imagine you could unwind a podocyte so that it was lying flat. Draw and label a diagram to show what such a cell would look like.**

REGULATION OF BODY FLUIDS IN ANIMALS: HOMEOSTASIS

The podocytes of the inner wall of the Bowman's capsule and the endothelium of the glomerulus forms an **endothelial capsular membrane**. This membrane consists of three parts listed below in the order in which substances from the blood pass through to reach the capsular space.

Endothelium of the glomerulus. A single layer of endothelial cells which contain pores between 5000–11 000 nm in diameter.

The basement membrane of the glomerulus. Consisting of fibrils arranged in a glycoprotein matrix, this extracellular membrane lies beneath the endothelium and contains no pores. It serves as the filter during the ultrafiltration process.

Inner wall of the Bowman's capsule. This is composed of **podocytes** with their footlike processes called **pedicels**. Notice that the pedicels are arranged parallel to the circumference of the capillaries in the glomerulus and cover the basement membrane except for the spaces between them called **filtration slits** (**slit pores**).

ANALYSIS

The process of ultrafiltration

This exercise will help you to make the link between a biological structure and the physical principles which make it work.

When blood enters the glomerulus, the high blood pressure forces water and dissolved components of the blood through the endothelial pores of the capillaries, the basement membrane and on through the filtration slits into the capsular space. The composition of plasma and the **filtrate** in the capsular space is given in Table 11.2.

Table 11.2 The mean composition of human plasma and filtrate

Molecule or ion	Approximate concentration/g dm^{-3}	
	Plasma	Filtrate
water	900.0	900.0
protein	80.0	0.0
glucose	1.0	1.0
amino acids	0.5	0.5
urea	0.3	0.3
inorganic ions	7.2	7.2

(a) Account for the differences in the composition of plasma and the filtrate.
(b) What other major component of the blood will not be present in the filtrate?
(c) The diameter of the efferent arteriole is smaller than that of the afferent arteriole. How will this affect the hydrostatic pressure of the glomerular blood compared with the hydrostatic pressure found in other capillaries in your body?

Tubular reabsorption

Ultrafiltration is so efficient that 15–20% of the water and solutes are removed from the plasma that flows through the glomerulus. The ultrafiltrate is produced at the staggering rate of approximately 125 cm^3 min^{-1} or 200 dm^3 day^{-1}. Clearly you do not pass this amount of urine each day, so during its passage through the rest of the nephron the original ultrafiltrate must be modified.

Reabsorption in the first convoluted tubule

This exercise gives practice in the interpretation of experimental data.

Look at Table 11.3: it shows the differences in the average daily amount of some important constituents of urine and the ultrafiltrate from which it was obtained.

Table 11.3   Average daily amounts of urine and ultrafiltrate constituents

Substance	Amount in ultrafiltrate	Amount in urine
sodium	550 g	5 g
potassium	27 g	2 g
calcium	5 g	0.2 g
ammonia	0.3 g	0.75 g
glucose	180 g	trace
urea	60 g	35 g
water	180 dm^3	1.5 dm^3

(a) What must be happening in the tubules of the nephron to produce the values in Table 11.3?

(b) Account for the values shown for ammonia.

(c) Calculate the concentration of each component of the urine in g dm^{-3}.

(d) Which two substances are most concentrated in the urine? Explain the value of their high concentration to normal body function.

Look at the electronmicrograph which shows the cells lining the first convoluted tubule (Fig 11.16).

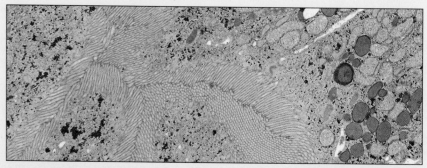

Fig 11.16   Electronmicrograph of cells from the first convoluted tubule showing microvilli on the inner side of the cell.

(e) How are these cells adapted for efficient absorption?

(f) What does the presence of large numbers of mitochondria in these cells suggest?

Frog nephrons are large and easy to work with for experimental purposes. Individual samples of fluid were obtained by micropuncture sampling at different places along the nephron tubule (Fig 11.17). The results of two experiments are plotted in Fig 11.18. The vertical axis represents the ratio of glucose concentration in tubular fluid (ultrafiltrate) to that in blood (plasma). Phlorizin is a substance that inhibits glucose uptake.

(g) What does a low ultrafiltrate : plasma ratio tell you about the relative concentrations of glucose in the tubular fluid and plasma?

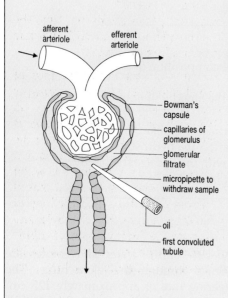

afferent arteriole

efferent arteriole

Bowman's capsule

capillaries of glomerulus

glomerular filtrate

micropipette to withdraw sample

oil

first convoluted tubule

REGULATION OF BODY FLUIDS IN ANIMALS: HOMEOSTASIS

(h) Where does the greatest reabsorption of glucose occur?
(i) Do the data support an active or passive reabsorption of glucose? Explain your answer.
(j) Suggest why the ultrafiltrate : plasma ratio of glucose increases in the first convoluted tubule treated with phlorizin.

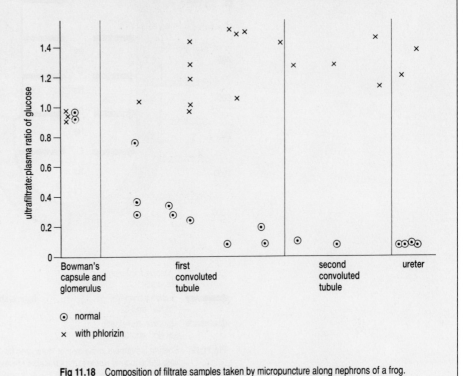

Fig 11.18   Composition of filtrate samples taken by micropuncture along nephrons of a frog.

The events in the first convoluted tubule can be summarised as follows.

- The ultrafiltrate which enters the first convoluted tubule is hypotonic to the blood plasma.
- Active reabsorption of glucose and Na⁺ ions occurs with virtually all the glucose and 67% of the Na⁺ ions being reabsorbed.
- Reabsorption of water and Cl⁻ ions occurs passively. By the time the filtrate reaches the loop of Henle it is considerably reduced in volume and is isotonic with the tissue fluid surrounding the tubule.

Of special interest in tubular reabsorption is the activity of the loop of Henle. It is the activity of this structure which allows a mammal to produce a urine with a more negative water potential than that of its blood, i.e. hypertonic urine. The thicker ascending limb of the loop of Henle is impermeable to water, but it actively removes sodium chloride from the ultrafiltrate into the tissues of the medulla (actually it is chloride ions that are removed, sodium ions follow by diffusion). As a result, the water potential of the tissues of the medulla becomes more negative and that of the ultrafiltrate becomes less negative. Having passed through the loop of Henle, the ultrafiltrate flows through the second convoluted tubule and then to the collecting duct. Since the walls of the second convoluted tubule and collecting duct are permeable, water passes from the ultrafiltrate to the tissues around the loop of Henle. Here it enters the capillaries which form the vasa recta and is carried away.

The activity of the loop of Henle is summarised in Fig 11.19: notice how a gradient of sodium ion concentration in the medulla ensures that water will leave the collecting duct along the whole of its length.

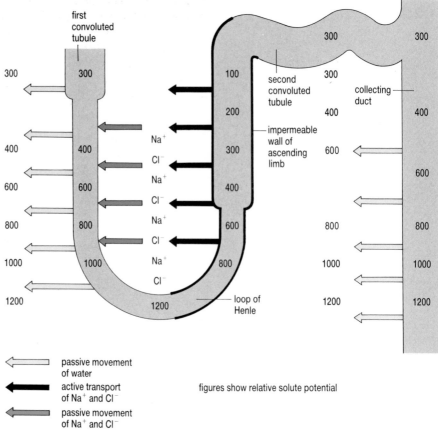

first
convoluted
tubule

Na+
Cl−
Na+
Cl−
Na+
Cl−
Na+
Cl−

second
convoluted
tubule

impermeable
wall of
ascending
limb

collecting
duct

loop of
Henle

 passive movement
of water

active transport
of Na+ and Cl−

passive movement
of Na+ and Cl−

figures show relative solute potential

Fig 11.19 Sodium movement in the loop of Henle and its effect on water reabsorption. The wall of the ascending limb of the loop of Henle is impermeable to water. All figures are in mOsmol dm^{-3}, a measure of concentration. Remember that as the concentration of the solution increases its water potential becomes more negative.

Tubular secretion

The final process controlling urine composition involves the addition of materials to the filtrate from the blood and occurs primarily in the second convoluted tubule. These secreted substances include K^+, H^+, NH_4^+, creatinine and drugs such as penicillin. Tubular secretion, therefore, has two principal functions

- it rids the body of certain materials
- it controls blood pH.

A blood pH of between 7.35 and 7.45 is maintained despite the fact that a normal diet provides more acid-producing than alkali-producing fluids. The second convoluted tubules secrete H^+ and ammonium (NH_4^+) ions into the filtrate, resulting in urine with a pH of about 6. The regulation of blood pH is summarised in Fig 11.20. The activities of filtration, reabsorption and secretion in the nephrons are summarised in Table 11.4.

12 Why is it important to control blood pH?

Control of urine volume and concentration

In some respects the kidney is a filter in which the ultrafiltrate is modified by active and passive reabsorption of key substances as it passes through the first convoluted tubule. However, you know that if you drink a large amount of liquid on a cold day you produce large quantities of pale coloured urine, whereas on a hot day, when perhaps you have not drunk very much, you produce a dark coloured urine in small quantities. Now imagine you were asked to drink 1 dm³ of distilled water while a friend drank 1 dm³ of salty water. You would produce a large amount of urine in about 2 hours whilst your friend would produce very little urine.

Clearly, the nephron is in some way able to control both the volume and concentration of the urine produced. This is, primarily, the function of two parts of the nephron – the loop of Henle and the second convoluted tubule – and the collecting duct.

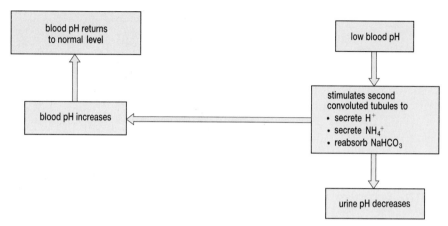

Fig 11.20 Regulation of blood pH. A constant concentration of hydrogen ions, ammonium ions and sodium hydrogencarbonate is necessary.

Table 11.4 Filtration, reabsorption and secretion

Region of nephron	Activity
Bowman's capsule	ultrafiltration of blood in the glomerulus under hydrostatic pressure produces an ultrafiltrate free of plasma proteins and cellular elements of the blood
First convoluted tubule	reabsorption of water by osmosis
	reabsorption of solutes such as Na^+, K^+, Cl^-, HCO_3^-, amino acids and glucose by a mixture of active and passive processes
Loop of Henle	active secretion of chloride ions from ultrafiltrate in ascending limb
Second convoluted tubule	reabsorption of Na^+
	facultative reabsorption of water under control of ADH
	secretion of H^+, NH_4^+, urea, creatinine and some drugs
Collecting duct	facultative reabsorption of water under control of ADH

 Mammals which live in deserts have very long loops of Henle whilst aquatic mammals, like beavers, have short loops of Henle. Explain the advantages of these adaptations.

Control of water reabsorption by ADH

The permeability of the walls of the second convoluted tubule and collecting duct to water is increased by **antidiuretic hormone (ADH)**. ADH is produced by the hypothalamus but secreted into the posterior lobe of the pituitary body where it is stored (Section 14.2). A rise in blood concentration (i.e. its water potential becomes more negative) is detected by osmoreceptors in the hypothalamus which send impulses to the posterior lobe of the pituitary gland (Fig 11.21). As a result, the posterior lobe of the pituitary gland releases more ADH into the blood which increases the permeability of the second convoluted tubule and collecting duct to water. More water passes to the medulla and a more concentrated (hypertonic)

Diabetes insipidus is a clinical condition in which insufficient ADH is released by the posterior lobe of the pituitary and which results in the continual production of large volumes of dilute urine, regardless of the water potential of the blood.

urine is produced. ADH also increases the permeability of the collecting duct to urea. Urea passes from the ultrafiltrate in the collecting duct to the medulla, making its water potential more negative and causing a greater loss of water from the descending limb of the loop of Henle.

A fall in blood concentration (i.e. its water potential becomes less negative) inhibits the release of ADH. As a result, the walls of the second convoluted tubule and collecting duct are impermeable to water, less water is reabsorbed and a hypotonic urine is produced.

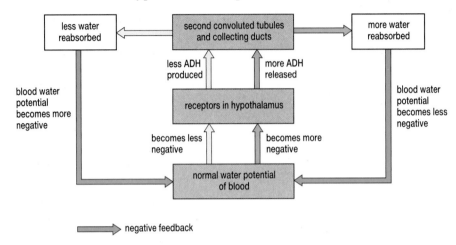

Fig 11.21 Control of blood water potential. By a negative feedback system, receptors in the hypothalamus regulate the release of ADH which controls the amount of water reabsorbed from the collecting ducts.

 Drugs, such as caffeine, and low temperatures inhibit the release of ADH. How will this affect the volume of urine produced?

Control of salt reabsorption by aldosterone

Aldosterone stimulates the active uptake of sodium ions from the ultrafiltrate in the second convoluted tubule. In doing so it has a secondary effect on water reabsorption. Control of its release is more complex than that of ADH.

A loss of sodium ions from the blood causes the blood's water potential to become less negative. As a result, water passes by osmosis to the tissues of the body and the volume of the blood gets less. A group of cells lying between the second convoluted tubule and the afferent arteriole, called the **juxtaglomerular complex**, is stimulated by a decrease in blood volume to release renin. Renin is an enzyme that acts on a blood-borne protein to form **angiotensin**, itself a hormone which stimulates the adrenal cortex to secrete **aldosterone** into the blood stream. On reaching the kidney, aldosterone stimulates the active uptake of sodium ions from the glomerular filtrate. This uptake is accompanied by an uptake of water and loss of potassium ions. Aldosterone also stimulates the brain to increase the sensation of thirst.

ANALYSIS

Urine production

Often in biology you need to consider several factors at the same time in order to provide a reasoned answer, a skill you can practise in this exercise.

Using human subjects, an investigation was carried out to determine the effect of drinking distilled water on the production of urine. At the start of the experimental period the subjects voided as much urine as possible. They then rapidly drank 800 cm^3 of distilled water. Urine was then collected at regular intervals and measurements made of the volume of each sample and its salt concentration. The results are given in Fig 11.22.

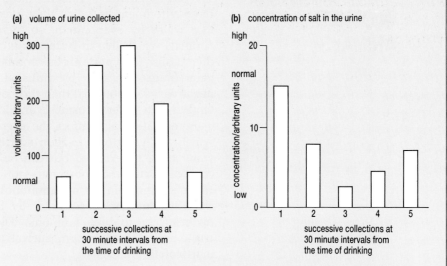

(a) volume of urine collected

(b) concentration of salt in the urine

Fig 11.22 Volume and salt concentration of urine collected at 30-minute intervals after drinking distilled water.

(a) Use your knowledge of kidney functioning to explain briefly
   (i) the change in the volume of urine collected between collections 1 and 2 and between collections 3 and 5.
   (ii) the change in concentration of salt in the urine between collections 1 and 3.

(b) (i) What results would you have expected if the person had drunk 300 cm³ of an isotonic saline solution instead of distilled water? Consider change in volume of urine collected and change in concentration of salt in the urine.
   (ii) State two variables which would need to be considered when comparing results from a number of people who had drunk similar volumes of similar solutions.
   (iii) Blood pH is around 7.4. Why is the pH of urine usually below this?

QUESTIONS

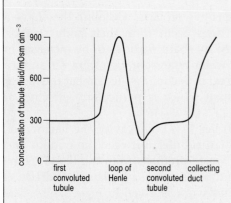

Fig 11.23 Changes in the concentration of fluid along the nephron (mOsm is a measure of concentration).

11.5 The graph (Fig 11.23) represents the concentration of fluid within various parts of a nephron and the collecting duct into which it empties.
   (a) Explain why the concentration of the fluid remains the same along the first convoluted tubule even though glucose, amino acids and ions are actively absorbed from it.
   (b) Explain the changes in concentration of fluid as it passes along the loop of Henle.
   (c) Explain what the changes in concentration of fluid within the second convoluted tubule and collecting duct show about their respective activities.

11.6 Devise a diagram to summarise the effects of ADH and aldosterone on urine composition.

11.7 Why are people who suffer from high blood pressure or from oedema given diuretics, drugs which decrease the permeability of the walls of the collecting duct?

11.4 OSMOREGULATION IN OTHER ANIMALS

The osmoregulatory problems faced by animals are largely dependent on the environment in which they live. Animals which live on land tend to lose water by evaporation, so their tissues become more concentrated. Animals which live in water of more negative water potential than their own tissues (a **hypertonic** environment) tend to lose water by osmosis and gain solutes, e.g. salt, by diffusion. Animals which live in water of less negative water potential than their own tissues (a **hypotonic** environment) tend to gain water by osmosis and lose salts by diffusion. The problems of osmoregulation also affect the excretion of nitrogenous waste. Animals which are hypertonic to their surroundings gain water by osmosis. This water must be excreted but can be used to dilute ammonia to harmless concentrations and most hypertonic animals excrete their nitrogenous waste as ammonia (i.e. are **ammonotelic**). Animals which are hypotonic to their surroundings lose water by osmosis and cannot continually lose further water to dilute ammonia. These animals convert their nitrogenous waste into a less toxic form, such as urea in **ureotelic** animals and uric acid in **uricotelic** animals.

 Are freshwater or salt water fish more likely to be ammonotelic? Explain your answer.

Fig 11.24   This Lugworm, *Arenicola marina* is an animal which is isotonic with its environment.

Fig 11.25   Woodworms lose almost no water in their faeces. They are thus able to survive in very dry conditions. This adult woodworm is just emerging from a hoie it has made to get out of the wood.

Insects

Like mammals, the majority of insects live on land. Although they have a waterproof epicuticle, evaporation does occur from their body surface, especially if the epicuticle is damaged by sharp particles or by increases in temperature. Insects inevitably lose water by evaporation from the surface of their gas exchange system, described in Section 9.4. To combat this water loss, insects have an excretory system which is particularly effective at conserving water.

The gut of an insect (Fig 11.26(a)) lies in a body cavity, the haemocoel, surrounded by blood. The **Malpighian tubules**, between the midgut and hindgut, are involved in removing waste products from the blood. Some species of insect have only two such tubules, others may have hundreds. A summary of how Malpighian tubules work is given in Fig 11.26(b). The cells of an insect's body produce uric acid, from surplus amino acids and nucleic acids, and secrete it into the haemocoel. Here it reacts with potassium hydrogencarbonate to form potassium urate (KHU). This salt is actively absorbed into the lumen of the distal end (furthest from the gut) of the Malpighian tubules, making the water potential of their contents more negative and causing the entry of water by osmosis. In the proximal (nearest the gut) part of the Malpighian tubules, a reaction occurs between the potassium urate, carbon dioxide and water. The hydrogencarbonate so formed is actively reabsorbed into the haemocoel, causing a fall in the pH of the contents of the lumen of the tubule. This fall in pH causes the uric acid to form crystals, in which state it is passed into the hindgut and mixed with the faeces. Almost no water has been lost with this 'urine' and further water is actively reabsorbed from the mixture of uric acid and faeces by **rectal glands** in the wall of the rectum.

(a)

(b)

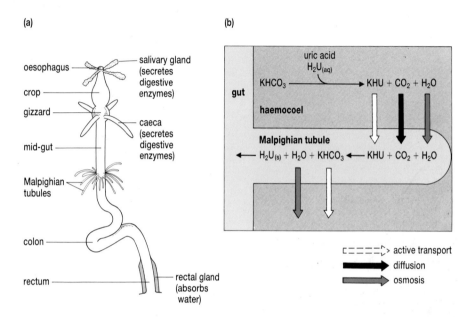

Fig 11.26 (a) The gut of an insect showing the position of Malpighian tubules.
(b) A mechanism for the excretion of uric acid by an insect's Malpighian tubules.

 16 Insects, such as dragonflies, have larvae which live in fresh water whilst the adults are free flying. How would you expect the urine of these larvae and adults to differ?

Fish

The water potential of tissues of Osteichthyes (bony fish) is more negative than that of freshwater streams, rivers and lakes but less negative than that of seawater. Although their outer covering of scales is relatively impermeable, the surface of their gills and of their guts is freely permeable to water and inorganic ions.

Freshwater bony fish excrete ammonia, actively absorb inorganic ions across their gills and produce large volumes of dilute urine. Marine bony fish excrete the less toxic nitrogenous waste trimethylamine oxide (TMO) which gives them their characteristic fishy smell, actively secrete inorganic ions across their gills and produce small volumes of urine. Unlike the freshwater bony fish, their kidneys lack glomeruli. As a result, no ultrafiltration occurs in their kidneys and urine is formed entirely by active secretion of salts and TMO with accompanying osmosis of water.

Some Osteichthyes change environment during their lives. Salmon (*Salmo salar*) hatch in freshwater, spend most of their lives in the sea but migrate back to freshwater to breed. Eels (*Anguilla vulgaris*) do exactly the opposite; they hatch in the sea, spend most of their lives in freshwater but migrate to the sea to breed. The gills of these animals actively absorb inorganic ions when they are in freshwater but actively excrete them when they are in sea water. Whether this involves different cells on their gills, or whether the same cells are able to operate in both directions, is not known. However, the switch does take several hours, during which time salmon and eels stop swimming.

QUESTIONS

11.8 Freshwater Osteichthyes have a large Bowman's capsule enclosing a glomerulus: marine Osteichthyes have no Bowman's capsule or a Bowman's capsule which is very small. Suggest how this difference might affect the volume and composition of urine produced by these two groups of animals. Explain your answer.

11.9 Chondrichthyes (cartilaginous fish) have kidney tubules with well-developed Bowman's capsules and excrete urea. The urea is retained in the bloodstream in such high concentrations that the water potential of their blood is almost the same as that of sea water.

(a) Explain how the above adaptations might enable Chondrichthyes to excrete large volumes of urine without being in danger of dehydrating.

(b) Chondrichthyes also possess a rectal gland which actively removes sodium ions from the blood and secretes them into the faeces. Explain the advantage to the Chondrichthyes of the activity of this gland.

11.10 How do insects produce concentrated urine and excrement?

11.5 CONTROL OF BLOOD GLUCOSE

Glucose in the blood provides the cells of a mammal's body with an energy source. If the concentration of blood glucose falls, cell respiration will be slowed and cells will die. Cells in the brain are particularly susceptible and a lack of glucose causes coma. On the other hand, too high a blood glucose concentration makes the water potential of blood more negative, causing cells to lose water by osmosis. A concentration of 80–100 mg of glucose per 100 cm^3 of blood is maintained in mammals by a homeostatic mechanism involving the pancreas and the liver.

The pancreas

As well as its digestive role, described in Section 8.4, the pancreas is an endocrine gland. Scattered among the cells which secrete digestive enzymes are groups of different cells, called **islets of Langerhans**. Within these are large **alpha cells** (α cells), which are sensitive to the concentration of blood glucose and secrete **glucagon**, and small **beta cells** (β cells), which are also sensitive to the concentration of blood glucose but secrete **insulin**. These two hormones have opposite effects on the blood glucose concentration, which are summarised in Fig 11.27.

Glucagon

Glucagon is a polypeptide, 29 amino acids long. Its secretion by the pancreas is stimulated by a fall in the blood glucose concentration (**hypoglycaemia**). As it passes in the blood through the liver, it binds to receptors on the surface membrane of liver cells and stimulates

- an increase in the rate of breakdown of glycogen to glucose
- an increase in the rate of conversion of amino acids and glycerol to glucose-6-phosphate (gluconeogenesis).

As a result, glucose and glucose-6-phosphate are produced and released into the blood. Other hormones, including adrenalin, also increase blood glucose concentration (see Chapter 14).

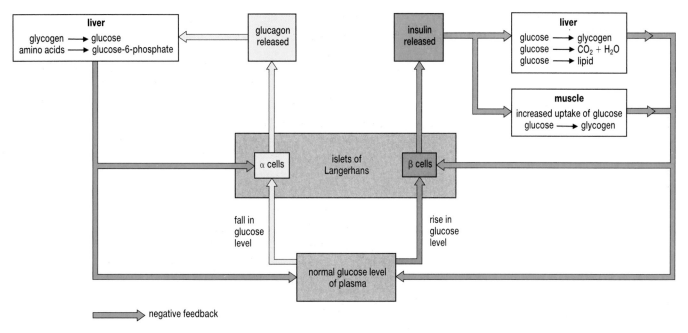

Fig 11.27   The dual control of blood glucose concentration. The release of glucagon from α cells and insulin from β cells is controlled by a negative feedback of glucose from the liver (and muscle) present in blood plasma.

Fig 11.28   Gary Mabbut, captain of Tottenham Hotspur, has diabetes mellitus, but clearly leads a normal active life!

REGULATION OF BODY FLUIDS IN ANIMALS: HOMEOSTASIS

Insulin

Insulin is a protein, 51 amino acids long. Its secretion is stimulated by a rise in blood glucose concentration (**hyperglycaemia**) and inhibited by hypoglycaemia. When bound to receptor sites on the surface membranes of cells throughout the body it stimulates

- an increase in the rate of cell respiration, using glucose as respiratory substrate
- an increase in the rate of conversion of glucose to glycogen (glycogenesis) in liver and muscle cells
- an increase in the rate of absorption of glucose by muscle cells
- conversion of glucose to fat in adipose cells.

As a result of insulin activity, glucose is removed from the blood and used or stored as glycogen or fat. Insulin is the only hormone which can cause this reduction in blood glucose concentration.

11.6 HOMEOSTATIC ROLE OF THE MAMMALIAN LIVER

Structure of the liver

The liver is the largest organ in a mammal's body, making up about 5% of its body mass, and is involved in many metabolic processes. The position of the human liver in shown in Fig 8.6. Like any organ in the body, the liver is supplied with oxygenated blood from an artery which arises from the aorta, in this case the hepatic artery. The liver also receives blood directly from the gut via the hepatic portal vein (see Fig 10.14). Within the liver, branches of these two vessels lead into channels, called **sinusoids**, which drain into branches of the hepatic vein. The cells of the liver, called **hepatocytes**, are arranged in single rows along these sinusoids (Fig 11.29). As blood flows through each sinusoid to the central branch of the hepatic vein, materials are removed from and added to it by the hepatocytes. Between rows of hepatocytes are small channels, called **canaliculi**, into which the hepatocytes secrete bile. Notice in Fig 11.29 that the sinusoids, carrying blood, are lined by flattened endothelial cells. Among these are **Kupffer cells** which ingest particulate matter, including foreign organisms, that have entered the blood via the intestines.

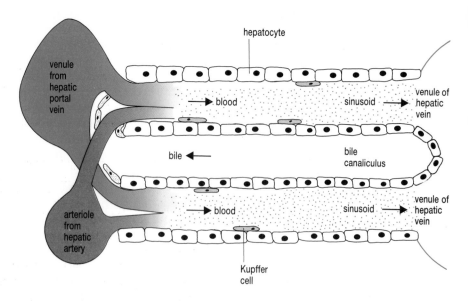

Fig 11.29   Structure of the mammalian liver. Materials circulating in the blood are conducted through the liver in the broad sinusoid channels, and exchanged with materials in the rows of hepatocyte cells which secrete bile into the bile canaliculus. The Kupffer cells ingest foreign particles.

Functions of the liver

Though relatively simple in appearance, the hepatocytes carry out most of the various functions of the liver. The major homeostatic functions of the liver are listed and explained below.

Carbohydrate metabolism

All the hexose sugars absorbed from the small intestine are converted to glucose in the liver. Under the influence of insulin, the hepatocytes convert glucose to glycogen, which they then store. Under the influence of glucagon, the hepatocytes break down stored glycogen to release glucose into the blood stream. The action of insulin and glucagon is described in Section 11.5.

Lipid metabolism

The liver of a human adult can store about 100 g of glycogen. Carbohydrate in excess of this amount is converted by the hepatocytes into lipids. The hepatocytes can also remove certain lipids from the blood, including cholesterol.

Protein metabolism

The liver produces a number of plasma proteins. The hepatocytes can also convert one amino acid to another, a process known as **transamination**. In this way, temporary dietary deficiencies of non-essential amino acids can be remedied. A major role of the liver is in the **deamination** of surplus amino acids, which cannot be stored. As its name suggests, deamination is the removal of the amine group (NH_2) from an amino acid. It must be removed before the carbohydrate residue can be stored. The amine group is removed with a hydrogen atom, so that the nitrogenous waste product ammonia is formed:

$$2NH_2CHRCOOH \quad + \quad O_2 \quad \rightarrow \quad 2CROCOOH \quad + \quad 2NH_3$$

$$\text{amino acid} \qquad \text{oxygen} \qquad \text{carboxylic acid} \quad \text{ammonia}$$

The bulk of this ammonia is converted by the hepatocytes into urea, which is then excreted in the urine. The production of urea occurs by a cyclic series of reactions called the **ornithine cycle**, which is summarised in Fig 11.30.

Storage function

Apart from their ability to store glycogen, hepatocytes store

- the lipid-soluble vitamins A, D, E and K
- the water-soluble vitamins of the B group and vitamin C
- the inorganic ions iron, potassium, copper, zinc and cobalt.

Formation and breakdown of erythrocytes

The red blood cells of a fetus are made by its liver. Only around the time of birth does this role transfer to the bone marrow. The breakdown of red blood cells by the Kupffer cells continues throughout life, however.

Production of bile

Bile is produced by the hepatocytes and released into the bile canaliculi (shown in Fig 11.29). Bile comprises mainly water with less than 1% of each of dissolved bile salts, inorganic salts, bile pigments and suspended cholesterol. From the canaliculi, the bile drains into the gall bladder and is released during digestion (see Section 8.3).

Breakdown of hormones

The liver breaks down some hormones very rapidly, e.g. testosterone; others it breaks down more slowly, e.g. insulin. All hormones are broken down by the liver, however.

Detoxification

The major toxin which is eliminated in the liver is ammonia produced from the deamination of surplus amino acids. The production of urea, a less harmful substance, is described in Fig 11.30. Other toxins which are made harmless include those released by pathogens, alcohol, nicotine and a variety of drugs. Excessive intake of certain toxins can cause fatal damage to the liver, e.g. **cirrhosis** is a fatal disease in which cells of the liver are damaged by excessive alcohol and is common among alcoholics.

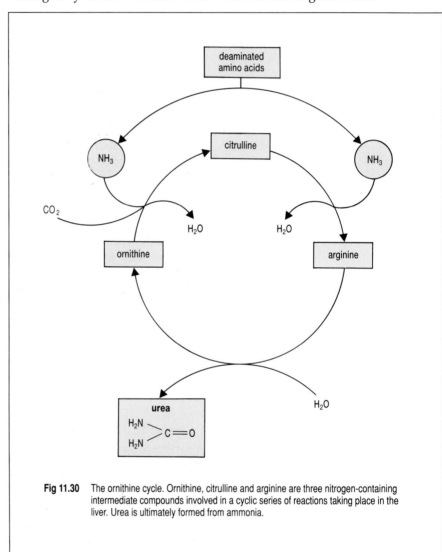

Fig 11.30 The ornithine cycle. Ornithine, citrulline and arginine are three nitrogen-containing intermediate compounds involved in a cyclic series of reactions taking place in the liver. Urea is ultimately formed from ammonia.

Heat release

Energy transfers in cells are not fully efficient and much energy is lost as heat. Some physiologists argue that the activity of the liver is so great that it must be a major source of heat in the body. More recently, other physiologists have questioned this view. Whilst the hormone adrenalin, secreted during times of stress (see Chapter 16) stimulates the breakdown of glucose in the liver, releasing waste heat, it is suggested that at other times as many liver reactions are endothermic as are exothermic, so that the liver might be thermally stable.

11.11 Summarise the interaction between the products and activities of the pancreas and liver.

11.12 In Britain, it is illegal to drive a motor car if the concentration of alcohol in the driver's blood is over 80 mg 100 cm^{-3}. Suspected motorists can elect to give either a blood sample or urine sample for analysis. The upper legal limit is 107 mg of alcohol 100 cm^{-3} of urine.
  (a) Use this information to describe the interaction between the liver and kidney in dealing with toxins.
  (b) Suggest why the upper legal limit is greater for urine than for blood.

SUMMARY

Cells in an animal's body function well only within a narrow range of conditions. Homeostasis is the process by which this narrow range of conditions is maintained by controlling the composition and properties of the blood. Most homeostatic control mechanisms involve negative feedback between an effector and a sensor: changes in the blood resulting from the activities of the effector are detected by the sensor which, at a threshold value, causes the effector to act in the opposite way from before.

Homoiothermic animals have physiological mechanisms which maintain their body temperature at about the optimum for their enzymes. Their skin is usually well insulated by fat, fur or feathers and its blood supply can be regulated. Stimulation by the heat gain centre of the hypothalamus of homoiotherms increases the rate of heat generation and decreases the rate of heat loss through the skin when the blood temperature falls below its optimum: stimulation by the heat loss centre of the hypothalamus has the reverse effect when the blood temperature is too high.

Poikilothermic animals depend on environmental heat to maintain their temperatures and lack physiological mechanisms to control temperature.

A mammal's blood water potential is regulated by its kidneys. The functional unit of a kidney is the nephron, consisting of glomerulus and kidney tubule. Ultrafiltration of blood in the glomerulus results in water and dissolved solutes passing into the Bowman's capsule of the kidney tubule. As it passes along the rest of the kidney tubule, selective reabsorption of water and solutes and active secretion of solutes determine the final composition of the ultrafiltrate that is excreted as urine. Antidiuretic hormone and aldosterone are among the hormones regulating selective reabsorption in the kidney tubules. In mammals and in fish, the structure of the nephron is adapted to the environment in which the animal lives. Relatively long loops of Henle are found in desert mammals and marine fish, relatively short loops of Henle and aglomerular nephrons are found in freshwater fish. The Malpighian tubules of insects enable them to produce crystalline urine in which very little water is lost.

Blood glucose concentration is regulated by the activities of the pancreas. Glucagon, released by the α cells of the pancreas, stimulates the liver to raise the blood glucose concentration. Insulin, released by the β cells of the pancreas, stimulates a lowering of blood glucose concentration through the activities of the liver, muscles and adipose tissues. Apart from its role in regulating glucose concentration, the mammalian liver has many other homeostatic functions. These include: conversion of carbohydrates to lipids; deamination of amino acids to form urea; storage of vitamins A, B, C, D, E and K; breakdown of erythrocytes, hormones and toxins.

Answers to Quick Questions: Chapter 11

1 Positive feedback causes the existing change to continue. For example a positive control system on a water bath would detect a temperature rise and cause it to rise still further by switching the heater on.
2 **(a)** return to pH 7.00 **(b)** become even more acidic.
3 Gain from sun by radiation; loss to snow by conduction and to air by convection, radiation and evaporation of water in lungs.
4 They bask in the sun on the banks when cold. Leave bank and go into the water when they are warm.
5 It is an endotherm but is absorbing heat from the environment by conduction and radiation.
6 Larger intake of energy foods needed.
7 Body more readily gains heat from, or loses heat to, environment.
8 More blood passes through connecting vein and less through skin capillaries in a given time. As a result less heat will be lost through skin.
9 Air trapped within hairs provides extra insulation.
10 Loop of Henle and collecting ducts lie in the medulla, Bowman's capsule, first and second convoluted tubules lie in the cortex.
11 See diagram

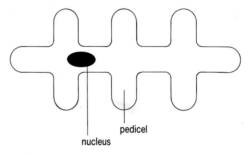

12 Body enzymes only work within a narrow range of pH.
13 Long loops of Henle absorb more water than short loops. Desert mammals must conserve water; freshwater mammals do not need to conserve water.
14 Increases volume of urine produced.
15 Freshwater fish which continually gain water from their surroundings; they produce copious quantities of urine which can dilute potentially toxic ammonia.
16 Larvae produce copious quantities of dilute urine; adults produce a little concentrated urine.

Chapter 12

TRANSPORT IN PLANTS

> **LEARNING OBJECTIVES**
>
> When you have studied this chapter you should be able to:
>
> 1. outline the need for transport systems in plants;
>
> 2. recognise the tissues which transport water, inorganic ions and organic compounds in flowering plants and describe their distribution in roots, stems and leaves;
>
> 3. describe the absorption of water and inorganic ions in roots and their transport within flowering plants;
>
> 4. interpret data relating to the loss of water from leaves;
>
> 5. describe and evaluate theories to explain the movement of organic molecules in flowering plants;
>
> 6. relate the structure of a number of plant transport tissues to their function.

Imagine a large tree during a warm summer's day. Its photosynthesising leaves need certain substrates to make organic compounds. In turn, these organic compounds serve as substrates for respiration or assimilation in other parts of the plant. The movement of metabolites in this plant is summarised in Fig 12.1.

Adequate exchange of gases can occur by diffusion since respiring and photosynthesising cells are only a short distance from the atmosphere. Gas exchange is described in further detail in Chapters 7 and 9.

Organic compounds made during photosynthesis (**photosynthate**) must be moved from leaves to other organs. Photosynthetic cells need water and inorganic ions, both of which are usually available only in the soil. Hence, movement of water and inorganic ions from the roots to sites of photosynthesis is essential. Diffusion is an adequate way of moving metabolites only if the distance from source to destination is short. Although the distance between a photosynthesising cell in a leaf and carbon dioxide in the atmosphere is short, this cell is a long way from the soil water and inorganic ions which the cell also needs for photosynthesis and assimilation. The diffusion of water (**osmosis**) and inorganic ions would not be sufficient to supply this photosynthetic cell's needs if it were more than a few centimetres from the soil. Similarly, although a root epidermal cell is in contact with soil water and soil air, it is a long way from photosynthesising cells in the leaves. The diffusion of sucrose, amino acids etc. over this distance would not be sufficient to meet the metabolic needs of this root epidermal cell. Plants can only grow to a height of more than a few centimetres if they have efficient transport systems that overcome the above problems.

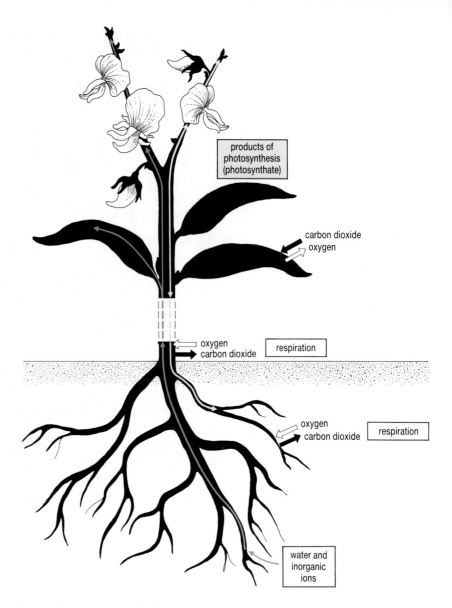

Fig 12.1 Movement of metabolites through the body of a flowering plant. Photosynthate moves from the leaves to other organs. Water and inorganic ions are transported from the roots. Oxygen and carbon dioxide can move by diffusion and are not transported large distances within the plant.

12.1 TRANSPORT SYSTEMS IN FLOWERING PLANTS

Unlike animals, plants do not transport water, dissolved inorganic ions and organic molecules together in a single transport system. Water and dissolved inorganic ions are transported upwards in a plant mainly in hollow tubes formed by dead cells in the **xylem**. Dissolved inorganic ions and organic molecules are transported upwards and downwards in the cytoplasm of specialised cells within the **phloem**.

Xylem

Xylem tissue of angiosperms (flowering plants) contains the two types of water-conducting cell, tracheids and vessels, shown in Figs 12.2 and 12.3. When mature, cells of both types are dead, consisting only of a cell wall. Water and dissolved inorganic ions can pass through these empty cells without being slowed down by cell contents. Lignin deposited in their secondary cell walls adds mechanical strength to both types of cell so that they give support to the plant (see Chapter 18) and are able to withstand the strong pressures which develop during water transport. Small areas occur over their walls which lack a secondary wall and consist of middle lamella and primary cell wall only: these **pits** allow rapid movement of water and dissolved inorganic ions from cell to cell. Xylem which develops just behind the growing tips of the stem and root is called **primary xylem**. As woody plants increase in girth, they produce rings of new xylem, called **secondary xylem**. The formation of secondary xylem is described in Section 21.5.

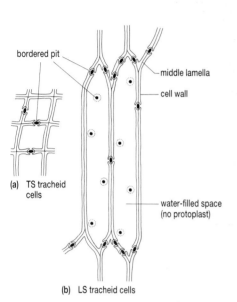

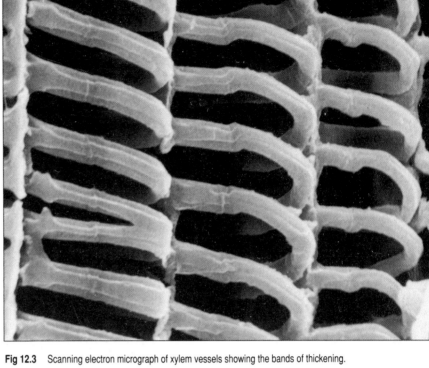

Fig 12.2 Tracheid cells. Pits through the cell walls allow rapid movement of water and dissolved ions from one cell to another. Lignin in the cell wall gives mechanical strength to the plant.

Fig 12.3 Scanning electron micrograph of xylem vessels showing the bands of thickening.

 The inside of a tree trunk may be many centimetres, indeed metres, from the atmosphere. Suggest why trees do not possess a special gas exchange system to supply the needs of this tissue.

A **tracheid**, shown in Fig 12.2, is an elongated cell with tapering ends. When compacted together in groups, these tapering ends overlap those of cells above and below, adding to the strength of this tissue. **Xylem vessels** are formed by differentiation of cells called **vessel elements**. Before the death of their contents, vessel elements grow end to end to form stacks. On the death of each cell, the end walls between individual elements disintegrate, forming a continuous vessel. This vessel is like a drainpipe and allows the unimpeded transport of relatively large volumes of water. The xylem vessels in Fig 12.3 have been lignified and show spiral thickening. **Pitted vessels** are uniformly lignified except at the small pores which are seen as pits, **reticulated vessels** are thickened by interconnecting bars of lignin. Vessels are clearly visible protruding from a piece of celery petiole which has just been bitten (Fig 12.4). The cell walls of these vessels are thickened by rings and spirals of lignin which allow them to stretch. Such vessels are only found in short-lived structures, such as the petiole (stalk) of a leaf, or in the growing tip of a root or shoot, where they form **protoxylem** (first-formed xylem). In more permanent plant structures, these vessels eventually collapse and are replaced by the pitted and reticulated vessels of the **metaxylem**.

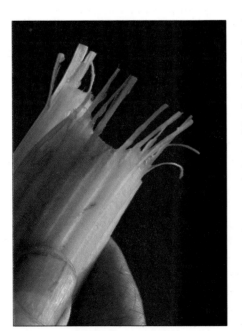

Fig 12.4 Xylem tissues are quite tough. They account for the stringyness of this celery.

 Why is it advantageous to the plant to replace protoxylem by metaxylem?

Phloem

Phloem tissue contains **sieve tube cells** which transport organic compounds. These cells lie end to end to form a continuous stack (Fig 12.5). The thin cellulose wall at the ends of these cells is perforated to form **sieve plates** that allow the cytoplasm from one cell to run into an adjacent cell. **Phloem protein** is present in the cytoplasm running through the sieve plates. Smaller strands of cytoplasm (plasmodesmata) run through the side walls of the sieve tube cells into the adjacent **companion cells**. These smaller companion cells do not transport organic materials but control the activity of the sieve tube cells. This is an important function since sieve tube cells lose their nucleus and cell organelles as they mature.

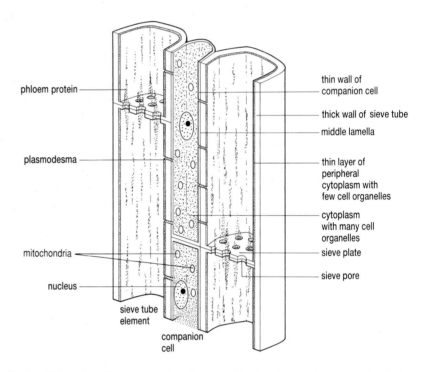

phloem protein

plasmodesma

mitochondria

nucleus

sieve tube element

companion cell

thin wall of companion cell

thick wall of sieve tube

middle lamella

thin layer of peripheral cytoplasm with few cell organelles

cytoplasm with many cell organelles

sieve plate

sieve pore

Fig 12.5 Phloem tissue. The wide sieve tube cells are joined by sieve plates which have some strands of phloem protein running through their pores to form stacks of cells, called sieve elements. The associated companion cells contain many cell organelles and are connected by plasmodesmata (singular plasmodesma) to the sieve tube elements.

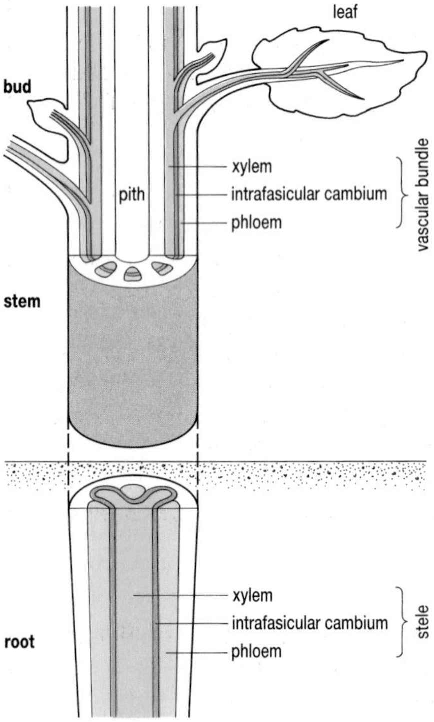

leaf

bud

pith

xylem

intrafasicular cambium

phloem

vascular bundle

stem

xylem

intrafasicular cambium

phloem

stele

root

Fig 12.6 The location of vascular tissue in the major organs of a flowering plant. In the stem of a dicotyledonous angiosperm, each vascular bundle comprises a zone of phloem tissue on the outside, xylem tissue on the inside, with cambium in between. The arrangement of vascular tissue in the root is slightly different as the xylem forms the central core of tissue.

Vascular tissue

Xylem and phloem cells form **vascular tissue** which is found throughout the body of a flowering plant (Fig 12.6). Other types of cell are also found in the vascular tissue. Between the xylem and phloem of dicotyledonous angiosperms is **intrafasicular cambium**, tissue which is able to divide to produce new xylem and phloem in plants, so increasing the diameter of their roots and stems. **Parenchyma cells**, which store food and allow lateral movement of solute through the vascular tissue, and cells which provide support, such as **fibres** and **sclereids**, are also found in vascular tissue. The vascular tissue is arranged in a plant in such a way that its cells provide maximum support for the plant organs. Vascular tissue forms a central **stele** in the root but peripheral **vascular bundles** in the stem (see Fig 18.21). In leaves, vascular tissue forms a central **midrib** from which smaller veins run through the blade. Section 18.5 describes the mechanical advantage of these arrangements.

Evidence that transport occurs through xylem and phloem

This exercise gives you practice in analysing and drawing conclusions from experimental results.

Read the following account of experiments on plant transport and then answer the questions at the end.

1. Ringing experiment

The active phloem of a woody stem is located on the inside of the bark; the xylem is located in the wood. Hence, removal of a complete strip of bark from around a woody stem, known as **ringing**, removes phloem but leaves xylem intact. Ringing experiments show that the movement of water and inorganic ions from roots to leaves is unimpeded by removal of the bark, but organic compounds cannot be transported past the region where bark has been removed. The more difficult operation of removing a section of xylem, leaving a water-filled cavity inside intact phloem, has also been performed. The transport of organic compounds is unimpeded by such rings and significant amounts of inorganic ions are also transported across them.

2. The use of ^{14}C-labelled sucrose

Radioactive isotopes have been used to further investigate the site of transport in plants. We must assume that plants transport sucrose molecules containing radioactive carbon atoms (^{14}C) in the same way as those containing only non-radioactive carbon atoms (^{12}C). Autoradiographs of sections through plant tissue show that radioactive sucrose is carried in the sieve tube cells of the phloem but not in the xylem.

The result of an experiment using autoradiographs of a whole plant is shown in Fig 12.7. It shows the movement of radioactive carbon. Used in conjunction with ringing experiments, this technique can show the effect of removing phloem on transport of fixed ^{14}C.

3. The use of radioactively labelled ions

If plant roots are treated with solutions containing the radioactive isotopes of inorganic ions (e.g., $^{42}K^+$, $^{32}PO_4^{3-}$, $^{82}Br^-$ and $^{24}Na^+$), subsequent autoradiographs show radioactivity in both the xylem and phloem. If xylem tissue and phloem tissue in the lower stem are separated by paraffined paper and a solution containing radioactive inorganic ions is applied to the roots, radioactivity is found in the lower stem only in the xylem. Higher up the stem, where the xylem and phloem have been left in contact, radioactivity is found in both xylem and phloem. Applications of solutions containing radioactive inorganic ions to the leaves result in radioactivity within the stem. However, radioactivity is now found in the phloem but not the xylem.

4. Experiments using aphid stylets

The use of feeding aphids enables an analysis of the contents of phloem sieve tube cells. These insects feed on plants by piercing a leaf or stem with their sharp pointed mouthparts, called **stylets**. When feeding aphids are anaesthetised and carefully severed from their stylets, sections through the plant show that the stylets always pierce an individual phloem sieve tube cell. For several days after a feeding aphid has been cut from its stylet, liquid droplets form at the severed end of the stylet and these can be collected using a fine capillary tube. Chemical analysis of this fluid shows it to be an alkaline watery solution

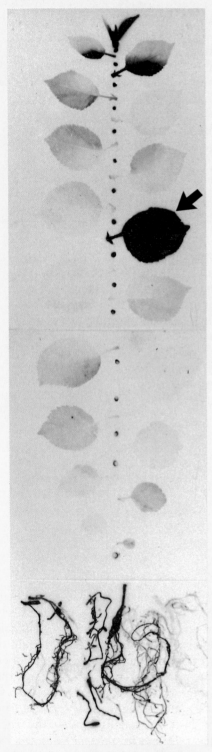

Fig 12.7 Autoradiograph of a rooted apple shoot supplied with $^{14}CO_2$ at the leaf marked by the arrow for 20 mins.

containing a mixture of organic compounds and inorganic ions. Up to about 90% of the organic solute is sugar (mainly sucrose) and up to about 12% is amino acids. ATP, proteins (including enzymes), hormones, alkaloids, vitamins and herbicides may also be found in lower concentrations.

(a) Which item(s) of experimental evidence support the following statements?
  (i) Photosynthate is transported in phloem but not in xylem.
  (ii) Inorganic ions are transported in both xylem and phloem.
  (iii) Transport of inorganic ions in xylem is unidirectional but in phloem is bidirectional.
  (iv) Transport of inorganic ions from the roots occurs initially in the xylem but lateral transfer to phloem tissue occurs.

The tissue behind developing buds contains no differentiated cells. As cells differentiate, phloem cells are produced before xylem cells.

(b) Suggest how inorganic ions from the root travel to the developing bud.
(c) Suggest how your explanation could be tested experimentally.

QUESTIONS

12.1 Members of the plant phylum Bryophyta (Chapter 26) lack xylem, phloem or any waterproof covering. The members of one group of bryophytes, the mosses, are generally less than 2 cm high; members of a second group of bryophytes, the liverworts, have a body which is thallus-shaped (flat). Use this information and your knowledge from reading Section 12.1 to suggest why bryophytes might be
(a) restricted to growing in damp environments
(b) able to grow successfully despite the absence of transport tissue.

12.2 Draw simple diagrams to show the position of xylem and phloem
(a) in transition from root to shoot
(b) in transition from shoot to petiole of a leaf.

12.2 TRANSPIRATION

Terrestrial organisms lose water by evaporation from the parts of their body in contact with air. Although most terrestrial plants have a relatively waterproof covering over their stems and leaves, water loss is inevitable through their permeable gas exchange surfaces (Section 7.2). The bulk of this water loss occurs by diffusion of water vapour through the open stomata of the leaves and is called **transpiration**. Water is lost through the stomata of leaves (Fig 12.8). Water in the walls of spongy mesophyll cells is in contact with air. Consequently, some water in the walls evaporates. The latent heat of vaporisation needed for this evaporation may cool the plant on a hot day. The resultant water vapour diffuses into the air spaces of the mesophyll. If the stomata are open and the air outside the leaf holds less water vapour than the air in these air spaces, a water potential gradient exists from inside to outside the leaf. Therefore, diffusion of water vapour from the leaf to the atmosphere occurs.

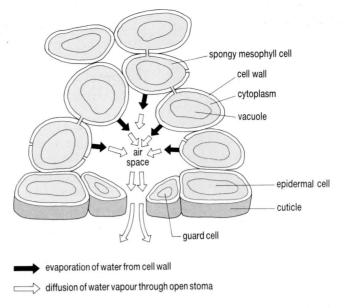

evaporation of water from cell wall

diffusion of water vapour through open stoma

Fig 12.8   Water loss through the stoma of a leaf. By the process of evaporation, water is lost from the cell wall and, as water vapour, it is lost from the leaf by diffusion through the open guard cells of a stoma.

Measuring transpiration rates

A **potometer** is used to measure transpiration rates (Fig 12.9). It contains a leafy shoot, the stem of which was cut and fixed into the potometer under water to stop any air entering its xylem vessels. All the joints are watertight. Assuming that the shoot does not use water for any metabolic process, the rate at which it takes up water from the potometer is the same as the rate of transpiration. This rate can be measured by following the movement of an air bubble along the graduated scale of the potometer over a measured time interval. The actual volume of water taken up can be found if the volume of the graduated tube corresponding to each division is determined.

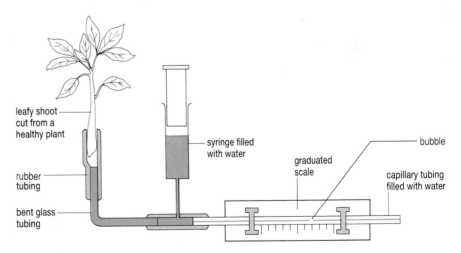

Fig 12.9   A simple potometer. This apparatus must be set up with watertight joint otherwise the bubble of air inside the capillary tube will move in the direction of the leak.

Environmental factors affecting transpiration rates

During transpiration, diffusion of water vapour occurs along a water potential gradient. Its rate is affected by

- the temperature, which affects the rate of random thermal movement of water molecules

Fig 12.10 The effects of transpiration. The water vapour produced by the transpiration of the plants has condensed on the glass.

- the difference in water potential between the atmosphere and leaf air spaces
- the state of the stomata (i.e. whether they are fully open).

Temperature

High temperatures increase the rate of transpiration in two ways.

1. Heat energy can provide the latent heat of vaporisation of water, causing an increase in the rate of evaporation of water from the walls of mesophyll cells.
2. Heat energy also increases the random thermal movement of molecules in the water vapour inside air spaces in the mesophyll.

Thus, high temperatures both increase the water potential gradient between the leaf air spaces and the atmosphere and increase the rate of movement of molecules. Water vapour will, therefore, diffuse out of the leaf faster.

Air conditions

The gradient in water potential between leaf air spaces and the atmosphere is also affected by the **humidity** of the air and by any **air movements** around the leaf. If the air is humid, the water potential gradient out of the leaf is less steep and diffusion will be slower than if the air is dry.

Plant leaves have a thin layer of still air, the **boundary layer**, around them. Since the air is still it remains humid; the boundary layer reduces the rate of diffusion. The thickness of the boundary layer depends upon the wind speed – the faster the wind blows the thinner the boundary layer.

 Explain why air movements increase the rate of transpiration.

ANALYSIS

Water loss from leaves

This exercise involves two important skills: experimental design and the ability to interpret experimental data.

Phaseolus has leaves with a thin cuticle and hairless epidermis. *Pelargonium* has leaves with a thicker cuticle and hairy epidermis. These plants were used to carry out an investigation into water loss from detached leaves by measuring changes in mass of leaves over several hours. The results of the investigations are shown in Fig 12.11.

(a) Describe briefly how you would carry out such an investigation.
(b) (i) Suggest why the percentage loss of mass from both sorts of leaf was rapid between X and Y.
   (ii) Why does the rate of loss decrease between Y and Z?
   (iii) Deduce, giving reasons, whether *Pelargonium* was Plant 1 or 2.

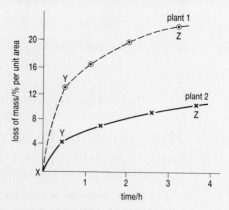

Fig 12.11 Water loss measurements on leaves of two plants.

Fig 12.12 Planting marram grass with an extensive underground rooting system, to stabilise sand dunes. This plant is a xerophyte with features which reduce the rate of transpiration. There are many hairs associated with sunken stomata on the inner side of the leaf which becomes rolled and needle-like.

USING XEROPHYTES

Except in the most humid environments, plants growing in the field are likely to suffer a deficit of water. Such water deficit causes stress, shown by reduction in the rate of photosynthesis and of growth. Plants which are able to tolerate water stress in extremely dry habitats are known as **xerophytes**. Some of the features which enable plants to tolerate these dry habitats are summarised in Table 12.1. Deserts are not the only dry habitats. Areas at high altitude, where water may be frozen for much of the year and where high winds increase evaporation rates, and areas with rapidly draining soil, such as sand dunes, are also inhabited by xerophytes. Marram grass is deliberately planted to stabilise sand dunes; an important part of maintaining coastal defences and conserving rare animals, e.g. natterjack toads.

Table 12.1 Summary of xerophytic features

Advantage of feature	Xerophytic feature	Example
trap water in rapidly draining soil	extensive shallow root network	marram grass (*Ammophila*)
	deep network of roots	date palm (*Phoenix* sp.)
reduce rate of evaporation	thick cuticle	rubber plant (*Ficus*)
	few stomata	holly (*Ilex*)
	stomata sunk into leaf	pine (*Pinus*)
	leaf covered by hairs	marram grass (*Ammophila*)
	leaf very small	all conifers
	leaf rolled with stomata on inside	marram grass (*Ammophila*)
	closing of stomata during day with fixation of carbon dioxide at night to form organic acids which are used as a carbon source during the day when the stomata are closed (crassulacean acid metabolism)	members of the Crassulaceae family
	increased level of abscisic acid which causes closure of stomata	most
water storage	water-storing parenchyma cells in stem	succulents, e.g. cacti
tolerance to water loss	lower solute potential of cytoplasm	wheat (*Triticum*)
	production of denaturation-resistant isozymes	resurrection plants, e.g. *Borya nitida*
	ability to rapidly repair structural damage to cell	resurrection plants, e.g. *Borya nitida*

Light and carbon dioxide

Transpiration occurs more quickly in bright light than in the dark and in an atmosphere with low carbon dioxide concentration than in one with high carbon dioxide concentration. To understand why, we need to look at the way in which stomata open and close.

Stomata

Look carefully at Fig 12.13 which shows the surface of a leaf. The rectangular shapes are cells of the epidermis: they lack chloroplasts and have a uniformly thick cellulose cell wall. The two sausage-shaped cells surrounding each stoma are **guard cells**. These cells possess chloroplasts and the cell wall on their inner curve is much thicker and less elastic than that over the rest of the cell. As a result, these cells expand irregularly when they are turgid. The stomatal pore opens when the guard cells are turgid and closes when the guard cells are flaccid (Fig 12.14). Guard cells become turgid when they gain water by osmosis and become flaccid when they lose water by osmosis. They gain water when the concentration of their

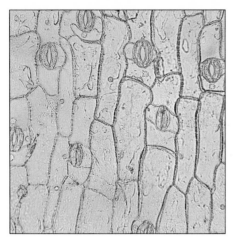

Fig 12.13 Micrograph of the lower surface of a leaf showing epithelial cells, stomata and their guard cells.

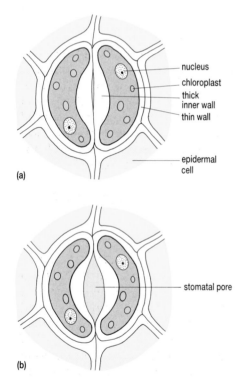

(a)

(b)

- nucleus
- chloroplast
- thick inner wall
- thin wall
- epidermal cell
- stomatal pore

Fig 12.14 The stoma is closed when guard cells are flaccid (**a**), but open when the guard cells are turgid (**b**).

contents rises (their water potential becomes more negative) and lose water when the concentration of their contents falls (their water potential becomes less negative). Various mechanisms have been proposed to explain these changes in the water potential of guard cells.

4 **What will happen to the rate of photosynthesis when the guard cells are flaccid?**

Even under constant environmental conditions, most plants open and close their stomata in an approximately 24-hour rhythm (**circadian rhythm**). Exactly what internal (endogenous) mechanism controls this rhythm is not known. However, under normal environmental conditions the endogenous circadian rhythm becomes synchronised with fluctuations in the environment. A typical pattern of stomatal behaviour during a 24-hour period of 12 hours darkness and 12 hours light is shown in Fig 12.15. The stomata begin to open just before the light period (**night-opening**) and may take up to three hours to open to their maximum size. They remain open for most of the light period (called **steady state opening**) until the onset of darkness when they close very rapidly. They then remain closed until just before daylight. In fact, steady state opening is a misnomer: in temperate regions periodic cloud cover results in temporary stomatal closure and even in the tropics midday closure is common.

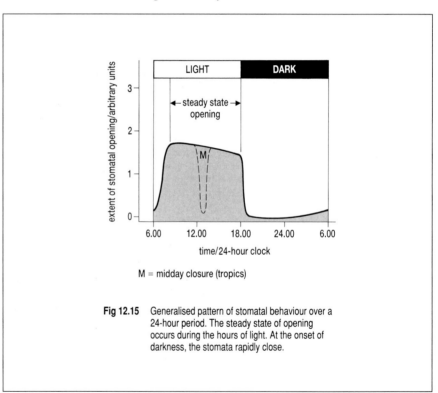

M = midday closure (tropics)

Fig 12.15 Generalised pattern of stomatal behaviour over a 24-hour period. The steady state of opening occurs during the hours of light. At the onset of darkness, the stomata rapidly close.

5 **What advantage do tropical plants gain by closing their stomata at midday? What is the price they pay?**

Temperature, carbon dioxide concentration (inside the leaf rather than in the atmosphere) and air humidity all affect the circadian stomatal rhythm. However, it is obvious from Fig 12.16 that another major environmental factor affecting the stomatal rhythm is light intensity. Light-sensitive pigment, thought to be similar to phytochrome (Section 17.4), is located within the guard cells. The pigment is most sensitive to blue light (of wavelength 464 nm) and stomata are able to respond to intensities of light which are too low for photosynthesis to occur.

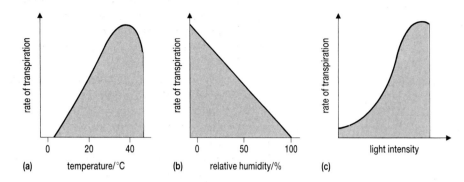

Fig 12.16 Graphs of rate of transpiration in different conditions.
  (a) Increase in temperature leads to an increase in rate of transpiration, but high temperatures reduce this rate.
  (b) As the relative humidity increases, the rate of transpiration decreases.
  (c) Within extremes of light intensity, the rate of transpiration increases as light intensity increases.

The control of stomatal closure

A hypothesis is an explanation which has not been disproved by direct experimental evidence.

Three hypotheses have been put forward to explain stomatal opening and closure.

Hypothesis 1

Since guard cells possess chloroplasts, it was once assumed that they could photosynthesise. An early theory to account for stomatal opening proposed that guard cells would photosynthesise during daylight to produce sugars. Since the presence of sugars within guard cells would make their water potential more negative, water would pass into guard cells by osmosis and cause the stomata to open. However, although the guard cells of most plants do contain chloroplasts, none has been shown to possess all the enzymes for the Calvin cycle needed to produce sugars (Chapter 7).

 Why does this observation allow us to reject this hypothesis?

Hypothesis 2

The chloroplasts of guard cells commonly store starch. In 1908, a diurnal rhythm in the starch content of guard cells was found which showed a correlation with stomatal movement. A theory to link these observations proposed that photosynthesis by mesophyll cells during daylight would remove carbon dioxide from air spaces within the leaf. Since carbon dioxide is an acidic gas, removal of carbon dioxide might raise the pH of the guard cells. The theory suggested that starch-hydrolysing enzymes in the guard cell might work better in such alkaline conditions and would bring about the conversion of starch to sugar. As with the earlier theory, accumulation of sugar would make the water potential of the guard cells more negative, causing a net influx of water into the guard cells and opening of the stomata. A starch-hydrolysing enzyme does indeed occur in guard cells. It is called starch phosphorylase and it catalyses a reversible reaction whose equilibrium point is affected by pH:

$$\text{starch} + \text{inorganic phosphate} \underset{\text{acidic}}{\overset{\text{alkaline}}{\rightleftharpoons}} \text{glucose-1-phosphate}$$

 Why do the observations on the following page invalidate this hypothesis?

Starch is insoluble whilst inorganic phosphate and glucose-1-phosphate are both soluble. Water potential depends upon the total number of molecules in solution not their size.

Hypothesis 3

The currently favoured theory to explain stomatal opening and closure is based on the observation that guard cells actively accumulate high concentrations of potassium ions (K^+) when stomata open. During stomatal closure, K^+ are actively removed from guard cells, reversing the process just described.

 How will these movements of K^+ ions cause the stomata to open and close? Think in terms of changing water potentials and osmosis.

Further studies suggest that the transport of K^+ ions utilises ATP generated by photosystem I of photosynthesis (Chapter 7), providing a role for the chloroplasts within the guard cell.

 Will this ATP be generated by cyclic or non-cyclic photophosphorylation? How could you test your answer given a solution containing intact chloroplasts isolated from guard cells?

An accumulation of K^+ ions in the guard cells might be expected to result in the development of a large potential difference across their surface membranes – the inside becomes more positive with respect to the outside. Such potential differences have not been found and current evidence suggests that there is no single mechanism by which electroneutrality is maintained. In onions *(Allium cepa)*, electroneutrality seems to be maintained by the active uptake of chloride ions (Cl^-) along with the K^+ ions. In other plants, malate accumulates in the cytoplasm of guard cells during stomatal opening. Since it is an anion, malate will balance the positively charged K^+ ions. However, the source of the malate is still uncertain. One possibility is that malate is produced from phosphoenolpyruvate via a series of reactions similar to those in the C_4 Hatch-Slack pathway (Fig 7.19).

QUESTION

12.3 **(a)** Explain what is meant by transpiration.
(b) Look at Fig 12.16 which shows the rate of transpiration of a plant under a number of different conditions. Explain the shape of each curve.

12.3 FROM ROOT TO LEAF

The water leaving a plant leaf by transpiration is replaced by water absorbed through the roots. In tall trees this means a movement of up to 100 m. This section examines how plants achieve this.

Movement of water through the leaf

Water lost from the leaf is replaced using water from the xylem in the leaf's vascular bundles. Look at Fig 12.17 which represents a portion of leaf between a stoma and the xylem in a vascular bundle. It shows three routes by which water can move from the xylem to the rest of the leaf.

Symplast route

Cells lining the air space around the stoma are separated from the xylem vessels by mesophyll cells. Plasmodesmata connect the cytoplasm of one mesophyll cell with that of another, forming a continuous network of cytoplasm with no intervening cell surface membranes. This network is called the **symplast**.

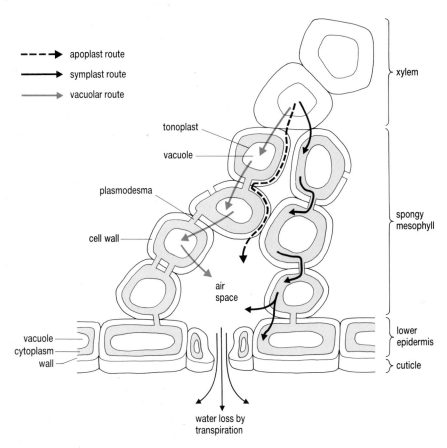

Fig 12.17 Water movement through the leaf. Water is transported by (i) apoplast route (cellulose cell walls) (ii) symplast route (cytoplasm in plasmodesmata) and (iii) vacuolar route (vacuoles and cell surface membranes).

Apoplast route

Air spaces around the mesophyll cells and within their cellulose cell walls form a second network, called the **apoplast**. Since there are no barriers to its movement, water can diffuse freely through the symplast and apoplast.

Vacuolar route

Water is also present in the vacuoles of mesophyll cells. To pass from cell to cell, this water must move by osmosis through the tonoplast (membrane surrounding the vacuole) and through the cell surface membranes separating the cells, which slows its movement.

Movement of water through the xylem in the stem

Three theories have been suggested to explain the movement of water through the stem xylem.

Capillarity theory

If one end of a glass capillary tube is placed in water, the water rises some way up the tube by **capillarity**. Xylem vessels have an internal diameter ranging from 20 μm to 400 μm: as they are stacked end to end (Fig 12.3), they form a capillary tube from root to leaf. Consequently, water will rise up the xylem from the root by capillarity. However, even in the finest vessels, it would only rise about 1 metre which is not far enough to reach

Fig 12.18 Xylem sap rises from the cut stump of a stem, demonstrating root pressure.

Many florists cut the bottom of flower stems to remove blocked xylem vessels and prolong the life of the flowers.

the leaves of even a small tree. So, another mechanism must also be involved.

Root pressure theory

If the stem of a well-watered plant is cut off, water continues to seep from the cut stump. By attaching the cut stump to tubing, as shown in Fig 12.18, the hydrostatic pressure forcing this water from the plant can be measured. This is called **root pressure**. Its value depends on the species of plant, the growing conditions and time of year, but root pressures of up to 200 kPa have been measured. Like capillarity, root pressure is not sufficient on its own to push water to leaves at the top of tall plants.

Transpiration-cohesion theory

Experimental results suggest that water is not pushed to the top of woody plants at all; it is pulled from above. If a column of water is pulled from above, it will be under tension (negative pressure) rather than under pressure. Such tensions have been found in xylem tissue causing the walls of xylem vessels to be pulled inwards when they are transporting water. (Although the reduction in diameter of a single vessel is very small, the reduction in diameter of a tree trunk is large and can be measured using an instrument called a dendrometer.) The pull from above is provided by water evaporating from mesophyll cells in the leaves, i.e. by transpiration. In fact, transpiring twigs have been shown to pull water through their xylem more effectively than a vacuum pump, i.e. they can raise a column of mercury to greater heights than would be supported by atmospheric pressure. This movement of water through the stem is called the **transpirational stream**.

 Suggest how the diameter of a deciduous tree growing in Britain might change during the course of a summer's day. Explain your suggestions.

However, this theory requires that xylem vessels contain unbroken columns of water. Water molecules certainly attract each other very strongly (see Section 1.2), but can they stick to each other so strongly that as molecules diffuse from xylem vessels in the leaf they pull a column of water molecules up the rest of the xylem? Calculations suggest that the tension of the contents of xylem vessels may be as high as 3 MPa. This would be sufficient to pull water to a height of about 150 metres. In other words, it seems that water molecules do stick to each other strongly enough to account for the transpiration stream.

The transpiration-cohesion theory is not without its flaws, however. Any break in the water column ought to stop the flow of water through the xylem. Such breaks in the column might be expected when the plant is shaken by the wind, yet movement through the xylem continues. Similarly, air bubbles introduced into a xylem vessel should stop the flow of water through it. In fact, the breakage of a single twig might be expected to let air into the xylem throughout the whole plant. Clearly, plants survive breakages of twigs. During winter when the contents of the xylem often freeze, dissolved air in the xylem sap is trapped as bubbles in the ice and should remain as bubbles when the sap thaws the following spring. Yet sap flow in the xylem continues in spring. It is thought possible that flow of sap continues around such air bubbles (and around older vessels which have become blocked by resins and gums) through the micropores of the cell walls and by lateral movement to unblocked vessels. Nonetheless, the transpiration–cohesion theory is the best explanation of water movement in large plants to date.

Uptake of water

So far we have accounted for the movement of water through leaves and up the xylem. We now need to know how water is taken up by roots and transported to the xylem in the first place.

The root anchors a plant in the soil and may act as a storage organ. Only a small part of the root, the **piliferous region**, allows water entry. Here the epidermis is freely permeable to water but selectively permeable to inorganic ions: it has **root hair cells** which increase the surface area in contact with the soil. The cell walls of most plant cells are filled with pores which lie between the fibres of cellulose. Water can diffuse from the soil into the walls of the epidermal cells of the root.

 How will this soil water in the walls of an epidermal cell enter its cytoplasm?

Transport of water across the root into the xylem

Transport of water across most of the root occurs in the same way as transport through the leaf, i.e. by the symplast, apoplast and vacuolar routes (Fig 12.19). Root hair cells are separated from the xylem cells in the central stele by several layers of parenchyma cells forming the **cortex**, a single layer of cells in the **endodermis** and several layers of cells forming the **pericycle**. Plasmodesmata connect the cytoplasm of one cell with that of another, forming a symplast in the root similar to that in the leaf. As water enters epidermal cells from the soil, their cell contents are diluted. As a result of this dilution, water diffuses from the cytoplasm of these cells to adjacent cells with cytoplasm of a more negative water potential. Water will continue to move across the root by diffusion through the symplast.

Spaces between root cells and within their cellulose cell walls form an apoplast in the root like that in the leaf. Water can freely diffuse through these spaces. Normally, the water potential of soil water will be less negative than that within the root, so water will diffuse down a water potential gradient until it reaches the cells of the endodermis. Water will also be pulled through the apoplast by the effects of the transpiration stream. Unlike cells in the rest of the root, the walls of endodermal cells have a **Casparian strip**, which completely encircles each cell (Fig 12.20). These Casparian strips contain suberin, a substance which is not permeable to water.

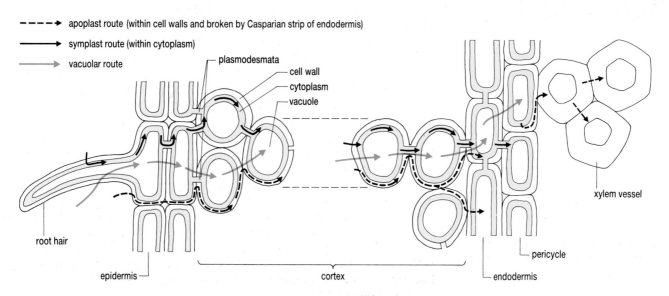

Fig 12.19   Horizontal section through a root. The arrows show three pathways by which water can cross the root (apoplast, symplast and vacuolar).

To travel beyond the strip, water must pass from the apoplast into the cytoplasm of the endodermal cells. This allows active control of the passage of water and any dissolved inorganic ions it contains into the xylem and is thought to be essential for the development of root pressure. From the cytoplasm of the endodermal cells, water may be actively secreted into the xylem or may be pulled through the apoplast by the transpiration stream.

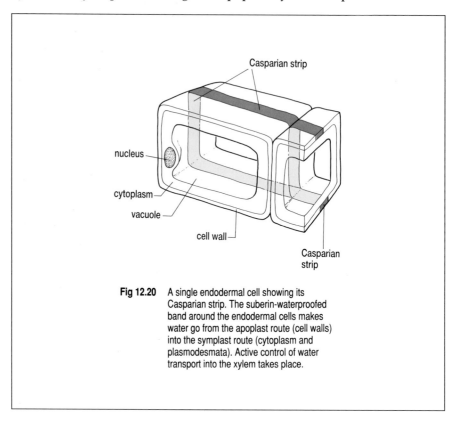

Fig 12.20 A single endodermal cell showing its Casparian strip. The suberin-waterproofed band around the endodermal cells makes water go from the apoplast route (cell walls) into the symplast route (cytoplasm and plasmodesmata). Active control of water transport into the xylem takes place.

QUESTIONS

12.4 (a) Distinguish between symplast and apoplast.
(b) Explain how water movement through the symplast and apoplast occurs.

12.5 When plant stems are cut, droplets of sap from the xylem leak from the cut stump. Suggest why these droplets form.

12.6 How is the transpiration stream in the stem brought about?

12.7 Explain the part played by the Casparian strip in the movement of water across a root.

12.8 The table below gives the relative rates of transpiration of four trees under identical conditions. Suggest reasons to account for the major differences shown.

Tree	Relative transpiration rate
spruce	100
pine	181
beech	286
oak	286

12.4 UPTAKE AND TRANSPORT OF INORGANIC IONS

The biosynthesis of proteins, nucleic acids and so on requires elements such as nitrogen, phosphorus and sulphur which are supplied to the plant in the form of inorganic ions, nitrates, phosphates and sulphates.

Uptake of inorganic ions and their transport across the root

Plants need a number of inorganic ions (sometimes called minerals) for their normal metabolism. The functions of the more important ions are summarised in Table 7.2.

Typically, inorganic ions are taken up from the soil water via the roots. Dissolved in soil water, inorganic ions pass into the walls of cells in the piliferous region of the root. Active uptake of inorganic ions into cells also occurs.

Once in the root, some inorganic ions are retained within the cytoplasm and vacuole or adsorbed to the cellulose walls of the root cells. Most are moved to the xylem, either through the symplast or apoplast route. To enter the symplast route, ions must be actively pumped across the surface membranes of the outer cells of the root. From the cytoplasm of these cells they may pass along the symplast during cytoplasmic streaming through the plasmodesmata. Passage of ions through the apoplast route is passive, partly by diffusion but mainly by being carried along in the water which is pulled through the root by transpiration. When they reach the Casparian strip in the wall of the endodermal cells, ions in the apoplast must cross a cell surface membrane into the cytoplasm of an endodermal cell. This is achieved by active transport. From endodermal cells, ions are actively transported into the xylem with ions that followed the symplast route.

 13 Suggest how the Casparian strip may ensure the selective absorption of ions into the stem.

Transport of ions from roots to aerial parts of plants

Water is moved from the roots through the xylem vessels by the pull of water lost from the leaves during transpiration. Inorganic ions in solution in this water are passively pulled along with it, a process called **mass flow**. Once they reach the organs where they will be used, inorganic ions pass from the xylem vessels along the apoplast of the leaf and are then taken up by cells of the leaf by active transport. Provided xylem and phloem cells are in contact, ions may also pass laterally into sieve tube elements, where they are carried in solution both up and down the stem.

SPOTLIGHT

FEEDING PLANTS

After World War II (1939–45) the Haber process, which converts gaseous nitrogen into ammonia, made nitrate fertilisers cheap and readily available. This led to a revolution in farming, particularly cereal production, which ensured that Britain could produce large quantities of food relatively cheaply. The ecological impact of such a change in farming practice has been enormous, especially when coupled to the extensive use of farm subsidies which allow farmers to produce grain which nobody wants to buy.

In commercial glasshouses, on a smaller scale, farmers can also feed plants by spraying nutrients onto their leaves – so called **foliar feeding**. Such feeds are expensive but they are invaluable for supplying high cost crops with a quick boost of nutrients when they need it most.

Even finer control of nutrient supply can be achieved using **hydroponics**. Here the plant roots grow directly in a solution of nutrients which is continually recycled. By checking the levels of nutrients in the circulating solution, the grower knows exactly how much the plants are absorbing and so which nutrients need to be added to the culture medium.

Fig 12.21   Plants growing under hydroponic culture. The plastic has been cut back to show the roots, normally it would cover them.

QUESTION

12.9 State which of the following you would expect to be slowed by cyanide and explain each of your answers (note cyanide stops ATP production)

(a) uptake of ions from the soil
(b) transport of ions in the apoplast of the root
(c) transport of ions through the symplast of the root
(d) transport of ions via the endodermis of the root.

12.5 TRANSLOCATION OF ORGANIC COMPOUNDS

Organic compounds are made in the leaves of a plant during photosynthesis. Most of these molecules are transported in aqueous solution from the leaves to areas called **sinks** where they are utilised or stored. These sinks include growing regions, such as new leaves, flowers and fruits, and stores, such as seeds, tap roots, stem tubers, corms and bulbs. Thus, organic molecules may be transported in phloem both upwards (to growing leaves and developing seeds) and downwards (to underground storage organs). This contrasts with transport in xylem, which occurs upwards from root to leaves only. Although amino acids and plant growth substances are carried by the phloem, up to 90% of the photosynthate is carried as sucrose. Sucrose is an ideal transport sugar: it is highly soluble yet being relatively unreactive plays little part in cell metabolism and is not used by phloem cells during its transit.

Although evidence that sugars are transported in the phloem stems from the 1930s, there is no agreement as to how this movement occurs. Plant biologists are in agreement about one thing, however: the rate of movement of sugars is much too fast to be the result of diffusion. A number of theories have been proposed to explain translocation in phloem. Many of them are not mutually exclusive and it is possible that translocation occurs differently in different plants. The difficulty of making measurements in translocating phloem tissue without disrupting the process itself has led to a lack of convincing evidence for, or against, the various theories.

Mass flow (pressure flow) hypothesis

This hypothesis, proposed in 1930 by Munch, depends on the movement of water through partially permeable membranes in response to differences in water potential and hydrostatic pressure.

Two imaginary cells, A and B, are shown in Fig 12.22: each is surrounded by a partially permeable membrane. Since each contains a sucrose solution, there is a tendency for water to enter these cells by osmosis and increase the volume of their contents. Cell A contains a more concentrated sucrose solution than cell B, i.e. its water potential is more negative.

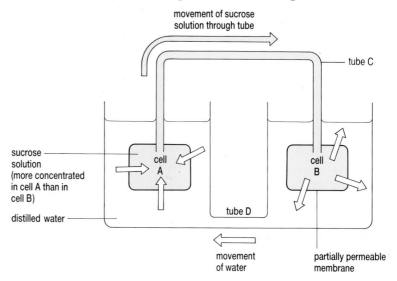

Fig 12.22   A model to illustrate Munch's mass flow hypothesis of translocation in phloem. In response to differences in water potential (caused by the sucrose solutions) water moves through partially permeable membranes; the resulting high hydrostatic pressure in cell A pushes sucrose through the system.

 14   Why will cell A take in water faster than cell B?

Hydrostatic pressure in cell A increases faster than in cell B pushing the sucrose solution through tube C to cell B. This represents the mass flow of solutes. Water will be forced out of cell B by the increase in hydrostatic pressure resulting from mass flow from A to B. Tube D keeps the water system open so that water forced out of B can diffuse back through the system into A. Now refer to Fig 12.23 which shows how this model can be applied to photosynthesising cells, phloem, xylem and sinks in flowering plants. The initial loading of sugars into the phloem and unloading from phloem to cells in the root is thought to be by active transport. Otherwise, movement of sugars would be passive, as in the model.

 15   Which parts of Fig 12.22 correspond to the parts shown in Fig 12.23?

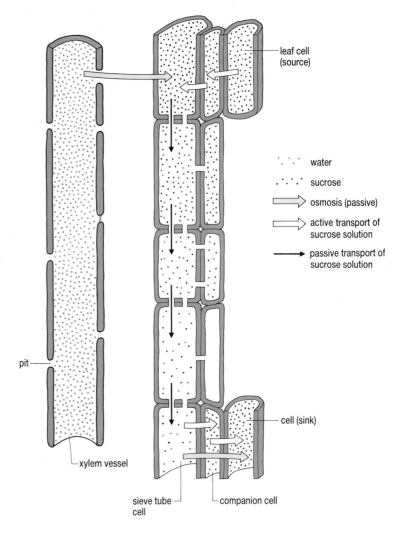

Fig 12.23 Mass flow in action within the essential parts of the vascular system – xylem vessel and sieve tube elements with companion cells.

Evidence to support Munch's hypothesis of mass flow includes the following observations.

- Labelled organic molecules applied to leaves are only transported out of the leaf in the phloem if the leaves are illuminated and, therefore, photosynthesising.
- When translocation occurs, hydrostatic pressures are found within phloem cells.

Evidence against includes other observations.

- The measured hydrostatic pressures are not as high as those calculated to be necessary to drive mass flow.
- The existence of sieve plates between sieve tube elements (see Fig 12.5) hinders mass flow.
- After the initial loading of sugars in the leaf, mass flow is passive. Since phloem tissue has been found to have a high turnover of ATP along its length, phloem is thought to play an active rather than passive role in translocation.

The electroosmosis hypothesis

This mechanism, proposed by Spanner, relies on the fact that

- ions (charged atoms, e.g. K^+) move in an electrical field to the pole with a charge opposite to their own, $+ \rightarrow -$, $- \rightarrow +$
- ions with a like charge repel each other
- ions in aqueous solution are surrounded by a shell of water, i.e. they are hydrated
- water, and its dissolved solutes, e.g. sucrose, which surrounds the hydrated ions is bound to the hydration shell by hydrogen bonds
- when hydrated ions move in an electrical field, water and dissolved solutes will follow such ions.

Electroosmosis is the movement of ions in an electrical field through a fixed, porous surface which is electrically charged, carrying with them water and any dissolved solute.

The sieve plates and phloem proteins (Fig 12.5) are normally negatively charged, thus forming a fixed, porous surface with an electrical charge. As mass flow occurs through the negatively charged sieve plates, anions will be repelled but cations will be able to pass through. When mass flow occurs downwards through the phloem, the repulsed anions will accumulate above the sieve plate so that the cell above the sieve plate will become negative with respect to that below. The sieve plate will now be a fixed porous surface within an electrical field, such as is needed for electroosmosis to occur. When a critical potential difference across the sieve plate is reached, protons (H^+ ions) surge from the wall of the upper cell into its cytoplasm, lowering its pH and making the cytoplasm above the sieve plate positively charged. The increased positive charge generated by the H^+ surge pushes other cations (mainly K^+), by electrical repulsion, through the sieve plate from the upper to the lower cell, i.e. electroosmosis occurs. This surge of hydrated potassium ions carries water molecules and dissolved solutes like sucrose across the sieve plate.

Using ATP from the mitochondria in the companion cell, proton pumps in the surface membrane of the sieve tube cell quickly pump protons out of its cytoplasm and back into its cell wall. Thus, the cell reverts to its original state and the whole process begins again. As a result, solutes would be moved through the phloem by mass flow as proposed by Munch, but the flow would be boosted at intervals by electroosmosis. Spanner's theory provides a role for the companion cells and sieve plates, whilst being compatible with Munch's mass flow theory. However, although high concentrations of K^+ ions have been found in the phloem sap there is no direct experimental evidence to support the theory of electroosmosis.

The cytoplasmic streaming hypothesis

The cytoplasm of plant cells is often observed to move around within the cell, a process called **streaming**. It has been proposed that solutes might be carried from one end of a sieve tube element to the other by streaming and then transferred across the sieve plate by active transport. Both streaming and transfer through the sieve plates would be energy-dependent, explaining the high turnover of ATP in phloem cells. A major problem with this hypothesis is that the observed rates of cytoplasmic streaming are too low to account for the observed rates of translocation.

A second hypothesis involving cytoplasmic streaming was proposed by Thaine in 1962. Using light microscopy, Thaine claimed to have seen cytoplasmic strands 1–7 µm thick passing through the sieve plates of living phloem cells and containing actively streaming cytoplasm. Using electron microscopy, he also demonstrated structures in the phloem of *Cucurbita* which he interpreted as being the cytoplasmic strands he had seen previously. Under the high resolution of the electron microscope, the strands appeared as tubules passing through, and supported by, the pores of sieve plates. Thaine proposed that solutes might be transported in one of two ways: in the streaming cytoplasm outside the transcellular tubules and in the streaming cytoplasm within the transcellular tubules. Thus, translocation could occur separately in both directions inside a single phloem cell.

The main criticism of Thaine's hypothesis is that all other research workers have failed to find the transcellular strands of tubules. Thaine suggested that this is because these structures are extremely fragile but the possibility that they are artefacts, caused by the severe treatment needed before cells can be examined using an electron microscope, cannot be excluded.

The current position

In conclusion, it can be said that none of the existing theories to explain translocation is entirely satisfactory and the mechanism of phloem translocation remains one of the unsolved problems of plant physiology. Part of the problem is that no-one is yet sure whether solutes are moved within the phloem by mass flow or whether they are transported independently, so there is uncertainty about the motive forces involved. Attempts to investigate phloem transport inevitably interfere with the phloem cells and the process of translocation itself. According to the method of preparation and species examined by microscopy, different structures are attributed to phloem cells. These alternative structures are consistent with some, but not all, the hypotheses to account for translocation. Similarly, physiological evidence does not clearly disprove or support any hypothesis.

QUESTIONS

12.10 Some electron microscope studies of sieve plate structure suggest that the pores in the sieve plate are empty. Explain how this observation would affect the feasibility of
(a) Munch's mass flow theory
(b) Thaine's theory of translocation.

12.11 It has not yet proved technically possible to dissect an intact phloem cell from vascular tissue and study translocation through it. Suppose this had been achieved and two different solutes had been found to be translocated in opposite directions at the same time. How might this finding affect each of the following theories of translocation
(a) Munch's theory
(b) the cytoplasmic streaming theory
(c) Thaine's theory?

12.12 Explain how xylem vessels and sieve tube elements are adapted for their functions.

SUMMARY

Flowering plants transport water and inorganic ions in a unidirectional flow through their xylem. They transport organic compounds, water and inorganic ions in a two-directional flow in their phloem. Xylem contains two types of cell, tracheids and vessels (found only in angiosperms). At maturity, these cells are dead. Phloem consists of two types of cell: sieve tube elements, through which transport occurs, and companion cells which control the activities of, and provide ATP to, the cytoplasm of the sieve tube elements.

Water evaporates from mesophyll cells into the air spaces in the leaves and is lost through the stomata during transpiration. As water molecules are lost, they are replaced by further water molecules which are pulled from the xylem. This process depends on cohesion of water molecules to one another by hydrogen bonding holding the water within the xylem as if it were a solid column. This transpirational pull is sufficient to move water within the xylem from the roots to the leaves. The rate of transpiration is increased by such factors as high temperatures, low humidity and wind currents. Plants which can control or reduce transpiration from their leaves are able to inhabit areas with a shortage of free water.

Water is absorbed from the soil by osmosis and travels across the root to the endodermis via the walls, plasmodesmata and vacuoles of root cells. The walls of the endodermal cells are impermeable to anything and so ensure that all water must pass through their cytoplasm where active secretion of water into the xylem occurs. Once inside the root, inorganic ions are carried in solution in the flow of water across the root, into the xylem and up to the leaves. Active transport explains the selective permeability of the root epidermal cells to ions and their secretion into the xylem by the endodermis.

The mechanism of transport within the phloem is poorly understood. The mass flow hypothesis postulates that the high hydrostatic pressures resulting from osmosis into parts of the plant producing sugars push water and dissolved solutes to regions with a lower hydrostatic pressure. The electroosmosis hypothesis postulates a mechanism in which an electrical imbalance between the cytoplasm of adjacent phloem cells pushes hydrated cations from cell to cell. Movement of these cations pulls water and dissolved solutes with them. Streaming of cytoplasm through transcellular strands may also be involved in transport through phloem.

Answers to Quick Questions: Chapter 12

1 Bulk of tree trunk is dead tissue which does not respire; living tissues form a thin band just inside the bark.
2 Metaxylem has pitted and reticulate vessels which increase mechanical strength of plant.
3 Air movements reduce the thickness of the boundary layer around the leaf. This increases the rate of diffusion of water vapour from the leaf.
4 Decrease
5 Prevent excessive water loss; a reduced rate of photosynthesis.
6 Guard cells cannot complete Calvin cycle which produces sugar; hypothesis depends on presence of sugars to make water potential of guard cells more negative.
7 In equation, number of glucose-1-phosphate molecules is equal to number of inorganic phosphate ions (being insoluble, starch is irrelevant). Water potential of guard cells would therefore be the same regardless of the pH.
8 Entry of K^+ into guard cells makes their water potential more negative, water enters by osmosis from surrounding cells which have a less negative water potential, guard cells become more turgid and stoma opens.
9 Cyclic photophosphorylation (only PSI involved). If cyclic expect no evolution of oxygen (and no reduced NADP); non-cyclic photophosphorylation produces both oxygen and reduced NADP.
10 Decrease towards middle of day; water in xylem under tension. Increase later in day as transpiration reduced by stomatal closure.
11 By osmosis through cell surface membrane provided that the

11 By osmosis through cell surface membrane provided that the cytoplasm has a more negative water potential than that of soil water.

12 Blocks movement of water in the apoplast route since suberin is impermeable.

13 Water containing dissolved ions has to enter the cytoplasm of the endodermal cell to avoid the Casparian strip: cell surface membrane has the opportunity to 'select' ions entering the cell during active transport.

14 Cell A contains more concentrated sucrose solution than cell B therefore A has a more negative water potential than B. The greater the water potential difference, the greater the rate of osmosis and this is true of cell A compared to B.

15 A = photosynthesising palisade cell (source)
    B = root cell (sink)
    tube C = phloem tissue
    tube D = xylem tissue.

Chapter 13

THE DEFENCE AGAINST DISEASE: IMMUNOLOGY

LEARNING OBJECTIVES

When you have studied this chapter you should be able to:

1. describe how human skin and mucous membranes help reduce the risk of infection;

2. describe the non-specific inflammatory response to infection and explain how it occurs;

3. define the term antigen and explain how lymphocytes are able to recognise specific antigens;

4. define the term antibody and explain how antibodies are produced;

5. describe the role of blood cells in humoral and cellular immunity;

6. discuss medical implications of the immune system.

Your body is an ideal 'incubator' for the growth of micro-organisms. It contains an abundant supply of water and nutrients and has a constant temperature of about 37 °C. There are countless millions of micro-organisms in the environment around us which can use this 'incubator': they include viruses, bacteria, protoctists and fungi and are present in the air we breathe, in the water and food we consume and on the objects we touch. However, relatively few of these micro-organisms are disease-causing, or **pathogenic**.

The human body has three lines of defence against pathogens. The first is to prevent their entry by defences in the **skin** and in the **mucous membranes** of the respiratory, digestive and urogenital tracts. If this fails, two internal defences exist: an initial **non-specific inflammatory response** to injury which, if unsuccessful, is followed by a second **specific immune response** (Fig 13.1).

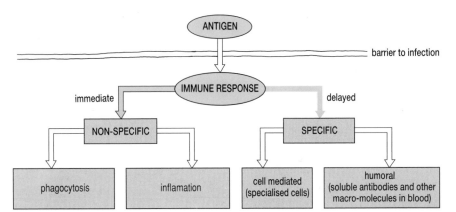

Fig 13.1 The two basic divisions of the immune response: non-specific and specific. An antigen is any molecule which the body recognises as being non-self, e.g. a protein in the coat of a virus.

Immunology is an area of intense scientific activity and there is still much to be learned. How far we have come in the last 50 years can be gauged by the following passage:

> Sir Winston Churchill, in hospital with an infection during World War II, is said to have looked at his hospital chart and asked his physician, 'What are these lymphocytes?' He was told, 'We don't know, Prime Minister.' 'Then why do you count them?' Churchill retorted.

Nature, 30 June 1988

13.1 BARRIERS TO ENTRY OF PATHOGENS

Intact skin is a barrier to the entry of most micro-organisms. Its outer surface, the epidermis, consists of layers of dead cells (Fig 11.7). These cells are dry and filled with a tough, indigestible protein called **keratin**. The lack of water and suitable food inhibits the growth of micro-organisms inside the epidermis. Further protection is given by the **sebum** secreted by the sebaceous glands of the skin. This oily secretion contains fatty acids which are toxic to many micro-organisms.

Areas not covered by skin have alternative defences, which are summarised in Fig 13.2. Tears, saliva and urine contain the enzyme **lysozyme**, which catalyses the hydrolysis of molecules within the cell wall of many bacteria. Mucus, secreted by cells lining the respiratory tract, traps micro-organisms and prevents them penetrating the underlying membranes. Cilia normally sweep this mucus up the respiratory tract into the pharynx, where it is coughed out or swallowed.

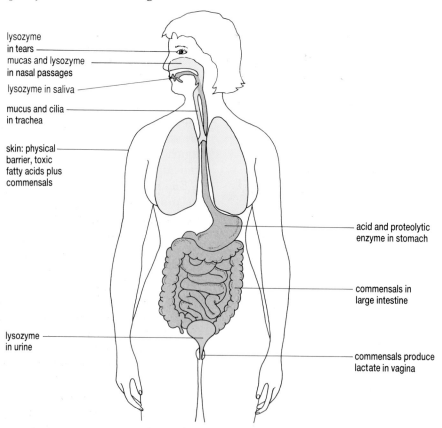

lysozyme in tears

mucas and lysozyme in nasal passages

lysozyme in saliva

mucus and cilia in trachea

skin: physical barrier, toxic fatty acids plus commensals

lysozyme in urine

acid and proteolytic enzyme in stomach

commensals in large intestine

commensals produce lactate in vagina

Fig 13.2 Surfaces which are exposed to pathogens are adapted to form barriers to microbial invasion.

 How would most micro-organisms be killed in the stomach?

THE DEFENCE AGAINST DISEASE: IMMUNOLOGY

Ironically, a vast number of bacteria help us prevent pathogens from entering our bodies. Millions of these bacteria live as harmless commensals in our large intestines, the so-called **gut flora**, where they utilise food which we have been unable to digest ourselves (Section 8.5). By filling all the available ecological niches in our large intestine, these bacteria competitively exclude other, harmful bacteria. (The concepts of ecological niches and competitive exclusion are described more fully in Section 28.5.) Other harmless bacteria live on the surface of the skin and in the vagina. The vagina secretes a carbohydrate on which these bacteria feed, producing lactate as a waste product. This makes the vagina acidic and discourages the growth of many pathogens.

Despite all these defences, mucous membranes are more susceptible to infection than the skin and many pathogens enter the body in this way. Once inside the body, these pathogens are much more difficult to combat: not only must the body distinguish them from its own cells, but it must also destroy them without damage to its own tissues.

QUESTIONS

13.1 What is a pathogen?

13.2 (a) Design and complete a table to summarise the way in which body surfaces reduce the likelihood of infection.

(b) Treatment with a wide spectrum antibiotic (which kills micro-organisms non-selectively) often results in the death of the body's normal gut and skin flora. Suggest why this might be dangerous.

(c) People who have an inadequate immune system, such as those born with severe combined immunodeficiency (SCID) or those who suffer acquired immune deficiency syndrome (AIDS), possess the first and second lines of defence against infection, but lack the final specific immune response. The longest that anyone is known to have lived with such a defective immune system is twelve years. What does this suggest about the relative importance of the three lines of defence against pathogens?

13.2 INFLAMMATORY RESPONSE: THE NON-SPECIFIC RESPONSE

Micro-organisms might gain entry to the body via a cut or similar injury to the skin and underlying tissues. Such injury leads to the formation of a blood clot (described in Chapter 10) which helps 'wall off' the site of injury and prevent entry of micro-organisms or their spread to the rest of the body. Injury also initiates the **inflammatory response**. This is a generalised response to injury and involves local inflammation at the site of injury and destruction of pathogens by the body's phagocytes. These effects are brought about by leucocytes which possess small vacuoles (**lysosomes**) of proteolytic enzymes (**lysozymes**) in their cytoplasm, making their cytoplasm appear granular (Fig 10.8). Because the inflammatory response cannot deal specifically with infection by novel organisms, it is referred to as the non-specific or non-adaptive immune response.

Histamine release

Basophil cells in the bloodstream (Fig 10.8) and **mast cells** (basophil-like cells that lie beneath the skin and around blood vessels) release histamine into the area of a wound. Histamine has no effect within the cytoplasm of basophil and mast cells. Once in the bloodstream, histamine has a number of effects.

- Histamine relaxes the smooth muscle of the arterioles leading to the wounded area, so increasing its blood flow.
- It causes the cells of the capillary walls to draw away from one another, so that the capillaries become leaky. As a result, more plasma escapes from the capillaries and collects in the site of the wound. This fluid causes the site of the wound to swell (oedema) and become warm.
- Histamine also increases the sensitivity of sensory neurones.

Some of the effects of histamine are enhanced by the simultaneous release of other compounds.

- **Prostaglandins** play an important part in causing inflammation and in triggering the sensation of pain: aspirin and many other pain killers work by blocking the synthesis of prostaglandins.
- **Bradykinins** are produced when blood clots or tissues are damaged. Bradykinins cause pain and help to increase the permeability of capillaries.

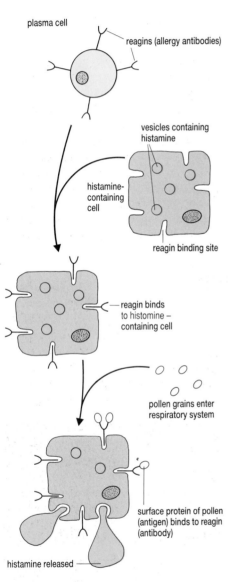

Fig 13.3   An allergic reaction occurs when appropriate plasma cells release antibodies against an allergen. Unlike normal antibodies, these antibodies become attached to cells which contain histamine. When subsequent allergen molecules attach to these antibodies, the carrying cells release histamine, causing local inflammation and symptoms of allergy.

Activity of phagocytes

Other chemicals released by damaged cells attract phagocytic white blood cells. The neutrophils are the first phagocytes to arrive at the site of infection but, with major infections, they are replaced by the longer-lived macrophages. Both types of phagocyte squeeze through the leaky capillary walls (**diapedesis**) and engulf bacteria and debris from damaged cells. Ingested cells are enclosed within a vacuole and enzyme-containing lysosomes fuse with the vacuole, releasing their digestive enzymes onto the engulfed cells. Thus, the phagocytes destroy bacteria in much the same way as an *Amoeba* feeds on bacteria (Fig 8.1). In doing so, they are often killed themselves. The whitish pus which may collect around a wound consists of living and dead phagocytes, together with cell debris and bacteria.

2 Histamine increases the permeability of blood capillaries. How will this assist phagocytosis?

Helping the phagocytes: complement proteins

Since phagocytes do not ingest the body's own living tissues, they must be able to recognise pathogens. In doing so, they are aided by a group of nine proteins, called complement proteins, which are produced by phagocytes themselves and by the liver. These complement proteins circulate in the blood where they are normally inactive. However, in the presence of bacteria and yeasts the complement proteins are activated by a mechanism involving antibodies (see Section 13.3). Activated complement proteins destroy invading microbes in four ways.

- Some complement proteins are deposited in a giant complex in the surface membrane of the foreign cell where they form pores in the membrane, leading to lysis of the foreign cell.
- Some complement proteins enhance inflammation by causing the release of histamine from mast cells and leucocytes.
- Some complement proteins attract phagocytes to the site of infection.
- Some complement proteins interact with receptors on phagocytes, promoting phagocytosis – the process of **opsonization** (Fig 13.4). Each fragment has two reactive surfaces. One of these reactive surfaces chemically combines with molecules on the outside of a bacterial or yeast cell: the other matches receptor sites on the membranes of the phagocytes. As the phagocytes bind to the complement proteins, they engulf the pathogen.

Killing larger pathogens

Phagocytes can only engulf pathogens smaller than themselves. Multicellular parasites, such as parasitic worms, cannot be engulfed. Three other types of granulocyte deal with these parasites: the basophils, eosinophils and mast cells. Like phagocytes, these cells contain small vacuoles (granules) in their cytoplasm. Unlike phagocytes, they release the contents of these granules outside the cell (**degranulation**). Eosinophils have surface receptors which bind to molecules on the surface of parasites; they then

complement protein

polypeptide fragments

attaches to receptor site on bacterial cell surface

bacterial cell

} mirror image of receptor site on surface membrane of phagocyte

Fig 13.4 In the presence of bacteria and fungi, complement proteins break down to form polypeptide fragments with two reactive surfaces. One of these chemically binds with molecules on the outside of the pathogen; the second fits a receptor site on one of the body's phagocytes.

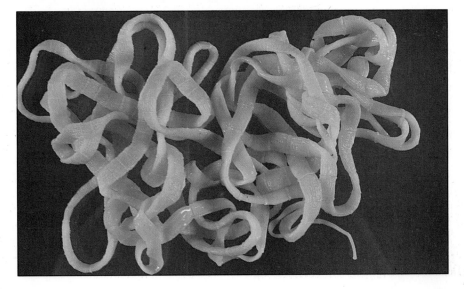

Fig 13.5 Large pathogens cannot be treated in the same way as microscopic ones! Each of the larger segments of this human tapeworm is about 1 cm long.

release an enzyme called **major basic protein (MBP)** which digests the parasite. Mast cells and basophils have a less direct effect on parasites. The contents of their granules stimulate other parts of the immune system, attracting cells or causing vasodilation and increased permeability of capillaries.

ANALYSIS

The order of battle

This exercise tests your ability to select and organise biological information logically.

Using the information contained in Section 13.2, which is summarised in Fig 13.6, rearrange the following list of events, which occur during a non-specific immune response, into their correct sequence.

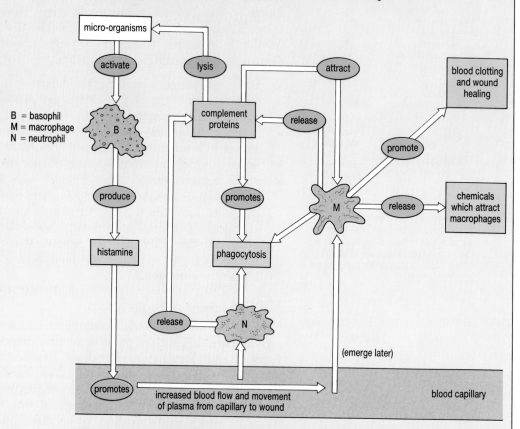

B = basophil
M = macrophage
N = neutrophil

Fig 13.6   A summary of the events occurring during the inflammatory (non-specific) response. Note that the antigen is first detected by a few macrophages and polymorphs present in the tissue. These release chemicals which recruit further cells from the blood system.

(a) Neutrophils and other polymorphs arrive on the scene and begin to engulf foreign material, damaged host cells, debris, etc.

(b) The presence of a micro-organism or other foreign substance (**antigen**) activates the few local phagocytes to release a range of active compounds, e.g. histamine.

(c) Macrophages arrive which, in addition to their phagocytic activity, release a range of substances which damage bacteria, inhibit virus replication and attract more macrophages.

(d) Histamine produces inflammation and increases the permeability of capillaries, allowing more phagocytic cells to emerge from the capillaries in the area of infection.

(e) Substances released by macrophages initiate blood clotting and wound healing.

The inflammatory response is effective only if

- the infection is localised
- the pathogen reproduces slowly
- the pathogen causes a certain amount of tissue damage.

If none of these is true, the pathogen may gain access to the blood stream, rapidly increase in number and endanger the whole body. If this happens, the final specific immune response is directed against the specific pathogen which is causing the infection.

QUESTIONS

13.3 In a few sentences, summarise the contribution made by macrophages to non-specific immunity.

13.4 Outline the part played by granulocytes in dealing with invading pathogens. Under what circumstances is the effect of granulocytes effective?

13.3 SPECIFIC IMMUNE RESPONSE

Any substance that triggers the immune response is called an **antigen**. Relatively small molecules, such as sugars, triglycerides, amino acids and small peptides, do not trigger the immune response. The major antigens are larger and more complex molecules, i.e. proteins, polysaccharides and glycoproteins. Although the basic units of these large molecules are the same throughout all forms of life (Chapter 1), the order in which they are assembled differs according to the genetic code of the organism concerned (Chapter 4).

How will different genetic codes affect the shape of the protein molecules in different organisms?

Self antigens occur on the surface membrane of cells throughout each individual's body. The molecules forming these self antigens are encoded by a series of genes called the **major histocompatibility complex (MHC)** and so are called **MHC proteins**. An individual does not produce antibodies against its own self antigens. **Non-self antigens** occur on the cell wall or surface membrane of all other organisms, including pathogens and cells from other humans (except an identical twin).

Suggest how your body might recognise non-self antigens.

B and T cells

The immune response depends on the activity of the two types of lymphocyte, called **B cells** and **T cells**. Both originate from **pluripotent stem cells** in the bone marrow. Early in embryological development, however, the newly formed T cells migrate to the **thymus** where they mature (hence the name, T cells). B cells probably mature in the bone marrow. Mature B and T cells have quite different functions. In response to infection, B cells divide many times by mitosis to form large **plasma cells**. These produce specific proteins, called **immunoglobulins**, which they secrete into the blood as antibodies. This type of immunity is called **humoral immunity**. T cells do not secrete antibodies; their specific proteins remain attached to their surface membrane. T cells bring about **cellular immunity**. Both types of immunity are described later in this Section.

Humoral
Another name for body fluid is humor, hence humoral means 'of a body fluid'.

AUTO-IMMUNITY

During embryological development, those plasma cells which produce antibodies against self-antigens are destroyed. As a result, a person's immune system does not normally produce antibodies against the body's own cells. However, this process sometimes goes wrong and the body starts to produce antibodies against some of its own antigens. The result is an **auto-immune disease** which destroys the type of cell that carries the appropriate antigen.

A number of common diseases result from auto-immunity. Insulin-dependent diabetes mellitus results from the body's inability to produce insulin (Chapter 11). This type of diabetes arises during childhood after the destruction by antibodies of all the insulin-producing β cells in the islets of Langerhans. Other auto-immune diseases include thyroiditis, pernicious anaemia, glomerulonephritis (the most common cause of kidney failure) and rheumatoid arthritis. As with allergies, there is no effective way to cure auto-immune diseases at present. Some relief can be given by replacement therapy, e.g. injections of insulin or blood transfusions. Suppression of the immune system unfortunately leaves the body unable to defend itself against day-to-day attacks by pathogens.

Recognition of non-self antigens

Both B cells and T cells have receptors on their surface membrane. The receptors on one B or T cell are able to combine with a specific antigen, in much the same way that an enzyme combines with its specific substrate (Chapter 5). Once it has combined with its specific antigen, the B cell produces its response (see Fig 13.7). **Antibodies** are immunoglobulins (**Ig**) which have been released by B cells; their structure is represented in Fig 13.8. Each antibody is a Y-shaped molecule made of four polypeptide chains: a large (heavy) pair and a small (light) pair. Each chain has a **constant region**, whose amino acid sequence is the same (or nearly the same) in each type of antibody, and a **variable region** which is different in each antibody. The nature of the heavy chains in the constant region gives rise to the different types of antibody (**isotypes**) listed in Table 13.1, e.g. α chains are present in the isotype **IgA**, γ chains in the isotype **IgG**, and so on.

 Which other class of molecules exhibits such a high degree of specificity? Suggest a mechanism to account for the specificity of antibodies.

A conservative estimate is that humans can produce several million different antibodies. This creates a problem since humans do not possess sufficient genes to encode this number of molecules. The enormous diversity of antibodies is generated by two mechanisms. Each variable region is encoded by a number of genes. As cells in the developing immune system divide, these genes cross over and genetically recombine (these terms are explained in Chapter 22). They also mutate at a rate which is much faster than in other cells, generating new antibody genes. As a result, each immune cell has its own combination of antibody genes and will produce a different antibody. Thus, there is a huge number of different types of immune cell within the body, each type bearing different surface antibodies and represented by only a few cells.

The surface receptors of T cells are chemically similar to immunoglobulins, but they have two polypeptide chains instead of four and are never released as antibodies. Like those of B cells, these receptors are almost infinitely variable and attach to specific antigens.

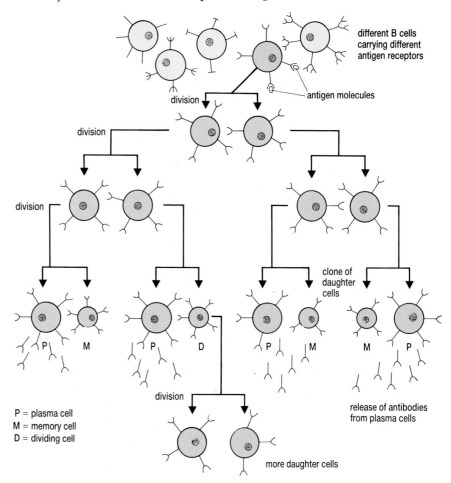

Fig 13.7 The blood contains a vast number of different types of B cell, each with different immunoglobulins on its surface membrane. When an antigen enters the bloodstream it binds to one or a few B cells causing them to divide rapidly to form a large **clone** of daughter cells. The majority of cells in this clone (P) will produce antibodies. Some will become long-lived memory cells (M) which remain dormant in the blood ready to respond to a repeated attack by this particular antigen. Some cells continue to divide (D) to produce more and more B cells.

P = plasma cell
M = memory cell
D = dividing cell

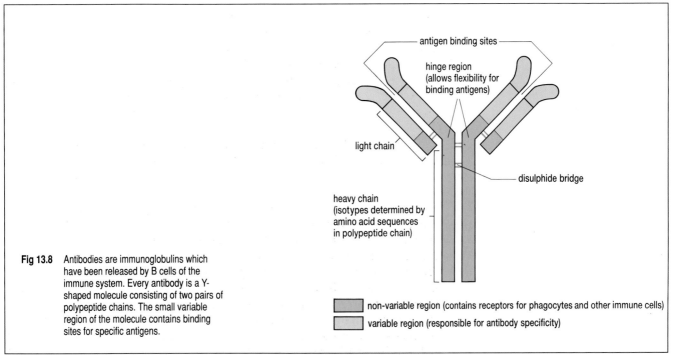

Fig 13.8 Antibodies are immunoglobulins which have been released by B cells of the immune system. Every antibody is a Y-shaped molecule consisting of two pairs of polypeptide chains. The small variable region of the molecule contains binding sites for specific antigens.

THE DEFENCE AGAINST DISEASE: IMMUNOLOGY

Humoral immunity

Humoral immunity is brought about by the activity of B cells. If a pathogen or its toxins enter the bloodstream they encounter numerous B cells. A few of these will carry the appropriate immunoglobulin on their surface membrane and will attach to antigens on the surface of the pathogen or toxin. This attachment has a dramatic effect on the B cell, summarised in Fig 13.7. Given the right signals from T helper cells (see later), the B cell divides rapidly to produce large numbers of genetically identical daughter cells, i.e. a **clone**. Some of these daughter cells develop into **plasma cells** which produce and secrete up to 2000 molecules of their specific antibody per second. Other daughter cells continue to divide, producing more plasma cells. Finally, some daughter cells develop into **memory cells** which remain in the circulation without secreting antibodies.

Antibodies

Circulating antibodies can be grouped into four categories (isotypes) according to the amino acid sequence of their heavy chains. These different isotypes have different effects on antigens and operate in different parts of the body. The isotypes, their effects and site of action are summarised in Table 13.1. These effects include the following.

- **Neutralisation.** The antibody may combine with the active part of a toxin molecule and prevent it from attaching to the surface of body cells. Similar neutralisation may inactivate viruses.
- **Precipitation.** The two binding sites of a single antibody molecule (Fig 13.8) may link with two molecules of antigen to produce the lattice-like immune complex shown in Fig 13.9. In this way soluble antigen molecules may be precipitated and are more easily ingested by phagocytes.
- **Agglutination.** This process is similar to precipitation except that the two binding sites of each antibody attach to antigens on two different pathogens. Agglutination of pathogens into an immune complex seems to help their ingestion by phagocytes.
- **Complement reactions.** The antibody-antigen complex on the surface of a pathogen may trigger a series of reactions in other blood proteins, called the complement system. Some of these proteins are enzymes that directly lyse the pathogen whilst others attract phagocytes (see Section 13.2).

Table 13.1   The main categories of immunoglobulins (isotypes) secreted by B lymphocytes. The antibodies are arranged in order of their concentration in the blood, starting with the highest (IgG has a concentration of about 1000 mg 100 cm^{-3}) and ending with the lowest (IgE has a concentration of about 0.003 mg 100 cm^{-3})

Immunoglobulin	Site of action	Position of antigen	Function (secreted as antibody)
IgG	blood, tissue fluid, across placenta	bacterial surface	causes lysis of cell
		bacteria, viruses	helps engulfment of cells
		viruses and toxins	causes neutralisation
IgA	blood, at epithelial surfaces to form secretions, e.g. saliva, tears and colostrum	viruses and toxin	causes neutralisation
		bacterial adhesions	blocks adhesion of bacteria
IgM	within blood vessels	bacterial surface	causes lysis of cell
IgD	attached to surface of mature B cells	unknown	unknown
IgE	in connective tissue beneath epithelia	allergens,	causes vascular changes
		parasitic worms	directly kills parasites

THE DEFENCE AGAINST DISEASE: IMMUNOLOGY

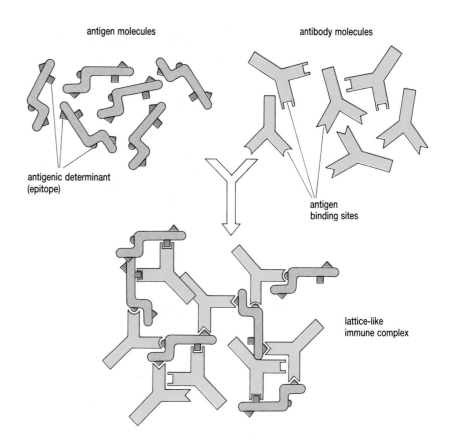

antigen molecules   antibody molecules

antigenic determinant
(epitope)

antigen
binding sites

lattice-like
immune complex

Fig 13.9   Antigen molecules may have more than one site to which antibodies attach, called epitopes. In the diagram, three specific antibodies have attached to the epitopes of one type of antigen molecule to form an insoluble immune complex.

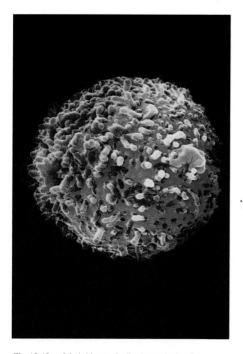

Fig 13.10   A hybridoma. Antibody-producing B lymphocyte cells are fused under laboratory cell culture conditions with some cancer cells. This hybrid clone of B cells grows well and the antibodies it produces are all identical – monoclonal antibodies.

SPOTLIGHT

MAGIC BULLETS?

If we inject a particular antigen into an animal, say a mouse, antibodies will be produced against that antigen by its plasma cells. However, since the antibodies are produced by plasma cells which are the descendants of many different B cells, they vary physically and chemically: they are not pure.

Immunologists thought they could get around this problem by isolating an individual B cell and growing it in tissue culture. The B cells would then divide to produce a clone of identical plasma cells all producing the same antibody. Unfortunately, B cells die after only a few days in cell culture. However, cancer cells grow indefinitely in cell culture. The trick immunologists learned was to fuse individual antibody-producing B lymphocytes with cancer cells. The resultant **hybridoma** grows well in tissue culture and produces antibodies. Since all the members of the resulting cell population are genetically identical, i.e. they form a clone, the antibodies which result are identical **monoclonal antibodies**.

Such monoclonal antibodies are useful in diagnosing allergies and diseases such as hepatitis and rabies. They may also prove useful in fighting diseases, for example cancer, where it may be possible to use them to deliver toxic drugs to specific cancer cells, the proverbial magic bullet!

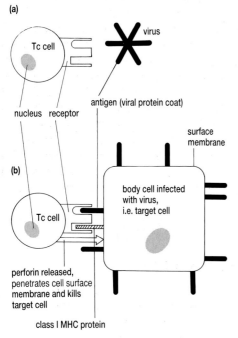

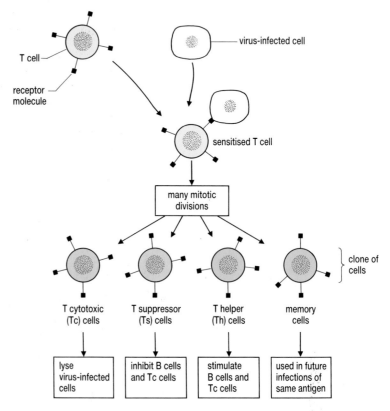

Cellular immunity

Cellular immunity is brought about by the activity of T cells. Like B cells, T cells have highly specific receptor molecules on their cell surface membrane which attach to antigens. Normally, they will only do so if the antigen lies on the surface membrane of one of the body's own cells next to a chemical marker. These chemical markers are examples of the major histocompatibility complex proteins (MHC proteins) described earlier. In humans they are sometimes called **human leucocyte antigens (HLA)**. Two types of these MHC proteins exist: class I proteins are found on the surface membrane of all body cells; class II proteins are found only on B cells and on a group of cells called **antigen presenting cells (APCs)**. T cells only recognise their specific antigen if it is on the surface of a human cell next to one of these two types of MHC proteins (Fig 13.11). Once they have attached to their specific antigen, T cells become sensitised and divide rapidly to form a clone (Fig 13.12). Four types of T cell are present in this clone.

Fig 13.11 **(a)** T cytotoxic (Tc) cells cannot destroy a lone virus and will not attach to antigens on the surface of a lone virus.
(b) A T cytotoxic cell only attaches to a viral antigen if it is on the surface membrane of an infected cell, lying next to a class I major histocompatibility complex (MHC) protein.

Fig 13.12 Sensitised T cells divide by mitosis to produce genetically identical daughter cells (a clone). Four major groups of T cells are present among this clone.

- **Memory cells** remain in the blood and are used in future infections of the same antigen.
- **T cytotoxic cells (Tc cells)** destroy their target cells. Their main targets are the body's own cells infected by viruses, which they destroy before the virus can proliferate. Tc cells are also involved in the destruction of cancer cells.

  Tc cells have surface receptors that bind to a protein on the outer coat of viruses. The Tc cell shown in Fig 13.11(a) does not attach to one of these proteins when it encounters a lone virus which it cannot kill. A Tc cell will only attach to one of these proteins if it is next to a class I MHC protein on the cell surface membrane of one of the body's own cells, as shown in Fig 13.11(b). Once attached to its target cell, a Tc cell releases a protein called **perforin**. Perforin incorporates itself in the membrane of the target cell (an infected body cell) producing pores of about

5 nm–20 nm in diameter. As a result, the target cell bursts. Thus, the Tc cell stops the production of new viruses by this body cell. Only certain antibodies from the B cells can destroy the virus itself.

- **T helper cells (Th cells)** stimulate the action of Tc cells and stimulate B cells to divide and produce plasma cells. The specific receptor sites on the cell surface membrane of Th cells only combine with their antigen if it lies next to a class II MHC protein on the cell surface membrane of an antigen presenting cell or a B cell. It is the Th cell which is inactivated by the human immunodeficiency virus (HIV), the organism which is thought to cause Acquired Immune Deficiency Syndrome (AIDS).
- **T suppressor cells (Ts cells)** inhibit the production of antibodies, probably by inhibiting the Th cells specific for the same antigen. By regulating the activity of Th cells, the body can control the sort of antibodies it produces. They probably also modify the action of Tc cells.

SPOTLIGHT

IMMUNE DEFICIENCY DISEASES

A person whose immune system produces no, or very few, cells is said to suffer an immune deficiency disease. Two such diseases were cited at the beginning of this chapter. **Severe combined immunodeficiency (SCID)** is an inherited condition which may occur in new-born babies. Provided they are breast-fed, these babies survive the first few months of life because they are protected by IgA antibodies in their mother's milk (a natural example of passive immunity). Unless they are then kept in a sterile environment, they are likely to die. In some of these children, bone marrow transplants have resulted in successful antibody production.

Fig 13.13 This boy suffers from SCID and is protected from viruses and bacteria by the 'bubble'.

Acquired immune deficiency syndrome (AIDS) is the result of infection by a virus, the human immunodeficiency virus (**HIV**) which attacks and inactivates T helper cells. Since the major destructive cells of the immune system, T cytotoxic cells and B cells, depend on stimulation by T helper cells for their activity, inactivation of T helper cells knocks out the whole immune system. Like SCID sufferers,

AIDS sufferers are likely to die from infections which non-sufferers normally overcome. HIV is not thought to be transmitted by casual skin contact but only by penetrative sexual intercourse with a carrier or injection of the blood of a carrier. However, the virus can infect the fetus of a pregnant woman. Unprotected sexual intercourse (intercourse without use of a barrier, such as a condom) with many partners and sharing needles to inject drugs intravenously increase the risk of HIV infection and are known as high risk activities.

ANALYSIS

The clonal selection hypothesis

The critical evaluation of experimental evidence is a key scientific skill which this exercise is designed to develop.

Lymphocytes undergo a developmental process either in the thymus gland (T cells) or elsewhere in the body (B cells) so that they become immunologically competent. This immunological competence is conferred, in the case of T cells, shortly before and for a few months after birth. During this developmental process each lymphocyte becomes committed to react with a particular antigen before ever being exposed to it. A cell expresses this commitment in the form of surface receptor proteins that specifically fit the antigen. The binding of antigen to receptors activates the cell, causing it to both multiply and mature. This is the basis of the clonal selection theory which postulates that the immune system is composed of millions of different families, or clones, of cells each consisting of T or B lymphocytes descended from a common ancestor. Since each ancestral cell is already genetically committed to make one particular antigen-specific receptor protein, all cells in a clone produced when B and T cell divide have the same antigen specificity.

Two rats, X and Y, were injected twice at 28-day intervals with polysaccharide antigens taken from different strains of pneumococcal bacteria. Rat X received antigens from Type I bacteria (Type I antigen), whereas rat Y received antigens from Type VIII bacteria (Type VIII

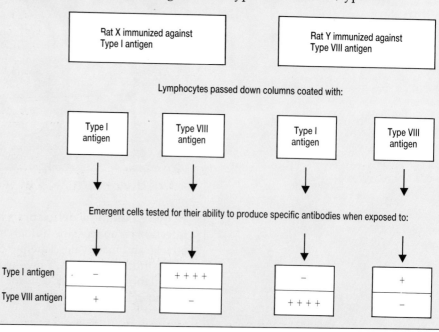

Fig 13.14   Experimental protocol and results. A column is a glass tube filled with beads covered in particular antigens.
− = no production,
+ = little production,
++ = much production.

THE DEFENCE AGAINST DISEASE: IMMUNOLOGY

antigen). Samples of lymphocytes were removed from the rats a week after the second antigen injection, and passed through columns of inert beads coated with either Type I or Type VIII antigens (see Fig 13.14). Cells emerging from each of the four columns were exposed separately to the two antigens *in vitro*, and their ability to synthesise antibodies binding specifically to each antigen was measured.

(a) Interpret the results of the above experiment in terms of clonal selection theory. Use Fig 13.7 to help you.

Cytokines

A number of cells, including B cells and sensitised T cells, release a group of proteins which act as chemical messengers, called **cytokines**. Many different types of cytokine have been found, including colony stimulating factor, interleukins and tumour necrosis factor. **Interferon**, which received much attention in the popular press during the late 1980s, is yet another type of cytokine. Cytokines are extremely potent messengers. For example, it has been found during *in vitro* experiments that 10^{-12} g of alpha interferon can protect one million cells from attack by ten million virus particles. Cytokines interact in complex ways and our knowledge of their effects is developing all the time. Some of their known effects are summarised in Table 13.2.

Table 13.2   Some of the known effects of cytokines. Immunologists are discovering other effects of cytokines and new cytokines all the time.

Name of cytokine	Some of the effects discovered
colony stimulating factor	stimulates growth and division of leucocytes
interferon	increases activity of Tc cells increases phagocytic activity of macrophage cells, forming 'angry macrophages' stimulates division by B cells stimulates antibody secretion by plasma cells stops cells growing protects cells from attack by viruses
interleukin	produces fever stimulates differentiation of leucocytes stimulates lymphocytes to destroy cancer cells stimulates release of prostaglandins is a hyperalgesic agent
tumour necrosis factor	stimulates growth of connective tissue cells helps formation of blood clot protects some cells against viral infection kills certain types of cancer cell

QUESTIONS

13.5 (a) What is (i) an antigen (ii) a self-antigen (iii) an antibody?
  (b) Describe two different ways in which antibodies might affect their targets.
  (c) Explain why the action of a B cell and of a T cell is limited to a particular antigen.

13.6 Which of the following statements refer(s) to (i) specific immunity (ii) non-specific immunity (iii) both?
  (a) Vertebrates have this type of immune system, but invertebrates do not.

QUESTIONS *continued*

(b) Is capable of distinguishing between self and foreign cells or macromolecules.

(c) Can distinguish between self and foreign material, but cannot distinguish between different types of bacteria or different virus particles.

(d) Repeated exposure to the same antigen does not produce an enhanced response.

(e) Retains a 'memory' of having encountered an antigen, which may persist for years.

(f) The principal humoral substance combines specifically with macromolecules with a 'foreign' structure (antigens).

(g) The principal cells eliminate foreign material by engulfing it (phagocytosis) and then digesting it with lysosomal enzymes.

13.7 Compare the action of a B cell and a T cytotoxic cell on its target.

13.8 Design and complete a flow-diagram to summarise the interaction of B cells and various T cells during the immune response.

13.4 IMMUNISATION

People suffer the symptoms associated with an infection when the causative pathogen has been able to multiply in their bodies, causing harm. Initially, their body lacked a large number of cells with the specific surface receptor for the antigens of that pathogen. In the early stages of the illness, the few B or T cells with the appropriate surface receptor were stimulated to divide and the humoral and cellular immune responses followed their course. People recover from an infection when they have produced sufficient antibodies to destroy the causative pathogen. Having once done so, memory cells of the particular antibody-producing cell remain in their bloodstream, ready to combat future invasions by the same pathogen; these people rarely suffer the same disease again.

Passive immunity

Nowadays, immunity to a disease can arise without ever having to suffer from the disease itself. **Immunisation** is the artificial induction of immunity to disease. In one method of immunisation, antibodies which have been made by another organism are passed into the body of a recipient where they react against their specific antigen. Because the recipient organism has not made these antibodies, this type of immunity is called **passive immunity**.

The venom of a king cobra contains a neurotoxin which causes human death within two hours of being bitten. The victim's life can be saved if he or she is injected with antivenin, which contains antibodies against neurotoxin. These antibodies are produced by injecting venom from king cobras into a horse. The injections begin with very dilute, non-lethal doses, and gradually increase in concentration. Eventually, blood taken from the horse contains a high concentration of antibodies against neurotoxin.

 How would an injection of serum containing these antibodies help someone suffering from a snake bite?

Since antibodies are themselves proteins, injected antibodies are recognised as non-self antigens and new antibodies may be produced against them. Therefore, protection may be only short-lived. However, immunisation of this type is vital if people have been bitten by a poisonous snake or rabid mammal or if they may have been exposed to infection by the bacterium which causes tetanus (lock-jaw).

Fig 13.15 Immunisation programmes can have an enormous impact on the health of a population. The life expectancy of these babies attending a post-natal clinic in Tanzania is greatly increased because they will receive innoculations against diseases which have been such a deadly scourge. For example, after a worldwide vaccination programme sponsored by the World Health Organization, smallpox was declared to have been eradicated in 1977.

THE DEFENCE AGAINST DISEASE: IMMUNOLOGY

Active immunity

Vaccinations are injections of antigens against which the recipient's body makes its own antibodies. Because the recipient makes antibodies, this is called **active immunity**. Active immunity is long-lived because, as well as antibodies, the recipient's body makes memory cells in the normal way. Before injection, the antigens are treated in a way which makes them relatively harmless. Such treated antigens are called **vaccines** (from *vacca* the Latin name for cow), in honour of Edward Jenner who, in 1798, published the results of a twenty-year study to establish a procedure for conferring immunity to smallpox by innoculating material from a cowpox pustule into the skin of a patient.

Three different kinds of vaccine are widely used

- pathogens which have been killed
- pathogens which have been weakened so that they no longer cause disease (**attenuated**)
- chemically modified toxins (**toxoids**).

Examples of these vaccines are given in Table 13.3. With developments in genetic engineering, new ways of producing vaccine are now used. One method is to synthesise the antigenic proteins which are recognised by the receptor sites of cells of the immune system. These can be injected without the need to isolate, purify and attenuate the pathogen itself. This method has been used with anthrax, a severe disease of farm livestock and of humans. A second method is to insert the genes for the antigens of a harmful pathogen into a relatively harmless organism, such as a bacterium like *Escherichia coli*. These 'designer microbes' would produce the appropriate antigen without causing debilitating disease and could be used for vaccination.

Table 13.3 Examples of vaccines which are used to confer immunity to specific diseases

Causative organism	Disease	Type of vaccine
virus	hepatitis B	viral antigen purified from blood of individuals with chronic hepatitis
	measles	attenuated strain
	polio	attenuated strain (Sabin oral vaccine) or formalin-fixed virus (Salk vaccine)
	rubella	attenuated strain
bacteria	cholera	killed pathogen
	diphtheria	toxoid
	tetanus	toxoid
	whooping cough	killed pathogen

SPOTLIGHT

A FORGOTTEN KILLER

In eighteenth century Britain, smallpox caused 19% of all deaths and 30% of all deaths in children aged under five. To combat this many parents resorted to a technique introduced from Asia which is a crude form of immunisation. The technique is based on the ancient observation that a person who had once recovered from a disease was either totally immune to that disease or may only suffer a mild second attack. The approach favoured by eighteenth-century parents is described in a letter written in Scotland in 1765.

I am assured that in some remote highland parts of this country it has been an old practice of parents whose children have not had the smallpox to watch for an opportunity of some child having a good mild smallpox that they may communicate the disease to their children by making them bed fellows to those in it and by tying worsted threads with the pocking material around their wrists.

A. Munro, *An Account of the Innoculation of Smallpox in Scotland,*
Edinburgh 1765

What panic this disease must have caused if parents were willing to take the fearful risk associated with this procedure in order to protect their children.

Modern immunisation methods still follow the ancient principle but with the safeguard of first rendering the disease-causing agent harmless. Use of such vaccines have now meant that smallpox, once a worldwide scourge, has now been eliminated entirely from the planet.

QUESTIONS

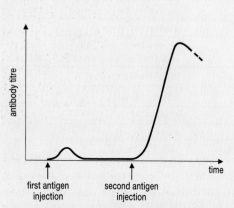

Fig 13.16 Increase of antibody titre after repeated injections with the same antigen.

13.9 The changes occurring in the number of antibodies (the antibody titre) in an individual injected with the same antigen on two separate occasions are shown in Fig 13.16. Explain the shape of the curve in terms of the clonal selection hypothesis.

13.10 Children in Britain are usually immunised against a number of infectious diseases such as tetanus and diphtheria. Immunisation normally takes the form of an injection of killed bacteria or virus particles, or some closely related organism that produces little or no symptoms of the disease. Immunisations are generally given on three or four occasions, initially separated by only a few months, but with a gap of perhaps years between the penultimate and the final dose. However, if a person immunised in childhood intends to travel as an adult to an area in which one of these diseases is common, they usually have another immunisation some weeks before departure.

Given the above information and your knowledge of B cells, answer the following questions.
(a) What is the rationale behind immunisation?
(b) Why is more than one immunisation usually given, at increasing intervals between injections?
(c) Why would an additional immunisation be valuable before travelling to a country where the disease is endemic?

13.11 One advantage of breast-feeding newborn babies is that breast milk contains antibodies to infectious diseases that the mother has either recovered from, or been immunised against. The antibodies can be absorbed by the baby and offer passive protection against those diseases. Why is this passive protection advantageous in the early weeks of life?

13.5 INFLUENZA: THE UNBEATABLE DISEASE?

We make antibodies against a pathogen when we recover from an illness caused by infection of that pathogen or after an appropriate vaccination. Production of sufficient antibody takes between several days and several months, depending on the pathogen. Once we have 'learned' to make a specific antibody we are immune to that disease. On subsequent infection, memory cells initiate antibody production so quickly that the pathogen is killed before it produces any symptoms. Thus, we are unlikely to suffer a disease more than once. This is untrue of a very common disease, influenza (flu).

Flu is the result of infection by orthomyxoviruses. The outer surface of these viruses is studded with the glycoprotein haemaglutinin with which the virus attaches to its target cells in the respiratory tract. These haemaglutinins are also the antigens against which our antibodies act. So, having survived one attack of flu, why are we not immune for the rest of our lives, as we would be to measles or polio? The answer lies in the ability of the genes of the influenza virus to mutate.

The viral genes which encode the haemaglutinins mutate so rapidly that there are likely to be ten new mutations in every million newly formed viruses. Whilst a single mutation is unlikely to cause a great change in the antigenic properties of the haemaglutinins, four or five may cause a big enough change to reduce the ability of the binding site of our existing flu memory cells to attach to this virus. Thus, the new virus can affect the body whilst new lines of B and T cells are formed. The result is that we suffer flu whilst the immune system learns to make new antibodies.

Fig 13.17 A source of protein, but also a potential health hazard.

Every ten to twenty years a new strain of the flu virus emerges which has totally different surface antigens. These antigens are not just the result of mutations of the old genes but are the result of entirely new genes. Never having encountered anything like these new virus strains, our existing B and T cells are totally inadequate. Whilst new lines of B and T cells are being selected, multiplication of the virus within our body cells goes unchecked. The virus multiplies so rapidly that people suffer severe symptoms and may even die. For example, an estimated twenty million people were killed by the flu pandemic of 1918.

The origin of these new genes has long been a puzzle. Their source has now been traced to influenza viruses in ducks. Experimental studies show that, although they can multiply in duck cells, human flu viruses are not transmitted from human to duck. For obvious reasons, experiments have not been done to test the transmission of duck flu viruses amongst humans. Pigs can become infected by both types of flu virus and transmit them not only from pig to pig but also back to their original hosts. Within a single

A **pandemic** is an epidemic affecting a wide geographical area.

cell of a pig which is infected with both duck and human flu viruses, new viruses are made containing genes from the duck flu virus and genes from the human flu virus. Some of these new viruses combine, from our point of view, the worst of the genes of both types: from the human flu virus, the genes which allow it to damage human cells; from the duck flu virus, the genes encoding the surface antigens of the duck virus. The result is a virus which causes flu in humans but which has surface antigens which humans have never before encountered.

Agricultural practices in which fresh animal faeces are used as fertiliser for fish ponds are common throughout the world. In some polyculture systems used in Hong Kong, Malaysia and Nepal, ducks are caged above pigs which are housed in pens directly above fish ponds. The pigs consume duck faeces and themselves defaecate directly into the fish pond. Such systems must present a potential human health hazard in bringing together the duck and human flu virus; indeed the most recent flu pandemic in 1968 was from one of these regions. In ignorance of this, such reputable organisations as the Food and Agriculture Organisation of the United Nations and the UK Overseas Development Administration have been recommending an extension of these farming systems elsewhere in Asia and into rural Africa.

QUESTION	13.12 A patron of a restaurant contracted food poisoning after eating food contaminated with bacteria. For some inexplicable reason, six weeks later this person went to the same restaurant. The food served was contaminated with the same bacteria. In addition, the diner was also sneezed upon by a waiter with influenza. This time the unfortunate person, though not succumbing to food poisoning did contract flu. Explain these observations and draw a graph of the antibody responses you would expect following the second meal.

13.6 BLOOD GROUPS

Normally we only produce antibodies if our B cells have encountered an appropriate antigen. This is not the case with some of the antibodies which give us our blood group; most of these are produced shortly after birth and thereafter throughout life. The antigens for the blood group antibodies are found on the cell surface membrane of erythrocytes, but we never produce antibodies against our own erythrocytes. If blood containing erythrocytes is mixed with an appropriate antibody solution, the erythrocytes are clumped together, or **agglutinated**. For this reason the antibodies are called **agglutinins** and the antigens **agglutinogens**. A number of blood group systems occur in humans, including the ABO, Rhesus, MN, P and Duffy systems.

The ABO system

Two agglutinogens, called A and B, may be present on the surface membrane of human erythrocytes. All of one person's erythrocytes will carry the same agglutinogens but these may be different from those of other people. Two agglutinins may be present in the plasma: anti-A and anti-B. The agglutinogens and agglutinins found in the blood of people of each blood group are listed in Table 13.4. Notice that the blood group name comes from their agglutinogens: cells with agglutinogen A belong to blood group A, those with agglutinogen B to blood group B and so on. The inheritance of the human ABO system is described in Chapter 22.

Normally, ABO blood groups are important only during blood transfusion. Care is taken to ensure that the millions of erythrocytes present in transfused blood match the agglutinins of the recipient. Agglutinins in transfused blood are so diluted within the recipient's body that their effect is minimal and they can be ignored. Some transfusions are safe whilst some will result in lethal agglutination of the recipient's erythrocytes (see Table 13.5).

Table 13.4   The antigens (agglutinogens) and antibodies (agglutinins) of the human ABO blood group system

Blood group name	Agglutinogens on all erythrocytes	Agglutinins in plasma
A	A	anti-B
B	B	anti-A
AB	both A and B	neither anti-A nor anti-B
O	neither A nor B	both anti-A and anti-B

Table 13.5   The agglutinogens on the surface membrane of transfused erythrocytes must be compatible with the agglutinins of the recipient's plasma

Blood group of transfused blood	Blood group of recipient			
	A	B	AB	O
A	safe	unsafe	safe	unsafe
B	unsafe	safe	safe	unsafe
AB	unsafe	unsafe	safe	unsafe
O	safe	safe	safe	safe

The Rhesus system

People who are Rhesus positive (Rh+ve) have agglutinogen D on the surface membrane of their erythrocytes: people without this agglutinogen are Rhesus negative (Rh-ve). Agglutinin against the D agglutinogen (anti-D) is never produced by Rh+ve people and is not normally produced by Rh-ve people. Anti-D is only produced by Rh-ve people whose blood has been contaminated by erythrocytes carrying agglutinogen-D, either during a blood transfusion or during delivery (when severe uterine contractions may force fetal blood cells across the placenta into the mother's blood).

Anti-D agglutinins will cause agglutination of any future transfused blood cells from an Rh+ve donor. They will also cross the placenta during any future pregnancies and agglutinate the erythrocytes of an Rh+ve fetus. The resulting **haemolytic disease of the newborn** can be eliminated by giving the fetus several blood transfusions as it develops in the uterus. An alternative solution is to inject an Rh-ve mother with anti-D immediately she has given birth to an Rh+ve baby. This injected anti-D will destroy the fetal cells before the mother has time to make her own antibodies.

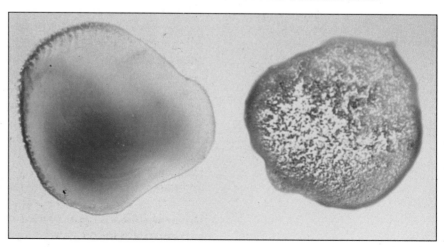

Fig 13.18   The Coombes test is used to identify blood groups. The blood droplet on the right of this test card shows agglutination.

13.13 During the nineteenth century, blood transfusions were given by direct person-to-person methods. Prior to the discovery of the ABO blood group system in 1900, many such blood transfusions resulted in the death of the recipient. Explain why a recipient of blood group A would have been killed by transfused blood from a donor of group B but not from a donor of group O.

13.14 **(a)** Suggest why transplanted organs are often rejected by their recipients.
   (b) Discuss the likely benefits and harm that may arise from giving immunosuppressant drugs to transplant patients.

SUMMARY

The body has three lines of defence against pathogens: the physical barriers of the skin and mucous membranes; the non-specific inflammatory response; the specific immune response.

The unbroken layer of dead cells of the skin's epidermis forms a barrier to the entry of pathogens. Fatty acids in the sebum inhibit the growth of many bacteria and fungi. Mucous membranes of the eyes and of the digestive, respiratory and urethral tracts release antibacterial substances. Micro-organisms in the respiratory tract are trapped in the mucus which is swept up into the throat by cilia.

Pathogens which enter the body are first attacked by the inflammatory response. This generalised response involves local inflammation at the site of infection and destruction of the pathogen by the body's phagocytes. Blood clots wall off the site of injury, preventing further spread of pathogens. Aided by the release of such chemicals as histamine and prostaglandins, neutrophils and macrophages move to the site of infection and engulf the invading cells.

If pathogens avoid the inflammatory response and spread to the rest of the body they are attacked by the immune response. This response is specific to each type of pathogen and involves the activity of the B and T lymphocytes.

Each B and T cell synthesises a particular immunoglobulin which it carries on its cell surface membrane. These immunoglobulins bind to part of the outer surface of a particular pathogen: in this way pathogens are recognised. Once a B or T cell has bound to its antigen it divides rapidly to produce vast numbers of genetically identical cells. The humoral response occurs when B cells divide to form plasma cells which secrete their immunoglobulin as antibodies. Circulating antibodies destroy their specific pathogens in one of four ways: direct neutralisation; precipitation; agglutination; complement reactions. T cells do not release their immunoglobulin molecules. They bind to their specific pathogens and either destroy them directly or promote their destruction by other cells.

Both B and T cells form memory cells that remain in circulation long after the infection is over. If the same pathogen reappears in the bloodstream, these memory cells are quickly activated so that the immune response occurs much faster than during the first infection.

Different antibodies are formed by rapid mutation and reassortment of a particular group of genes during development. Destruction of inap-

propriate combinations ensures that the body's own cells are not normally attacked. Human blood groups form a special case of inherited antibody production.

Certain medical disorders of the immune system are common. Allergies occur when some people produce antibodies against harmless particles, such as pollen. Auto-immune diseases result when the immune system destroys some of the body's own cells. Immune deficiency diseases occur when part of the immune system is inactivated so that trivial diseases cannot be overcome.

Answers to Quick Questions: Chapter 13

1   By the acidic conditions which would denature the bacterial enzymes.
2   Phagocytes will find it easier to leave the blood capillaries and so gain access to the site of infection.
3   Different genetic codes will produce different sequences of amino acids and, so, differently shaped protein molecules.
4   By recognising their shape as being foreign.
5   Enzymes; a lock and key mechanism between variable regions of antibody and antigen similar to that found in enzymes.
6   The antibodies would bind to, and so inactivate, the neurotoxins contained in the venom.

3.1 **(a)** The graph shows the changes in pressure which take place in the left side of a mammalian heart during one complete cardiac cycle, i.e. one complete heart beat.

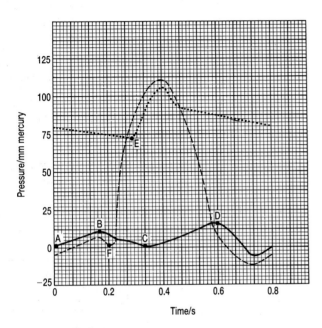

(i) During which part of the cardiac cycle is the pressure in the left ventricle at a maximum? (1)

(ii) From which vessel does blood flow directly into the left atrium? (1)

(iii) Pressure in the left atrium increases between points C and D due to the inflow of blood. Explain the increase in atrial pressure between A and B. (1)

(iv) Name the valve which closes at point F. (1)

(v) Which valve must open to allow pressure to increase in the aorta after point E? (1)

(vi) Assuming that the cardiac cycle remains constant, calculate, from the data, the heart rate as beats per minute. (1)

(b) The diagram represents part of the circulatory system of a fish.

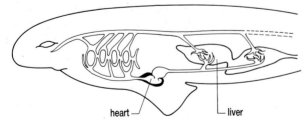

(i) The pattern of blood circulation in a fish is described as single. What is meant by this term 'single'? (1)

(ii) Show, by means of arrows, the direction of blood flow from the heart to the liver. (1)

(iii) Give one way by which the heart of an amphibian differs from that of a fish. (1)

(iv) How does a portal vein differ from other veins in its general position within a circulatory system? (1)

(SEB 1988)

3.2 Two leafy shoots, **A** and **B** were placed in water. A ring of bark and phloem was removed from shoot **B** in the position shown. A drop of sucrose solution labelled with ^{14}C was applied to the upper surface of one leaf of each shoot. After 24 hours an autoradiograph was developed from each shoot. The diagrams show the shoots and their autoradiographs.

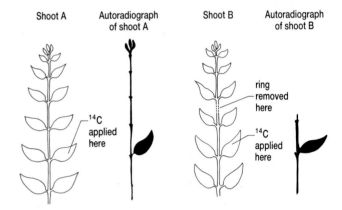

(a) Why was sucrose used in the experiment in preference to glucose? (1)

(b) Give **one** piece of evidence which shows that the labelled sucrose was transported in the phloem. (1)

(c) Give **one** piece of evidence which shows that the labelled sucrose was not transported in the xylem. (1)

(d) Account for the distribution of ^{14}C in shoot **A**. (2)

(AEB 1989)

3.3 Porometers can be used to measure air flow through a leaf under different environmental conditions. Figure 1 shows a simple porometer attached to the underside of a leaf.

To obtain results, the time is taken for the meniscus to travel from X to Y.

(a) (i) Explain how the meniscus would be set at X at the beginning of the experiment.

(ii) Describe two different ways in which the apparatus might be modified to give a faster flow of fluid from X to Y. (3)

The apparatus shown in Figure 1 was used with detached leaves to obtain the results shown in Figure 2.

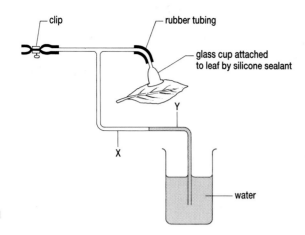

Fig 1

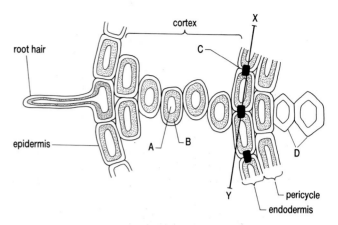

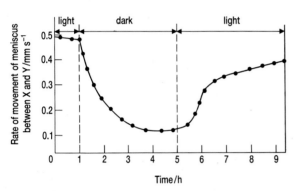

Fig 2

(b) Using your knowledge of leaves, interpret the changing shape of the graph between
   (i)   0 and 1 hour
   (ii) 1 and 5 hours
   (iii) 5 and 9 hours   (5)
(c) (i)   Suggest **one** objection to the use of this apparatus for gathering information about the response of leaves to periods of light and darkness.
   (ii) Would similar results have been obtained if the porometer had been attached to the upper surface of the leaf. Explain your answer.   (2)
(d) Describe in outline how you would use this experimental method to investigate the effect of different wavelengths of light on the porosity of detached leaves.   (4)

(JMB 1988)

3.4 The diagram represents part of a transverse section across a young root. The relative distance across the cortex has been shortened for simplicity.
   (a) (i)   Name the features labelled A, B, C and D.
      (ii) State **two** advantages to the plant of root hairs.
      (iii) Make a diagram of an endodermal cell as it would appear in a tangential longitudinal section along X-Y. Shade and label feature C on your diagram.
      (iv) What is the function of feature C?   (9)

(b) (i)   It is believed that water can move across the cortex along the cellulose cell walls. Suggest how this movement is maintained.
   (ii) Name the process by which water moves from a cortical cell into a cell of the pericycle.   (3)
(c) Consider the two adjacent cortical cells represented below.

	Cell P		Cell Q	
Ψ_{cell}	$= -1200$ kPa	Ψ_{cell}	$= -800$ kPa
Ψ_{p}	$= 800$ kPA	Ψ_{p}	$= 600$ kPa
Ψ_{s}	$= -2000$ kPa	Ψ_{s}	$= -1400$ kPa

   (i)   Which cell has the higher water potential?
   (ii) In which direction will water move?
   (iii) State the water potential of the cells at equilibrium.
   (iv) Calculate the pressure potential of cell Q at equilibrium.   (4)

(WJEC 1988)

3.5 This question refers to the flowering plant.
   (a) (i)   Draw a labelled diagram to show the distribution of tissues in a herbaceous stem.
      (ii) Relate the structure of the xylem to its function of water transport.   (3)
   (b) Describe briefly **one** experiment to show that water is transported in the xylem rather than the phloem of flowering plants.   (7)
   (c) Explain concisely the 'root pressure' and 'cohesion-tension' hypotheses which have been proposed to account for the transport of water from roots to leaves. For **each** hypothesis, outline the supporting experimental evidence.   (10)

(JMB 1987)

3.6 The table at the top of the next page gives raw data of pulse rates of 25 students.

Pulse rate of 25 students/beats min⁻¹

70	82	87	75	77
83	88	66	80	92
78	96	85	71	82
76	79	81	72	81
64	80	90	78	74

(a) Copy and complete the table below. The third line has been done for you. (2)

Pulse rate/beats min⁻¹	Tally	Number of students
60–64		
65–69		
70–74	/ / / /	4
75–79		
80–84		
85–89		
90–94		
95–99		

(b) (i) Construct a histogram of the data in your table. (5)

(ii) What is the modal class in your histogram? (1)

(c) The diagram shows the sequence and relative duration of atrial and ventricular systole and diastole through two cardiac cycles.

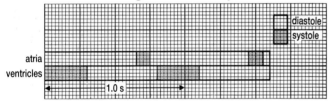

(i) Calculate the pulse rate. (1)

(ii) What is the duration of atrial systole? (1)

(iii) What is happening to the cuspid valves, the atria and the blood in the heart during ventricular diastole? (3)

(iv) What is the function of atrial systole? (2)

(AEB 1989)

3.7 **(a)** What is meant by the term *homeostasis*? (4)

(b) Describe the role in homeostasis of:

(i) the kidneys,

(ii) the pituitary gland,

(iii) the pancreas. (16)

(UCLES 1988)

3.8 **(a)** Describe the role of the mammalian kidney in the following processes.

(i) Osmoregulation

(ii) Excretion. (14)

(b) Explain the significance of the following as excretory products.

(i) Ethanol in yeast

(ii) Ammonia in freshwater teleosts. (6)

(ULSEB 1989)

3.9 The figure below shows the progress of a patient who developed acute kidney failure after emergency heart surgery. The operation was carried out on Thursday, 18th October. Blood dialysis began on the Saturday afterwards and was needed over the next 13 days. As a result of treatment the patient eventually recovered full health.

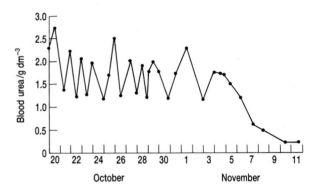

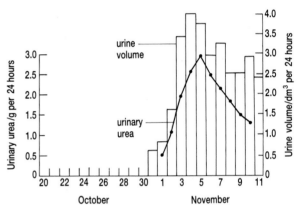

(a) (i) Using this information, state how many times the patient received dialysis.

(ii) Explain how you arrived at this answer. (2)

(b) On what date did urine production start again? (1)

(c) The patient's mass fell by 7 kg between 20th October and 11th November. Suggest **two** possible reasons for this change in mass. (2)

The efficiency of the working of the kidneys can be judged from renal:plasma ratios. For urea, for example, this is calculated as the concentration of urea in urine divided by the concentration of urea in the blood. The table shows how this was calculated for November 5th.

Date	Urea concentration in urine (g dm⁻³)	Blood urea	Renal:plasma ratio for urea
November 5th	$\frac{2.9}{3.8} = 0.76$	1.5	0.5
November 10th			

(d) (i) Complete the table to show the missing values for November 10th.

(ii) On which of these two days were the kidneys working more efficiently? Explain you answer. (4)

(JMB 1989)

3.10 The table shows the concentration of some of the substances present in the blood plasma, the glomerular filtrate of the kidney and normal urine.

Substance	Concentration (g/100 cm³)		
	Blood plasma	Glomerular filtrate	Urine
Urea	0.03	0.03	2.00
Protein	7.00	0.00	0.00
Glucose	0.10	0.10	0.00
Sodium	0.32	0.33	0.60

(a) Use your knowledge of the structure and working of the kidney to explain:
  (i) The difference in composition between the plasma and the glomerular filtrate.
  (ii) The difference in composition between the glomerular filtrate and the urine. (5)
(b) In a person with diabetes mellitus, how might the concentration of glucose in the three fluids differ from the figures given in the table? Explain your answer. (2)
(c) How, without using commercial methods, would you test a sample of urine for the presence of glucose? (2)
(d) In an artificial kidney the patient's blood flows on one side of a dialysing membrane, with a dialysing fluid on the other. Explain why the dialysing fluid is made up of various salts in aqueous solution at the same concentration as that of the plasma of a normal person. (1)

(AEB 1989)

3.11 Describe how excessive water loss is
  (i) prevented by the mammalian kidney, (13)
  (ii) minimised in terrestrial insects. (7)

(WJEC 1990)

3.12 Consider a guard cell from the leaf of a flowering plant and a cell from the proximal tubule of a mammalian kidney.
  (a) Explain briefly the functions of **each** of these cells in the life of the organism. (4)
  (b) State **four** ways in which the two cells differ in structure and relate these differences to the activities carried out by the cells. (10)
  (c) Describe how **each** of these cells is provided with nutrients and oxygen. (6)

(JMB 1989)

3.13 The diagram shows a human kidney tubule.

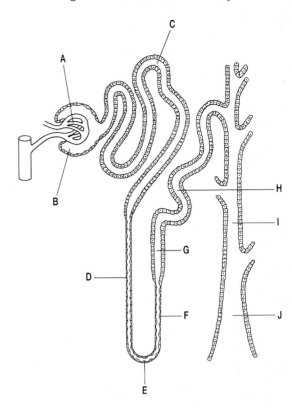

(a) Label each of the parts **A, B, C, E, H,** and **I**. (6)
(b) Under normal conditions where, in **A, B, C, D, E, F, G, H, I** and **J**, would you expect to find
  (i) fluid especially hypertonic to blood plasma?
  (ii) fluid isotonic with blood plasma?
  (iii) podocytes?
  (iv) the countercurrent multiplier mechanism operating?
  (v) active transport of Cl⁻ out of the tubule?
  (vi) glucose absorbed from the lumen of the tubule? (6)
(c) Name **one** condition under which you would expect to find **H** almost impermeable to water. (1)
(d) What would stimulate **H** to increase its permeability? (1)

(AEB 1987)

3.14 Water loss from the leaves of two different species of plants, *Species A* and *Species B*, was compared using two experimental methods.

Experiment 1
Leafy shoots were taken, one from each plant. They were weighed, hung in air and re-weighed at 15-minute intervals. The results are shown in the table (– means that a result is not available).

Time after start of experiment/minutes	Mass of leafy shoot/g	
	Species A	Species B
0	210.0	240.0
15	195.3	–
30	184.4	–
45	176.4	–
60	170.1	–
75	166.3	213.1

(a) (i) Plot a graph of mass against time for *Species A*.

(ii) Account for the shape of the curve for *Species A*. (5)

(b) (i) Calculate the percentage change in mass for each species after 75 minutes.

(ii) Explain which of these two species you would expect to be more successful in an arid habitat. (3)

Experiment 2
Different leafy shoots from each species were placed in a simple potometer, as shown below.

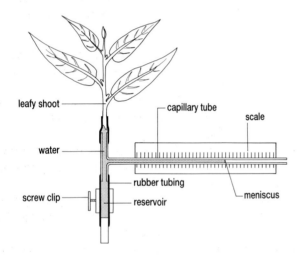

leafy shoot
capillary tube
scale
water
rubber tubing
screw clip
reservoir
meniscus

(c) (i) State **two** precautions that should be taken when using a potometer, to ensure that the results obtained are comparable for the two different species.

(ii) What **three** measurements are required to calculate the rate of water taken up in $cm^3min^{-1}g^{-1}$ in *Experiment 2*?

(iii) Give **one** reason why the values obtained for water loss in *Experiment 2* were likely to have been higher than those from *Experiment 1* for the same species. (6)

(JMB 1990)

Answer guideline for Question 3.1
This question tests a number of skills. Its clear division into small sections, each scoring only one mark, makes it easy to spot which skills are being tested.

(a) (i) ventricular systole;

(ii) pulmonary vein;

(iii) muscle of atrium contracts/atrial systole;

(iv) bicuspid/mitral (because the pressure in the ventricle becomes greater than that in the atrium);

(v) semi-lunar valves at base of aorta;

(vi) one beat in 0.8 seconds gives 75 beats per minute;

(b) (i) blood goes once round body for each passage through heart/heart to gills to body tissues to heart;

(ii) your arrows should go forwards from the heart, through the gills, backwards towards the liver and down into the liver;

(iii) the fish's heart has two chambers, the amphibian's has three (two atria);

(iv) veins usually empty into another vein or into heart, portal vein empties into liver.

Theme 4

COORDINATION AND MOVEMENT

Theme 1 describes how a single cell can control its activities by 'switching on and off' the genes which carry the code for appropriate enzymes. Multicellular organisms might consist of billions of cells. If they are to function efficiently, the activities of these cells must be coordinated so that they work together.

Coordination between cells is by chemicals which are secreted by one cell, bind to receptors on the surface membrane of a target cell and affect the activities of the target cell. The following chapters show how this relatively simple concept underlies complex patterns of control and coordination.

- *Hormones* – coordination by chemicals which are released into the blood of animals and so bring about slow and widespread effects throughout their bodies.
- *Nerves and nervous systems* – instead of releasing their chemical into the blood, nerve cells grow to their target cell and release their chemical directly onto its surface.
- *Receptors and effectors* – the way in which animals detect stimuli in their environment, convert the stimulis into a nerve impulse and produce appropriate responses.
- *Control and coordination in plants* – chemicals called plant growth substances interact to control the activities of plants during their germination, growth, reproduction and ageing.
- *Support systems and movement* – movement and locomotion is at the heart of most responses to environmental stimuli.

Chapter 14

CONTROL AND COORDINATION IN ANIMALS: HORMONES

LEARNING OBJECTIVES

When you have studied this chapter you should be able to:

1. explain the terms endocrine gland and hormone;

2. compare the effects of hormones and nerves in stimulating their targets;

3. explain the specificity of hormones;

4. describe the position of the major mammalian endocrine glands, the hormones they secrete and the effects of these hormones;

5. use one example to show how hormones interact;

6. outline the role of hormones in controlling metamorphosis in insects and amphibians.

Most cells within the body of a multicellular animal are specialised to perform only one function. Groups of these cells work with other cells in tissues, organs and systems. In carrying out their functions, the cells in tissues and organs are coordinated so that they cooperate. Animals have two coordinating systems, the **endocrine** system and the **nervous** system. These systems operate in a similar way: a cell of the endocrine system or nervous system secretes a chemical coordinator which affects the activities of a target cell. The pathway of these chemical coordinators differs in the endocrine and nervous systems (Fig 14.1). Chemical coordinators of the

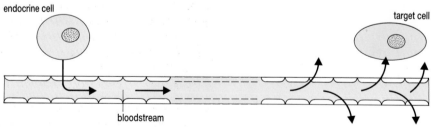

endocrine cell

target cell

bloodstream

pathway of chemical coordinator in endocrine system

nerve cell

target cell

Fig 14.1 Comparison of the pathway of chemical coordinator in the endocrine and nervous systems.

pathway of chemical coordinator in nervous system

⟶ movement of chemical coordinator

endocrine system, called **hormones**, are secreted into the bloodstream and so reach all parts of the animal's body: chemical coordinators in the nervous system, called **neurotransmitters**, are secreted directly onto the target cell. As a consequence of these different pathways to the target tissues, there are differences in the nature of the endocrine and nervous systems. Table 14.1 summarises the major differences between the ways in which these systems work.

Table 14.1 Comparison of the endocrine and nervous systems

Feature	Endocrine system	Nervous system
speed of effect	generally slow	generally rapid
duration of effect	generally long-lasting	generally short-lived
localisation of effect	secreted into blood, so effect may be widespread	secreted onto target cell, so effect very localised
nature of chemical coordinator	many different types, each affecting different specific tissue	very few types, which are secreted only onto target tissue

The endocrine and nervous systems interact in controlling the internal environment of an animal: in fact some nerve cells secrete hormones into the bloodstream (so-called **neurosecretion**) and noradrenaline is released both into the bloodstream by endocrine cells and directly onto target cells by nerve cells. For convenience, the nervous system is described in Chapter 15. This chapter describes the endocrine system, the activities of specific endocrine glands and some of their hormones. Because of their effects on other organs, the action of some animal hormones is described elsewhere in this book. Hormones involved in controlling the release of digestive juices are described in Section 8.3, those involved in controlling the sugar, inorganic ion and water content of the blood in Sections 11.4 and 11.5 and those involved in reproduction in Section 20.3.

14.1 AN OVERVIEW OF THE ENDOCRINE SYSTEM

A **gland** is a tissue which produces and secretes mixtures of chemicals. Many glands release their secretions through ducts, e.g. the salivary glands release saliva through salivary ducts (see Section 8.3). These glands are called **exocrine glands**.

Endocrine glands release hormones directly into the bloodstream. Since they lack ducts, they are often termed **ductless glands**. Look at Fig 14.2: it shows the positions of the major endocrine glands in a human. Notice that some of these glands are organs which have other functions, e.g. the pancreas, ovaries and testes. In these organs, small groups of hormone-secreting cells occur among the cells with other functions which make up the bulk of the organ.

Hormones are released as a result of the following stimuli.

- Direct nervous stimulation of the appropriate gland, e.g. secretion of adrenaline from the adrenal medulla occurs following stimulation by the sympathetic nervous system (see Section 15.6).
- The presence of particular compounds in the blood. Such compounds may be:
  1. metabolites, e.g. the release of insulin from the pancreas is stimulated by high concentrations of glucose in the blood;
  2. other hormones secreted by the anterior lobe of the pituitary gland.

Negative feedback loops characterise the latter two methods of controlling endocrine secretion. A simplified summary of such a negative feedback loop is shown in Fig 14.3.

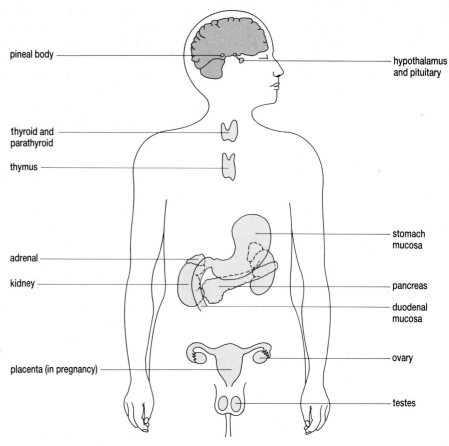

Fig 14.2 Positions of the major human endocrine glands.

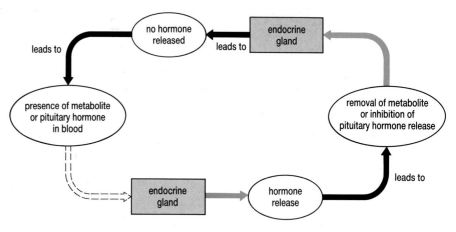

Fig 14.3 Negative feedback controls the secretion of many endocrine glands.

The nature and action of hormones

Hormones fall into one of four groups of chemical compounds: amines (derivatives of amino acids), polypeptides of less than 100 amino acids, proteins and steroids. A common feature of these compounds is that their molecules can be carried freely in the bloodstream. With such chemical heterogeneity, compounds are classed as hormones if they

Hormone activity was first demonstrated in 1905 by Bayliss and Starling, who were working with the digestive hormone secretin.

- are released into the bloodstream by a ductless gland
- are carried by the bloodstream to a target organ
- stimulate a change in the activity of the target organ.

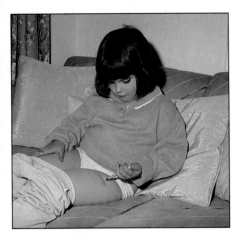

Fig 14.4   Artificial insulin taken by injection. Those people not able to produce their own must be provided with adequate supplies to regulate their blood sugar levels.

Because they are carried in the bloodstream, hormones reach every cell in the body. Their effect on each cell is dependent on the presence of specific **receptor sites** on the cell surface membrane: if a cell lacks an appropriate receptor site it will not respond to the hormone. These sites may be either protein or lipoprotein molecules and are probably similar to the other receptor sites on cell surface membranes. Once attached to its specific receptor, a hormone may affect its target cell in one of several ways. A hormone may do the following

- change the permeability of the cell surface membrane, e.g. insulin is thought to increase the activity of the carrier molecule which transports glucose across cell surface membranes;
- cause the release of a second internal messenger which is chemically linked to the receptor site, e.g. adrenaline is thought to activate the enzyme adenyl cyclase in the surface membrane of a target cell. (Adenyl cyclase catalyses the formation of an internal messenger compound called cyclic adenosine monophosphate (cAMP) which then produces the appropriate response within the cytoplasm of the target cell (Fig 14.5).
- pass through the cell surface membrane as a hormone–receptor complex which itself acts as an internal messenger molecule, e.g. oestrogen (Fig 14.6).

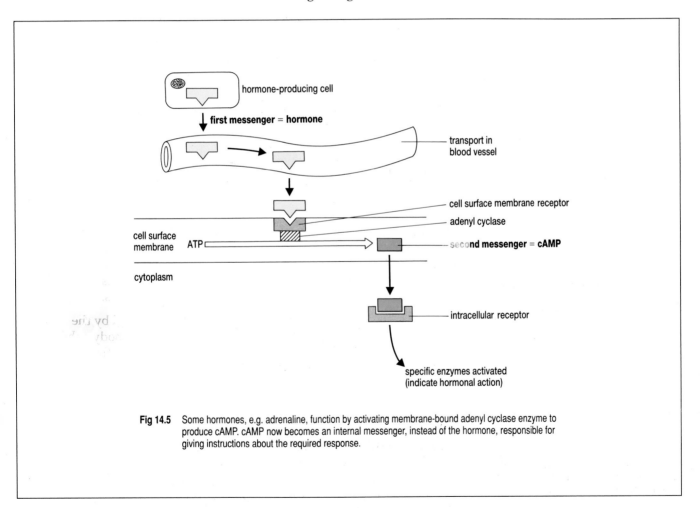

Fig 14.5   Some hormones, e.g. adrenaline, function by activating membrane-bound adenyl cyclase enzyme to produce cAMP. cAMP now becomes an internal messenger, instead of the hormone, responsible for giving instructions about the required response.

The internal messengers that are stimulated by hormones often affect enzymes in the cytoplasm, e.g. thyroxine affects enzymes involved in ATP synthesis inside mitochondria. Many hormones, especially the steroids such as oestrogen, affect the transcription of DNA in the nucleus of the receptor cell (Fig 14.6).

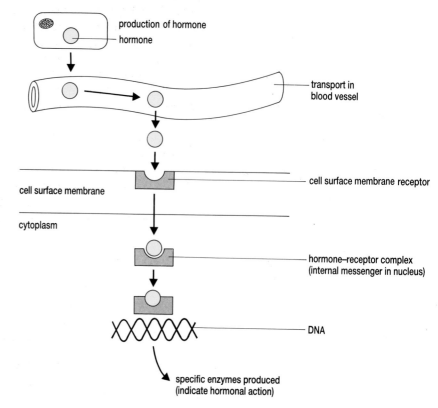

production of hormone

hormone

transport in blood vessel

cell surface membrane

cell surface membrane receptor

cytoplasm

hormone–receptor complex (internal messenger in nucleus)

DNA

specific enzymes produced (indicate hormonal action)

Fig 14.6 Some hormones, e.g. oestrogen, link with an intracellular receptor and this complex becomes the internal messenger for the target cell. Many steroid hormones are known to affect DNA transcription.

QUESTIONS

14.1 Define the terms endocrine gland and hormone.

14.2 (a) How does an endocrine gland differ from an exocrine gland?
(b) How does a cell in an endocrine gland differ from a neurosecretory cell?

14.3 Imagine you are in a 'house of horrors' at a pleasure park. Much of your sense of fear is caused by the effects of the hormone adrenaline. Once outside in the daylight, your symptoms of fear quickly vanish as you joke about your experiences with friends. Use the information in Table 14.1 to compare the effects of adrenaline with those of other hormones and of nerves.

14.4 The hormones insulin and glucagon are both secreted by the pancreas. Insulin affects cells in many organs of the body, while glucagon affects only cells of the liver. Suggest a hypothesis to account for the different specificity of these two hormones.

14.2 THE HYPOTHALAMUS AND PITUITARY GLAND

The hormones of the pituitary gland regulate so many activities that it is sometimes called the master gland. It is in fact surprisingly small, only about 1.3 cm in diameter in humans, and lies at the base of the brain (Fig 14.2). In reality the function of the pituitary is so closely linked to the hypothalamus that we could often consider them together.

Hypothalamus

The hypothalamus consists largely of nervous tissue and is part of the underside of the brain (see Section 15.5). The pituitary gland is attached to its underside. The hypothalamus has a number of functions, among which is the monitoring of the levels of certain metabolites and hormones in the blood. In response to changes in these levels, two groups of **neurosecretory cells** within the hypothalamus influence the activity of the pituitary gland. Cells of one group (neurosecretory cells (1) in Fig 14.7) release transmitter substances from the ends of their axons into the blood vessel connecting the hypothalamus with the anterior lobe of the pituitary. These transmitter substances (called releasing factors and release-inhibiting factors) control hormone secretion from this lobe of the pituitary gland. The second group of neurosecretory cells (neurosecretory cells (2) in Fig 14.7) release transmitter substances from the ends of their axons directly into the blood vessels within the posterior lobe of the pituitary, where these transmitter substances are then stored. When appropriate, other neurosecretory cells in the second group stimulate the posterior lobe of the pituitary to release the stored transmitter substances as hormone. Oxytocin and antidiuretic hormone are produced and released in this way.

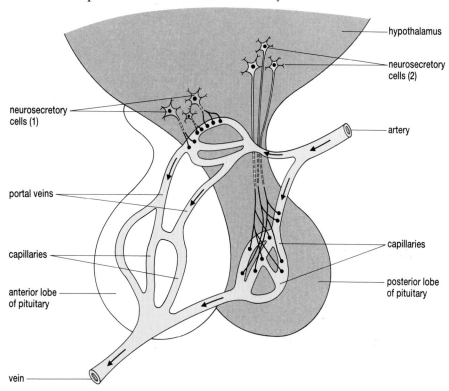

Fig 14.7   The relationship between the hypothalamus and pituitary gland. Notice that the neurosecretory cells (1) secrete into the capillaries connecting the hypothalamus with the anterior lobe of the pituitary but the neurosecretory cells (2) secrete directly into the capillaries of the posterior lobe.

 How does the control of hormone release from the anterior and posterior pituitary differ?

Pituitary gland

The pituitary is, in fact, two glands (Fig 14.7). The **posterior lobe** develops as an outgrowth of the hypothalamus and is stimulated by nerve cells from the hypothalamus. The **anterior lobe** develops as an outgrowth of glandular tissue from the roof of the buccal cavity: its connection with the hypothalamus is via a complex of blood vessels. The effects of hormones produced by the pituitary gland are summarised in Tables 14.2 and 14.3.

The anterior lobe of the pituitary is under the control of the hypothalamus and produces and stores the six hormones listed in Table 14.2.

The control of anterior pituitary hormones is well illustrated by **thyroid stimulating hormone** (TSH) which controls the rate of iodine uptake and formation of thyroxine by the thyroid gland. Its secretion is controlled by **thyrotrophin releasing factor** (TRF) in a negative feedback loop, summarised in Fig 14.8.

Table 14.2   Anterior pituitary hormones, their principal actions and associated hypothalamic regulating factors

Hormone	Principal actions	Associated hypothalamic regulating factors
growth hormone	growth of body cells	growth hormone releasing factor
	protein anabolism	growth hormone release-inhibiting factor
thyroid stimulating hormone (TSH)	controls secretion of hormones by thyroid gland	thyrotrophin releasing factor (TRF)
adrenocorticotrophic hormone (ACTH)	controls secretion of some hormones by adrenal cortex	adrenotrophin releasing factor (ARF)
follicle stimulating hormone (FSH)	in female, initiates development of ova and induces ovarian secretion of oestrogens	gonadotrophin releasing factor (GnRF)
	in male, stimulates testes to produce sperm	
luteinizing hormone (LH)	in female, together with oestrogens stimulates ovulation and formation of progesterone-producing corpus luteum; prepares uterus for implantation and mammary glands to secrete milk	gonadotrophin releasing factor (GnRF)
	in male, stimulates interstitial cells in testes to develop and produce testosterone	
prolactin	together with other hormones initiates and maintains milk secretion by the mammary glands	prolactin releasing factor (PRF)
		prolactin inhibiting factor (PIF)

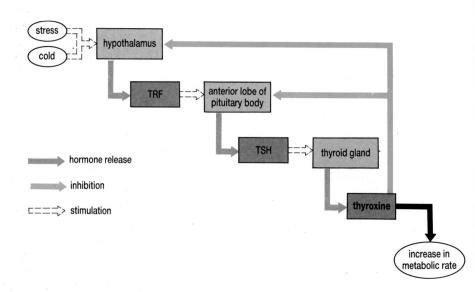

Fig 14.8   Interaction between the hypothalamus, anterior lobe of the pituitary and thyroid gland in controlling metabolic rate (TRF = thyrotrophin releasing factor, TSH = thyroid stimulating hormone).

Table 14.3 Posterior pituitary hormones, their principal actions and control of secretion

Hormone	Principal actions	Control of secretion
oxytocin	stimulates contraction of smooth muscle cells of pregnant uterus during labour and stimulates contraction of contractile cells of mammary glands for milk ejection	neurosecretory cells of hypothalamus secrete oxytocin in response to uterine distension and stimulation of nipples
antidiuretic hormone (ADH)	decreases urine volume	neurosecretory cells of hypothalamus secrete ADH in response to low water concentration of the blood, pain, stress, acetylcholine, and some drugs
	raises blood pressure by constricting arteries during severe haemorrhage	
		alcohol inhibits secretion

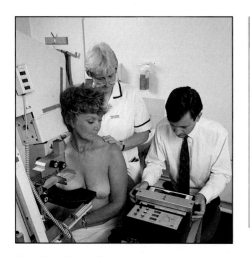

Fig 14.9 A mammogram is used to screen for breast cancer. Areas of increased density in the breast tissue are revealed as opacities on the X-ray film.

The posterior lobe of the pituitary does not produce any hormones. Instead, it stores the hormones antidiuretic hormone (ADH) and oxytocin which are produced by neurosecretory cells in the hypothalamus. Attached to carrier protein molecules, these hormones are passed down the axons of the neurosecretory cells and secreted directly into capillary beds in the posterior lobe of the pituitary. Under nervous stimulation from the hypothalamus, these hormones are released from these capillaries into the general circulation. ADH and oxytocin are short-chain polypeptides with a very similar structure. However, their effects are quite different (Table 14.3).

 What do the properties of ADH and oxytocin suggest about the nature of the cell surface receptors to which these hormones attach?

QUESTIONS

14.5 Explain the biological importance of the relationship between the hypothalamus and the pituitary gland.

14.6 Refer back to Table 14.2 and Fig 14.3. Choose one substance, other than TRF, which is released by the hypothalamus and which affects the release of a hormone from the pituitary gland. Construct a diagram, similar to Fig 14.8, to show how the release of your chosen hormone is controlled.

Whilst the hypothalamus and pituitary control so many body activities, a large number of other endocrine glands exist throughout the human body. The activities of some of these are described in this section. Rather than looking at these as a list, try to think of the way they interact and how their secretions are controlled.

Thyroid gland

The two large lobes of this gland lie like a bow tie on each side of the larynx (Fig 14.2). Under the influence of TSH from the anterior pituitary, these lobes produce the three hormones listed in Table 14.4. **Triiodothyronine** (T_3) and **thyroxine** (T_4) contain iodide ions, acute dietary shortage of which leads to an enlargement of the thyroid gland (**simple goitre**). They affect many processes in the body concerned with the rate of metabolism. **Calcitonin**, produced in different cells from those that produce T_3 and T_4, is released if the concentration of calcium ions in the blood rises above 2.5 mmol dm^{-3} and causes a reduction in the calcium ion concentration of the blood. In this way, it interacts with parathormone from the parathyroid gland in the control of calcium metabolism (Fig 14.10).

Table 14.4 Thyroid gland hormones, their principal actions and control of secretion

Hormone	Principal actions	Control of secretion
thyroxine (T_4)	regulates metabolism, growth and development and activity of nervous system	thyrotrophin releasing factor (TRF) is released from hypothalamus in response to low thyroid hormone levels, low metabolic rate, cold, pregnancy and high altitudes
		TRF secretion is inhibited in response to high thyroid hormone levels, high metabolic rate, high levels of oestrogens and androgens and by ageing
triiodothyronine (T_3)	same as above	same as above
calcitonin	lowers blood levels of calcium by accelerating calcium absorption by bones	high blood calcium levels stimulate secretion
		low levels inhibit secretion

SPOTLIGHT

ABNORMALITIES OF THE THYROID GLAND

Underactivity of the thyroid results in the secretion of abnormally low amounts of thyroid hormones (**hypothyroidism**). This condition is more serious during childhood since the reduction in metabolic rate leads to retarded physical and mental growth (**cretinism**). If, after a normal childhood, the condition occurs during adulthood, the reduction in metabolic rate leads to mental and physical sluggishness. This condition, known as **myxoedema**, leads to obesity and swelling of the thyroid. Thyroxine can be administered orally to sufferers of hypothyroidism.

Overactivity of the thyroid leads to abnormally high levels of thyroid hormones in the blood (**hyperthyroidism**). An increased metabolic rate results which is characterised by a high pulse rate (tachycardia), high breathing rate and high body temperature. Heart failure may result in extreme cases. Treatment includes both surgical removal of part of the thyroid or its destruction by the administration of radioactive iodine.

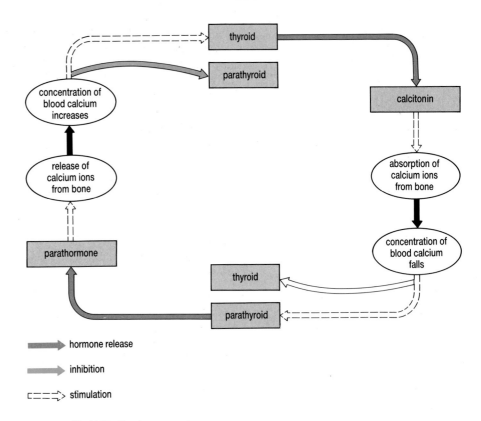

Fig 14.10 Parathormone and calcitonin interact in controlling calcium levels in the blood.

Parathyroid glands

During the early days of surgery to remove part of the thyroid gland, a number of patients suffered disruption to their calcium metabolism and died within 48 hours of their operation. Investigation of these deaths led to the discovery of the parathyroids, embedded in the thyroid like four spots on a bow tie. Their only hormone, **parathormone**, increases the calcium ion concentration of the blood by stimulating its rate of absorption by the kidney and the gut wall and by stimulating its release from bone. Oversecretion of this hormone results in excessive removal of calcium ions from the skeleton: undersecretion results in a low calcium concentration in the blood, leading to nervous disorders and prolonged contraction of muscles (tetany). Fig 14.10 shows how parathormone acts antagonistically to calcitonin in controlling blood calcium levels.

ANALYSIS

Calcium and hormones

This exercise enables you to practise the skills of assimilating information and then applying that information to a problem.

Calcium is an important constituent of teeth and bones. Insufficient calcium in the diet leads to a variety of disorders of which perhaps the most notable is rickets in young children. Rickets is a condition in which the protein of new bone fails to mineralise by deposition of calcium phosphate. As a consequence, the growing bones become deformed. Rickets was commonly associated with the poor industrial areas of temperate climates and was particularly prevalent in the industrial areas of Britain during the late nineteenth and early twentieth centuries. It is extremely rare in socially deprived communities living in subtropical or tropical countries.

Another possible consequence of severe calcium deficiency (hypocalcaemia) is that of spontaneous muscular spasm or calcaemic

---

tetany, since calcium is of great importance in muscular contraction and in the regulation of nerve impulse transmission across synapses.

Calcium is also an essential cofactor in the clotting of blood. It occurs in blood plasma at a concentration of about 0.1 mg cm^{-3} and deviates only by about 5% either way from this norm. Of the free calcium present in blood, about half is in a free, ionised and chemically active state and the remainder is bound to plasma proteins.

Regulation of blood calcium level is therefore extremely important and almost exclusively hormonal, since the cells of the endocrine glands involved (i.e. the thyroids and parathyroids) are themselves sensitive to the calcium levels circulating in the blood: the parathyroids secrete a polypeptide hormone, parathormone, and the thyroid a polypeptide hormone, calcitonin.

(a) Study Fig 14.11 which summarises the results of an experiment involving parathormone (PTH). Account for the changes in the blood serum and urine levels of calcium ions and of phosphate ions.

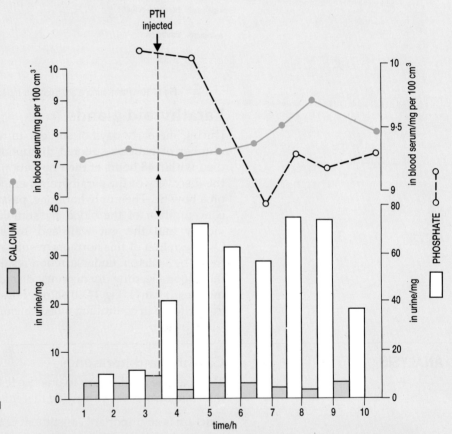

Fig 14.11 The effect of parathormone on Ca^{2+} and PO_4^{3-} levels in urine and blood serum.

(b) Sometimes a growth or tumour may develop on the parathyroid gland, with the result that large amounts of parathormone are manufactured and released into the blood, giving a permanently elevated level of parathormone. When this situation lasts for many months, what effect do you think it is likely to have on
- the level of Ca^{2+} ions in the blood?
- the constitution of the bones?
- the levels of Ca^{2+} ions and PO_4^{3-} ions excreted in the urine?

You can assume that the thyroid will not be capable of producing enough calcitonin to counteract the abnormal amount of parathormone produced by a substantial growth of this kind.

Adrenal glands

Like the pituitary gland, the adrenal glands consist of tissue from two different origins, working independently. The outer **cortex** develops from the same embryological tissue as the kidneys and gonads. It makes up about 80% of each adrenal gland and is essential for life. The inner **medulla** is derived from nervous tissue and is not essential for life.

Adrenal cortex

Under the influence of adrenocorticotrophic hormone (ACTH) from the pituitary, the adrenal cortex produces a number of steroid hormones, all derived from cholesterol (see Section 1.5). Steroid hormones from the adrenal cortex are called **corticoids**, and are divided into two groups.

- **Mineralocorticoids** control the metabolism of inorganic ions (minerals)
- **Glucocorticoids** control glucose metabolism

It is thought that, being lipid-soluble, corticoids diffuse through the surface membrane of their target cells and attach to receptors in the cytoplasm. These complexes migrate to the nucleus and attach to specific areas of a chromosome, inhibiting or stimulating the transcription of genes. The effects of **aldosterone**, one of the mineralocorticoids, are described in Section 11.3.

Adrenal medulla

Adrenaline and noradrenaline are released only during times of excitement, fear or stress. Their effects, shown in Table 14.5, are almost identical, are widespread throughout the body and help the body to prepare for action. Noradrenaline is also secreted by nerve cells of the sympathetic nervous system (see Section 15.6).

Table 14.5   The effects of adrenaline and noradrenaline

Effect on body	Advantage during stress
pupil of eyes dilate	increases visual sensitivity, especially to movement
sensory threshold lowered	perception of external stimuli is faster
mental awareness increased	'decisions' made faster
bronchioles dilated	more air inhaled into lungs
heart rate and stroke volume increased	increases delivery of metabolites to cells and removal of wastes from cells
vasoconstriction of most arteries and arterioles	increases blood pressure
vasodilation of vessels to brain and muscles (adrenaline only)	increases supply of metabolites and removal of waste products
glycogen converted to glucose in liver	more glucose available for cell respiration
peristalsis and digestion inhibited	allows blood to be diverted to more life-saving processes
hair raised ('goose pimples' in humans)	makes furry animals appear larger, which may deter attackers

 3 **In what ways does the action of hormones of the adrenal medulla differ from that of most other hormones?**

Pancreas

The role of the pancreatic hormones **insulin** and **glucagon** is described in Section 11.5. Both are polypeptides and work antagonistically to regulate blood glucose concentration. Insulin is secreted at times when the blood glucose concentration is high and stimulates the uptake of glucose and its storage as glycogen. Glucagon is released when the blood glucose concentration is low and stimulates the breakdown of glycogen to release glucose.

 4   **Suggest how the effect of adrenaline on glucose production from glycogen (Table 14.5) might be of advantage to a mammal.**

The gut

The initial release of gastric juices into the stomach is stimulated by a nervous reflex. Later secretion is stimulated by the hormone **gastrin**, itself released by cells lining the stomach wall. The release of bile and pancreatic juices into the duodenum is stimulated largely by hormonal action. **Secretin** and **cholecystokinin-pancreozymin** are released by the lining of the duodenum and stimulate the release of pancreatic juice and bile, respectively. The control of gut secretion is further described in Chapter 8.

SPOTLIGHT

GASTRIN, BACTERIA AND DUODENAL ULCERS

You may be surprised to learn that one of the best-selling drugs in the world is for the treatment of intestinal ulcers. Duodenal ulcers are thought to occur when the stomach produces too much acid. Many doctors believe that stress and diet are part of the cause of increased acid production. New evidence suggests that a bacterium, *Helicobacter pylori*, may also be involved.

H. pylori lives in the lower part (antrum) of the stomach. Cells lining this part of the stomach release the hormone gastrin, which stimulates the stomach to secrete acid. When the acidity in the stomach reaches its optimum level, the cells that produce gastrin are 'switched off'.

The acid in the stomach kills most bacteria. *H. pylori* can survive in the stomach because it neutralises the stomach's acid by converting urea (present in very small amounts in gastric juice) into alkaline ammonia. When the ammonia neutralises the stomach's acid, the gastrin-producing cells continue to release gastrin and overproduction of acid results.

A Scottish research team led by Kenneth McColl found they could kill the *H. pylori* in their patients' stomachs using a cocktail of antibiotics and a bismuth compound. The stomachs of treated patients produced less gastrin and were less acidic than those with *H. pylori* infections. Treated patients remained free of duodenal ulcers for longer periods than untreated patients.

H. pylori can be spread from one person to another by close contact. If the Scottish research team turn out to be right, it is possible that a duodenal ulcer may be something you catch!

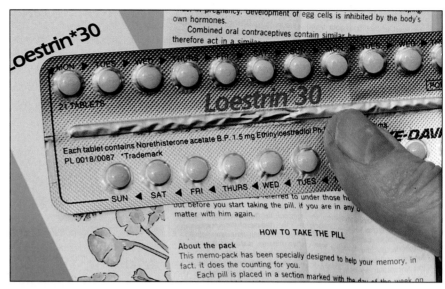

Fig 14.12 Contraceptive pills, an example of an artificial hormone which has tremendous impact on our lives.

Gonads

The female and male gonads are described in more detail in Section 20.1. Each produces gametes, egg cells or sperm, from puberty onwards. In the female, under the influence of **follicle stimulating hormone**, small follicles within the ovary develop. These contain a developing egg cell and also secrete **oestrogen**, a mixture of hormones which stimulate the development of secondary sexual characteristics of the female, including growth of the uterine lining. After the release of an egg cell from a ripe follicle (ovulation) and under the influence of **luteinising hormone**, the follicle secretes **progesterone** which prepares the body for pregnancy. The complex interaction between these hormones in bringing about the menstrual cycle is described in Chapter 20. Hormonal control of male development is somewhat simpler. In males, luteinising hormone stimulates groups of cells in the testes to produce male hormones, called **androgens**. The major androgen is **testosterone** which initiates and maintains the secondary sexual characteristics of the male and stimulates sperm production.

 Both male and female athletes have been known to take extra testosterone. Suggest the advantages and disadvantages of this practice which is banned by the International Amateur Athletics Federation.

Kidneys

When the hydrostatic pressure of its blood supply is low, the kidney produces a hormone called **renin**. This hormone stimulates the conversion of an inactive plasma protein, **angiotensinogen**, into active **angiotensin**. Because it causes generalised vasoconstriction, angiotensin raises the arterial blood pressure. It also stimulates the adrenal cortex to release aldosterone which, in turn, stimulates the uptake of sodium and chloride ions by the kidney tubules.

 Why will aldosterone help to increase blood pressure?

Pineal gland

The pineal gland is only about 8 mm in length and 4 mm wide. Its function is largely unknown, though it is thought to be affected by light. It secretes **melatonin** which is thought to influence physiological processes that are affected by time, e.g. sleep. Tumours of the pineal gland result in early sexual development in males, suggesting that melatonin might influence the timing of puberty.

PROSTAGLANDINS AND ENDORPHINS

Prostaglandins

These are fatty acids which are produced by many, if not most, cells of the body. The Nobel Prize was awarded in 1983 to the discoverers of prostaglandins (Sune Bergstrom, Bengt Samuelsson and John Vane), and research into their function is still in its early stages. Among the effects of prostaglandins are

- initiation of contraction of the uterus
- aggregation of thrombocytes during blood clotting (see Section 10.2)
- promotion of inflammation in response to injury or infection (see Section 13.2)
- constriction of the bronchi.

Aspirin (acetylsalicylic acid) is commonly used as a pain reliever. Its main effect in the body is to inhibit the production of prostaglandins, suggesting that some prostaglandins must result in pain. The value of aspirin as a pain reliever for menstrual pains can be explained from what we already know about prostaglandins. Just before childbirth, and to a lesser extent during menstrual flow, prostaglandin secretion by the uterine lining causes the muscles of the uterus to contract. A cramp is experienced if excessive contraction of these muscles squeezes off the blood supply to the uterus. By inhibiting prostaglandin synthesis, aspirin can reduce the severity of these cramps.

Endorphins

Pain is usually associated with some injury to the body. However, pain itself is a sensation. Like all sensations, pain is actually perceived by special cells in the brain. Opiates, such as opium and morphine, have been used as pain killers for centuries. In the early 1970s it was discovered that opiates bind to the surface membrane of certain nerve cells within the brain. It seemed unlikely that brain cells would have evolved receptor sites for these plant substances. Further research showed that the receptor sites were for a group of chemicals secreted in the body called **endorphins** (endogenous morphine-like compounds). All endorphins discovered so far are peptides and have a similar effect to opiates, i.e. they block the sensation of pain even though tissue damage continues. Further research is needed to find the normal role of these endorphins, but it is possible that they allow us to function during an emergency by ignoring pain until we are safe.

CONTROL AND COORDINATION IN ANIMALS: HORMONES

14.7 Suggest why radioactive iodine affects cells of the thyroid but not cells in other tissues when administered to a sufferer of hyper-thyroidism.

14.8 Copy the table below and correctly fill in the spaces labelled A – H.

Name of gland	Position of gland	Name of hormone	Effect of hormone
A	B	C	stimulates the release of thyroxine
D	E	thyroxine	F
G	adrenal medulla	H	stimulates the conversion of glycogen to glucose

14.9 The hormone adrenaline is supposed to prepare the body for 'fight or flight'. Explain this statement. If it helps, think about what might happen to your body if you saw a car coming towards you too quickly to stop before you had crossed the road.

14.4 HORMONES IN OTHER ANIMALS

Hormones play a central role in controlling the life cycle of many groups of animals. To illustrate this we will consider two groups, insects and amphibians.

Insect hormones

Many insects have a complex life cycle (see Fig 27.5). Adult females lay eggs which develop into worm-like larvae. These larvae grow, eventually forming pupae in which the larval body form is broken down and reorganised into that of an adult. The body of an insect, including that of a larva, is enclosed within a fairly rigid exoskeleton. Although it allows some expansion of the body, a larva must lose its exoskeleton (moult) if it is to grow. A series of moults each followed by rapid growth leads to discontinuous growth, described in Chapter 21. The timing of moulting (ecdysis) is controlled by the insect's brain, which produces a peptide called **brain hormone**. Release of this hormone stimulates the prothoracic gland to release a steroid moulting hormone (**ecdysone**). As with many

Fig 14.13 An axolotl from the lakes in Mexio. This amphibian retains larval characteristics, such as the gills visible in the photograph, permanently. Only rarely do they mature into the adult salamander.

mammalian hormones, ecdysone exerts its effect on DNA, promoting the transcription of a number of genes which bring about moulting.

When it moults, a larva either becomes a new, larger larva or it develops into a pupa. Which of these occurs depends on the concentration of a second hormone, juvenile hormone (**neotonin**), released from two regions behind the brain called the **corpora allata**. During the early larval stages (**instars**), the corpora allata release relatively large amounts of neotonin and a larger larva is formed after each moult. As the larva matures, the brain inhibits release of neotonin. When the levels of neotonin are at a threshold low level, a pupa is formed after the next moult.

Amphibian hormones

Amphibians, such as frogs, also have a complex life cycle (see Section 21.2). A female frog lays eggs which develop into tadpoles. Growth of the tadpole ultimately results in a land-dwelling frog. The change in body form from tadpole to miniature frog is triggered by a sudden release of **thyroxine** from the tadpole's thyroid gland. Like humans, tadpoles need iodine to manufacture thyroxine. Tadpoles which are supplied with adequate food but kept in iodine-deficient water do not metamorphose but grow into giant tadpoles.

QUESTIONS

14.10 As grain pests become increasingly resistant to insecticides, agricultural research scientists have studied the effect of a family of chemicals which mimic the effect of insect juvenile hormones, called JH analogues. Unlike insecticides, these JH analogues are thought to be harmless to other forms of life.
(a) Explain the likely effects of JH analogues on insects.
(b) Suggest why the effects of JH analogues have proved effective in controlling populations of cockroaches and cat fleas but not of insects whose larvae are pests of stored food.

14.11 What would be the effect of administering thyroxine to an axolotl?

SUMMARY

Hormones are chemicals that are released by ductless glands directly into the bloodstream and stimulate a change in their target organs. Although they reach every cell in the body, hormones specifically attach to receptor molecules present only on the surface membrane of their target cells. Ductless glands which secrete hormones are called endocrine glands.

Typically, hormones bring about effects in the body that are slow to develop, widespread and long-lasting. Control of hormone secretion is typically by negative feedback in response to the concentration of either a specific metabolite or of another hormone in the bloodstream. Adrenaline is a notable exception, causing rapid and short-lived responses to stress.

The pituitary is the major endocrine gland of mammals. The anterior lobe of the pituitary is under the chemical control of the hypothalamus: it produces a number of hormones which themselves control growth, activity of the sexual organs and the release of hormones by the gonads and adrenal glands. The posterior lobe of the pituitary does not produce hormones but stores the hormones oxytocin and antidiuretic hormone (ADH) which are produced by the hypothalamus.

Other mammalian endocrine glands include the thyroid, parathyroids, adrenal cortex, adrenal medulla, pancreas, gut, gonads and kidneys. Each produces hormone(s) with a specific role in metabolism.

Two little-known groups of chemicals have hormone-like effects. Prostaglandins are fatty acids produced by most cells in the body. Among their effects are promotion of inflammation and blood clotting during infection. Endorphins are peptides which block pain.

A number of hormones are centrally involved in the control of metamorphosis in the life cycle of insects. Moulting (ecdysis) is controlled by brain hormone and ecdysone; development from a larva into an adult is controlled by juvenile hormone (neotonin). The hormone thyroxine, which regulates metabolic rate in mammals, stimulates metamorphosis in amphibians.

Answers to Quick Questions: Chapter 14

1   Anterior lobe is controlled by hormones, posterior lobe by nervous stimulation.
2   Although similar in structure, the cell-surface receptors are specific to each hormone.
3   Their effects are widespread, appear quickly but are short-lived.
4   Response to stress, fighting or running away, increases rate of respiration in muscle cells. Adrenaline stimulates release of glucose, respiratory substrate used by muscles, into bloodstream from glycogen stores.
5   Major advantage – testosterone stimulates muscle development. It also inhibits ovulation, which may occur at an inconvenient time. Major disadvantage – is lethal in fairly low doses; also causes females to develop male characteristics.
6   Vasoconstriction causes same volume of blood to flow into a smaller diameter bore. Increased uptake of ions causes the blood's water potential to become more negative. As a result, water passes from tissues into blood by osmosis: its volume increases and hence so does its pressure.

Chapter 15

CONTROL AND COORDINATION IN ANIMALS: NERVES AND THE NERVOUS SYSTEM

LEARNING OBJECTIVES

When you have studied this chapter you should be able to:

1. describe the structure of receptor, effector and relay neurones and explain how they may connect in an unconditioned reflex arc;

2. outline the way in which a conditioned reflex is formed;

3. explain how a nerve impulse is initiated and transmitted along unmyelinated and myelinated neurones;

4. describe the structure and functioning of a synapse and the role of different neurotransmitter substances;

5. outline the structure and functions of the major parts of the central nervous system and peripheral nervous system;

6. distinguish between the effects of the sympathetic and parasympathetic branches of the autonomic nervous system.

Consider for a moment what is going on in your mind as you read this page. You receive sensory information about the page from your eyes and use reading skills which are stored as memory in your brain to convert it into some sort of meaning. You will, hopefully, store memories of some of this information in your brain and relate it to other information which is already stored there. At the same time, your brain receives information

Fig 15.1 This driver is not only steering the car, watching the road, using foot pedals but also listening to instructions. We can all do several things at the same time, but an added stress, such as being late, can overload the system!

from touch receptors as your clothing, the chair in which you are sitting and the pages of the book rub against your skin. You control movements of your muscles as you turn over pages, shift position in your chair and make notes of what you are reading. You may also be aware of the stereo playing in the background as you read, of the smell of a meal which is being prepared and have time for the occasional day-dream. All this awareness is the result of millions and millions of separate nerve cells which transmit brief electrical signals around your body. This chapter examines the structure of these nerve cells, the way in which they transmit their electrical signals and the way in which they are organised within the nervous system.

15.1 NEURONES

Nerve cells (neurones)

Nerve impulses are transmitted by highly specialised cells called **neurones** which are linked together to form nervous pathways. The structure of three types of neurone is shown in Fig 15.2 and their functions summarised in Table 15.1. All neurones have a **cell body** which contains the bulk of the cell organelles and has a number of cytoplasmic extensions. The number of such extensions provides one basis for classifying neurones

- **unipolar neurones** have only one extension
- **bipolar neurones** have two extensions
- **multipolar neurones** have more than two extensions.

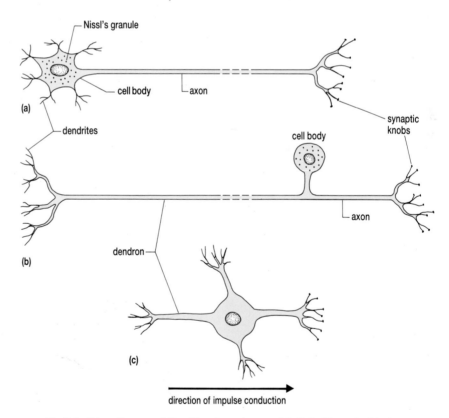

Fig 15.2 Schematic representation of three types of neurone (**a**) effector (**b**) receptor (**c**) relay.

Cytoplasmic extensions of the cell body only transmit impulses in one direction. Extensions which transmit impulses to the cell body are called **dendrons** or, if they are very small and numerous, **dendrites**. Extensions which transmit impulses away from the cell body are called **axons**; they are usually thinner than dendrons and may be up to several metres long.

Table 15.1   Functions performed by neurones

Function	Part of neurone which performs this function
receive input information from receptors	**Dendrites**, fine cytoplasmic strands, receive stimuli and convert them to electrical signals. In some neurones, the dendrites collect into one or more larger dendrons.
transmit electrical signals ensuring they do not fade in intensity as they travel	**Axons**, cytoplasmic extensions which may be several metres long, carry signals to the target.
transmit a signal to the target organs	**Synaptic knobs**, microscopic, bulb-like endings of the axons, release chemical transmitters onto the target cell across a gap of about 20 nm called the **synapse**.
coordination of metabolic activities of cell	**Cell body**, which contains nucleus, mitochondria, ribosomes, endoplasmic reticulum, etc. Many have densely staining rows of endoplasmic reticulum, called **Nissl's granules**.

 1 State whether each cell in Fig 15.2 is unipolar, bipolar or multi-polar.

Neurones can also be classified on the basis of their function.

- **Receptor neurones** (also called sensory or afferent neurones) transmit impulses from receptors to the central nervous system (brain and spinal cord).
- **Effector neurones** (also called motor or efferent neurones) transmit impulses from the central nervous system to muscles and glands.
- **Relay neurones** link receptor and effector neurones in nervous pathways within the central nervous system. (These are also known as intermediate, association and internuncial neurones.)

Satellite cells

These cells cannot conduct impulses but surround neurones within nervous tissue. In the brain and spinal cord, satellite cells called **neuroglia** are more numerous than the neurones. Some of these neuroglia form membranes, others are phagocytic and yet others are thought to be involved in memory. Outside the brain and spinal cord, the most common satellite cell is the **Schwann cell**. As the axons and dendrons of many mammalian neurones develop, Schwann cells wrap around them. Look at Fig 15.3 which shows a transverse section across such an axon. Notice how many layers of the Schwann cell's surface membrane are wrapped around the axon. The membranes of Schwann cells are largely composed of lipid and they do not contain the protein channels which, as we will see below, are essential to the transmission of nerve impulses. Consequently, the layers of membrane around the neurone effectively insulate it against 'electrical leakage'. Since the lipid in the membrane of Schwann cells is mainly myelin, neurones which are enclosed by Schwann cells are said to be **myelinated**. Between adjacent Schwann cells of a myelinated axon there are small gaps called **nodes of Ranvier** (Fig 15.4).

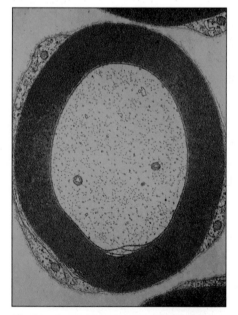

Fig 15.3   A transverse section through a myelinated axon. The myelin in the surface membrane of the Schwann cell, visible as the black concentric rings outside the axon, insulates against loss of ions.

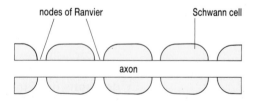

nodes of Ranvier   Schwann cell

axon

Fig 15.4   Part of a myelinated axon showing nodes of Ranvier, i.e. the gaps between adjacent Schwann cells.

15.1 Draw a myelinated effector neurone and annotate it to show the function of its parts.

15.2 Compare the structure of a neurone with that of the live wire which supplies electricity to a refrigerator.

15.2 SIMPLE NERVOUS PATHWAYS

This section looks at how nervous systems work. We shall start with a simple system as illustrated by **reflexes**.

Unconditioned reflex

Unconditioned reflexes are inborn and unlearned. They result in a fixed response (**unconditioned response**) always being produced following a particular stimulus (**unconditioned stimulus**). For example, you jerk your knee if the tendon of your thigh muscle is suddenly stretched, you blink your eyelids if an object flies near your face and you salivate if you taste food. The nervous pathway controlling an unconditioned reflex is called a **reflex arc** and is the simplest nervous pathway in a mammal's body. The reflex arc controlling the knee-jerk reflex is represented in Fig 15.5(a). It involves only two neurones: a receptor neurone detects stretching of the thigh muscle tendon produced, for example, by tapping it with a hammer; an effector neurone stimulates contraction of a cell in the thigh muscle. Hundreds of these arcs result in the contraction of sufficient cells in the thigh muscle to lift the lower leg when its tendon over the knee is tapped.

The reflex arc shown in Fig 15.5(b) has three neurones: a receptor neurone, an effector neurone and a relay neurone. Notice that the central nervous system (brain and spinal cord) is involved in both reflex arcs, irrespective of whether a relay neurone is involved. The part of the central nervous system involved in both the reflex arcs shown in Fig 15.5 is the spinal cord, so these reflexes are called **spinal reflexes**. The contraction of the circular muscle of the iris to close the pupil in bright light is a **cranial reflex**, involving the brain and not the spinal cord.

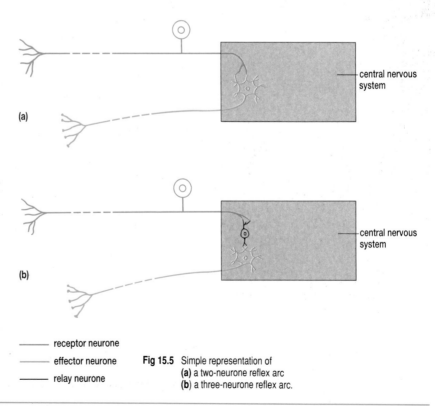

(a)

(b)

— receptor neurone
— effector neurone
— relay neurone

Fig 15.5 Simple representation of
(a) a two-neurone reflex arc
(b) a three-neurone reflex arc.

central nervous system

central nervous system

Conditioned reflex

Unlike an unconditioned reflex, a conditioned reflex is learned. Learning involves areas of the brain, called **association areas**. A conditioned reflex occurs when a new stimulus (**conditioned stimulus**) is presented at the same time as the unconditioned stimulus. If these two stimuli are presented together often enough, an animal will eventually produce the unconditioned response even if the conditioned stimulus is given alone. The response is now said to be a **conditioned response**. For example, the Russian scientist Pavlov rang a bell at the same time as feeding his dogs. Initially, the dogs salivated (unconditioned response) at the sight of food (unconditioned stimulus). However, after a while the dogs would salivate if the bell was rung but no food was presented. Salivation is now a conditioned response to a conditioned stimulus (ringing the bell). Salivation became the conditioned response to a conditioned stimulus because certain areas within the dog's brain associated the unconditioned and conditioned stimuli. A new nervous pathway (Fig 15.6) had developed in which an association area of the dog's brain had linked two stimuli and elicited the same response to them. The brain, and its role in learning, is described more fully in Section 15.5.

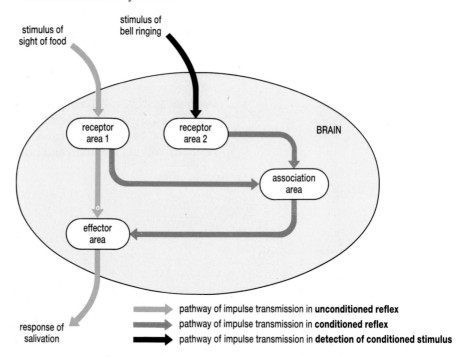

Fig 15.6 The nervous pathway involved in a conditioned reflex.

Associating an unconditioned and conditioned stimulus need not be the result of a deliberate act performed by a research biologist, and it occurs in animals other than dogs. Vomiting is an unconditioned response to the presence of poison in the stomach. It has obvious survival value and is controlled by an unconditioned reflex arc. Some people who have suffered a particularly bad attack of food poisoning find that, for a long time afterwards, the sight of the offending meal makes them feel sick. They would normally vomit only in response to a poison in the stomach, but now they feel as if they might vomit when given a totally different stimulus. To an unconditioned reflex (vomiting in response to poison in the stomach) has developed a conditioned reflex (vomiting in response to the sight of a particular meal).

15.3 Summarise the properties of a simple reflex and outline its nervous control.

15.4 Many cat owners find that their cats appear at their feet within seconds of starting to use a can-opener in their kitchen. Suggest how this response has been learned by the cats.

15.3 THE ELECTRICAL NATURE OF THE NERVE IMPULSE

The speed with which reflex actions occur suggests that they must involve an active process rather than a passive diffusion of some chemical. A long series of experiments starting in the last century showed us that the impulses going down neurones are electrochemical in nature. The key to understanding the nerve impulse is the transport processes that occur in cell membranes, particularly the sodium pumps. These are discussed in Chapter 2. You may need to review that chapter as you work through the tutorial: The nerve impulse.

TUTORIAL

The nerve impulse

The data in Table 15.2 show the concentration of some major ions in the cytoplasm of an axon and in the fluid around the axon. The data show an imbalance in the concentration of these ions on each side of the surface membrane of the axon.

(a) Compare and contrast the distribution of positively charged and negatively charged ions
   (i) within the cytoplasm of the axon
   (ii) in the fluid around the axon
   (iii) between the cytoplasm within and the fluid around the axon.

Table 15.2 The distribution of ions inside and outside the axon of a typical mammalian neurone

Ion	Concentration / mmol dm^{-3}	
	in cytoplasm of axon	in fluid around axon
chloride (Cl$^-$)	4	120
organic anions (e.g. proteins)	163	29
potassium (K$^+$)	155	4
sodium (Na$^+$)	12	145

The imbalance in the concentration of organic ions is relatively easy to explain: their molecules are synthesised inside the cell and are too large to diffuse through the cell surface membrane. All the inorganic ions shown in Table 15.2 are small enough to diffuse through the surface membrane of the neurone. However, it is the activity of the sodium pumps that maintains the imbalance in these inorganic ions. As sodium ions diffuse into the cytoplasm of the axon, the sodium pumps use energy from the hydrolysis of ATP to carry them back out again.

(b) How do you account for the higher concentration of potassium ions inside the axon than outside?

As a result of this ionic imbalance, there is a voltage (or potential difference) across the surface membrane: the membrane is said to be **polarised**. A cathode ray oscilloscope and electrodes (Fig 15.7) can be used to measure this potential difference at a point (P) along an axon. For a resting neurone, i.e. one not conducting an impulse, this potential difference is called the **resting potential**. Values between −40 mV and −120 mV for the resting potential have been measured in neurones from different animals (the 'minus' sign indicates that the inside of the axon is

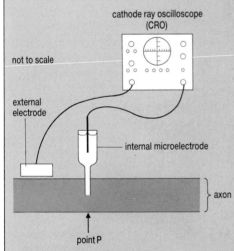

cathode ray oscilloscope
(CRO)

not to scale

external electrode

internal microelectrode

axon

point P

Fig 15.7 The potential difference across the cell surface membrane of an axon is transmitted via electrodes to be measured by a cathode ray oscilloscope (CRO).

A television receiver displays electrical signals as a picture on a fluorescent screen; a CRO displays very small signals as a single line against a grid background to measure their duration and strength.

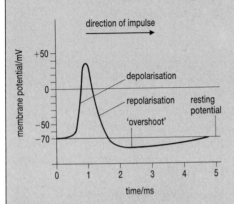

Fig 15.8 An action potential trace as seen on the CRO screen.

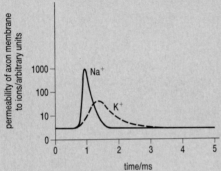

Fig 15.9 Changes in the permeability of the axon's cell surface membrane to Na⁺ and K⁺ during an action potential.

negative with respect to the outside of the cell.) Calculations using the data in Table 15.2 suggest that a potential difference of −90 mV would be measured for this axon.

If an electrical stimulus is now applied to one end of the axon, the potential difference at P is found to change. Shortly after the stimulus has been applied, the inside suddenly becomes positive with respect to the outside. The surface membrane is said to have been **depolarised**. The potential then gradually returns to its normal resting value; the surface membrane is **repolarised**. These changes, called an **action potential**, are summarised in Fig 15.8, which shows the sort of trace you would observe on the screen of the cathode ray oscilloscope.

(c) During depolarisation, what is the change in the potential difference across the axon's surface membrane?

In an unmyelinated axon, the action potential is accompanied by dramatic changes in the permeability of the cell surface membrane to sodium and potassium ions shown in Fig 15.9. When the axon potential reaches a certain value, called the **action potential threshold**, the permeability to sodium ions at that point of the axon suddenly increases about one thousandfold. This allows the rapid diffusion of sodium ions into the axon.

(d) What effect will this sudden influx of sodium ions have on the membrane potential of the axon?

(e) How is this change at the ionic level reflected in the trace on the cathode ray oscilloscope?

After about 0.3 ms, the potential reaches a value at which the sodium ions are in equilibrium and the inward diffusion of sodium ions slows down. Meanwhile the permeability of the surface membrane to potassium ions has gradually increased about thirtyfold. As a result, potassium ions begin to diffuse out of the axon.

(f) How will this affect the membrane potential of the axon?

The potential at which potassium ions are in equilibrium is lower than the resting potential of the axon, explaining why the axon membrane potential actually reaches a value below the resting potential of the cell. Eventually, the resting potential is restored. Over the first millisecond, this return to the resting potential is due to changes in the permeability of the cell surface membrane to potassium ions and not to the effects of the sodium pumps, which act much more slowly. The sodium pumps do restore the resting concentrations of sodium and potassium ions, but this takes about 50 ms.

Propagation of an impulse

Look at Fig 15.10 which shows the changes in charge and the movement of ions when a nerve impulse begins.

- Fig 15.10(a) represents part of an unmyelinated axon at rest. The plus and minus signs represent the overall charges on each side of the cell surface membrane which contribute to the resting potential of this axon.
- When a nerve impulse reaches any point on the axon, an action potential occurs and the balance of charges changes to that shown in Fig 15.10(b). The arrows in Fig 15.10(b) show how local electrical currents occur when adjacent portions of the surface membrane have a different potential.

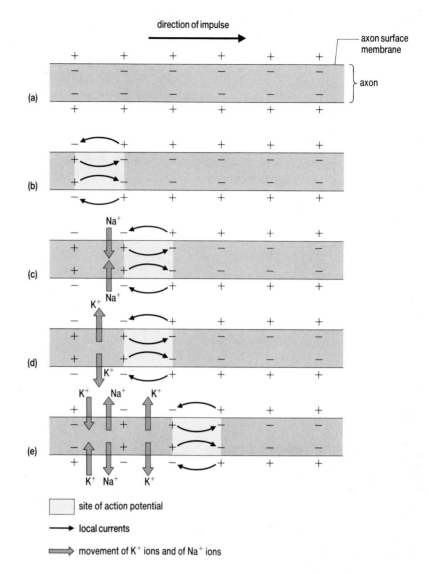

Fig 15.10 The balance of charges and movement of ions during the propagation of a nerve impulse.
(a) Part of the axon of a resting neurone.
(b) Stimulation of the axon causes an action potential and local currents are set up on the axon membrane.
(c) Once the action potential threshold has been reached, sodium ions diffuse into the adjacent portion of the axon and a new action potential occurs.
(d) Potassium ions diffuse out behind the impulse as it progresses along the axon and this portion of surface membrane is capable of depolarisation again.
(e) Sodium ions are pumped from the cytoplasm of the axon while potassium ions are pumped back again, restoring the original ion concentrations.

- These local currents change the potential of the surface membrane at the next point in the direction of the impulse transmission until it reaches its action potential threshold: an increase in the permeability of this portion of the cell surface membrane to sodium ions results and sodium ions rapidly diffuse into the axon so that an action potential occurs here (Fig 15.10(c)).
- Local currents are set up between this new portion of axon and the portion immediately next to it in the direction of the nerve impulse transmission and so the impulse progresses along the axon.
- Behind the impulse, the outward diffusion of potassium ions causes the axon to become repolarised (Fig 15.10(d)), allowing another impulse to be passed .
- Eventually the activity of the sodium pumps restores the original ionic balance (Fig 15.10(e)).

3 **What keeps the action potential moving in one direction only?**

Impulse propagation in myelinated axons

Relatively few ions pass through the myelin sheath except at the nodes of Ranvier, which are about 1 mm apart. Since they are the result of movement of ions, action potentials can only occur at the nodes of Ranvier in myelinated axons. Hence, the action potentials jump from node to node, increasing the speed of impulse transmission in myelinated axons (see Fig 15.11). This type of impulse propagation is called **saltatory** (after the Latin *saltare*, to jump).

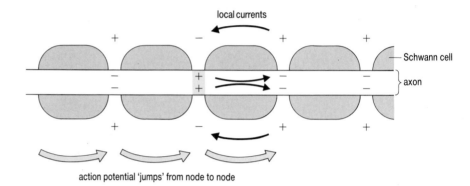

Fig 15.11 Saltatory propagation of an impulse in a myelinated axon.

Refractory period

There is a time immediately following an action potential when the cell surface membrane of an axon cannot change its permeability to sodium ions, no matter how intense the stimulus. This is the **absolute refractory period** which lasts for about 1 ms. The **relative refractory period**, during which time new action potentials occur only if the intensity of stimulation is much greater than the action potential threshold, lasts for up to about 10 ms after this.

The refractory period has two important consequences.

- It ensures that an impulse can only flow in one direction along an axon, since the portion of axon behind the impulse cannot be depolarised again.
- It limits the frequency with which successive impulses can pass along axons.

ANALYSIS

The speed of propagation of nerve impulses

This exercise gives practice in data analysis.

Look at the data in the table.

Neurone	Diameter/μm	Myelinated or unmyelinated	Speed of impulse transmission/m s^{-1}
cat axon	1.0	unmyelinated	3.0
cat axon	10.0	myelinated	50.0
frog axon	10.0	myelinated	30.0
squid axon	1000.0	unmyelinated	30.0

(a) Describe the relationship between the properties of these neurones and the speed with which they conduct impulses.

(b) Suggest an explanation of these data.

Fig 15.12 A squid can eject a jet of water to escape from predators by rapidly contracting the muscles around its mantle cavity.

When a squid is attacked it escapes by darting backwards and ejecting a cloud of inky pigment which confuses its enemy as it makes good its escape. This swift backwards movement is produced by contraction of the muscles surrounding the water-filled mantle cavity. The water in the mantle cavity is squirted out of a small opening, effectively providing the squid with rocket propulsion. The impulses needed to make the muscles contract originate from the stellate ganglion, a group of nerve cells near the front end of the squid. The muscles at the rear of the mantle cavity contract before those at the front end.

(c) Explain why the muscles must contract in this sequence to be effective in jet propulsion.

(d) Suggest two explanations of rapid transmission of the impulse to the more distant muscles but slower transmission to the muscles near the stellate ganglion.

(e) How could you test your explanations given in (d)?

QUESTIONS

15.5 Explain the meaning of
  (a) polarised membrane
  (b) resting potential.

15.6 Describe how the resting potential of a neurone is maintained.

15.7 Describe the electrical changes which can be measured during an action potential and explain their occurrence.

15.8 Electrical currents fade as they pass along a wire. Nerve impulses do not fade as they pass along neurones. Use your knowledge of the propagation of nerve impulses to suggest why.

15.9 Action potentials are often described as being all-or-nothing. Suggest an explanation for the meaning of this term.

15.4 THE SYNAPSE

A **synapse** is the junction between two neurones (Fig 15.13): some neurones in your brain may form several hundred synapses. Neurones do not actually touch their target cells. Instead there is a gap, called the **synaptic cleft**, of up to 20 nm between the first (**pre-synaptic**) neurone and the second (**post-synaptic**) neurone. Although the synaptic cleft is bridged electrically in some neurones, in the vast majority of synapses the pre-synaptic neurone communicates with the post-synaptic neurone by releasing a hormone-like **neurotransmitter substance**. The special synapse between an effector neurone and a muscle cell, called a **neuromuscular junction**, is described in Section 15.6.

Structure of a synapse

The axons of neurones end in small bulbous **synaptic knobs**. The surface membrane of the synaptic knob, known as the **pre-synaptic membrane**, is separated from the surface membrane of a dendrite on its target cell, the **post-synaptic membrane**, by the synaptic cleft (Fig 15.13). The post-synaptic membrane has numerous channels through which movement of specific ions can be controlled. It also has large protein molecules on its surface which act as receptor sites for the neurotransmitter substances released by the synaptic knob. Prior to its release, the neurotransmitter substance is located in small membrane-bound structures in the cytoplasm of the synaptic knob, called **synaptic vesicles**. Although a number of neurotransmitter substances exist, the two main chemical transmitters in mammals are **acetylcholine** and **noradrenaline**. Neurones which release acetylcholine are termed **cholinergic** neurones; those which release noradrenaline are termed **adrenergic** neurones. Examples of other neuro-transmitter substances are given in Table 15.3.

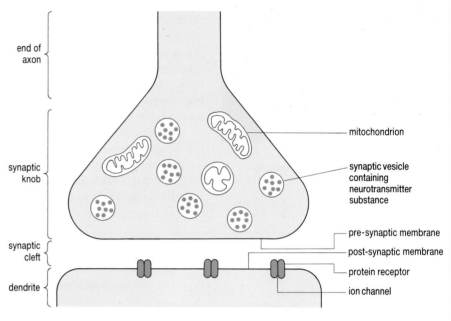

Fig 15.13 The structure of a synapse.

Table 15.3 Neurotransmitter substances in the mammalian nervous system

Neurotransmitter	Site of action	Function	Effects of certain drugs
acetylcholine	throughout the nervous system	excitation or inhibition	atropine blocks action in parasympathetic nervous system
			botulinum toxin prevents release
			curare blocks action on muscles
			nicotine mimics action
			strychnine stops action of acetylcholinesterase
noradrenaline	in sympathetic nervous system	excitation	amphetamines stimulate release
			imipramine inhibits reabsorption
			reserpine reduces storage in synaptic knob
dopamine	brain	excitation	chlorpromazine blocks receptors of post-synaptic membranes
serotonin	brain	excitation	lysergic acid diethylamide (LSD) affects synapses

TUTORIAL

Synaptic transmission

The first step in transmitting a nerve impulse across a synapse involves calcium ions. The extracellular concentration of calcium ions is ten thousand times greater than their concentration inside the synaptic knob.

When a nerve impulse reaches a synaptic knob it causes an increase in the permeability of the pre-synaptic membrane to calcium ions (Fig 15.14(a)).

(a) Explain why calcium ions will now diffuse into the synaptic knob.

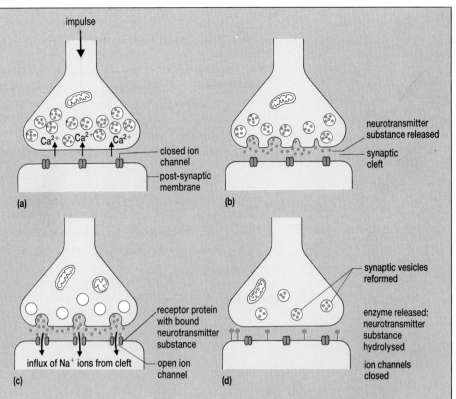

Fig 15.14 Synaptic transmission at an excitatory synapse is a sequence of events involving Ca^{2+} ions, a neurotransmitter substance, Na^+ ions and receptor sites on the post-synaptic membrane.

The resulting influx of calcium ions causes the synaptic vesicles to fuse with the pre-synaptic membrane and discharge their contents into the synaptic cleft (Fig 15.14(b)).

(b) What do you think will now happen to the calcium ions?

The released neurotransmitter substance diffuses across the synaptic cleft and attaches to its specific receptor sites on the post-synaptic membrane. The resulting change in configuration of these receptor sites affects the post-synaptic membrane's ion channels mentioned earlier. At **excitatory synapses**, these changes in configuration cause sodium ion channels to open so that they allow an influx of sodium ions into the post-synaptic neurone (Fig 15.14(c)).

(c) What will be the effect of this rapid influx of sodium ions on the potential difference across the post-synaptic membrane?

At **inhibitory synapses**, the configurational change opens channels which are specific to chloride and potassium ions. The outward movement of potassium ions and the inward movement of chloride ions increases the polarisation of the post-synaptic membrane so it is, therefore, more difficult for the action potential threshold of the post-synaptic membrane to be exceeded and less likely that a new action potential will be created.

Whether a synapse is excitatory or inhibitory depends on the nature of the receptor sites on the post-synaptic membrane, rather than on the nature of the neurotransmitter substance. For example, acetylcholine has an inhibitory effect on heart muscle and gut muscle, but an excitatory effect on skeletal muscle.

Having affected the permeability of the post-synaptic membrane, the neurotransmitter substance is almost immediately lost from the synaptic cleft, by rapid diffusion out of the cleft and hydrolysis by enzymes. In cholinergic synapses, for example, the enzyme **acetylcholinesterase**, located on the post-synaptic membrane, catalyses the hydrolysis of acetylcholine into choline and ethanoic acid. These compounds are reabsorbed by the pre-synaptic membrane where they are recoupled and stored inside the synaptic vesicles, ready to be used again (Fig 15.14(d)).

(d) What is the advantage of destroying acetylcholine almost as soon as it reaches the post-synaptic membrane?

Depolarisation of the post-synaptic membrane of an excitatory synapse only occurs when sufficient neurotransmitter substance has accumulated at the post-synaptic membrane. This phenomenon is called **summation** and can occur in two ways.

1. **Spatial summation:**   Simultaneous stimulation of several synaptic knobs may cause depolarisation of the post-synaptic membrane even though each synaptic knob may release insufficient neurotransmitter substance to cause depolarisation on its own.
2. **Temporal summation:**   Repeated stimulation of the same synaptic knob may occur until sufficient neurotransmitter substance has been released to cause depolarisation.

Thus, impulses can be generated in a post-synaptic neurone either by weak stimulation by many pre-synaptic neurones or by the repeated stimulation by one pre-synaptic neurone.

SPOTLIGHT

NEUROTOXINS

Continued transmission of nerve impulses between nerve cells and between nerve cells and muscles depends upon the enzyme acetylcholinesterase. This hydrolyses acetylcholine, the molecule which carries the nerve impulse across the synapses. Any substance which inhibits acetylcholinesterase will therefore inhibit the hydrolysis of acetylcholine which will, in turn, inhibit the functioning of the nervous system. This is the basis on which nerve gases like Sarin work. The muscles of a person affected by the nerve gas rapidly enter a permanently contracted state with death following, from asphyxiation, soon after.

It was research on nerve gases during World War II that led to the development of organophosphorus insecticides like parathion. These excellent insecticides are highly toxic to insect pests and break down rapidly in the environment so there is no accumulation of the compounds in food chains (see Chapter 30); a major advantage over insecticides such as DDT.

However, these organophosphorus insecticides are also toxic to humans and there are several thousands of deaths each year from insecticide poisoning, especially in countries where safety regulations are poorly implemented.

15.10 Produce a flow chart to show what effect the arrival of a nerve impulse has on a synaptic knob.

15.11 Explain how a weak stimulus applied to a single axon may not produce an action potential in a post-synaptic neurone whereas several weak stimuli applied to the axon might.

15.5 THE CENTRAL NERVOUS SYSTEM (CNS)

The central nervous system consists of the brain and spinal cord. These are surrounded by three protective membranes, called **meninges**. A space between the inner two meninges is filled with **cerebrospinal fluid (CSF)**, which is circulated by the action of ciliated cells and acts as a transport system within the CNS. Exchange of metabolites between the CSF and capillaries of the blood circulatory system occurs at thin membranes in the brain, called **choroid plexuses**.

The spinal cord

The spinal cord is a hollow tube passing from the lower abdomen to the brain. It has a central **spinal canal** filled with cerebrospinal fluid, around which is a region of cell bodies, unmyelinated fibres and synapses called the **grey matter**. The outer **white matter** contains myelinated axons which form tracts between the grey matter and the brain. **Ascending tracts** carry receptor impulses from the spinal cord to the brain and **descending tracts** carry effector impulses from the brain to appropriate parts of the spinal cord. It is the colour of the myelin surrounding the axons in these tracts which gives this white matter its name.

Spinal nerves occur at intervals along the spinal cord. Each spinal nerve is a mixed nerve (contains both receptor and effector neurones) but just outside the spinal cord the receptor neurones and effector neurones are separated into the dorsal and ventral roots, respectively. The cell bodies of the receptor neurones lie in a swelling on the dorsal root, called the **dorsal root ganglion**. The structure of the spinal cord and the position of neurones within it are shown in Fig 15.15.

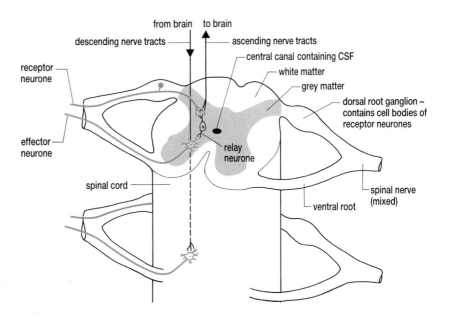

Fig 15.15 A transverse section through the spinal cord at the point of entry of a spinal nerve. The coloured lines represent the position of neurones.

The brain

Like the spinal cord, the brain is made up of grey and white matter. The grey matter, located in the outer cortex of the brain, consists of groups of cell bodies and synapses called **centres**. The association areas involved in conditioned reflexes (see Section 15.2) are centres of this type. Cells within these centres may form several hundred synapses, allowing a wide variety of responses to a given stimulus. The white matter is situated mainly in the inner part of the brain and contains myelinated nerve fibres, the **nerve tracts,** which carry impulses between centres.

During its development, the brain grows as a swelling of the spinal cord and differentiates into three distinct regions: the **forebrain, midbrain** and **hindbrain**. This division is difficult to see in humans, where the forebrain (especially the cerebrum) is very large and folded backwards to completely cover the midbrain and hindbrain (Fig 15.16). Table 15.4 summarises the functions of the major parts of the brain.

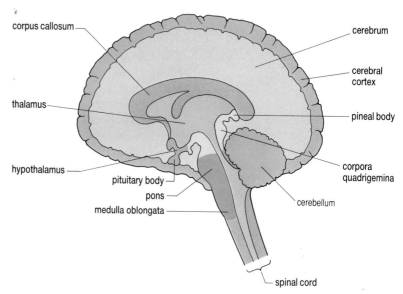

Fig 15.16 A vertical section through the human brain. The cerebral cortex is shown disproportionately thick – otherwise it would not be visible in the diagram.

The **medulla oblongata** (also called the medulla) controls the activity of many internal actions. Five of the cranial nerves originate from the medulla. It is via one of these, the **vagus nerve**, that the medulla controls autonomic activities, such as heart rate, blood pressure and peristalsis of the gut. The ascending and descending nerve tracts from the spinal cord cross over from one side to another in the medulla, e.g. the left-hand spinal nerve tracts cross to the right side of the brain.

4   Explain why damage to the right-hand side of the brain produced, for example, by a stroke often leads to paralysis of the left-hand side of the body.

The **cerebellum** of the hindbrain, consisting of two cerebellar hemispheres, controls fine muscular movements. During early learning periods, the cerebrum controls the cerebellum, resulting in jerky and clumsy movements. Smooth, reflex movement only occurs when the cerebellum alone controls this movement. You only have to watch a child learning to walk or remember how you felt when you learned to play a musical instrument or drive a car to realise the difference between cerebral and cerebellar control of movement.

Table 15.4 Functions of the major parts of the mammalian brain

Embryonic division of brain	Region of brain	Functions of this region of brain
hindbrain	medulla oblongata	contains reflex centres for control of heart rate, blood pressure, peristalsis (including vomiting), swallowing, saliva production, breathing rate, coughing and sneezing
	pons	contains centres which relay impulses to the cerebellum
	cerebellum	receives information from the muscles and the ears about posture and balance
		coordinates smooth movement
midbrain	corpora quadrigemina	controls visual and auditory reflexes
forebrain	cerebrum	controls the body's voluntary behaviour, learning, reasoning, personality, memory
	corpus callosum	connects the left and right cerebral hemispheres, allowing the two sides of the cerebrum to communicate
	thalamus	processes all sensory impulses before relaying them to the appropriate part of the brain
		perception of pain and of pleasure
	hypothalamus	coordination centre of the autonomic nervous system
		controls hunger, thirst, aggressive behaviour and reproductive behaviour
		monitors blood
		regulates pituitary gland

The cortex of the **cerebrum** is only about 3 mm thick but controls most of the functions which give us our self-awareness. Decision-making, sensory awareness, memory, intelligence, speech and learning are all functions of the cerebral cortex. Experiments during the 1950s and 1960s, in which the cerebrums of volunteers who were undergoing surgery for removal of a brain tumour were electrically or chemically stimulated, showed **localisation of function** within the cerebrum (Fig 15.17).

 5 **The cortex of the cerebrum is highly folded. Suggest an advantage of this.**

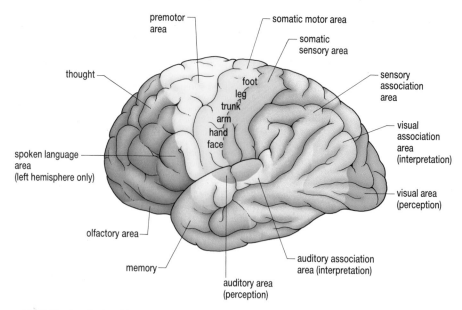

Fig 15.17 Localisation of function in the cortex of the left cerebral hemisphere. The right cerebral hemisphere has similar areas except that those concerned with speech are absent.

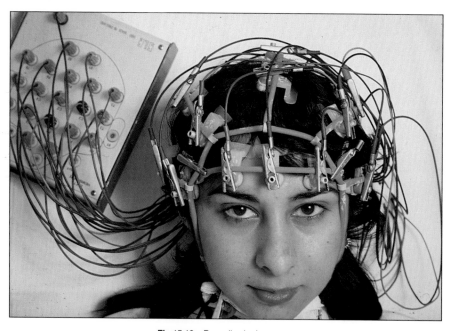

Fig 15.18 Recording brain waves.

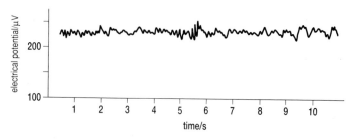

Fig 15.19 Brain waves – traces on an EEG recorded during the early sleep period.

15.12 An unconditioned reflex causes us to drop hot objects which we have picked up. If the object is valuable to us we may overcome this reflex until we have a safe place to put the object down. Suggest how synapses of relay neurones in ascending and descending tracts of our spinal cords allow us to overcome this reflex.

15.13 What are meninges? Suggest why the disease meningitis often has serious consequences.

15.14 Describe the contents of
(a) white matter
(b) grey matter.

15.15 In a simple experiment, a blindfolded person picked up an object in the left hand.
(a) Describe the parts of the nervous system through which impulses from the subject's fingers must pass if he or she is to describe the object to the person conducting the experiment.
(b) In the early treatment of epilepsy, the corpus callosum of the sufferer was surgically cut. Use the information in Table 15.4 to suggest why a person with a cut corpus callosum would not be able to describe the object in the experiment above.

15.6 THE PERIPHERAL NERVOUS SYSTEM (PNS)

KEY CONCEPT

CNS has two parts – the brain and the spinal cord.
PNS has two parts – somatic nervous system for voluntary action and autonomic nervous system for involuntary action.

The peripheral nervous system consists of the nerves which run to and from the central nervous system. The **somatic** part of the peripheral nervous system coordinates voluntary activities; the **autonomic** part of the peripheral nervous system coordinates involuntary activities.

Nerves

A nerve is not the same as a nerve cell (neurone): nerves contain bundles of neurones. Within the nerve shown in Fig 15.20, you can see many individual axons which have been cut in transverse section. The axons are surrounded by connective tissue, called **endoneurium**. Small bundles of axons are surrounded by sheaths, of connective tissue, called **perineurium**, which are formed as ingrowths of the sheath around the entire nerve, the **epineurium**.

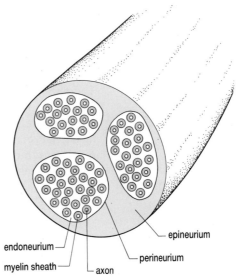

endoneurium
myelin sheath
axon
perineurium
epineurium

Fig 15.20   A transverse section through a myelinated nerve – several small bundles, each containing many axons are encased by a protective epineurium, the outer coat of the nerve.

Fig 15.21 Jacqueline du Pré suffered from MS, but still gave performances in the periods between loss of muscle coordination (remission).

MULTIPLE SCLEROSIS (MS)

This disease is caused by the progressive destruction of the myelin sheaths of neurones. The sheaths deteriorate to **scleroses**, which are hardened scars or plaques. The destruction of myelin sheaths interferes with the transmission of impulses from one neurone to another, literally short-circuiting conduction pathways.

The result of decay of the myelin sheaths is a progressive loss of function, the nature of which depends upon the region of the CNS affected. Death occurs anywhere between 7 and 30 years after the onset of the disease, usually as a result of an infection produced by a loss of motor activity. For example, if stimulation of the bladder is reduced the bladder wall muscles never constrict fully. This means the bladder never fully empties and the stagnant urine provides an ideal environment for bacterial growth. The resulting infection may then spread up the ureter to the kidney.

The cause of MS is unknown. However, much evidence points to it being the result of a viral infection which causes an autoimmune response (see Section 13.3). Viruses may trigger the destruction of nerve cells by the body's own antibodies and by killer T cells. Like all other demyelinating diseases, MS is, unfortunately, currently incurable. Treatment is aimed at alleviating symptoms and complications and at slowing down the progress of the disease.

All nerves run to or from the central nervous system. **Cranial nerves** run to or from the brain; **spinal nerves** run to or from the spinal cord at regular intervals along its length. Each nerve branches many times, forming smaller and smaller nerves which eventually supply individual organs or tissues of the body. Nerves are of three types.

1. **Mixed nerves** contain both receptor and effector neurones, e.g. all spinal nerves and some cranial nerves.
2. **Receptor nerves** (also called sensory nerves) contain only receptor neurones, e.g. some cranial nerves.
3. **Effector nerves**, (also called motor nerves) contain only effector neurones, e.g. some cranial nerves.

The optic nerve, which carries impulses from the retina of the eye to the brain, is a receptor nerve; the hypoglossal nerve, which carries impulses from the brain to control movements of the tongue, is an effector nerve; the facial nerve is a mixed nerve carrying impulses from the brain to the muscles which control facial expression and salivation and impulses relating to taste from the tongue to the brain.

The autonomic nervous system

The autonomic nervous system is that part of the peripheral nervous system which controls the activities of internal organs, such as the sweat glands and muscles of the gut, blood vessels and reproductive tracts.

MEDITATION

Activity of the internal organs is usually considered involuntary, although deep meditation has been shown to have an effect on their functioning. Perhaps, since overall control of the autonomic nervous system is carried out by centres in the hypothalamus and medulla oblongata, some control by the cerebrum is possible if people are trained. After all, most of us learn to control voluntarily the sphincter muscles of our bladder and anus, which are under the control of the autonomic nervous system.

Fig 15.22 Deep meditation has been shown to have an effect on the functioning of internal organs controlled by the autonomic nervous system.

The autonomic nervous system can be considered in two parts:

1. **sympathetic nervous system**
2. **parasympathetic nervous system**.

Both contain only effector neurones which connect the central nervous system to the effector organs. Receptor neurones from the internal organs to the central nervous system are not part of the autonomic nervous system.

In both sympathetic and parasympathetic branches, unmyelinated **preganglionic neurones** leave the central nervous system and synapse inside a ganglion with one or more unmyelinated **postganglionic** neurones. The postganglionic neurones innervate the effector organ. In the sympathetic branch of the autonomic nervous system, the synapses between preganglionic neurones and postganglionic neurones are very near the central nervous system; in the parasympathetic branch, these synapses are very close to, or inside, the effector organ. This gives rise to the structural differences between the two systems summarised in Fig 15.23.

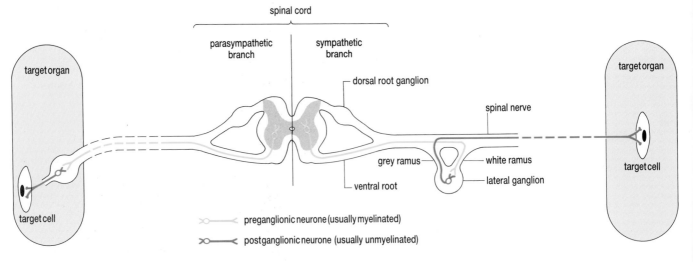

Fig 15.23 The structural features of the sympathetic and parasympathetic branches of the autonomic nervous system.

Most internal organs are supplied by both sympathetic and parasympathetic neurones. Table 15.5 shows how stimulation of an organ by these two branches of the autonomic nervous system has opposite effects. Generally, stimulation by the sympathetic system has an excitatory effect and stimulation by the parasympathetic has an inhibitory effect. These differences are the result of different neurotransmitter substances secreted by the postganglionic neurones. Like most neurones in the peripheral nervous system, postganglionic neurones of the parasympathetic system secrete acetylcholine. In contrast, sympathetic postganglionic neurones secrete noradrenaline. As a result, stimulation by the sympathetic branch of the autonomic system has similar effects to those of adrenaline.

QUESTIONS

15.16 Describe one structural difference and one functional difference between the sympathetic and parasympathetic branches of the autonomic nervous system.

15.17 Complete the diagram (Fig 15.24) to show the position of
(a) a receptor neurone
(b) a relay neurone
(c) an effector neurone of the somatic nervous system
(d) a preganglionic fibre of the sympathetic nervous system.

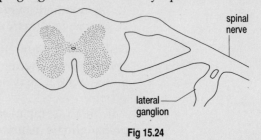

Fig 15.24

15.18 Suggest how will the interaction of the parasympathetic and sympathetic nervous systems might maintain the human body in different states of alertness from very relaxed to very frightened.

Table 15.5 The opposing effects of the sympathetic and parasympathetic branches of the autonomic nervous system

Organ or tissue	Effect of sympathetic stimulation (neurotransmitter is noradrenaline)	Effect of parasympathetic stimulation (neurotransmitter is acetylcholine)
iris of eye	dilation of pupil	constriction of pupil
tear gland	none	stimulates secretion of tears
salivary gland	inhibits saliva secretion	stimulates saliva secretion
intercostal muscles	increases breathing rate	decreases breathing rate
bronchi and bronchioles	dilation	constriction
heart	increases heart rate and stroke volume	decreases heart rate and stroke volume
blood vessels	increases blood pressure	decreases blood pressure
urinary system	decreases urine output	increases urine output
	contracts sphincter of bladder	relaxes sphincter of bladder
	relaxes wall of bladder	contracts wall of bladder
reproductive system	stimulates ejaculation	stimulates erection of penis and clitoris
gut	inhibits peristalsis	stimulates peristalsis
	inhibits secretion of gastric juice	stimulates secretion of gastric juice
	contracts anal sphincter	relaxes anal sphincter
skin	contracts erector muscles of hairs	no effect on erector muscles of hairs
	stimulates sweat production	no effect on sweat production
	stimulates vasoconstriction	stimulates vasodilation

SUMMARY

Nerves are made from bundles of individual nerve cells (neurones) surrounded by connective tissue. Receptor nerves (sensory nerves) contain only neurones that transmit impulses to the brain or spinal cord; effector nerves (motor nerves) contain only neurones that transmit impulses to muscles and glands; mixed nerves contain both types of neurone.

Individual neurones have a cell body which contains a nucleus, one or more dendrons (with smaller dendrites) which receive stimuli, and one or more axons which transmit impulses to the target organ. Myelinated neurones have an insulating layer of Schwann cells wrapped around their dendrons and axons: as a result they carry impulses faster than unmyelinated neurones of the same diameter. In the simplest inborn response to a stimulus (an unconditioned reflex), neurones are connected together in a simple reflex arc. In such an arc, receptor neurones carry impulses from a receptor organ to the brain or spinal cord where relay neurones pass the impulses to effector neurones. The latter carry the impulses from the brain or spinal cord to the target organ so that a response is made. Simple learned actions, called conditioned reflexes, arise when new pathways develop among the relay neurones within the brain and spinal cord.

A nerve impulse passes along a neurone as an electrochemical wave. Sodium–potassium pumps are normally active within the membrane of a neurone. As a result, there is an imbalance of sodium and potassium ions between its cytoplasm and the surrounding tissue fluid, producing a potential difference (the resting potential) across its surface membrane. As a suitable impulse passes along a neurone, the sodium-potassium pumps along its surface membrane momentarily stop working (depolarisation) and an influx of sodium ions causes a reversal of the resting potential (called the action potential) at that point in the neurone. Following an action potential, the sodium-potassium pumps recover and quickly restore the original imbalance of ions (repolarisation). Since during this recovery (refractory period) the membrane cannot be depolarised again, there is a maximum rate at which neurones can transmit impulses. The short refactory period ensures that an impulse does not travel backwards along a neurone.

An action potential occurs at one point of the cell at a time. As an action potential occurs it depolarises the adjacent surface membrane causing a new action potential in that membrane: as a result an impulse is self-propagating. In myelinated neurones, depolarisation can only occur at the short gaps between adjacent Schwann cells, called the nodes of Ranvier. Consequently, action potentials jump from node to node (a saltatory impulse).

Neurones are separated from each other and from their target cells by a short gap, called a synapse. The pre-synaptic neurone ends in small swellings known as synaptic knobs (or motor end plates in effector neurones). When depolarised by an impulse, the synaptic knobs release chemical transmitter onto their post-synaptic neurone, causing its depolarisation. Acetylcholine is the chemical transmitter released by most neurones. Other chemical transmitters include noradrenaline, dopamine and serotonin.

The nervous system can be considered in two parts. The central nervous system consists of the brain and spinal cord; the peripheral nervous system consists of nerves. The central nervous system is surrounded by protective membranes, called meninges. It consists of unmyelinated relay cells (the grey matter) and the ends of myelinated receptor and effector neurones (the white matter). The brain is essential to life. The medulla oblongata controls the vital involuntary activities of many internal organs; the cerebellum controls balance and fine muscle movements; the cerebrum controls conscious thought and memory.

The somatic part of the peripheral nervous system contains both receptor and effector nerves: it is involved in voluntary activities, such as those involving muscles which move the skeleton. The autonomic part of the peripheral nervous system contains only effector nerves: it coordinates involuntary activities, such as those of the gut, reproductive tract and sweat glands. Neurones of the sympathetic branch of the autonomic nervous system release noradrenaline as their transmitter substance whereas those of the parasympathetic branch release acetylcholine. The sympathetic and parasympathetic nerves of the autonomic nervous system have opposing effects on their target organs.

Answers to Quick Questions: Chapter 15

1   Effector neurone is unipolar; receptor neurone is bipolar; relay neurone is multipolar.

2   Inborn (i.e. unlearned) mechanisms that cause rapid, protective movements.

3   Membrane cannot immediately be depolarised again, therefore only the cell surface membrane in front of the action potential can be depolarised.

4   Ascending and descending nerve tracts of the spinal cord cross over from left to right and vice versa in the medulla oblongata.

5   Cortex contains the multipolar cells. Folding increases the surface area available for these cells, so more synapses and more complex (intelligent) behaviour result.

Chapter 16

CONTROL AND COORDINATION IN ANIMALS: RECEPTORS AND EFFECTORS

LEARNING OBJECTIVES

When you have studied this chapter you should be able to:

1. classify receptors according to their position and the type of stimulus to which they respond;

2. explain the terms generator potential, convergence and adaptation in relation to receptors;

3. describe the structure of the mechanoreceptors in the mammalian ear and outline their involvement in the transduction of sound, gravity and movement into nerve impulses;

4. describe the structure and function of accessory structures of the mammalian ear which are involved in the reception and amplification of sound;

5. describe the structure of the mammalian eye and explain how it detects light of different intensities and wavelengths;

6. recognise involuntary and voluntary muscle tissue and explain how contraction of the latter is brought about.

Just as a computer is of no use unless it has an input device (e.g. a keyboard) and output device (e.g. a printer), a system of nerves is of value only if it connects **receptors**, i.e. cells which can detect stimuli, to **effectors**, i.e. cells which bring about a response to those stimuli. This chapter describes the activity of some important receptors and effectors.

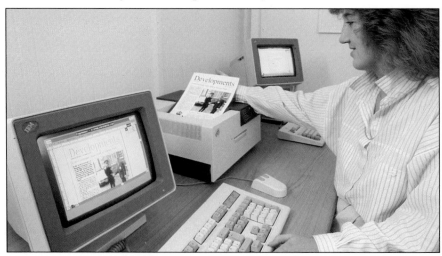

Fig 16.1 Communication networks need input and output devices. This system has only two input devices, the mouse and the keyboard. The human 'system' has many more.

All receptors have the same basic function: to convert energy in one form, light, sound and so on, into the electrochemical energy of an action potential which can then propagate through the nervous system. This conversion process is called **transduction**. Given that animals respond to a wide range of stimuli, it is not surprising to find that there are many types or receptor.

Types of receptor

The simplest form of receptor is a single neurone whose dendron is sensitive to a single type of stimulus and whose axon carries an impulse to the central nervous system. Such a receptor cell is called a **primary receptor cell**. Temperature and pressure receptors in the skin are primary receptor cells. The more complex **secondary receptor cells** are epithelial cells that are adapted to detect a particular stimulus and communicate with a receptor neurone, via a synapse. Taste buds are composed of secondary receptor cells. **Sense organs** are complex receptors, consisting of many receptor cells, receptor neurones and other supportive tissues. The mammalian ear and eye are examples of complex sense organs, described in Sections 16.2 and 16.3, respectively.

Receptors are often classified according to their position. **Exteroceptors** are sensitive to stimuli outside the body whereas **interoceptors** are sensitive to stimuli inside the body. **Proprioceptors** are a particular type of interoceptor which are sensitive to stimuli concerned with the relative position and state of contraction of muscles. A more useful classification is based on the type of stimulus energy detected by each receptor. Receptors are energy transducers, i.e. they convert energy from one form to another. Whilst a receptor can only be stimulated by one specific form of energy, they all convert their stimulus energy into the electrical energy of a nerve impulse. Table 16.1 summarises how receptors can be classified according to the type of stimulus energy to which they respond.

Table 16.1 Classification of receptors

Class of receptor	Type of stimulus energy	Examples of stimuli detected
chemoreceptor	chemical	pH of blood; smell; taste
electroreceptor	electromagnetic	electrical fields (mainly fish)
mechanoreceptor	mechanical	gravity; movement; pressure; tension
photoreceptor	electromagnetic	visible light; UV light (insects)
thermoreceptor	thermal	temperature differences

 How would you classify a smell receptor?

Electrical changes in receptors

Like neurones, receptor cells have a resting potential, i.e. a voltage across their surface membrane. The resting potential of receptor cells is produced by the activity of sodium pumps, in an identical way to that of neurones (see Section 15.3). A specific stimulus causes the local breakdown of the sodium pump mechanism and a rapid exchange of ions across the surface membrane of the receptor cell, known as **depolarisation**. This depolarisation of the receptor cell's surface membrane leads to a change in voltage, called a **generator potential**. The magnitude of this generator potential varies according to the strength of the stimulus, e.g. light pressure may result in only a small generator potential in a mechanoreceptor but heavy pressure will result in a large generator potential.

 How does a generator potential differ from an action potential?

In many ways, generator potentials in receptor cells are like action potentials in neurones. However, unlike generator potentials, the action potentials in receptor and effector neurones do not vary in magnitude. Instead they follow the **all-or-nothing** law: stimulations of low magnitude do not cause an action potential in a neurone (nothing) but stimulations at or above a threshold magnitude always produce an identical action potential (all). Thus, a receptor cell may be affected by a stimulus of low intensity but fail to cause an action potential in its receptor neurone so that no impulse will pass to the central nervous system. In many sense organs, cells are arranged in such a way that several receptor cells synapse with a single receptor neurone. This is called **convergence** and is shown in Fig 16.2. Whilst the effect of a stimulus on a single receptor cell in Fig 16.2 might not be great enough to cause an action potential in the receptor neurone, the stimulation of several receptor cells will. Rod cells in the retina (see Section 16.3) synapse with receptor neurones in this way, allowing increased visual sensitivity. This cumulative effect of stimulating the receptor neurone with several receptors is called **summation** and is similar to summation in synapses, described in Section 15.4.

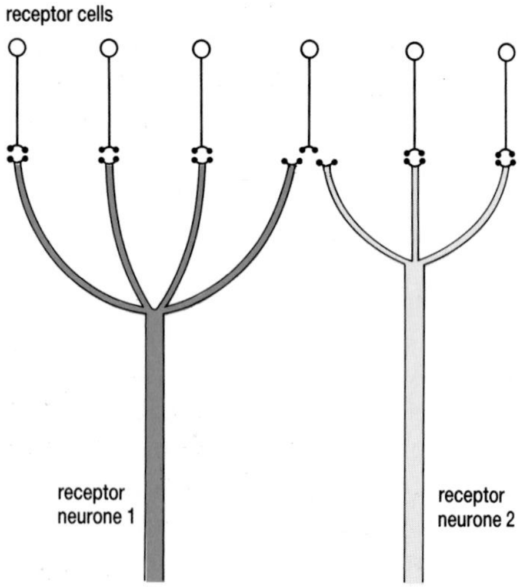

 3 Suggest a chemical explanation of convergence.

Exposure to a constant, intense stimulus causes a lessening of the response in most receptor cells. This is called **adaptation**. For example, when you get into a bath of hot water you might initially feel it is too hot. However, after a short time the water feels comfortable even though the temperature of the water is the same. Temperature receptor cells in your skin have become adapted. The mechanism by which receptor cells become adapted is not fully understood but it is thought that the cell surface membrane of the receptor cell becomes progressively less permeable to ions. An advantage of adaptation is that it frees the central nervous system of irrelevant information, allowing full awareness of stimuli with greater survival value.

receptor cells

receptor neurone 1

receptor neurone 2

Fig 16.2 Convergence: a number of receptor cells synapse with a single receptor neurone.

QUESTIONS

16.1 What are the basic elements of the system which allows an animal to perceive a stimulus?

16.2 Explain the term generator potential and outline how it is produced in a receptor cell.

16.3 Use Fig 16.2 to suggest why convergence of rod cells in the retina of your eye reduces your ability to read small print in dim light.

16.2 THE MAMMALIAN EAR

The mammalian ear responds to sound, gravity and movement of the head. Although sound, gravity and movement seem unrelated stimuli, they are all forms of mechanical energy. Mechanoreceptors in three different parts of the ear convert this mechanical energy into impulses which pass to the brain in receptor neurones in the auditory nerve. The basic structure of these mechanoreceptors is remarkably similar and is summarised in Fig 16.3. Each consists of a mound of receptor cells with cilium-like projections (often called sensory hairs) embedded in a second structure. Relative movement of the sensory mound and the structure in which its projections are embedded results in stretching of these projections, producing a generator potential in the receptor cell.

Structure of the ear

The ear consists of the three chambers shown in Fig 16.4. The **outer ear** consists of a **pinna**, which is the flap of skin-covered elastic cartilage that can be seen protruding from the head of a mammal. This pinna collects and

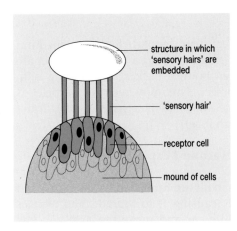

structure in which 'sensory hairs' are embedded

'sensory hair'

receptor cell

mound of cells

Fig 16.3 General structure of a mechanoreceptor from the mammalian ear.

Fig 16.5 The fennec or desert fox has very large ears. This animal hunts at night for lizards, birds and insects.

funnels sound waves down the **ear canal** (external auditory meatus) to the **tympanic membrane** (ear drum). In many mammals the pinna can be moved, allowing the mammal to locate sounds without moving its head. Most humans are able to move their pinnae only slightly: location of sound depends on the difference in the time when sound reaches each ear.

 Suggest how the fennec fox benefits from having ears that are cone-shaped.

The **middle ear** is an air-filled cavity in the skull. It contains three small bones called **ossicles** which transmit vibrations of the tympanic membrane to the membrane covering a small oval hole in the skull, the **oval window**. These ossicles act like levers, magnifying the vibrations of the tympanic membrane over twenty times. Since the middle ear is air-filled, damage would occur if the air pressures on each side of the tympanic membrane were very different. This is prevented by a connection between the middle ear and the pharynx called the **Eustachian tube**. Swallowing and yawning open the Eustachian tube allowing air to move into or out of the middle ear.

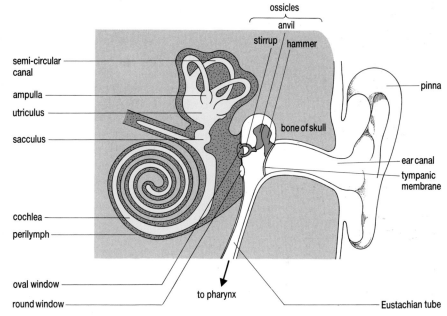

Fig 16.4 Vertical section through a human ear.

 Explain why an infection which blocked the Eustachian tube could cause ear ache.

The **inner ear** is a bony cavity in the skull which is filled with a fluid, called **perilymph**. The whole cavity is lined by membrane which forms the **oval window** and **round window** where it passes over holes in the bony cavity. Inside this cavity is a group of membrane-bound structures containing **endolymph** and mechanoreceptors

- the **cochlea** contains sound receptors
- the **semi-circular canals** contain receptors for movement of the head
- the **sacculus** and **utriculus** contain gravity receptors.

Pressure changes in the perilymph caused by movement of the oval window are balanced by the opposite movement of the round window.

Perception of sound

Sound occurs because a vibrating object disturbs particles in the medium surrounding it. The particles of the medium around a vibrating object are first pushed by the vibration and then bounce back after colliding with particles further from the vibrating object. This movement of particles is called a **sound wave**. The disturbed medium may be a gas (e.g. air), liquid (e.g. water) or solid (e.g. a wall in your house).

 6 **Could sounds travel through a vacuum? Explain your answer.**

In the mammalian ear, sound waves in air are funnelled by the pinna and ear canal to the tympanic membrane, causing it to vibrate. The ossicles carry these vibrations across the middle ear and make the oval window vibrate. Vibrations of the oval window cause vibrations in the perilymph of the inner ear. Since a greater force is required to disturb the perilymph than is required to disturb air, it is important that the ossicles act as levers, magnifying the vibrations of the tympanic membrane. Thus, vibrations in the air have been converted into vibrations of perilymph:

vibrations of air → vibrations of tympanic membrane → vibration of ossicles → vibrations of oval window → vibrations of perilymph.

The oval window lies at one end of a hairpin canal running above and below the cochlea. Look at Fig 16.6 which represents an imaginary uncoiled cochlea (in humans, this would be 3–4 cm long). You can see the perilymph-filled canal running from the oval window along the top of the cochlea, around its end and then below the cochlea to the round window. The upper part of the canal is the **vestibular canal** and the lower part is the **tympanic canal**. The arrows in Fig 16.6 show how vibrations of the perilymph, which fills these canals, run from the oval window to the round window.

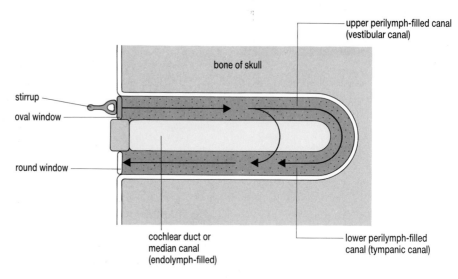

Fig 16.6 An imaginary uncoiled cochlea. The arrows show the path of vibrations in the perilymph.

As you can see in Fig 16.8, the cavity within the cochlea, the **median canal**, is filled with endolymph. The median canal is separated from the upper vestibular canal by the **Reissner's membrane** and from the lower tympanic canal by the **basilar membrane**. Running along the whole length of the basilar membrane is a mound of receptor cells, called the **organ of Corti**. These receptor cells have sensory projections which are embedded in a third membrane, the **tectorial membrane**, which runs along the cochlea like a rigid shelf. If the tectorial membrane and basilar membrane move apart, sensory projections are stretched and a generator potential occurs in the appropriate receptor cell. This causes a release of neurotransmitter substance across synapses to receptor neurones which transmit impulses to the brain.

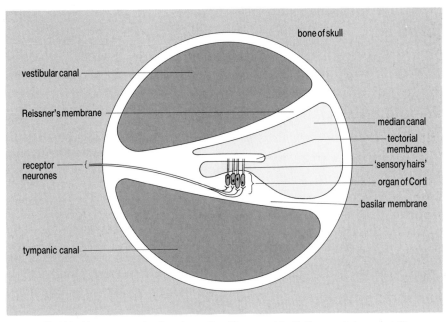

Fig 16.8   Transverse section through a cochlea. The organ of Corti is the mound of receptor cells.

Fig 16.9 shows what is thought to happen when a sound wave causes vibrations of the oval window. Pressure waves pass along the perilymph in the vestibular canal and at some point along the cochlea cause the Reissner's membrane to vibrate. Vibrations of the Reissner's membrane cause the endolymph in the median canal to vibrate which, in turn, causes the basilar membrane to vibrate.

 How will the vibration (mechanical energy) of membranes be converted into the electrochemical energy of an action potential in the receptor neurones to the brain?

The brain interprets the stimuli as sound. Since the basilar membrane is thinnest near the oval window, it is caused to vibrate by sounds of higher pitch (frequency) than the thicker end away from the oval window. The brain is able to determine the pitch of each sound according to the source of the impulses from the cochlea. In the meantime, vibrations of the basilar membrane have also caused vibrations in the perilymph filling the tympanic canal and in the round window. Sound waves which are of too long a wavelength to vibrate Reissner's membrane pass all the way along the vestibular canal, around the apex of the cochlea (called the **helicotrema**) and back along the tympanic canal where they cause vibrations in the round window.

 Will you be able to 'hear' these wavelengths?

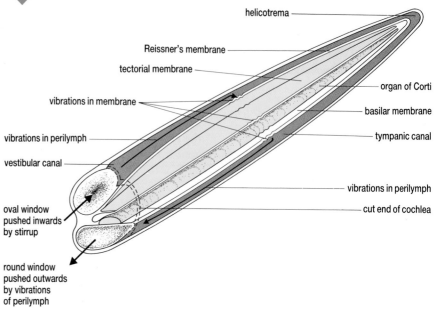

Fig 16.9 Stereogram of an uncoiled cochlea showing how vibrations of the basilar membrane are caused. Notice that the inward movement of the oval window results in a compensatory outward movement of the round window.

Thus, **pitch** of a sound is perceived by the brain depending on the position of the receptor cells being stretched. **Intensity** or loudness of a sound depends on the presence of receptor cells with different thresholds at each point along the organ of Corti. A quiet sound stimulates only a few receptor cells whilst a loud sound of the same frequency stimulates more receptor cells. This is an example of spatial summation (see Section 15.4). **Tone** is related to the number of waves of different frequency which make up a sound. Secondary waves, called **harmonics**, give a particular sound a distinctive quality which is characteristic of a particular musical instrument or human voice. By controlling the presence of harmonics, musicians are able to reproduce the voice of many instruments using electronic synthesizers.

Perception of gravity

Like the cochlea, the utriculus and sacculus are filled with endolymph. Each has a mound of receptor cells, called the **macula**, on the inside of its wall. The sensory projections of these receptor cells are embedded in a mass of jelly-like material containing small chalk crystals, called **otoliths**. These otoliths are affected by gravity. When the head is upside down, the otoliths of the utriculus fall away from the macula, as shown in Fig 16.10. The resultant pulling of their sensory projections results in a generator potential in the receptor cells and the initiation of an impulse along receptor neurones in the auditory nerve. When the head is on one side, only some of the sensory projections are stretched. When the head is in its normal position, no sensory projections are stretched. The corresponding pattern of impulses along receptor neurones in the auditory nerve are interpreted by the brain accordingly, providing information about the position of the head.

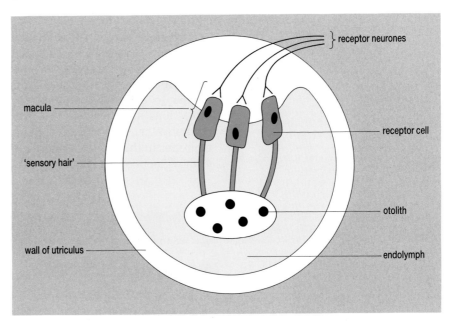

Fig 16.10 Section through an inverted utriculus. Gravity causes the small chalk crystals, otoliths, to fall away from the macula; the sensory hairs are stretched and the receptor cells are depolarised.

Perception of movement of the head

The semi-circular canals are arranged so that each is at right-angles to the other two. This arrangement enables movement of the head in any direction to be detected. At one point along each canal is a small swelling, the **ampulla**, containing a mound of receptor cells, called the **crista**. The receptor cells of the crista have sensory projections embedded in a jelly-like structure, called the **cupula**. A section through a single ampulla is shown in Fig 16.11.

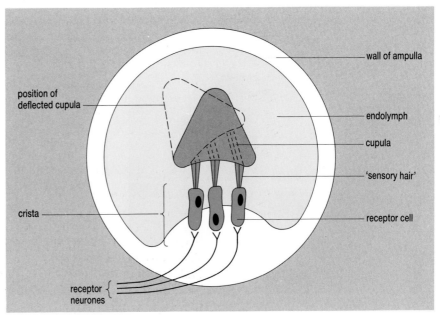

Fig 16.11 Section through a single ampulla of a semi-circular canal. Any movement of the jelly-like cupula is detected by the receptor cells of the crista.

Imagine what happens if you drop an upright glass of water. Long before it hits the ground, water seems to rise from the top of the glass. This is because the water does not begin to fall at exactly the same moment as the glass that contains it. Instead, there is a short time period during which the glass is falling but the water is stationary. This tendency for the water to resist changes in its motion, from resting to moving, is called its **inertia**. Your inertia causes you to overbalance when standing in a train which suddenly moves forwards; the inertia of cutlery and crockery causes them to stay on a table when someone who is skilled in the trick pulls a table cloth from under them. The endolymph in the semi-circular canals has inertia. When a semi-circular canal and cupula move, the inertia of the endolymph causes it to stay motionless for a short time. This causes the cupula to be pushed to one side (see Fig 16.11). As a result, sensory projections at one side of the crista are stretched, a generator potential occurs in their receptor cells and impulses pass to the brain along receptor neurones in the vestibular nerve. The brain is able to interpret these stimuli as movement of the head.

NOISE POLLUTION

Whilst we are only too aware of visual pollution, how many of us think of excess sound as a source of pollution? The ear is a sensitive detector of changes in air pressure: that is what sound is. Sound is measured using a unit called the **bel**, named after Alexander Graham Bell the inventor of the telephone. The bel scale is logarithmic. That means an increase of 1 bel unit indicates a tenfold increase in sound energy. The bel is a rather large unit and so the decibel is generally used for measuring sound. The threshold of hearing is 0.1 decibels.

A whisper at 3 decibels is 10 000 times louder than a sound at 0.1 decibels.

As the decibels increase, the sound energy increases rapidly. Sound can damage the delicate structures in the cochlea or even rupture the tympanic membrane. For example, when exposed to music at 122 decibels, a level similar to that found in the average discotheque, guinea pigs had widespread and irreversible damage to the receptor cells in their cochleas.

Fig 16.12 Both these environments can cause permanent ear damage!

16.4 The receptor regions of the ear each contain a mound of receptor cells whose sensory projections are embedded in a second structure. Copy Table 16.2 and complete the spaces labelled A – I.

Table 16.2 Perception in the ear

Receptor area	Stimulus to which receptor responds	Mound of receptor cells	Structure in which sensory projections are embedded
ampulla	A	B	C
cochlea	D	E	F
utriculus	G	H	I

16.5 (a) List the sequence of vibrations of structures within the ear following vibrations of the ear drum.

(b) Which of these vibrations can the ear detect?

16.6 (a) Describe how the ear is able to detect sounds of different pitch.

(b) Suggest how very loud noises might permanently damage a person's hearing.

16.7 Use your understanding of the functioning of the ear to attempt to explain the following.

(a) (i) Most people feel a pain in their ears when they skin dive at a depth of 5 metres.

(ii) Experienced skin divers overcome this pain by pinching their nose and blowing air into their closed nostrils.

(b) (i) After spinning around for a short time, people feel dizzy when they stop.

(ii) During this dizziness, they seem to be spinning in the opposite direction to the way they were actually spinning before.

16.3 THE MAMMALIAN EYE

Light is a form of electromagnetic radiation (see Fig 7.2) emitted and absorbed in discrete packets called **quanta** or **photons**. Quanta of light in the wavelength range 400–700 nm produce photochemical changes within receptor cells of the eye, enabling us to see. The mammalian eye contains structures which

- control the amount of light which can enter
- bend (refract) the light to a focus
- transduce the light energy into nerve impulses.

Structure of the eye

A section through a mammalian eyeball is shown in Fig 16.13. Each eyeball is held in a protective bony socket of the skull, called an **orbit**. Two antagonistic pairs of **rectus muscles** and one antagonistic pair of **oblique muscles** bring about movement of each eyeball. These muscles attach to the tough outer layer of the eyeball, the **sclera**. The sclera is white because of the numerous collagen fibres within it. However, these collagen fibres are normally absent in the sclera covering the front of the eyeball, forming a transparent **cornea** through which light can pass.

Beneath the sclera is a layer of tissue containing blood vessels and melanin-containing cells, the **choroid**. At the front of the eyeball, this choroid layer forms the **ciliary body**, with **suspensory ligaments** that support the lens, and the pigmented iris, which controls the amount of light entering the eye. Beneath the choroid layer is the layer of light-

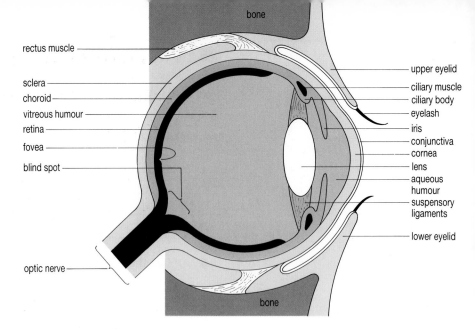

Fig 16.13 Section through a mammalian eyeball.

sensitive cells, the **retina**. Receptor neurones which synapse with light-sensitive cells leave the centre of the eyeball causing a break in the retina, appropriately called the **blind spot**.

The ciliary body and lens separate the eyeball into two chambers. The anterior chamber is filled with watery **aqueous humour**, which is secreted by the ciliary body. The posterior chamber is filled with a jelly-like **vitreous humour**. Both humours exert a pressure on the eyeball. Table 16.3 summarises the major functions of these structures within the eye.

Table 16.3   Functions of the major parts of the mammalian eye

Part of eyeball	Function
conjunctiva	protects the cornea at the front of the eyeball against friction
sclera	protects eyeball against mechanical damage
	allows attachment of eye muscles
cornea	allows passage of light
	refracts light
choroid	contains blood vessels supplying the retina
	prevents reflection of light within the eye
ciliary body	supports the lens
	contains muscles which enable the lens to change shape
	secretes aqueous humour
iris	controls the size of the pupil
	pigment reduces passage of light
lens	focuses, by changing shape
retina	contains light-sensitive cells; rods and cones
fovea (yellow spot)	part of the retina which contains only cones
blind spot	region where optic nerve leaves inside of eyeball and which contains no light-sensitive cells
vitreous humour	supports the lens
	helps to keep the shape of the eyeball
aqueous humour	maintains the shape of the anterior chamber of the eyeball

SPOTLIGHT

CATARACTS

The commonest cause of blindness, cataracts, is the clouding of the lens so that it becomes milky. This prevents light passing through and so reduces vision. It can now easily be treated by removing the lens and replacing it with an implanted artificial lens or spectacles.

Focusing of light rays onto the retina

Light rays from an object spread out in all directions (diverge). If we are very close to the object, a cone of divergent light enters our eyes (Fig 16.14(a)). If we are a very long way from the object, however, the angle of the cone of light which enters our eyes is so small that the light rays appear parallel (Fig 16.14(b)). In both cases, light rays from the object must be bent (**refracted**) to produce a sharp point on the retina if they are to be seen clearly. This refraction of light to produce a sharp image is called **focusing**.

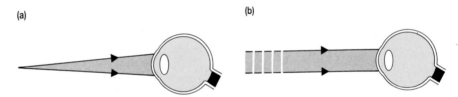

(a)   (b)

Fig 16.14   A cone of light rays from **(a)** a near object and **(b)** a distant object.

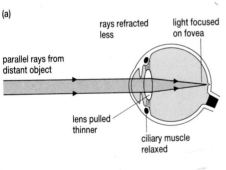

(a)
rays refracted less
light focused on fovea
parallel rays from distant object
lens pulled thinner
ciliary muscle relaxed

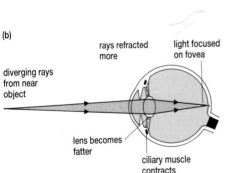

(b)
rays refracted more
light focused on fovea
diverging rays from near object
lens becomes fatter
ciliary muscle contracts

 Which set of light rays in Fig 16.14 will have to be refracted most to bring them into focus?

Light is refracted when it passes through transparent media of different densities. In the mammalian eye the conjunctiva, cornea, aqueous humour, lens and vitreous humour have different densities from each other, hence light is refracted at the junction (interface) of each of these media. Most refraction occurs at the air–conjunctiva–cornea interfaces. Since they are more strongly diverging, light rays from near objects need more refraction to focus them on the retina than do light rays from distant objects. The mammalian eye is able to change the degree of refraction of light by changing the shape of the lens (the conjunctiva, cornea, aqueous humour and vitreous humour all have a fixed refractive power). Light from a distant object is focused by a long thin lens (Fig 16.15(a)), whilst light from a near object is focused by a short fat lens (Fig 16.15(b)). The change in shape of the lens is brought about by the action of **ciliary muscles**, which are arranged circularly in the ciliary body. When the ciliary muscles contract they close the aperture around the lens. As a result the tension in the **suspensory ligaments** is lessened. Because the lens is normally elastic, it becomes short and fat when the tension is lessened: its ability to bend light is now at a maximum. When the ciliary muscles relax, the aperture around the lens increases in diameter and the tension in the suspensory ligaments is increased. The suspensory ligaments now pull on the lens, making it long and thin: in this condition (called the unaccommodated state) its ability to bend light is at a minimum. The process of focusing objects at different distances from the eye is called **accommodation**: Fig 16.15 shows how near and far accommodation are brought about.

ANALYSIS

Sight and age

This exercise will enable you to practise the analysis of numerical data.

The data in Table 16.4 show the mean refractive power of human eyes at various ages. The refractive power is measured in diopters, which is calculated as the reciprocal of the distance in metres between the eye and the closest object that can be focused. So, if the nearest object you could see was 2 metres away, this would give a refractive power of 1/2 dioptre. If it was 0.5 metres away the refractive power would be $1/0.5 = 2$ dioptres.

Table 16.4   Mean refractive power of human eyes at different ages

age/years	10	20	33	45	69
mean refractive power/diopters	14.0	9.0	4.5	1.0	0.3

(a) Calculate the nearest distance at which an object could be focused by a 20-year-old.
(b) Describe what the data show about the ability of humans to focus on near objects as they age. What must reading glasses do to light rays to overcome this problem?
(c) Explain how each of the following phenomena might affect the refractive power of human eyes.
- As we age, chemical changes occur in our lenses which make the lens less elastic.
- The unaccommodated lens of an infant is about 3.3 mm thick whilst that of a 70-year-old is about 5.0 mm thick. Since they attach to the front of the lens, this increase in lens thickness stretches the suspensory ligaments.

Control of the amount of light entering the eye

If too much light enters the eye, the light-sensitive cells may be overstimulated, resulting in painful dazzling, or even damaged. If too little light enters the eye, light-sensitive cells may not be stimulated at all. By controlling the diameter of the pupil, the amount of light entering the eye can be controlled. Just as with the iris diaphragm of a camera, reducing the diameter of the pupil improves our **depth of focus**. This is a second advantage of being able to control the size of our pupil.

Changes in the diameter of the pupil are normally controlled by unconditioned reflexes (see Section 15.1) in response to the number of light-sensitive cells stimulated by light, but the pupils are also enlarged by the action of adrenaline (see Chapter 14). The iris has muscles arranged in two layers: circular muscles and radial muscles. These two sets of muscles are antagonistic, i.e. when one set of muscles contracts the other relaxes.

- In bright light, contraction of the circular muscles and relaxation of the radial muscles causes the pupil to become smaller.
- In dim light, relaxation of the circular muscles and contraction of the radial muscles causes the pupil to become larger.

A detached retina will result in lack of vision but newly developed, surgical methods using a laser beam can now be used to weld the retina back in place.

 Many nocturnal mammals have a reflective layer, the tapetum, lining the back of the eye rather than a pigmented layer. Suggest the value of this arrangement.

Perception of light by the retina

The retina contains over one hundred million light-sensitive cells and the neurones with which they synapse. The arrangement of these cells is represented in Fig 16.16(a): notice that light must pass through the neurones before striking the light-sensitive cells. The two types of light-sensitive cell, **rods** and **cones**, are similar in structure (Figs 16.16(b) and (c)). Each contains a pigment within membranous vesicles in its **outer segment**. A narrow constriction, containing a pair of cilia, connects the outer segment to an **inner segment** in which are the cell's nucleus and such organelles as mitochondria and ribosomes. An extension of the inner segment synapses with a bipolar neurone. The major structural and functional differences between rods and cones are summarised in Table 16.5.

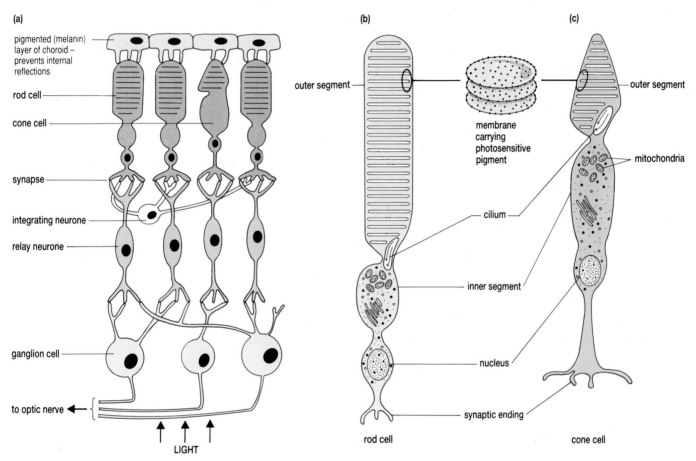

Fig 16.16 (a) Arrangement of cells within a mammalian retina
(b) structure of rod cell
(c) structure of cone cell.

Table 16.5   Structural and functional differences between rods and cones

Feature	Rods	Cones
frequency in single retina	about 120×10^6	about 6×10^6
distribution	evenly spread throughout retina except at fovea	tightly packed at fovea (about 5×10^4 mm^{-2})
shape of outer segment	rod-shaped	cone-shaped
light-sensitive pigment	single pigment	three pigments (but each cone contains only one)
	affected by light of low intensities	affected only by light of high intensities
	unable to distinguish colour	able to distinguish colour
synapse with bipolar cell	synapse in groups (retinal convergence)	synapse individually

Photoreception in rods

Vesicles in the outer segment of rod cells contain the light-sensitive pigment **rhodopsin** (visual purple). This pigment is formed from a lipoprotein, called **opsin**, and **retinene,** a light-absorbing derivative of vitamin A. When one photon of light is absorbed by a molecule of rhodopsin, a change occurs in the retinene molecule which results in the breakdown of rhodopsin to its constituent parts. This process is called **bleaching** and results in a generator potential within the rod cell. If sufficiently large, this generator potential results in an impulse in the receptor neurone leading to the brain.

Before it can be stimulated again, the rod cell must resynthesise its rhodopsin. It does this using ATP from its mitochondria to rejoin the retinene and opsin. In normal daylight conditions, most of the rhodopsin in the rods is bleached: the eyes are said to be **light-adapted**. Exposure to dim light has little effect on the rods of a light-adapted eye and continues to have little effect until the rods have resynthesised more rhodopsin. When sufficient rhodopsin has been resynthesised to enable us to see, the eye is said to be **dark-adapted**. It takes about 30 minutes in complete darkness for human eyes to become fully dark-adapted.

 Why should pilots waiting to fly at night not be exposed to bright lights?

Colour vision in cones

Cones work in a similar way to rods except that their pigment is **iodopsin**. A greater amount of light is needed to break down iodopsin, explaining why cones are not operative in dim light. It is thought that there are three different types of iodopsin, each mainly sensitive to one of the primary colours of light (Fig 16.17). Each cone cell contains only one of these types of iodopsin and so is mainly affected by red, green or blue light. The **trichromatic theory** of colour vision suggests that different colours are interpreted by the brain according to the way in which these different types of cone cell are stimulated (Table 16.6).

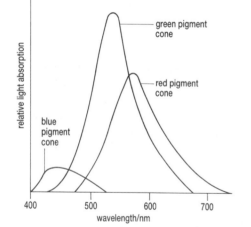

Table 16.6 **Perception of different colours by the brain according to the trichromatic theory of colour vision**

Light stimulates			Colour perceived
red cones	green cones	blue cones	
✔	✕	✕	red
✕	✔	✕	green
✕	✕	✔	blue
✔	✔	✕	orange/yellow
✕	✔	✔	cyan
✔	✕	✔	magenta
✔	✔	✔	white

There is an overlap in the range of light absorbed by the three cone pigments, particularly by the red and green cones (Fig 16.17). Thus, in the absence of red cones, red light will still be detected by the green cones. However, in the absence of red cones, the brain will be unable to distinguish between red light and green light. This condition is known as **red–green colour blindness** and can equally result from an absence of green cones. The presence and extent of colour blindness is determined by using a series of cards with ingenious patterns of coloured dots (Ishihara test cards). Red–green colour blindness results from the inheritance of a defective allele of the colour-vision gene which is carried on the X chromosome (see Section 22.5).

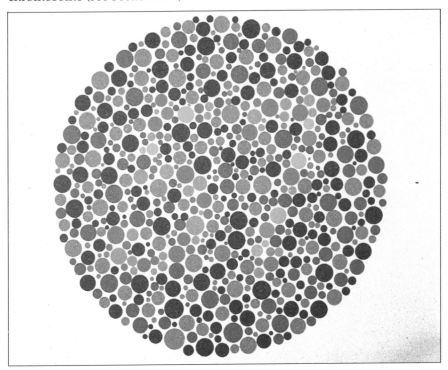

Fig 16.18 An Ishihara test chart for red-green colour blindness. Those with red-green deficiencies will see the number 35, not 57.

Perception of distance and size

During normal vision in bright light, impulses pass along several million receptor neurones from the fovea of both eyes to the visual cortex of the cerebrum. The pathway of most of the neurones from the retina of both eyes to the visual cortex is shown in Fig 16.19. Neurones from the left side of each eye pass to the visual cortex in the left cerebral hemisphere; neurones from the right side of each eye pass to the visual cortex in the right cerebral hemisphere.

In humans, both eyes point forwards and see much the same visual field. Because they are differently placed, each eye views the same visual field from a slightly different angle. This is called **binocular vision**. These different images of the left of the field of view are integrated in the visual cortex of the left cerebral hemisphere to produce a single perceived image; those from the right of the field of view are integrated in the visual cortex of the right cerebral hemisphere. During this integration differences between the overlapping fields of view, together with memories of learned experiences, enable us to judge distance.

Our ability to perceive size is related to our judgment of distance (Fig 16.20). A distant object appears smaller than a close object. We have learned to compensate for distances when assessing size, so we know that a distant elephant is actually larger than a mouse which is close to us.

Animals which need to judge distance and size accurately in striking prey or grasping branches usually have forwardly directed eyes, for example humans. Animals which do not need to judge distances accurately but need to see predators approaching from any direction usually have eyes placed on the sides of the head, e.g. rabbits.

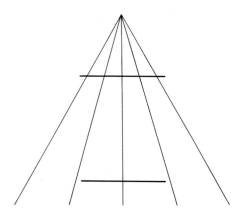

Fig 16.20 Is the upper line really longer than the lower line? This figure contains visual clues which encourage us to perceive distance. As a result, we perceive the upper horizontal line as further away and thus, longer than the lower line even though they are exactly the same length.

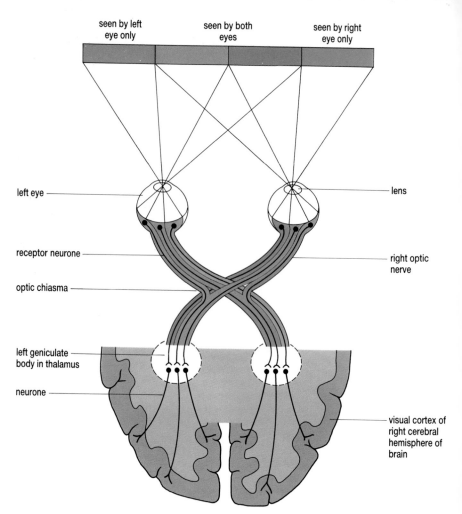

Fig 16.19 Pathway of neurones from the retina of both eyes to the visual cortex of the cerebrum in humans. Integration of both fields of view gives binocular vision.

QUESTIONS

16.8 The image of part of your field of view lands on the blind spot, where there are no light-sensitive cells. Suggest why you are unaware of this blind area
(a) when viewing objects with both eyes open
(b) when viewing objects with one eye closed.

16.9 Use Fig 16.16(a) to explain why visual acuity is much greater at the fovea than at the periphery of the retina.

16.10 (a) The retina of most mammals contains only rod cells. Discuss the significance of this observation.
(b) When sitting in a dimly lit room objects often appear slightly fuzzy and colours are more difficult to distinguish. Explain these observations.

16.11 A pigeon has eyes which are on the sides of its head. Use a sketch drawing to show
(a) the pigeon's field of view
(b) the field of view over which a pigeon will be able to judge size and distance
(c) that a pigeon loses sight of its food as it picks it up in its beak.

16.12 Consider a man sitting in a dim room looking at his watch. To check the time he looks out through the window at the distant church clock which is brightly illuminated by the sun. He then glances back at his watch. Outline the changes occurring in this man's eyes and explain their significance.

Effector neurones stimulate their target cells, effectors, into action. Endocrine glands and muscles are effectors. Endocrine glands are described in Chapter 14.

Muscle types

Three types of muscle tissue occur in mammals: cardiac muscle, involuntary muscle and voluntary muscle.

SPOTLIGHT

MUSCLE TYPES AND SLIMMERS

Voluntary muscle makes up the largest mass of tissue in the human body. It contains two different types of muscle fibre.

Type 1 muscle fibres use fatty acids as their energy source. Since type 1 fibres contain a store of fat, they fatigue fairly slowly.

Type 2 muscle fibres use glucose as their energy source, not fat. As they contain very little stored carbohydrate, type 2 muscle fibres tend to tire much faster than type 1 fibres.

By analysing the composition of the thigh muscle (quadriceps) of male volunteers, research workers at the London Hospital Medical College have been able to determine the proportion of type 1 and type 2 fibres in human muscle tissue. They found that slim men had a high proportion of type 1 fibres whilst fatter men had a higher proportion of type 2 fibres. They also found that fat men (higher proportion of type 2 fibres) metabolised less fat during periods of standard exercise than lean men (high proportion of type 1 fibres).

If all obese people have a greater proportion of type 2 fibres in their muscle, they would break down less fat than lean people when working. This might explain why some people grow fat whilst others stay slim. The finding that type 2 muscle fibres tire more easily than the type 1 fibres might also explain why doctors are often unsuccessful when trying to persuade obese people to take more exercise.

Cardiac muscle

It is found only in the heart and is described in Chapter 10.

Involuntary muscle

This occurs in the iris and ciliary body of the eye and also lines the tubular structures within the body, i.e. gut, reproductive ducts, urinary ducts and blood vessels. It is made of spindle-shaped cells bound together by collagen fibres (Fig 16.21(a)). The contractions of involuntary muscle are usually slow and rhythmical, such as those causing peristalsis in the gut, and are not normally under voluntary control.

Voluntary muscle

In mammals it is attached by **tendons** to bones in the skeleton. Its contraction is normally a voluntary action and causes movement of one of the bones to which it is attached (see Chapter 18). An individual muscle is made of hundreds of **muscle fibres** (Fig 16.21(b)), each of which may be several centimetres long. The surface membrane of each muscle fibre, the **sarcolemma**, encloses cytoplasm (**sarcoplasm**) containing many nuclei and a membrane-bound network, the **sarcoplasmic reticulum**. Within the sarcoplasm are many parallel **myofibrils** which have characteristic striations along their length.

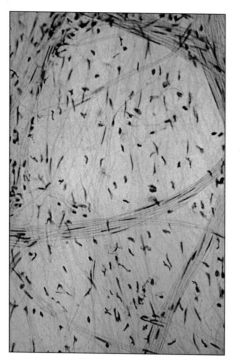

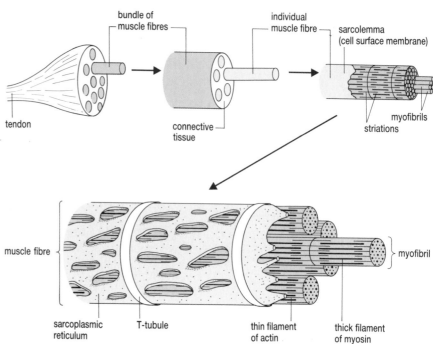

Fig 16.21　(a) *(left)* Involuntary muscle showing spindle-shaped, unstriated muscle cells. (b) *(right)* Structure of a voluntary muscle.

ANALYSIS

The structure of striated muscle

This is an exercise in the interpretation of electronmicrographs.

The striations of voluntary muscle are caused by interdigitating filaments of two types of protein: **myosin** and **actin**. Myosin filaments are thicker than actin filaments, causing the darker (i.e. pink) striation in the myofibril, shown in the electronmicrograph below. Fig 16.23 represents the distribution of actin and myosin filaments within a myofibril such as the one in Fig 16.22.

Fig 16.22　A false colour transmission electron-micrograph of voluntary muscle. Using this technique, those areas which would be light in a conventional black and white electronmicrograph look blue and the dark areas look pink.

(a) Using the electronmicrograph (Fig 16.22) and Fig 16.23 account for
- the light appearance of the **I** band
- the very dark region at each side of the **A** band
- the lighter **H** zone.

(b) What do the **Z** and **M** lines represent?

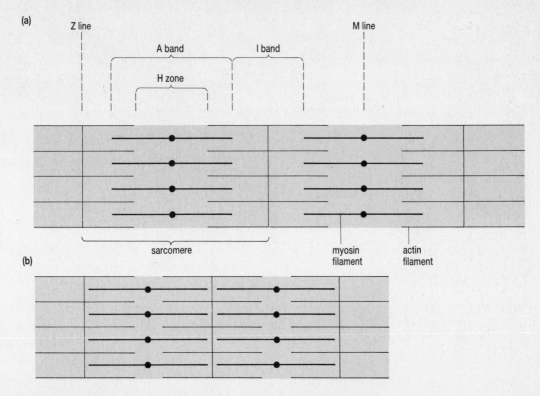

Fig 16.23 The ultrastructure of the parallel myofibrils in a length of voluntary muscle showing the distribution of actin and myosin protein filaments in **(a)** relaxed muscle **(b)** contracted muscle.

During contraction of a muscle fibre, each sarcomere shortens by up to 30% of its length. It does so because the actin filaments slide over the myosin filaments further into the A band, in much the same way as you can push the middle finger of your right hand between the second and third fingers of your left hand.

(c) Account for the following as the sarcomere contracts
- the **A** band stays the same width
- the **H** zone is shorter
- the **I** band is shorter.

Stimulation of muscle fibres: the neuromuscular junction

Unlike cardiac muscle, voluntary muscle fibres will not contract of their own accord but must be stimulated by effector neurones. Each effector neurone stimulates a group of muscle fibres causing them to contract together. The point where an effector neurone stimulates a muscle fibre is called a **neuromuscular junction** (Fig 16.24). It works in much the same way as a synapse (see Section 15.4). When an impulse reaches the neuromuscular junction, synaptic vesicles in its motor end plate fuse with the pre-synaptic membrane and release acetylcholine. The acetylcholine diffuses into the sarcolemma and changes its permeability to sodium ions. As a result of an influx of sodium ions into the sarcoplasm, the sarcolemma is depolarised and an action potential occurs in the muscle fibre. The muscle fibre contracts as the action potential is propagated along its length.

The sarcolemma contains the enzyme acetylcholinesterase which rapidly hydrolyses acetylcholine. In this way overstimulation of the muscle fibre is prevented and the motor end plate reverts to its normal resting state.

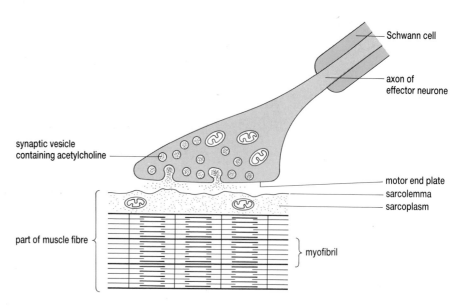

Fig 16.24 A neuromuscular junction.

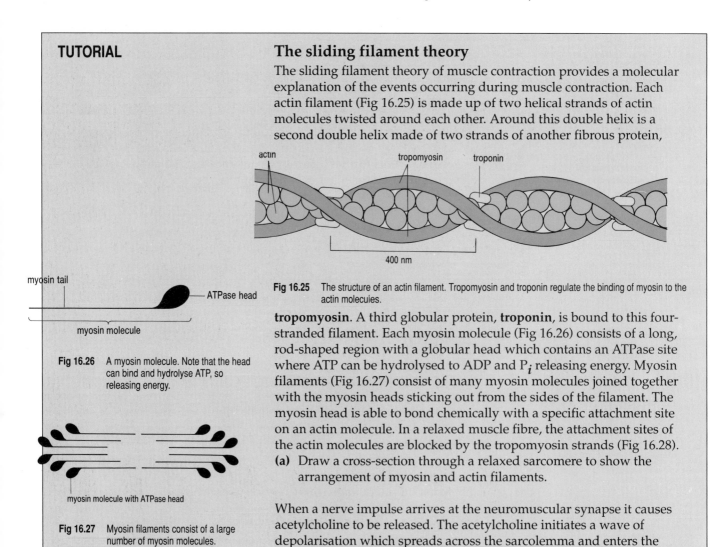

TUTORIAL

The sliding filament theory

The sliding filament theory of muscle contraction provides a molecular explanation of the events occurring during muscle contraction. Each actin filament (Fig 16.25) is made up of two helical strands of actin molecules twisted around each other. Around this double helix is a second double helix made of two strands of another fibrous protein,

Fig 16.25 The structure of an actin filament. Tropomyosin and troponin regulate the binding of myosin to the actin molecules.

Fig 16.26 A myosin molecule. Note that the head can bind and hydrolyse ATP, so releasing energy.

Fig 16.27 Myosin filaments consist of a large number of myosin molecules.

tropomyosin. A third globular protein, **troponin**, is bound to this four-stranded filament. Each myosin molecule (Fig 16.26) consists of a long, rod-shaped region with a globular head which contains an ATPase site where ATP can be hydrolysed to ADP and P_i releasing energy. Myosin filaments (Fig 16.27) consist of many myosin molecules joined together with the myosin heads sticking out from the sides of the filament. The myosin head is able to bond chemically with a specific attachment site on an actin molecule. In a relaxed muscle fibre, the attachment sites of the actin molecules are blocked by the tropomyosin strands (Fig 16.28).

(a) Draw a cross-section through a relaxed sarcomere to show the arrangement of myosin and actin filaments.

When a nerve impulse arrives at the neuromuscular synapse it causes acetylcholine to be released. The acetylcholine initiates a wave of depolarisation which spreads across the sarcolemma and enters the

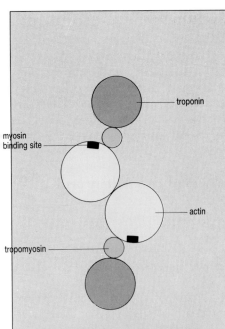

Fig 16.28 Transverse section of an actin filament showing arrangement of troponin and tropomyosin in a relaxed muscle filament. Notice how the tropomyosin blocks the myosin binding site on the actin molecule.

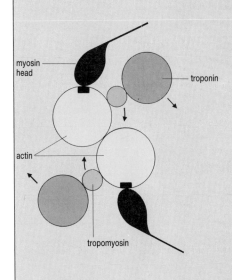

Fig 16.29 Transverse section showing formation of actomyosin cross-bridges.

T tubules and sarcoplasmic reticulum. This stimulates the sarcoplasmic reticulum to release calcium ions from storage into the sarcoplasm. The calcium ions activate myosin which hydrolyses the ATP attached to the ATPase site, so 'energising' the myosin head. Calcium ions also bind to the troponin in the actin filaments and cause a change in its shape. In changing shape, the troponin pushes the tropomyosin away from the attachment site of the actin molecules. The myosin heads can now become attached to the actin filaments, forming an **actomyosin cross-bridge** (Fig 16.29).

As actomyosin cross-bridges are formed (Fig 16.30(a)), the energised myosin heads rotate and change their angle. As the angle of the actomyosin cross-bridge changes, the myosin head pulls the actin filament over itself towards the centre of the sarcomere (Fig 16.30(b)). The myosin head, thus, acts like your arm as you close a sliding door. When rotation of the myosin head is complete the myosin head disengages from the actin filament and rotates back to its relaxed position (Fig 16.30(c)). This disengagement depends on the binding of ATP to the ATPase site on the myosin head. The ATP is now hydrolysed by the ATPase site and new actomyosin cross-bridges form. The energised myosin head rotates again, further pulling the actin filament towards the centre of the sarcomere. Since all the filaments in a sarcomere are behaving in this way, the sarcomere shortens.

(b) The amount of ATP present in a muscle cell is quite small. Using this information, explain the process of *rigor mortis* during which the muscles of a dead person enter a state of contraction.

After passage of the action potential, calcium ions are actively pumped from the sarcoplasm back to the sarcoplasmic reticulum and T system. Calcium ions dissociate from the troponin allowing tropomyosin to block the attachment site of the actin once again, so that the actomyosin cross-bridges cannot reform. Consequently, the sarcomere reverts to its normal resting tension.

Creatine phosphate is used to regenerate ATP ready for the next contraction. Much later, the creatine phosphate is itself resynthesised using ATP from the breakdown of fatty acids or glycogen.

(c) Why do you think it might be advantageous for creatine phosphate to be used to resynthesise ATP rather than, say, glucose?

Very active muscles respire anaerobically to produce lactate. Once the muscle contraction has stopped, some lactate is oxidised to yield energy for the synthesis of ATP, which is then used to convert the remaining lactate to glycogen. The oxygen needed for this processing of lactate is called the **oxygen debt**.

Fig 16.30 Longitudinal section of part of an actin filament. Movement of the myosin heads causes the actin filament to slide towards the centre of each sacromere.

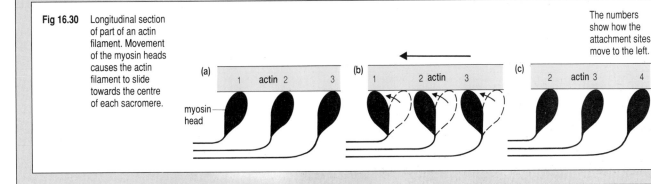

LAW OF ALL-OR-NOTHING

Action potentials and muscle fibre contractions are either 'on' or 'off'; there is no intermediate state.

Properties of muscle contraction

A resting muscle fibre will only contract if stimulated at or above a threshold level. But if stimulated at or above its threshold, a vertebrate muscle fibre fully contracts. This is known as the **all-or-nothing response** of the muscle fibre. Our ability to alter the degree of contraction of an entire muscle depends on the number of individual muscle fibres which are contracted and not the degree of contraction of each fibre. After its contraction, there follows a **refractory period**, during which the muscle fibre cannot respond to a further stimulus.

13 What do you think is happening during the refractory period?

If a muscle is stimulated, there is a short **latent period** before it contracts. A contraction then rapidly occurs and a force develops, after which the muscle returns to its relaxed state. These changes, known as a **muscle twitch** can be recorded using appropriate apparatus and a trace of the changes made as shown in Fig 16.31.

If a second stimulus is applied to a contracting muscle, a second contraction occurs as more muscle fibres contract. This second response develops a greater force than the first (Fig 16.32). This effect is called **mechanical summation**. Rapid stimulations of a muscle cause a smooth prolonged contraction. This is called **tetany** and is shown in Fig 16.33. Eventually, the muscle becomes **fatigued** and the contraction diminishes.

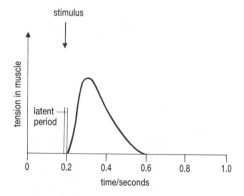

Fig 16.31   Recorded trace of a muscle twitch following a single stimulus.

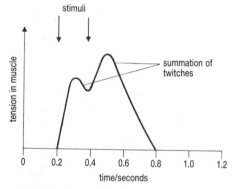

Fig 16.32   Mechanical summation: the second response develops a greater force than the first muscle contraction.

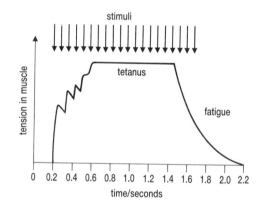

Fig 16.33   Frequent stimulation of a muscle causes tetany.

QUESTIONS

16.13 Copy Table 16.7 and complete it (labels A–L) to compare the three types of muscle.

Table 16.7   Comparison of muscle types

Feature	cardiac muscle	involuntary muscle	voluntary muscle
location	A	B	C
shape of cells	D	E	F
presence of striations	G	H	I
need for nervous stimulation	J	K	L

16.14 Distinguish between each of the following levels of muscle organisation: myofilaments, myofibrils, muscle fibres and muscle.

16.15 Discuss the functions of myosin, actin, troponin and tropomyosin in producing and controlling the contraction of muscle fibres.

SUMMARY

Receptors are cells which are sensitive to a single type of stimulus. They all convert (transduce) the energy of their stimulus into the electrochemical energy of a nerve impulse. Provided a specific stimulus has enough energy, it causes depolarisation of the surface membrane of its specific receptor, i.e. a generator potential. Stimulation of a receptor neurone by a receptor causes an impulse to pass along the neurone: the repeated stimulation of a neurone by a single receptor and the simultaneous stimulation of a neurone by several receptors (convergence) may help weak stimuli to be translated into impulses in receptor neurones.

Exteroceptors are sensitive to stimuli outside the body; interceptors are sensitive to stimuli inside the body; proprioceptors are a particular type of interoceptor sensitive to the state of contraction of the body's muscles.

Mechanoreceptors convert mechanical energy into generator potentials. Within the inner part of the mammalian ear there are mounds of mechanoreceptors with hair-like projections embedded in movable structures. Pulling of these 'hairs' causes a generator potential in their receptor cells and then impulses pass along the auditory nerve to the brain. In the cochlea, movement of the basilar and tectorial membranes pulls on mechanoreceptors in the organ of Corti: this is interpreted as sound. The air-filled external and middle parts of the ear serve to funnel sound waves and to amplify and transmit their vibrations to the inner ear, respectively. In the sacculus and utriculus, movement of chalky otoliths pulls on mechanoreceptors in the macula: this is interpreted as position with respect to gravity. Within the semi-circular canals, movement of a jelly-like cupula pulls on mechanoreceptors in the crista: this is interpreted as movement of the head.

Light receptors convert light energy into generator potentials. Two types of light receptor are found in the retina of the mammalian eye. Rods are very sensitive to light and to movement but insensitive to colour: they are most common at the periphery of the retina. Cones are less sensitive to light and movement than rods but are sensitive to colour: they are densely packed at the centre of the retina (the fovea) but absent at its periphery. Light receptors have a pigment located in their internal membranes: rod cells have the pigment rhodopsin; cone cells have the pigment iodopsin. Bleaching of these pigments by light causes a generator potential in the receptor cells and then impulses pass along the optic nerve to the brain. The brain uses cues from mechanoreceptors in the ear and from memory cells to interpret these impulses.

The retina is supplied with blood by the choroid layer of the eye. This layer also contains melanin, a dark pigment which absorbs light and so prevents its internal reflection within the eye. At the front of the eye, the choroid forms an iris diaphragm, which controls the amount of light entering the eyeball, and a ciliary body, which suspends the lens and causes it to change shape (and hence focal length). The outer layer of the eye, the sclera, is a protective layer. At the front of the eyeball the sclera forms a transparent cornea. The shape of the eyeball is maintained by two fluids, the anterior aqueous humour and the posterior vitreous humour. Light rays are bent (refracted) at the interface of each of the different transparent layers of the eye. Only by changing the shape of the lens can focusing of objects at different distances from the eye (accommodation) be achieved.

Muscles are effectors, i.e. they convert action potentials into reactions. Mammals have three types of muscle tissue. Involuntary muscle occurs in the eyeball and around internal tubular structures such as the gut and urogenital tracts. This tissue has spindle-shaped cells which lack striations in their cytoplasm. Cardiac muscle is found only in the myocardium of the heart. It has cylindrical cells with striations in their cytoplasm: the cells form interconnecting columns. Unlike other types of muscle tissue, cardiac muscle needs no stimulation to contract, i.e. it is myogenic. Voluntary muscle is attached by tendons to the bones of the skeleton. Its cells, or fibres, may be several centimetres long, contain several nuclei and have cytoplasm which is striated.

Striations in muscle cells result from filaments of two types of protein: actin and myosin. These filaments are arranged in repeated units called sarcomeres. Within each sarcomere, the actin and myosin overlap along part of their length. When stimulated by an effector neurone, calcium ions flood into the cytoplasm of a voluntary muscle cell and cause changes in the actin and myosin. As a result, actin and myosin filaments slide over each other increasing the region of overlap. Repeated in the sarcomeres along the whole of a muscle cell, this causes a muscle cell to contract.

Muscle cell contraction depends on stimulation at or above a threshold energy level. Stimulated at or above this threshold, a muscle cell always fully contracts – the all-or-none law. Different degrees of muscle contraction result from the full contraction of different numbers of muscle cells. Like neurones, muscle cells have a short recovery (refractory) period immediately after responding to stimulation, during which they cannot be restimulated. Repeated stimulation of muscle tissue causes smooth, prolonged contraction of the entire muscle, known as tetany. Eventually, voluntary muscles tire as they run out of ATP; this is known as fatigue and is associated with high concentrations of lactate produced during anaerobic respiration.

Answers to Quick Questions: Chapter 16

1 Chemoreceptor; exteroceptor.
2 Depolarisation results from a specific stimulus, not from stimulation by a chemical transmitter substance. The action potential follows an all-or-nothing law; the generator potential is proportional to the size of the stimulus.
3 Only when many synaptic knobs release their chemical transmitter simultaneously does its concentration reach a threshold at which depolarisation of the post-synaptic membrane occurs.
4 Aperture of the cone has a relatively large area which collects sound waves. Apex of the cone acts as a funnel, directing and concentrating sound waves to ear drum.
5 Pressures on either side of the ear drum cannot be equilibrated, causing the ear drum to be pushed outwards or inwards.
6 No – sound is transmitted as vibrations of molecules: a true vacuum contains no molecules.
7 Vibrations stretch projections of receptor cells in the organ of Corti, resulting in a change of permeability of their surface membranes. An influx of sodium ions causes a generator potential in these receptor cells, an electrochemical event. Resulting release of synaptic transmitter from receptor cells to receptor neurones causes electrochemical impulse to pass to brain.
8 No – they will not cause vibrations of tectorial and basilar membranes, so no stretching of receptor cell projections occurs.

9   Their pressure maintains the shape of the eyeball; their transparency enables light to pass to the retina; their difference in density from the cornea and lens results in additional interfaces for the refraction of light.

10 Those in Fig 16.14(a) – the near object – which are diverging more.

11 Light will be reflected back off the tapetum so that it passes twice through the light-sensitive cells, thereby increasing the size of the generator potential developed in the cells.

12 Even a short exposure to bright light would cause the rhodopsin in the rods to break down so destroying the pilot's night vision.

13 The sarcolemma is repolarising.

Chapter 17

CONTROL AND COORDINATION IN PLANTS

LEARNING OBJECTIVES

When you have studied this chapter you should be able to:

1. outline the control of rapid leaf movements in the Sensitive plant and Venus flytrap;

2. name the major classes of plant growth substances in flowering plants and outline the way in which they interact to control the plant's life cycle;

3. review the way in which plant growth substances are thought to control geotropism and phototropism;

4. interpret experiments on tropisms;

5. define photoperiodism and name a number of photoperiodic responses in plants;

6. summarise current understanding about the role of phytochromes in photoperiodism;

7. realise the limitations in our knowledge of the regulation of plant growth.

As far as we know, plants do not possess sense organs comparable with those, for example, in a mammal. However, plants do detect environmental stimuli that are vital for their survival: these include the direction of gravity; the direction, intensity and duration of light; touch; changes in temperature. Plants also respond to these environmental stimuli although, in comparison with many animal responses, plant responses are usually slow. Slow responses are common in animals when they are controlled by hormones. Most plant responses are controlled by hormone-like chemical coordinators. Since these usually exert their effect by controlling plant growth, they are commonly called **plant growth substances**. It is the interactions of these substances which control plant development. Although the growth response is generally slow, plants do show some rapid responses to plant growth substances. For example, changes in the membrane potential of bean roots can be measured within one minute of auxin application (auxins are described in Section 17.2).

17.1 CONTROL OF RAPID RESPONSES IN PLANTS

Not all growth responses in plants are slow nor are they all controlled by plant growth substances. Leaf closure of the Sensitive plant, *Mimosa pudica*, and the Venus flytrap, *Dionaea muscipula*, occurs very quickly. In both cases, leaf closure results from a change in the size of epidermal cells as they gain or lose water by osmosis.

All cells have active sodium pumps in their surface membranes (see Chapter 2). The activity of these sodium pumps results in a potential

Fig 17.1 *Mimosa pudica*, the Sensitive plant. When the leaves are touched they immediately collapse and close up.

difference (voltage) across cell surface membranes so that the inside of the membrane is negatively charged with respect to the outside. This is called a **resting potential**. In ways which are poorly understood, touching an appropriate part of the leaf of *M. pudica* or *D. muscipula* causes depolarisation of the surface membrane of cells in its epidermis, resulting in leaf movements.

Control of movement in the Sensitive plant (*Mimosa pudica*)

When an individual leaflet of *M. pudica* is touched, depolarisation of the surface membranes of its leaf cells occurs, resulting in an action potential like that in neurones (see Section 15.2). This action potential is propagated through surrounding phloem cells to a specialised group of parenchyma cells at the base of the leaf called **motor cells**.

The motor cells respond to depolarisation by actively pumping potassium ions from their cytoplasm into the extracellular spaces around themselves. As the concentration of potassium ions within the motor cells falls, their water potential becomes less negative and they lose water by osmosis (see Chapter 2). Since the motor cells have unusually elastic walls, they shrink when they lose water and pull the leaflets upwards (Fig 17.3). Recovery of the normal leaf position occurs when the movement of potassium ions is reversed. Motor cells are also thought to release an, as yet unidentified, chemical transmitter which is thought to move via the xylem to adjacent leaflets where it initiates a new action potential.

Control of movement in the Venus flytrap (*Dionaea muscipula*)

Touching the sensory projections on the surface of the special 'trapping leaves' of *D. muscipula* causes action potentials in the cells of its upper epidermis. These action potentials are transmitted to cells in the lower epidermis of the leaf, which respond by actively pumping hydrogen ions into their cellulose walls. Activated by the resulting acidic conditions, enzymes within the walls loosen its cellulose fibres. Since wall pressure stops osmosis into a turgid plant cell (see Chapter 2), loosening the fibres in its wall allows an epidermal cell to absorb more water from the extracellular spaces. Affected cells expand by about 25% of their volume as this water enters. Since only the lower epidermis expands, the whole leaf is pushed closed (Fig 17.4). The activity of hydrogen ion pumps in the lower epidermal cells requires so much energy that over 30% of the ATP in the entire leaf is used in the seconds that it takes to close the trap.

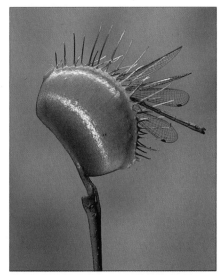

Fig 17.2 *Dionaea muscipula*, the Venus flytrap. The fly is trapped in a cage of spines and is digested by enzymes secreted by the leaf.

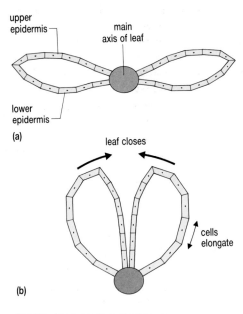

Fig 17.4 Diagrammatic section through
(a) an open leaf
(b) a closed leaf of a Venus flytrap plant.

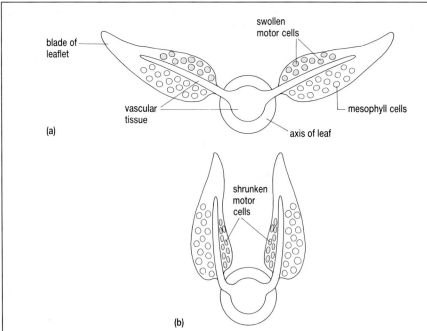

Fig 17.3 Diagrammatic transverse section through the main leaf axis and leaflet of *Mimosa pudica*
(a) with open leaves
(b) with closed leaves.

QUESTIONS

17.1 How is the response to touch of epidermal cells of *M. pudica* and *D. muscipula*
(a) similar to
(b) different from, that of mammalian receptor cells (see Chapter 16)?

17.2 Suggest why rapid leaf movements are advantageous to *M. pudica* and *D. muscipula*.

17.2 PLANT GROWTH SUBSTANCES

Plant responses are coordinated by chemicals which do not always move from their sites of production. Hence they should not be called hormones. Because they usually affect some aspect of growth, they are called plant growth substances.

During one investigation, 1300 kg of peas yielded only 9 mg of abscisic acid.

The slow responses of plants to light, gravity etc. result from cell growth. Growth of plants is coordinated by **plant growth substances**: compounds produced in one more or less precise region of a plant and usually transported to other parts of the plant where they affect cell division, cell elongation or cell differentiation.

The study of plant growth substances has been hampered by a number of experimental difficulties. For example, the yield of growth substances in plant extracts is very low and these extracts are not always pure. It also appears that several biological effects are stimulated by more than one plant growth substance. As a result, the nature of their mode of action remains controversial. However, five major groups of plant growth substances have been identified: abscisins, auxins, cytokinins, ethene and gibberellins. The major effects with which they are attributed are summarised in Table 17.1.

Seed dormancy and germination

It might seem that, on a warm day, a mature seed inside a juicy fruit has ideal conditions for its germination. In spite of this, most seeds remain dormant inside their fruit and for some time after the fruit has been eaten or decomposed. Seed dormancy and germination are controlled by the levels of abscisic acid and gibberellins in the seed.

Abscisic acid is produced by a maturing seed and causes a reduction in the water content from about 85% to 10% of its fresh mass. The continued

Some seeds do germinate inside their fruits. They are called **viviparous embryos**. Such viviparity is common in cultivated grapefruit, in which the roots emerging from the seed are pickled in the acidic flesh of the grapefruit, and in certain mutants of maize. Study of these viviparous embryos has provided clues about the control of desiccation and dormancy in seeds. In most mutants, viviparity is associated with lack of **abscisic acid (ABA)**.

presence of abscisic acid in the dried seeds inhibits their development. These seeds will not germinate until the concentration of abscisic acid has fallen. In some desert plants, a heavy fall of rain washes the abscisic acid out of seeds which are in the soil. In temperate plants, exposure to prolonged cold (**vernalisation**) induces the destruction of abscisic acid. As its concentration of abscisic acid falls, a seed absorbs water and its concentration of **gibberellins** increases. The presence of gibberellins causes breakdown of stored food within the seed, enabling the embryo to grow.

 Why is the process of vernalisation advantageous to a temperate plant?

Table 17.1 The major effects of plant growth substances

Class of plant growth substance	Site of production	Major effects on plant growth
abscisins (e.g. abscisic acid)	widespread	promote stomatal closure under water stress
		stimulate loss of water during final stages of seed development
		prolong seed dormancy
		stimulate abscission of leaves, flowers and fruits
		slow growth of most plant organs
auxins (e.g. indole acetic acid)	growing tips of stem and root (apical meristems)	stimulate elongation of cells in shoots
		inhibit elongation of cells in roots
		inhibit growth of lateral shoots (apical dominance)
		stimulate growth of lateral roots
		stimulate retention of ovary (fruit set) and its development into a fruit
		inhibit fall of leaves and fruits
		stimulate production of ethene by seeds within fruit
		stimulate production of vascular tissue (high concentrations)
cytokinins (e.g. kinetin)	actively dividing tissues (meristematic tissue)	promote cell division
		promote sprouting of lateral buds
		inhibit senescence of leaves
		inhibit dormancy of seeds
ethene	most plant organs	stimulates ripening of fruit
		stimulates abscission of fruits and flowers
		inhibits the polar flow of auxin
gibberellins (e.g. gibberellic acid)	apical leaves, buds, seeds, root tips	stimulate elongation of stems of dwarf varieties
		affect flowering
		stimulate fruit development
		stimulate germination of seeds
		stimulate growth of buds

Both abscisic acid and gibberellins seem to exert their effect by switching genes on or off (see Chapter 4). Abscisic acid is thought to stimulate the transcription of certain genes whose protein products accumulate during desiccation: it is the accumulation of these proteins which appears to protect plant tissues from death during desiccation. Gibberellins are thought to stimulate the transcription of genes which carry the genetic code for enzymes that are needed to hydrolyse the food reserves in the cotyledon or endosperm (see Chapter 20), enabling the embryo to grow.

Growth of the seedling

Seeds often germinate within soil. If they are to survive, their shoots must grow upwards and their roots downwards. Once the shoot has emerged from the soil, growth towards incident light enables its leaves to photosynthesise. Growth with respect to gravity is **geotropism**; growth with respect to light is **phototropism**. One theory to explain these tropisms, the **Cholodny–Went theory**, suggests that both are controlled by **auxins**, of which **indole-3-acetic acid (IAA)** is the most active. Actively dividing cells form a plant tissue called **meristematic tissue**. Auxins are produced by such tissue and move to other regions in a one-directional (polar) flow, either by diffusion between cells or in the phloem. The effect of IAA on elongation of cells behind the growing tip of a root and shoot is shown in Fig 17.5.

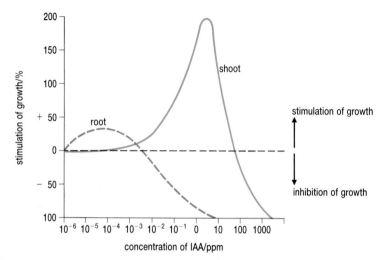

Fig 17.5   The effect of IAA concentration on elongation of cells in the region behind the growing tip of a root and shoot. Notice that the axis for IAA concentration has a logarithmic scale.

 Summarise the effects of IAA on root and shoot elongation.

Geotropism

Auxins are produced by the tip of a shoot and root (and root cap) and move into the rest of the plant. In a vertical root or shoot, the concentration of auxins reaching growing cells is equal all around the plant. However, if the root or shoot is horizontal, a greater concentration of auxins is found in the lower side than in the upper side (Fig 17.6). According to the Cholodny–Went hypothesis, this unequal distribution in shoots results from lateral transport of auxins as they move backwards through the stem, i.e. the upper side loses auxins to the lower side. There is some controversy about the cause of the unequal distribution of auxins in roots, e.g. there is evidence that auxins reach the root from the stem and that additional auxins are manufactured in the lower part of a horizontal root.

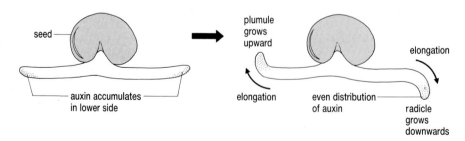

Fig 17.6   Geotropism. Root tips are positively geotropic and shoot tips grow away from the pull of gravity. Note it is suggested that auxins alone may not control geotropism.

 What effect will this unequal distribution of auxins have on horizontal shoots and on horizontal roots?

CONTROL AND COORDINATION IN PLANTS

Recent observations have shown certain problems with the Cholodny–Went explanation of geotropism. Although radioactively labelled auxins have been found to be redistributed in shoots and roots, the resulting differences in concentration are insufficient to explain the different growth rates of the upper and lower side of a horizontal shoot. Much greater differences between the upper and lower sides of shoots and roots have been found in the concentration of gibberellins: in horizontal sunflower shoots, the gibberellin concentration of the lower side has been found to be ten times higher than that of the upper side. Possibly, gibberellins might contribute to geotropic growth. The results of other investigations suggest that abscisic acid or ethene might be the growth inhibiting substance in roots.

Problems such as these have led some plant physiologists to suggest that other growth-controlling factors might be more important than plant growth substances. Gradients in several properties, including pH, electrical potential, water potential, rate of respiration and enzyme activity, are known to exist between the upper and lower sides of horizontal shoots and roots. Furthermore, a mechanism to explain these differences has been widely accepted. This mechanism, the **statolith theory**, suggests that starch granules within the cytoplasm fall to the lower side of cells under the influence of gravity. Here, these granules cause pressure on a membrane or membrane system, resulting in the physiological differences observed.

Phototropism

Although a mature root seldom emerges from the soil, shoots usually do. Most shoots grow towards light (are **positively phototropic**) whilst roots are either negatively phototropic or unaffected by light.

Most research into phototropism has involved the use of coleoptiles (see Section 21.4) of cereals (Fig 17.7). Light from one direction (unidirectional light) causes illumination gradients which are detected by a photoreceptor pigment (probably a flavine) within coleoptiles. According to the Cholodny–Went hypothesis, this illumination gradient results in the lateral movement of auxins from the illuminated to the shaded side of the coleoptile. Exactly how this lateral movement happens is not known, but it is known that light neither inactivates nor destroys auxins. Since elongation of cells in a shoot is stimulated by relatively high concentrations of auxins, an accumulation of auxins in the shaded cells will cause the coleoptile to bend towards the light (Fig 17.8). Possibly, the lack of response to light shown by most roots is because they lack the necessary photoreceptor pigment.

Fig 17.7 Cereal coleoptiles. Although they are the protective cover for the emerging leaf at germination, coleoptiles behave as stems and are used in auxin experiments.

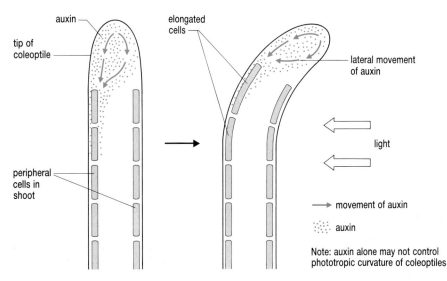

Fig 17.8 A mechanism to explain positive phototropism in shoots. Cells elongate on the shaded side where the concentration of auxin is higher.

As with the Cholodny–Went hypothesis of geotropism, the above explanation of phototropism is not without its critics. Experiments have shown that phototropic curvature of coleoptiles can occur in the apparent absence of the lateral transport of auxins. In the hypocotyl of sunflower seedlings it appears that lateral movement of auxins does not occur but lateral movement of an abscisin does occur. However, the movement of abscisins is towards the illuminated side: perhaps abscisins inhibit elongation or inhibit transport of auxin down the illuminated side.

ANALYSIS

Phototropism

The following exercise is intended to help you to practise the important skill of drawing conclusions and making predictions from experimental results. It is important that you only use the evidence which you have available from the experiments.

Fig 17.9 summarises a number of experiments that were performed to investigate the control of tropisms in plants. A number of spaces in the figure have been left blank.

- Complete Fig 17.9 by drawing the result you would expect or by suggesting an explanation, as appropriate (A, B, C and D). The first line has been done as an example.

Name of investigator (date)	Experimental procedure	Observed results	Suggested explanation
Darwin (1880)	unilateral light — intact oat coleoptile 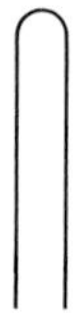	coleoptile bends towards light 	The coleoptile is positively phototropic. It bends towards the light by unequal elongation of the region just behind the tip.
	unilateral light — lightproof cover is placed over intact tip of coleoptile 	A	Light is perceived by the tip of the coleoptile.
Boysen–Jensen (1913)	unilateral light — mica inserted into shaded side 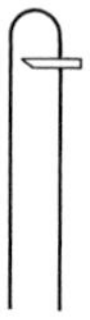	no elongation or bending 	B
Paal (1919)	 darkness; tips removed and replaced but displaced to one side 	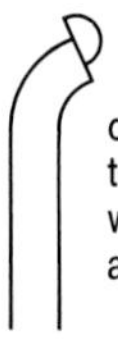 coleoptiles bend towards side where tip is absent 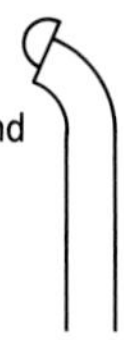	C
Briggs (1957)	unilateral light — thin glass plate separates the two sides of the coleoptile 	elongation but no bending 	D

Fig 17.9 Experiments used by eminent plant scientists to investigate phototropism in coleoptiles.

CONTROL AND COORDINATION IN PLANTS

17.3 PATTERNS OF VEGETATIVE DEVELOPMENT

To someone who lives in Britain, the trees in Fig 17.10 are instantly recognisable by their size, shape and branching pattern. The roots of these trees have characteristic patterns of growth too. Whilst these species-specific differences are inherited, growth pattern is controlled by a balance of plant growth substances.

Fig 17.10   The branching system of a tree gives it a very distinct appearance. How many different species of tree can you see in this photograph?

Stem growth and branching

Stem height is controlled by gibberellins and auxins which stimulate cell elongation. Dwarf varieties of plants are often genetic mutants which cannot synthesise gibberellins.

 What do you think would happen to such a dwarf mutant if you applied gibberellins to the stem?

Many gardeners know that by removing the top from a plant they can make it grow bushier. The reason for this is that the uppermost (apical) bud inhibits the growth of lateral buds, a process called **apical dominance**. The apical meristem, which is a site of auxin production, is located within the apical bud (Fig 17.11). As auxins move downwards to other parts of the plant, they stimulate elongation of the cells just behind the apical meristem but inhibit the growth of lateral buds. As they move through a plant, auxins are eventually degraded by enzymes so that buds which are a long way from the apical meristem are relatively unaffected by its auxins. Any auxins which reach the roots stimulate growth of lateral roots.

 How could a commercial grower exploit the properties of auxins?

Root meristems are thought to produce very little auxin, but do produce **cytokinins**. These growth substances are transported upwards through the roots into the shoot system, where they stimulate the growth of lateral buds. Thus, stem branching results from the antagonistic effects of auxins from the shoot apex and cytokinins from the root apex. There is a high concentration of auxins and low level of cytokinins at the top of the shoot (Fig 17.11), so growth of lateral buds is inhibited. In comparison, the lowest branches are far from the source of auxins but nearer the source of cytokinins, so they will sprout. Somewhere in between, the balance of auxins and cytokinins is just sufficient to inhibit the growth of lateral buds.

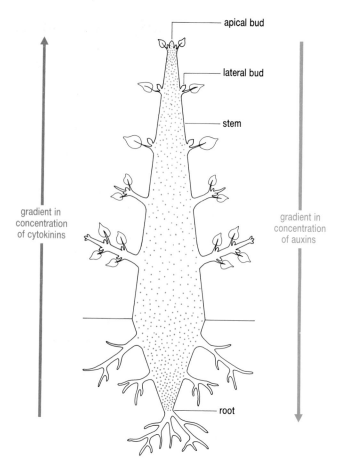

- apical bud
- lateral bud
- stem
- root

gradient in concentration of cytokinins

gradient in concentration of auxins

Fig 17.11 Diagrammatic summary of the interplay between auxins from the shoot apex and cytokinins from the root apex.

 6 Why, when the shoot grows just a few more millimetres, will these lateral buds begin to sprout? What effect will the small amount of auxins reaching the roots have?

Through this interaction of auxins and cytokinins, the root and shoot system help to regulate each other's growth. A large root system produces more cytokinins than a small root system. Therefore, more cytokinins reach the shoot system and stimulate the growth of lateral buds into branches. On the other hand, a slowly growing root system produces less cytokinins so there is less stimulation of lateral buds of the shoot system. At the same time, a vigorously growing and branching stem produces more auxins, which stimulate branching of the root.

7 Suggest why such a system of dual control is advantageous to the plant.

Differentiation of vascular tissue

Relatively high concentrations of auxins help to stimulate new cells to form the tissues of xylem and phloem. High concentrations of auxins are found just behind the apical meristems and in the region of sprouting buds. It is in these regions that new vascular tissue develops, helping to interconnect the older parts with the new parts of the plant as they grow.

Notice once again how the effect of the auxin depends upon

- its concentration
- its site of action.

There are two extra factors to be considered. Firstly, higher concentrations of auxins are needed to stimulate development of vascular tissue than those which inhibit the growth of lateral buds. Secondly, buds which are dormant can be kept dormant by low auxin concentrations. However, once the genes which control sprouting have been switched on, the new cells formed during sprouting are not inhibited by auxins and produce auxins themselves.

Flowering and development of fruit

Flowering occurs at a specific time of year in most species of plants. It is thought that a hormone, called **florigen**, controls the development of flowers, although this hormone has never been isolated and its existence is doubted by many plant physiologists. The onset of flowering is affected by the relative proportions of daylight and darkness during each day (**the photoperiod**). A possible mechanism to account for photoperiodism is described in Section 17.4.

 Why is it important for a plant to control the time of its flowering?

After fertilisation, some parts of the flower grow to produce a fruit (see Chapter 20). In many plants, the auxins released by germinating pollen grains and by the style and ovary of the pollinated flower prevent abscission of the ovary. This is called **fruit set**. The further growth of these fruits is controlled by auxins released by the developing seeds which they enclose. Under auxin stimulation, cells of the ovary divide and enlarge as they store a variety of food materials. In plants of the genus *Prunus*, the same effects are brought about by gibberellins released by their developing seeds.

Plant fruits are adapted to aid dispersal of seed. Some fruits aid dispersal when they are eaten by animals which deposit the undigested seeds in their faeces. Unripe fruits of this kind are often green, hard and distasteful: as they mature, they ripen, becoming brightly coloured, soft and sweet. Ripening is stimulated by **ethene**, a plant growth substance which is thought to be made by the cells of the fruit in response to a surge of auxin production by mature seeds. Ethene is a gas and so diffuses from a ripe fruit into the air. When fruit growers pack fruit together in containers, the ethene released by one ripe fruit stimulates the ripening of others, explaining why all the fruit in one container ripens within a few days.

Senescence and dormancy

Temperate regions have distinct seasons. In many countries the season autumn is descriptively named fall. It is during this season that fruits which have not been eaten fall from the parent plant and the leaves of deciduous plants change colour and fall from the plant. Both fruits and leaves undergo rapid ageing, called **senescence**, prior to being shed from the plant. Fruit fall and leaf fall is called **abscission**. At the point of abscission (**the abscission zone**), the middle lamellae of a group of cells breaks down and these cells separate to form an upper **abscission layer** and a lower **protective layer**. The plant organ is now held to the plant only by vascular tissue. Mechanical action, such as the effects of wind, eventually breaks this vascular tissue so that the fruit or leaf falls from the plant.

Fig 17.12   Plant growth substances are widely used in the fruit industry. For example, bananas are shipped around the world from their origin to their point of sale. To avoid loss in transit, bananas are picked and shipped when they are unripe. Exposing them to ethene in storage will ripen the bananas. Onto the growing crop of various commercially important fruits other plant growth regulators are applied to control fruit size, shape and to prevent early abscission.

Senescence and abscission are controlled by a balance of several growth substances, the secretion of which is thought to be controlled by daylength (see Section 17.4). Both auxins and cytokinins help to prevent senescence, keeping leaves and fruits healthy. During the autumn, plants produce less auxin and cytokinin so their inhibition of senescence becomes lessened. At the same time, abscisic acid is produced which promotes formation of the abscission layer. In fruits, the release of ethene during ripening stimulates the release of abscisic acid; in leaves it is not clear what stimulates the release of abscisic acid but a response to photoperiod is probably involved.

Abscisic acid also slows growth of most plant organs and induces dormancy (see Table 17.1). When abscisic acid contributes to senescence and abscission in fruit and leaves it also stimulates seeds and buds to become dormant and the plant to stop growing. Dormancy enables trees to survive adverse winter conditions with little risk of damage to growing organs or of excessive water loss from leaves which would not photosynthesise effectively because of low temperatures and poor light.

CONTROL AND COORDINATION IN PLANTS

USING PLANT GROWTH SUBSTANCES

Plant growth regulators are powerful tools which growers can use to increase the productivity of their crops. For example, yields of wheat have been greatly increased by selective breeding and the application of nitrogen fertilisers. However, too much nitrogen can weaken the stem making the crop more susceptible to fungal attack and allowing the crop to be beaten down by the wind and rain (lodging). One way to counteract this is to spray the crop with a growth retardant, e.g. chlormequat chloride. This shortens the internodes between the leaf joints and so stiffens the straw. Plant growth substances can also be used to dwarf indoor pot plants. For example, potted chrysanthemum plants are sprayed with daminozide.

Fig 17.13 The wheat on the left-hand side of this field received applications of chlormequat to reduce growth. With shorter internodes the crop will be less likely to fall over, or lodge.

ANALYSIS

Abscission

This exercise will help you develop skills in interpreting experimental data.

Two leaf blades were removed from a healthy intact plant. One of the petioles was left exposed while the other was treated with a plant growth substance. Two weeks later a longitudinal section through the petiole stumps and stem was taken. The tissue distribution of this section is shown in Fig 17.14.

petiole treated with plant growth substance

(a) (i) Describe three differences between the treated and untreated parts of the plant which are apparent from Fig 17.14.
    (ii) Copy Fig 17.14 and mark clearly with crosses the position of two areas where actively dividing cells would be present.
(b) (i) Can you suggest what method should be used to apply the plant growth substance?
    (ii) Describe a suitable control for this investigation.

CONTROL AND COORDINATION IN PLANTS

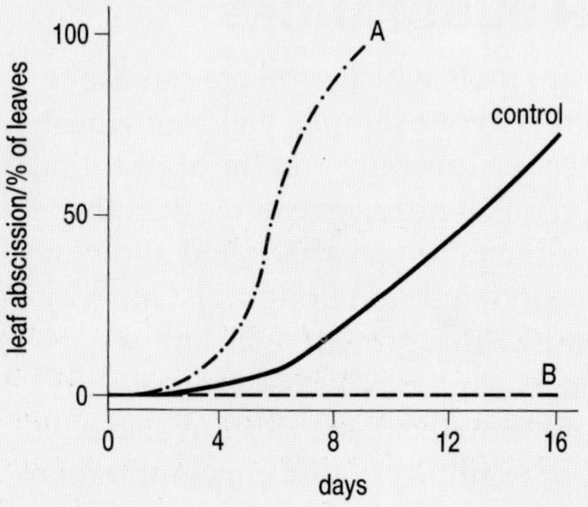

Fig 17.15

(c) The effect of two plant growth substances A and B, on abscission of leaves in similar plants was measured. The results are shown in Fig 17.15.
   (i) From the results shown in Fig 17.15, which of the two hormones do you think was applied to the cut petiole of the plant shown in Fig 17.14? Explain your answer.
   (ii) Give the name of a hormone which could be hormone A.

QUESTIONS

17.3 In what ways are plant growth substances similar to mammalian hormones (see Chapter 14)? In what ways do they differ?

17.4 Use the information in Sections 17.2 and 17.3 to suggest the effect of each of the following horticultural practices.
   (a) The cut end of a stem cutting is dipped into a synthetic auxin powder before being planted in sand or compost.
   (b) Synthetic auxins are commonly sprayed on grass lawns to remove broad-leaved weeds.
   (c) Synthetic auxins and extracted gibberellins may be sprayed on flowering fruit trees.
   (d) Abscisic acid may be sprayed on fruit trees before cropping.
   (e) Cytokinins are often sprayed on lettuce and cabbage after they have been harvested.

17.4 PHOTOPERIODISM

Each plant species flowers and produces fruit at a particular time of year. For plants which grow on the floor of deciduous woodlands, it is advantageous to flower early in the year before the canopy of leaves develops and reduces light intensity. For plants in temperate regions, it is advantageous to produce fruit at a time when photosynthesis is rapid but before the adverse conditions of winter. Of the environmental factors which can vary throughout the year in temperate regions, daylength (**photoperiod**) is the most reliable indicator of season. Long days always accompany summer, short days always accompany winter.

Mechanism for detecting daylength: phytochrome

The fact that plants are able to respond to daylength indicates that they must possess a mechanism that measures time, and a mechanism that detects light. It is thought that most organisms have a mechanism that measures time, called a **biological clock**. Though the way in which this timing mechanism might work is unknown, it works independently of environmental variables to produce **endogenous rhythms** in the behaviour of organisms. Often, these endogenous rhythms are synchronised with the day–night cycle to produce circadian rhythms (*circa* means about, *diem* means day). Movement of leaves of wood sorrel (*Oxalis* sp.) and runner bean (*Phaseolus* sp.), opening and closing of tulip (*Tulipa* sp.) flowers and fixation of carbon dioxide by the air plant (*Bryophyllum* sp.) all show circadian rhythms.

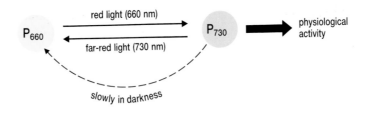

Fig 17.16   The effect of red light and far-red light on phytochrome.

Synchronisation of endogenous rhythms with daylight is possible because the leaves (including cotyledons) of plants contain minute quantities of a light-sensitive pigment, called **phytochrome**. A molecule of phytochrome can exist in two different forms. One of these, P_R also called P_{660}, maximally absorbs red light of mean wavelength 660 nm; the other, P_{FR} or P_{730}, maximally absorbs far-red light of mean wavelength 730 nm. Absorption of light by one form of phytochrome causes its rapid conversion to the other; slow conversion of P_{730} to P_{660} also occurs in darkness (Fig 17.16). Since daylight contains more red light than far-red light, more of the phytochrome will exist as P_{730} than as P_{660} during the day. After sunset P_{730} will be converted slowly to P_{660}.

The two forms of phytochrome have different physiological effects. In most plants, P_{730} is considered to be the active form, i.e. a suitable concentration of P_{730} can stimulate or inhibit physiological processes on which P_{660} has no influence. These physiological processes are, thus, stimulated by exposure to red light and inhibited by exposure to far-red light (Table 17.2).

Table 17.2   Stimulation of physiological processes by light

Process	Effect of far-red light and of darkness	Effect of red light and of white light
conversion of phytochrome	promotes P_{730} to P_{660}	promotes P_{660} to P_{730}
germination of small seeds	inhibits	promotes
flowering in long-day plants	inhibits	promotes
flowering in short-day plants	promotes	inhibits
leaf expansion	inhibits	promotes
growth of internodes	promotes	inhibits
chloroplast development	inhibits	promotes
chlorophyll synthesis	inhibits	promotes

Flower induction

Flowering plants are sensitive to daylength. Flowering, leaf fall from deciduous trees (abscission), formation of winter buds in temperate trees and shrubs, and formation of perennating organs, e.g. tubers, are affected by daylength. One of the most widely studied effects of daylength is on flowering: flowering plants are often classed according to the photoperiod in which they flower. **Day neutral plants**, such as snapdragon (*Antirrhinum* sp.), flower whenever they have grown sufficiently, irrespective of daylength. **Long day plants**, such as poppy (*Papaver* sp.), flower only if the daylength exceeds some critical value. **Short day plants**, such as the commercial chrysanthemums, flower only if the daylength is less than some critical value.

ANALYSIS

The control of flower induction

This exercise will help you to understand how flowering is controlled.

Flower induction is related to the type of phytochrome present in the flower.

Spinach (*Spinacia oleracae*) is a long day plant that flowers only if the daylength is greater than a critical period of about 14 hours whilst cocklebur (*Xanthium pennsylvanicum*) is a short day plant that flowers only if the daylength is less than a critical period of 15.5 hours. However, both plants will flower with a photoperiod of 15 hours light and 9 hours darkness.

(a) If spinach and cocklebur were under a lighting regime of 8 hours daylight and 16 hours darkness, which would you expect to flower?

If cocklebur plants were grown under a light regime of 8 hours light and 16 hours dark but half way through the dark period the plants were given a flash of white light, different results would be obtained.

(b) During the light period, what form of phytochrome would predominate in the cocklebur plant?

(c) What would happen to this form of phytochrome once the dark period started?

(d) What would happen to this phytochrome when the cocklebur plants were exposed to the brief flash of white light (look at Fig 17.16).

(e) Cocklebur plants treated in this way fail to flower. Suggest a hypothesis to account for this result.

(f) If we repeated this experiment with spinach it would flower. Explain why.

(g) Which appeared to be the significant factor in controlling flowering; the length of daylight or the length of darkness?

(h) If the dark period were interrupted by a brief flash of red light followed by a flash of far-red light, cocklebur would flower. Explain this result.

Although flowers are produced at the shoot tips, it is exposure of the leaves to light which results in flowering. In fact, exposure of just one leaf to the appropriate lighting regime will result in flowering. Together with evidence that the flowering stimulus can be transmitted from one plant to another across grafts, this suggests that a hormone passes from the stimulated leaves to the shoot tip to bring about flowering. This hypothetical flowering hormone has been called **florigen** but has not yet been found.

SPOTLIGHT

FLOWERS OUT OF SEASON

Using knowledge about the photoperiodic response of plants enables growers to bring plants to the consumer when they will command the highest price, for example at Christmas (Fig 17.17). Thus chrysanthemums, short day plants which usually flower in the autumn as the days shorten, can be prevented from flowering by maintaining them in a long day lighting regime. A few weeks before Christmas, flowering can be initiated by providing a short day lighting regime in the glasshouse. Other plants, for example hyacinths, which would normally only grow in the spring are treated by cooling, simulating the effect of winter. This breaks the plant's dormancy so that when you plant the bulbs in the late autumn they flower at Christmas time.

Fig 17.17 Some flowering plants, like these Poinsettia, are sometimes induced to flower at the wrong time of year by exposure to long dark periods. They are short day plants.

The relationship between phytochromes and other plant growth substances

All the major events in the life cycle of a flowering plant, such as dormancy of buds and seeds, growth of buds and flowers, leaf fall and germination of seeds, occur during a specific season. In fact, the survival of a plant depends on its ability to restrict each stage of its growth to a specific season. It is likely that seasonal variables, such as temperature or photoperiod, must be involved in controlling seasonal growth in plants. However, each stage of growth is also controlled by plant growth substances (see Table 17.1). Gibberellic acid promotes flowering in some long day plants and inhibits flowering in some short day plants. This resembles the effect of P_{730}.

 What does this suggest might be the relationship between phytochrome, gibberellins and other plant growth substances?

The nature of this relationship is just emerging, thanks to genetic engineeering techniques. Plant growth substances are produced by enzyme-controlled reactions within plant cells. Being proteins, enzymes are made when an appropriate DNA sequence (gene) in the cell nucleus is transcribed to messenger RNA. The transcription of many genes is itself controlled by regulatory sequences of DNA (see Section 4.1). Gene transfer experiments carried out during the mid-1980s have shown that some of these regulatory DNA sequences are affected by light; they are called **light-responsive elements** (**LRE**s). In a way which is not yet understood, it is thought that phytochrome pigment is responsible for the effect of light on these LREs. Possibly, the phytochrome activates other molecules that influence the LRE; only further research will tell.

To date, investigation of LREs has been restricted to their effect on the production of such enzymes as ribulose 1,5-bisphosphate carboxylase. This enzyme is involved in carbon dioxide fixation during photosynthesis. The transcription of its gene is controlled by an LRE and is stimulated by red light. LREs have also been shown to be capable of affecting **transgenes** (genes transferred from one species to another), so there seems little reason to doubt that they might control the genes for growth substance production.

Even if this turns out to be true, it is unlikely to be the whole story. The action of LREs does not explain how exposure of only one of a plant's leaves to an appropriate light regime will induce flowering by the whole plant. Not all long day plants or short day plants react in the same way to gibberellic acid, so the association between photoperiod and growth substance is not universal. It has long been known that gibberellins can be used to stimulate flowering in plants which normally require a period of chilling (**vernalisation**) before they flower. Thus, it seems likely that exposure to low temperatures might also induce gibberellic acid secretion in some plants. Unfortunately, our understanding of the control of plant growth and development is far from complete.

QUESTIONS

17.5 Imagine a seed which germinates in deep soil beneath other plants. Its plumule grows through the soil and between the shoots of other plants until it reaches daylight. (Chlorophyll absorbs white light but emits far-red light; white light contains more red light than it does far-red light). Name the form of phytochrome in this plant and suggest its effect on the plant's growth
(a) during seed germination
(b) during growth of the plumule through the soil

(c) during growth of the plumule between the other plants

(d) once the plumule has grown above the other plants.

17.6 Look at Fig 17.18 which shows the day/night regimes used to grow different samples of one plant species. The effect of each regime on flowering is also given.

(a) Use the data from regimes 1 and 2 to deduce the photoperiodic group to which this species of plant belongs.

(b) Explain the result of regimes 3 and 4.

(c) Suggest the result of regime 5 and explain your answer.

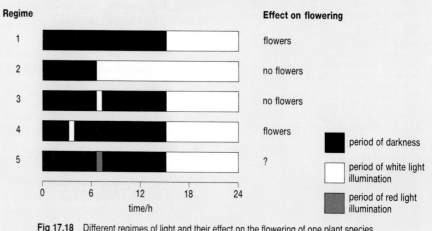

Fig 17.18 Different regimes of light and their effect on the flowering of one plant species.

SUMMARY

Most plant responses are slow and involve the differential growth of organs. Exceptions are the rapid movements of the Sensitive plant, *Mimosa pudica*, and the Venus flytrap, *Dionaea muscipula*. In these plants, depolarisation of cell surface membranes results in osmotic changes inside crucial leaf cells. As a result of water movements, rapid leaf movements occur.

Responses associated with growth are controlled by plant growth substances. These chemicals are produced by one or more regions of plant, usually the growing regions, and are transported to other tissues where they affect growth. Depending on the target organ, each plant growth substance has a number of effects. They also interact in the control of certain responses.

Seed dormancy is broken when a decrease in the concentration of abscisic acid and an increase in the concentration of gibberellins occurs in the seed tissues. As the stem of the seedling grows, its length is controlled by gibberellins, but its response to light and gravity is controlled by auxins, such as indole-3-acetic acid (IAA). IAA also controls the geotropic growth of the root and the formation of vascular tissue in roots and shoots. Along with cytokinins, IAA determines the branching pattern of the root and shoot systems.

After fertilisation in the ovules, auxins prevent the ovary from falling from the plant (fruit set) and stimulate its development into a fruit. Ripening of succulent fruits is stimulated by ethene. Abscission of the ripe fruit and of leaves is controlled by a balance of several growth substances; principally, low concentrations of auxins and cytokinins and high concentrations of abscisic acid.

Leaves possess minute quantities of a light-sensitive pigment called phytochrome. Phytochrome exists in two forms. When exposed to darkness or far-red light (wavelength 730 nm), phytochrome is converted to P_{660}; when exposed to white or red light (wavelength 660 nm), phytochrome is converted to P_{730}. In most plants, P_{730} is physiologically active and stimulates such activities as chlorophyll development and leaf expansion. Phytochrome is involved in the ability of plants to synchronise their activities with daylength (photoperiodism). Plants which are induced to flower by P_{730} are termed long day plants and flower only when exposed to long periods of illumination. Plants which are stimulated to flower by P_{660} are termed short day plants and flower only when exposed to long periods of uninterrupted darkness or far red light. In a way which is not understood, it is thought that phytochrome affects the transcription of critical genes.

Answers to Quick Questions: Chapter 17

1   Cold shock occurs in winter so that growth resumes in spring when temperatures, light availability and availability of (unfrozen) water increase.
2   IAA inhibits elongation of root except at relatively low concentrations and stimulates growth of shoot only at relatively high concentrations.
3   Horizontal shoots will grow upwards and horizontal roots will grow downwards.
4   Internodes would increase – the stem would get longer.
5   Cuttings dipped in an auxin mixture root faster. Removal of apical buds produces attractive bushy plants.
6   The balance is tipped in favour of the cytokinins (whose source is now nearer as the apical meristem grows further away). The small amount of auxin reaching the roots will stimulate branching of the root.
7   Root system grows as shoot system that it supports grows.
8   Energy (from sunlight) is needed for fruit and seed production; pollination and seed dispersal may depend on animals that are active only at certain times of year.
9   P_{730} may stimulate release of gibberellins and influence secretion of other plant growth substances.

Chapter 18

SUPPORT SYSTEMS AND MOVEMENT

LEARNING OBJECTIVES

When you have studied this chapter you should be able to:

1. distinguish between movement and locomotion;

2. explain the terms kinesis, taxis, nasty and tropism and describe examples of each;

3. describe how cells move using pseudopodia and undulipodia;

4. explain why muscular movement requires a skeleton and compare hydrostatic skeletons, exoskeletons and endoskeletons;

5. describe the major skeletal tissues in mammals and outline their distribution;

6. outline the main features of a mammal's skeleton and explain how it is adapted to support the animal's weight and allow locomotion;

7. describe the major plant skeletal tissues and account for their distribution.

The molecules from which cytoplasm is made continually move about (see Chapter 2). Indeed, such vital processes as cell respiration (see Chapter 6), photosynthesis (see Chapter 7) and nerve impulse transmission (see Chapter 15) depend on the movement of particles within the cytoplasm. In biology, the term movement is usually applied to active movement of the whole organism (called **locomotion**) or of its parts.

Fig 18.1 This seasonal migration of huge numbers of wildebeest in Africa ensures their survival. How?

Active movement involves expenditure of energy by the organism; the random drifting of organisms in the currents of air or water in which they live is not active movement. Locomotion is often described as if organisms had a purpose. Thus, it is said that animals move from place to place to obtain food, find a mate, escape predators or disperse away from crowded conditions. Whilst humans have purpose in their behaviour, do animals? It is wrong to interpret animal behaviour as though the animal had *human* purpose. So rather than ask questions such as 'Why do animals move from place to place?' it is better to ask 'What advantage does an animal gain in moving from place to place?'

18.1 TYPES OF MOVEMENT

Most movement and locomotion is related to stimuli from the environment, and we can classify movement in terms of the type of response elicited by the stimuli.

Kinesis

An unfavourable environment causes many organisms to move faster and change their direction more often. Such a non-directional response to an unfavourable stimulus is called a **kinesis** (plural, kineses). Woodlice exhibit a kinesis in response to humidity; the drier the air the faster they move and the more often they change direction. As a result of this kinesis, woodlice are likely to move away from dry conditions and come to rest in humid conditions.

 What advantage will the woodlice gain as a result of this behaviour?

Taxis

Unlike kineses, some locomotion is directional, either towards or away from the appropriate stimulus. This behaviour is called a **taxis** (plural, taxes). Many photosynthesising protoctists, such as *Euglena*, swim towards light; they are **positively phototactic**. Many carnivorous animals move towards chemicals emitted by their prey; they are **positively chemotactic**.

 How would you describe the behaviour of insects moving away from an insect repellent?

SPOTLIGHT

A PHENOMENAL SENSE OF SMELL

The female silkworm moth attracts males by emitting minute quantities of the chemical bombykol from glands at the tip of her abdomen. The male silkworm moth starts seaching for the females when they are immersed in as few as 14 000 molecules of bombykol per cm^3 of air. The male detects the molecules using some 10 000 distinctive sensory hairs on each of its two feathery antennae (Fig 18.2). Each hair has one or two receptor cells with neurones that lead inwards to nerve centres in the brain. Only a single molecule of bombykol is required to activate a receptor cell, and the cells respond to virtually nothing else other than bombykol. When about 200 cells in each antenna are activated per second, the male moth makes a chemotactic response, flying up the bombykol gradient centred on the tip of the female's abdomen.

Fig 18.2 The adult male silkworm moth showing its very large antennae.

Plant responses

Some plants show nastic movement. A **nasty** is a non-directional response of part of a plant in response to a stimulus, i.e. it involves movement not locomotion. The direction of the response is determined by the structure of the plant organ, not by the external stimulus. The rapid leaf movements of the Venus flytrap and mimosa plants in response to touch (see Section 17.1) are examples of nasties.

A **tropism** is a directional movement of part of a plant in response to a unidirectional stimulus. A tropism is caused by unequal growth of a plant organ so that the organ becomes positioned in a direction that is related to the direction of the stimulus. If growth is towards the stimulus, the movement is described as **positively orthotropic**; if directed away from the stimulus, as **negatively orthotropic**; if at an angle to the stimulus, as **plagiotropic**; if at a right angle to the stimulus, as **diatropic**. Plants show tropisms with respect to such stimuli as light, gravity, water, chemicals and touch. The control of two types of tropism, phototropism and geotropism, is described in Chapter 17.

QUESTIONS

18.1 Explain the difference between movement and locomotion.

18.2 Copy the table below and complete the second column by adding your own definition of each type of movement.

Type of movement	Definition	Example
kinesis		
nasty		
taxis		
tropism		

Add each of the following examples of movement to the appropriate row in the third column of your table.
(a) The Livingstone daisy (*Dorotheanthus* sp.) closes its flowers at night.
(b) Male mosquitoes fly towards the high-pitched whine of a female mosquito in flight.
(c) A free-living platyhelminth (*Polycelis nigra*) turns less frequently in shaded conditions than in brightly lit conditions.
(d) The pollen tube of a flowering plant grows towards a chemical secreted by the ovule and away from air.

Locomotion occurs among the 'adult' members of three kingdoms: Prokaryotae (bacteria), Protoctista and Animalia. Locomotion is achieved in all these kingdoms by one of a few methods: by pseudopodia (amoeboid movement), by undulipodia (or flagella in prokaryotic organisms) and by muscles. Although the gametes of some plants and fungi show locomotion, adult plants and fungi do not.

Locomotion using pseudopodia (amoeboid movement)

A **pseudopodium** is a temporary cytoplasmic projection from a cell. Pseudopodia are formed by members of the protozoan phylum Rhizopoda, e.g. *Amoeba*. The cytoplasm of these cells is differentiated to form two distinct regions

- an outer layer of **ectoplasm** – viscous cytoplasm called **plasmagel**
- **endoplasm** – fluid cytoplasm called **plasmasol**.

During amoeboid movement, endoplasm has been observed to flow in the direction of movement into one or more **pseudopodia**. At the tips of the pseudopodia, endoplasm is converted to ectoplasm, the process of **gelation** (Fig 18.3). At the trailing edge, ectoplasm is converted to endoplasm, the process of **solation**.

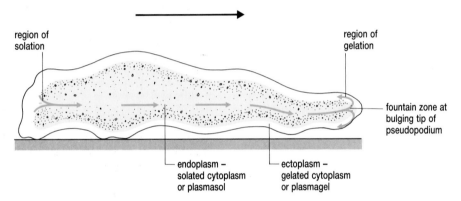

region of solation

region of gelation

fountain zone at bulging tip of pseudopodium

endoplasm – solated cytoplasm or plasmasol

ectoplasm – gelated cytoplasm or plasmagel

Fig 18.3 Changes in the cytoplasm of a cell moving by pseudopodia. At the bulging front tip of the pseupodium, endoplasm gelates to ectoplasm. At the trailing edge, in the region of solation, ectoplasm is converted to endoplasm.

The outstanding problem with amoeboid motion is to explain the origin of the motive force which causes the endoplasm to flow forwards. Is the endoplasm being pushed forwards by a contraction of the ectoplasm at the trailing edge of the animal, like squeezing toothpaste from a tube, or is the endoplasm being pulled forwards by some mechanism in the advancing pseudopodia?

It has recently been discovered that actin and myosin, the proteins involved in muscle contraction, are present in all eukaryotic cells. Actin is embedded within the surface membrane and myosin is free in cytoplasm. Perhaps amoeboid movement occurs with shearing movements between the actin and myosin in much the same way as occurs during muscle contraction (see Section 16.4).

 Which cells in your body show amoeboid movement?

Locomotion using undulipodia

Undulipodia are permanent membrane-covered extensions at the surface of eukaryotic cells. They are of two types, commonly called cilia and eukaryotic flagella. Both have a diameter of around 0.2 μm but most cilia are about 10 μm long whereas eukaryotic flagella are about 100 μm long. Cilia occur in groups, often covering the entire cell surface, whilst eukaryotic flagella are often single or in pairs. Both cilia and eukaryotic flagella arise from **basal bodies (kinetosomes)** which, in cilia, form an interconnecting network (Fig 18.4).

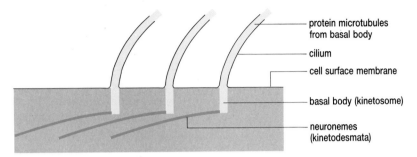

- protein microtubules from basal body
- cilium
- cell surface membrane
- basal body (kinetosome)
- neuronemes (kinetodesmata)

Fig 18.4   Cilia — a type of undulipodia, about 0.2 μm diameter and 10 μm long. At the base of each cilium is a kinetosome which is interconnected with others by neuronemes.

Many prokaryotic cells also have long, thin, movable extensions which, although called flagella, are not undulipodia. The major differences between prokaryotic flagella and eukaryotic undulipodia are summarised in Table 18.1.

Table 18.1   The major differences between prokaryotic flagella and eukaryotic undulipodia

Prokaryotic flagellum	Undulipodium
extracellular	intracellular, so covered by cell surface membrane
not composed of microtubules	composed of microtubules
9-fold symmetry absent	9-fold symmetry present
single globular protein, called flagellin	contain many different proteins, the most common of which is tubulin

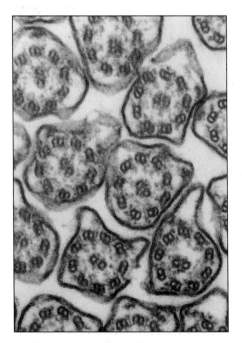

Fig 18.5   Electronmicrograph of a transverse section of undulipodia showing the characteristic 9 +2 arrangement of microtubules.

Cilia and eukaryotic flagella have an identical internal structure (Fig 18.5). Within the cell surface membrane is an **axoneme**, consisting of a pair of central filaments surrounded by nine pairs of peripheral filaments (the so-called **9 + 2 arrangement**). Each peripheral filament consists of two microtubules (called A and B) made of the protein **tubulin**. During movement, ATP is hydrolysed and the energy which it releases is used to slide the microtubules over each other. It is thought that this sliding occurs in five of the filaments on one side of the undulipodium, causing it to bend. The other four microtubules slide slightly later, causing bending in the opposite direction. The two central ones may be involved in transmitting impulses from the basal body.

Sliding of the microtubules produces different actions in cilia and eukaryotic flagella. A single eukaryotic flagellum produces a series of waves from its basal body to its tip (Fig 18.6(a)). Usually, the wave is in one plane but may be spiral, in which case the cell spins as it moves forwards, e.g. *Euglena*. The eukaryotic flagellum is normally at the back of the cell and pushes it forwards, rather like a tadpole's tail. In some cells it is at the front of the cell, e.g. *Euglena*, and works to pull the cell forwards.

A beat of a cilium has two distinct phases (Fig 18.6(b)), which are similar to the two phases of the breast stroke in human swimming. During the **effective stroke**, the cilium is held fairly rigidly and moved rapidly through about 180°. It is this stroke which causes movement. During the **recovery stroke**, the cilium is flexible and slowly returned to its starting position with as little drag as possible. The basal bodies of the numerous cilia covering a cell surface are connected by strands called **neuronemes**, or kinetodesmata (see Fig 18.4). These coordinate the cilia so that they beat in

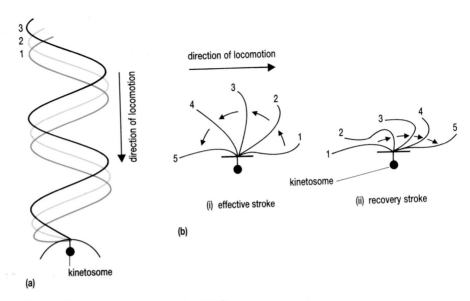

Fig 18.6 The sequence of movements of undulipodia.
(a) A single eukaryotic flagellum makes a series of waves from the kinetosome to the tip; these are sometimes spiral.
(b) Cilia beat in two phases. (i) Effective or power stroke with undulipodium moving from position 1, through 2, 3, 4 to finish at position 5. (ii) The recovery stroke then follows through positions 1 to 5.

a synchronised manner. Each cilium beats slightly before the cilium to one side and slightly later than the cilium to the other side, creating a **metachronal rhythm.**

The forces generated by eukaryotic flagella and cilia are weak. These organelles are generally used for locomotion only in small organisms, such as protozoa, or for single-celled gametes. Although capable of muscular swimming, many free-living flatworms normally move over surfaces using cilia. They are probably the largest organisms to use ciliary locomotion: their large surface area to volume ratio and secretion of mucus undoubtedly help them to do so. Cilia are commonly used to create currents over the surface of larger animals. The gills of mussels and the trachea and oviducts of mammals are lined by beating cilia.

Locomotion using muscles

The structure and physiology of striated muscle is described in Chapter 16. When the cells (fibres) within a muscle contract, the muscle itself shortens: the greater the number of cells which contract the greater the degree of shortening of the muscle. Muscles cannot expand to return to their normal position, however. To return to its resting position, a relaxed muscle must be pulled by the contraction of a second muscle. Thus, in mammals, muscles must always work in pairs. Since the muscles within a pair work against each other, one relaxing and being expanded as the other contracts and shortens, these pairs are called **antagonistic pairs**. If you bend your elbow, you have contracted a muscle (the biceps) and relaxed its antagonist (the triceps) (Fig 18.7). To straighten your arm again, you must relax the biceps and contract the triceps.

At each of its ends your biceps is connected to the outer membrane of bones in your skeleton (see Fig 18.7). These attachments are by **tendons**. One of these tendons, the **origin**, attaches to a part of your skeleton which does not move when you contract your biceps; the other tendon, the **insertion**, attaches to a movable bone. Contraction of your biceps pulls on the **radius** of your lower arm, decreasing the angle of your elbow (**flexion**). Contraction of the triceps pulls in the opposite direction on the **ulna** of your lower arm, increasing the angle of the elbow (**extension**). To allow these movements, one muscle must relax as the other contracts. You can demonstrate this by making an effort to contract both biceps and triceps at the same time.

4 **Why is it important that tendons are inelastic whilst ligaments, which join bones together at joints, are elastic?**

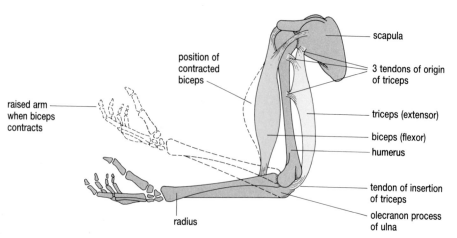

Fig 18.7 Muscular movement of the human forearm showing the antagonistic pair of muscles. When the triceps relaxes, the biceps contracts and the forearm is raised.

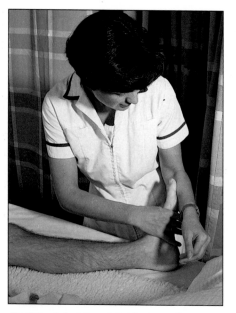

Fig 18.8 A physiotherapist helping to exercise a muscle, atrophied by inactivity caused by injury.

MUSCLES AND EXERCISE

Exercise makes muscles more efficient, larger and stronger. This enlargement is not the result of an increase in the number of muscle cells. Rather, each individual cell increases in diameter as the number of myofibrils per cell increases, so increasing the total force that a muscle can exert as it contracts.

A muscle that is not used **atrophies** or shrinks. This may occur because of a disease like poliomyelitis or multiple sclerosis where the nerves supplying the muscles are damaged, or an injury like a broken leg which means the limb cannot be used.

Locomotive forces

Locomotion is movement of the whole body from one place to another. Contraction of your biceps caused movement of part of your body; it did not move your whole body. To cause locomotion, muscle contraction must cause part of an organism's body to push against something which has resistance to movement. The effect of contraction of the blocks of muscles (**myotemes**) on the right side of a fish are shown in Fig 18.9(a). Movement of the tail exerts a force (thrust) against the water. The thrust of the tail against the water causes the water to exert a force in exactly the opposite direction, called the **reaction**. (Note that forces have a direction as well as a magnitude, i.e. they are **vectors**.) The reaction has two components, which are shown. The forward component pushes the fish forwards and the lateral component pushes the fish sideways. As the tail moves from side to side, the lateral components tend to cancel each other out so that the remaining forward component pushes the fish forwards.

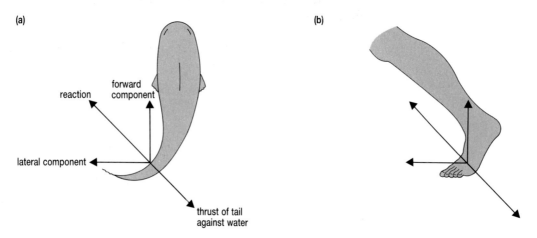

(a)

reaction

forward component

lateral component

thrust of tail against water

(b)

Fig 18.9   Forces acting on **(a)** a fish as it swims **(b)** a human foot during walking.

QUESTIONS

18.3 Explain the difference between plasmasol and plasmagel and explain how these two forms of cytoplasm allow locomotion in *Amoeba*.

18.4 How do cilia and eukaryotic flagella differ in structure and function?

18.5 Explain why muscles usually occur in antagonistic pairs.

18.6 Look at Fig 18.9(b). It shows a force diagram of a human foot during walking (similar to the locomotive forces acting on a fish as it swims). Describe Fig 18.9(b) in words.

18.3 ANIMAL SKELETONS

Aquatic organisms are supported by the water in which they live. Only the smallest animals are sufficiently well supported by water to need no further internal supporting system. A skeleton is a supporting system, which helps to give shape to an animal. More importantly, a skeleton provides the rigid framework against which muscles can work to cause locomotion. In some animals the skeleton also acts like an armour, giving protection against attacks by predators. Three major types of skeleton occur in the animal kingdom

- hydrostatic skeletons
- exoskeletons
- endoskeletons.

Hydrostatic skeletons

Because water is incompressible, it can form a rigid medium against which muscles can contract. Such hydrostatic skeletons are found only in soft-bodied animals. The body of an earthworm can be considered as a cylinder of watery fluid (coelomic fluid) surrounded by two layers of antagonistic muscles, the circular and longitudinal muscles.

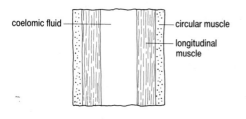

coelomic fluid — circular muscle
— longitudinal muscle

>
> **5**
>
> **What will happen to the shape of the segment in Fig 18.10 when**
> **(a) the longitudinal muscles contract and the circular ones relax**
> **(b) the circular muscles contract while the longitudinal ones relax?**

Fig 18.10   The circular and longitudinal muscles in one segment of an earthworm's body are antagonistic; while one contracts the other relaxes.

Short, fat segments push against the sides of an earthworm's burrow and act as an anchor. By contracting its circular muscles, an earthworm can elongate its body forwards from such an anchor point; by contracting its longitudinal muscles, an earthworm can pull its body forwards to such an anchor point. A simplified representation of such movement is shown in Fig 18.11. Normally, the earthworm's body is divided into several waves of such contraction.

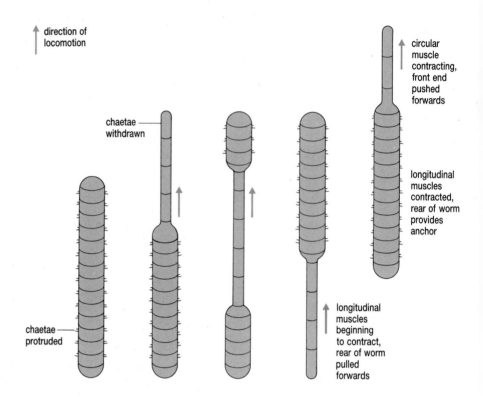

direction of locomotion

circular muscle contracting, front end pushed forwards

chaetae withdrawn

longitudinal muscles contracted, rear of worm provides anchor

chaetae protruded

longitudinal muscles beginning to contract, rear of worm pulled forwards

6 **Suggest a role for the chaetae.**

Exoskeletons

Exoskeletons are found only among the arthropods. The exoskeleton, or **cuticle**, has three layers which are secreted by the epidermal cells beneath it. The outermost layer is a waxy waterproofing layer, called the **epicuticle**. Beneath this are two layers containing the polysaccharide **chitin**. The outer of these layers, the **exocuticle** is made rigid by the presence of tanned proteins and, sometimes, by various salts. The inner **endocuticle** is untanned and remains flexible.

7 **Suggest some advantages and disadvantages of an exoskeleton.**

Movement of an arthropod's body is possible because the cuticle is segmented; movable joints occur between adjacent segments (Fig 18.13). Flexibility is possible because of the absence of the rigid exocuticle. Movement is brought about by one pair of antagonistic muscles at each joint. Like the human elbow joint, one muscle acts as a flexor and one as an extensor. The tendons of these muscles attach to projections on the inside of the cuticle, called **apodemes**. Although each joint can be moved in one plane only, successive joints in an arthropod's limb articulate at different angles, giving the whole limb great flexibility.

8 **What will happen to the joint in Fig 18.13 if muscle B contracts whilst muscle A relaxes?**

Fig 18.12 The exoskeleton can be extremely strong and difficult to break into, but then it is also relatively heavy.

SUPPORT SYSTEMS AND MOVEMENT

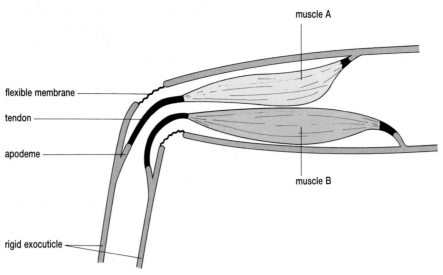

Fig 18.13 The joint of an arthropod. The ends of each flexor and extensor, the muscle pair at a joint, are attached to apodemes. A flexible membrane connects each segment of the cuticle and the rigid exocuticle is missing at the joint allowing movement.

Whilst providing support, protection against mechanical damage and protection against water loss, an exoskeleton restricts growth of the tissues inside it. Arthropods must periodically lose their exoskeleton, a process called moulting, or **ecdysis**. During ecdysis, a new soft cuticle forms beneath the old. When the old is shed, the arthropod uses air or water to expand its body before the new cuticle hardens. After hardening the arthropod reduces its body back to its normal size, leaving a gap within the cuticle into which it can grow. Before the new cuticle has hardened, an arthropod is unable to move or support its own weight. It is also vulnerable to predators and many arthropods hide away during this time. The energy cost of shedding and rebuilding the entire skeleton must place a considerable strain on arthropods. The hormonal control of ecdysis is described in Chapter 14.

Endoskeletons

Endoskeletons are internal, i.e. inside the soft tissues. They are found in radiolarians (protozoa), cephalopods (a class of the phylum Mollusca) and vertebrates (phylum Chordata).

 Suggest the advantages and disadvantages of endoskeletons.

Vertebrate skeletal tissue consists of living cells surrounded by a non-living matrix. During early embryological growth, the skeleton of vertebrates is laid down as cartilage. In one class, the Chondrichthyes, e.g. sharks and rays, cartilage forms the entire adult skeleton as well. In all other classes, the skeleton is converted into bone by a process called **ossification**, although some cartilage remains, principally at the joints.

QUESTIONS	
	18.7 Explain why animals need skeletons.
	18.8 Compare and contrast the structure and function of hydrostatic skeletons, exoskeletons and endoskeletons.

18.4 MAMMALIAN SKELETON

The bulk of the skeleton is formed of bone tissue. At the joints, cartilage protects the bone tissue from damage caused by friction as the bones move, and elastic **ligaments** hold bones together. **Tendons** hold muscles to individual bones. All these tissues consist of living cells which are surrounded by a non-living matrix which they secrete. Such tissues are often called **connective tissues**. Table 18.2 shows the types of cell and nature of the matrix in each of these tissues.

Table 18.2   A comparison of the main skeletal tissues

Tissue	Cell type	Matrix secreted by these cells
bone	osteoblasts	protein (osein)
		calcium salts (mainly phosphate)
cartilage	chondroblasts	protein (chondrin)
	fibroblasts	protein (collagen and elastin to different extents)
ligaments (yellow elastic connective tissue)	fibroblasts	protein (mainly elastin)
	mast cells	protein (mucin and chondrin)
tendons (white fibrous connective tissue)	fibroblasts	protein (mainly collagen)
	mast cells	protein (mucin and chondrin)

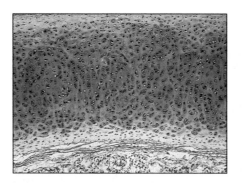

Fig 18.14   Light micrograph of hyaline cartilage, showing matrix and chondroblasts.

Skeletal tissues

Cartilage tissue is shown in Fig 18.14. The chondroblasts have become trapped in spaces (called **lacunae**) within the matrix: in this condition they are termed **chondrocytes**. **Compact bone** is a type of bone tissue which is found in long bones (Fig 18.15). The dark, spider-like objects are **osteoblasts**, the bone-secreting cells. They are trapped in lacunae within the hardened bone matrix but communicate one with another via the fine cytoplasmic extensions running within spaces in the bone matrix, called **canaliculi** (Fig 18.16). The osteoblasts are arranged in concentric rings, called **lamellae**, around a central hole, the **Haversian canal**, together forming a **Haversian system**. Blood vessels and lymph vessels run within the Haversian canals and capillaries branch from them through the canaliculi of the Haversian system they supply. The Haversian canals also contain nerves.

A second type of bone tissue, **spongy** or **cancellate bone**, also occurs within a long bone. This has less inorganic material in its matrix than compact bone and is arranged in a network of thin struts, called **trabeculae**. The space between these struts is filled with soft tissue, called **marrow**. Marrow which appears red is coloured by red blood cells: marrow which appears yellow is coloured by fat cells.

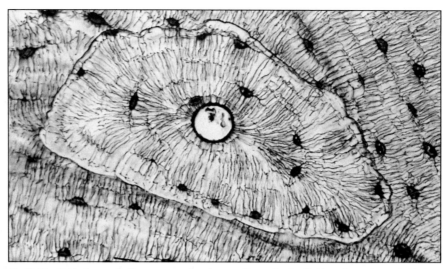

Fig 18.15 Light micrograph of transverse section of compact bone, showing the Haversian system, osteoblasts and canaliculi.

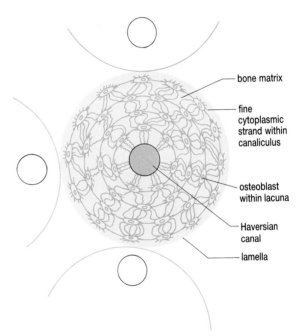

bone matrix

fine cytoplasmic strand within canaliculus

osteoblast within lacuna

Haversian canal

lamella

Fig 18.16 The structure of compact bone. The hard bone matrix contains osteoblasts (bone-secreting cells) arranged in the concentric circles of a Haversian system. Each Haversian canal contains blood vessels, lymph and nerves. Fine cytoplasmic strands within the canaliculi link together the osteoblasts.

 10 In addition to support and protection, give two other functions of long bones in mammals.

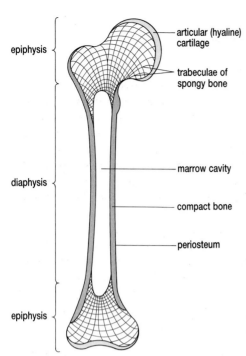

epiphysis

articular (hyaline) cartilage

trabeculae of spongy bone

diaphysis

marrow cavity

compact bone

periosteum

epiphysis

Fig 18.17 A vertical section through a long bone, such as the femur. The shaft contains soft marrow making it a light but strong structure. At the ends, hyaline cartilage protects the articulating surfaces and the trabeculae are arranged along the lines of stress.

Bone structure

A vertical section through a long bone, such as the femur is shown in Fig 18.17. Its shaft (**diaphysis**) is hollow, consisting of an outer cylinder of compact bone around a cavity containing soft marrow. This arrangement is strong but light (engineers use hollow steel cylinders in construction). Each end of the bone (**epiphysis**) is made from spongy bone surrounded by a thin layer of compact bone and hyaline cartilage. The trabeculae of the spongy bone are arranged along the lines of stress which the bone experiences: changes in the stress on a bone result in a new arrangement of trabeculae in line with these stresses. In this way, the strength of the bone is maximised. The epiphyseal cartilage protects the underlying bone from damage whilst being easily replaced itself. Muscle tendons attach to the membrane surrounding the entire bone (**periosteum**).

Joints (arthroses)

There are three types of joints (Fig 18.18)

- synarthroses – rigid joints, e.g. in the skull
- amphiarthroses – slight movable joints, e.g. intervertebral discs
- diarthroses – fully movable joints, e.g. elbow and hip.

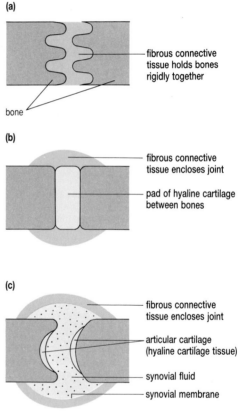

(a)

fibrous connective tissue holds bones rigidly together

bone

(b)

fibrous connective tissue encloses joint

pad of hyaline cartilage between bones

(c)

fibrous connective tissue encloses joint

articular cartilage (hyaline cartilage tissue)

synovial fluid

synovial membrane

Fig 18.18 Simplified structure of mammalian joints
(a) an immovable synarthrose
(b) a slightly movable amphiarthrose
(c) a freely movable diarthrose (synovial joint).

During embryological development, some developing bones fuse together, so that no movement is possible. Such joints are termed **synarthroses** and occur in rigid, protective parts of the skeleton, such as the sutures in the cranium, joints between the individual bones of the pelvic girdle and between the vertebrae in the sacrum.

SUPPORT SYSTEMS AND MOVEMENT

11 Why does the cranium have joints?

Like the cranium, the vertebrae which make up the backbone protect delicate nervous tissue. They must also allow movement, otherwise locomotion in most vertebrates would be severely restricted. Slightly movable joints, **amphiarthroses**, are present between the bodies of adjacent vertebrae. A pad of fibrous cartilage, **the intervertebral disc**, is held in place by a capsule of connective tissue which attaches to the periosteum of the vertebrae. Displacements of these discs, through poor posture, asymmetric backbones or injury, are known as slipped discs. They are painful because the displaced cartilage presses against nervous tissue of the spinal cord or spinal nerves.

Freely movable joints, **diarthroses**, are present between the transverse processes of adjacent vertebrae and between most other bones of the vertebrate skeleton. The fibrous capsule of these joints is lined by **synovial membrane** which secretes **synovial fluid**. This fluid lubricates movement of the joint so that the bones, lined by protective hyaline cartilage, do not rub directly against each other. Diarthroses are more commonly called synovial joints, because of the presence of this fluid.

12 Suggest the essential properties of materials used for artificial joints.

Plan of the skeleton

The size and shape of a mammal's bones are relatively constant within a species and are the result of natural selection (Section 23.3). In spite of their different ways of life, the arrangement of bones within the skeletons of different species of mammal is strikingly similar. The distinctive features of the human skeleton (Fig 18.19) are related to three aspects of human

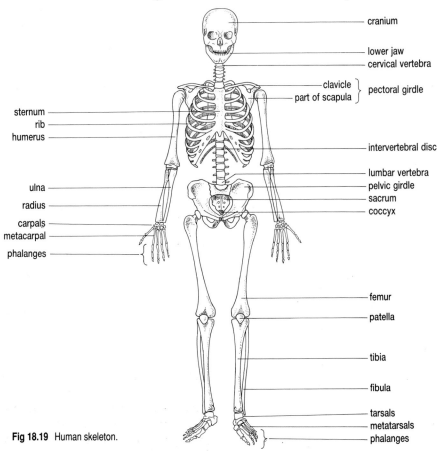

Fig 18.19 Human skeleton.

evolution. The first is our ancestry; the ancestors of humans, and all the other primates of 60 to 70 million years ago, were arboreal animals, rather like a modern lemur. Swinging from tree to tree requires great mobility of the shoulder, an ability to twist the forearm and to flex the digits of the hands. These movements also require precise coordination, as does the ability to judge distance when making grasping movements. The second important aspect of our evolution is our bipedalism, stimulating changes in our vertebral column, pelvic girdle and legs. The third is the increase in the size of our brains which accompanied our use of tools.

The skeleton performs a number of functions, summarised in Table 18.3. It is commonly considered in two parts: the axial skeleton and appendicular skeleton. The **axial skeleton** comprises the cranium and lower jaw, vertebral column and rib cage. It forms a central axis on which the body is arranged. The **appendicular skeleton** comprises the pectoral girdle, pelvic girdle and limbs.

Table 18.3 Functions of the mammalian skeleton

Function	Part of skeleton involved
support	All bones within the skeleton provide anchorage and give shape to the body.
protection	Cranium protects the brain; vertebral column protects the spinal cord; rib cage protects the heart and lungs; pelvic girdle protects the female reproductive organs.
movement	Muscles are attached to the periosteum of all bones. Muscle contraction causes movement of bones except in the synarthroses of the cranium, sacrum and pelvic girdle. Specialised movement includes mastication, breathing and transmission of sound waves through the middle ear.
production of blood cells	Erythrocytes are manufactured in the marrow of all the bones of young mammals and the flat bones (cranium, ribs, sternum, vertebrae and girdles) of adults. Leucocytes (other than T lymphocytes) are produced by the marrow of all mammals, including adults.
storage	The matrix of all bones contains calcium salts which can be released into the bloodstream. The process is controlled by parathormone.

QUESTIONS

18.9 Use Figs 18.17 and 18.19 to explain how the human femur is adapted to support the weight of the body and allow movement.

18.10 Distinguish between spongy and compact bone in terms of general appearance, location and function.

18.5 PLANT SKELETONS

Although its structure is stabilised by a cytoskeleton of protein tubules (see Section 3.3), cytoplasm has a semi-fluid consistency, like a jelly. Just as a large jelly is not strong enough to maintain its shape, a large group of cells cannot maintain its shape without further support. Hence the advantage of skeletons.

 How will a plant cell function as a hydrostatic skeleton?

The mechanical strength of turgid cells is sufficient to support small plants, such as mosses. Turgid cells, however, are not strong enough to support larger plants. Terrestrial plants gain support in a similar way to animals, by the production of specialised supporting tissues which are placed at strategic points around the plant.

 How can some large aquatic plants survive without specialised supporting tissues?

Supporting tissue in plants

The structure of a number of plant supporting tissues is summarised in Fig 18.20. **Parenchyma** is one of the least structurally specialised plant tissues. The cells are roughly spherical and have a thin wall of cellulose (Fig 18.20(a)). Parenchyma is often called packing tissue, because it seems to fill the spaces between more specialised cells. When turgid, parenchyma cells become tightly packed and rigid. Turgid parenchyma cells provide the main support in herbaceous (non-woody) stems.

 What will happen to a non-woody plant when its parenchyma cells lose their turgidity, e.g. on a hot, dry day?

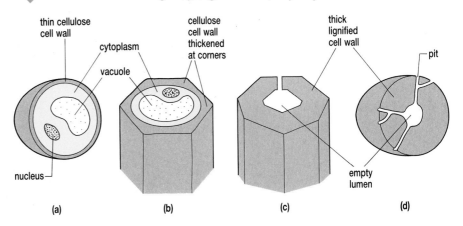

Fig 18.20 The structure of a number of plant supporting tissues
(a) parenchyma (b) collenchyma (c) sclerenchyma fibre (d) sclereid.

Specialised supporting tissues contain cells in which the cell walls have been thickened by the addition of extra cellulose or a stronger substance. For example, the walls of **collenchyma cells** are strengthened by the deposition of extra cellulose, particularly at the corners of the cells (Fig 18.20(b)). Being alive, these cells can grow as the organ around them grows. Collenchyma is, therefore, an important supporting tissue in growing stems and leaves and is often found around the outside of a plant organ, just under the epidermis.

There are two types of **sclerenchyma** cell:

- **fibres** (Fig 18.20(c))
- **sclereids** (Fig 18.20(d)).

After depositing a primary wall of cellulose, the living protoplasts of both types of cell secrete **lignin**, a substance which is much stronger than cellulose. In mature cells, the protoplasts die, leaving an empty cell wall with small holes, **pits**, through which groups of plasmodesmata had run from the once-living cells. Fibres are usually collected into strands or sheets which enhance the strength of individual cells. Unlike fibres, sclereids are often found singly or in small groups; the grittiness of the fruit of pears is caused by such groups. More typically, they form the tough shells of nuts and stones of some fruits.

Xylem tissue is the main supporting tissue in woody stems in which the majority of tissue is xylem. Xylem contains tracheids and vessels, both of which are dead and have lignified walls. The structure of xylem vessels and tracheids is described in some detail in Chapter 12.

Distribution of skeletal tissues in plants

Plant skeletal tissues do not need to be arranged in a way which allows movement. They are distributed in a plant in such a way that they confer maximum support to each organ. Collenchyma and sclerenchyma fibres are often associated with the transport tissues in vasular bundles. These are arranged in a central stele in roots, where they exert a force which counteracts the pull of the shoots as they are blown from side to side; in bundles around the periphery of stems and leaf petioles, where they resist the compression and extension as these organs bend; in the branching veins of leaves, where they support the delicate lamina.

Collenchyma and sclerenchyma fibres are often found in a layer beneath the epidermis of plant organs where they form a hollow cylinder of supporting tissue. The distribution of plant supporting tissues in the root and stem of an annual dicotyledonous plant is summarised in Fig 18.21. Perennial plants grow from year to year. The stems and roots of these plants increase in diameter so that they support the additional weight of new tissue. Such **secondary thickening** is described in Section 21.5.

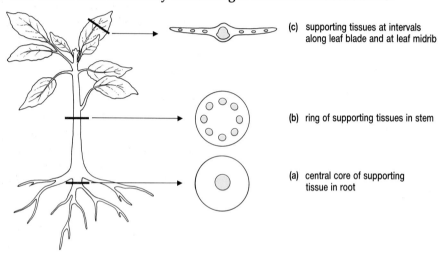

Fig 18.21 The distribution of supporting tissues in (a) the root (b) the stem and (c) the leaf of a generalised dicotyledonous plant. The supporting tissues are shown in blue.

(c) supporting tissues at intervals along leaf blade and at leaf midrib

(b) ring of supporting tissues in stem

(a) central core of supporting tissue in root

QUESTIONS

18.11 Suggest why the stems of herbaceous plants collapse in dry conditions whilst those of woody plants do not.

18.12 Locate the supporting tissues in Fig 18.21 and relate their position to the forces acting on each plant organ.

SUMMARY

Movement is an energy-consuming process in which parts of an organism's body change their relative position. Locomotion occurs when the whole body changes position.

The movement of organisms is usually a reaction to environmental stimuli. Non-directional responses are those which are not related to the direction from which the stimulus comes, only to its intensity. Non-directional movement of part of the body is called a nastic response: a non-directional change in the rate and direction of locomotion is called a kinesis.

SUPPORT SYSTEMS AND MOVEMENT

Locomotion towards or away from a stimulus is called a taxis. The differential growth of plants in response to a unidirectional stimulus is called a tropism.

Single cells are able to move using pseudopodia or undulipodia (flagella and cilia). During amoeboid movement, a flow of cytoplasm occurs from the posterior end of the cell into one or more advancing pseudopodia. The flow of cytoplasm is maintained by solation of rigid ectoplasm at the rear of the cell and gelation of liquid endoplasm at the front of the pseudopodium. Undulipodia arise from a basal body (kinetosome) and have a pair of internal protein filaments surrounded by nine pairs of peripheral filaments (9 + 2 arrangement). Sliding of the filaments over each other causes the undulipodium to bend. Flagella are usually long and occur in small numbers. Waves of bending along their length pull or push their cell along. Cilia are short and numerous. Their activities are integrated by a network connecting their kinetosomes, so that cilia beat in a coordinated metachronal rhythm. Each cilium is rigid during its effective or power stroke but flexible during its recovery stroke.

Only small organisms can use cilia for locomotion: large animals rely on muscles. Muscles work in antagonistic pairs; when one contracts the other relaxes and vice versa. Locomotion occurs because muscle contractions push part of the body against air, water or a solid surface. The resistance to this thrust causes a force in the opposite direction, pushing the animal forwards.

Muscles are useful only if they work against an incompressible framework, i.e. a skeleton. Fluid (environmental water, blood or coelomic fluid) within soft-bodied animals serves as a hydrostatic skeleton. Arthropods have a rigid exoskeleton to which the muscles attach internally. As well as providing a rigid framework for locomotion, an exoskeleton gives protection against mechanical damage: it does, however, restrict growth except during moults.

Chordates have an internal skeleton (endoskeleton). Chondrichthyes have a skeleton entirely made of cartilage. Other chordates have a skeleton in which most of the cartilage changes to bone during development. In mammals, spongy bone is filled with blood cavities; compact bone has fewer blood spaces and is much stronger. Compact bone is found around the periphery of the shaft (diaphysis) of long bones where it resists compression and tension; spongy bone is found at the ends (epiphysis) of long bones where it resists stress.

The main axis of the mammalian skeleton is the axial skeleton. It consists of skull, vertebral column and rib cage. The pectoral girdle, pelvic girdle and limbs form the appendicular skeleton. Mammals have a pentadactyl limb (five digits) which has been modified during evolution to perform a number of different tasks. Where individual bones meet, a joint is formed. Rigid joints are found in protective structures such as the skull. The cartilage of intervertebral discs protects the bone surfaces against friction during slight movement; the cartilage and fluid of synovial joints enables movement that is restricted only by the shape of the bones.

Although they do not show locomotion, plants have skeletal tissues which support their mass. In non-woody plants, turgid parenchyma cells provide most of the skeletal support. Specialised supporting cells have walls which are thickened by cellulose, e.g. collenchyma and phloem, and lignin, e.g. sclerenchyma and xylem. Specialised supporting tissues are arranged in such a way that they provide maximum support for the plant's organs.

They form a central stele in roots, a peripheral cylinder in stems and a branched network in leaves. Plants which live for many years increase the amount of supporting tissue during secondary thickening.

Answers to Quick Questions: Chapter 18

1 Water loss is less rapid in humid air than in dry air, so the behaviour will reduce the rate of dehydration.
2 Negative chemotaxis.
3 Monocytes and all classes of granulocyte.
4 The elasticity of ligaments allows bones to move relative to each other at joints. However, if tendons were elastic some of the mechanical force developed by muscle contraction would be used to stretch the tendon rather than moving the bone.
5 In (a) the segment becomes shorter and thicker, in (b) it becomes longer and thinner.
6 Anchor the segment against the substratum.
7 Advantages include protection of the body against water loss, friction and other mechanical damage. Disadvantages include its weight and its restriction of growth.
8 The joint will flex.
9 Advantages include lack of restriction of growth and lightness. Disadvantages include lack of protection for body organs.
10 (i) Production of blood cells – erythrocytes, granulocytes and monocytes. (ii) Store of calcium (needed for muscle contraction and blood clotting).
11 The cranium forms as individual bones in the fetus which then grow together and fuse.
12 Light, rigid, durable, inert.
13 Its vacuole contains an incompressible aqueous solution.
14 Density of water provides support, especially if surface area to volume ratio of plant is large.
15 The cells will become flaccid and the plant will droop (wilting).

4.1 When an oat grain germinates, its shoot is enclosed in a sheath of tissue known as coleoptile. Lengths of coleoptile floating in sucrose solution will continue to grow for some time. Different concentrations of auxin solution added to the sucrose will stimulate growth of the coleoptile to a greater or lesser extent.

(a) Given a solution containing 100 parts per million (ppm) of auxin, describe how you would prepare a range of auxin solutions, each 1/100 of the concentration of the previous one. (2)

(b) How would you determine the effect of each of these solutions on the growth of the coleoptiles? (3)

(c) What control would it be necessary to set up and why? (2)

(d) Draw two graph axes and label them to show how you present the result of this investigation. (1)

Oak apples are plant galls. They are the swollen buds, 2–3 cm in diameter, that result from the feeding activity of a small wasp inside the bud.

(e) Outline how you would determine the approximate concentration of auxin in oak apples using the results you would have obtained from (a) to (d) above. (3)

(AEB 1990)

4.2 The diagram below shows part of two nerve fibres and a synapse. The figures indicate the value in mV of the potential across the membrane between the cytoplasm of the fibres and the extracellular fluid at intervals along each fibre.

(a) (i) Draw a circle round *one* region of the diagram where an action potential exists. Explain your choice. (2)

(ii) By means of an arrow on the diagram, indicate the direction in which action potentials would normally travel along these fibres. Explain your choice of direction. (2)

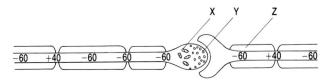

(b) Identify structures X and Y, and state how each is involved in the transmission of nerve impulses. X (2) Y (2)

(c) (i) What is the major chemical constituent of structure Z? (1)

(ii) State *two* effects of structure Z on the transmission of action potentials. (2)

(ULSEB 1989)

4.3 (a) What are the principle effects of **each** of the following plant hormones on plant tissue?
(i) auxin
(ii) gibberellin
(iii) cytokinin
(iv) abscisic acid. (10)

(b) Explain how the balance between **two** or **more** of the hormones listed in (a) controls
(i) seed dormancy and germination,
(ii) leaf senescence and abscission. (5)

(c) What is a bioassay? Describe how the auxin concentration of a solution could be determined by bioassay. (5)

(JMB 1987)

4.4 Explain each of the following events as precisely as you can.

(a) A puppy salivates at the sight of a tin opener. (4)

(b) Shining a torch into one eye causes both pupils to be constricted. (4)

(c) Green tomatoes ripen after a few days when enclosed in a polythene bag with a ripe banana and kept in a warm place. (4)

(d) Blood coagulates following a cut in the skin. (4)

(e) Tissue is rejected following a transplant in a mammal. (4)

(JMB 1988)

4.5 The diagram shows the fine structure of a junction between a nerve ending and a striated muscle fibre as seen in section.

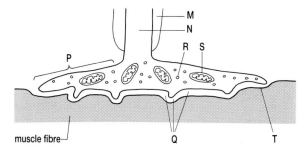

(a) Name the parts labelled **M, N, P, Q, R** and **T**. (3)

(b) (i) Describe the roles of **M, R** and **S** in the nerve impulse transmission.

(ii) Suggest how the role of **S** could be verified. (5)

(c) Before stimulation, protein filaments in part of the muscle fibre appeared as follows:

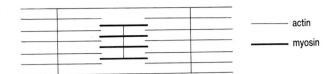
— actin
— myosin

(i) Using the same scale draw an **accurate** diagram showing the appearance of the same fibre after it has been stimulated sufficiently to contract.

(ii) Briefly explain what has happened to cause the effect shown in your diagram. (4)

(d) The following traces were obtained from an isolated leg muscle of a frog when subjected to a series of stimulations via its nerve.

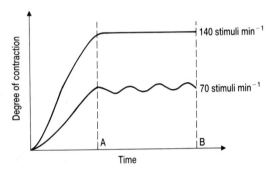

140 stimuli min^{-1}

70 stimuli min^{-1}

Degree of contraction

Time

A   B

(i) Explain the difference in shape of the two curves between A and B.

(ii) State two experimental conditions under which this nerve-muscle preparation should be kept while the experiment was proceeding. (4)

(WJEC 1987)

4.6 **(a)** Describe the structure of each of the following tissues, indicating in each case how structure is related to function.
(i) Parenchyma
(ii) Collenchyma
(iii) Sclerenchyma (12)

(b) Compare the distribution of tissues in a dicotyledonous stem and root in relation to the mechanical functions of the stem and root. (8)

(ULSEB 1991)

4.7 **(a)** State **three** functions of cartilage in a fully formed mammalian skeleton, giving **one** example illustrating each function. (3)

(b) The drawings (not to the same scale) show the skeletons of a human arm and a bird wing. Both these left fore-limbs have the same basic structure, although each is adapted to its particular function.

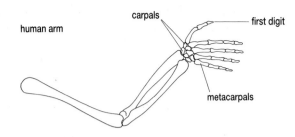

human arm

carpals

first digit

metacarpals

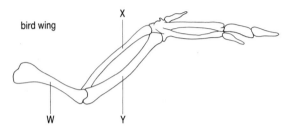

bird wing

X

W   Y

(i) Using your knowledge of the structure of the human arm, suggest the names of the bones labelled W, X and Y in the bird wing. (3)

(ii) Describe **three** ways, visible in the drawings, in which the carpals, metacarpals and digits of the bird wing differ from those of the human arm. (3)

(c) The diagram represents a vertical section through part of the leg of an insect.
(i) Name the substance which forms the exoskeleton.
(ii) Draw and label on the diagram
1. a flexor muscle,
2. an extensor muscle. (3)

(d) Briefly explain how an insect walks. (3)

(e) The hollow tubular exoskeleton, made from a substance less dense than bone, is very efficient for small animals. Suggest **two** reasons why it would be less efficient in a large animal such as a bird. (2)

(WJEC 1988)

4.8 **(a)** Explain what is meant by the term 'negative feedback'? (4)

(b) Describe the effects of the following mammalian hormones.
(i) Insulin
(ii) Adrenaline. (8)

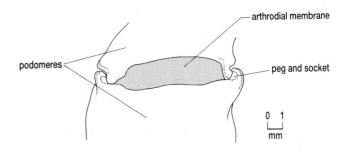

(AEB 1986)

(c) Describe how hormones from the ovary and pituitary gland interact to control the human menstrual cycle. (8)

(ULSEB 1990)

4.9 (a) Give an illustrated account of the supporting tissue in
    (i) the stem of a herbaceous flowering plant (6)
    (ii) a mammalian limb. (6)
(b) Compare the ways in which these tissues provide support in these structures and the amount of such tissues in relation to their function. (8)

(WJEC 1990)

4.10 Plants and animals respond to changes in their environments through systems of receptors and effectors. (3)
(a) Compare the ways in which angiosperm and mammalian receptors are linked to effectors.
(b) The relative daylength (the photoperiod) may influence the activities of flowering plants, particularly the flowering process itself.
    (i) What part of the plant is involved in the perception of the photoperiod? What is the evidence for this?
    (ii) Different species of flowering plant respond differently to changes in the photoperiod. How are these different species categorised and what is the stimulus which leads to flowering in each category? How does this influence the time of flowering? (9)
(c) Describe the role of the following structures in the formation and/or detection of sharp images in the mammalian eye
    (i) cornea,
    (ii) lens and ciliary body,
    (iii) retina,
    (iv) choroid. (8)

(NISEC 1988)

4.11 (a) Name an animal with a hydrostatic skeleton and explain how its skeleton helps its locomotion. (6)
(b) Give **four** differences between arthropod and mammal skeletons. (4)
(c) The diagram shows the surface of an arthropod limb joint.
    (i) Explain how this joint allows movement in one plane. (4)
    (ii) Compare and contrast structure and movement in this joint and in a human finger joint. (6)

4.12 (a) The diagram shows interaction between the hypothalamus, pituitary and thyroid glands. Arrows indicate probable pathways of direct influence.

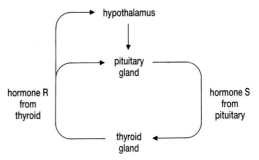

    (i) Give the names of the hormones **R** and **S**.
    (ii) State the effect of hormone **S** on the thyroid.
    (iii) State the effect of hormone **R** on the hypothalamus.
    (iv) Describe the control of the pituitary by the hypothalamus in this situation.
    (v) What is this kind of control of the pituitary by the thyroid called? (8)
(b) Read the following passage and then answer the questions below. Water soluble hormones and steroid hormones both alter cell function by activation of genes but by somewhat different mechanisms. Since steroid hormones are lipid-soluble, they easily pass through the plasma membrane of the target cell. Upon entering the cell, a steroid hormone binds to a protein receptor site in the cytoplasm and the hormonal-receptor complex is translocated into the nucleus of the cell. The complex interacts with specific genes of the nucleus DNA and activates them to form the enzymes necessary to alter cell function in a specific way.
    (i) Explain what is meant by a target cell.
    (ii) Why should lipid-soluble hormones pass through plasma membranes easily?
    (iii) Suggest a mechanism and a pathway by which the hormonal-receptor complex may be translocated.
    (iv) Describe a way in which the complex may alter cell function. (9)
(c) Explain clearly the part played by hormones

in regulating the concentration of calcium in the blood. (8)

4.13 This question is about a veterinary investigation of a sick cat, thought to be suffering from a malfunction of the thyroid gland.

(a) Give **one** symptom that might have led the vet to suspect a thyroid hormone deficiency. (1)

A sample of tissue was taken from the thyroid of the sick cat and a stained section prepared. This was compared with a similar section from a healthy cat. Drawings of parts of these sections are shown below.

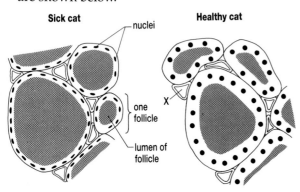

(b) Suggest a function for the lumen of the follicle. (1)

(c) (i) Name the tubular structure labelled **X**.
    (ii) Give **one** reason why there should be many such tubular structures (**X**) in thyroid tissue. (2)

(d) Describe **two** ways, apparent from the drawings, in which the thyroid of the sick cat differed from that of the healthy cat. (2)

(e) The investigator thought that the cat might be suffering as a result of a lack of iodine in its diet. Why should this element be needed for healthy thyroid function? (1)

The investigator arranged for both cats to receive radioactive iodine compounds and then measured the levels of radioactivity in their blood. These levels were never high enough to cause harmful effects. The results are shown in the graph.

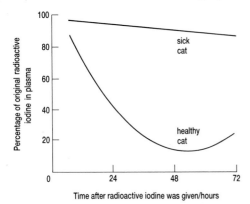

(f) (i) Do these results support the hypothesis that a dietary deficiency of iodine was the cause of the illness? Explain your answer.
    (ii) Suggest why the curve for the healthy cat rose after about 48 hours.
    (iii) Suggest a treatment which might enable the sick cat to recover. (4)

(JMB 1990)

4.14 Birds are thought to have evolved from prehistoric reptiles. The diagrams compare three structures in prehistoric reptiles and present-day birds.

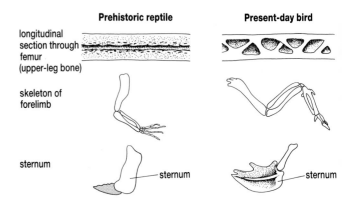

For each structure, describe how it might have been modified in the course of evolution. In each case, suggest how the change from the condition in the prehistoric reptile to that in the bird is an adaptation to flight.

(a) Bone (2)
(b) Forelimb (2)
(c) Sternum (2)

(AEB 1990)

4.15 (a) What is meant in plant biology by the terms:
    (i) *tactic movements;*
    (ii) *nastic movements;*
    (iii) *tropic movements*? (3, 3, 6)

(b) Explain the mechanisms involved that result in the shoot of a potted plant on a window sill growing to face the light. (10)

(c) What do you understand by 'short day plants' (SDP) and 'long day plants' (LDP)? How is flowering controlled in such plants? (8)

(UODLE 1987)

4.16 (a) Make a fully-labelled diagram to show the principal muscles and bones involved in movement of the fore-limb of a **named** mammal. (8)

(b) (i) The most widely accepted theory of muscle contraction is termed '*the sliding filament theory*'. Describe, with special reference to the micro-anatomy of striated muscle, how this theory helps to explain muscle contraction.

(ii) In order to initiate contraction, a nerve impulse must arrive at the muscle. Explain how potassium and sodium ions are important in this process. (16, 6)

(UODLE 1987)

Answer Guidelines for Question 4.1

This question tests the ability to design experiments. By splitting the question into five parts, the examiner has helped us structure our answer into the five stages of the experiment.

(a) Pipette 1 cm^3 of auxin into 99 cm^3 of water; mix thoroughly by shaking; repeat with each new solution to obtain required range; *(maximum 2 marks)*

(b) Accurately measure the cut pieces of coleoptile; use ten or more pieces per sample; put each sample into sucrose solution with one of the auxin solutions added and leave for specified time; ensure conditions (e.g. temp and darkness) same for all samples; measure final length of each piece. *(maximum 3 marks)*

(c) put sample into sucrose with no added auxin solution; comparison will show that sucrose itself has no stimulatory effect;

(d) percentage change in length on *y*-axis and concentration of auxin in ppm on *x*-axis;

(e) grind measured mass of oak apples; in specified volume of water; repeat exactly the experiment given in **(b)** above; compare percentage change in length with graph from **(d)** above. *(maximum 3 marks)*

Theme 5

REPRODUCTION, GROWTH AND DEVELOPMENT

Organisms do not live for ever. Some trees are known to be thousands of years old; the oldest authenticated age lived by any mammal was 122 years 237 days by a human; many insects live only for a few months: eventually all organisms die.

Life on Earth continues because organisms reproduce before they die. However, few organisms can reproduce when they are first formed. They need time to get bigger and to accumulate the energy stores needed for reproduction. Most also need to develop the biological structures by which reproduction is eventually achieved.

We will examine these processes in the chapters of this theme:

- *Reproductive strategies* – a comparison of asexual and sexual reproduction.
- *Sexual reproduction in mammals and in flowering plants.*
- *Patterns of growth and development.*

(below left) The plantlets of this plant eventually fall from its leaves and grow independently.

(below right) A cloud of pollen grains: an external sign of the reproductive cycle of this catkin.

Chapter 19

REPRODUCTIVE STRATEGIES

LEARNING OBJECTIVES

When you have studied this chapter you should be able to:

1. distinguish between asexual reproduction and sexual reproduction and compare their products;

2. give examples of asexual reproduction by fission, budding, sporulation, fragmentation and parthenogenesis;

3. outline a variety of methods of vegetative propagation in flowering plants;

4. outline gametogenesis in mammals and flowering plants and explain the terms isogamy, anisogamy and oogamy;

5. compare internal and external fertilisation;

6. explain the differences between inbreeding and outbreeding and describe mechanisms which promote outbreeding.

Even if they are not killed by predators, parasites or environmental hazards, organisms do not live indefinitely. Instead, they seem to have a programmed life expectancy which varies from species to species. After a certain time, organisms function less efficiently than is needed to keep them alive; they age (the process of **senescence**) and die. The world continues to be inhabited by living organisms only because they reproduce before they die. Unlike biological processes such as feeding or locomotion, reproduction is of no apparent survival value to an organism: it will still die. Whatever its biological value, organisms expend a great deal of energy during reproduction; in fact, some organisms expend so much energy that they die during the process of reproduction itself!

Fig 19.1 This animal can live for well over 100 years.

19.1 REPRODUCTIVE STRATEGIES: AN OVERVIEW

Although there seems to be a bewildering array of reproductive structures and behaviours, organisms reproduce in one of two ways.

- **Asexual reproduction** is the production of new individuals by division of the genetic material and cytoplasm of a single parent cell.
- **Sexual reproduction** is the production of new individuals by the fusion of two nuclei which are usually contained in specialised sex cells (**gametes**). The cell resulting from this fusion is called a **zygote**.

 Why will asexual reproduction be a particularly advantageous reproductive strategy for a solitary organism colonising a new environment?

ANALYSIS

Reproductive myths

Asexual reproduction differs from sexual reproduction because it does not involve any fusion of nuclei. It is often said that other differences exist between these two types of reproduction, making one more advantageous than the other under certain environmental conditions. Some of these comparisons are worth questioning.

This exercise helps you to structure questions in challenging commonly held views.

Myth 1   Gametes are always produced by meiosis.

The amount of genetic material (DNA) in a zygote is double that of the gametes, i.e.

$$\text{gamete 1} \; + \; \text{gamete 2} \; = \; \text{zygote}$$
$$n \qquad\qquad n \qquad\qquad 2n$$

(a) Why must meiosis occur at some stage in the life cycle of a sexually reproducing organism?

The life cycles of two sexually reproducing organisms, a mammal and a flowering plant, are shown in Fig 19.2.

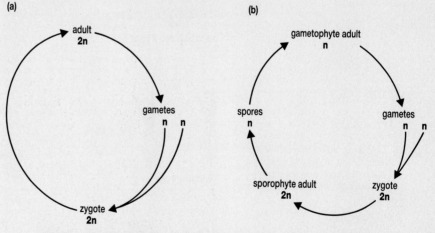

Fig 19.2   Life cycles of sexually reproducing organisms **(a)** a mammal and **(b)** a flowering plant.

(b) Where does meiosis occur in each of the life cycles?
(c) Are the plant gametes produced by mitosis or meiosis?
(d) Is it true that gametes are always produced by meiosis?

Myth 2 Asexual reproduction always involves mitosis.

(e) What is the fundamental difference between meiosis and mitosis?

Look again at the plant's life cycle in Fig 19.2(b).
(f) What form of cell division is involved in producing the adult sporophyte from the zygote? Is this sexual or asexual reproduction?
(g) What form of cell division is involved in producing the spores from the sporophyte? Is this asexual or sexual reproduction?
(h) Does asexual reproduction always involve mitosis?

Myth 3 Organisms produced by asexual reproduction are genetically identical.

Mitosis involves copying the genetic information contained in one nucleus into two daughter nuclei. Unless a mutation occurs, the products of mitosis will be genetically identical; such groups of genetically identical offspring form a **clone**.
(i) Will the haploid spores produced by the sporophyte in Fig 19.2(b) be genetically identical?
(j) Can asexual reproduction therefore produce genetically different offspring?

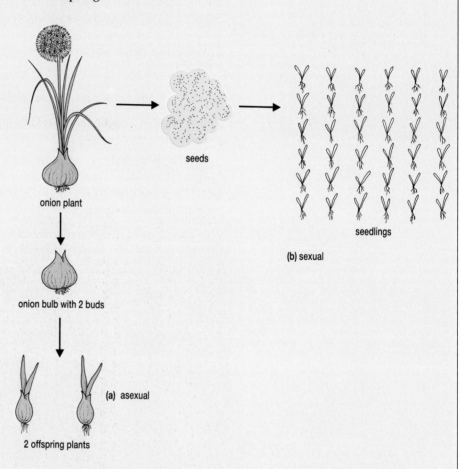

seeds

onion plant

seedlings

(b) sexual

onion bulb with 2 buds

(a) asexual

Fig 19.3 Onions can reproduce asexually and sexually. In a two-year life cycle, an onion may produce **(a)** two offspring asexually and **(b)** hundreds of offspring sexually.

2 offspring plants

Myth 4 Asexual reproduction **always** results in a more rapid increase in population size than sexual reproduction.

Certainly this would be true of a bacterium under ideal conditions, dividing every 20 minutes or so by binary fission. However, when conditions are not ideal the rate of fission will be less than this.

(k) Assume that a single bacterium divides once a month. How many bacteria will there be at the end of one year?

(l) An oyster can produce over a million eggs in one year. Will the bacterium or the oyster produce more offspring?

(m) Onions can reproduce asexually by producing new bulbs and sexually by producing seeds (Fig 19.3). Which method will produce more offspring per plant in one year?

(n) What assumption are you making about the survival of the seeds?

QUESTIONS

19.1 Identify the major difference between sexual and asexual reproduction.

19.2 In some species of bacteria, individual cells join together by a conjugation tube and transfer part of their circular genome to the other cell involved in the conjugation. The recipient cell incorporates this new DNA into its own genome. Is this sexual reproduction? Explain your answer.

19.2 ASEXUAL REPRODUCTION

During asexual reproduction, offspring are produced by a single parent without the need for gamete fusion. The five major methods of asexual reproduction are

- fission
- sporulation
- budding
- regeneration and fragmentation
- vegetative propagation.

Fission

This method occurs only in prokaryotes and in protoctists (e.g. *Amoeba*). It involves replication of the cell's DNA, division of the nucleoplasm (prokaryotes) or nucleus (protoctists) and division of the cytoplasm. Typically, cells divide into two (see Fig 19.4), a process called **binary**

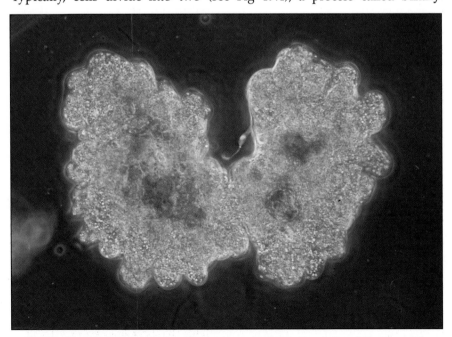

Fig 19.4 Binary fission in *Amoeba*. In most prokaryotes and protoctists, the cytoplasm divides transversely.

REPRODUCTIVE STRATEGIES

Fig 19.5   A female *Anopheles* taking a meal of blood.

fission. Certain stages in the life cycle of *Plasmodium vivax*, the species of protozoan which causes benign tertian malaria in humans, undergo **multiple fission**. In this process, repeated division of the nucleus is eventually followed by division of the cytoplasm to produce many new cells. Malaria is transmitted by female mosquitoes of the genus *Anopheles* when they bite a human to obtain a blood meal. Cells of *P. vivax*, known as **sporozoites**, are injected into human blood in the saliva of an infected female mosquito (Fig 19.6). After about half an hour, they develop into spherical **schizonts** inside cells in the human host's liver. Here multiple fission (**schizogony**) occurs resulting in up to 1000 offspring. These offspring, called **merozoites**, are released into the bloodstream and infect the host's erythrocytes, where they undergo further schizogony to produce up to 24 further merozoites each. Within a few weeks of infection by a single sporozoite, several million merozoites will be present in the host's bloodstream. The cyclic release of merozoites coincides with repeated bouts of fever, the sweating which results attracting the female mosquitos to the infected person.

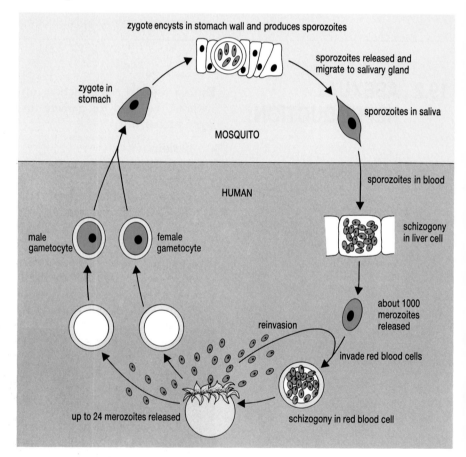

Fig 19.6   Simplified life cycle of the malarial parasite *Plasmodium vivax* to show the occurrence of schizogony.

2   **What is the advantage of schizogony to *Plasmodium vivax*?**

Sporulation

This involves the production of asexual **spores** which are unicellular structures produced by mitosis or meiosis in a parent cell. They are often dormant and have a protective outer coat which enables them to resist adverse environmental conditions. Typically, asexual spores are small and light, with minimal food stores. Once released, they are easily carried by currents of air, allowing rapid dispersal to new habitats. However, asexual

REPRODUCTIVE STRATEGIES

KEY CONCEPT

Pollen grains are spores not gametes. Whereas the microspores are dispersed, the megaspores are retained within the body of the adult sporophyte.

spores may be active and motile. For example, some protoctistan spores, called **zoospores**, have undulipodia with which they swim, while the spores of one phylum of protozoa, the sporozoans, are naked cells which move by amoeboid movement.

Asexual spores occur most commonly in fungi and plants, where they are produced in structures called **sporangia** (see Chapter 26). A single sporangium usually produces large numbers of spores. In plants, spores are produced by meiosis in the sporangia of the sporophyte adult. In flowering plants (angiosperms), two types of sporangia occur (see Section 20.5). Microsporangia (pollen sacs) produce microspores (pollen grains) and megasporangia (ovules) produce megaspores (embryo sacs).

As well as producing spores asexually, some fungi produce spores as a result of sexual reproduction (see Chapter 26). Such **zygospores** are larger than asexual spores and contain food stores for the subsequent growth of the embryo.

 3 What are the two main functions of asexual spores?

SPOTLIGHT

ANTHRAX: A PERSISTENT KILLER

Spores provide a mechanism which enables unicellular organisms to survive hostile environmental conditions. For example, some bacteria, particularly in the genus *Bacillus*, are able to enclose the contents of their cells within structures called endospores (**Note: These are not reproductive structures**). The walls of such spores can be resistant to prolonged boiling and exposure to disinfectants. Destruction of such bacterial endospores on, say, surgical instruments, is achieved by autoclaving them at a temperature of 121 °C for up to one hour. However, destroying endospores in the environment is much more of a problem. For example, as a result of biological warfare experiments during World War II, the soil of the Scottish island of Gruinard became contaminated with *Bacillus anthracis*, the causative agent of anthrax. Endospores of this bacterium isolated from the soil on this island were still viable and capable of causing infection forty years after the experiments ended. The contaminated soil was eventually fumigated with the sterilising agent, formaldehyde, which destroys the endospores.

Fig 19.7 The island of Gruinard was used during the second world war (1939–45) to investigate *Bacillus anthracis* as a biological warfare agent. The bacterium causes anthrax, a disease known to kill people, cattle and other livestock. These technicians are spraying on decontaminant.

Fig 19.8 The buds of these *Hydra* catch and digest their own food long before they break off from the parent.

Budding

The term bud has two meanings in biology.

1. Flowering plants form buds, which are dormant, compact shoots which grow at the tips of stems and in the axils of leaves. Whilst asexual reproduction may occur if two buds grow into shoots (see vegetative propagation later in this chapter), these buds are not organs of reproduction.
2. The term bud is also used to refer to an outgrowth of a parent organism which eventually constricts its point of attachment to the parent and is released as an independent individual. This is a form of asexual reproduction.

Yeast cells reproduce by budding. In these unicellular fungi, budding is like binary fission except that the two cells produced are of unequal size. Among multicellular organisms, buds develop as fully differentiated multicellular structures before they detach from the parent. Among the animal kingdom, budding is restricted to relatively simple groups. Cnidarians (Chapter 27) have relatively unspecialised interstitial cells which divide to replace more specialised cells which are lost. Mitotic division of these cells produces buds, like the ones shown in Fig 19.8.

4 **Suggest what the biological advantage(s) of budding might be.**

Regeneration and fragmentation

Many organisms are able to replace parts of their body which are lost during accidents. This is known as **regeneration**. Although they appear sluggish and innocuous, starfish are voracious predators of corals and bivalve molluscs, e.g. the crown of thorns starfish (*Acanthaster planci*) has destroyed large parts of the Great Barrier Reef, costing the Australian authorities about A\$3 million between 1985 and 1988. In many parts of the world, oyster fishermen protected their beds of oysters by catching starfish and chopping them into pieces which they threw back into the sea. Unfortunately each fragment regenerated a whole new starfish; the starfish populations exploded. The fishermen had, inadvertently, caused asexual reproduction of the starfish.

Some organisms frequently and spontaneously divide into fragments which then regenerate. This natural process of **fragmentation** occurs in many fungi, filamentous algae, sponges, cnidarians and platyhelminths.

Vegetative propagation

Vegetative propagation occurs in flowering plants. It is similar to budding in that a relatively large, fully differentiated structure becomes detached from the parent plant's body to become an independent plant. These structures may arise from stems, leaves, buds or roots. They often contain stored food from the parent plant's photosynthesis and can remain dormant in the soil from one growing season to the next as **perennating organs** (Table 19.1). During periods of adverse environmental conditions, the aerial parts of the parent plant die, leaving only the dormant perennating organ. During the next growing season, buds in these perennating organs grow into new shoots. The shoots photosynthesise, producing food which is stored in a new perennating organ, and flower. If only one bud grows from each perennating organ, only one new perennating organ will be formed during the next year; no reproduction occurs. Reproduction occurs if more than one bud grows from each perennating organ, so that two or more new shoots are formed leading to the production of two or more new perennating organs.

Table 19.1 Perennating organs through which vegetative propagation is achieved. Note that new plants only arise if more than one bud develops into a new shoot.

Name	Example	Description	How reproduction is achieved
bulb	onion, daffodil	short, triangular stem with swollen leaf bases	buds in axils of leaf bases grow into individual shoots which each produce a newbulb
corm	crocus	swollen base of stem, surrounded by dry scale leaves	buds in axils of scale leaves grow into individual shoots which each produce a new corm
rhizome	iris	underground, horizontal stem, often swollen with food reserves	buds develop to produce new lateral rhizomes or new vertical shoots
runner	strawberry	lateral stem growing away from parent plant	adventitious roots develop where runner touches ground and a new plant develops from these points
tuber: root	dahlia	swollen adventitious roots of a single parent plant	buds at base of old stems develop to produce new shoot from each tuber
tuber: stem	potato	tip of slender rhizome becomes swollen with stored food	buds on swollen stem tip develop to produce new shoots

 5 **Are perennating organs produced by mitosis or meiosis?**

Because vegetative propagation in plants involves mitosis, the cloned offspring are usually genetically identical to the parent. This is useful if the parent had some desirable quality, such as a particular pattern or colour of flowers. Many commercial plants are produced by splitting a perennating organ, e.g. horticulturists split onion sets and cut rhizomes into small lengths, each carrying a bud.

SPOTLIGHT

TOTIPOTENCY AND CLONING

A favourite idea among science fiction writers and horror film directors is that it may be possible to take a highly differentiated cell, say one from the intestine, and use that cell to manufacture a new human being by cloning. However, such a cell would have to be **totipotent**, that is have the capacity to proceed through all the stages of development and thus produce a normal adult. This has not yet been achieved with humans, but is already common practice with plants.

In the 1950s, Frederick Steward demonstrated that highly differentiated phloem cells in the roots of carrots were totipotent. Using a single phloem cell grown in nutritive medium, Steward was able to produce a complete carrot from a single phloem cell. Provided the necessary nutrients and growth substances are present in the growth medium, a mound of cells, called a **callus**, develops. From these, individual plants can be grown. Using this technique, now called **cloning**, it is possible to obtain a large number of identical plants from the somatic cells of a single plant. The impact of this work on plant breeding has been enormous. For example, almost all commercially available orchids are produced by cloning. The technique is particularly important with sexually reproducing organisms because sexual reproduction disrupts useful combinations of genes during the process of meiosis. Cloning retains these combinations of genes intact and is used commercially, for example, to maintain genetically identical collections (**lines**) of trees.

Fig 19.9 These orchid plants have been produced by artificially cloning cells from the parent plant on an agar-based medium containing nutrients and growth substances.

QUESTIONS

19.3 Compare and contrast multiple fission and sporulation.

19.4 Tap roots, such as carrots, are perennating organs. Are they also methods of vegetative propagation? Justify your answer.

19.3 SEXUAL REPRODUCTION

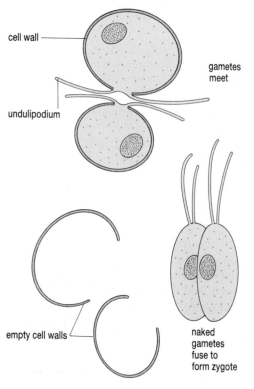

Fig 19.10 Isogametes of the photosynthetic protoctist *Chlamydomonas* swim around before joining together at their flagellar ends during fertilisation.

cell wall

undulipodium

empty cell walls

gametes meet

naked gametes fuse to form zygote

Sexual reproduction involves the formation of a diploid nucleus from two haploid nuclei. The haploid nuclei are usually inside special cells, called **gametes**; fusion of the gametes is called **fertilisation** and the diploid product of this fusion is the **zygote**.

Types of gamete

For fertilisation to occur, at least one of the gametes must be motile. In many motile protoctists the gametes are indistinguishable from each other and from the vegetative cell. The production of such identical gametes is called **isogamy**. The **isogametes** of the photosynthetic protoctist *Chlamydomonas* in Fig 19.10 swim freely until they come together in pairs during fertilisation. The gametes of another photosynthetic protoctist, *Spirogyra*, are identical in appearance but only one moves: the other remains within the wall of the parent cell (Fig 19.11). Such gamete production is called **anisogamy** and the gametes are **anisogametes**, also called **heterogametes**. In higher organisms the non-motile (**female**) gamete is swollen with reserves of food whereas the motile (**male**) gamete is small. Such differentiation of gametes is called **oogamy** after the female egg cell (**ovum**). Some oogamous species are **monoecious** (Greek for one house), meaning that individuals can produce both male and female gametes. Such individuals are also known as **hermaphrodite**. Other species are **dioecious** (Greek, meaning two houses): some individuals produce only female gametes and others only male gametes.

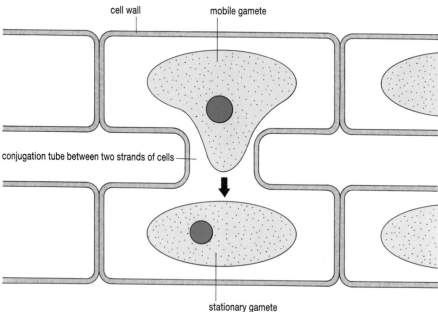

cell wall   mobile gamete

conjugation tube between two strands of cells

stationary gamete

Fig 19.11 One of the gametes of *Spirogyra* moves from the parent cell and joins with a physically similar non-motile gamete.

 Classify humans on the basis of the types of gamete we produce.

Isogamy	the production of identical gametes, e.g. *Chlamydomonas* (Fig 19.10).
Anisogamy	the production of physically identical gametes, only one of which moves to fertilise the other, e.g. *Spirogyra* (Fig 19.11).
Oogamy	the production of non-identical gametes. The larger female gamete, the ovum, is usually non-motile whereas the smaller male gametes are motile, e.g. mammals, flowering plants.
Dioecious	the condition in which a single individual produces only male or female gametes during its life cycle, e.g. hazel, malarial mosquito.
Monoecious	the condition in which a single individual produces both male and female gametes at some stage of its life cycle, e.g. rose, earthworms.
Hermaphrodite	• A plant species in which male and female gametes are produced in the same flower of a single individual (compare monoecious). • An animal which produces both male and female sex organs.

Gametogenesis

In all sexually reproducing organisms, meiosis must occur at some stage of the life cycle. Table 19.2 summarises the main stages of gamete production (**gametogenesis**) in mammals and plants. Whilst the stages are superficially similar, you should remember that the life cycle of plants involves two generations of adults, so called **alternation of generations** (Fig 19.2(b)).

Table 19.2 Gametogenesis in mammals and flowering plants. Whilst they appear similar, plants have two adults in their life cycle, and their gametes are naked nuclei.

Diagrammatic summary	Mammals		Flowering plants	
	Male	Female	Male	Female
2n	primordial germ cell	primordial germ cell	anther tissue	ovary tissue
mitosis 2n 2n	spermatogonia	oogonia	none	none
growth 2n	primary spermatocyte	primary oocyte	microspore mother cell	megaspore mother cell
meiosis I n n	secondary spermatocyte	secondary oocyte and polar body	none	none
meiosis II n n	spermatid	none	microspore (pollen grain)	megaspore (embryo sac)
differentiation n	sperm	ovum plus polar body	generative nucleus and pollen tube nucleus	none
			male gametes	ovum nucleus, embryo sac nuclei, polar nuclei etc.

Fertilisation

There are two types of fertilisation

- **External fertilisation** occurs outside the body, with the gametes being shed directly into the environment.

- **Internal fertilisation** occurs when the male gametes are placed inside the female's body and then move through the female's reproductive tract to the female gametes.

 Why is external fertilisation common only in aquatic environments?

Successful external fertilisation is more likely if large numbers of gametes are present in the water at the same time. Millions of egg cells and sperm may be released. Spawning among sessile (non-motile) organisms is often synchronised with a major environmental variable, such as a full moon. Sessile or sluggish animals often release **pheromones** as they spawn. A pheromone is a chemical which is released from the body of one animal and affects the behaviour of another animal. The pheromone released by a female mussel as she spawns stimulates the surrounding males to release their sperm, thus synchronising spawning. In actively swimming animals, complex courtship rituals, such as the zig-zag dance of the male stickleback (Fig 19.12), ensure that their gametes are released very close to each other.

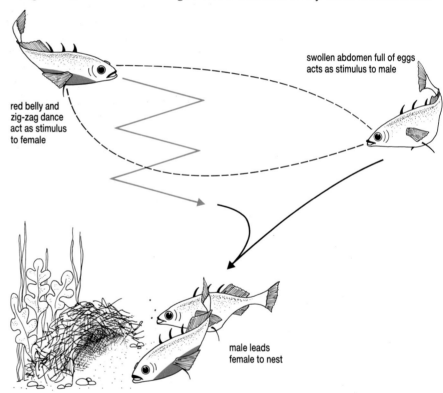

swollen abdomen full of eggs
acts as stimulus to male

red belly and
zig-zag dance
act as stimulus
to female

male leads
female to nest

Fig 19.12 A male stickleback defends his territory against intruders. During the mating season, the absence of red coloration and the swollen belly of a gravid female act as a stimulus for a series of reflexes by which the male eventually leads the female to a nest which he has built. Once the female has laid her eggs in the nest, the male drives her off, ejaculates sperm onto the eggs and stays to guard the zygotes.

In spite of the synchronisation of spawning, many gametes are not fertilised, representing an enormous waste of energy. The probability of fertilisation is greater with **internal fertilisation**. Transfer of the male gametes from the male's reproductive organs to the female's reproductive organs occurs in a variety of ways. The male gametes of angiosperms are the nuclei which grow down the pollen tube of a germinated pollen grain after its transfer to the female part of a flower (see Chapter 20).

Among many animal groups, the male packages the sperm into **spermatophores** which may be deposited on the ground to be picked up by the female (e.g. scorpions), transferred to the genital opening of the female using modified limbs (e.g. spiders and cephalopod molluscs, such as squid) or even injected hypodermically into the female (e.g. some leeches). Mammals release free sperm directly into the female's reproductive tract during copulation.

 In non-vascular plants, such as bryophytes, the male gametes swim through a film of water, attracted by the release of chemical stimulants (often monosaccharides or disaccharides) to the female reproductive organs. Is this internal or external fertilisation? Explain your answer.

Transfer of sperm does not ensure fertilisation of gametes. Fertilisation will only occur if the female gametes are mature at a time when there are viable male gametes to fertilise them. Many animals, such as snails and bats, store sperm for up to several months, ensuring that there is always a ready supply of sperm whenever ovulation occurs. More typically, animals and plants only transfer male gametes at certain times of the year, when environmental cues have caused sexual maturation of both sexes. In rabbits, copulation itself is the trigger for the release of egg cells. All these strategies help to ensure that a majority of the egg cells are fertilised.

Female animals whose egg cells are fertilised inside their bodies produce far fewer egg cells than those which spawn. The energy saved is often put to other uses which increase the survival chances of the fertilised egg cells. The largest egg cell known is that of the ostrich; its size is the result of a large food store. Female mammals do not produce such large egg cells but keep the zygote inside their body during its embryonic development. Most mammals also care for their young for some time after their birth, further increasing their chances of survival.

 What is the price mammals pay in return for this increased probability of survival for their young?

Inbreeding and outbreeding

Since they produce only one type of gamete, dioecious organisms are obligate **outbreeders**, i.e. their gametes must be fertilised by those of another organism. Those monoecious organisms which produce both male and female gametes at the same time may fertilise their own gametes. Such **inbreeding** may be advantageous in organisms which are solitary, such as tapeworms inside the gut of their host. Because fertilisation of their haploid gametes is at random, inbreeders may still produce different genetic variants among their diploid offspring. Inbreeding does have a disadvantage, however. If a monoecious parent is heterozygous for a harmful allele of one of its genes, half of its gametes are likely to carry the harmful allele (see Chapter 22). As a result of random fertilisation amongst these gametes, one quarter of its offspring are likely to be homozygous for the harmful allele (see the Hardy-Weinberg equation in Chapter 23), which may result in their premature death.

The Double Cross

Inbreds   Inbreds

← Seed Parent
Pollen Parent →

Crop

Fig 19.13   **(a)** *(above)* An example of hybrid vigour in maize. The cob in the middle (labelled crop) is the hybrid produced by crossing the two parent plants. Notice that the hybrid not only has much bigger cobs – but it also has larger seeds.

     (b) *(right)* The effects of inbreeding depression in maize over eight generations. The plant on the left is a very vigorous hybrid: the seven plants on the right show the steady decline in vigour as a result of inbreeding.

BREEDING CEREALS

Some cereals, for example wheat, are natural inbreeders. This means that such plants will be genetically similar to each other. The advantage of this to the farmer is that it provides a uniform crop in which, for example, all the plants are of a similar height which aids mechanical harvesting. However, the cereal crop maize is an outbreeder. This means that individual maize plants will be genetically different from each other and so will not provide a uniform crop. Forcing the maize to inbreed, you would produce genetically uniform plants but they would suffer from inbreeding depression (Fig 19.13(a)). However, if you now cross two of these inbred plants the offspring produced are extremely large, a phenomenon known as hybrid vigour (Fig 19.13(b)). This is the basis of breeding the so-called F_1 hybrid plants which are becoming increasingly popular with vegetable growers.

Further information on plant breeding is given in Geoff Hayward's book in this series, *Applied Genetics*.

Mechanisms which prevent inbreeding among monoecious organisms include the **temporal separation** of male and female gamete production. Many organisms develop mature female gametes before they produce male gametes (**protogyny**) or mature male gametes, which they shed, before they produce female gametes (**protandry**). Such differential sexual maturity may occur within one reproductive season, as is the case with many species of flowering plants, or within the organism's lifetime, as is the case in many species of snail.

Self-sterility

Even if pollen (which contains the male gamete) is transferred to the female part of a flower, sometimes it will not germinate. Male and female gametes can, thus, never meet. Such self-sterility is an inherited trait and explains why certain varieties of apples, pears and cherries will not set fruit unless another tree of the same variety grows nearby.

Fig 19.14 Female aphids can reproduce sexually or asexually. During spring and early summer they reproduce parthenogenetically, giving birth to live young, which are all female. The development of their ovaries is so rapid that many young are already pregnant when they are born. During autumn the females reproduce sexually, their egg cells being fertilised by sperm from the males.

Parthenogenesis

Even though parthenogenesis involves only one parent and no fusion of gametes, it has been included in this section on sexual reproduction because sexual structures are involved.

The females of some animals can reproduce by the process of **parthenogenesis**. Haploid egg cells are produced in their ovaries, in the usual way. The egg cell then divides and develops into an adult without ever having been fertilised. In some animals the resulting adult is haploid, like the unfertilised egg cell. Male honeybees (drones) are parthenogenetically produced haploid individuals; female honeybees (queens and workers) are diploid individuals which develop from fertilised egg cells.

In other groups of animals, a duplication of chromosomes occurs within the egg cell without the cytoplasm dividing, so restoring the diploid number. The resulting offspring are all female. This occurs, for example, in aphids in spring when food is abundant, and produces more diploid females (Fig 19.14). In the autumn, their egg cells are fertilised by males and they reproduce sexually in the normal way. Populations of the whiptail lizard contain no males at all. The all-female populations reproduce solely by parthenogenesis.

QUESTIONS

19.5 Distinguish between isogamy, anisogamy and oogamy.

19.6 In angiosperms and in mammals, fertilisation occurs inside the body of the female.
(a) What are the advantage(s) of such internal fertilisation over external fertilisation?
(b) Suggest disadvantages of internal fertilisation over external fertilisation.

19.7 What are the potential advantages and disadvantages of being hermaphrodite?

19.8 Devise a diagram to summarise the annual life cycle of an aphid.

SUMMARY

Organisms reproduce either asexually or sexually. Asexual reproduction occurs when the genetic material and cytoplasm of a single parent divides to form new individuals. Except for spore production by the sporophyte generation of plants and some protoctists, such division is by mitosis and the offspring are genetically identical. A group of genetically identical individuals is called a clone.

Sexual reproduction occurs when two haploid gametes fuse at fertilisation to form new diploid individuals. The random fusion of gametes always results in genetic variation among the offspring.

Asexual reproduction is usually confined to groups of organisms with relatively unspecialised tissues. It may occur by fission, sporulation, budding, or fragmentation. Fission occurs in unicellular organisms and involves replication of the cell's genetic material followed by division of the parent nucleus and cytoplasm to form new cells. Sporulation may occur in unicellular and multicellular organisms and produces unicellular offspring which are dormant and able to withstand adverse conditions. Budding involves division of the parent to form a miniature offspring as an outgrowth of itself. Eventually the bud detaches and becomes independent. Organisms which reproduce by fragmentation spontaneously divide into small pieces, each of which regenerates a whole organism.

A special type of asexual reproduction, called vegetative propagation, occurs in flowering plants. Individual plant organs become swollen with stored food and then remain dormant during adverse conditions. Growth of one or more buds on this perennating organ results in the production of new offspring.

For sexual reproduction to occur, two single-celled gametes fuse to form a zygote at fertilisation. Isogamous fertilisation involves fusion of two identical gametes; anisogamous fertilisation involves fusion of gametes which appear similar but behave differently; oogamous fertilisation involves fusion of one large, non-motile gamete and one small, motile gamete.

Fertilisation may occur outside the bodies of the parents (external fertilisation) or inside the body of one of the parents (internal fertilisation). Internal fertilisation has two major advantages over external fertilisation: fewer gametes need be produced and fertilisation can be accomplished without the need for water.

Organisms which reproduce sexually may be able to produce both male and female gametes (monoecious or hermaphrodite) or produce only one type of gamete (dioecious). Dioecious organisms are obligate outbreeders since they can never fertilise their own gametes. Theoretically, it is possible for monoecious organisms to fertilise their own gametes. Such inbreeding may result in offspring which inherit pairs of disadvantageous alleles of genes from their single parent.

Mechanisms exist to prevent inbreeding. These include chemical self-sterility and the development of sexually mature male and female structures on the same individual at different times.

In some groups of animals, haploid egg cells may develop without the need for fertilisation. Such asexual reproduction involving a single gamete is called parthenogenesis. In honeybees, the resulting adult is a haploid male. In aphids, chromosome duplication restores the diploid chromosome number and the offspring are all female.

Answers to Quick Questions: Chapter 19

1 Because it does not require the presence of another member of the same species.
2 By increasing the number of infective cells, schizogony increases the probability that the parasite will be (a) picked up from an infected person by a female mosquito and (b) transmitted to a new host by that mosquito.
3 (i) Provide a dormant, resting stage which enables an organism to survive periods of unfavourable environmental conditions.
   (ii) Dispersal.
4 Allows a rapid increase in numbers when conditions are favourable.
5 Mitosis.
6 Dioecious, oogamous.
7 Gametes dehydrate unless they are surrounded by water. Movement of single cells can only occur in water.
8 Although the male gamete is shed into the environment (a feature of external fertilisation) the female gamete is kept within the female reproductive organ. Like many classifications, using terms such as internal and external fertilisation is oversimplistic.
9 Mammals must expend more energy during care of their young, so fewer young can be produced per unit of resource, and they are also vulnerable to predators during this stage.

Chapter 20

SEXUAL REPRODUCTION IN MAMMALS AND IN FLOWERING PLANTS

LEARNING OBJECTIVES

When you have studied this chapter you should be able to:

1. describe the reproductive system of human males and females;

2. explain the hormonal control of human sperm and egg cell production;

3. describe the changes which occur to human reproductive systems during copulation;

4. explain how fertilisation of a human egg cell occurs;

5. interpret the structure of entomophilous and anemophilous flowers and describe their structure using floral diagrams and half-flowers;

6. explain how entomophilous and anemophilous flowers are adapted for pollination;

7. describe how pollination leads to fertilisation;

8. explain the formation of seeds and fruit and explain how fruits aid seed dispersal.

Fig 20.1 Most mammals show seasonal sexual activity.

All mammals are unisexual. Each has male or female **primary sex organs (gonads)**, which produce both gametes and sex hormones, and a number of **accessory sex organs**, which are involved with the storage and transport of gametes. The female gamete is the egg cell or **ovum**, the male gamete is the spermatozoon, or **sperm**. In most species of mammal, egg cells and sperm are only produced at certain times of the year. Human reproduction is an exception; sperm production in mature males is virtually continuous whilst mature females ovulate roughly once each month. The following account of mammalian reproduction concentrates on human reproduction.

It is tempting to regard plants as having a similar sexual pattern to the one we know best, i.e. our own. The sexual cycle of plants, called **alternation of generations**, is very different from our own, however. Sexual reproduction is carried out by single-sexed haploid **gametophytes**. The progeny of these haploid organisms are diploid individuals which are released inside seeds. These diploid plants are called **sporophytes** because, as adults, they reproduce asexually by producing haploid spores. Successful spores eventually germinate to produce new gametophytes.

20.1 THE HUMAN MALE REPRODUCTIVE SYSTEM

A human male has two gonads, which produce sperm, and a number of accessory structures which store the sperm, produce secretions which activate and nourish the sperm and conduct the sperm to the female reproductive system. These are shown in Fig 20.2(a).

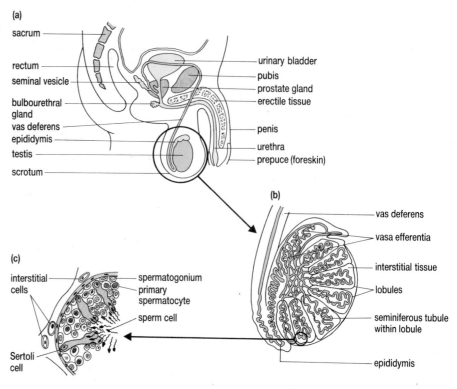

Fig 20.2 (a) The male reproductive system consists of primary sex organs, the testes, and a number of secondary sex organs.
(b) Each testis contains over 100 m of coiled seminiferous tubules.
(c) A single seminiferous tubule continuously produces sperm cells whilst the surrounding interstitial cells produce male hormones.

Testes

The male gonads are called **testes** (singular **testis**). They produce the male gametes (**spermatozoa**, or sperm) and male hormones (**androgens**). Whilst the testes are formed in the abdomen of a male fetus, by the time of birth they have descended into a sac of skin, called the **scrotum**. The temperature within the scrotum is about 3 °C lower than that of the abdomen. This lower temperature is essential for successful sperm production.

Each testis is an oval structure about 5 cm long. Its 300 or so lobules each contain up to four highly coiled, sperm-producing **seminiferous tubules**; around these are **interstitial cells**, which produce androgens (Fig 20.2(b)). Lining the walls of the seminiferous tubules is a layer of **spermatogonia**, forming the **germinal epithelium**. Some of these spermatogonia divide by mitosis to produce more spermatogonia, others produce sperm. Between the sperm-producing cells are large **Sertoli cells** which nourish the spermatids as they mature into spermatozoa (Fig 20.2(c)). Sperm production begins at puberty and a healthy adult male produces several hundred million sperm each day.

A mature spermatozoon is shown in Fig 20.3. Its **head** is almost filled by a haploid nucleus, in front of which is a specialised lysosome, the **acrosome**. This contains enzymes which are needed to digest the protective layer around an egg cell if the sperm is to enter its cytoplasm. Behind the head is the **midpiece**, which is packed with mitochondria arranged in a spiral.

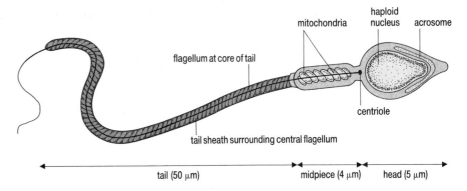

Fig 20.3 A mature sperm is well adapted to deliver its haploid nucleus to the egg cell. Its tail uses energy from the mitochondria to swim at a rate of 4 mm h⁻¹; proteolytic enzymes in its acrosome digest an opening through the protective outer layers of the egg cell.

 Suggest the function of the mitochondria in a sperm cell.

Male accessory sex organs

The accessory organs are known as secondary sexual structures. They include the structures into which the seminiferous tubules empty their contents. Within each testis, the seminiferous tubules merge together to form a further network of tubes, the **vasa efferentia**. These in turn empty into a network of tubes lying just outside the testis, the **epididymis**. Finally, the epididymis empties into a single **vas deferens**. The two vasa deferentia (singular of vas deferens) leave the scrotum and join the urethra, just below the bladder. The urethra passes to the tip of the **penis** and is shared by sperm (during ejaculation) and urine (during urination). Ducts from three glands also join the urethra at about the same point as its junction with the vasa deferentia: a pair of **seminal vesicles**; a single, diffuse **prostate gland**; and a pair of **bulbourethral glands**. Secretions from these glands make the semen alkaline, aiding survival of sperm in the acidic conditions of the vagina. The seminal vesicle secretions also contain fructose, an energy source for swimming sperm.

> ### KEY TERM
>
> Vasectomy involves severing the vasa deferentia so preventing sperm mixing with the rest of the semen.

20.2 THE HUMAN FEMALE REPRODUCTIVE SYSTEM

A human female has two gonads, which produce egg cells (**ova**), and a number of secondary sex organs which allow sperm deposition, fertilisation and development of a fertilised egg cell. These are shown in Fig 20.4.

Ovaries

The female gonads are called **ovaries**. They produce ova (egg cells), and female hormones. They are suspended by ligaments on either side of the abdominal cavity. Each ovary is an oval structure about 3 cm long. It is largely composed of connective tissue around which is a protective capsule. Unlike males, gamete production starts in the early months of fetal development. Beneath the protective capsule of a fetal ovary is a germinal epithelium, consisting of **oogonia**. These divide by mitosis to produce **primary oocytes** so that, by the third month of fetal development, no oogonia are left. Meiosis starts in the primary oocytes but is suspended during the first prophase. At birth, each ovary contains about one million oocytes, each surrounded by a group of smaller epithelial cells to form a **primary follicle**. The majority of these die so that only about 400 000 remain at puberty. By the time of the menopause (45–50 years) only about 10 000 follicles remain.

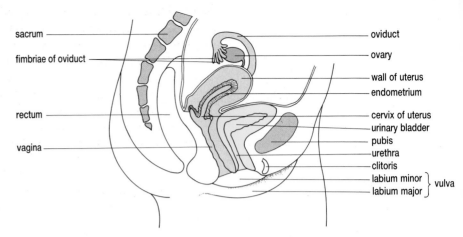

Fig 20.4 The female reproductive system consists of primary sex organs, the ovaries, and a number of secondary sex organs.

◆**2** **Is the loss of follicles between the onset of puberty at, say, 11 years of age and 50 years attributable to the release (ovulation) of an oocyte (egg cell) each month?**

After puberty, hormones released by the pituitary gland stimulate the further development of some of these follicles. Each month, several primary follicles begin to develop, although only one usually completes development. Under hormonal stimulation, the primary oocyte completes the first division of meiosis to become a **secondary oocyte** and a smaller polar body (Table 19.2). At the same time, the smaller cells around the follicle multiply so that the follicle increases in size from less than 1 mm to over 10 mm in diameter. Fluid-filled cavities now form between the follicle cells, and the follicle cells begin to produce oestrogen. The mature follicle, known as a **Graafian follicle**, presses against the surface of the ovary. At **ovulation** it bursts through the ovary wall releasing not an egg cell but the secondary oocyte: the second meiotic division to produce an ovum only occurs if a sperm penetrates this secondary oocyte. Some follicle cells leave the ovary with the secondary oocyte and help the fimbriae of the oviduct entrap the oocyte (see below). Most follicle cells remain inside the ovary, where they become large and glandular to form the **corpus luteum**. One complete series of stages in the development of primary follicle to corpus luteum is represented in Fig 20.5.

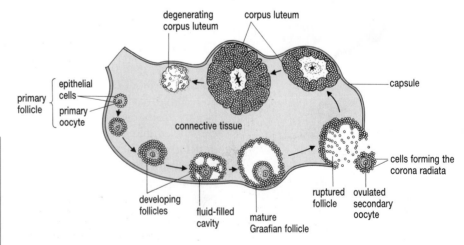

Fig 20.5 This vertical section through an ovary represents the stages of development of a single follicle. Of the 400 000 primary follicles in the ovaries of an adolescent girl, only about 450 will undergo this development.

SEXUAL REPRODUCTION IN MAMMALS AND IN FLOWERING PLANTS

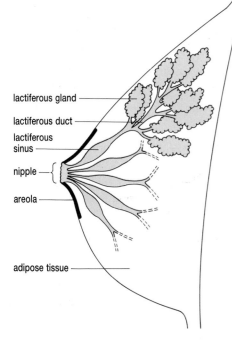

lactiferous gland

lactiferous duct

lactiferous sinus

nipple

areola

adipose tissue

Fig 20.6 Vertical section through a mature mammary gland. The suckling of a baby stimulates the release of milk from the lactiferous ducts.

It has been known for a male goat to have a mastectomy: the removal of his mammary glands was necessary as they gave milk painfully and had enlarged so much that they interferred with normal functioning of his male sex organs.

Female accessory sex organs

Each ovary is adjacent to an **oviduct**, more commonly called a **uterine tube** or **Fallopian tube** in humans. The open end of this tube is fringed with finger-like **fimbriae**. These are lined by cilia whose movement traps the released secondary oocyte. The oviducts lead into the **uterus**. The bulk of the uterus wall consists of smooth muscle (the **myometrium**). The rhythmic contraction of these muscles expels the fetus from the uterus at birth (and causes the regular pains called contractions). The inner layer of the uterus is the **endometrium**. Hormones released from the follicle and corpus luteum cause an increase in the thickness and vascularisation of the endometrium. If a woman becomes pregnant, the endometrium contributes the mother's part of the placenta; if not, the thickened endometrium will disintegrate and be lost as the **menstrual flow** (period).

A narrow ring of muscle, the **cervix**, closes the lower end of the uterus. This muscle holds a developing fetus within the uterus, expanding only during labour. Beyond the cervix is a muscular tube, the **vagina**. This opens to the outside via the fleshy **labia** of the **vulva**, the only part of the female reproductive system which is external to the body. Enclosed within the labia is the **clitoris**, a small organ of sensitive, erectile tissue.

Although not strictly a part of the female reproductive system, the breasts (**mammary glands**) will be described. Like all secondary sexual characteristics, the mammary glands develop after puberty under the influence of sex hormones. Each mammary gland is composed largely of fat within which are a number of milk-secreting **lactiferous glands** (Fig 20.6). Under the influence of the hormone **prolactin**, these glands produce milk, which is stored in **lactiferous ducts**. Stored milk is released (let down) via the nipple under the influence of the hormone **oxytocin**. Males have undeveloped mammary glands; in some mammalian species both females and males produce milk, e.g. goats.

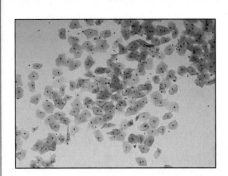

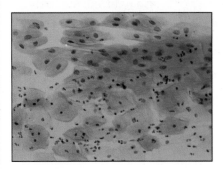

CERVICAL CANCER

One of the commonest forms of cancer, cervical cancer, starts with a change in the shape, growth and number of cervical cells called **cervical dysplasia**. Though not cancerous in themselves the abnormal cells associated with cervical dysplasia tend to become malignant.

Early diagnosis of cervical dysplasia is achieved in more than 90% of cases using the **smear test**. A few cells are removed from the cervix with a swab, via the vagina. The cells are then stained and examined microscopically (Fig 20.7). Abnormal cells have a characteristic appearance. If detected they can be removed using a laser. All women should have regular cervical smear tests every 3–5 years.

Fig 20.7 **(a)** Normal and **(b)** abnormal cervical smears as viewed under the optical microscope. Note the relatively large blue-staining nuclei at the top of (b). These cells are cancerous.

20.1 By reference to the male and female reproductive system, distinguish between primary sex organs, accessory sex organs and secondary sexual characteristics.

20.2 Using Fig 20.2 and Table 19.2, summarise the stages in the development of sperm.

20.3 THE HORMONAL CONTROL OF HUMAN REPRODUCTION

The control of the mammalian reproductive system is complex not only because of the range of hormones involved but also because of differences between species and sexes of mammals. Whilst we will again concentrate on the control of human reproduction, you should not think this is the whole story of the control of reproduction in mammals. For example one obvious distinction would be between continuous breeders, e.g. humans and other mammals, which are seasonal breeders, e.g. deer. Another difference would be between female mammals which have an oestrous cycle involving regular ovulation, e.g. humans and others where ovulation is induced by copulation, e.g. rabbits.

 What activities must be coordinated if sperm and egg cell are to come together at the same time?

Male hormones

Sexual maturity begins at puberty when the hypothalamus stimulates the anterior pituitary gland to release two hormones: **follicle stimulating hormone (FSH)** and **luteinising hormone (LH)**. FSH stimulates sperm production in the seminiferous tubules; LH stimulates the interstitial cells of the testes to produce **testosterone**. FSH and LH derive their names from the effects they have in females; in males, LH is often referred to by its alternative name of **interstitial cell stimulating hormone (ICSH)**. Testosterone stimulates development of the male accessory sex organs, which grow throughout adolescence. It also controls development of **secondary sexual characteristics**, such as facial and pubic hair, breaking of the voice and muscle development, as well as stimulating sexual drive (**libido**). A negative feedback loop (Chapter 14) between the testes, pituitary gland and hypothalamus regulates the rate of sperm production and the concentration of testosterone in the blood.

ANALYSIS

Male sex hormones

This is a data analysis and interpretation exercise.

If a cockerel is castrated it becomes a capon and loses its characteristic wattles and combs; its aggressive behaviour is also reduced. Similarly castrating bulls and stallions makes them less aggressive and easier to handle. However injections of testicular extract into capons restores both their secondary sexual (male) characteristics and their aggressiveness.

In men, at least five sex hormones are produced in the testes, as shown in Table 20.1.
(a) Which of these male sex hormones appears to be the most important? Explain your answer.
(b) Which hormones have the greatest potency?
(c) Which has the least potency? Can you suggest why?

(d) Androgens (male sex hormones) are responsible for stimulating the development and activity of male secondary sex characteristics. List as many male secondary sexual characteristics in mammals as you can which are likely to be induced by androgen activity.

(e) When X-rays are administered in controlled doses to the testes of rats, the seminiferous tubules are destroyed but secondary sex characteristics are unaffected. Where does this suggest male sex hormones are made within the testes?

Table 20.1 Analysis of male sex hormones

Hormone	Blood plasma levels /μg 100 cm^{-3}	Rate of testicular secretion /mg 24 h^{-1}	Androgenic potency / % restoration of combs in capons
androstenedione	0.1	No data	25
dehydroepiandrosterone	0.5	0.15	10
stanalone	0.05	No data	100
oestrogen	0.003	0.03	0
testosterone	0.7	6.2	100

Female hormones

Gamete production in females is stimulated by FSH, in much the same way as sperm production in males. Whilst males release FSH continuously, females do not. Thus, egg cells are not continuously produced. It is easy to see the biological advantages of this. The development of fertilised egg cells occurs within the female's uterus, which must be prepared for pregnancy. The size of the uterus limits the number of fertilised egg cells which can be supported during pregnancy. Furthermore, in many mammalian species, birth of the young coincides with the time of year when food availability and climate are most suitable for their survival. The females of some mammalian species produce egg cells only once or twice per year. Only at the time of ovulation are they receptive to the male, when they are said to be in **oestrous**, or on heat. The interaction of hormones regulating their behaviour is called the **oestrous cycle**. Egg cell production in human females (and in other primates) is not seasonal, but occurs approximately each month. The cycle in humans is called the **menstrual cycle** (after the Latin *mensis*, meaning month).

The menstrual cycle

A cycle has no beginning or end. For convenience, a description of the menstrual cycle will begin with the development of a follicle within the ovary (Fig 20.8). Numbers in the text refer to the stages labelled (1) – (12) in Fig 20.8.

At the start of the menstrual cycle, gonadotrophin releasing hormone (GnRH) is released by the hypothalamus and stimulates the anterior pituitary to release FSH (1). FSH circulates in the blood to the ovaries where it has two major effects. The first is to stimulate the development of a number of primary follicles (2). For reasons not yet understood, most of these stop growing and disintegrate, so that only one usually finishes development.

 What is the second effect of FSH? (Look at (3) in Fig 20.8)

Oestrogen has three main effects on the menstrual cycle. Firstly, it promotes further growth of the follicle. Secondly, it promotes growth of the endometrium (4), causing it to increase in thickness by about 0.5 mm.

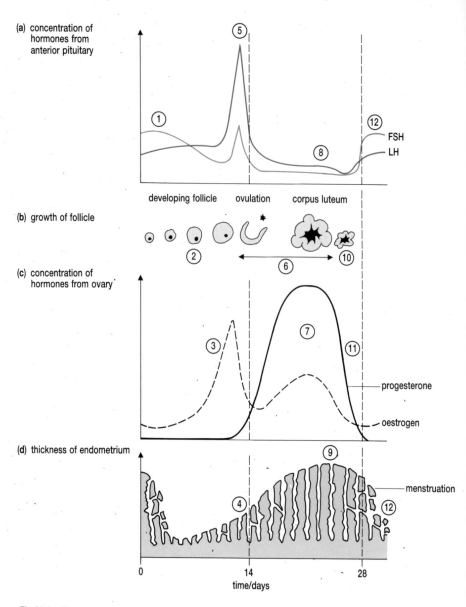

(a) concentration of hormones from anterior pituitary

FSH

LH

developing follicle　ovulation　corpus luteum

(b) growth of follicle

(c) concentration of hormones from ovary

progesterone

oestrogen

(d) thickness of endometrium

menstruation

0　　　14　　　28

time/days

Fig 20.8 The human menstrual cycle involves changes in the blood concentration of hormones from the anterior pituitary and ovary, in the activity of follicles in the ovary and in the development of the endometrium. The encircled numbers indicate steps which are referred to in the accompanying text.

Thirdly, it inhibits the further secretion of FSH by the pituitary whilst stimulating the pituitary to release LH. The surge of LH in the blood (5) at around day 12 has several effects. It stimulates

- the resumption of meiosis I in the primary oocyte, so that a secondary oocyte and polar body are formed
- the growth of the follicle, terminating in ovulation
- the development of follicle cells which remain in the ovary to form a **corpus luteum (6)**.

The corpus luteum secretes both oestrogen and progesterone (7). These hormones inhibit further release of FSH and LH by the pituitary (8), so that no further follicles develop. Progesterone and oestrogen stimulate the further growth of the endometrium and its blood supply, so that it eventually reaches a thickness of about 5 mm (9).

The corpus luteum has a limited life span. Unless stimulated by chemicals released by a developing embryo, it starts to degenerate after about seven days (10). As it degenerates, the corpus luteum stops secreting oestrogen and progesterone (11).

5 What are the two effects of this reduction in oestrogen and progesterone secretions by the corpus luteum? (Hint – look at points (12) in Fig 20.8.)

A developing embryo releases the hormone **human chorionic gonadotrophin (HCG)** which has a similar effect to LH and prolongs the life of the corpus luteum well into pregnancy. By the third month of pregnancy, the placenta secretes sufficient progesterone to maintain the endometrium and inhibit further FSH secretion.

6 Why must FSH secretions be inhibited if pregnancy is to be maintained?

You have doubtless realised that the interaction of hormones controlling the menstrual cycle is complex. A summary is given in Fig 20.9. The majority of women who have read the above account of the menstrual cycle will realise that it is incomplete. The hormones regulating the cycle have effects on the breasts, abdomen and brain which have not been described. The resulting enlargement and tenderness of the breasts, abdominal swelling, painful cramps of the uterus and changes of mood can be severe. In addition, only a minority of women have a 28-day menstrual cycle; 28 days is the mean.

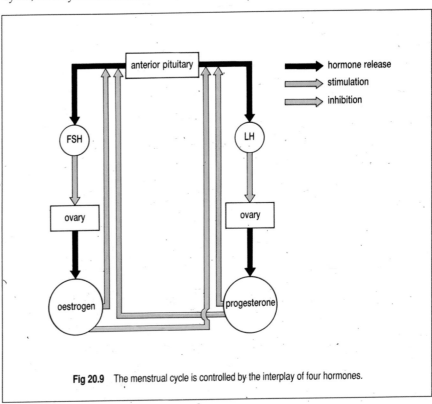

Fig 20.9 The menstrual cycle is controlled by the interplay of four hormones.

ANALYSIS

The oestrous cycles of ewes and rats

This exercise involves the interpretation of data.

The menstrual cycle is found only in human females and some other primates. If fertilisation does not occur the endometrium breaks down at the end of each monthly cycle, accompanied by bleeding (a period). A more typical mammalian **oestrous cycle** is shown by the ewe, (Fig 20.10(a)). During oestrus the female becomes receptive to the advances of the male and her basal body temperature rises, i.e. she is on heat.

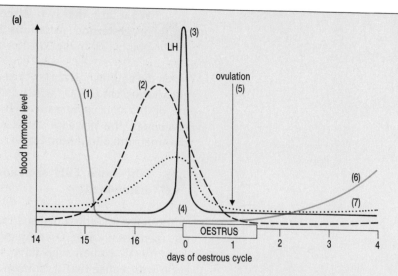

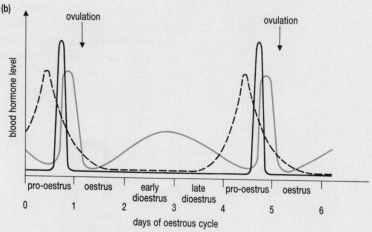

Fig 20.10 The oestrous cycle of **(a)** ewes and **(b)** rats. (Whilst rats do produce FSH, levels of this hormone have been left off graph (b) for clarity.)

(a) Can the human menstrual cycle be regarded as a special modification of the mammalian oestrous cycle? Why?

(b) How does the LH peak (3) of the ewe differ from that seen in the human menstrual cycle?

(c) How could changes in the FSH level account for the oestrogen peak?

(d) How could changes in the activity of the corpora lutea account for the drop in progesterone level at (1) and the rise in level at (6)?

The hormonal changes in the oestrous cycle of a rat are shown in Fig 20.10(b).

(e) Given that the rat ovulates like women and ewes, how does the rat's oestrous cycle differ from that of ewes and women?

QUESTIONS

20.3 Explain how the male and female reproductive systems are adapted for the production and release of gametes.

20.4 FSH and LH are hormones released by both females and males.
  (a) What different effects do these hormones have in the two sexes?
  (b) An ideal male contraceptive might be one that would suppress release of FSH but not of LH. Discuss why.

Copulation in humans

Copulation is also known as **coitus** or **sexual intercourse**. During copulation, the male ejaculates sperm into the vagina of the female. To enable this to happen changes must occur in both male and female reproductive systems.

In its normal condition, the penis is flaccid. Nervous stimulation causes dilation of the arteries carrying blood into the penis and constriction of the veins carrying blood away from it. The result is that blood accumulates in the columns of spongy tissue so that the penis increases in size and becomes firm. This is called an **erection**. In this condition, the penis can be inserted into the vagina of a receptive female. Nervous stimulation in the female causes increased blood flow to the vagina, labia and clitoris. As a result, the vagina enlarges and secretes mucus, which is necessary if insertion of the penis is to be comfortable for both partners. The clitoris, which is embryologically similar to the penis, becomes erect.

After insertion of the penis into the vagina, body movements of both male and female stimulate sensitive touch receptors of the penis, vagina and clitoris. In both partners, this stimulation may result in peristaltic contractions of involuntary muscles encircling the reproductive tracts, a feeling of intense pleasure and release, known as **orgasm**. In males, orgasm involves peristaltic contraction of the smooth muscle surrounding the epididymes, vasa deferentia and urethra. At the same time the seminal vesicles, prostate gland and bulbourethral glands release their secretions. The result is the release of 2 to 6 cm^3 of semen containing about 100 million sperm cm^{-3}, called **ejaculation**. In females, orgasm involves peristaltic contractions of muscles lining the vagina and uterus.

 Biologically, what is (a) the function and (b) the advantage of coitus in mammals?

Fertilisation

Fertilisation is the fusion of two haploid gamete nuclei to form a diploid **zygote** nucleus. However, neither secondary oocyte nor sperm lives very long. After ovulation, the secondary oocyte is wafted into the uterine tube by the action of cilia lining the fimbriae. The uterine tube is lined with cilia and capable of contractions, both of which move the oocyte along. The secondary oocyte remains viable for up to 24 hours, by which time it has only been moved a short distance along the uterine tube. Fertilisation must therefore occur towards the upper end of the uterine tube.

Under ideal conditions, sperm remain viable for up to two days. During this time they must reach the upper part of the uterine tube. Exactly how they accomplish this is not clear. Although they can swim, their arrival in the uterine tube is much earlier than could be accounted for by their swimming ability. In addition, they spend periods of time resting in groups. It is likely that the contractions of the female reproductive tract assist sperm passage. Prostate secretions in the semen not only contain enzymes which stimulate sperm to swim but also prostaglandins which further stimulate contractions of the uterus. However it occurs, only a few hundred of the millions of sperm ejaculated reach the site of fertilisation in the uterine tube; the rest die en route.

ANALYSIS

Male fertility

This is an exercise in numerical analysis.

A male ejaculation has a mean volume of 3.4 cm^3, about 10% of which will consist of spermatozoa giving an average concentration of 100 000 sperms mm^{-3}.

(a) Calculate the total number of sperm in a single ejaculation.

(b) The distance from the point of ejaculation in the vagina to the upper end of the oviduct is about 19 cm. How many sperm lengths is this? The dimensions of a sperm are given in Fig 20.3.

(c) Sperm have been shown to cover this distance in three hours. If a sperm swam this distance, how many times its own length would it have to swim each second? Do you think it likely sperm could swim this fast?

(d) If there are less than about 20 000 sperm mm^{-3} of ejaculate fertilisation is unlikely to occur. Suggest why?

Male fertility can be assessed using the **nomogram** shown in Fig 20.11. This takes into account the number of sperm, their motility and appearance. Abnormal sperm, i.e. those without normal heads, do not survive well in the female reproduction tract.

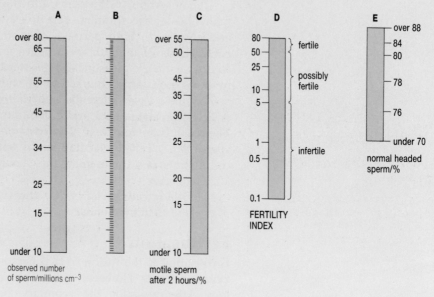

Fig 20.11 A nomogram for the assessment of male fertility.

To use the nomogram:

• draw a straight line between the observed number of sperm (scale A) and the percentage of sperm motile after 2 hours (scale C);

• from the intersection of this line with scale B, draw another straight line to scale E;

• read off the fertility index where this second line crosses scale D. This gives a relative assessment of fertility.

(e) Upon analysis a semen sample is found to contain 25 million sperm cm^{-3}, 35% of which are motile after 2 hours and 84% of which have normal heads. Assess the fertility of this man.

Look at Fig 20.12. It shows a human secondary oocyte (often referred to as the egg cell) surrounded by sperm. You will recall that the structure ovulated is actually the secondary oocyte, surrounded by a number of follicle cells. These follicle cells can be seen forming the outer **corona radiata** around the large oocyte. Between this layer and the secondary oocyte is a clear, proteinaceous membrane, the **zona pellucida**. Proteolytic enzymes, released from the acrosomes of the sperm, digest the corona radiata and zona pellucida. It is unlikely that a single sperm would release

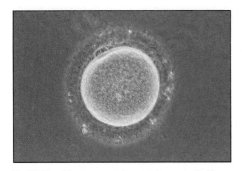

Fig 20.12 A human secondary oocyte surrounded by sperm.

sufficient enzyme to digest these layers so that fertilisation could not occur unless a larger number of sperm were present. Eventually one sperm wriggles through the weakened outer layers and touches the surface membrane of the secondary oocyte. The surface membranes fuse and the head of the sperm is taken into the cytoplasm of the secondary oocyte. This triggers two important events. Firstly, vesicles near the surface membrane of the secondary oocyte release membranous material into the zona pellucida which prevents the entry of further sperm. Secondly, the oocyte undergoes its second meiotic division to produce the female gamete. The sperm head absorbs water and swells, so that its chromosomes spill into the ovum. Fusion of the pronuclei of the sperm and ovum produces a diploid **zygote** from which the embryo develops.

 All the mitochondria in a human body were inherited from one parent. Which one? Explain your answer.

Pregnancy

The zygote begins to divide by mitosis as it is moved along the uterine tube (Fig 20.13). In humans, about four days after fertilisation it has formed a hollow ball of cells, called the **blastocyst**. The outer layer of the blastocyst is the **trophoblast**. This embeds in the endometrium in a process called **implantation**. Villi from the trophoblast increase its surface area for absorption of nutrients and anchor the blastocyst firmly in the endometrium. Later in development, the trophoblast forms part of the **chorion**. This structure secretes chorionic gonadotrophin (HCG) which maintains the corpus luteum. As this hormone travels in the blood stream, some is excreted in the urine of the mother; most pregnancy tests are assays for this hormone. A mass of cells within the blastocyst will form the embryo and its associated membranes. Details of embryological development are described in Chapter 21.

Absorbing nutrients from the cells of the endometrium will not support the embryo for longer than a week or two. During this time the placenta develops. It consists of blood-filled cavities in the endometrium and membranes of the embryo. Although embryonic and maternal tissues remain distinct, diffusion of metabolites occurs through the membranes of the placenta. The structure of the placenta is further described in Section 21.3.

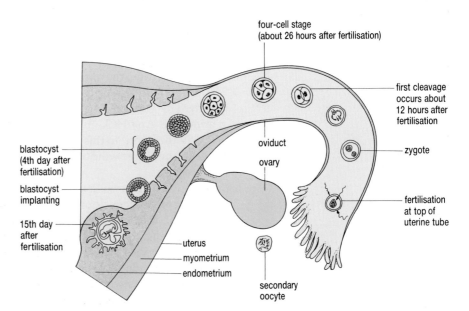

Fig 20.13 During the first weeks after fertilisation the zygote is moved along the uterine tube. During this process it divides by mitosis so that a hollow ball of cells, the blastocyst, finally enters the uterus and embeds in the endometrium.

four-cell stage (about 26 hours after fertilisation)

first cleavage occurs about 12 hours after fertilisation

zygote

fertilisation at top of uterine tube

oviduct

ovary

blastocyst (4th day after fertilisation)

blastocyst implanting

15th day after fertilisation

uterus

myometrium

endometrium

secondary oocyte

PREGNANCY TESTING

A woman excretes a hormone called human chorionic gonadotropin (HCG) immediately after conception. The presence of this hormone can be detected using antibodies in a simple pen-like tester in a simple, speedy and reliable test devised by a British biotechnology company (Fig 20.14(a)). The woman removes the top of the tester to reveal a wick which she places in her stream of urine first thing in the morning.

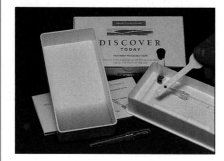

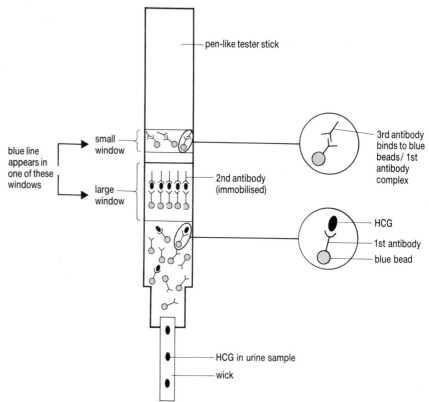

Fig 20.14 **(a)** *(left)* A pregnancy test to detect HCG, a hormone produced immediately a woman is pregnant. **(b)** *(right)* Diagrammatic section through tester.

The urine moves up the wick until it reaches the first layer shown in Fig 20.14(b). This consists of a mobile layer of tiny blue latex beads. Attached to each bead is an HCG antibody. If there is any HCG in the woman's urine, the antibodies and the blue latex beads will become attached to it. The urine then carries the blue latex beads to the second zone. This contains another HCG antibody, but this time the antibody is immobile. Any HCG in the urine, plus the blue latex beads attached to it, will then bind to this immobile antibody. This will appear as a blue line in the larger window of the stick which shows the woman that she is pregnant.

If the woman does not have any HCG in her urine, i.e. she is not pregnant, the blue latex beads swept up with the urine will become attached in a third zone, to a line of immobilised antibodies which bind to the antibodies on the blue latex beads. This produces a blue line in the small window of the test stick. This gives the woman confidence that the test has worked and that she is not pregnant.

Birth

In humans, delivery of the developed fetus occurs after about 38 weeks of development. By the time of delivery, a fetus has normally moved within the uterus so that it is head downwards. Delivery of a fetus which is the other way around (breach delivery) is hazardous for the fetus and, to some extent, the mother.

In response to a signal from the fetus, the fetal membranes and endometrium secrete prostaglandins. These hormones cause intense (and often painful) contractions of the muscles of the uterus wall. As these muscles contract, they push the head of the fetus against the cervix, making it dilate. Stretch receptors in the cervix send impulses to the mother's hypothalamus which triggers the release of oxytocin from her posterior pituitary. The combined effect of prostaglandins and oxytocin causes continued contractions of the uterine wall until the baby is finally pushed through the cervix and vagina. Continued uterine contractions result in shrinkage of the uterus. During these contractions, the placenta becomes detached from the uterus and is pushed through the vagina as the afterbirth. This is the final stage of delivery.

 What other effect will the release of oxytocin have on the mother?

Further prostaglandin release causes constriction of the blood vessels in the umbilical cord. This shuts off the baby's blood flow into the placenta and a new individual takes its first breath.

QUESTIONS

20.5 Look at Fig 20.15 which summarises hormonal interactions which can occur in a female human. From the list of hormones given, select the one hormone which is most important in each of the eight pathways, (1) to (8), shown.

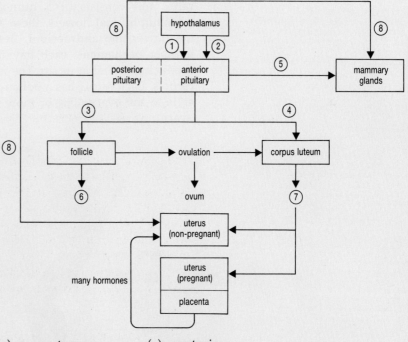

Fig 20.15 Hormonal interactions associated with reproduction in a female human.

(a) progesterones   (e) oxytocin
(b) oestrogens   (f) prolactin
(c) LH   (g) FSH releasing factor
(d) LH releasing factor (f) FSH

20.6 Summarise the major events of each phase of the menstrual cycle.

20.5 SEXUAL REPRODUCTION IN FLOWERING PLANTS

The flowering plants which surround us, whether grasses, border plants or trees, are diploid sporophyte adults. When cells in their reproductive tissue undergo meiosis, the resulting haploid cells are spores, not gametes (as they would be in our gonads). The spores of a flowering plant are produced within its flowers. One type of spore, a tough, drought-resistant **pollen grain**, is shed and develops into a male gametophyte only if it falls onto a suitable part of the flower of a member of the same species. A second type of spore, the **ovule**, is kept within the flower where it germinates to produce a female gametophyte. The dominant sporophyte *appears* to undergo sexual reproduction only because the female gameto-phyte is inside its flowers and never leads an independent existence.

Flower structure

Flowers develop singly or in groups, called an **inflorescence**. The swollen tip of the stem which supports the flower is called a **receptacle**.

Complete flowers

A complete flower consists of four sets (**whorls**) of modified leaves: calyx, corolla, androecium and gynaecium. These can be represented in a **floral diagram** such as the one shown in Fig 20.16. A second way in which floral structure may be represented is a **half-flower**; this represents the parts of a flower which would be seen if a flower were cut vertically through its middle and viewed from one side. Whilst a half-flower does not convey the same information as a floral diagram regarding the number of components in each whorl, it does show their structure in an easily recognisable way.

A half-flower of apple is shown in Fig 20.17. The outer **calyx** consists of **sepals**. These are often small, green and leaf-like, but in many flowers are indistinguishable from the petals. The sepals surround the flower bud, protecting the other whorls as they develop. The **corolla** lies within the calyx and consists of a number of **petals**. In most insect-pollinated (**entomophilous**) flowers, these are large and coloured. The male part of the flower, the **androecium**, lies just inside the petals. It consists of a number of **stamens**, each having a long **filament** supporting a pollen-producing **anther** at its tip. The female **gynaecium** consists of one or more **carpels**. Each carpel has a pollen-receiving **stigma** which is connected by a **style** to the **ovary**. One or more female gametophytes develop inside the ovary.

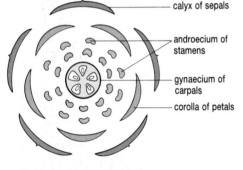

Fig 20.16 A floral diagram representing a horizontal section through an apple flower.

- calyx of sepals
- androecium of stamens
- gynaecium of carpals
- corolla of petals

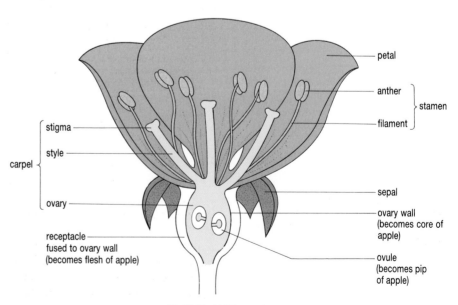

Fig 20.17 Half-flower of apple.

- petal
- anther
- stamen
- filament
- stigma
- style
- carpel
- ovary
- sepal
- ovary wall (becomes core of apple)
- receptacle fused to ovary wall (becomes flesh of apple)
- ovule (becomes pip of apple)

Drawing floral diagrams

This exercise involves accurately observing, interpreting and recording information in the form of a diagram. The photograph, Fig 20.18, shows a vertical section of an entomophilous flower. Use it to construct

(a) a labelled half-flower drawing for this species
(b) a half-flower drawing

for this species.

Fig 20.18

Incomplete flowers

Incomplete flowers lack one or more of the four floral parts described. Some plants produce male flowers which lack a gynaecium or female flowers which lack an androecium. Cucumber plants are monoecious (produce male and female flowers on the same plant) whilst willows are dioecious (produce either male or female flowers, but not both). Wind-pollinated (anemophilous) plants often lack sepals and petals. The grass flower, shown in Fig 20.19, is enclosed by two modified leaves, called **pales**.

 How is the grass flower adapted for wind pollination?

KEY CONCEPT

The pollen grain is the microspore and the embryo sac is the female gametophyte of flowering plants. They are produced within the stamens and carpels, respectively.

Pollen production

Each anther has either two or four cylindrical lobes, according to the species. Within each lobe is an outer layer which protects and nourishes the developing pollen, and an inner region of **pollen mother cells**, which produce the pollen. Each diploid pollen mother cell divides once by meiosis to produce four haploid microspores, the pollen grains. Each pollen grain divides once by mitosis to produce a small **generative cell** and a larger **tube cell**. You can see the nucleus of these cells in the pollen grain in Fig 20.20; also visible is the thick, resistant coat which each pollen grain secretes around itself.

SEXUAL REPRODUCTION IN MAMMALS AND IN FLOWERING PLANTS

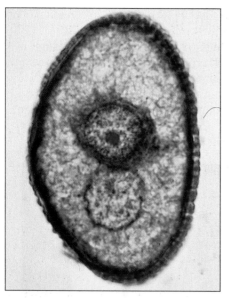

Fig 20.20 Micrograph of transverse section of pollen grain.

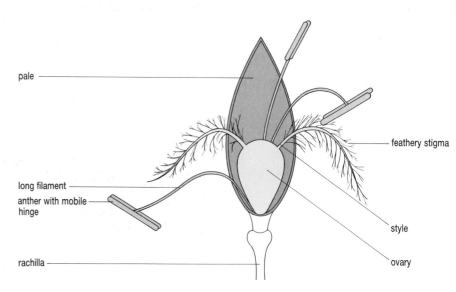

pale

feathery stigma

long filament

anther with mobile hinge

style

rachilla

ovary

Fig 20.19 Half-flower of a grass plant.

When ripe, the anthers split along their length and open out. The pollen of entomophilous flowers remains loosely attached to the inside walls of the lobes whilst the pollen of anemophilous flowers spills out into the air. It is the pollen released from wind-pollinated plants, especially grasses, which causes such misery to sufferers of hay fever.

SPOTLIGHT

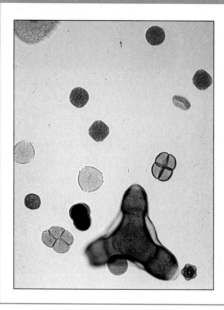

CLIMATIC CHANGE

The climate of the British Isles has undergone dramatic changes over the last 100 000 years. These changes can be followed using pollen analysis.

The tough outer coat of a pollen grain survives intact for many thousands of years when entombed in peat. The pattern of surface sculpturing on the pollen grain is species-specific. By taking a core of peat, cutting it into slices, ageing each slice (for example by carbon dating) and then extracting the pollen from each slice, we can work out which plants were growing when the peat was laid down. By correlating the plant species with the climates they grow under today we can work out what the climate was like in the past.

Fig 20.21 Light micrograph of pollen grains from a number of different plant species. Using the pattern of sculpturing on the outside of the pollen grains, botanists can recognise pollen which has been preserved for thousands of years in peat bogs. Pollen grains are different colours, although in this photograph they have been stained which, in this case, accounts for the variation in colour.

Embryo sac formation

Depending on the species, one or more **ovules** develops inside the ovary; for simplicity, only one ovule is shown in Fig 20.22. Each ovule has a mass of parenchyma tissue, the **nucellus**, which contains a single, large **megaspore mother cell**. Around the nucellus lie one or more layers of cells, the **integuments**. As these grow, they almost completely enclose the nucellus, leaving only a microscopic pore, the **micropyle**, at its tip. Production of the embryo sac is a little more complex than production of pollen and is represented in Fig 20.23. A megaspore mother cell divides by

(a)

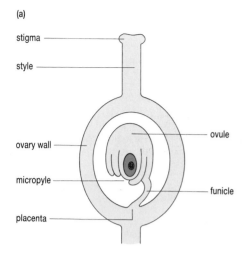

stigma

style

ovary wall

micropyle

placenta

ovule

funicle

(b)

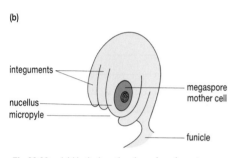

integuments

nucellus
micropyle

megaspore
mother cell

funicle

Fig 20.22 (a) Vertical section through an immature ovary, containing a single ovule. (b) An immature ovule.

meiosis to produce four haploid cells, the **megaspores**. Only one of these megaspores survives and grows until it almost fills the nucellus. The nucleus of this megaspore divides three times by mitosis, forming a cell with eight haploid nuclei. The cytoplasm then divides into seven, not eight, cells: three small uninucleate cells at one end, called the **antipodal cells**; one in the middle formed by the fusion of the two polar nuclei, and three at the other end made up of two **synergids** and the ovum. In some plants fusion of the two polar nuclei occurs very late. This seven-celled **embryo sac**, situated near the micropyle, is the female gametophyte.

Pollination

Pollination occurs when a ripe pollen grain lands on the stigma of a plant of the same species. **Self-pollination** occurs when pollen is transferred from anther to stigma of the same plant: **cross-pollination** occurs when pollen is transferred from the anther of one plant to the stigma of another plant. During cross-pollination, pollen may be carried from one flower to another by wind or by other animals, notably insects. Some of the ways in which flowers are adapted to one or other of these methods are summarised in Table 20.2.

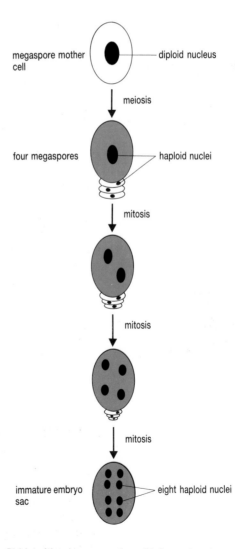

megaspore mother cell — diploid nucleus

meiosis

four megaspores — haploid nuclei

mitosis

mitosis

mitosis

immature embryo sac — eight haploid nuclei

Fig 20.23 Division of the megaspore mother cell to form an immature embryo sac.

Table 20.2   Comparison of wind-pollinated and insect-pollinated flowers

Feature	Wind-pollinated flowers (anemophilous flowers)	Insect-pollinated flowers (entomophilous flowers)
sepals	often absent	may be indistinguishable from petals
petals	small, inconspicuous	large and conspicuous through colours and patterns visible in UV light
		may be arranged in such a way that they provide a landing platform for insects, a lever mechanism to push the anthers onto an insect's body may mimic the female insect so that the male insect attempts to mate with the flower
odours	usually absent	odour attracts specific pollinators, e.g. smell of rotting meat attracts carrion feeders and musky smells attract night-flying moths
nectar	absent	nectar is secreted by nectaries often situated at the bottom of a tube of petals which ensure the pollinator must partly enter the flower
stamens	long filaments protrude from flower, anthers are often hinged	enclosed within flower
pollen grains	light and powdery, often with air bladders	no air bladders
		surface slightly sticky
carpels	stigmas protrude from flower and are often 'feathery' and sticky	stigma enclosed within flower
		stigma is flat, lobed and sticky

ANALYSIS

Ensuring cross-pollination

An important skill for you to develop, and one you can practise in this exercise, is the ability to relate structure to function.

Primrose populations are polymorphic for flower structure. Some plants produce pin-eyed flowers and others produce thrum-eyed flowers. The structure of these flowers is shown in Fig 20.24(a).

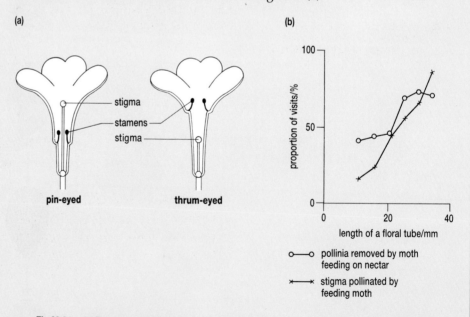

Fig 20.24   (a) Two types of primrose flower (b) Visits of moth on orchid (*Plantanthera bifolia*) (Pollinia are sticky sacs of pollen grains, typically found in orchids.)

(a) What name would be given to this type of diagram representing flower structure?

(b) Describe the differences between thrum-eyed and pin-eyed flowers.

(c) Bees insert their long proboscises into both types of flower to collect nectar from the base of the corolla. Explain how the flower polymorphism will ensure cross-pollination among primroses.

The flowers of many orchids have long, tube-like spurs at the base of which they secrete nectar. Flowers of the white-flowered lesser butterfly orchid *(Platanthera bifolia)* are pollinated by night-flying moths. These moths insert their proboscises into the spurs to obtain nectar. In doing so the tiny pollen-containing sacs (pollinia) of the flower may attach to their eyes.
(d) Suggest how pollination is achieved in *P. bifolia*.
(e) Account for the information shown in Fig 20.24(b).

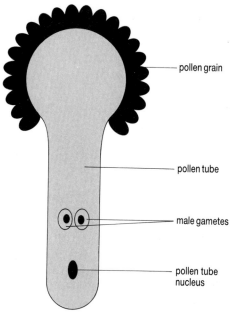

Fig 20.25 The tube cell nucleus moves down the pollen tube as it organises its growth. The generative cell, which divides to form the two male gametes, follows close behind.

Fertilisation

Once ripe pollen lands on a stigma of the same species of plant, it absorbs water, swells and splits open. A **pollen tube** is formed which digests its way through the tissues of the style and ovary wall as it grows towards the embryo sac. The tube nucleus is always found at the tip of the tube and organises its growth. As the generative cell follows the tube cell nucleus down the pollen tube, it divides by mitosis to produce the two male gametes which can be seen in Fig 20.25. This structure represents the male gametophyte stage of the flowering plant's life cycle.

11 **What type of cell division produces the male gametes?**

When it reaches the ovule, the pollen tube grows through the micropyle and nucellus (see Fig 20.26). As soon as it penetrates the embryo sac, the tip of the pollen tube and tube nucleus disintegrate, so that only the two male gametes enter the embryo sac. A **double fertilisation**, unique to flowering plants, now occurs. One of the male gametes fuses with the female gamete to form a diploid zygote. The other male gamete fuses with both polar nuclei to form a triploid **primary endosperm cell**. The other five cells of the embryo sac disintegrate.

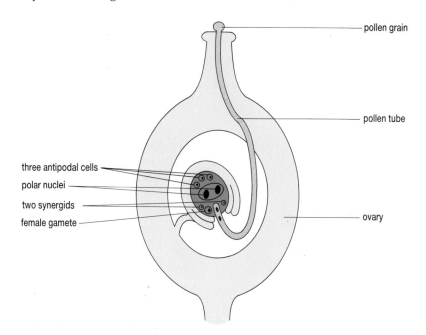

Fig 20.26 Following pollination, a pollen tube grows through the style to the embryo sac. Fertilisation occurs when a male gamete moves through this tube to fuse with the female gamete. A second male gamete fuses with both polar nuclei.

SEXUAL REPRODUCTION IN MAMMALS AND IN FLOWERING PLANTS

20.7 Name the four whorls within a complete flower, the components of each whorl and one function of each component.

20.8 Outline how a two-celled pollen grain is formed.

20.9 Describe the structure of an embryo sac and explain how it is formed.

20.10 How could you determine whether a flower was insect-pollinated or wind pollinated?

20.11 Explain the difference between pollination and fertilisation.

20.12 What is meant by the term double fertilisation in flowering plants? Suggest why the term double fertilisation is a misnomer.

20.6 DEVELOPMENT OF SEEDS AND FRUIT

Following fertilisation the embryo sac develops into a seed, the integuments into a seed coat and the ovary wall into a fruit (Fig 20.27).

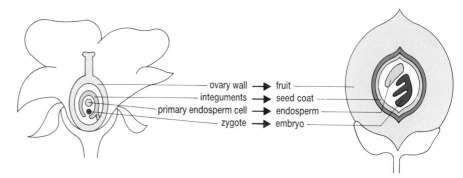

Fig 20.27 The flower of a fertilised ovule eventually forms a fruit. The ovary wall and integuments of the parent sporophyte form the fruit and seed coat, respectively. The female gametophyte produces the endosperm and the zygote forms the embryo.

Seed formation

Within the embryo sac, two distinct processes occur. The triploid primary endosperm cell divides by mitosis, absorbing nutrients from the parent plant to form a mass of food tissue, the **endosperm** (Fig 20.28(a)). After endosperm formation is well under way, the diploid zygote starts to divide by mitosis to form an **embryo**. A typical embryo has three parts: the embryonic shoot, or **plumule**; the embryonic root, or **radicle**; the seed leaves, or **cotyledons**. The number of cotyledons forms the basis for the two major divisions of the angiosperms: the **dicotyledons** and **monocotyledons** (see Chapter 26). In dicotyledonous plants, most of the food is usually transferred from the endosperm to both cotyledons before the seed is mature. In these **non-endospermic seeds** virtually no endosperm remains and the embryo virtually fills the seed (Fig 20.28(b)). In contrast, seeds of monocotyledons plants are endospermic.

 How does an endospermic seed differ from a non-endospermic one?

As these changes occur within the embryo sac, the integuments of the ovule become lignified to form a protective seed coat (**testa**). Only the **micropyle** perforates this seed coat. During the final stages of seed formation, the water content of the seed falls from about 85% to 10% of fresh weight and the seed becomes dormant. This dehydration is controlled by abscisic acid (Chapter 17).

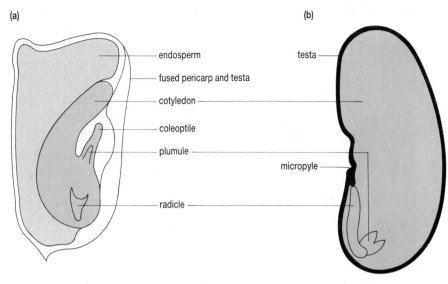

Fig 20.28 **(a)** An endospermic maize *(Zea mays)* seed **(b)** Broad bean *(Vicia faba)*, a non-endospermic seed.

 Why is dehydration essential to seed dormancy?

Fruit formation

Developing seeds produce hormones which stimulate the development of the ovary to form a fruit (Chapter 17). A fruit protects the seeds within it and aids their dispersal. In **dry fruits**, the fruit dries out at the same time as the seeds. Table 20.3 summarises some of the ways in which dry fruits aid dispersal of the seeds they contain. **Succulent fruits** develop a fleshy food store which is generally eaten by animals. Unaffected by the acid or enzymes in the animal's stomach (in fact, such treatment is sometimes necessary for germination to occur), their seeds are deposited with the animal's faeces some distance from the parent plant. In **true fruits** it is the ovary which swells with stored food; in **false fruits** some other part of the flower swells with food, e.g. the receptacle forms the flesh of an apple fruit. Under the influence of growth substances, notably ethene, succulent fruits change in colour and taste as they ripen (Chapter 17).

> **Ethene** was previously called ethylene: this hydrocarbon ($CH_2=CH_2$) is one of the fractions obtained from crude petroleum. As the polymer, polythene, it is used widely in many thermoplastic products.

Table 20.3   Fruits are adapted to ensure dispersal of the seeds they contain

Dispersal mechanism	Description of fruit	Example
wind	very light in weight; surface extensions catch the wind	dandelion, sycamore
explosion	two compartments which violently split open as they dry out	gorse, broom
pepper pot	holes through which seeds are released when fruit is shaken by the wind	poppy
animal dispersal	adhesive hairs, spines or hooks which adhere to feathers and fur	cleavers (a bur fruit)
	eaten but seed is deposited without damage in animal's faeces	blackberry, apples

CALVARIAS AND DODOS

Calvaria trees grow only on the island of Mauritius in the Indian Ocean. They produce a large fruit rather like a peach. Formerly one of the most common trees on Mauritius, only a few extremely old trees survive, in spite of the fact that these survivors produce perfectly normal fruit each year.

It is now believed that calvaria trees depended for seed dispersal and germination exclusively on the dodo. As you may know, dodos were flightless birds about the same size as turkeys. They fed mainly on fruits and seeds. Mariners in the sixteenth and seventeenth centuries found dodos an easily caught source of fresh meat after their long sea journeys and, by 1681, had hunted them to extinction. Without the physical and chemical treatment of passing through a dodo's digestive system, seeds of the calvaria tree failed to germinate, explaining their demise. Using this information, a few calvaria trees have been cultivated in nurseries after their seeds had been force-fed to domestic turkeys.

Fig 20.29 A calvaria tree. It grows only on the island of Mauritius and only a few very old trees remain. The succulent fruit it produces were probably eaten by dodos and passage through their digestion tract aided seed germination.

QUESTIONS

20.13 A fruit containing endospermic seeds contains the tissue of two spore-producing plants and one gamete-producing plant. Explain this statement.

20.14 Choose a number of fruits which you commonly eat, e.g. a bean with pod, a raspberry, a plum and a tomato. Identify the ovary wall, seed coat and embryo plant in each.

20.15 Compare and contrast the processes of gametogenesis and fertilisation in mammals and flowering plants.
Note, the term 'compare and contrast…' means that you need to discuss both the similarities and differences between two processes or structures. In an examination you would not gain marks for giving two separate accounts. You would gain marks for picking out individual differences and similarities, e.g. X has… whereas Y has….

SUMMARY

Humans have primary sex organs, which produce gametes, accessory sex organs, which nourish and transport the gametes, and secondary sexual characteristics, which affect sexual behaviour.

In human males, the primary sex organs are the testes. Under the control of hormones from the anterior pituitary, these produce sperm as well as the sex hormone testosterone. The male accessory sex organs transport sperm to the female's reproductive tract and release fluids, nutrients and factors which stimulate sperm to swim.

In human females, the primary sex organs are the ovaries. Under the control of hormones from the anterior pituitary, these produce secondary oocytes and the hormones oestrogen and progesterone. The accessory sex organs carry sperm to the secondary oocyte and receive and nourish an embryo during its early development. In human females, hormone production, release of the secondary oocyte (ovulation) and development of the lining of the uterus form a sequence of events called the menstrual cycle. This cycle is controlled by a complex interaction of hormones from the hypothalamus, anterior pituitary body and ovaries.

During copulation, the erect penis of the male deposits semen (sperm plus fluids) in the vagina of the female. Aided by the female's reproductive tract, the sperm swim into the Fallopian tube, where fertilisation usually takes place. Before fertilisation, the secondary oocyte is surrounded by two barriers which must be digested by enzymes released from the acrosomes of the sperm. Following digestion of these layers, only one sperm enters the secondary oocyte, stimulating its division to form an ovum. Fertilisation occurs when the pronuclei of the ovum and sperm fuse.

Following fertilisation, the zygote divides by mitosis to form a ball of cells called a blastocyst. Contractions of the Fallopian tube push the developing blastocyst into the uterus where it implants into the thickened uterine lining. Implantation and the sudden release of a hormone, chorionic gonadotrophin, by the embryo prevent further menstrual cycles.

During pregnancy, the fetus is nourished by the placenta which is formed from both fetal and maternal tissue. After about 38 weeks, prostaglandins released by the fetus initiate contraction of the uterine muscles, resulting in birth.

The reproductive parts of flowering plants are born in whorls of modified leaves in an inflorescence (flower). The calyx consists of sepals which protect the flower when in bud, the corolla consists of petals, the androecium consists of a number of pollen-producing stamens and the gynaecium consists of one or more 'female' carpels.

Pollen grains are produced on the stamens in pollen sacs called anthers. When ripe, these pollen sacs split open and release their pollen grains. Depending on the species of plant, these pollen grains may be carried to another flower by wind or by animals, notably insects. In insect-pollinated flowers, the androecium and gynaecium are usually contained within the flower, surrounded by large and brightly coloured petals. In wind-pollinated flowers, the androecium and feathery gynaecium often protrude from the flower, beyond the small and inconspicuous petals. If the pollen grains land on the stigma of a suitable carpel, they germinate to form a pollen tube containing two nuclei: a pollen-tube nucleus and a generative nucleus which will divide to form two male gametes.

The carpels contain ovules within which are formed embryo sacs. A germinating pollen grain produces a tube which digests its way through the carpel to the embryo sac where a double fertilisation occurs. One male gamete fuses with a haploid ovum to form a diploid zygote. The second male gamete fuses with two other nuclei (polar nuclei) within the embryo sac to form a triploid primary endosperm cell.

Following this double fertilisation, the primary endosperm cell divides to form a food store, the endosperm. Using this food store, the diploid zygote divides to form an embryo plant. Other changes around the embryo plant result in the formation of a seed surrounded by a fruit. Dispersal of the seed, containing the embryo plant, is aided by the fruit. Succulent fruits attract animals which eat them, depositing the unharmed seeds in their faeces. Dry fruits may stick to the fur of animals or help in wind dispersal of the seeds they contain.

Answers to Quick Questions: Chapter 20

1 To provide ATP needed for movement of the sperm's tail.
2 No. Assuming a menstrual cycle of 4 weeks, i.e. 13 cycles per year, ovulation during this period would lead to the loss of only $13 \times 39 = 507$ eggs.
3 (i) Reproductive cycles so egg cells and sperm are mature at the same time.
   (ii) Release of egg cell and sperm.
4 Stimulates the secretion of oestrogen.
5 (i) Breakdown of the uterine lining, i.e. menstruation.
   (ii) FSH secretion starts to increase.
6 Essential to prevent other follicles developing, becoming fertilised and implanted in the uterus so placing a huge physiological burden on the mother.
7 (a) To place sperm in the female genital tract.
   (b) Increases the probability of fertilisation.
8 Mother. Only the head of the sperm, which contains no mitochondria, penetrates the egg cell. (Some human features are throught to be controlled by mitochondrial DNA.)
9 Stimulates growth of lactiferous glands.
10 A list is given in Table 20.2.
11 Mitosis
12 In an endospermic seed, stored food remains as endosperm and is not absorbed by the cotyledons.
13 It prevents enzymic hydrolysis of stored food and enables the seed to survive a long period of dormancy.

Chapter 21

PATTERNS OF GROWTH AND DEVELOPMENT

LEARNING OBJECTIVES

When you have studied this chapter you should be able to:

1. define growth and review ways in which it may be measured;

2. distinguish between continuous and discontinuous growth and between primary and secondary growth;

3. define development and suggest how it may be controlled;

4. outline the growth and development of a human fetus;

5. interpret data relating to growth of humans to adulthood and decline during ageing;

6. outline the events which occur as a seed germinates;

7. describe primary and secondary growth of the stems and roots of dicotyledonous angiosperms.

A human being is a kind of miracle. It begins as a single cell but the adult body contains many millions and millions of cells of different kinds. So during its existence from a single cell to an adult human being both growth and development, the specialisation of cells to perform specific functions, have taken place. It is these twin processes of growth and development which are discussed in this chapter.

21.1 DEFINING GROWTH

As organisms grow they increase in size or, simply, they get bigger. Although we may use these terms in everyday speech, as biologists we should use the terms size and bigger with care since they do not accurately or completely describe growth.

Measuring growth

Biologists use the term size in a number of ways.

1. A linear dimension such as height or circumference. We measure changes in the circumference of a baby's head or in a child's height when we measure its growth. However, a child's party balloon increases in circumference and height when inflated with air but it has not grown. Birds that are able to inflate bags of skin around their necks do not grow when they inflate them.

2. Mass. Many people regularly check their size by measuring their own mass (incorrectly called weight). However, a bucket which is full of water has a greater mass than when it is empty but it does not grow when it is filled up. A pelican appears to increase in mass when it fills its beak with fish, but it has not grown.

Thus changes in height, circumference or mass may be poor indicators of growth. This makes it difficult to define growth. The best definition of biological growth must involve a genuine increase in the amount of cytoplasm within an organism. The amount of cytoplasm is usually best measured as mass. The most common substance within an organism is water. The mass of an organism together with the water it contains is the **wet mass**.

1 Why would a change in wet mass be a poor indication of growth?

A better indication of growth is an increase in the amount of substances other than water in the cytoplasm. This is the **dry mass** of an organism. The dry mass is found by heating dead organisms in an oven at a temperature which is high enough to evaporate all their water but not so high that it burns their carbohydrates, lipids and proteins; 100 °C is a convenient temperature. While the organisms are heated in this way, they are regularly weighed. Once their mass no longer changes it is assumed that all their water has been evaporated. In this way, the organisms have been dried to constant mass.

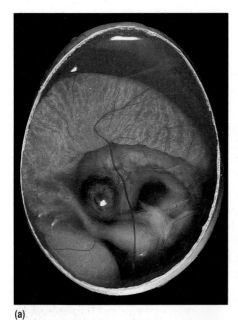

(a)

(b)

You will have realised that finding the dry mass of an organism kills it. You could not use dry mass to measure the growth of a single organism over a period of time, instead you would have to use a large number of organisms and measure the mean dry mass of samples taken from them. Even so, it is often undesirable or unethical to kill organisms during the investigation of their growth. It is in these circumstances that we should have to use one of the other, less reliable, indicators of growth, such as height, length or wet mass.

(c)

Fig 21.1 Growth and development of a chicken.
(a) a 10-day-old embryo inside the egg
(b) a chick 12 hours after hatching
(c) a chick one day after hatching
with the adult female.

Patterns of growth

Growth is affected by

- environmental factors, such as food availability, temperature and light intensity
- an organism's genotype.

This interaction between genotype and environment means that rates and patterns of growth will differ both between species and between individuals of the same species. You only have to look at your friends to see this.

Continuous and discontinuous growth

If increases in growth per unit time are plotted as a graph, we obtain a **growth curve**. Two contrasting growth curves are shown in Fig 21.2. In (a), even though the rate of growth changes, the organism does not stop growing between birth and reaching its maximum size; this is called **continuous growth**. In (b), the organism shows a number of periods of extremely rapid growth followed by periods in which very little, if any, growth occurred; this is called **discontinuous growth**. Most organisms undergo continuous growth; the human growth curves shown in Fig 21.11 are examples of continuous growth.

Discontinuous growth is characteristic of arthropods, which have exoskeletons. The body of an arthropod can grow until it is pressed tightly against its own hard, tanned exoskeleton. No further growth is possible unless the animal sheds its exoskeleton (**ecdysis**). What normally happens at ecdysis is that an arthropod loses its exoskeleton, expands its body, e.g. by taking in air, and forms a new exoskeleton. After the exoskeleton has hardened, the animal's body reduces to its normal size, leaving a little room within the new skeleton into which it can grow. Formation and hardening of the exoskeleton take from hours to several days, depending on the class of arthropods. The hormonal control of ecdysis and the production of new exoskeletons is described in Section 14.4.

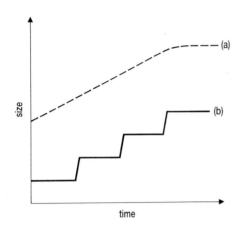

Fig 21.2 Growth curves showing (a) continuous growth and (b) discontinuous growth.

Fig 21.3 An insect undergoing ecdysis.

An arthropod's growth could be followed by measuring the length of a part of its exoskeleton. With time it would be recorded that the arthropod had rapidly increased in length during the spurts on the graph and that no further increase in length occurred until the next moult. If wet mass of the animal were to be measured, the story would be different. The arthropod's mass would actually decrease as it moulted, stay the same or slightly decrease as the new exoskeleton was formed from existing food reserves and then increase between moults as the animal grew to fill the space inside its new skeleton.

Has length or wet mass been used as a measure of growth in Fig 21.2? (Hint – think about the nature of the animal represented in curve (b)).

Primary and secondary growth

Organisms grow as new cells are produced by mitosis and then enlarge. In some animal and plant phyla, all the cells within an organism are capable of dividing: these organisms can usually reproduce asexually by fragmentation (Section 19.2).

In the so-called higher animal and plant phyla, many cells are so specialised that they have lost the ability to divide. In flowering plants, tissue which is capable of dividing by mitosis is called **meristematic tissue**. It is found in only a few places, such as the **apical meristem** at the tips of the shoot, root and buds, and in the **cambium** between the xylem and phloem of the vascular bundles.

Primary growth is an increase in length of the stem or root following division of cells in the apical meristems. The cells so formed increase in length and differentiate to form primary tissue, e.g. primary xylem and phloem.

Secondary growth is an increase in the diameter of the stem or root following division of cells in the vascular cambium. The cells so formed increase in length and differentiate to form secondary tissue, e.g. secondary xylem and phloem. Primary and secondary growth are described more fully in Section 21.5.

QUESTIONS

21.1 Define biological growth and explain why it may be difficult to measure experimentally.

21.2 (a) Distinguish between continuous and discontinuous growth.
   (b) Arthropods have a rigid 'shell' which results in discontinuous growth. Most molluscs also have a rigid shell yet, under suitable conditions, their growth is continuous.
      (i) Suggest what these suitable conditions might be.
      (ii) Suggest an explanation to account for the ability of these molluscs to grow continuously.

21.3 What is meant by secondary growth?

21.2 DEVELOPMENTAL STAGES

The single-celled zygote of an animal or plant eventually produces an adult consisting of millions of cells. This process is not simply one of cell division or the adult would be a mass of identical cells. As the cells divide, they **differentiate**, i.e. they specialise to carry out particular functions. Thus, **development** involves an increase in complexity as well as an increase in size. What causes cells to differentiate during development?

ANALYSIS

Differentiation

Stating hypotheses accurately and using data to test hypotheses are important skills for a biologist to learn. You can practise these skills in this exercise.

A zygote contains copies of all the genes which are needed to form a complete adult organism. During its development, the zygote divides by mitosis. Chapter 4 describes how mitosis produces genetically identical daughter cells. Therefore, all the cells within an animal or

---

plant should contain copies of every gene needed to form a complete adult. Look at Fig 21.4 which summarises an experiment using the African clawed toad, *Xenopus laevis*, to investigate whether this is true. *Xenopus* was used in the experiment because amphibia produce large egg cells which are easier to manipulate after fertilisation than the smaller ones of mammals.

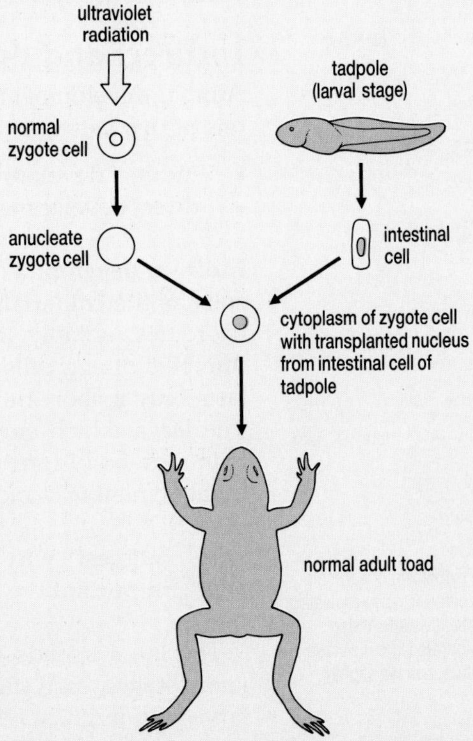

Fig 21.4 In a classic experiment, nuclei from intestinal cells of a tadpole were transplanted into zygotes whose nuclei had been destroyed by ultraviolet radiation. Surviving zygotes developed into normal toads, showing the intestinal cells contained all the genes needed for development of an entire individual.

(a) State, concisely, the hypothesis being tested in this experiment.
(b) How and why was the original nucleus of the zygote destroyed?
(c) Why was only the nucleus and not the cytoplasm of the intestinal cell transferred to the anucleated zygote?
(d) What can you conclude from this experiment with respect to your original hypothesis?
(e) Would you be justified in extending your conclusions from this experiment to another species, say a frog? Justify your answer.

Gene regulation during development

Although genes may be present in a cell, they do not exert an effect unless they are transcribed. When a gene is transcribed, its base sequence is copied onto a strand of messenger RNA, which moves to the ribosomes in the cytoplasm. Chapter 4 describes how the transcription of genes may be regulated by substances within the cytoplasm. These substances bind with the chromosomes, blocking the transcription of some genes and promoting the transcription of others.

3 ▶ **List three such gene-regulating substances.**

Female gametes are nearly always bigger than male gametes since their cytoplasm contains stored food for the early development of the zygote. Also present in the cytoplasm of female gametes are **gene-regulating substances**. Fig 21.5 demonstrates how these gene-regulating substances

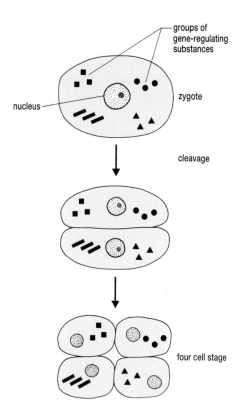

groups of gene-regulating substances

nucleus

zygote

cleavage

four cell stage

Fig 21.5 Different parts of the cytoplasm of a female gamete may contain different gene-regulating molecules. After mitotic divisions, these molecules may be separated into individual cells. Each cell will, thus, behave slightly differently.

might become confined to separate cells during early division of the zygote. As a result, these cells will develop differently from others not containing the gene-regulating substance, for example they are likely to produce different molecules. The presence of such molecules on their surface or secreted into their immediate surroundings is thought to influence the transcription of DNA in their neighbouring cells. Similarly in adults, newly-formed cells are thought to differentiate under the influence of the cells surrounding them.

Indirect and direct development

Animal development from an egg to a sexually mature adult may exhibit one of two patterns

- indirect development
- direct development.

Indirect development occurs in most of the invertebrate animal phyla, such as the cnidarians, molluscs, arthropods and echinoderms, and a few chordates, notably the amphibia. Typically, these animals produce vast numbers of egg cells with a small amount of stored food (**yolk**). The yolk lasts only a short time, during which each embryo develops into a **larva**. The larva which emerges from these egg cells is quite unlike the adult animal in body form or in lifestyle. In sessile animals, the larva is capable of movement or is moved by water currents, ensuring dispersal of young.

 Suggest why this dispersal function of these larvae is so important.

The larva spends much of its time feeding. There is often a series of larval stages, each stage feeding on a different type of food. Eventually, the larvae undergo a revolution in body form, called **metamorphosis**, and become sexually mature adults. In the life cycle of an amphibian such as the African clawed toad, *Xenopus laevis*, the larval form lasts only a few weeks, the adult is the main stage of the life cycle (Fig 21.4). In contrast, mayflies spend most of their lives as aquatic larvae: the adults live only the few days needed to reproduce; they do not even feed. An insect life cycle is shown in Fig 27.19.

 In addition to reproduction what is the other function of the adult in the mayfly life cycle?

Fig 21.6 Zooplankton contains millions of larvae of sessile marine organisms.

PATTERNS OF GROWTH AND DEVELOPMENT

Direct development is typical of most vertebrates. The life cycle of these animals lacks a larval form; the animal which hatches from the egg or is born live from its mother is a miniature, sexually immature version of the adult. Apart from the growth of sex organs and secondary sexual characteristics (Section 20.1), these young do not alter their body form or pattern of life as they mature. The growth of an embryo to produce a miniature version of the adult needs a considerable amount of food. Animals have evolved two ways of dealing with this food demand. In some, notably reptiles and birds, the females produce large egg cells containing large amounts of yolk.

Other animals, notably placental mammals, have relatively little or no yolk in their egg cells. Instead, the females keep the developing embryo within their bodies. In both cases, the supply of food for the developing embryo makes large demands on the mothers, which produce fewer egg cells.

The largest birds' eggs are those of the ostrich. An ostrich egg has a mass of 1.0 to 1.5 kg; eggs of the extinct *Aepyornis*, or elephant bird, of Madagascar had a mass of up to 11.5 kg!

QUESTIONS

21.4 Explain the difference between growth and development.

21.5 After fertilisation, a frog's egg cell forms a pigmented area of cytoplasm called the grey crescent. In the laboratory, fertilised egg cells can be made to divide so that, although division of the nucleus and cytoplasm proceeds normally, one daughter cell receives all the grey crescent and the other daughter cell receives none. If separated, the daughter cell with the grey crescent develops into a normal tadpole whilst the other daughter cell forms an undifferentiated mass of cells which soon dies. Suggest an explanation for these observations.

21.6 Plants have two adults in their life cycle. The asexual spores of the diploid adult sporophyte develop into a haploid gametophyte which bears little resemblance to the sporophyte. Sexual reproduction by the gametophyte produces a zygote which develops into a sporophyte. Does this life cycle demonstrate complete development or incomplete development? Justify your answer.

21.7 Discuss the advantages and disadvantages of indirect and direct development.

21.3 GROWTH AND DEVELOPMENT IN HUMANS

Compared with other animals, humans have a very long period of growth and development. This involves both an increase in cell number (**growth**) and cell differentiation, i.e. an increase in the number of different types of cell (**development**).

Fetal growth and development

Section 20.4 describes development of a human zygote during the seven days or so after fertilisation. After about four days a hollow ball of cells, called the **blastocyst**, has formed. After seven days this blastocyst has embedded itself (**implanted**) in the endometrium of the mother's uterus: Fig 21.7 shows a section through such a blastocyst. Already at this stage, cells have differentiated. The inner cell mass contains cells which will form the embryo and its membrane-bound cavities. The outer layer of cells, or trophoblast, forms the **chorion**, which releases enzymes that digest the endometrium and prolong the life of the corpus luteum. Even within the chorion, differentiation has occurred; implantation always occurs on the side containing the inner cell mass.

The inner cell mass contains two cavities, the **amniotic cavity** and **yolk sac**. Separating these cavities are two layers of cells (the **embryonic disc**) which will form the embryo.

- Nearest the amniotic cavity is **ectoderm**; as well as the epidermis, glands and hair of the skin, this layer of cells will form the inner ear, the lens of the eye and the entire nervous system of the embryo.
- Nearest the yolk sac is **endoderm**; these cells will eventually form the linings of the digestive and respiratory tracts, the liver and the pancreas.

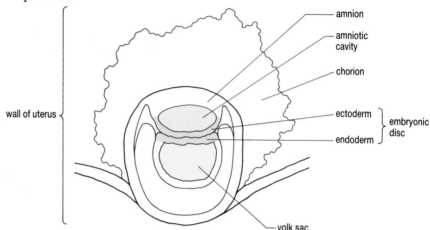

Fig 21.7 Section through a human blastocyst which is implanted in the mother's uterus 7 days after fertilisation. The embryonic disc, here made of two layers, will develop into the embryo. The fluid filled amniotic cavity has formed.

Shortly after implantation, the two layers of the embryonic disc split apart and a depression, the **primitive streak**, develops in the ectoderm. Ectodermal cells along the primitive streak migrate into the newly formed space between endoderm and ectoderm. These cells form the **mesoderm** (Fig 21.9(a)) which will eventually form most of the internal systems, such as musculo-skeletal, circulatory, excretory and reproductive systems, as well as the dermis of the skin. Initially, some of these mesodermal cells form a hollow tube, the **notochord**, which is a distinctive feature of all chordate animals. The formation of the notochord stimulates the ectoderm to form the **neural tube**, the forerunner of the spinal cord and brain.

BIRTH DEFECTS

Development is a complex process during which functional or structural abnormalities may arise. In the early 1960s a medical centre in the United States studied 6000 children from the fourth month of fetal life through the first year after birth and recorded all malformations. Of live births (by far the majority of abnormal embryos never reach full term), 4% had severe disorders such as heart defects, mental defects and imperfect closure of the neural tube resulting in a neural canal open to the exterior of the body, the condition known as **spina bifida**.

Birth defects generally result from three conditions

Fig 21.8 The carbon monoxide inhaled during smoking reduces the growth rate of this woman's fetus. The inhaled nicotine may cause it to develop abnormally.

- defective genes, e.g. cystic fibrosis
- abnormal arrangements of chromosomes, e.g. Down's syndrome
- **teratogens**, environmental agents which act either directly or indirectly on the embryo and produce malformation. Teratogens are responsible for 80% of the birth (congenital) defects in children. Teratogens include drugs, e.g. thalidomide, alcohol and nicotine, diseases, e.g. German measles (rubella) and X-rays.

During the third and fourth week, the embryo curls ventrally (Fig 21.9 (b)). As it does so, part of the empty yolk sac becomes pinched off to form the gut. Most of the major tissues are now present in the four-week-old embryo; the rest of embryonic existence involves the growth and further development of these tissues.

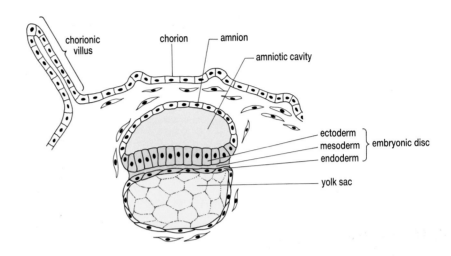

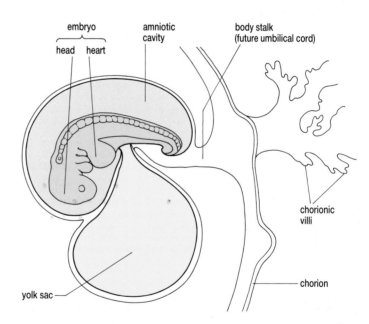

Fig 21.9 The later stages of human embryonic development:
(a) mesoderm has developed; later this forms the spinal cord and brain.
(b) 3–4 weeks after fertilisation the embryo is visible lying inside the amniotic cavity. The chorion is joined with the endometrium (lining of the uterus).

(b)

The placenta

When the four-week-old embryo curls ventrally, the amnion also curls until it almost completely encloses the embryo (see Fig 21.10). The **amniotic fluid** secreted by this membrane bathes the developing embryo, protecting it from mechanical damage. This fluid is only lost when the amnion ruptures just before birth ('waters breaking'). The part of the embryo which is not enclosed within the amnion is the **umbilical cord** by which the embryo is attached to the placenta. The **placenta** is formed from the tissue of the mother's endometrium and from the tissues of the fetal chorion. Two **umbilical arteries** carry fetal blood into capillary networks in the chorionic villi. These villi are surrounded in the endometrium by the membrane-bound lakes of maternal blood which are shown in Fig 21.10. Rapid exchange of metabolites occurs along concentration gradients between the two blood systems. This exchange includes the diffusion of respiratory gases and waste products, such as urea, and the active transport of nutrients. Unless the fetal and maternal membranes are damaged, as often happens during delivery, exchange of cells and maternal antibodies does not occur.

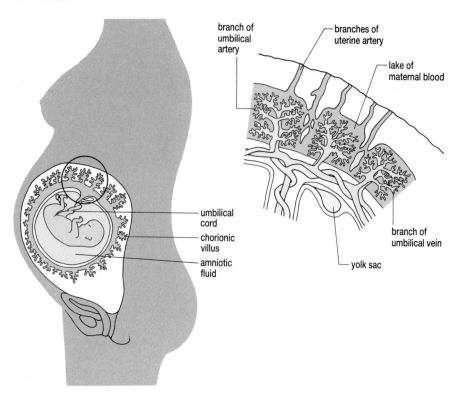

Fig 21.10 A 12-week fetus within the uterus of its mother is bathed by amniotic fluid. The structure of the placenta, which nourishes the fetus and holds it in place, is shown in greater detail.

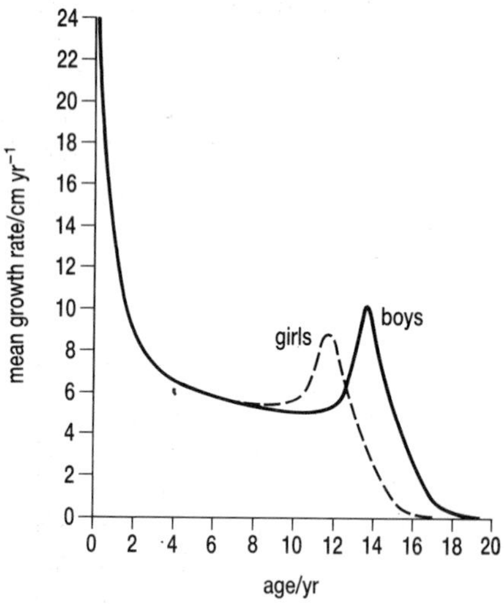

Fig 21.11 The human growth rate changes throughout life and is different in adolescent males and females. The spurt of growth for girls starts and finishes at an earlier age than for boys.

Growth after birth

The rate of increase of human height from birth to adulthood is not constant throughout life (see Fig 21.11) and is different for females and males. The rate of growth is highest immediately after birth and steadily declines until late childhood. At **puberty**, the time at which sexual maturity is reached, the growth rate of both sexes increases and then declines to zero. In females, the growth spurt starts and ends earlier than in males, who subsequently reach a greater adult height than females. An individual's rate of growth is affected by its genotype, its diet and the concentration of growth hormones in its blood. The growth spurts are the result of the sex hormones oestrogen and testosterone which are released at puberty. The effect of these hormones on growth and sexual development is described in Chapters 14 and 20.

GROWTH HORMONE

Growth hormone or somatrophin, produced by the anterior pituitary, is essential for growth. Its principal action is on the skeleton and skeletal muscles, acting to increase their growth and maintain their size once growth is achieved. Growth hormone causes cells to grow and multiply by directly increasing the rate at which amino acids

PATTERNS OF GROWTH AND DEVELOPMENT

enter cells and are built up into proteins. Unfortunately some children suffer from a lack of growth hormone and their growth is therefore stunted.

This condition can be treated by injecting growth hormone, but the problem is that this is both difficult and expensive to produce in large quantities. Originally the growth hormone was extracted from pituitary glands removed from dead people. Genetic engineers have now been able to place the human gene which encodes growth hormone into bacteria which will, hopefully, ensure more plentiful supplies in the future.

Different parts of the human body grow at different rates (see Fig 21.12). The curve for the growth of the body correlates closely with the curve of growth rate in Fig 21.11. This is not surprising since they are virtually measuring the same thing. The head reaches its adult size well before adulthood, making the head of a fetus and young child seem out of proportion with its body. This is further demonstrated in Fig 21.13, which also shows changes in the relative size of the genitals, limbs and body musculature throughout life.

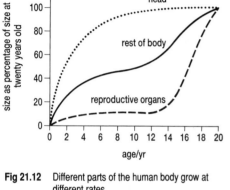

Fig 21.12   Different parts of the human body grow at different rates.

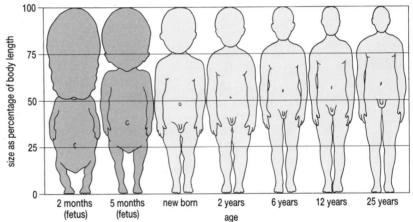

Fig 21.13   The same information in Fig 21.12 in a more easily visualised form.

> **6**   Using Figs 21.12 and 21.13, state which parts of the body grow fastest: (a) between childhood and puberty; (b) after puberty.

Human development reaches its peak at about the same time as growth stops. During childhood, skills involving coordination and dexterity improve. At the same time, a child learns a body of knowledge and its ability to manipulate and communicate that knowledge improves. After a certain age the body begins to deteriorate in a process called **ageing**. A number of physiological activities decrease with age (see Fig 21.14). Also, during ageing

- the intervertebral discs wear and people lose height
- calcium is lost from the bones, which become brittle
- muscles decrease in size and lose their tone, causing a body stoop
- skin loses its elasticity and becomes wrinkled
- hair becomes white and may be lost from certain parts of the body

- changes in the lens of the eye and its supporting ligaments result in long-sight
- hearing is lost at the extreme pitches of sound
- deterioration of the nervous system results in loss of memory (especially short-term memory) and an increase in reaction time
- reduction of sex hormone secretion and loss of libido occurs
- menopause occurs in women; reduction of sperm production occurs in men.

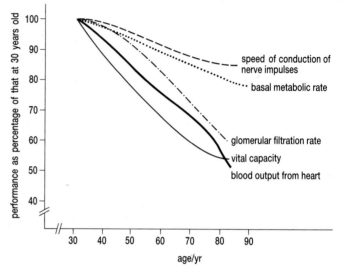

Fig 21.14 Despite a healthy life-style, human bodily functions inevitably decline with age.

QUESTIONS

21.8 **(a)** Explain why it is clear that the cells in a human blastocyst have already differentiated.

(b) Human identical twins arise when a single embryo divides into two during its development. After the formation of the primitive streak, identical twin formation is not possible. Suggest why.

21.9 Which of the human systems is the first to develop in a fetus? Suggest an advantage of this.

21.10 List the functions of the placenta. Suggest why a fetus is seldom affected if its mother suffers a bacterial infection but may be affected if she suffers a viral infection.

21.4 GERMINATION OF PLANT SEEDS

Seeds contain an embryo plant, consisting of a root (**radicle**), shoot (**plumule**) and one or two seed leaves (**cotyledons**), together with a food store. The formation and structure of seeds is described more fully in Section 20.6. During germination, the food store is used up to enable the radicle to grow into the soil, where it absorbs water and inorganic ions, and the plumule to grow towards light and begin photosynthesis. Germination is over once the miniature plant (**seedling**) is able to photosynthesise.

Since their food store is hydrolysed and its carbohydrates used in aerobic respiration, all seeds require water, oxygen and a suitable temperature to germinate. However, most newly mature seeds will not germinate immediately even given these conditions. Instead, they remain alive but dormant for long after their fruits have been removed. One of the mechanisms controlling seed dormancy is described in Section 17.2.

Fig 21.15 Seeds need to be moved to a suitable place for germination.

Dormancy is broken in a number of different ways, depending on the seeds. Among these are

- disruption of the seed coat by passage through the gut of an animal
- exposure to cold (in temperate plants)
- removal of substances from the seed coat by rainfall in desert plants
- exposure to light.

Water uptake

Germination begins when the seed takes up water, often doubling its mass in a few hours. Water enters the seed as it is attracted by the hydrophilic groups of molecules within the seed, notably membrane-bound proteins. Since the seed coat is impermeable, water enters the seed through its micropyle. As a result of water entry, **cell elongation** occurs within the seed and enzymes become functional.

Mobilisation of food stores

Seeds of different plants have food stores of a different composition. All seeds contain some carbohydrates, lipids and proteins, however. During germination, these are hydrolysed in much the same way as they are hydrolysed in the guts of animals (Chapter 8). The food stores are hydrolysed by enzymes then transported and utilised as shown in Fig 21.17.

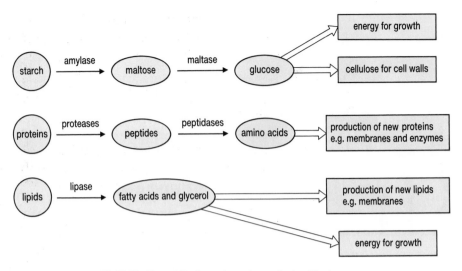

Fig 21.17 The mobilisation and use of a seed's stored food reserves.

Germination

This exercise develops two commonly assessed examination skills: the ability to translate information from one form to another and the ability to interpret data.

Table 21.1 Changes in the dry mass of the endosperm and the embryo of a germinating maize seed.

Time / days	Dry mass / mg g^{-1}	
	endosperm	embryo
0	200	2
1	189	3
2	188	5
3	155	9
4	115	15
5	84	23

Much of the food store is used by the embryo in building up new tissues and some is used in respiration. Table 21.1 gives some data on the changes occurring in a germinating maize seed.

(a) Why was the dry mass and not the wet mass of the endosperm and embryo used?

(b) Plot the data on a single set of axes.

(c) Describe the changes which occurred in the dry mass of the endosperm and embryo. Make your answer quantitative, e.g. the mass of endosperm decreased by $x\%$ in y days.

(d) What happened to the overall dry mass of the seed?

(e) Account for the changes you have described.

In another experiment two barley seeds, one of which had been boiled, were cut in half and placed in wells cut in the surface of agar jelly which contained 1% starch (Fig 21.18). After two days the plate was flooded with iodine solution which was then washed off.

(f) With the aid of a sketch, describe the results you would expect from this experiment.

(g) Explain the results you have depicted.

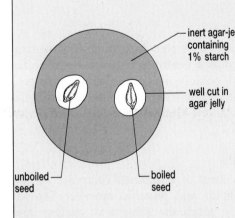

inert agar-jelly containing 1% starch

well cut in agar jelly

unboiled seed

boiled seed

Growth of radicle

Uptake of water causes the cells within the radicle to elongate. As it does so it pushes its way through the micropyle. Roots are positively geotropic (Section 17.2) so the radicle grows downwards into the soil and begins to absorb water and inorganic ions. The tip of the root, containing its apical meristem, is protected from damage as it grows through the soil by a collection of loose cells called the **root cap**.

Growth of plumule

Elongation of cells in the plumule causes it to grow out of the soil. Seeds differ in the way in which the plumule grows. One difference is found between the seed of monocotyledons and dicotyledons. In monocotyledons, the plumule is enclosed within a sheath, the **coleoptile**. After growth of the radicle into the soil, this sheath grows and emerges above the ground; only then does the plumule increase in length, growing through the protective coleoptile. Dicotyledons lack a protective coleoptile; their plumules often protect their delicate tips by curling into a hook shape as they grow through the soil (Fig 21.19).

A second difference relates to whether or not the cotyledon(s) appear above ground or remain below ground which, in turn, depends on whether the part of the embryo just below the cotyledon (the **hypocotyl**) or the part of the embryo just above the cotyledon (the **epicotyl**) elongates (Fig 21.19). In the broad bean, the epicotyl elongates and the cotyledon(s) remain below ground (Fig 21.19(a)). This is called **hypogeal germination**. In the French bean and in sunflowers (both dicotyledons), the hypocotyl elongates and the cotyledons emerge from the soil, enclosing and protecting the plumule (Fig 21.19(b)). This is called **epigeal germination**.

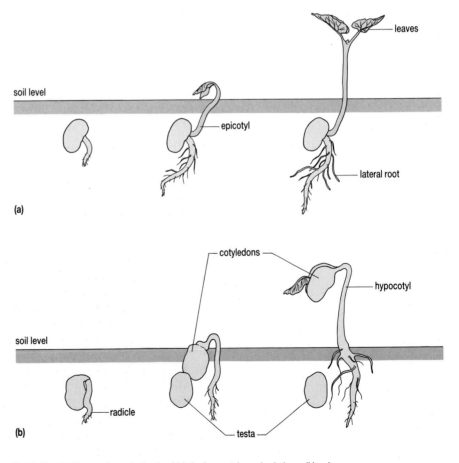

Fig 21.19   (a) Hypogeal germination in which the hypocotyl remains below soil level.
(b) Epigeal germination in which the hypocotyl emerges above the soil.

Writing continuous prose

The ability to draw together information from many different areas of biology to write an essay is an important skill. This exercise will give you an opportunity to practise this skill. You should first analyse the question to identify the key issues. For example, which areas do you need to bring together to answer the questions below? Next draw up a plan, putting down all the points you think you might need to cover and then selecting and organising them into a logical pattern. The marks allotted to each question give you a clue as to how many points you need to include. Finally, write your essay. The more you practise this essential skill, the better you will become.

(a) Name the principal food reserves of seeds. State their location within the seed and their form and location within the cells.

(6 marks)

(b) Describe how the food reserves in seeds are mobilised and the factors which initiate and control this process.

(6 marks)

(c) Given that a seed is planted at a depth of two centimetres in the soil and has started to germinate, describe the growth of the seedling and its responses to the environment up to the time it commences photosynthesis.

(8 marks)

21.11 List the stages which occur during the germination of a seed.

21.12 Name two ways in which the germination of a maize seed differs from the germination of a sunflower seed.

21.5 PRIMARY AND SECONDARY GROWTH IN PLANTS

Since they contain a large number of different types of cells, growth in plants must involve both cell division and cell differentiation. It is convenient to divide plant development into two stages: primary and secondary growth.

Primary growth

Two processes result in an increase in the length of a root and shoot

- production of new cells by mitosis
- elongation of these cells as they differentiate.

The hormonal control of cell differentiation is described in Chapter 17. Few cells in a mature plant retain the ability to divide by mitosis. Those that do are called **meristematic tissue**.

Primary growth in stems

Stems develop from dividing cells in the stem tip, called the **apical meristem** (Fig 21.20). Within this apical meristem are two regions, an outer **tunica** and an inner **corpus**. Division of cells in the tunica causes growth of the epidermis around the stem. Division of cells in the corpus causes growth of the internal tissues enclosed by the epidermis. Both tunica and corpus are involved in the production of regularly spaced swellings at the stem tip, called **leaf primordia**. These grow to form leaves so rapidly that they form an **apical bud** around the stem tip. Lower down the stem, these primordia produce leaves. Enclosed within the **axil** of each leaf (the angle between the leaf and the stem) is a small **axillary bud primordium** (Fig 21.20). Release of auxin by the apical meristem inhibits cell division in the axillary bud primordia closest to it (Chapter 17). Growth of the more distant axillary bud primordia produces lateral stems in much the same way as growth of the main stem.

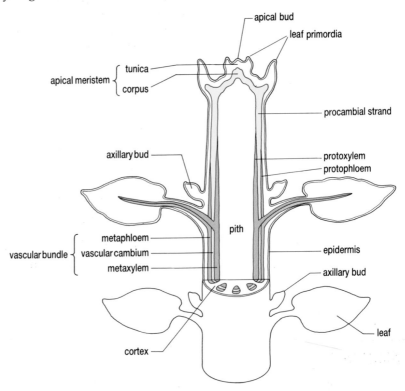

Fig 21.20 Cell division in the apical meristem results in growth of stems. Most plant tissues develop directly; xylem and phloem develop from procambial strands.

 What would happen to the axillary bud primordia if the apical bud were removed?

As new cells are produced by the tunica and corpus they elongate. Most cells formed by the corpus also vacuolate. These eventually form the tissues **parenchyma**, **collenchyma** and **sclerenchyma** which make up the bulk of the primary stem (the outer **cortex** and inner **pith**). A few elongate but do not vacuolate. These form **procambial strands**, i.e. tightly packed bundles of cells which retain the ability to divide. When the cells in these procambial strands differentiate they form vascular tissue; those on the outside of each bundle produce **protophloem** and those on the inside of each bundle produce **protoxylem**. Later still, when the stem has stopped elongating, further differentiation of cells in the procambial strand produces **metaphloem** and **metaxylem**. Cells of the latter have walls with complex patterns of lignification which prevent further elongation.

 What is the main function of this lignification?

The procambial strands of some plants change entirely into metaphloem and metaxylem. These are called **closed vascular bundles** and lack the potential for future growth. In many dicotyledons, remnants of the original procambial tissue remain between the metaxylem and metaphloem as **vascular cambium** (also called intrafascicular cambium). These are called **open vascular bundles** and retain the ability for secondary growth. Primary growth and differentiation in stems is summarised in Fig 21.20. The structure and function of plant tissues is described in Chapters 12 and 18.

487

Primary growth in roots

Like the stem, cell division in the root occurs at an apical meristem. Division of the root meristem produces three to four regions of cells. Two separate nomenclature systems are used to refer to the tissues in these regions. The alternative names for these tissues (called **histones**), and the outcome of their development, are summarised in Table 21.2.

Table 21.2 Two systems exist for naming tissues in the root meristem

Name of tissue	Alternative name of tissue	Region of plant formed by division of this tissue
calyptrogen	calyptrogen	root cap
dermatogen	protoderm	epidermis (and root cap in some plants)
periblem	ground meristem	cortex and endodermis
plerome	procambial strand	phloem, xylem and pericycle

The pattern of differentiation within the root is shown in Fig 21.21. **Protoderm** (dermatogen) produces the outer epidermis. Unlike the epidermis of the stem, that near the root tip has no waterproof cuticle so is able to absorb (or lose) water. Epidermal cells in this region also produce elongated root hairs (Chapter 12). Cells of the **ground meristem** (periblem) differentiate to form parenchyma of the cortex. These normally store food which is manufactured by the leaves during photosynthesis. The innermost layer of the cortex differentiates to form the **endodermis**. These cells have an impermeable strip of suberin, called the **Casparian strip**, across their radial walls which helps control the passage of water across the root (Chapter 12).

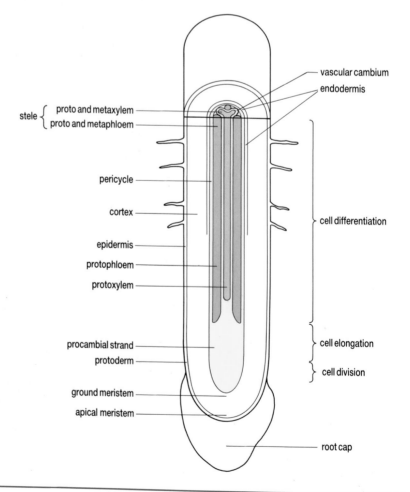

Fig 21.21 Cell division in the apical meristem results in growth of roots. Several groups of tissues in the root apex (histones) give rise to the mature tissues of the root.

PATTERNS OF GROWTH AND DEVELOPMENT

The **procambial strand** (plerome) differentiates to form the central vascular bundle (**stele**) of the root. Strips of protophloem are first formed around the outside of the procambial strand followed by alternating strips of protoxylem. Metaxylem later develops in the centre of the vascular bundle, so that the mature vascular bundle looks like a star of metaxylem with phloem between its arms (Fig 21.21). Some cambial tissue remains between the xylem and phloem (vascular cambium) which allows secondary growth. Some cambial tissue also remains in the very outer layer of the procambial strand, forming the **pericycle**. Division of the pericycle may later produce lateral roots. (Notice how lateral roots are produced from within the root whilst lateral stems arise from the stem's periphery.)

Secondary growth

Most conifers and many dicotyledonous angiosperms last for several years, i.e. are **perennial**. Since these plants grow in length each year, their mass increases beyond the limits that can be supported by their primary structure alone. Such plants overcome this problem by increasing their girth as their length increases. This secondary growth, producing secondary tissues, is often called **secondary thickening**. Most of the primary tissues in a plant have become so specialised that they have lost the ability to divide. Only a few cells, located in **lateral meristems**, retain the ability to divide and produce new tissues during secondary thickening.

 Why do few annual plants need secondary thickening?

Secondary growth in stems

The lateral meristems of a stem are the **vascular cambium** between the primary xylem and primary phloem of open vascular bundles (see above). As secondary thickening starts, parenchyma cells between the vascular cambium become meristematic, so that there is a complete ring of cambium (sometimes called interfascicular cambium) around the stem (Fig 21.22(a)). The division of these cells produces more cambium cells and cells of the secondary tissues.

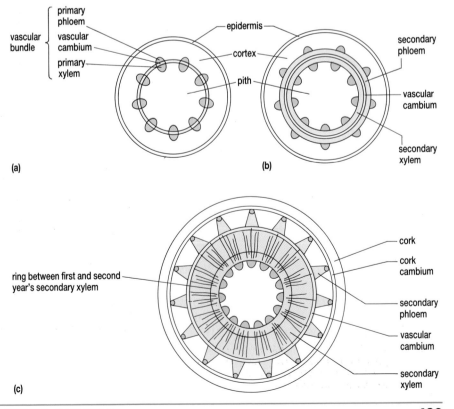

Fig 21.22 Transverse sections of stem to show stages in the secondary growth of
(a) a young stem
(b) an intermediate stage
(c) a 2-year-old stem.

Vascular cambium actually contains two types of cells which are arranged with their long sides at right angles to each other.

- **Fusiform initials** are cambial cells with their long sides running up and down the stem. When they divide they form **secondary xylem** to the inside and **secondary phloem** to the outside.
- **Ray initials** are cambial cells with their long sides running around the stem. When they divide they form tracts of parenchyma cells called **rays**. These rays run radially through the secondary xylem and secondary phloem, carrying water and nutrients between the two. Ray initials divide more in the secondary phloem than in the secondary xylem, giving rise to the wedge-shaped phloem tissue shown in Fig 21.22(c).

As they mature and die, secondary xylem cells secrete thick walls. These cells form the wood which fills a woody stem and contributes most of its strength. Young xylem cells are found towards the outside of the stem and transport water and inorganic ions. Older xylem cells are found towards the middle of the stem and contribute little towards transport since their lumens are filled by waste substances.

 What is the main function of this older xylem tissue in a tree?

The increase in girth of the stem strains the cortex and epidermis, which split and disintegrate. Before this happens, parenchyma cells within the cortex become meristematic and form a ring of phellogen (**cork cambium**). Division of these cambial cells produces a new tissue, called **periderm**. The parts of the periderm, together with their functions, are summarised in Table 21.3. The periderm and phloem are commonly called bark. Complete removal of this bark, known as **ringing**, severs the phloem tissue so that substances formed during photosynthesis cannot be passed to the root.

 Why will this cause the plant to die?

Some conifers, such as the giant sequoia tree, *Sequoiadendron giganteum*, are over 300 years old; their height is over 80 metres and their girth is over 24 metres.

Table 21.3   Components of the periderm and their functions.

Name of tissue	Alternative name	Function
cork	phellem	dead cells with suberin in walls reduce water loss
		protects against entry of pathogens
		allows gas exchange through lenticels
cork cambium	phellogen	divides to form the periderm
secondary cortex	phelloderm	inner layer of parenchyma cells

ANALYSIS

Ageing trees

This exercise will help you develop observation and interpretation skills.

In temperate latitudes, division of the cambial cells stops during the cold winter months. Cell division resumes in spring and new cambium, xylem, phloem and ray cells are formed. As with all newly formed plant cells, these elongate by swelling with water whilst their cell walls are still soft. Water is usually most abundant in spring, so cells formed at this time swell more before they stop growth than those formed during the dryer summer. As a result, spring wood can be clearly distinguished from summer wood because it has a less compact appearance, peppered with the lumens of large xylem vessels. The pattern of alternating spring

PATTERNS OF GROWTH AND DEVELOPMENT

and summer xylem cells produces the familiar **annual rings** by which we can age trees.

Look at Fig 21.23 which shows a transverse section through a woody stem.

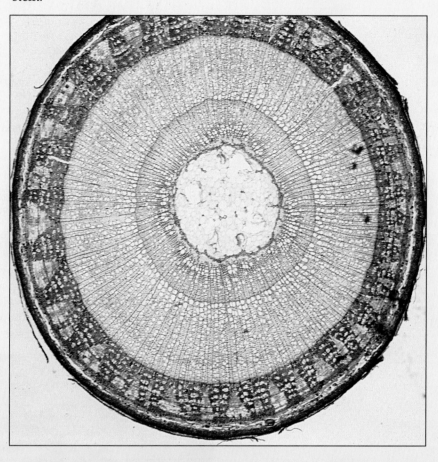

Fig 21.23 Transverse section through a woody stem.

(a) Locate the following tissues: periderm, secondary phloem, vascular ray of phloem, secondary xylem, vascular ray of xylem, spring wood and summer wood.
(b) How old is this woody stem? Justify your answer.

Secondary growth in roots

A young and old root are compared in Fig 21.24. The lateral meristems of the root are the vascular cambium within the stele and the pericycle around the stele. As cells in the vascular cambium divide they produce cells which differentiate into vascular tissue as well as more cambium cells. Those daughter cells formed on the inside of the cambium develop into new **secondary xylem**, those formed on the outside of the cambium develop into new **secondary phloem**. Since the centre of the root is already filled by primary xylem, the growth of secondary xylem pushes the tissue of the cambium, phloem, cortex and epidermis outwards. An increase in diameter of the root thus occurs.

Just as with the stem, secondary thickening of the root strains the outer tissues. Since cells in the epidermis and cortex have lost the ability to divide, they cannot respond to increases in growth of the central tissues. Eventually they split and disintegrate. This does not leave the phloem unprotected, however. Before this happens, division of the pericycle produces **phellogen** (cork cambium). This tissue divides to form periderm, in much the same way as in the stem (Table 21.3).

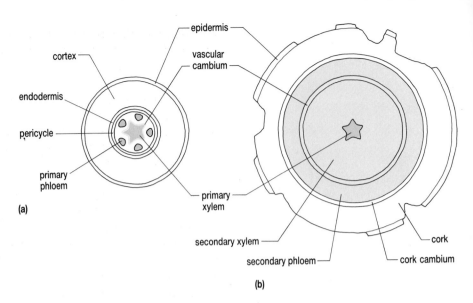

Fig 21.24 **(a)** The primary structure of a young root compared with **(b)** the secondary structure of an older root in which secondary thickening has taken place.

QUESTION

21.13 Look at Fig 21.25 which shows a transverse section through a young root where it is forming a lateral root.
   (a) Locate the following primary tissues: epidermis, cortex, pericycle, endodermis, xylem and phloem.
   (b) Explain how the formation of the lateral root differs from the formation of a lateral stem.

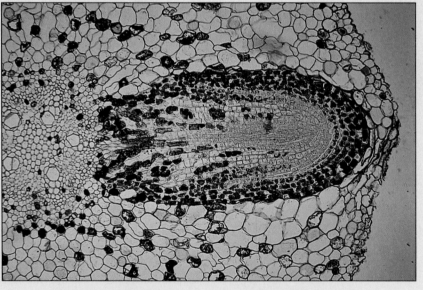

Fig 21.25 Transverse section through a young root with a developing lateral root.

SUMMARY

Biological growth involves an increase in the amount of cytoplasm within an organism. In multicellular organisms, it is usually accompanied by an increase in cell number. The best measure of an increase in cytoplasm is the dry mass. Since this involves killing experimental organisms, less ideal measures of growth, such as wet mass or length, are often used.

Growth may be continuous or discontinuous. Although its rate may vary throughout life, growth does not cease from birth to the adulthood of an

PATTERNS OF GROWTH AND DEVELOPMENT

organism showing continuous growth. Discontinuous growth is characteristic of arthropods whose rigid exoskeletons restrict growth except during moults (ecdysis).

In plants, an increase in length is called primary growth. It occurs as cells in the apical meristems divide and their daughter cells elongate. An increase in the girth of a plant is called secondary growth and occurs mainly in plants which survive more than one reproductive season. It occurs as cells in the cambium divide.

Growth is usually accompanied by an increase in complexity as newly formed cells differentiate to perform different functions. This increase in complexity is called development. Since all the vegetative cells in the body of a multicellular organism contain an identical set of genes, differentiation of cells occurs through activation and repression of genes. One of the major influences on cell differentiation in animals is the presence of regulatory substances in the cytoplasm of the egg cell which are separated unequally into specific daughter cells during cell division. The development of the daughter cells is greatly influenced by which substance it receives from the cytoplasm of the egg cell.

Animals undergo either direct or indirect development. Egg cells with relatively little yolk hatch into small, immature feeding stages, called larvae. After a period of feeding and growth, these larvae metamorphose into a different adult form. Egg cells with large amounts of yolk or those that are nourished within the mother's body usually produce young which are immature, miniature versions of the adult. Subsequent development does not radically change their body form.

A fertilised human egg cell undergoes cell division without cell growth to form a hollow ball, the blastula. This embeds into the lining of the mother's uterus. An inner mass of cells contains the embryonic disc which will form the embryo. The embryonic disc develops three layers: ectoderm, which will give rise to the epidermis and nervous system; mesoderm, which produces musculo-skeletal, circulatory, excretory and reproductive systems; endoderm, which produces the linings of the digestive and respiratory tracts. Throughout the whole of embryological development, metabolites are exchanged between fetal and maternal blood across the placenta.

After birth, humans show continuous growth. The general growth rate decreases from birth, showing a growth spurt during adolescence. Different parts of the body do not grow at the same rate, the head reaching its adult size earlier than the limbs or genitals. During development, cells begin to die and eventually ageing and death of the individual occurs.

Seed germination begins when a seed takes up water through its micropyle, causing cell elongation and mobilisation of enzymes. As a result, stored food is hydrolysed and used for cell division and elongation. Protected by its root cap, the embryonic root, or radicle, grows downwards into the soil and begins to absorb water and inorganic ions. The embryonic shoot, or plumule, grows upwards out of the soil. In monocotyledonous plants a sheath, called the coleoptile, protects the plumule during its passage through the soil. In plants which show epigeal germination, the cotyledons protect the plumule and emerge above the soil. In plants which show hypogeal germination, the cotyledons stay below the ground.

Primary growth of plants occurs as cells in the apical meristem divide and their daughter cells elongate and differentiate to form mature plant tissues. Mature tissues which retain the ability to divide form a tissue called cambium which is found in the vascular bundles in stems and roots and in the pericycle of roots.

Further division of the cambium produces secondary tissues, resulting in secondary thickening in both roots and shoots. As they increase in girth, their epidermis ruptures and is replaced by cork.

Answers to Quick Questions: Chapter 21

1 An organism's water content changes after taking up water (e.g. drinking) or losing water (e.g. by evaporation or in urine).
2 Length – there is no increase between moults.
3 See Chapter 4.
4 The adults are sessile so they cannot move to colonise new habitats. Without a dispersal stage, overcrowding and severe competition would occur among adults.
5 Dispersal.
6 (a) arms and trunk (b) legs.
7 Axillary buds would start to develop.
8 Support.
9 They do not grow sufficiently large to warrant the development of the more robust support provided by secondary thickening.
10 As it no longer conducts, the older xylem tissue's remaining contribution is to support the plant.
11 Roots would have no source of respiratory substrate, hence root cells would die so killing the whole plant.

5.1 The graphs show growth curves for a potato plant grown from a tuber and for a human.

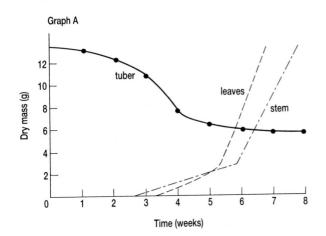

Graph A

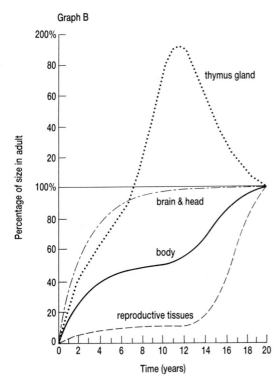

Graph B

(a) Suggest **one** reason in **each** case why the units of growth shown were chosen for
  (i) Graph A,
  (ii) Graph B. (4)
(b) (i) Describe the phases of growth for the stem of the potato plant.
  (ii) Explain the shape of the curve for the tuber of the potato plant. (4)
(c) (i) State the age range at which the human reproductive organs are growing most rapidly.
  (ii) If the ratio of brain size to body size is 1:24 in the human adult, estimate from

the graph the ratio in a 10-year child. (3)
(d) (i) Describe the growth of the human thymus gland.
  (ii) What does this pattern of growth suggest about the function of the thymus gland? (2)
(e) Many insects increase in length rapidly at the time of moulting, but show little change in length between moults. On the axes below, sketch the graph you would expect for an insect which moulted five times before reaching adult size. (2)

(WJEC 1987)

5.2 **(a)** Describe, with the aid of annotated diagrams, the essential features of the process of meiosis in a cell whose diploid number of chromosomes is 4. (9)
(b) As a result of meiosis, non-identical haploid daughter cells are produced.
  (i) State the significance of the fact that they are haploid.
  (ii) With reference to the description given in part **(a)**, account for the variation in the daughter cells. (5)
(c) Outline the events leading to and including fertilisation in:
  (i) a flowering plant after a pollen grain has landed on the stigma
  (ii) a mammal after the release of spermatozoa in the female genital tract. (6)

(NISEC 1989)

5.3 This question refers to the mammal.
(a) Describe
  (i) spermatogenesis,
  (ii) the production of semen,
  (iii) the fertilisation of the egg,
  (iv) the development of the fertilised egg up to its implantation. (14)
(b) What hormonal changes take place during gestation and what are their effects? (6)

(JMB 1987)

5.4 **(a)** What is meant by the term 'growth'? (4)
(b) Explain how each of the following takes place.
  (i) Increase in girth of a flowering plant stem
  (ii) Increase in size of a mammal after birth. (12)

(c) Comment on the importance of the larval stage in the life cycle of many insects. (4)

(ULSEB 1988)

5.5 The diagrams below show stages in the life cycle of a flowering plant.

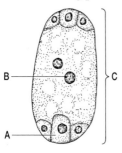
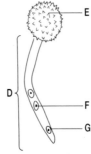

(a) Complete the table below by choosing the letter from the diagrams which refers to each of the stages given. (4)

Stage in life cycle	Letter
Female gametophyte	
Tube nucleus	
Female gamete	
Male gamete	

(b) (i) State *one* function of D. (1)
(ii) How has structure D enabled flowering plants to adapt to terrestrial life? (2)
(c) Comment on the surface structure of E. (2)
(d) Suggest *two* ways in which self fertilisation may be avoided in flowering plants. (2)

(ULSEB 1989)

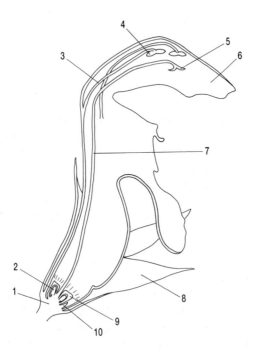

5.6 (a) The diagram shows a half flower.
(i) Name parts 1 to 10. (5)

(ii) Describe **five** visible features in the flower in the diagram which you would expect to be different or absent in wind-pollinated flowers. (5)

(b) Explain where each of the following **four** parts of an angiosperm flower develops:
(i) microspore
(ii) megaspore
(iii) male gamete
(iv) female gamete. (8)

(c) *Zostera* is a marine angiosperm whose flowers are *submerged*. Give **two** ways in which you would expect its pollen grains to differ from those of terrestrial insect-pollinated angiosperms. (2)

5.7 After fertilisation, developing mammals and flowering plants are protected and nourished. How is this achieved in
(a) mammals until birth, (10)
(b) flowering plants until photosynthesis begins? (10)

(JMB 1988)

5.8 The following diagram represents a vertical section through the carpel and ovule of a flower such as buttercup. Pollination has occurred.

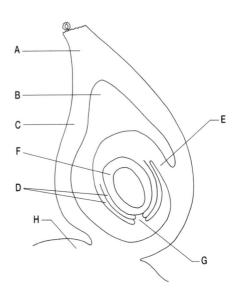

(a) Give the names of the structures labelled **A-H**. (4)
(b) Suppose that fertilisation has just occurred. On the diagram draw in clearly and label the following structures:
(i) the zygote cell;
(ii) two synergid cells;
(iii) the endosperm nucleus;
(iv) the antipodal cells;
(v) the pollen tube. (5)
(c) Explain concisely what is meant by *cross pollination*. (2)

(d) How is the endosperm formed? (2)

(e) What part could *chemotropism* play in the processes of fertilisation in flowering plants? (2)

(f) What is the significance of the spiny nature of the pollen grain wall? (1)

(g) What do the structures labelled **D** on the diagram eventually form? (1)

(UODLES 1988)

5.9 Compare the processes of sexual reproduction in a flowering plant and in a mammal. (20)

(ULSEB 1988)

5.10 (a) Why is a sexual phase important in the life cycle of different organisms? (4)

(b) With reference to the life cycles of a moss and a human, describe and comment on the following.
(i) The occurrence of haploidy and diploidy
(ii) The transfer of gametes. (12)

(c) Describe *two* roles of the placenta in a mammal. (4)

(ULSEB 1989)

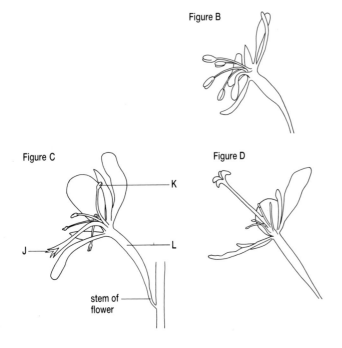

Figure B

Figure C

Figure D

K

L

J

stem of flower

5.11 (a) Distinguish between the processes of *pollination* and *fertilisation* in a flowering plant. (2)

The drawings above show the detailed structure of individual dissected flowers taken at different stages from a common flowering plant. The drawings are all to the same scale.

(b) The key separates this species from three other flowering plants of similar appearance.

1. Petals fused into a corolla tube	Species **W**
Petals free and not fused into a corolla tube	2
2. Ovary inferior (flower parts inserted above ovary)	Species **X**
Ovary superior (flower parts inserted below ovary)	3
3. Petals 4 in number, approximately equal in length to sepals	Species **Y**
Petals 5 in number, at least three times as long as sepals	Species **Z**

Identify the common flowering plant as species **w, x, y,** or **z**. (2)

(c) Describe **one** piece of evidence that this species is a member of the class Dicotyledones. (1)

(d) Name structures **J, K** and **L** In Figure **C**. (3)

(e) Describe the changes in appearance of the stigma and style as the flower matures. (2)

(f) A flower in which the male parts mature before the female parts is described as *protandrous*. Give the evidence from the drawings that this plant is protandrous. (3)

(g) The diagram shows the results of a study of the behaviour of bees visiting this species of plant.

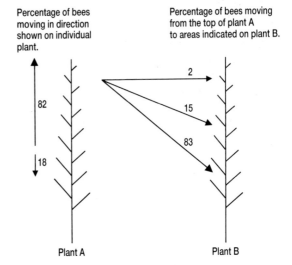

Percentage of bees moving in direction shown on individual plant.

Percentage of bees moving from the top of plant A to areas indicated on plant B.

82

18

2

15

83

Plant A

Plant B

Using the information in this diagram and your answers to questions **(e)** and **(f)**, describe how cross-pollination from plant to plant is achieved in this species. (6)

(h) Describe **four** ways in which the flowers of this plant might be expected to differ from the flowers of a typical wind-pollinated plant. (4)

(AEB 1990)

5.12 An investigation was carried out into the relationship between the birth weight of human babies and the concentrations in the mother's placenta of the metallic elements lead and zinc. The results of the investigation are given in the table below. Figures for metal concentrations are the means (averages) of many measurements, given together with their standard deviations.

(a) (i) Plot these data in the form of histograms using graph paper. Do not include the standard deviations   (5)

(ii) Describe the relationships between placental lead and zinc concentration and birth weight, as shown by the histograms. Suggest reasons for these relation ships.   (6)

Birth weight range in kg	Placental concentration in arbitrary units			
	Lead		Zinc	
	Mean	Standard deviation	Mean	Standard deviation
2.00–2.49	31	0.15	33	4.45
2.50–2.99	22	0.92	41	13.55
3.00–3.49	13	0.46	60	15.10
3.50–3.99	12	0.42	58	14.52
4.00–4.49	11	0.38	51	15.29
4.50–4.99	13	0.15	52	6.39

(b) (i) Which *two* ranges of birth weights appear to show the lowest overall variability in placental metal concentration?   (2)

(ii) Suggest *one* reason why these ranges show the lowest variability.   (1)

(c) Explain how the efficiency of the human placenta is increased by the presence of

(i) placental villi   (2)

(ii) maternal and fetal blood flow in opposite directions.   (2)

(d) How many metal ions such as lead and zinc enter the body of a pregnant woman from the environment, and reach the placenta?   (2)

(ULSEB 1989)

Answer Guidelines to 5.1

(a) **(i)** the amount of water in these plant organs is variable and it is easy with plant organs to take samples and dry to constant weight;

(ii) it is unacceptable to harm humans, percentage of size in adult was chosen because individual parts of the body start off at different sizes;

(b) **(i)** steady increase in growth from day two to end of day 5; faster growth from day 5;

(ii) the mass decreases as stored food is used for the growth of the stem and leaves; the rate of decrease slows after day 4 as the leaves begin to photosynthesise and so stop using stores from tuber;

(c) **(i)** 14–19 years;

(ii) at ten years old brain is almost adult size and body is about 50% of adult size: ratio is therefore about 1:12;

(d) **(i)** increases from birth to almost twice adult size at 12 years and then decreases to adult size by 20 years;

(ii) *suggests* it is particularly important during childhood/is associated with growth during childhood;

(e) step-like appearance with 6 horizontal lines and five vertical lines; slight growth between moults.

Theme 6

DIVERSITY

Variety's the very spice of life,
That gives it all its flavour.
William Cowper (1731–1800)

Whether or not variety is the spice of life, it is certainly a feature of life. There are countless varieties of organisms on Earth and most of us learn to recognise only a few of them. Even within one small group of organisms, such as the members of your own family, there are major differences in appearance, chemical make-up and behaviour.

This theme examines how this variety comes about and how biologists attempt to make order out of it.

- *Inheritance* – the basic principles which explain why offspring differ from each other and from their parents.
- *Evolution* – the way in which the genetic make-up of whole populations changes.
- *Species* – the way in which biologists believe evolution leads to formation of new species and the principles by which biologists classify these species to show their evolutionary relationships.
- *Prokaryotae and Protoctista* – the basic features of two kingdoms of organisms, both of which contain single-celled organisms.
- *Fungi and Plantae* – the basic features of yeasts, moulds and plants.
- *Animalia* – the basic features of members of the animal kingdom.

(Above) A mosaic of different flowers and stem lengths.
(Below) The diversity of coat colour is only a tiny fraction of the inherited difference in this litter.

Chapter 22

INHERITANCE

LEARNING OBJECTIVES

When you have studied this chapter you should be able to:

1. define the terms genotype and phenotype and explain the relationship between them;

2. explain the terms gene, allele and locus and relate them to the structure of DNA and its control of cell activity;

3. use simple genetics diagrams to explain the inheritance of up to two features;

4. define the terms dominant, recessive, linked loci, unlinked loci and sex linkage and recognise their occurrence from inheritance patterns;

5. calculate a cross over value from given data and use it to produce a chromosome map.

An organism's **phenotype**, its physical and chemical characteristics, is the result of the interaction between its **genotype** (the genetic information inherited from its parents) and the environment in which development occurs (Fig 22.1). The members of any group, e.g. people, all have different phenotypes. Many of those differences in these phenotypes have been inherited. The study of inheritance is known as **genetics**.

Inherited differences occur between the individuals of most populations. The grass plants in a playing field are no more identical than the people who walk over them; we are simply not good at recognising individual grass plants. Experimental genetics has been dominated by work on organisms which are easy to keep in laboratories and which have a fairly short life cycle (so that many generations can be produced in a short time) and short maturation times (so that the characteristics of the offspring can be seen a short time after the parents have reproduced). On the other hand, the study of human inheritance depends on inferences which can be made from family pedigrees or, more recently, from biochemical studies on human DNA.

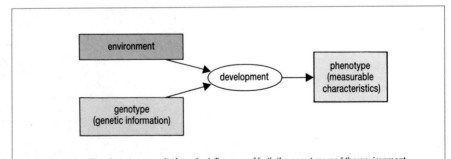

Fig 22.1 The phenotype results from the influences of both the genotype and the environment.

It is important to understand the terms and notations used by geneticists before proceeding. Revision of meiosis is advised (see Section 4.7) since it is the events which occur during cell division which largely determine patterns of inheritance.

Loci, genes and alleles

The way in which the nucleic acid in a cell controls the cell's activities is described in Chapter 4. Most of this nucleic acid is found in strands called **chromosomes**. Some segments of the nucleic acid in a chromosome carry a code which specifies the amino acid sequence of polypeptides and proteins which the cell manufactures. These specific lengths of nucleic acid are called **genes**. Within each species, each gene occupies a fixed position on a particular chromosome. This place on a chromosome, which is the normal site of a particular gene, is called a gene **locus**. Like all prokaryotes, cells of the bacterium *Escherichia coli* have only one chromosome. At the same place along the chromosome in each *E. coli* cell there is the locus of the β-galactosidase gene described in Fig 4.6. The fruit fly *Drosophila melanogaster* (Fig 22.2) has two copies of four different chromosomes. At the same point on each copy of chromosome 2 is the locus of a gene which controls the wing length of the adult: this is the locus of the wing length gene.

Chromosomes contain DNA and protein. The DNA of a bacterium does not contain protein so, strictly, is not a chromosome. It is called a chromosome in this chapter for simplicity and because it has the same function as a true chromosome.

 1 **How many genes controlling wing length will (a) a gut cell (b) a spermatozoan from *Drosophila* contain?**

Fig 22.2 *Drosophila melanogaster*, the fruit fly, is widely used for genetics studies. It has only eight chromosomes (diploid number) and has a short life cycle of about 10–14 days.

Section 4.2 describes how chromosomes are copied before cell division occurs. Sometimes the base sequence of a nucleic acid is replicated wrongly during this process. Such **mutations** result in new nucleic acid codes. When translated, these new codes may result in the production of a different protein. This new protein may be so different from the normal protein that it is functionless and its possessor dies. Alternatively, it may be a usable protein which affects the phenotype in a different way from the protein produced when the original nucleic acid code is translated. The possessors of such mutant codes may survive and have offspring, in which case the mutant codes are inherited. It is the combination of mutant and 'normal' codes within an organism's genotype which causes inherited variation amongst its offspring. The different nucleic acid codes which may be found at a particular locus are called **alleles** of that particular gene. At the wing length locus on chromosome 2 of *D. melanogaster* described above, there may be an allele for long wings or an allele for vestigial (short) wings. The relationship between a locus, gene and allele is summarised in Fig 22.3.

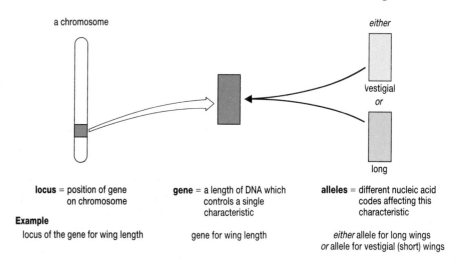

locus = position of gene on chromosome

Example
locus of the gene for wing length

gene = a length of DNA which controls a single characteristic

gene for wing length

alleles = different nucleic acid codes affecting this characteristic

either allele for long wings
or allele for vestigial (short) wings

Fig 22.3   The relationship between locus, gene and allele is shown in this diagram of one chromosome.

2 Use the term gene or allele to complete the phrases: (a) the ... for brown hair (b) the ... for hair colour.

Fig 22.4   Red-eyed wild-type and the white-eyed mutant of *Drosophila melanogaster*.

QUESTIONS

22.1 Which of the following human phenotypic features result from (i) the effects of the genotype (ii) the effects of the environment (iii) the combined effects of both?
(a) A missing top finger joint following an industrial accident.
(b) Polydactyly (more than five digits on the hands or feet).
(c) Size of the person.

22.2 Seed coat colour in peas may be green or yellow. Are the nucleic acid codes controlling these alternative seed colours (a) loci (b) genes or (c) alleles? Explain your answer.

Life cycles

When the gamete nuclei fuse during fertilisation, a zygote nucleus is produced which contains two copies of each chromosome, one copy from each gamete nucleus. These chromosome pairs are called **homologous chromosomes** and a cell which contains pairs of homologous chromosomes is called **diploid**. In a diplontic life cycle, diploid zygote cells divide by mitosis, giving rise to diploid adults; in a haplontic life cycle, diploid zygote cells divide by meiosis, giving rise to adults with only one copy of each chromosome in each cell (i.e. **haploid**). In the former case, a division to reduce the chromosome complement of cells must occur prior to gamete production, otherwise the number of chromosomes would double at each fertilisation. The behaviour of chromosomes in each of these two life cycles is summarised in Fig 22.5.

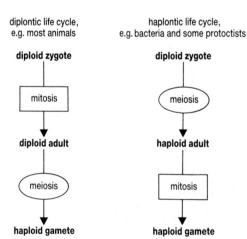

Fig 22.5   The timing of mitosis and meiosis during two different life cycles. In both cases the gamete nuclei are haploid and the zygote nuclei resulting from their fusion are diploid.

 Human males have two types of sex chromosome: one X chromosome and one Y chromosome. Which sex chromosome will their sperm carry?

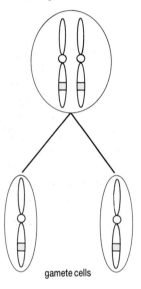

parent cell with a pair of
homologous chromosomes

gamete cells

Fig 22.6 A diploid zygote cell has two copies of each
locus. After meiosis, the gamete cells have
only one copy of each locus.

Homologous chromosomes	Chromosomes which carry the same loci and which pair with each other during the first division of meiosis.
Diploid	1. A cell which has two sets of homologous chromosomes, e.g. a human somatic cell (a cell not destined to become a gamete) has 46 chromosomes consisting of 23 pairs of homologous chromosomes. 2. An individual which has two chromosome sets in each of its somatic cells.
Haploid	1. A cell having one chromosome set, e.g. a human sperm or egg cell contains only 23 chromosomes, one from each homologous pair. 2. An organism composed of cells which contain only one set of chromosomes, e.g. the fungus *Sordaria* is a haploid organism.

Homozygosity and heterozygosity

Each chromosome in one gamete has the same loci as its homologous chromosome in the gamete with which it will pair. After fertilisation, every diploid zygote will, therefore, have two copies of each locus. The products of meiosis of a diploid cell with only one pair of homologous chromosomes are represented in Fig 22.6. Being diploid, the parent cell has two copies of the locus whose position is marked. The haploid gametes have only one copy of this locus, however.

If the alleles of the gene found at the locus which is marked in Fig 22.6 are the same on both copies of the diploid cell's homologous chromosomes, the cell's genotype for this particular characteristic is described as **homozygous**. If the alleles of the gene found at this locus are different on the two chromosomes, the cell's genotype is termed **heterozygous**. There is a difference between the gametes produced by a cell which is homozygous at one locus and those produced by a cell which is heterozygous at one locus (Fig 22.7). In the former, all the gametes have the same genotype: in the latter, gametes with two different genotypes are produced in equal numbers.

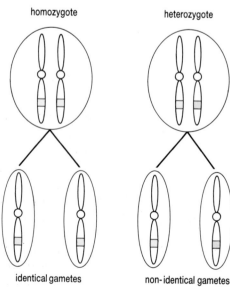

homozygote   heterozygote

identical gametes   non-identical gametes

Fig 22.7 Heterozygous cells produce different
gametes whereas homozygous cells produce
identical gametes.

How many copies of each of your genes will be present in (a) your liver cells (b) your gametes?

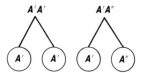

genotype of diploid cells

genotype of haploid gametes

Fig 22.8 The information in Fig 22.7 is more simply represented using letters to denote alleles without drawing entire chromosomes.

Producing genetics diagrams

Patterns of inheritance are best represented using diagrams to show the genotypes of parents, gametes and offspring. Diagrams like the one in Fig 22.7 could be used but would take too long to draw. Since the only part of the chromosome which is of interest is the locus where the gene controlling the characteristic under discussion is found, it is easier to represent this by a letter. Look at Fig 22.8 which shows how the information in Fig 22.7 can be represented using the letters *A'* and *A"* to represent the differently coloured alleles. Note the labels on the left of the diagram and how the gametes are encircled: these are conventions which should be followed.

QUESTION

22.3 Using the symbols *P'* and *P"* to represent two alleles of the same gene, write the genotype of
(a) a homozygote
(b) a heterozygote
(c) a haploid cell
(d) the gametes produced by a heterozygote.

22.2 MONOHYBRID INHERITANCE

The inheritance of characteristics which are controlled by different alleles of a single gene is called **monohybrid inheritance**. Seed coat colour in peas is an example of monohybrid inheritance.

Dominant alleles

All the alleles of a particular gene carry the DNA code for a protein or polypeptide. Some of these proteins may be completely functionless, however, and will produce no discernible effect on the phenotype.

Flower colour in the edible garden pea plant (*Pisum sativum*) is controlled by alleles of a single gene. One allele of the gene for flower colour carries a DNA code for a particular enzyme. This enzyme is involved in a biochemical reaction within the cells of a pea plant's flowers, resulting in a purple pigment being produced (Fig 22.10). It may, therefore, be called the purple-flower allele. A second allele of this gene carries a different DNA code. The resulting protein is not a functional enzyme in the reaction

KEY CONCEPT

A monohybrid cross involves the inheritance of genes at only one locus.

normally leading to the production of purple pigment. If a plant is homozygous for the latter allele its flowers are white, so it may be called the white-flower allele. Purple flowers will be produced even if a plant has only one purple-flower allele in its genotype.

5 What does this suggest about the amount of the flower-colour enzyme produced from the purple-flower allele?

Fig 22.9 Purple and white peas of the type used in experiments. The plants are self-fertile, so steps have to be taken to ensure cross pollination.

Fig 22.10 Many alleles exert their effect on the phenotype through coding for enzymes. Here an allele of the flower-colour gene encodes an enzyme which converts a colourless precursor into the purple-flower pigment. If this allele were to undergo a mutation such that the new allele no longer coded for a functional enzyme, no purple pigment would be made and the flower would appear colourless (white).

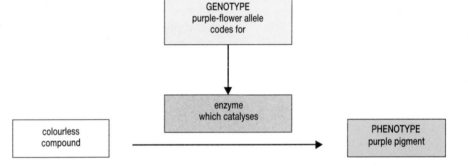

The effect of the white-flower allele on the phenotype of the pea plant has been masked by the effect of the purple-flower allele. The white-flower allele is **recessive** to the purple-flower allele and the purple-flower allele is **dominant** to the white-flower allele. Note that it is not correct to describe the white-flower allele as recessive without naming the allele to which it is recessive. As you will learn later in this chapter, it is quite possible for an allele to be recessive to a second allele but dominant to a third allele of the same gene.

KEY TERMS

Dominant allele	An allele whose effect is expressed in the phenotype even in the presence of a recessive allele; thus if allele *A* is dominant to allele *a*, then *AA* and *Aa* have the same phenotype.
Recessive allele	An allele whose effect is expressed in the phenotype of a diploid organism only in the presence of another identical allele.

Representing alleles

There are a number of short-hand ways of representing the alleles of a gene. One way is to use an upper case, or capital, letter for a dominant allele and the corresponding lower case letter for a recessive allele. Since the dominant allele for flower colour in peas produces purple flowers, the letter 'p' would be used. The dominant, purple-flower allele would, thus, be represented *P* and the recessive white-flower allele would be represented *p*. These symbols are used in Fig 22.11 to show the results of a **cross** between a homozygous purple-flowered plant with a homozygous white-flowered plant.

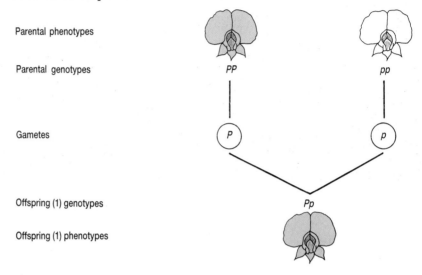

Parental phenotypes	
Parental genotypes	*PP* *pp*
Gametes	*P* *p*
Offspring (1) genotypes	*Pp*
Offspring (1) phenotypes	

Fig 22.11   A cross between a homozygous purple-flowered and a homozygous white-flowered pea plant. The offspring (1) are all purple-flowered.

6 **What is the genotype of a homozygous purple-flowered plant?**

SPOTLIGHT

PLANT AND ANIMAL BREEDING

This chapter describes the basic genetics rules which are used in breeding better strains of plants and animals. Such selective breeding has resulted in enormous increases in the yield from practically all agricultural plants and animals in the last hundred years. This application of genetics in selective breeding programmes will continue to be essential if we are going to feed an ever increasing world population.

Representing crosses

*   The first line of Fig 22.11 represents the phenotype of the parent plants. Diagrams have been used because they may help you to understand the genetic process better; you could have written 'purple-flowered' and 'white-flowered' instead.
*   The second line of the genetics diagram shows the genotype of these plants. It is important that you have been told what *P* and *p* represent (see earlier) otherwise you would not be able to understand these genotypes.
*   The third line in Fig 22.11 shows the genotype of the gametes which

The terms F₁ and F₂ are sometimes used for offspring, but should only be used when both the original parents were homozygous. F_1 (first filial generation) specifically refers to the offspring of homozygous parents, F_2 to the offspring produced by crossing those F_1 individuals.

will be produced from each flower. Since both parents are homozygous they can produce only one type of gamete.

- The fourth line shows that, after fertilisation, all the offspring (offspring (1)) will have the genotype **Pp**, i.e. will be heterozygotes.
- The fifth line shows the phenotype of the offspring (1), in this case purple-flowered.

A diagram representing any monohybrid cross between a homozygous dominant individual and a homozygous recessive individual will always be the same as Fig 22.11, the only difference being the symbols used (which you must always explain before you use them).

Look at Fig 22.12 which is a little more complicated; it represents a cross between two of the offspring (1) from the previous cross. There is no need to explain the symbols again since the same characteristic is being represented.

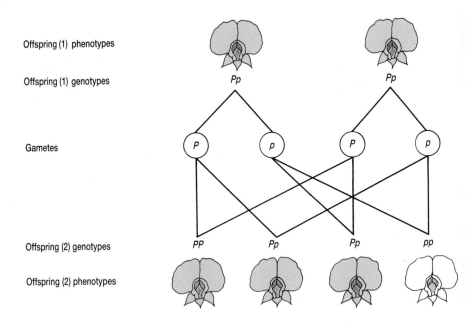

Offspring (1) phenotypes

Offspring (1) genotypes — *Pp* — *Pp*

Gametes — *P* *p* *P* *p*

Offspring (2) genotypes — *PP* *Pp* *Pp* *pp*

Offspring (2) phenotypes

Fig 22.12   A cross between two heterozygous purple-flowered pea plants.

- The first and second lines of the diagram represent the phenotype and genotype of the offspring (1) plants.
- The third line represents the gametes which each plant can produce. Since they are both heterozygous, they will each produce some gametes carrying the purple-flower allele *and* some gametes carrying the white-flower allele.

 Why will they probably make these gametes in equal abundance?

- The fourth line of Fig 22.12 shows the ways in which these gametes may combine to form the genotype of the offspring (2) plants.

The first gamete of the plant on the left (**P**) has been joined with the first gamete of the plant on the right (also **P**) to make the first offspring (2) genotype (**PP**). The second genotype on the offspring (2) line has been made by joining the same gamete with the second allele (**p**) of the plant on the right, and so on. If you check, you will find that each gamete from one plant has been combined with each gamete from the other plant *once only* and that the gametes of one plant have not been combined.

The resulting offspring (2) ratio of phenotypes is 3:1. This ratio is characteristic of monohybrid inheritance if large enough numbers of offspring are produced.

8 Will the offspring of actual crosses always be in the exact ratio of 3:1?

Diamond checkerboards and Punnet squares

Figure 22.13 shows a different way of representing the same information in Fig 22.12. The genotype of the gametes is shown on either side of a diamond checkerboard. The contents of each small diamond are produced by combining the gametes along the edge of the diamond checkerboard. Each small diamond therefore represents the genotype of one of the offspring (2) plants. The phenotype of these plants has been shown in diagrammatic form but could just as well have been written in words. This diamond-shaped checkerboard has been used because it shows the genotypes of the gametes produced by each parent. An alternative is to represent exactly the same information in a square (called a Punnet square) rather than a diamond. It does not matter whether the format of Figs 22.12 and 22.13 or a Punnet square is chosen to represent genetics crosses. As long as symbols are explained, and each line of the diagram is correctly labelled, the important thing is to choose the way which is easiest to use and which leads to fewest mistakes.

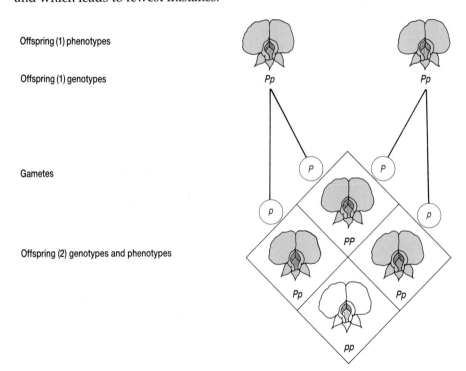

Offspring (1) phenotypes

Offspring (1) genotypes

Gametes

Offspring (2) genotypes and phenotypes

Fig 22.13 The same cross as Fig 22.12 using a diamond checkerboard. The offspring (2) are purple-flowered and white-flowered in the ratio 3:1.

ANALYSIS

The inheritance of leaf spotting in Blue-eyed Mary

This problem is intended to help you understand how to answer questions about genetics.

The plant Blue-eyed Mary grows wild in British Columbia in western Canada. One plant in a wild population was observed to have blotched leaves. This plant was allowed to pollinate itself (**selfed**). The seeds were collected and grown into progeny. One randomly selected (but typical) leaf from each of the progeny is shown in Fig 22.14.
(a) What is the ratio of blotched to unblotched leaves?
(b) Formulate a genetics hypothesis to explain how this ratio might be

---

produced. Explain all the symbols you use and show all genotypic and phenotypic classes.

(c) What was the genotype of the original plant?

(d) How would you test your hypothesis? Give details of the crosses you would make and show, using genetics diagrams, the results you would expect.

Make sure that you lay out your diagrams correctly.

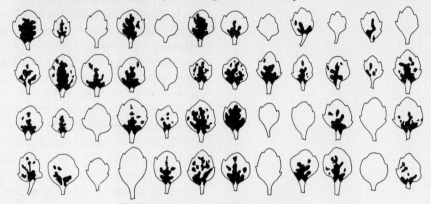

Fig 22.14 Leaves from Blue-eyed Mary plants.

Detecting heterozygotes – the test cross

If one allele of a gene controlling a character is dominant to a second allele, the homozygous dominant individual appears identical to the heterozygous individual. There may be times when breeders need to know whether an individual showing the effect of a dominant allele is homozygous or heterozygous, e.g. to know whether a purple-flowered pea plant was homozygous or heterozygous. This cannot, as yet, be determined by simple tests on the pea plant. It can be determined by breeding from the pea plant and making deductions about its genotype from the phenotype of its offspring.

To determine whether an individual showing the effect of a dominant allele is homozygous or heterozygous for that gene, it must be mated with

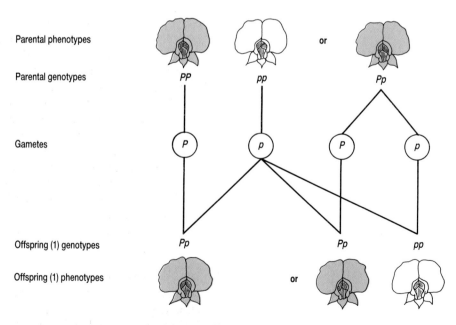

Fig 22.15 A test cross to detect whether a purple-flowered pea plant is homozygous or heterozygous. A heterozygote produces some white-flowered offspring.

a number of homozygous recessive individuals. A homozygous recessive individual is used because all its gametes must contain the recessive allele of the gene, which will not have an effect on the phenotype of the offspring. In this way the phenotype of the offspring will immediately lead to deductions about the genotype of the unknown parent. If all the offspring show the effect of the dominant allele it is reasonable to assume that the unknown parent was homozygous dominant. If even only one of the offspring shows the effect of the recessive allele in its phenotype the unknown parent must have been heterozygous. Using a number of homozygous recessive mates in a test cross increases the number of offspring which can be produced during the experiment and hence the chance of one of the offspring inheriting any recessive allele from the parent under test. This technique to determine an individual's genotype is represented in Fig 22.15 and is known as a **test cross**.

 Why is the homozygous recessive genotype used in performing a test cross?

QUESTIONS

N.B. When constructing diagrams to represent genetics crosses, take care to ensure that

- parents and offspring are always diploid but that gametes are always haploid;
- you put a circle around the symbols representing each gamete;
- you make a point of explaining each line of your diagram by labelling it Parental phenotypes, Parental genotypes, gametes, Offspring (1) genotypes, Offspring (1) phenotypes, gametes, Offspring (2) genotypes and Offspring (2) phenotypes, as appropriate.

Without these labels, genetics diagrams, and hence their use as explanations, are incomplete.

22.4 Use the symbols for the flower-colour alleles in pea plants given in the text to make genetics diagrams to show the genotype of the offspring of a white-flowered pea plant crossed with
 (a) a homozygous purple-flowered pea plant
 (b) a heterozygous purple-flowered pea plant
How do your genetics diagrams verify the test cross technique?

22.5 The fruit fly (*Drosophila melanogaster*) lays its eggs in rotting fruit and is commonly found around dustbins in warm weather. The length of its life cycle depends on the ambient temperature, but is about two weeks. It has been used in genetics laboratories around the world since about 1905. A gene controlling wing length has two alleles. The dominant allele for long wings can be denoted *L* and the recessive allele for vestigial (short) wings can be denoted *l*. Use these symbols in a diagram to show that, when a homozygous long-winged fly mates with a vestigial-winged fly and the offspring (1) are left to interbreed, the phenotypes of the offspring (2) occur in the ratio of 3:1.

22.6 Holstein-Friesian cattle are popular since they have a very high milk yield. Homozygous Holstein-Friesians have a black and white pattern in their coats. There is a recessive allele of the gene for coat colour which, in the homozygous condition, produces a red and white coat colour (the 'red' is really a brown colour). Red and white cattle are barred from pedigree registration, so it is important to know that any bulls used as sperm donors by artificial insemination organisations are homozygous for the dominant allele of the coat colour gene.
 (a) Devise appropriate symbols to denote the alleles controlling coat colour in Holstein-Friesian cattle.
 (b) Use these symbols in genetics diagrams to show how the genotype of a black and white bull could be determined.

Codominant alleles

Sometimes both the alleles of one gene result in the production of a functional protein. Thus, both will produce an effect on the phenotype. Such alleles are termed **codominant alleles**.

Human blood groups are caused by the presence of various molecules

BLOOD GROUPS

A blood group is a characteristic recognised by the presence of antibodies (agglutinins) in the blood and antigens (agglutinogens) on the surface of erythrocytes. In addition to the ABO blood group system in humans there are many others, e.g. Rhesus (Rh), MN, Duffys. In cattle there are over 200 different blood group systems.

(called agglutinogens) on the cell surface membrane of erythrocytes (Fig 22.16). For reasons which are not fully understood, there are usually proteins in the plasma (agglutinins) which specifically combine with the appropriate agglutinogens causing the erythrocytes to stick together (agglutination). The resulting blockage of blood vessels may result in death. The agglutinins are only produced against foreign agglutinogens and not against selfproteins (see Chapter 13).

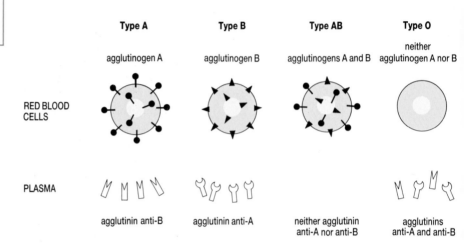

Fig 22.16 Agglutinogens and agglutinins involved in the ABO blood grouping system.

The MN blood group system is controlled by a single gene. One allele of the MN blood group gene results in the production of protein M on the erythrocytes and a second allele results in the formation of protein N on the erythrocytes. Since neither is recessive, the upper and lower case system of representing dominant and recessive alleles used so far is inappropriate. Instead, an upper case letter is used to denote the gene (in this case G to represent group, for example) and an upper case superscript to represent the alleles of this gene. Hence the allele causing production of protein M is represented G^M and the allele causing the production of protein N is represented G^N. A cross between a homozygous person of blood group M with a homozygous person of blood group N is represented in Fig 22.17; all the offspring (1) are heterozygous, blood group MN.

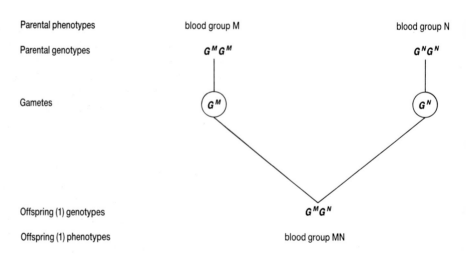

Fig 22.17 A cross between a person of blood group M and a person of blood group N; an example of codominant alleles.

22.7 **(a)** Continue the cross shown in Fig 22.17 to show that the offspring of a couple who both have the blood group MN will occur in the ratio of 1:2:1.

(b) Does this mean that if this couple have four children the genotype of one child will be G^MG^M, two children will be G^MG^N and the fourth child will be G^NG^N? Explain your answer.

22.8 The coat colour of shorthorn cattle is controlled by a gene with two alleles. Cattle which are homozygous for the red-coat allele have red coats; those homozygous for the white-coat allele have white coats. Heterozygotes have a coat in which half the hairs are red and half are white. The resulting colour is called roan.

(a) Using the rules described for alleles controlling the MN blood group system, devise suitable symbols to represent the red-coat allele and white-coat allele.

(b) Draw a fully labelled genetics diagram to show the expected ratio of coat-colour phenotypes in the calves produced by mating a number of roan cows with (i) a roan bull (ii) a white bull.

22.9 Human erythrocytes are round, biconcave discs that contain haemoglobin in their cytoplasm. Some people produce an abnormal haemoglobin which results in their erythrocytes easily becoming mis-shaped. Since the affected cells become sickle-shaped in low oxygen tensions, the condition is known as sickle-cell anaemia (see Section 23.5). Unaffected individuals are homozygous for the allele Hb^A; people who are homozygous for the sickle-cell allele Hb^S suffer from sickle-cell anaemia. Half the haemoglobin in the erythrocytes of heterozygotes is normal and half is abnormal. Heterozygotes do not usually show symptoms of sickle-cell anaemia.

(a) Use the symbols to represent the genotype of a heterozygote.

(b) Draw a fully labelled genetics diagram to show the phenotypes which might be expected among the offspring of a couple who are both heterozygotes.

(c) Explain why children with full sickle-cell anaemia are likely only when both parents are heterozygotes.

Flower colour in the snapdragon plant (*Antirrhinum* sp.) is controlled by two alleles of a flower-colour gene. Individuals which are homozygous for the red-flower allele have red flowers and those which are homozygous for the white-flower allele have white flowers. Heterozygotes have pink flowers; a flower colour which is intermediate between red and white. The cross represented in Fig 22.18 is similar to that for the inheritance of flower colour in pea plants (see Fig 22.13). The difference lies in the phenotypes of the offspring (1).

This pattern of inheritance can be explained if we assume that the presence of a red-flower allele in the genotype results in the formation of a certain amount of the enzyme which controls the production of flower pigment within each cell. The amount of enzyme then determines the amount of red pigment which is produced. The presence of two red-flower alleles in the cells of a homozygote might result in the production of twice as much flower-colour enzyme as the presence of only one red-flower allele in the cells of a heterozygote; hence twice as much pigment will be produced by the homozygote as by the heterozygote.

KEY TERM

CODOMINANCE

The situation in which a heterozygote shows the phenotypic effects of both alleles equally.

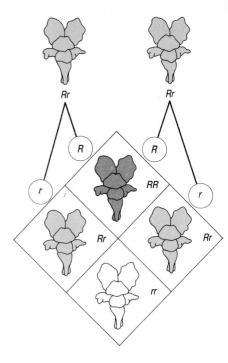

Parental phenotypes

Parental genotypes   Rr   Rr

Gametes   R   R

    r   RR   r

Offspring (1) genotypes and phenotypes   Rr   Rr

    rr

Fig 22.18   A cross between two pink-flowered snapdragons; an example of incompletely dominant alleles.

QUESTIONS

22.10 Draw genetics diagrams to show the ratio of phenotypes among the offspring of a pink-flowered snapdragon crossed with
(a) a white-flowered snapdragon
(b) a red-flowered snapdragon.

22.11 Thalassemia is a form of anaemia which is common in human populations around the Mediterranean Sea. It is inherited as a result of two codominant alleles. Individuals who are homozygous for the thalassemia allele suffer severe anaemia which may be fatal during childhood (thalassemia major); those who are homozygous for the normal allele show no anaemia; heterozygotes show a mild form of anaemia (thalassemia minor).
(a) Devise symbols to denote the two alleles.
(b) Use these symbols in genetics diagrams to represent the genotype and phenotype of the offspring produced by a person suffering from thalassemia minor who mates with (i) a similar sufferer (ii) a person who shows no anaemia.

22.12 Normal human brain cells produce an enzyme which regulates the metabolism of lipids in their surface membranes. Homozygotes for a defective allele synthesise very small amounts of this enzyme, resulting in an accumulation of lipids in the brain. This causes progressive mental retardation, blindness and poor motor control (symptoms of Tay-Sachs disease). Death usually occurs before the age of seven.
(a) Heterozygotes have a blood concentration of this enzyme which is about halfway between that of a normal person and that of a Tay-Sachs sufferer. Suggest why.
(b) Devise symbols to represent the normal and Tay-Sachs alleles and use them in genetics diagrams to show that Tay-Sachs children can only be born to two heterozygous parents.
(c) Suggest how the medical profession can minimise the risk of couples having children with Tay-Sachs disease.

Multiple alleles

In all the examples given so far there have been only two alleles of each gene. Often a gene may have more than two alleles. For example, the gene controlling the human ABO blood group has three alleles. Each results in the production of a different agglutinogen on the erythrocyte membrane (Table 22.1).

Table 22.2 shows the genotypes which are possible using all combinations of these three blood group alleles. Note that although there are three alleles, only two can be present in any one genotype. Each of the genotypes shown in Table 22.2 can exist and these combinations result in the blood groups which we observe in human populations.

Table 22.1   The alleles controlling the human ABO blood group system

Allele of blood group gene	Property	Production of agglutinogen on erythrocyte	
		agglutinogen A	agglutinogen B
I^A	codominant to I^B but dominant to I^O	yes	no
I^B	codominant to I^A but dominant to I^O	no	yes
I^O	recessive to both I^A and I^B	no	no

Table 22.2   Human ABO blood group genotypes and their resulting phenotypes
(Notice that, although there are three alleles, only two can be present in a person's genotype. Can you explain why?)

Genotype	Phenotype (blood group)
$I^A I^A$; $I^A I^O$	group A
$I^B I^B$; $I^B I^O$	group B
$I^A I^B$	group AB
$I^O I^O$	group O

SPOTLIGHT

MENDEL

During the middle of the nineteenth century an Austrian monk, Gregor Mendel, carried out a series of experiments in his monastery garden. He planted edible pea seeds, transferred pollen from one chosen plant to another, collected their seeds and replanted them. By choosing one character at a time, Mendel was able to deduce how inheritance occurred and produced a number of laws of inheritance. These laws can be restated in modern terms.

Mendel's First Law of Segregation
The phenotypic characteristics of an organism are determined by pairs of genes in its genotype. Only one of a pair of these genes can be present in a single gamete.

Mendel's Second Law of Independent Assortment
The alleles of unlinked genes segregate independently during meiosis.

Mendel's achievements are remarkable if you consider that no one was to see a chromosome for another 50 years, the processes involved in cell division were unknown and the discovery of the nature of the gene was about 100 years into the future.

22.13 A woman of blood group A claims that she was given the wrong baby when she left a maternity hospital. The baby has blood group O and her husband has blood group AB.

(a) Is the woman's claim justified?

(To answer this question you must work out the possible genotypes of the parents and explore the genotypes of the offspring that *might* be produced by all possible crosses between them, using clearly explained genetics diagrams.)

(b) Would it have made a difference to your answer if her husband had been blood group B? Again, the best way to justify your answer is to use clearly explained genetics diagrams.

22.3 DIHYBRID INHERITANCE

KEY CONCEPT

A dihybrid cross involves the inheritance of genes at two loci.

In examining monohybrid inheritance, all the loci except the one controlling the characteristic under discussion were ignored. This made our study of genetics much simpler. **Dihybrid inheritance** is concerned with the phenotypic patterns produced by the inheritance of genes at **two loci**. In interpreting patterns of dihybrid inheritance, all loci except the two being studied are ignored.

Genes occupying the two loci involved in dihybrid inheritance may control two different characteristics in the phenotype or they may act together to control the same characteristic. In either case, the two loci may be situated on different chromosomes, in which case they are said to be **unlinked**, or on the same chromosome, when they are said to be **linked**.

Dihybrid inheritance with unlinked loci

Look at Fig 22.19; it shows what happens to two pairs of homologous chromosomes during meiosis. One locus is shown on each of the pairs of chromosomes. It is not always convenient to draw the chromosomes, but it helps to remember what the symbols in genetics diagrams represent. Only one member of each pair of chromosomes from the diploid parent cell is present in the haploid gamete cell. The movement of one pair of chromosomes during meiosis has no effect on the movement of other pairs, i.e. they segregate independently (**independent assortment**). As a result, there are four combinations of chromosomes (and therefore of alleles shown in the diagram) possible.

Question 22.5 concerns the inheritance of wing length in *Drosophila melanogaster* which should be attempted before reading on. One of the loci controlling wing length in *Drosophila melanogaster* is on chromosome 2. A locus controlling body colour lies on chromosome 3; its gene has a dominant grey-body allele (*G*) and a recessive ebony-body allele (*g*). As they are on different chromosomes, these two loci are obviously not linked. The progeny resulting from a cross between a homozygous long-winged, grey-bodied fly and a vestigial-winged, ebony-bodied fly in which the offspring (1) were allowed to interbreed are shown in Fig 22.20.

As can be seen in Fig 22.20, all of the offspring (1) show the features of the dominant alleles. Just as a cross between a homozygous long-winged fly with a vestigial-winged fly would produce an offspring (1) generation which was all long-winged (Question 22.5), so the adults in this cross do the same: consideration of the alleles controlling body colour has had no effect on the inheritance of wing length. During gamete formation, the alleles in these offspring (1) individuals show independent assortment. Thus, each heterozygous offspring (1) individual produces equal numbers of four haploid combinations of alleles: *LG*, *Lg*, *lG* and *lg*.

Random fertilisation among these gametes will result in the sixteen different genotypes among the offspring (2) shown in the diagram. With

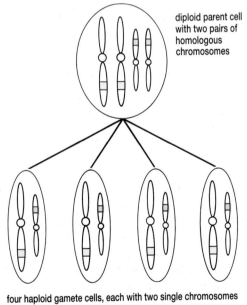

diploid parent cell with two pairs of homologous chromosomes

four haploid gamete cells, each with two single chromosomes

Fig 22.19 Four different gametes are produced by a cell which is heterozygous at two loci.

such a complicated diagram, use of the diamond checkerboard or Punnet square becomes essential.

Many of the genotypes in the offspring (2) in Fig 22.20 give rise to the same phenotypes, e.g. *LLGG*, *LLGg*, *LIGG* and *LLGg* all result in a phenotype of long wings and grey bodies. In fact, the sixteen different genotypes shown in the diagram result in only four phenotypes: long wings and grey bodies; long wings and ebony bodies; vestigial wings and grey bodies; vestigial wings and ebony bodies. If a large enough number of offspring is produced in the offspring (2), the ratio of these phenotypes will be **9:3:3:1**. This ratio is characteristic of the offspring (2) progeny of a dihybrid cross with no linkage.

A 9:3:3:1 ratio is characteristic of the offspring (2) generation of a dihybrid cross with no linkage.

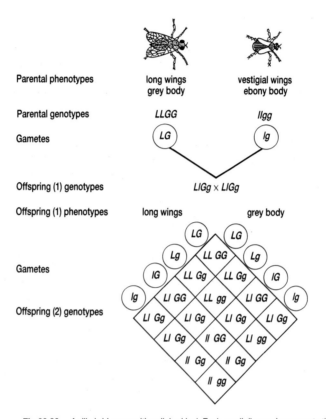

Parental phenotypes	long wings grey body	vestigial wings ebony body
Parental genotypes	*LLGG*	*llgg*
Gametes	*LG*	*lg*

Offspring (1) genotypes   *LlGg* × *LlGg*

Offspring (1) phenotypes   long wings   grey body

Fig 22.20 A dihybrid cross with unlinked loci. Each small diamond represents the genotype of an offspring that could be formed from this cross. The ratio of offspring (2) phenotypes is the characteristic 9:3:3:1.

Parental phenotype	Number of progeny			
	dark short	dark long	albino short	albino long
(a) Dark, short × dark, short	89	31	29	11
(b) Dark, short × dark, long	18	19	0	0
(c) Dark, short × albino, short	20	0	21	0
(d) Albino, short × albino, short	0	0	28	9
(e) Dark, long × dark, long	0	32	0	10
(f) Dark, short × dark, short	46	16	0	0
(g) Dark, short × dark, long	30	31	9	11

KEY CONCEPT

Linked loci occur on the same chromosome; unlinked loci are on different chromosomes.

Dihybrid inheritance with linked loci

A single chromosome carries the genetic code controlling the production of many proteins, i.e. it has many loci. When chromosomes migrate during cell division the linked loci on one chromosome will be moved together. Linked loci will, therefore, be inherited together unless pieces of chromosome break and join onto a different chromosome (**crossing over**).

Inheritance with no cross overs

The loci controlling shape of pollen grains and flower colour in sweet peas are linked. The pollen shape allele for long pollen grains, *L*, is dominant over that for round pollen grains, *l*, and the flower colour allele for purple flowers, *P*, is dominant over that for red flowers, *p*. These linked loci are shown on the chromosomes of the sweet pea plant in Fig 22.21. A cross between a parent which is heterozygous at both loci and a parent which is

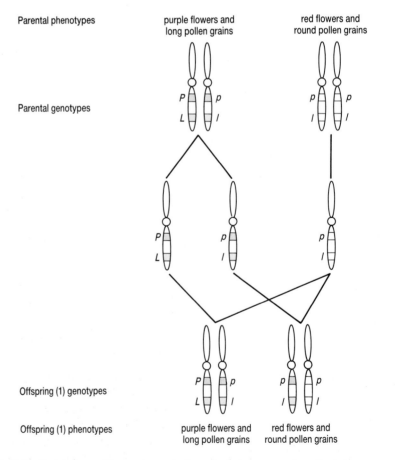

Fig 22.21 The inheritance of flower colour and pollen grain size in sweet pea plants. The loci are linked.

homozygous recessive at both loci is represented. Notice that linkage results in the offspring (2) showing exactly the same combination of characters that the parents had, i.e. purple flowers *and* long pollen grains or red flowers *and* round pollen grains.

 What would have been the result if these loci had not been linked?

Drawing chromosomes takes time and it is much easier to use symbols. Linkage can be represented using brackets around the linked alleles. The chromosomes in Fig 22.21 can thus be represented (*PL*) and (*pl*) and a diploid genotype can be represented as (*PL*)(*pl*).

 Redraw Fig 22.21 using these symbols: the parental genotypes are (*PL*)(*pl*) and (*pl*)(*pl*).

Inheritance with cross overs

At some stage during the cell cycle described in Fig 4.12, each chromosome replicates itself. Each chromosome then consists of two copies, called **chromatids**. When homologous chromosomes pair together during the first prophase of meiosis, they have already made copies of themselves to produce chromatids. The chromosomes pair together very precisely with homologous loci opposite each other. Very often the chromatids of homologous chromosomes intertwine, producing the X-shaped figures called **chiasmata** (singular chiasma) shown in Fig 22.22.

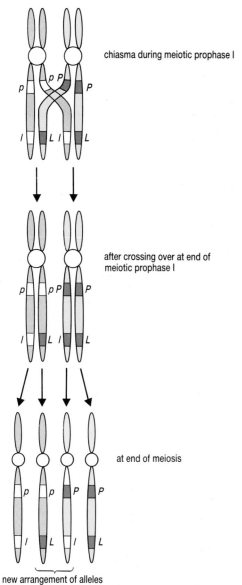

chiasma during meiotic prophase I

after crossing over at end of meiotic prophase I

at end of meiosis

new arrangement of alleles

Fig 22.23 Crossing over of chromosomes during meiosis produces new combinations of linked genes.

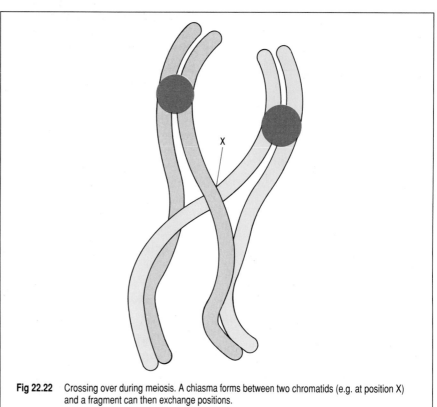

Fig 22.22 Crossing over during meiosis. A chiasma forms between two chromatids (e.g. at position X) and a fragment can then exchange positions.

If a breakage occurs at point X on the two chromatids in Fig 22.22 and it is followed by the two chromatid fragments joining the opposite chromosome, a genetic exchange, called **crossing over**, will occur. After meiosis has been completed, two daughter cells will contain a copy of one of the original chromatids and two daughter cells will contain these new chromatids.

Crossing over between the chromosomes of a sweet pea plant carrying the loci for flower colour and pollen shape is represented in Fig 22.23. You can see that two new genetic combinations appear among the gametes which were not produced previously, i.e. (*Pl*) and (*pL*), which might result in new genetic combinations among the offspring (plants with purple flowers and short pollen grains or with red flowers and long pollen grains). These new phenotypic combinations are called **recombinants**.

 Which genotype could be seen as a recombinant in Fig 22.23: (*PL*)(*pl*), (*Pl*)(*pl*) or (*Pl*) (*PL*)?

KEY TERMS	
Crossing over	The exchange of corresponding parts between homologous chromosomes by breakage and reunion.
Recombinants	Offspring with new phenotypic combinations as a result of new genetic combinations arising from crossing over during gamete production.

QUESTION

22.16 Write the genotypes of the gametes which would be formed by a parent of genotype *HhTt*
  (a) if the loci are unlinked
  (b) if the loci are linked (*hT*) and (*Ht*) and no crossing over occurs
  (c) if the loci are similarly linked and crossing over occurs.
  Comment on the frequency of the different types of gamete.

Cross over value and genetic mapping

Crossing over between linked loci does not happen in every cell which undergoes meiosis. Consequently, there will always be more offspring with the parental combination of characters than there are with new combinations. Crossing over between loci is more likely the further apart the loci are on the chromosome, so that loci which are adjacent are seldom separated. The frequency of crossing over between loci can therefore indicate how close the loci are to each other on their chromosome. The only way to tell how often crossing over has occurred is from the proportion of recombinants among the offspring. This leads to a value called the **cross over value (COV)**, defined as

$$\text{cross over value} = \frac{\text{number of recombinant offspring}}{\text{total number of offspring}} \times 100\ \%$$

 Theoretically the cross over value can never be above 50%. Can you suggest why?

ANALYSIS

Genetics maps

This exercise will give practice in creating and interpreting genetic maps.

Genetics maps are made using the cross over values found from repeated genetics crosses. Whilst it has no physical value, 1 **map unit** is defined as a 1% cross over value. If two loci A and B are found to have a cross over value of 20%, they are said to be 20 map units apart. If a third locus, C, has a cross over value with B of 15%, it is 15 map units from B.

In Fig 22.24 the cross over value between A and C must be either 35% (if locus B lies between locus A and C) or 5% (if locus C lies between locus A and B). Only by finding the cross over value from genetics experiments can the position of locus C be determined. Analyses like these have allowed chromosome maps to be deduced for many species of bacteria and other organisms commonly used in genetics, such as *Drosophila*.

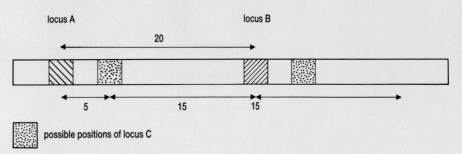

Fig 22.24 A hypothetical chromosome map constructed from the experimental results of the cross over values of three loci, A, B and C.

(a) Construct a chromosome map from the following cross over values (COV) between loci A, B, C and D as shown in the margin:

Loci	COV
A-B	15%
B-C	5%
C-D	20%
A-C	10%
B-D	25%

The mutant genes producing ebony bodies (*g*) and curled wings (*w*) in *Drosophila* lie 20.7 map units apart on chromosome 3. The alleles dominant to these mutant alleles are denoted *G* and *W* respectively for grey body and straight wings.

(b) Draw a pair of chromosomes to show the genotype (*GW*)(*gw*).

(c) What would be the proportion of different phenotypes among the offspring (1) of a cross between a female of genotype (*GW*)(*gw*) and a male of genotype (*gw*)(*gw*) if no cross overs occurred?

(d) In an actual cross between a grey-bodied, straight-winged female of genotype (*GW*)(*gw*) and a male of genotype (*gw*)(*gw*), the following offspring were found:

Grey bodies and straight wings	40
Grey bodies and curled wings	13
Ebony bodies and curled wings	38
Ebony bodies and straight wings	9

Use these data to calculate the cross over value between these loci and suggest why your value is not identical with the value of 20.7 given above.

22.4 POLYGENIC INHERITANCE

Many inherited characteristics are not controlled by genes at one locus, but rather by genes at many loci, located on several chromosomes. In this pattern of **polygenic inheritance** the different loci all contribute to the final phenotype. The simplest example of polygenic inheritance arises when only two loci are involved.

14 Stem length in pea plants and human height are inherited characteristics. Is each the result of monohybrid or polygenic inheritance?

The colour of the iris in human eyes is controlled by at least two loci. The genes at both loci have dominant alleles which cause the production of melanin in the iris; this is the same brown pigment which colours hair and skin. The recessive alleles at each locus result in no pigment production. The iris contains two layers of pigment, one in front of the other. If there is

Human skin colour is thought to be controlled by alleles at up to 40 loci. This is an example of polygenic inheritance.

no melanin in either layer (the albino condition), the iris appears red because the blood vessels within it can be seen. More commonly, the rear layer of the iris contains pigment. If the front layer contains little or no pigment, the iris appears blue. As progressively more melanin is added to the front layer the iris appears deeper blue or green, various shades of brown and finally almost black.

Table 22.3 shows how a range of eye colours is produced: the symbols *D* and *E* represent the alleles at each locus which result in melanin being deposited in the front layer of the iris and *d* and *e* represent the alleles which result in no melanin deposition.

Table 22.3 **Human eye colour results from the action of two unlinked loci controlling the deposition of melanin in the front layer of the iris. Alleles *D* and *E* are dominant to alleles *d* and *e* respectively**

Number of D and E alleles in genotype	Examples of genotype	Colour of iris
4	*DDEE*	dark brown or black
3	*DDEe* *DdEE*	medium brown
2	*DDee* *ddEE* *DdEe*	light brown
1	*Ddee* *ddEe*	deep blue or green
0	*ddee*	light blue

QUESTION

22.17 Draw a genetics diagram to show the number and expected frequency of different eye-colour phenotypes among the offspring of two parents with light brown eyes of genotype *Dd Ee*. (Refer to Table 22.3 for eye-colour phenotypes.)

22.5 SEX DETERMINATION AND SEX LINKAGE

Although regarded as a single characteristic, there are many physical and chemical differences between different sexes. The **sex chromosomes** are involved in controlling these differences. The remaining chromosomes which are not involved in sex determination are called **autosomes**; humans have 22 pairs of autosomes.

In some organisms, for example grasshoppers, a diploid male has only one sex chromosome whereas a female has two. In most diploid organisms both sexes have two sex chromosomes. Unlike the members of a pair of autosomes, however, these two sex chromosomes do not have the same loci and one is much smaller than the other (Fig 22.25). Both sex chromosomes in female mammals are the large **X chromosome**. The male has only one of these chromosomes with a much smaller **Y chromosome**. The sex chromosome genotype of a female mammal is thus *XX* and of a male mammal *XY*. Because all of the egg cells produced by a female mammal will have the *X* chromosome, she is called the **homogametic sex**. The male, half of whose sperm will carry the *X* chromosome and half the *Y* chromosome, is the **heterogametic sex**. In mammals, it is the sperm which determines the sex of the zygote. The male is not always the heterogametic sex, for example in most bird species the male is homogametic and it is the female which is the heterogametic sex. Equal proportions of female and male offspring might be expected from any mating (Fig 22.26).

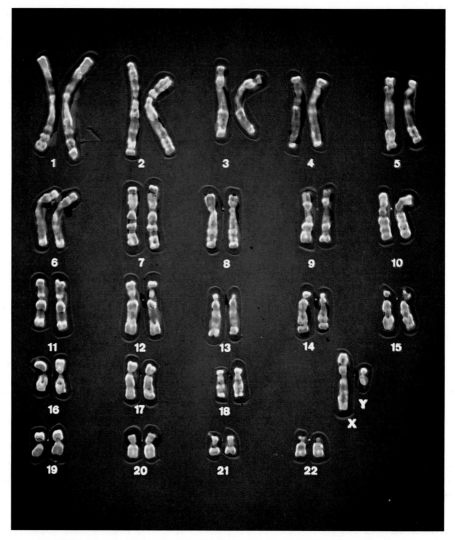

Fig 22.25   A human male karyotype showing all 44 autosomes plus the X and Y sex chromosomes.

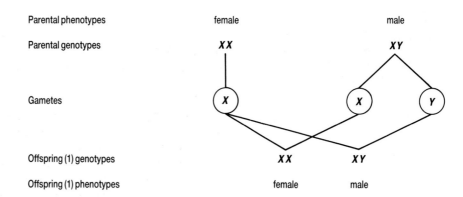

Fig 22.26   Inheritance of sex in humans as shown by X and Y, the sex chromosomes. Equal proportions of males and females are expected from parents.

Sex linkage

Not all the loci on the sex chromosomes are concerned with sexual characteristics. For example, the larger X chromosome in humans carries loci controlling the production of factor VIII, required for normal blood clotting, (see Section 10.2) and the ability to distinguish the colours red and green. The shorter Y chromosome in humans carries neither of these loci, so if a male inherits a recessive allele of these genes on the X chromosome from his mother he will suffer from haemophilia or red-green colour blindness. A female inheriting a recessive allele of one of these genes would not suffer either of these complaints provided her other X chromosome carried the normal, dominant allele of the gene (Fig 22.27).

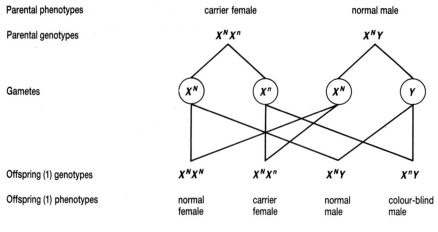

key:

X^N = X chromosome with allele for normal colour vision

X^n = X chromosome with allele for red-green colour blindness

Y = Y chromosome which has no colour-vision locus

Fig 22.27 Sex-linked inheritance of red-green colour blindness in humans. The male Y chromosome does not have a colour vision locus so a male need only have one affected X chromosome to have impaired colour vision, explaining why most red-green colour blind people are male.

15 Why do the sons of a haemophiliac man never inherit his haemophilia?

SPOTLIGHT

Fig 22.28 A calico (tortoiseshell) cat; a female genetic mosaic. Some cells determine orange fur and others determine black fur since one of the X chromosomes is inactivated. The gene for coat colour is on the X chromosome. (The white fur pattern seen in this cat is controlled by another, unlinked, gene.)

INACTIVATED X CHROMOSOME AND CALICO CATS

Although all the cells in the body of a female mammal carry two X chromosomes, during an early stage of development one of them becomes inactive and coils up to form a tight mass, called a **Barr body**. This Barr body can be seen using an optical microscope. Which of the X chromosomes becomes inactive in each embryonic cell seems to be a chance event, but all of its daughter cells have the same inactivated chromosome. This means that female mammals are a genetic mosaic, with patches of cells with one X chromosome inactivated, separated by patches of cells with the other X chromosome inactivated.

The effects of this are clearly seen in some female cats. One of the genes controlling coat colour in the domestic cat is situated on the X chromosome. It has two alleles; C^O and C^B. C^O results in orange fur and C^B results in black fur. Males, with only one X chromosome, will be either orange or black but heterozygous females, with one X chromosome carrying C^O and the other X chromosome carrying C^B, have patches of black and orange fur (the calico or tortoiseshell pattern). This is shown in Fig 22.28.

22.6 PLEIOTROPY

Throughout this chapter, alleles of genes have been described as though they had only one effect on the phenotype of their possessor. This is often not the case. **Pleiotropic genes** affect two or more, apparently unrelated, aspects of the phenotype.

Wild mice have grey-coloured fur (known as agouti) controlled by a recessive gene y. A dominant mutant allele of this gene, Y, causes the coat colour to be yellow. All mice with yellow fur are heterozygous (Fig 22.29) because the homozygous YY condition results in pre-natal death .

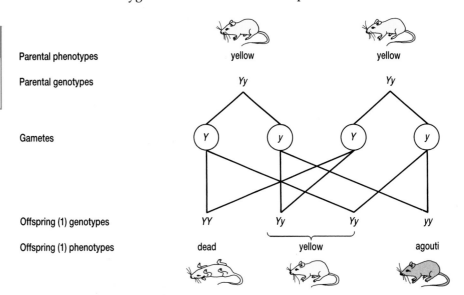

Fig 22.29 Inheritance of yellow coat colour in mice.

The homozygous yellow mice die because the Y allele has an effect on the phenotype other than that on coat colour. Also notice that, although the Y allele is dominant in its effect on coat colour, it is recessive in its lethal effect (since only the homozygote YY dies). With pleiotropic genes the terms dominant and recessive refer to the different effects of the alleles on the phenotype. When discussing genes with only one known effect on the phenotype, however, it is possible to refer to the alleles as dominant or recessive, as has been done throughout this chapter.

QUESTIONS

22.18 Palomino horses are golden in colour. Unfortunately for horse breeders they do not breed true. A series of matings between palomino horses produced 20 palomino, 10 cream-coloured and 10 chestnut (red-brown) foals.
  (a) This coat colour is controlled by a single gene. Are the alleles of this gene (i) dominant and recessive or (ii) codominant? Justify your answer.
  (b) Devise symbols to represent the coat-colour alleles and draw a genetics diagram to represent a cross between two palomino horses.

22.19 Refer to spotlight: 'Inactivated X chromosome and calico cats'. Devise suitable symbols and use them to draw a genetics diagram to represent the phenotypes which might be expected if an orange male cat mated with a calico female cat. Explain why it is not possible to have a calico male kitten among the litter.

22.20 Humans with blood group O are more likely to have a duodenal ulcer and humans with blood group A are more likely to suffer stomach cancer than humans with any other

---

ABO blood group. What does this suggest about the effects of the alleles of the human ABO blood group genes?

22.21 The allele for round seeds in peas (*W*) is usually described as dominant over the allele for wrinkled seeds (*w*). The table shows some properties of round and wrinkled pea seeds.

Genotype	Seed appearance	Starch content	Sugar content
WW	round	high	low
Ww	round	intermediate	intermediate
ww	wrinkled	low	high

(a) What does this show about the number of ways in which the round-seed and wrinkled-seed alleles affect the phenotype? What name is given to this property of alleles?

(b) Do these data confirm that the round-seed allele is dominant to the wrinkled-seed allele? Explain your answer.

SUMMARY

Every organism has visible or chemical characteristics, referred to as its phenotype. Organisms in a population may show variation in their phenotypes because of the effects of the environment and because of differences in their genetic make-up, or genotype.

The genotype of an organism depends on the nucleic acid in the chromosomes of its cells. This DNA is arranged along chromosomes as a series of segments, each of which controls the production of a polypeptide or protein. Each of these segments is called a gene and its specific location on a chromosome is called the locus of that gene. Changes in DNA structure may result in slight differences in the base sequence of some genes. If these changes result in the production of polypeptides or proteins which have a different effect on the characteristic controlled by that gene, they may be inherited. These alternative base sequences which control a single characteristic are called alleles of that gene. The flower-colour gene in peas has two alleles; one causes the production of purple flowers and the other causes the production of white flowers.

Every diploid adult has two copies of the gene controlling each characteristic, one on each of the chromosomes in a homologous pair. If both of these are identical alleles of that gene, the cell is said to be homozygous; if the two alleles are different, the cell is said to be heterozygous. During gamete production, haploid cells are formed which possess only one gene controlling each characteristic. Heterozygous cells always produce gametes with different genotypes whilst homozygous cells always produce gametes with an identical genotype.

If an allele of a gene always has an effect on the phenotype of its possessor, it is said to be dominant. If an allele of a gene only has an effect on the phenotype if its possessor is homozygous, the allele is said to be recessive. Strictly, a dominant allele is dominant to a recessive allele and a recessive allele is recessive to a dominant allele. Some alleles of the same gene exert their effect on the phenotype, even if the individual is heterozygous. Such alleles are termed codominant.

The phenotype of an organism which is heterozygous for a gene which has a completely dominant allele and a completely recessive allele is identical to that of the homozygous dominant genotype. To determine whether an organism with such a phenotype is homozygous or heterozygous a test cross must be performed. A test cross involves crossing the organism of unknown genotype with an individual known to be homozygous recessive. The presence of any offspring showing the recessive characteristic confirms that the unknown genotype must be heterozygous.

Inheritance of a characteristic controlled by alleles of a single gene is called monohybrid inheritance. Dihybrid inheritance involves the inheritance of characteristics which are controlled by the alleles of two genes. If these genes occur on separate chromosomes, they are said to be unlinked and they segregate independently of each other during gamete formation. Genes which are on the same chromosome are said to be linked. During gamete formation, linked genes are inherited together, unless chromosome breakage occurs during formation of a chiasma. As a result of chiasmata, genetic exchange between homologous chromosomes occurs, a process known as crossing over. Crossing over can be recognised because it results in new combinations of characteristics amongst the offspring of a cross, called recombinants. The percentage of recombinants among the total number of offspring is the cross over value. Because crossing over is more likely between genes which are far apart on their chromosome than between genes which are close together on their chromosome, the cross over value allows the production of genetic maps of chromosomes to be made.

Sex is a complex characteristic and is controlled by gene(s) on the sex chromosomes; other chromosomes are called autosomes to distinguish them from these sex chromosomes. In humans, two types of sex chromosomes occur: males have one X chromosome and one Y chromosome whereas females have two X chromosomes. Non-sexual characteristics controlled by genes on the sex chromosomes are called sex-linked characteristics. Human males have only one copy of each gene present on the X chromosome and so recessive sex-linked characteristics are more commonly seen in males than in females.

Many genes have more than one effect on the phenotype, a phenomenon known as pleiotropy. Sometimes one effect of an allele of such a gene may be dominant whilst another effect of the same allele may be recessive.

In discussing inheritance patterns, alleles of genes may be represented using symbols. Usually, dominant alleles are represented using upper case letters, e.g. A, and recessive alleles using the corresponding lower case letters, e.g. a. Codominant alleles, or genes with more than two alleles, are represented using an upper case letter to represent the gene and a superscript to represent the alleles, e.g. A^B, A^C, A^D and so on. The details of crosses between individual organisms are labelled using a standard format:

> Parental phenotypes
> Parental genotypes
> Gametes
> Offspring (1) genotypes
> Offspring (1) phenotypes

It is usually convenient to use a diamond-shaped checkerboard or the Punnet square to denote the offspring resulting from the random fusion of gametes during fertilisation. The ratio of phenotypes amongst the offspring of crosses is usually diagnostic in determining the pattern of inheritance and genotypes of individuals.

Chapter 22: Answers to Quick Questions

1 (a) 2 (b) 1
2 Allele for brown hair; gene for hair colour.
3 Each carries only one sex chromosome (is haploid). On average, half will carry the X chromosome and half will carry the Y chromosome.
4 (a) 2 (b) 1
5 One purple-flower allele 'produces' sufficient enzyme to produce the same phenotype as two purple-flower alleles.
6 *PP*
7 Because the chromosome assortment during meiosis is a random process.
8 No. Fusion of gametes occurs at random, so only in very large samples of offspring will an exact ratio of 3:1 be common.
9 Its gametes carry only the recessive allele which will not mask the effect on the offspring's phenotype of any allele from the parent of unknown genotype.
10 A ratio of 1:1:1:1, i.e. 1 purple-flowered, long-grained; 1 purple-flowered, round-grained; 1 red-flowered, long-grained; 1 red-flowered, round-grained.
11

Parental phenotypes	purple-flowered long pollen grains	×	red-flowered round pollen grains
Parental genotypes	(*PL*)(*pl*)	×	(*pl*)(*pl*)
Gametes	(*PL*)(*pl*)		(*pl*)
Offspring (1) genotypes	(*PL*)(*pl*)		(*pl*)(*pl*)
Offspring (1) phenotypes	purple-flowered long pollen grains		red-flowered round pollen grains

12 (*Pl*)(*pl*) would appear as the recombinant with purple flowers and round pollen grains. Although a cross over is needed to produce the genotype (*Pl*)(*PL*), the *l* allele will be masked by the *L* allele.
13 Every cross over results in two recombinant chromatids and two parental chromatids.
14 Stem length in peas is monohybrid; height in humans is polygenic.
15 His sons must inherit his Y chromosome. The haemophilia gene is carried on the X chromosome which the sons inherit from their mother.

Chapter 23

GENETIC CHANGES IN POPULATIONS: EVOLUTION

LEARNING OBJECTIVES

When you have studied this chapter you should be able to:

1. define the terms gene pool and allele frequency;

2. explain the Hardy-Weinberg equation and use it to calculate allele frequencies and genotype frequencies from data concerning the phenotypes within a population;

3. explain that, in its broadest sense, evolution is a change in allele frequency within a population and describe how such changes may result from mutation, genetic drift, migration, non-random mating and natural selection;

4. illustrate various types of gene mutation and chromosome mutation;

5. use examples from natural populations to explain the terms: stabilising selection, directional selection, disruptive selection, biological fitness, selection coefficient, transient polymorphism and stable polymorphism;

6. compare Darwin's theory of evolution through natural selection with a modern neo-Darwinian theory.

During the 1950s a number of chemical compounds were thought to be a solution to many human ills. It was hoped that DDT, a synthetic compound that kills insects, might end the starvation caused by locusts eating human crops and that antibiotics, substances which in small amounts inhibit the growth of bacteria, might cure all bacterial diseases. For a time these substances were effective but, after years of use, insect populations were found which were no longer affected by DDT and disease-causing bacteria

Fig 23.1 DDT brought enormous benefits to human beings, but then nature caught up with us!

were found to be immune to commonly used antibiotics. The resistance of these organisms is inherited, i.e. it is controlled by genes. A locust population which is resistant to DDT is, therefore, genetically different from a locust population which is susceptible to DDT. Since there were no DDT-resistant locust populations when DDT was first used, a genetic change must have occurred in these locust populations. **Population genetics** is the study of such genetic changes in populations. In its broadest sense, **evolution** is said to occur when a genetic change occurs in a population.

 Why didn't DDT wipe out all locust populations?

23.1 POPULATION GENETICS

The previous chapter dealt with how to represent the genotype of individual organisms using symbols. To represent the genotype of a diploid organism you needed to describe two genes. For example, the genotype of a pea plant which is heterozygous for long pollen grains can be represented as *Ll*, where *L* represents the dominant long-pollen-grain allele of the controlling gene and *l* represents the recessive round pollen-grain allele of the gene. To represent the genotype for pollen shape of 1000 pea plants in a population you would need to represent 2000 genes (two genes for each of 1000 organisms). The **gene pool** is the sum total of the genes in a population. Although the term gene pool is used in this way to represent the sum total of the genes in a population, we are usually only concerned with one phenotypic character at a time. Thus, when considering the inheritance of pollen shape in the pea plant population described above, the 2000 alleles represent the gene pool for that character.

Fig 23.2 This crowd carries part of the human gene pool.

Allele frequency

If all the pea plants in a population of 1000 individuals had the genotype *Ll*, there would be 1000 *L* alleles and 1000 *l* alleles in the gene pool of 2000. The frequency of the *L* allele is $1000/2000 = 50\%$ and for the *l* allele is $1000/2000 = 50\%$. Statisticians do not usually refer to frequencies as percentage values but rather as decimal values. The decimal equivalent of 50% is 0.5; the alleles *L* and *l*, therefore, each have a frequency of 0.5.

A general formula can be used to represent the frequencies in a population of two alleles of a single gene. If one allele is dominant to the other, the symbols *p* and *q* are used to represent the frequency with which the dominant and recessive alleles appear in a gene pool. This formula is:

GENETIC CHANGES IN POPULATIONS: EVOLUTION

Frequency of	+	Frequency of	=	Gene pool
dominant allele		recessive allele		
p	+	q	=	1

 If the frequency of the recessive allele of a gene is 0.4, what is the frequency of its dominant allele?

The Hardy-Weinberg equation

Whilst the allele frequency is an essential concept for analysing genetic change, it is not something which can be measured directly in populations of diploid organisms. What can be seen or measured is the **phenotype** of each individual in the population.

In 1908 an English mathematician, G. H. Hardy, and a German physician, W. Weinberg, independently developed a way to relate allele frequencies, genotype frequencies and phenotype frequencies using a formula which now bears their names.

KEY CONCEPT

THE HARDY-WEINBERG EQUATION

Frequency of homozygous dominant individuals	+	Frequency of heterozygous individuals	+	Frequency of homozygous recessive individuals	=	Total population
p^2	+	$2pq$	+	q^2	=	1

Using the Hardy-Weinberg equation

To use this equation to calculate allele frequencies, you need to know the frequency of at least one of the genotypes. The homozygous dominant individuals cannot be distinguished from the heterozygous individuals but homozygous recessive individuals can be identified. To calculate allele frequencies, you should start with the frequency of the homozygous recessive individuals.

Suppose there are 150 plants which produce long pollen grains and 50 plants which produce round pollen grains in a population of peas. The frequency of the long-pollen-grain and round-pollen-grain alleles of the pollen-shape gene can be found by calculating the frequency of homozygous recessive individuals, i.e. those plants which produce round pollen grains.

Number of plants which produce round pollen grains   = 50
Number of plants in population   = 200
Therefore, frequency of plants producing round
    pollen grains   = 50/200 = 0.25
∴ frequency of homozygous recessive individuals ($= q^2$) = 0.25
∴ frequency of the recessive allele   ($= q$) = √0.25 = 0.50

Using the Hardy-Weinberg equation has enabled us to find the frequency of the recessive allele (q). It is now an easy matter to find the frequency of the dominant allele (p):

Since $p + q = 1$,
frequency of the dominant allele   ($= p$) = 1 − 0.50 = 0.50

The frequency of all the pollen grain **genotypes** can then be calculated using the Hardy-Weinberg equation:

Frequency of **LL** genotype $(= p^2) = 0.50^2 = 0.25$
Frequency of **Ll** genotype $(= 2pq) = 2 \times 0.50 \times 0.50 = 0.50$
Frequency of **ll** genotype $(= q^2) = 0.50^2 = 0.25$
(Note that the frequencies of these three genotypes add up to 1).

Using these frequencies, you would, therefore, expect 50 **LL** individuals, 100 **Ll** individuals and 50 **ll** individuals among the 200 pea plants.

QUESTIONS

23.1 A population of *Drosophila* contains 64 long-winged flies and 36 vestigial-winged flies.
 (a) What is the frequency of vestigial-winged flies (q^2) in this population?
 (b) What is the frequency of the vestigial-wings allele (q)?
 (c) What is the frequency of the long-wings allele ($p = 1 - q$)?
 (d) What is the frequency of the homozygous long-winged flies (p^2)?
 (e) What proportion of the long-winged flies are homozygous and what proportion are heterozygous?

23.2 Humans suffering from insulin-dependent diabetes mellitus cannot secrete the hormone insulin. This disorder is inherited as a recessive allele at a single locus. If the frequency of this allele (q) in a human population is 0.07, calculate the frequency of
 (a) the normal allele
 (b) people who suffer diabetes
 (c) heterozygous carriers of the diabetes allele.

23.2 HARDY-WEINBERG PRINCIPLE OF EQUILIBRIUM POPULATIONS

Hardy and Weinberg each developed their equation whilst considering **equilibrium populations** in which no genetic change occurred, i.e. where the frequency of alleles and of genotypes remained constant from generation to generation. If allele frequencies do not change, evolution does not occur. Whilst you will not be expected to reproduce the mathematical proof which Hardy and Weinberg used, it is important that you are familiar with their principle. They found that the frequency of alleles of a gene in populations would only stay constant if a number of conditions were met. The more important of these conditions are:

- the population must be extremely large
- there must be no migration into (immigration) or out of (emigration) the population
- mating between individuals must be completely at random, e.g. there must be no tendency for certain genotypes to mate together either because of geographical proximity or mate selection
- there should be no genetic mutations
- all genotypes must be equally fertile.

 Which, if any, of these conditions are likely to be met by a natural population?

Provided all these conditions were met, the frequency of a dominant allele and recessive allele of a gene would be p and q respectively in every generation and the frequency of different genotypes would always follow the Hardy-Weinberg equation.

As you read through the list of conditions you might have realised that few, if any, populations satisfy every condition, so that few populations are in equilibrium. The importance of the Hardy-Weinberg principle is that it points to the causes of genetic change in populations, i.e. it shows the processes of evolution. In this section, each of Hardy and Weinberg's conditions, except mutation, will be examined in more detail, leading to a better understanding of how evolution within a population occurs. Chapter 24 discusses how the evolution of new species may occur.

Population size

The Hardy-Weinberg equation can be used to calculate the number of individuals in a population which carry a particular allele of a gene (Questions 23.1 and 23.2). Table 23.1 shows the number of organisms carrying a particular recessive allele in one population of 50 individuals and a second population of 5000 individuals. The frequency of the recessive allele (q) is 0.01 in both populations. In the smaller population only one individual carries the recessive allele (a). If this individual were to die before reproducing, the a allele would be lost from the population. It is less likely that the 100 individuals carrying the a allele in the second population would all die before reproducing and, hence, it is less likely that the a allele would be lost. Such random fluctuation in allele frequency is called **genetic drift**.

Table 23.1 The Hardy-Weinberg equation has been used to calculate the number of individuals carrying the a allele in two populations of different sizes. The a allele has a frequency of 0.01 in both cases. (Adjustment has been made to ensure whole numbers of individuals)

Population size	Number of individuals with each genotype			Number of individuals with the a allele
	AA (p^2)	Aa $(2pq)$	aa (q^2)	
50	49	1	0	1
5000	4900	99	1	100

Small, natural populations are fairly rare but there are well-known instances where genetic drift has played an important part in evolution. Occasionally, populations become very small, a phenomenon known as a **population bottleneck**, perhaps as a result of habitat destruction, drastic climatic change or human activities. Just by chance the population of individuals which survive a population bottleneck may have different allele frequencies from the original population so that, when their reproduction leads to a large population again, a permanent change in allele frequencies occurs. The human population of the northern hemisphere is thought to have been reduced to several pockets of a hundred or so individuals during the Ice Ages, so genetic drift may have played an important part in human evolution. The northern elephant seal (*Mirounga angustirostris*), shown in Fig 23.3, was hunted for its oil to such an extent

Fig 23.3 Northern elephant seals. These animals, found only off the west coast of North America, are the survivors of a species that was nearly extinct last century. Measurements of the genetic variation of this surviving population shows it to be less variable than expected.

that by the 1890s only about 20 individuals survived. There are now about 30 000 descendants of these elephant seals and biochemical analysis shows much less variation between individuals than would be expected of a sexually reproducing population.

Fig 23.4   Cheetah, an endangered species. Careful selection of breeding pairs is made by humans to avoid inbreeding to avoid lowering even more the genetic variation in zoo populations of cheetahs.

CONSERVING CHEETAHS

If we use electrophoresis to examine the variability of protein structure in populations of cheetahs we find it is very low compared with, say, populations of the domestic cat. This means that the amount of genetic variation in cheetah populations is also very low, i.e. if you were to select two cheetahs at random they are likely to be genetically very similar. This suggests that at sometime in the past cheetahs have gone through a population bottleneck and that all the cheetahs alive in Africa today are descended from a relatively small ancestral population. This lack of genetic variability presents problems for captive breeding programmes with this endangered species since crossing genetically similar individuals can lead to **inbreeding depression**, something already evident even in wild cheetah populations. Cheetahs used in captive breeding programmes have, therefore, to be matched carefully, using pedigree records and biochemical analysis, to minimise the effects of inbreeding depression.

A special case of genetic drift, the **founder effect,** is seen when an individual or small group of individuals leaves a large population and migrates to a new environment where it is reproductively isolated. Again, these small groups may have allele frequencies which are different from the larger parent population. The founder effect is thought to have played an important part in the evolution of island populations. It has also been important in some human populations. Religious persecution during the eighteenth century caused a number of migrations. Two groups, the Amish from Switzerland and the Dunkers from Germany, independently migrated to Pennsylvania where they have remained reproductively isolated from the rest of the American population. Both these groups have an incidence of certain genetic traits which is much higher than the rest of America, presumably because one or two individuals in the original migratory party happened to carry a rare allele.

 What would happen to the frequency of one of these genetic traits if the religious group were to allow free marriage with the American population?

ANALYSIS

Genetic drift

Modelling is an important technique in genetics. This exercise will give you some insight into the nature of the process.

You can mimic the effects of genetic drift using a diagram like the one on the next page. Each of six individuals in a population produces two identical offspring. Half of these die each generation; the population size stays the same, i.e. the total number of individuals producing offspring remains unchanged.

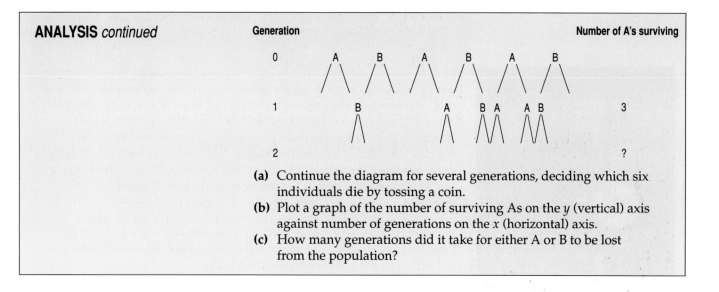

(a) Continue the diagram for several generations, deciding which six individuals die by tossing a coin.
(b) Plot a graph of the number of surviving As on the y (vertical) axis against number of generations on the x (horizontal) axis.
(c) How many generations did it take for either A or B to be lost from the population?

Migration

Some populations show predictable seasonal movements between breeding grounds and winter refuges. Since the whole population moves together, there is not likely to be any effect on allele frequencies within the population. If some individuals move from one population into another, as young baboons often do, alleles of genes are lost from one population and gained by another. Such **gene flow** may have two effects. On an irregular basis, the loss of alleles from one population and their gain by another will change the gene pool of both. On a regular basis, gene flow will ensure that all the populations of one species share a common gene pool, so that its disruption will lead to the accumulation of genetic differences between the populations (see Chapter 24).

 How will rates of gene flow be affected by (a) geographical separation (b) mobility of an organism?

Random mating

In many animal and plant species, fertilisation occurs when gametes are simultaneously released into the water surrounding hundreds or thousands of organisms. In the Great Barrier Reef off the eastern coast of Australia, the release of egg cells and sperm by many species of coral is synchronised by the phase of the moon. On the fourth and fifth night after the full moons of November and December, the sea is clouded by millions of egg cells and sperm from these corals. Under these conditions, fertilisation must almost be a random process; only distance between corals is likely to affect the likelihood of egg cell and sperm meeting.

In many animal species, however, even close proximity does not ensure reproductive success, since individuals are selective when mating.

Fig 23.5 The colour attracts a mate, but also attracts predators.

Consequently, egg cells can be regarded as a commodity in short supply whilst sperm are not. We might, thus, expect males to compete with each other for access to the females and their egg cells. Males with particular 'strong' features would mate successfully whilst those with 'weak' features would not: these male features would be subject to natural selection. Similarly, females that mated with 'strong' males would have more offspring than those that mated with 'weak' males: female choice would also be subjected to natural selection, a process called **sexual selection**.

In fact, sexual selection does occur and accounts for such features as large body size in male elephant seals, large antlers of red deer and the red belly of a male three-spined stickleback.

Competition for a mate often results in ritualised displays, such as those of the male peacock (shown on the front cover of this book) and male bowerbird. Even in relatively inconspicuous animals, sexual selection occurs. For example, female *Drosophila* mate only with a male that performs a suitable ritualised dance. Sexual selection also occurs in humans: statistical data show that human females are more likely to marry someone of the same ethnic group, height, IQ and even of the same length of finger. Sexual selection appears to act on some strange characteristics!

Different fertility of genotypes

Unless mutations occur, the offspring of an organism which reproduces by mitosis will be genetically identical to each other and to their parent. As you saw in Chapter 22, the offspring of organisms which reproduce by meiosis are genetically different from each other and from their parents, even when considering a single phenotypic character. In some cases, such as the case of Tay-Sachs disease in humans or yellow coat colour in mice, certain genotypes are harmful to their possessors.

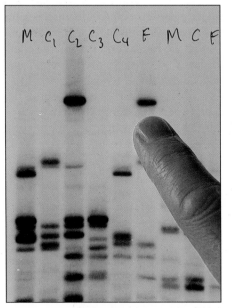

Fig 23.6 A genetic fingerprint – each of us has a unique genetic fingerprint, except for identical twins.

AND YOU THOUGHT YOU WERE ONLY ONE IN A MILLION?

Diploid human cells have 23 pairs of chromosomes. Haploid egg cells and sperm contain only one copy from each pair. During gamete production, the separation of chromosomes in one pair does not affect the way that members of other chromosome pairs separate (**independent assortment**). As a result, 2^{23} different chromosome combinations can occur during meiosis when one chromosome is taken from each pair ($2^{23} \approx 8 \times 10^6$). Since both parents can produce 2^{23} different chromosome combinations in their gametes and fertilisation of the mother's egg cell by the father's sperm may occur at random, $2^{23} \times 2^{23}$ possible chromosome combinations may occur among the zygotes of one human couple ($2^{23} \times 2^{23} = 2^{46} \approx 70 \times 10^{12}$). Crossing over, which results in new combinations of alleles on chromosomes, and mutations increase the potential number of different offspring which one couple may have still further.

If the possession of one allele of a gene makes its carrier likely to have fewer offspring than the carrier of a different allele of the same gene, the frequency of the harmful allele will be reduced in the next generation. This change in allele frequency resulting from differential fertility is known as **natural selection** and the reproductive success of a phenotype is its **fitness**. Notice that organisms do not need to die before reproducing in order for their alleles to be poorly represented in succeeding generations, they simply need to produce fewer offspring than fitter individuals. Because it is the main agent causing genetic changes in populations, natural selection will be examined in more detail in Section 23.4.

QUESTIONS

23.3 List five factors which may change the frequency of the alleles of a single gene in a population.

23.4 Like human speech, many bird species have local dialects in their song. A female is more likely to mate with a male which sings the same dialect as her own father. What effect might this have on variation in song dialects?

23.3 MUTATION

A **mutation** is a change in the genetic code of an organism. It may involve a change in

- the structure of DNA
- the amount of DNA.

Gene mutations

A gene can be described as a length of DNA which carries the genetic code for the production of a particular polypeptide or protein. Section 4.1 describes how the code is made of nucleotide bases which, in groups of three, cause individual amino acids to be incorporated into a developing polypeptide. If the nucleotide base sequence is incorrectly copied during DNA replication, a new amino acid sequence might result. The effect of several such gene mutations, using English words rather than 'DNA words', is shown in Table 23.2. Notice how in some cases a nonsense message is produced whilst in other cases a message with a new sense is produced by these mutations. In DNA mutations, these changes would result in non-functional polypeptides or in polypeptides with a different function.

Table 23.2   Three-letter English words have been used to represent the nucleotide base triplets which make up the DNA code. The effect of gene mutations can be seen on the sense of the new sentences

Type of mutation	Triplet code						
normal code (no mutation)	The	old	man	saw	the	cat	
base added (frame-shift insertion)	The	old	man	asa	wth	eca	t
base lost (frame-shift deletion)	The	olm	ans	awt	hec	at	
bases repeated (duplication)	The	old	mol	dma	nsa	wth	eca t
bases turned around (inversion)	The	old	man	was	the	cat	
base copied wrongly (substitution)	The	old	man	saw	the	car	

 Will all mutations produce new amino acid sequences?

Chromosome mutations

Section 22.3 describes how chromosomes may break during prophase 1 of meiosis and how this may lead to new combinations of linked genes. Other consequences may arise from chromosome breakage, however. One chromosome may lose the broken fragment and the genes it carries. This change in the amount of DNA is a chromosome mutation, referred to as a **deletion**. The broken fragment from a chromosome may join another similar chromosome (**translocation**) or may invert and then join its original chromosome again (**inversion**). The figures on the left of Fig 23.7 summarise these chromosome mutations: the letters in each figure represent the position of genes along the segment of chromosome shown. Sometimes chromosome mutations affect the pairing of homologous chromosomes during the first prophase of meiosis such that chromosome loops, visible under the microscope, occur. The figures on the right of Fig 23.7 explain how some of these loops occur as the mutated chromosomes pair with their normal homologous chromosomes.

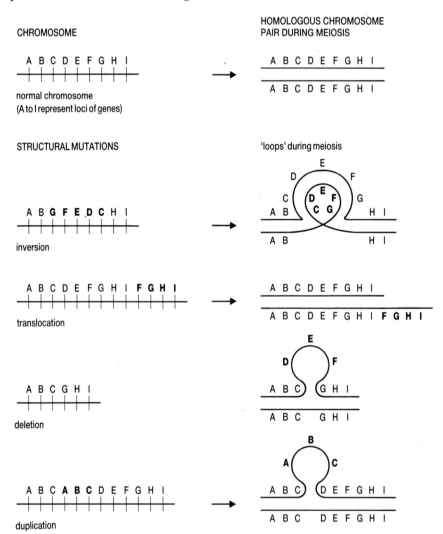

Fig 23.7 Structural mutation in chromosomes. Homologous chromosomes come together in pairs during meiosis at which time differences in structure are often visible as loops.

Other errors during cell division may result in the loss or gain of entire chromosomes. The failure of one homologous pair of chromosomes to separate during the first anaphase of meiosis (**chromosome non-disjunction**) gives rise to some gametes which lack an entire chromosome and some which have two copies of an entire chromosome (Fig 23.8).

GENETIC CHANGES IN POPULATIONS: EVOLUTION

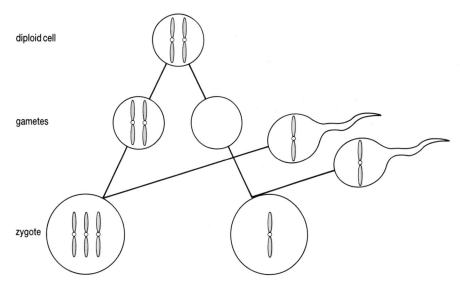

diploid cell

gametes

zygote

Fig 23.8 The effect of chromosomes failing to separate during egg cell production. One zygote will have an additional chromosome and another zygote will be missing one chromosome.

Following fertilisation, zygotes which lack an entire chromosome usually die. Zygotes with an extra chromosome often survive but may show a range of abnormalities. Some human conditions resulting from chromosome non-disjunction are shown in Table 23.3.

Table 23.3   Some human disorders caused by chromosome non-disjunction

Name of condition	Symptoms	Chromosome abnormality
Down's syndrome	mental retardation, stunted physical growth, reduced resistance to disease, congenital heart abnormalities	additional copy of chromosome 21
Klinefelter's syndrome	sterile male with breast development, little facial hair	additional X chromosome (genotype XXY)
Turner's syndrome	female with poorly developed sexual characteristics	deleted X chromosome (genotype $X–$)

Polyploidy

This is the condition where a cell has three or more times the haploid number of chromosomes, giving rise to triploid cells (three times the haploid number), tetraploid cells (four times the haploid number) and so on.

ANALYSIS

Karyotypes

This exercise helps you to practise the skills of observation and interpretation.

Chromosomal mutations can be detected by using karyotype analysis. Cells, usually white blood cells or fetal cells obtained by amniocentesis or chorionic villus sampling (CVS), are stimulated to undergo mitosis. The drug colchicine inhibits the formation of the mitotic spindle and is used to produce a cell which is full of contracted chromosomes. The cells and their contents are then stained and photographed under a microscope. The photograph is then cut up and the homologous chromosomes arranged to form a karyotype.

Look at Fig 23.9. This shows the chromosomal constitution of a human individual as seen during one of the stages of mitosis in a lymphocyte. In normal humans $2n = 46$. The chromosomal constitution shown here has several abnormalities.

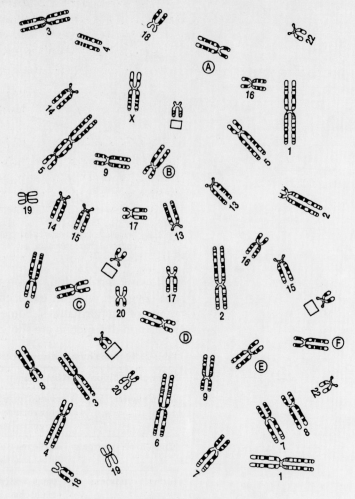

Fig 23.9 A human karyotype prepared for analysis.

The preparation of material for karyotype analysis may be done as follows.

- <u>Heparin</u> (or citrate) an anticoagulant, is added to a sample of blood which is then incubated for 48 hours.
- <u>Spindle inhibitor</u> is added and, several hours later, the white blood cells are centrifuged out.
- A squash preparation of the cells is made and stained. The chromosomes show a characteristic banding on examination under a miscroscope.

(a) (i) Suggest briefly the reason for using each of the two substances underlined in the description given.

  (ii) Why does each chromosome appear as a double structure?

(b) These questions concern the six largest chromosome pairs, numbered 1 to 6 in Fig 23.9.

  (i) One of the number 6 homologues has had an addition to it from another of the large chromosomes. Identify the chromosome from which the addition came.

  (ii) Which chromosome has become a dicentric (with two centromeres) by addition? Where did the added portion come from?

(c) This question concerns the pairs of medium-sized chromosomes, numbers 10 to 12. Instead of a number, these are shown on the diagram by a circle containing a letter. Identify the homologous pairs.

(d) Four of the small chromosomes have been identified with a box. On the basis of these four chromosomes, what would be the sex and condition of this individual?

 Suggest why a karyotype analysis for chromosome abnormality is not routinely carried out on all pregnant women.

The evolutionary importance of mutation

Mutations will only have evolutionary importance if they occur during the production of gametes since mutations in somatic cells are not inherited. However, the incidence of mutation is itself inherited. Crossing over, resulting from chromosome breakage, never occurs in male *Drosophila*. Some unknown mechanism must operate to reduce the likelihood of chromosome breakage in these males. The rate at which genes mutate is different for individual species and for each of the loci in one cell. It has been estimated that there is a mean natural mutation rate of about 10^{-5} per locus per generation for animals and plants and between 10^{-8} to 10^{-9} for micro-organisms such as bacteria. Note that a mutation rate of 10^{-5} per locus per generation means that one mutation will occur at this locus in every 10^5 or 100 000 cell divisions. Consequently, in a population of 1×10^6 individuals, about 10 offspring will carry a new mutation at a given locus; if the population size is a more realistic 1000 to 10 000 individuals, an offspring with this particular mutation will occur only once in a hundred or so generations. Some **mutagenic agents**, such as ionising radiation and mustard gas (a chemical used during World War I), increase the rate of mutation.

 Suggest some other mutagenic agents.

Many mutations will change the genetic code to such an extent that a non-functional protein is produced or future cell division is impossible and, thus, have a great effect on the phenotype. Others, called **neutral mutations**, have a negligible effect. A small proportion of mutations must result in the synthesis of a new protein which functions more effectively than the old.

Since the natural rate of mutation is only about 10^{-5} to 10^{-9}, changes in allele frequencies caused by mutation will be very small, probably only about 0.0001 per locus per generation at most. The evolutionary importance of mutation is that it is a source of genetic variation (in asexually reproducing organisms, the only source of genetic variation) on which natural selection can act.

QUESTIONS

23.5 (a) What is meant by a (i) gene mutation (ii) chromosome mutation?

(b) What is the importance of mutations in (i) asexually reproducing populations (ii) sexually reproducing populations?

23.6 (a) With respect to sex chromosomes, what gametes would be produced by non-disjunction during meiosis I of spermatogenesis?

(b) Describe the genotypes and phenotypes of the resultant offspring if these abnormal sperm were to fertilise a normal egg cell.

23.4 DARWINISM AND NEO-DARWINISM

Charles Darwin was not the first biologist to suggest that evolution had occurred or to suggest a mechanism by which it had occurred. Darwin is remembered not only because his theory forms the basis of our theory of evolution today but also because the evidence he collected over twenty six years forced people to accept that evolution had occurred. He published this evidence in a book called *On the Origin of Species by Means of Natural Selection* in 1859. Darwin knew that populations which reproduce sexually show great **variation** in phenotype. He also knew that they are capable of producing far more offspring than ever survive, leading to a **struggle for existence**, i.e. competition between members of the species (**intraspecific competition**, described further in Section 28.2) for scarce environmental resources. Darwin argued that those organisms which possess harmful alleles of genes are more likely to die (are less fit) than others. As a result of their death, they will not reproduce so their harmful genes will not be passed on to the next generation. He called this process **natural selection**. Since natural selection is the result of a struggle for existence, selection may not occur in environments where there is little or no competition.

 Suggest some other *natural* processes which could be important selective agents.

Industrial melanism

There were no data from studies of natural populations to support Darwin's theory until almost 100 years after the publication of *On the Origin of Species*. Then in the 1950s, studies by H. B. D. Kettlewell of British populations of the peppered moth (*Biston betularia*) supplied such data. There are two genetically determined patterns of *B. betularia* (Fig 23.10): a light-coloured peppered pattern (*B. betularia typica*) and a uniformly dark pattern (*B. betularia carbonaria*).

Fig 23.10 Two morphs of the peppered moth, *Biston betularia*.

The records of moth collectors show that the first dark form was found in Manchester in 1849 but that by the end of the nineteenth century 98% of *B. betularia* in Manchester were of the *carbonaria* form. Clearly, evolution (a change in allele frequency) had occurred. By releasing moths in various locations and investigating what happened to them, Kettlewell was able to show that more moths were eaten by birds if they were conspicuous against their background than if they blended in with their background (i.e. were **cryptic**). Thus, more individuals of the *carbonaria* form survived in northern industrial areas than survived in unpolluted south-western areas where they were very conspicuous on tree trunks speckled by lichens. In

GENETIC CHANGES IN POPULATIONS: EVOLUTION

Fecundity means the number of offspring produced. A frog produces hundreds of egg cells, each a potential tadpole to develop into a frog. Contrast this with a kangaroo normally producing only one offspring at a time.

industrial areas the reverse was true; the *typica* form was conspicuous against the blackened, lichen-free tree trunks and had little chance of surviving. This process, whereby species have become blacker since industrialisation, is called **industrial melanism** (after the pigment, melanin, which produces the blackness).

Neo-Darwinism

The developments in genetics, molecular biology, palaeontology (study of fossils) and ecology during the twentieth century have resulted in a modification of Darwin's theory. This modification of Darwin's theory, **neo-Darwinism** (*neo* means new), may be stated as the theory of organic evolution resulting from the differential fertility of organisms with different phenotypes, and hence genotypes, within a particular environment. This new theory incorporates the realisation that natural selection will work even if organisms with harmful alleles of genes survive, provided their fecundity is less than others with advantageous alleles of the same genes, and recognises the importance of different alleles of one gene. The modern theory also incorporates the discovery that selective advantages and disadvantages of an allele of a particular gene relate to one specific environment and one specific time only, i.e. an allele is not advantageous *per se* but only under certain circumstances.

Even now, the neo-Darwinian theory is the subject of debate between scientists and is still open to further modification. For example, one belief strongly held by evolutionists, that all genetic mutations are either selected for or against, was questioned during the 1970s when Professor Mootoo Kimura proposed his **neutral mutation theory**. This theory resulted from studies of proteins (the products of genes) using the technique of electrophoresis outlined in Chapter 1. These studies showed much more variation in the amino acid composition of individual proteins than had been expected. Since such variation must result from different nucleotide sequences in the genes controlling these proteins, Kimura questioned whether different nucleotide sequences made much difference to the survival of their possessors. All theories must be questioned as new evidence is brought to light and scientists should always keep their minds open to change.

 10 Suggest why Kimura called his theory the neutral mutation theory.

QUESTIONS

23.7 What is meant by biological fitness?

23.8 List the observations and conclusions on which Darwin based his theory of natural selection.

23.9 How does neo-Darwinism differ from Darwinism?

23.5 TYPES OF NATURAL SELECTION

An organism's phenotype is the result of an interaction between its genotype and the environment in which it develops and grows. Edible pea plants may be tall or dwarf. The allele for tallness is dominant to that for dwarfness. On measuring a large number of tall pea plants, the results would show they were not all the same height. Some pea seeds with an allele for tallness might germinate in an area of soil which is too dry or which has a harmful pH so that they cannot grow very well. The heights of the tall pea plants would be distributed as shown in Fig 23.11, if the sample were large enough. The vertical line represents the height which you would expect from their genotype alone. The area around this line represents the variation in size caused by the environmental conditions in which the pea

---

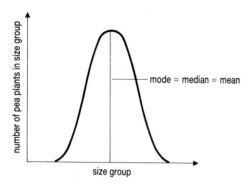

Fig 23.11 If the sample is large enough the frequency of the phenotype for height of tall pea plants follows a normal distribution.

plants grow. The graph shows a **normal distribution** in which the height with the greatest frequency (**mode**), the middle height of the range (**median**) and the arithmetic average height (**mean**) are all the same.

If the heights of a large sample of adult humans of one sex were measured, they would show a normal distribution (though the mean height would be different for females and males). Unlike height in pea plants, human height is controlled by many genes (**polygenic inheritance**). Look at Fig 23.12 which shows the distribution of genotypes that would be expected in a population if one characteristic were controlled by (a) one gene, (b) two genes and (c) many genes, each having two codominant alleles. Human height fits the third graph. Graphs like these are used in describing the types of natural selection: they help to show how the genotype and the environment interact to produce curves of this general shape.

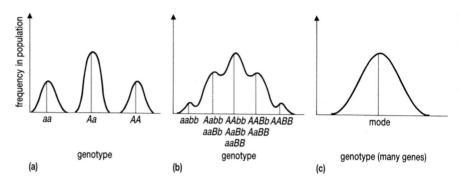

Fig 23.12 The graphs show the distribution of genotypes which would be expected if **(a)** one, **(b)** two or **(c)** many genes, each with two alleles showing codominance, controlled a single phenotypic character.

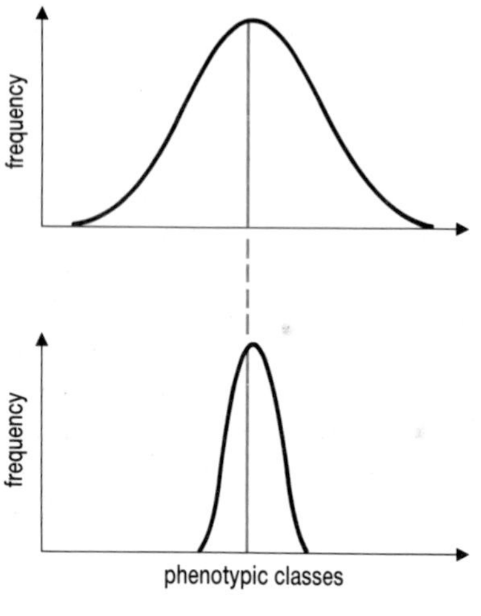

Stabilising selection

The fossil record of sharks suggests that they have the same structure today as they did tens of millions of years ago. This would suggest that their environment has been remarkably stable during that time and that the sharks have been well adapted to it. Under such stable conditions, most variations from the norm, arising through mutation or recombination of existing alleles of genes, are likely to be harmful. The organisms most likely to reproduce successfully are those with a phenotype which is close to the average for the population. The variants are less likely to reproduce and so are at a selective disadvantage compared with the norm. This is **stabilising selection**; variants at the extremes of the range are eliminated (Fig 23.13).

Stabilising selection probably occurs on all characteristics in stable populations. Stabilising selection might have acted on a prehistoric human population in which some individuals carried the sickle-cell condition, in an environment where malaria was endemic and without modern medical help. Sickling of red blood cells results from a single amino acid substitution in the haemoglobin-A polypeptide, caused by a codominant allele Hb^S (see Chapter 22). Homozygous individuals, Hb^SHb^S, might have died in infancy from severe anaemia whilst normal homozygotes, Hb^AHb^A, would not have suffered anaemia but would have been susceptible to malaria. The heterozygoytes, Hb^AHb^S, would suffer a mild anaemia but would have been resistant to malaria. In areas where malaria is endemic it would have been the heterozygote which would have most offspring, i.e. was fittest, and both homozygotes would have been at a selective disadvantage (Fig 23.14). This is known as **heterozygote advantage**. Such stabilising selection acting on prehistoric human populations might explain the high incidence of the sickle-cell allele in parts of the world where malaria is endemic today.

Fig 23.13 The effect of stabilising selection is to reduce the variation around the mode of the distribution range.

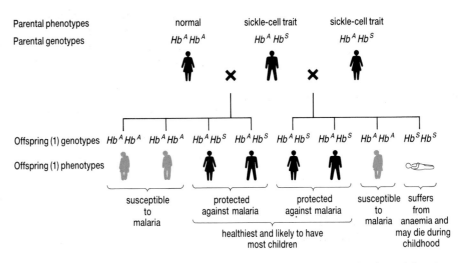

Parental phenotypes | normal | sickle-cell trait | sickle-cell trait

Parental genotypes: $Hb^A Hb^A$ × $Hb^A Hb^S$ × $Hb^A Hb^S$

Offspring (1) genotypes: $Hb^A Hb^A$ $Hb^A Hb^A$ $Hb^A Hb^S$ $Hb^A Hb^S$ $Hb^A Hb^S$ $Hb^A Hb^S$ $Hb^A Hb^A$ $Hb^S Hb^S$

Offspring (1) phenotypes:

susceptible to malaria | protected against malaria | protected against malaria | susceptible to malaria | suffers from anaemia and may die during childhood

healthiest and likely to have most children

Fig 23.14   Stabilising selection acting on sickle-cell anaemia. Heterozygotes are protected against malaria and those homozygous for normal red blood cells are susceptible to malaria.

 11 Would homozygotes be at an advantage in modern populations which have migrated to parts of the world that are free of malaria?

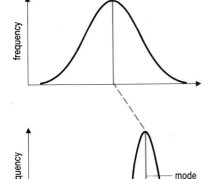

Fig 23.15   The effect of directional selection is to move the distribution of the phenotype, so that the mode coincides with a new environmental optimum.

Directional selection

Industrial melanism in populations of the peppered moth (*Biston betularia*) has been described (in Section 23.4). Before the industrial revolution changed the environment in Britain, *B. betularia typica* was well adapted to its environment and the occasional *B. betularia carbonaria* which turned up as a result of mutation was at a selective disadvantage. In other words, stabilising selection of body colour would have been occurring. The environmental effects of the industrial revolution reversed these selective advantages, so that *B. betularia carbonaria* became the optimum phenotype. When environmental change favours a new phenotype, **directional selection**, represented in Fig 23.15, occurs.

Similar examples of directional selection have occurred in populations of bacteria since the introduction of antibiotics. Presumably, mutations which would make bacteria resistant to antibiotics have occurred many times during evolution. Only when bacterial populations were being exposed to antibiotics did these mutant genes confer an advantage on their possessors. If one resistant bacterium arose by mutation in a population of one million cells, this bacterium would be the sole survivor after the population was exposed to antibiotic. The reproduction of this bacterium would produce a population of antibiotic-resistant bacteria and the frequency of the resistant allele would change from 0.000001 (one resistant cell in a population of one million cells) to 1.0 (one cell in a population of one cell) in a single generation.

The change of frequency of the *carbonaria* allele in Manchester populations of *B. betularia* from 1849 was more gradual than that of antibiotic resistance in bacteria however, and for decades both *typica* and *carbonaria* moths were found side by side. The existence of two body forms (**morphs**) in a population is known as **polymorphism**. A polymorphism which results from directional selection changing the optimum phenotype of a population in response to environmental change is known as **transient polymorphism**.

 12 The *carbonaria* form of *B. betularia* has declined in industrial areas in recent years. Suggest why.

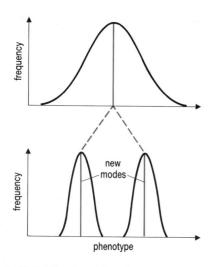

Fig 23.16 The effect of disruptive selection is to produce a bimodal population. Each mode has a different appearance and is a distinct morph of that species. This is the beginning of balanced polymorphism.

Disruptive selection

Sometimes selection will favour phenotypes towards the extremes of the range of phenotypic variation. Such selection is the reverse of stabilising selection and is called **disruptive selection**. The effect of disruptive selection is to eliminate phenotypes at the very extremes *and* the middle of the range, producing the **bimodal** distribution shown in Fig 23.16. Each of the modes in Fig 23.16 represents a distinct morph of that species. Unlike the transient polymorphism in *B. betularia*, the polymorphism resulting from disruptive selection is likely to be stable for as long as the environmental factors causing the disruptive selection last. This type of polymorphism is called **balanced polymorphism**.

Banding patterns in *Cepaea nemoralis*

Look at Fig 23.17 which shows a number of snails belonging to the species *Cepaea nemoralis*. Shells in the photograph show variation in background colour of the shell; yellow, pink and brown. Some snails show a variety of banding patterns. Fossil *C. nemoralis* shells indicate that this polymorphism is thousands of years old.

Fig 23.17 *Cepaea nemoralis*, a snail which shows balanced polymorphism of shell colour and banding.

These colour and banding patterns have been extensively studied throughout Europe. They are inherited, with the shell-colour gene and the banding-pattern gene being so closely linked that crossing over very rarely occurs (forming a **supergene**). The colour and banding patterns affect concealment and, in some areas of Britain, snails with conspicuous shell patterns are more likely to be eaten by predators, such as the song thrush (*Turdus philomelos*). Thus, yellow unbanded shells are at an advantage in grasslands but at a disadvantage among the reddish-brown leaf litter in beech woodlands. Banded shells and brown shells, which are at a disadvantage in grasslands, are at an advantage in woodlands where they are less conspicuous than the yellow-shelled snails.

 Explain the presence of both yellow and brown morphs in a single *C. nemoralis* population using the information in Fig 23.16.

Over large areas of France and Spain, the survival value of shell colour and pattern morphs depends not on their concealment from predators but on their ability to reflect the sun's heat. On sunny mountain sides, banded or brown shells absorb more heat than unbanded yellow shells and so brown or banded snails are more likely to die from overheating. In spite of

GENETIC CHANGES IN POPULATIONS: EVOLUTION

all the research on populations of *C. nemoralis* there is still no explanation for the distribution of colour and banding patterns found over large areas of southern England, known as **area effects**, showing the complexity of factors controlling this single balanced polymorphism.

Strength of natural selection

Natural selection occurs because different phenotypes within a population have different reproductive success (**biological fitness**). Conventionally, the fittest phenotype for a particular characteristic is given a **fitness value** of 1. Because selection is acting against the other phenotypes of that characteristic, they will have a fitness value which is less than 1. These fitness values can be estimated from analyses of survival rates of different phenotypes from birth, e.g. examining hospital records to determine the effect of live birth weight on survival in humans. They can also be estimated by releasing known ratios of different phenotypes into the environment and sampling those that survive, as Kettlewell did with inconspicuously marked *Biston betularia* morphs.

 Why was it important that the released moths were marked inconspicuously?

These estimated fitness values indicate the strength of natural selection against each phenotype, called **the selection coefficient**. Table 23.4 shows the estimated strength of natural selection acting against phenotypes in a number of naturally occurring populations. You will notice that some of the selection coefficients in Table 23.4 show a wide range of values; the value for stabilising selection on shell size in dogwhelks having the largest range of 0–91%. You may think that this range of values indicates the results of different investigations. This is not the case, however. The range of values arises because natural selection is not a constant force. The selective advantage of a particular phenotype may vary throughout its life, as the young organism grows, ages, escapes being eaten, finds mates and so on. A selective advantage may also be different in different parts of the species' geographic range. Whilst variation in the size of dogwhelk shells is much less in older groups than in younger groups on rocky seashores which are exposed to strong waves (the result of stabilising selection), there is less difference between the age groups on seashores which are very sheltered from waves.

Table 23.4   The strengths of natural selection measured in different populations

Organism	Selected phenotype	Strength of selection/%
common bent grass *Agrostis tenuis*	copper-tolerant plants on old mine tips	54–65
peppered moth *Biston betularia*	melanic form in industrial areas of Britain	5–35
land snail *Cepaea nemoralis*	lack of bands on shell in woodlands	19
dogwhelk *Nucella lapillus*	shell size on exposed shores	0–91
human *Homo sapiens*	birth weight and survival	3

The strength of selection

This exercise gives practice in the collection and interpretation of experimental data.

In some populations of *Biston betularia* and *Cepaea nemoralis*, more of the conspicuous morphs are eaten by birds than the less conspicuous (cryptic) morphs. Natural selection thus occurs.

You can test this natural selection on a lawn or playing field near where you live. Use pastry which has been coloured by culinary dyes to make two groups of 'caterpillars'. Make one group green so that it is cryptic against the grass background and the other group a contrasting colour so that it is conspicuous.

Scatter equal numbers of these 'caterpillars' on the lawn or playing field and let the birds feed. After feeding, count the number of each type of 'caterpillar' which is left.
(a) Are there more of the green ones compared with the other colour?
(b) Does the relative proportion of green and non-green 'caterpillars' at the start of the experiment affect the final result?

Mix a little vinegar with the pastry of the non-green 'caterpillars' to make them distasteful.
(c) Does this affect the survival of these 'caterpillars'?
(d) If so, does it make any difference if only some of the conspicuous 'caterpillars' have vinegar and others (like mimics in nature) do not?

Artificial selection

Humans have encouraged differential fertility among domesticated animals and plants for thousands of years. Crop plants which produced the largest seeds, cattle which produced most milk, birds with unusual and attractive plumage and so on were used for breeding purposes. At each generation, those individuals which best showed the desired characteristics were used for breeding; those with these characteristics only poorly developed were not used for breeding. Such **artificial selection** has resulted in the many different breeds of dogs, some of which are shown in Fig 23.18. Although the observable differences between these dogs is greater than between some species, all these dogs belong to the same species, *Canis familiaris*.

Artificial selection continues today, in efforts to further improve organisms for human purposes. The use of machines to harvest crops has led to artificial selection of plants for uniform size, and plants with both a high crop yield and resistance to various diseases are still being bred. Recent medical evidence linking the incidence of heart attacks in humans with a diet rich in animal fat has led to the selection of animals which grow quickly to produce meat with little fat.

Fig 23.18 Breeds of dog, all members of the same species, *Canis familiaris*, produced by many generations of selective breeding by humans.

23.10 The variety of plant which produces the cream-coloured beans used to make canned baked beans will not grow in the cold British climate. The variety of bean plant which will grow in Britain produces beans speckled with colour which some people do not like to eat. Suggest how plant breeders could produce a plant which would produce cream-coloured beans in Britain.

QUESTIONS *continued*

Fig 23.19 *Echinochloa oryzoides* mimics rice.

Organisms which closely resemble some other organism or inanimate object are called **mimics**.

23.11 Rice is grown in paddy fields throughout the world. In many paddy fields, weeds are removed by hand when the rice plants are quite young. A species of barnyard grass called *Echinochloa oryzoides* (previously called *E. phyllopogan*) is a weed which was first found in rice paddies throughout Asia. It has since spread from Asia to paddy fields in California and Australia. Like rice plants, but unlike other species of *Echinochloa*, *E. oryzoides* does not have drooping leaves or red leaf bases and flowers early in the season.

(a) What is the advantage to *E. oryzoides* of looking like rice plants?

(b) Suggest how this rice mimic might have evolved.

23.12 (a) What is the importance of genetic variation in natural selection?

(b) Use Chapters 22 and 23 to review the source of genetic variation in populations.

SUMMARY

The sum total of all the genes which control the phenotypic characteristics of a given population is called the gene pool of that population. Usually, only one characteristic is studied at a time, in which case the term gene pool is applied to the gene which controls this characteristic. If any gene has more than one allele, the frequency of each allele within the gene pool is called the allele frequency. For a gene with only two alleles A and a, the frequency of allele A in the gene pool is defined as p and the frequency of allele a in the population is defined as q. Using these allele frequencies ($p + q = 1$), the frequencies of the genotypes AA, Aa and aa in a population will be p^2, $2pq$ and q^2 respectively. These frequencies must equal unity, therefore $p^2 + 2pq + q^2 = 1$, this is the Hardy-Weinberg equation.

The Hardy-Weinberg equation was developed whilst considering populations in which no change in allele frequency occurred. When permanent changes in allele frequency occur, evolution of that population is said to have occurred. Such changes in allele frequency may arise as a result of mutation. Mutations are random, undirected changes in DNA composition. They may involve changes in the sequence of bases coding for amino acids (gene mutation) or in the structure or amount of DNA in the cell (chromosome mutation). Whilst there is a natural mutation rate, certain chemical compounds or forms of energy cause an increase in this rate. These are collectively called mutagenic agents. In asexually reproducing populations, mutation provides the only source of new genetic information and is the only source of variation. Changes in the allele frequency of a population may also arise as a result of random fluctuations in small populations (genetic drift), migration, selective mating and the differential fertility of organisms within the population (natural selection).

Natural selection is thought to be the most important agent causing changes in allele frequency, and hence evolution, in most populations. Natural selection operates because, within a given environment and at a given time, individuals showing one phenotype of a particular characteristic tend to have more offspring than those with a different phenotype of that characteristic. Since the phenotype is largely affected by the genotype, the allele causing the favourable phenotype is passed on to more individuals in the offspring generation than the allele of the same gene

gene causing the less favourable phenotype. Over several generations, this may result in a marked change in allele frequency for that gene.

Individuals which, as a result of their phenotype, have more offspring are said to be at a selective advantage over others with a different phenotype. They are also said to be biological fittest. If the fittest phenotype lies in the middle (mode) of the range of phenotypes for that characteristic, selection acts against those phenotypes at the extremes of the range: this is called stabilising selection. If the fittest phenotype occurs at one end of the range of phenotypes, selection acts against other phenotypes until the favoured phenotype becomes the mode of a new distribution: this is called directional selection. If different phenotypes are fittest under different circumstances, selection will produce a bi- or polymodal distribution: this is called disruptive selection. When two or more phenotypes exist in a population, a polymorphism is said to occur. During directional selection a transient polymorphism occurs whilst the range of phenotypes changes to the new favoured mode. Disruptive selection produces stable polymorphisms which may last for hundreds or thousands of years.

Chapter 23: Answers to Quick Questions

1 Some individuals in the population must have already had the gene for DDT resistance. They survived and their offspring inherited their resistance.
2 0.6 (the sum of the two frequencies must be 1).
3 Many populations are large but the other conditions are unlikely to be met.
4 It would decrease as more of the recessive alleles would be present in heterozygotes than in homozygotes.
5 (a) decrease (b) increase rates of gene flow.
6 No, some amino acids are coded by many base triplets (see Table 4.1).
7 It is a rare condition; it is expensive; not all women would opt for an abortion if chromosome abnormalities were found.
8 Radioactivity, e.g. α particles; chemical carcinogens such as benzene.
9 Predation, parasitism, herbivory, the weather, speed of current for aquatic organisms.
10 A change in the nucleotide sequence of a gene is called a mutation. If most of these changes have no effect on the survival or fecundity of their possessor, they are selectively neutral (neither selected for nor selected against). Hence, neutral mutation theory.
11 No, they would suffer mild anaemia and resistance to malaria would be irrelevant.
12 Control of air pollution has enabled lichens to grow on the tree trunks, putting melanics at a selective disadvantage.
13 Disruptive selection is occurring where brown shells are at an advantage in one part of the habitat and yellow shells in another part of the habitat.
14 They had to be marked so that they could be recognised when recaptured. Conspicuous marks might have affected their survival compared to unmarked moths (e.g. they might be more likely to be eaten by visually hunting birds).

Chapter **24**

SPECIES: THEIR FORMATION AND CLASSIFICATION

LEARNING OBJECTIVES

When you have studied this chapter you should be able to:

1. define the term species;

2. explain how new species may arise by phyletic or divergent speciation;

3. discuss the conditions needed for allopatric and sympatric speciation;

4. describe the two-kingdom and five-kingdom natural classification of species;

5. use correctly the terms kingdom, phylum, class, order, family, genus and species;

6. use the binomial nomenclature system and explain some of the rules which govern it;

7. devise a dichotomous identification key using an artificial classification scheme.

Long before having arrived at this part of my work, a crowd of difficulties will have occurred to the reader. Some of them are so grave that to this day I cannot reflect on them without being staggered; but to my best judgment, the greater number are only apparent, and those that are real are not, I think, fatal to my theory.

C. Darwin, *On the Origin of Species by Means of Natural Selection*

The above extract from Charles Darwin's *On the Origin of Species by Means of Natural Selection* shows that he anticipated some of the objections which would be raised to his theory of the mechanism by which the evolution of new species occurs. Since its publication in 1859, Darwin's theory has been repeatedly challenged, not only by religious fundamentalists but also by scientists. New discoveries, such as the principles of genetics, the apparently random variation of proteins within a species, and new fossil finds, cannot be ignored by scientists. The scientific debate which follows these discoveries is about the mechanism of evolution and not about whether evolution has occurred. Like the scientific theories of gravity or of chemical energetics, the majority of scientists regard evolution as a 'fact'.

24.1 THE NATURE OF SPECIES

The evolution of Darwinism

The process of natural selection was first described by Charles Darwin and Alfred Russell Wallace at a meeting of the Linnaean Society in 1858. The two men had arrived at their theory in different ways. Wallace had been impressed by the variety within and between butterfly populations in the tropical rain forests of the Malaya archipelago and rapidly came to the conclusion that natural selection must be the cause. Darwin had been impressed by the variety of tortoise, mockingbird and finch populations between the islands of the Galapagos archipelago (*galapagos* is Spanish for tortoise), which he visited as ship's naturalist between 1831 and 1836. For the 20 years after his return to England he collected evidence to support his ideas of evolution and was pushed into publication by the threat that Wallace would beat him to it. Darwin published his book, *On the Origin of Species by Means of Natural Selection*, in 1859 in which he expanded the theory and presented the evidence in its support which he had collected. It was not until 1959, that H. B. D. Kettlewell published the first field evidence to support the theory of natural selection in populations of the peppered moth, *Biston betularia*.

The really contentious issue that Darwin raised was that species were not immutable. In other words, new species could evolve from old ones: they did not have to be created by a divine agent. In fear of the great opposition from the Establishment, including the Church, Darwin modified his theory as each edition of *On the Origin of Species* was published. Studies since 1959 have shown that Darwin's ideas were slightly wrong. Darwin's theory involved survival of the fittest; the idea that only those organisms which were best suited to their environment survived to reproduce. In neo-Darwinism (*neo* means new) unfit individuals do not need to die, simply to leave fewer offspring than those which are better adapted to their environment. Darwin's ideas largely remain however, and through the effect his theory of natural selection has had on political, religious and social beliefs, he may be regarded as one of the most influential people who has ever lived.

Fig 24.1 A coelocanth. This bony fish was regarded as being extinct until 1938 when one was caught off the coast of south-east Africa. No evolution of this species is thought to have taken place in 60 million years because it is so similar to fossil coelocanths found in rocks 60 million years old.

Definition of a species

Members of the same species usually have distinctive features by which that species can be recognised; to see one lion is to recognise all other lions. These similarities also extend to the chemical level. In sexually reproducing populations, however, there is considerable variation between members of a species. It is, therefore, difficult to define a species in terms of chemical or physical similarities.

In their natural environment, members of one species do not usually breed with members of another species. In cases where they do, such as the

SPECIES: THEIR FORMATION AND CLASSIFICATION

breeding of domesticated asses (*Equus hemionus*) and horses (*Equus equus*) to produce mules, the offspring are usually sterile. Because two, small interbreeding populations of one species (**demes**) may live in distant environments, the chance that individuals from the two environments will mate is small. They are potentially capable of doing so, however. These considerations give us two definitions of species.

Definition 1. A group of organisms which are actively or potentially capable of interbreeding to produce fertile offspring.

Definition 2. A group of organisms which show a close similarity in physical (morphological), biochemical and ecological characteristics and life history.

 What problem does the first definition of species raise with respect to asexually reproducing organisms?

Fig 24.2 Beetles. All have thickened forewings which cover the membraneous hind wing. Otherwise beetles come in very many different shapes and sizes!

The formation of subspecies

As far as is known, the Lava Heron (*Butorides sundevalli*) shown in Fig 24.3 has restricted geographical distribution. It is found only on the Galapagos Islands off the coast of Ecuador. Most species are distributed over a wider geographical range. If there are different environmental variables in two parts of this range the two demes living in these regions might be subjected to different selection pressures. As described in Chapter 23, such selection pressures resulted in phenotypic differences between *Biston betularia* populations in polluted and unpolluted areas of Britain. Populations of one species which show phenotypic differences associated with different environments are called **subspecies**.

A subspecies is a group of populations of a species, inhabiting a geographical subdivision of the range of the species and differing taxonomically from other populations of the species. A trinomial nomenclature system is used for subspecies, e.g. *Hoplitis producta gracilis*, *H. producta subgracilis*.

Fig 24.3 The Lava Heron *(Butorides sundevalli)* has a restricted geographical distribution.

Clines

This exercise will help you to develop skills in formulating and testing hypotheses.

A **cline** is a gradual change in the frequency of phenotypes over the geographical range of a species.

Some white clover plants (*Trifolium repens*) inherit genes which enable them to produce hydrogen cyanide when their tissues are damaged. Tissue damage may be the result of herbivores, such as slugs and snails, or of frost damage. The frequency of cyanogenic clover plants in populations throughout Europe and isotherms of mean January temperatures are shown in Fig 24.4.

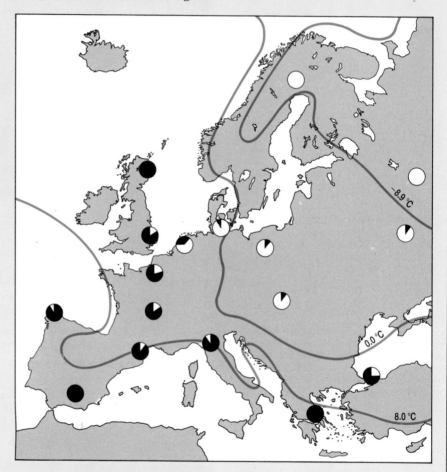

Fig 24.4 The frequency of cyanogenic forms of white clover in different parts of Europe. A solid black circle indicates 100% cyanogenic form; a white circle indicates 0% cyanogenic form. Blue lines are isotherms for the mean January temperature.

(a) How does the frequency of cyanogenesis (the ability to produce cyanide) vary over Europe? Is this a cline?

(b) How is the frequency of cyanogenic clover plants correlated with mean January temperatures?

(c) Suggest a hypothesis to account for the frequency of cyanogenic plants (i) in warm southern areas (ii) in cold northern areas.

(d) How could you test your hypothesis?

Ring species

Sometimes a cline may encircle a region, its two ends occupying the same area. Such a circular cline exists in the *Larus* (gull) population which encircles the north pole at latitudes between 50°N and 80°N. Differences in size, colour of legs, beak and wings have led to the recognition of a number of different subspecies of the *Larus* gull, but each subspecies can successfully interbreed with those to either side of it in the cline. The habitat of seven of these groups of gull which make up the circular cline or **ring species** is shown in Fig 24.5. Notice that the two ends of the cline meet in the British Isles, Netherlands and Germany. Here the gulls behave as the two distinct species shown in Fig 24.6. The herring gull (*Larus argentatus*) has a greyer plumage, yellow legs and is non-migratory; the lesser black-backed gull (*Larus fuscus*) has grey to black plumage, pink legs and migrates in winter. The two gulls also have different calls and, although they may nest side by side, they rarely interbreed. These gulls demonstrate the inadequacies of our definition of the term species, since they may be simultaneously regarded as different subspecies of one species or as different species.

Fig 24.6 Herring gull (*Larus argentatus*) and lesser black-backed gull (*Larus fuscus*) nest side by side; they can, but rarely do, interbreed. Are they different subspecies or different species?

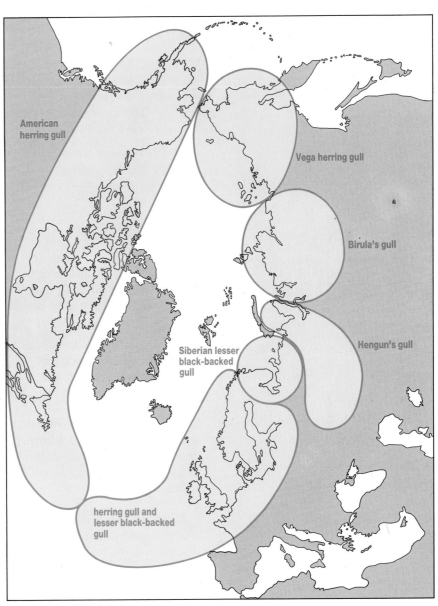

Fig 24.5 The distribution of different species of gull (*Larus*) in the northern latitudes forms a ring species. The different species form a circular cline around the North Pole.

 Suggest why taxonomists do not put different breeds of dog into separate subspecies of *Canis familiaris*.

QUESTIONS

24.1 (a) Explain what is meant by (i) a species (ii) a subspecies.

(b) How could you test whether two animals belonged to subspecies of one species or to different species (i) if both were alive (ii) if both were preserved dead specimens?

24.2 Explain why naming the two gulls shown in Fig 24.6 presents problems to biologists.

24.2 SPECIATION: THE FORMATION OF NEW SPECIES

How can one species evolve from another? This is the process of **speciation** that Darwin called 'the mystery of mysteries'. A number of mechanisms of speciation have been proposed including

- phyletic speciation
- divergent speciation which may arise through
   allopatric speciation
   sympatric speciation
   interspecific hybridisation and polyploidy.

These are discussed below. However, since speciation usually takes so long compared to a human lifetime, they are difficult to test. Nonetheless, the processes which cause evolution are still acting in the world today: the key is to spot populations in which evolution is occurring rapidly and study them.

 Give two examples of populations where evolution is occurring rapidly today.

Phyletic speciation

Biologists assume that species are well-adapted to the environment in which they live. In a well-adapted population, stabilising selection will ensure that the optimum phenotype is maintained from generation to generation. If the environment changes, directional selection will begin to act on the populations which live there. Directional selection following the changes in the environment caused by industrialisation resulted in the spread of melanic insects, such as *Biston betularia*. If widespread environmental changes occur over a long time, directional selection will act on a large number of inherited characteristics. Unless the widespread environmental changes result in a species becoming extinct, directional selection will cause the species as a whole to change, so that it can be recognised as quite different from the ancestral species. This evolution of the whole species is termed **phyletic speciation**.

Since the ancestral species may be extinct, it is impossible to test whether the new species and ancestral species could interbreed. It may be clear from fossil records, however, that the differences between the old and new species are much greater than those which occur within a single species. A new species may, thus, be regarded as having arisen. Whilst phyletic speciation may account for the existence of species which are different from those which existed in fossil records, it would not lead to an increase in the number of species.

 Why will phyletic speciation not lead to an increase in the number of species over time?

Divergent speciation

This mechanism of speciation results in an increase in the number of species (Fig 24.7). Imagine how this process might work using the *Larus* populations shown in Fig 24.5.

Since the *Larus* populations in Fig 24.5 form a continuous series of interbreeding populations around the North Pole, alleles of the genes present in the *L. argentatus* (herring gull) population in Britain may eventually pass to the *L. fuscus* (lesser black-backed gull) population. These gulls could, therefore, be regarded as a subspecies of a single species. Imagine that the *Larus* populations of North America disappeared. This would break the circular cline connecting *L. argentatus* and *L. fuscus* so that gene flow between the two would stop. The frequency of particular alleles and genotypes in each of these populations might gradually change, exaggerating the differences which already exist between them. We would now confidently refer to them as distinct species. Our confusion was removed as soon as the populations were genetically isolated from each other.

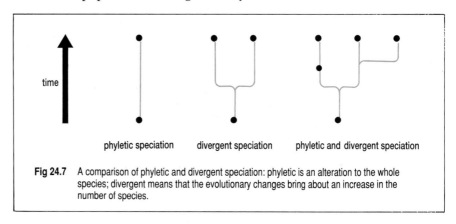

Fig 24.7 A comparison of phyletic and divergent speciation: phyletic is an alteration to the whole species; divergent means that the evolutionary changes bring about an increase in the number of species.

Genetic isolation is the central requirement in the theory of divergent speciation. If populations of one species are genetically isolated for long enough for them to accumulate different allele frequencies, they might eventually behave as different species. The mechanisms which maintain genetic isolation between populations of one species are termed **reproductive isolating mechanisms**. A summary of these isolating mechanisms is given in Table 24.1.

Three different types of divergent speciation are discussed below, i.e. allopatric speciation, sympatric speciation and interspecific hybridisation.

Table 24.1 Reproductive isolating mechanisms between populations of one species

Time that barrier is effective	Nature of barrier	Notes
before mating	geographical isolation	populations inhabit different environments
	ecological isolation	populations use different local habitats within one environment
	temporal isolation	populations occupy the same area but are active or reproduce at different times
	behavioural isolation	courtship displays among animals usually stop mating between species
	mechanical isolation	reproductive parts may not fit each other
after mating	gamete incompatibility	tissues and secretions of female may kill sperm or pollen of the wrong species
	hybrid inviability	hybrids fail to develop to maturity
	hybrid infertility	hybrids are formed but, as adults, are sterile

---

ANALYSIS

The evolution of copper tolerance in *Agrostis tenuis*

This exercise helps you to develop skills of hypothesis formation.

Plants of *Agrostis tenuis* (common bent grass) which grow in unpolluted soil are poisoned by high concentrations of copper. Plants which grow on the waste tips around copper mines possess an allele of a gene which makes them tolerant to these high copper concentrations.

(a) Suggest how the copper-tolerant allele might have originally arisen.

(b) Explain how the copper-tolerant populations of *Agrostis tenuis* probably arose.

(c) Suggest why there are few copper-tolerant individuals of *Agrostis tenuis* in unpolluted soil.

(d) Copper-tolerant *Agrostis tenuis* plants have an earlier flowering time than copper-susceptible plants. Explain why this might lead to the formation of a new species.

Allopatric speciation

Genetic isolation is much easier to imagine in populations which are physically separated from one another by barriers which they cannot cross. The different populations may be subjected to different natural selection pressures depending on the geography or climate of their environment, so that they may become increasingly different from each other as time goes on. The sea slows the movement of many species of animals and plants. If populations living on islands are separated for long enough, they may form new species which, even when they meet on the same island, will no longer interbreed. Such geographical isolation is thought to explain the variety of closely related species on neighbouring islands, such as the finch species on the Galapagos archipelago and the *Drosophila* species on the volcanic islands of the Hawaiian archipelago.

Evidence from a number of sources shows that the continents which now exist have not always appeared as they do today. At one time all the land is thought to have occurred in a single large mass, called Pangaea. This is thought to have broken up into two parts, northern Laurasia and southern Gondwanaland, which in turn split up into the continents we know. The Atlantic Ocean widens by a few centimetres each year as **continental drift**

SPECIES: THEIR FORMATION AND CLASSIFICATION

continues today. The existence of ancient Pangaea is supported by the presence of fossilised remains of the same extinct species found in present-day Antarctica, Australia, South America and South Africa. Such fossilised remains suggest that these continents were once populated by large populations of the same organisms.

 Explain how the break up of Pangaea may account for the differences in the animals inhabiting these southern continents today.

Sympatric speciation

Provided the isolating mechanisms are strong enough to prevent interbreeding between two subspecies, speciation may occur without geographical isolation, i.e. sympatrically. There are two common sub-species of house mouse in Europe, a southern, light-bellied *Mus musculus musculus* and a northern, dark-bellied *M. musculus domesticus*. These subspecies have been found to be interfertile in the laboratory but in the areas of Europe where they occur together there is very little interbreeding. If sufficient genetic differences accumulate between these two groups of mice they may one day be classified as different species.

Interspecific hybridisation and polyploidy

Mules, the offspring of a cross between a horse and an ass, are sterile because the chromosomes from the two different parents cannot pair together during meiosis in the mules' sex organs. Such sterility may be overcome if interspecific hybrids double their chromosome number, becoming polyploids, when their chromosomes will be able to pair again during meiosis. Doubling of chromosome numbers happens naturally but its likelihood can be increased by exposing organisms to certain chemicals. It is much more common in plants than in animals, possibly because the sex chromosomes of animals still cannot segregate properly in polyploid cells. Polyploidy has resulted in the formation of new species of primula (*Primula kewensis*), bread wheat (*Triticum aestivum*), macaroni wheat (*Triticum durum*) and hemp (*Galeopsis tetrahit*). A new species of cord grass (*Spartina townsendii*) is thought to have originated in Southampton Water around 1870 (Fig 24.8) where the native British species (*S. maritima*) hybridised with an introduced American species (*S. alterniflora*).

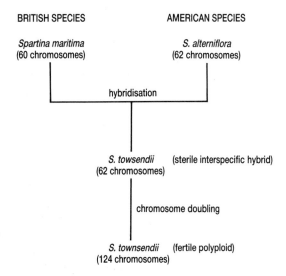

Fig 24.8   The immediate formation of a polyploid new species of cord grass (*Spartina townsendii*) by interspecific hybridisation.

Fig 24.9 Not only does this cultivated blackberry produce larger fruit than the wild plant, it also has no thorns. Both features are clearly advantageous to the fruit grower.

POLYPLOIDY AND PLANT BREEDING

Many of the food plants that we eat are polypoids. For example, common bread wheat is a hexaploid. Artificial induction of polyploidy has now been tried with many crops. Often, such polyploids are found to grow more vigorously than diploids or they produce larger fruit, tubers and so on. For example, wild blackberries are usually diploid (14 chromosomes), while cultivated blackberries are tetraploid (28 chromosomes), hexaploid (42 chromosomes) or octoploid (56 chromosomes) and have larger fruit than their wild counterparts. Cultivated strawberries are usually octoploid.

The rate of speciation

The rate at which new species are formed seems to be different for different groups of organisms. *Spartina townsendii*, which arose as a new species instantaneously as a result of polyploidy, had an exceptionally fast rate, but millions of years are not always needed for new species to evolve.

The moth genus *Hedylepta* is found only in the Hawaiian Islands. It feeds mainly on palms. Five species of this moth feed on banana and are different in many morphological ways from the palm-feeding moths. Banana was introduced into Hawaii by the Polynesians only about 1000 years ago, so the banana-eating species of *Hedylepta* must have evolved from palm-eaters within that time.

 Is this an example of divergent or phyletic speciation?

There are five species of cichlid fish which live only in Lake Nabugabo, Uganda. This lake is thought to have formed when Lake Victoria shrank less than 4000 years ago. Similarly, Lake Lanao, a lake in the Philippines which is about 10 000 years old, contains 14 species of cyprinid fish which are found nowhere else. It is hard to explain the occurrence of these species other than by divergent speciation from an ancestral species which was present in each lake when it was formed. This explosive speciation is called **adaptive radiation**.

Isolation need not always result in new species, however. Many populations which were isolated for between 20 000 years and two million years during the Pleistocene Ice Ages did not form different species.

When divergent speciation in a new habitat occurs, this is described as adaptive radiation.

24.3 Explain the difference between selection and speciation.

24.4 Complete Table 24.2 to show the differences between phyletic and divergent speciation. The first line has been done as an example.

Table 24.2

Property	Type of speciation	
	Phyletic	Divergent
occurs in response to a change in the environment	✔	✗
results in an increase in the number of species		
depends on an unoccupied microhabitat (or niche) within the environment		
the whole species changes		
results from directional selection		
results from disruptive selection		

24.5 Table 24.3 shows the number of chromosomes in the cells of several species of dock plant (*Rumex*). Suggest how these species might have evolved.

Table 24.3   Number of chromosomes of some species of the genus *Rumex*

English name	Species of dock plant (*Rumex*)	Number of chromosomes
red-veined dock	*R. anguineus*	20
broad-leaved dock	*R. obtusifolius*	40
curled dock	*R. crispus*	60
great water dock	*R. hydrolapathum*	200

24.3 THE BIOLOGICAL CLASSIFICATION OF SPECIES

Whether one accepts that modern species have evolved from ancestral species or that they arose through a Special Creation, there is a vast number of species on Earth. Fig 24.10 shows an animal which you might never have seen before. Your immediate reaction might be to ask what species it is. Its name may mean nothing to you, so you may then ask about its similarity to other species to which it may be related. In this process you are trying to find a name for this new animal and to classify it with other animals with which you are more familiar. To be told that the animal in Fig 24.10 is a sea

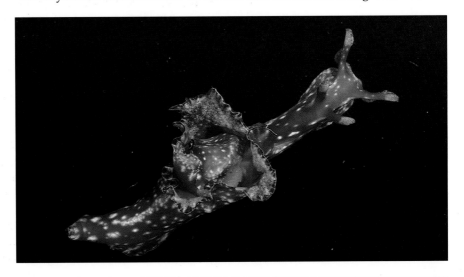

Fig 24.10   What is this organism?

hare (*Aplysia*), which is related to slugs and snails, somehow satisfies our curiosity.

Artificial and natural classifications

We all tend to classify things which we encounter around us. A young child might classify objects by putting them in its mouth to see whether they taste good or bad. You will have classified this book, not only as a textbook and as a biology book, but also as a book which is clear or complicated, well presented or dull and so on. If organisms are grouped according to the way they affect people, edible or poisonous, domesticated or wild, weed or crop, we make an **artificial classification**. This artificial classification is based on a few, superficial characteristics. (A superficial characteristic is one which can easily be seen at a glance.)

Most existing biological classifications of organisms are based on attempts to show evolutionary relationships between species (**phylogeny**) based on such evidence as physical characteristics, palaeontology, embryology, immunology, biochemistry, cell structure, ecology, behaviour and reproductive strategies. Classification systems which reflect an attempt to show evolutionary origins and relationships are termed **natural classifications**.

 Is the classification of our plant food into fruit and vegetables a natural or an artificial classification?

Convergent evolution, analogy and homology

Some biological structures which appear similar are thought to have evolved independently in different biological groups (**convergent evolution**). Such structures, which include the wings of birds and insects or the piercing and sucking mouthparts of bugs and mosquitoes, are referred to as **analogous structures**. If our classification is to link organisms with common ancestors, these analogous structures must be given less importance than features which are homologous. **Homologous structures**, like the wing of a bat, flipper of a seal and foreleg of a horse, have the same basic structure but different functions. Distinguishing between analogous and homologous structures is not as easy as it sounds. The mammals form a biological group. It is widely believed that the features used to classify mammals evolved independently in three different groups of therapsid reptiles, so that the mammals represent a level of organisation (a **grade**) rather than a group with a recent common ancestor (a **clade**).

 Are the eyes of a bee and the eyes of a cat analogous or homologous structures?

Taxonomy

Species are grouped together in larger inclusive sets called **categories**, e.g. genus and family are both categories. These categories are ranked, in order of decreasing size, **kingdom**, **phylum**, **class**, **order**, **family**, **genus** and **species**.

Each set of organisms within a category, e.g. *Drosophila* and Hominidae, is called a **taxon**. Taxonomy is the study of these taxa (plural of taxon). The major categories into which species are grouped are shown in Table 24.4, which also shows how a number of familiar animals are classified. The taxa which are written in colour show the similarities between the classification of each animal and that of humans. The greater the number of coloured groups in the table, the more closely related that animal is to humans.

> **KEY CONCEPT**
>
> **Analogous** structures have the same function but not the same biological structure, e.g. wings of birds, bats and insects.
> **Homologous** structures have the same underlying biological structure although they may perform different functions, e.g. foreleg of horse, arm of human and flipper of seal.

SPECIES: THEIR FORMATION AND CLASSIFICATION

Table 24.4 The major categories used in classification. The names in blue show those taxa which cats (*Felix* sp.) and fruit fies (*Drosophila* sp.) share with humans (*Homo* sp.)

Category	Human	Cat	Fruit fly
kingdom	Animalia	Animalia	Animalia
phylum	Chordata	Chordata	Arthropoda
class	Mammalia	Mammalia	Insecta
order	Primata	Carnivora	Diptera
family	Hominidae	Felidae	Drosophilidae
genus	*Homo*	*Felix*	*Drosophila*
species	*sapiens*	*felis*	*melanogaster*

ANALYSIS

Why should anyone care about taxonomy?

This analysis will develop your skills of comprehension.

Biological knowledge and understanding changes rapidly. To keep up to date you need to read and abstract information from scientific journals and periodicals, such as *New Scientist* and *Scientific American*. This exercise will give you some practice in this essential skill. Read the passage and then answer the questions relating to it.

Although another profession is usually accorded the distinction of being the oldest, there is reason to believe that the honour belongs to taxonomy (*Genesis* 2:20). Naming plant and animal species is still as important to modern biology as it would have been to early humans, because without a unique and universally applied name it is impossible to communicate information about an organism.

At the same time as naming an organism, a taxonomist is required to describe it formally (to characterise it morphologically) so that others may recognise it, and distinguish it from the multitude of similar organisms. In well-known groups, such as mammals and birds, it is usually enough to point out the features that separate it from its relatives.

For groups with large numbers of undescribed or poorly known species, however, the description of a new taxon is generally included as part of a monograph dealing with all of a group of organisms in a particular region – the beetles of Costa Rica, for instance. Though lacking in general appeal, these specialist treatises are the foundations upon which all popular guides to plants and animals are built.

Simply describing vast numbers of organisms would create an unmanageable, and largely useless, mountain of information. To handle such quantities of data, taxonomists must classify species. The earliest systems of classification, some of which are still used by forest-dwelling peoples, were simple, utilitarian schemes. Plants may have been classified as edible, medicinal or poisonous, for example. An early textbook of entomology classified insects as 'those bestowed by the Creator as a pestilence upon mankind' and 'those bestowed by the Creator for the benefit of mankind'.

Over 250 years ago Linnaeus proposed a system for classifying plants. His work is still valid today.

Such classifications have a critical short-coming: they tell nothing more about the organisms than the information that was originally used to classify them. Modern taxonomy is no longer so anthropocentric. Biologists now seek instead to classify organisms in a hierarchical manner that reflects their evolutionary relationships. Probably the most profound change in taxonomic practice since the days of Linnaeus (his *Systema naturea* was published in 1735) has come in the past 20 years, with the development of better techniques for working out those relationships.

Reasoned classifications are not mere filing systems for data, they are highly informative. Animals sharing a recent common ancestor are likely to share many features. So, species placed in a single genus on the basis of a few morphological characteristics generally resemble each other in a variety of other ways – genetically, in their behaviour and their physiology, for instance.

A classification has two important facets: it allows us to make generalisations about groups of organisms; and it enables us to predict some of the biological characteristics of a newly discovered species.

Economic entomologists looking for, say, a parasitic wasp that might help to control a leafhopper pest, can disregard many of the wasps because they can predict, from what they know of similar wasps, that they will not attack leafhoppers.

Phylogenetic classification also provides an evolutionary perspective to underpin other biological studies. To understand why a particular organism does what it does the way it does is not enough to comprehend the forces of natural selection that are working on it. It is also necessary to be aware of the constraints that a species is likely to have inherited as a result of its evolutionary history. We can deduce some measure of these constraints from studies of related, but more ancestral, species.

(a) Why do taxonomists classify organisms?
(b) How do modern methods of phylogenetic classification differ from the classification used by primitive people and early naturalists?
(c) Why will organisms which share a recent common ancestor have many features in common?
(d) How can classification be used as a predictive tool?
(e) Why do you think taxonomy is important?

Taxonomic systems

Taxonomists might disagree about how organisms should be classified for a number of reasons. Usually, the controversy concerns one of the smaller taxa, such as the family or genus. Some taxonomists, for example, would combine all cats (except the cheetah which, alone among the cats, has non-retractable claws) into a single genus called *Felis*. Other taxonomists would split them into a genus of smaller cats (*Felis*), a genus of larger cats (*Panthera*) and a genus of bob-tailed cats (*Lynx*). Difficulty may also arise whenever a new species is found when it must, if appropriate, be placed in an existing genus. Sometimes difficulty arises when it is discovered that a species had already been named before its currently accepted name was allocated. The dogwhelk (*Nucella lapillus*) was placed in the genus *Purpura* during the 1920s, *Thais* during the 1960s and *Nucella* during the 1970s as a result of disagreements between taxonomists.

Controversy sometimes occurs over large taxa. For example, there is disagreement about the number of **kingdoms**. This might surprise you since generally it is obvious what is an animal and what is a plant. This may be true with the large animals and plants but is not necessarily true for microscopic organisms. For example, bacteria were discovered long after the animal and plant kingdoms had been named. Because bacterial cells have a cell wall, they were placed in the plant kingdom. Studies with electron microscopes, however, showed that the eukaryotic structure and organisation of plant and animal cells is quite similar but totally different from that of prokaryotic bacterial cells. A new kingdom for the bacteria was therefore created. A similar problem over the validity of kingdoms is shown by single-celled organisms in the genus *Euglena*. Individuals have chloroplasts, which is a plant feature (Fig 24.11). Like animals, they do not have cell walls outside their surface membranes and are able to change shape. To help overcome problems such as these, all eukaryotic unicells are now placed in a new kingdom (called **Protoctista**).

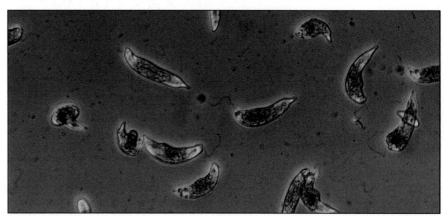

Fig 24.11 *Euglena* sp. is an organism which is plant-like because it has chloroplasts, but animal-like since it does not have a cell wall and changes its shape.

Fungi have also been classed as plants because they have a cell wall. However, apart from the presence of a cell wall they show very few similarities to plants. They never have chloroplasts (and so have a method of nutrition more like that of some bacteria and animals) and, unlike both animals and plants, often have bodies which are not made up from separate cells. The fact that their cell wall is sometimes made of chitin, a polysaccharide also found in insect cuticles, and not cellulose makes the link with plants even more doubtful. Fungi are now given a kingdom of their own. Lichens, which are a mutualistic relationship between a fungus and an alga, cause problems whichever system of classification is used.

Table 24.5 shows the two most commonly used groupings of organisms; the two-kingdom and five-kingdom schemes. Animal and plant groups which appear in different kingdoms are highlighted in colour, taxa over which there is no controversy are not shown. (Note: you should check your syllabus to see which of these classifications you are expected to use.)

Table 24.5   Conflicts between the two-kingdom and five-kingdom classifications

Two-kingdom classification		Five-kingdom classification	
Kingdom	**Some major groups**	**Kingdom**	**Some major groups**
Plantae	bacteria blue-green algae algae fungi	Prokaryotae	bacteria blue-green algae*
		Protoctista	protozoa algae
Animalia	protozoa	Fungi Plantae Animalia	

(* renamed blue-green bacteria)

9 Does the natural classification of organisms prove evolution has occurred?

SPOTLIGHT

THE BURGESS SHALE FAUNA

Classifying fossils can present real problems. Often, there is only a faint imprint in a rock from which to reconstruct the physical characteristics of the organism. The problem is even more complex with the animals found in the Burgess Shale, a middle-Cambrian rock formation in British Columbia, Canada. Quite simply the beautiful fossils from this formation look like nothing else on Earth. For example, consider the bizarre animal called *Hallucigenia sparsa* (Fig 24.12) which has seven pairs of walking legs and seven tentacle-like structures coming out of its back. Such an animal presents us with a clear problem: how to classify organisms when they do not resemble anything else which is familiar. How on earth is *Hallucigenia sparsa* related to other known living organisms?

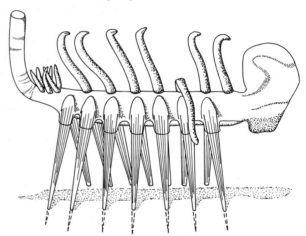

Fig 24.12 *Hallucigenia sparsa*, a fossil from The Burgess Shale; but to what is it related?

Nomenclature

The name which you would use for a particular animal or plant depends on your mother tongue and the area in which you were brought up. However, such a variety of names would cause confusion among biologists throughout the world. For example, the common two-spot ladybird found in English gardens has over 30 different names. This problem is overcome by reference to the names of the taxa used in classification. In whatever way biologists refer to a 'man' or a 'woman' in their mother tongue, every biologist in the world would understand the description 'Animalia Chordata Mammalia Primata Hominidae *Homo sapiens*'. Most of these names are not helpful because they are shared by the other animals in Table 24.4. In fact, only the name of the genus and species need be used to allow every known species to be individually named. This is known as the **binomial** (literally, two name) system.

10 Suggest why Latin names are commonly used to name organisms.

In the past the name of the person who discovered the new species was used as the root of the species name (e.g. *Spartina townsendii* is named after a person called Townsend). Often the name describes a feature of the species (e.g. *Adalia bipunctata*, the two-spot ladybird).

There are certain rules that are used in the biological binomial system.

- The name of the genus (generic name) is written with an upper case (capital) letter and that of the species (trivial name) with a lower case letter.
- The whole binomial must be written in italics or underlined if italics are not possible, such as in handwritten work, e.g. *Bellis perennis* or <u>Bellis</u> <u>perennis.</u>
- In scientific research papers the name of the person who discovered and named the organism is also included after the species name, though not written in italics. Names derived from those given by Linnaeus are denoted by 'L'. written after the species name, e.g. *Bellis perennis* L.
- The names are chosen by the taxonomist who discovers and names the species. If a new species is similar to others in an existing genus, it will be placed in that genus and the discoverer need only invent a new species name. If no existing genus exists, the discoverer must invent a new genus and species name (but must, of course, decide which existing family it must be placed in).
- Preference is usually given to the earliest name assigned to a species. In the nineteenth century an English clergyman found a fossil which he classified in a new genus *Hyracotherium* (hyrax-like animal). Independently, fossils of the same animal were found in North America where, because of other younger fossils in the same area, it was obvious that this animal was an early ancestor of the horse. Its American discoverers classified the fossil to a new genus *Eohippus* (dawn-horse). Although the latter is more helpful in describing the evolutionary importance of this fossil, the earlier name takes preference. The 'correct' name, therefore, gives 'wrong' information about the animal.

  An earlier name may be changed because it is 'wrong'. When Pere Armand David, a French missionary, first described the giant panda in 1869, he thought it was a bear and called it *Ursus melanoleucus* (black-and-white bear). Later examination of the skeleton of the giant panda showed it to be more closely related to racoons than to bears. The giant panda was renamed *Ailuropoda melanoleuca* (the new genus name reflects the giant panda's association with the raccoons; the species name was changed so that it grammatically fitted its new Latin genus name).

- The generic name may be used alone, but the species name must never be used by itself, e.g. write *Bellis*, but never *perennis* by itself.
- The name of the species must be written out in full when used for the first time, e.g. *Bellis perennis*, but provided there is no ambiguity, thereafter the name may be abbreviated, e.g. *B. perennis*.
- Many organisms have an English name (or other common name of the native language), which is not written in italics, e.g. common daisy.

KEY CONCEPT

Each species has its own unique name of two words; the binomial system of classification.

QUESTIONS

24.6 (a) Distinguish between a natural classification and an artificial classification.

(b) The books in your school or college library are probably arranged in the Dewey decimal classification system. Is this an artificial or natural classification? Explain your answer.

(c) Geologists classify rocks into three groups.

   1. Igneous   formed when molten magma below the Earth's crust rises to the Earth's surface, e.g. through a volcano, cools and hardens.

   2. Metamorphic formed when existing rocks are changed by exposure to high temperatures and pressures.

SPECIES: THEIR FORMATION AND CLASSIFICATION

3. Sedimentary   formed when particles, carried away from their parent rock by water and glaciers, are deposited and cemented together.

Is this an artificial or natural classification? Justify your answer.

24.7 Rearrange the list of biological categories in the list on the left into their correct sequence and then match them with appropriate taxon names from the list on the right:

Category	Taxon
kingdom	*lupus*
genus	Animalia
phylum	Mammalia
species	Carnivora
class	*Canis*
order	Canidae
family	Chordata

Your answer gives you the name of the wolf.

24.8 Explain why the biological name for the dogwhelk has the form *Nucella lapillus* L. and one of the barnacle species which it eats has the form *Elminius modestus* Darwin.

24.4 IDENTIFICATION KEYS

When encountering an organism for the first time you probably want to name it. Only after this (if at all) will you want to learn about its relationship with other organisms. Identification keys help you quickly to name an organism. The best keys do so by using a number of easily seen external features, i.e. they are based on an artificial classification.

Constructing and using dichotomous keys

To make a key for the animals shown in Fig 24.13 we need to choose some external features by which we can separate the animals into groups. Absolute size is not a good feature to use since organisms grow throughout life but the relative size of one of their parts (e.g. antennae one-third the

Fig 24.13 Organisms for use in constructing and using dichotomous keys. A is an earthworm; B is a slug; C is a fish; D is a fly; E is a lizard.

length of the body) may be useful. A number of species show polymorphism for colour patterns on their bodies (see Section 23.5) so colour is not a good feature to use in a key. Keys which are easiest to use start by sorting the organisms to be identified into two groups using a small number of features.

One way in which the animals in Fig 24.13 can be separated into two groups is by the presence or absence of legs.

The first line of a key might be

1. Legs absent 2
   Legs present ....

The number '2' at the end of the first line tells the user to refer to the second part of the key. A number cannot be given to the second line at this stage, but it will be given one later.

The animals labelled A, B and C in Fig 24.13 have no legs.

The second part of a key must use other features to sort them into two groups. Animal A has clear segments on the outside of its body which B and C do not have; animal B has antennae which A and C do not have; animal C has scales, fins, a tail and well-developed eyes which A and B do not have.

Any of these can be used in the second part of the key, e.g.

2. Body with fins and covered by scales C
   Body without fins or scales 3

Part 3 of the key must now separate animals A and B, e.g.

3. Definite head with two antennae .... B
   No head or antennae .... A

Now that A, B and C have been dealt with in the key, we can return to animals D and E. The next part of the key will be numbered '4'. Animal D has a body which shows clear segmentation and is divided into head, thorax and abdomen; it has compound eyes; its has antennae; it has wings; it has six walking legs. Animal E does not have a body showing segmentation, compound eyes, antennae or wings. It has four walking legs, a tail, a body covered by scales and teeth.

Any of these could be used in the key, e.g.

4. Walking legs 6, wings 2 .... D
   Walking legs 4, wings absent .... E

The whole key would, therefore, be

1. Legs absent 2
   Legs present 4
2. Body with fins and covered by scales C
   Body without fins or scales 3
3. Definite head with two antennae .... B
   No head or antennae .... A
4. Walking legs 6, wings 2 .... D
   Walking legs 4, wings absent .... E

The final construction of the key for the organisms is shown in Fig 24.14. Each point in the key has split the organisms into two groups. **Dichotomous keys** like this, based on identifiable features, are the easiest keys to construct and to use to distinguish different organisms.

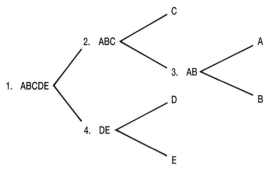

Fig 24.14 The dichotomous pattern of the key to separate the animals labelled A, B, C, D and E shown in Fig 24.13.

QUESTIONS

24.9 Construct a dichotomous key to distinguish the animals in Fig 24.15.

24.10 Collect leaves from the trees in your local area. Use an identification key to identify each tree if you are not sure of its name. Make your own dichotomous key to distinguish the leaves.

Fig 24.15 Organisms for use in question 24.9. A is a crab; B is a grasshopper; C is a water flea; D is a spider.

SUMMARY

A species is defined as a natural group of organisms which is actually or potentially capable of interbreeding to produce fertile offspring. Members of one species are physically and chemically more similar to each other than they are to other species. A geographically isolated population of one species is called a deme. Sometimes there is a gradation in allele fre-

SPECIES: THEIR FORMATION AND CLASSIFICATION

quencies between different demes of one species, called a cline. If a cline encircles a geographical area so that populations at its end inhabit the same habitat, a ring species is formed.

The formation of a new species from an existing one is called speciation. During phyletic speciation, an ancestral species changes over time to produce a single different modern species. Such a pattern of speciation is usually brought about by directional selection following changes in the environment. Since the ancestral species is usually extinct, identification of the two species must be on physical features alone. During divergent speciation an ancestral species gives rise to two or more modern species. For this to happen, gene flow between groups of a single population must stop, i.e. they must be genetically isolated from each other.

Genetic isolation may result from a number of isolating mechanisms. Geographical isolation, ecological isolation, temporal (time) isolation, behavioural isolation and mechanical isolation resulting from incompatibility of reproductive parts all reduce the chances of organisms in different populations from mating. Physiological incompatibility of gametes, inviability of hybrids and hybrid sterility all operate after mating to reduce the chances of successful reproduction. Allopatric speciation occurs when gene flow between populations is prevented by geographical isolation and is common in island populations. Sympatric speciation occurs without geographical isolation: one of the other isolating mechanisms must operate to reduce gene flow between such populations whilst they diverge genetically.

Species sometimes arise in a single generation: a new species is formed when an individual forms multiple copies of the chromosome complement of an existing species (polyploidy). Such polyploidy is more common in plants than in animals and is known to have occurred naturally to form new species within the past 100 years. Some animal species have been known to arise naturally within the past 1000 to 10 000 years.

Species are classified into groups, called categories, of increasing order, i.e. species, genus, family, order, class, phylum and kingdom. Each group of organisms within a category is called a taxon and the study of their classification is called taxonomy. In placing groups of organisms into taxa, taxonomists attempt to show their evolutionary relationships. This is called a natural classification: evolutionary relationships are reconstructed using chemical as well as developmental, physical, behavioural and ecological features. In contrast, artificial classifications are based on a few superficial features. There is considerable dispute regarding the appropriate grouping of organisms in natural classification systems. An old two-kingdom system has been superseded by a five-kingdom system: the kingdoms are Prokaryotae, Protoctista, Fungi, Plantae and Animalia.

Species are named using the name of their genus and species; the binomial system. By international convention, both names are written in italics, or underlined when handwritten, and the genus name takes an upper case initial letter while the species takes a lower case initial letter. The name of the discoverer of each species is recorded after its binomial.

Identification keys are useful for identifying unknown organisms. They rely on a few, easy-to-see features, although absolute size and colour are not used since they vary within a species. The most widely used key, the dichotomous key, repeatedly separates organisms into two groups until individual species have been described.

Chapter 24: Answers to Quick Questions

1 Asexually reproducing organisms do not breed with each other.

2 The different breeds are not populations in a geographical subdivision of the range of the species.

3 Insects and other pests subject to spraying by pesticides. Rats becoming resistant to warfarin. Bacteria becoming resistant to antibiotics.

4 One species will be replaced by another species.

5 Different groups of animals will have become isolated and followed different evolutionary pathways on the different fragments of Pangaea.

6 Divergent; there are five banana-eating species.

7 Artificial. The term vegetables has no biological meaning; such vegetables as tomatoes and capsicums are actually fruits (swollen ovaries containing seeds).

8 Analogous – they have the same function but very different structures.

9 No. The theory of evolution is one of the bases for our natural classification – taxonomists are trying to reflect probable evolutionary relationships when they classify organisms.

10 It is a language that is not used today and so cannot be corrupted.

Chapter 25

PROKARYOTAE AND PROTOCTISTA

LEARNING OBJECTIVES

When you have studied this chapter you should be able to:

1. describe the ultrastructure of prokaryotic cells and compare it with that of eukaryotic cells;

2. explain that bacteria are the only prokaryotic organisms and form the kingdom Prokaryotae;

3. describe the range of forms which are found in bacteria and the major phyla of bacteria;

4. outline the way in which the physiology of bacteria helps to explain the range of habitats which they colonise;

5. explain the classification of the kingdom Protoctista;

6. outline the major features of a number of protoctist phyla;

7. discuss the nature of viruses.

All taxa have an individual name which may be anglicised. In the following three chapters both the classical and English common names are given, e.g. Mammalia and mammals.

This chapter begins the survey of the diversity of life by considering organisms which are, on the face of it, extremely simple. However, further into the chapter the complexity of even single-celled organisms should be appreciated. In addition, the fundamental differences which exist between prokaryotes and eukaryotes should be noted. All organisms do not have life cycles similar to humans (*Homo sapiens*).

25.1 THE PROKARYOTIC CELL

Except in the kingdom Prokaryotae, all organisms have eukaryotic cells. Eukaryotic cells are so called because they have a nucleus which is surrounded by a nuclear envelope (from Greek *eu* meaning true and *karyon* meaning kernel or nucleus): they are described in Chapter 3. In contrast, prokaryotic cells (from Greek *pro* meaning before) lack a nucleus. In the transmission electronmicrograph of a bacterium shown in Fig 25.1, the light

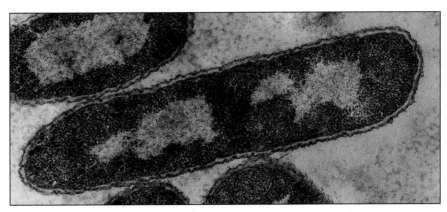

Fig 25.1 Transmission electronmicrograph of a bacterial cell. The light area in the cytoplasm, the nucleoid, contains the DNA of the bacterial cell.

area contains the cell's DNA. Within this light area, called the **nucleoid**, the DNA is present in a circular structure called a **genophore**. As well as a membrane-bound nucleus, eukaryotic cells have other complex cell organelles, such as mitochondria and chloroplasts. Prokaryotic cells do not have these complex cell organelles: instead, enzymes which catalyse the reactions of aerobic respiration and photosynthesis are bound to their cell surface membrane. Most prokaryotic cells are very small, less than 2 µm wide and 5 µm long: that is less than three millionths the size of a pin head.

1 Why have details of the internal structure of prokaryotic cells only become known in the last 40 years or so?

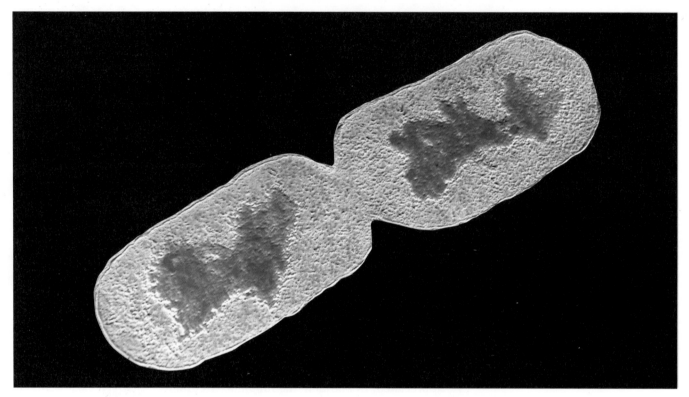

Fig 25.2   Reproduction in prokaryotes is by binary fission as seen in this scanning electronmicrograph of bacteria. The red area is the DNA.

Prokaryotic cells reproduce by binary fission (Fig 25.2). During fission, the nucleoid attaches to parts of the cell surface membrane so that, as the cell grows in volume, the cell surface membranes grow apart from each other and separate the nucleoid into two masses. Eventually, the cytoplasm of the cell may divide into two. Under ideal conditions, a prokaryotic cell may divide into two every 20 minutes (see Chapter 21).

2 How does binary fission in prokaryotes differ from mitosis in eukaryotes?

Some endospores have been isolated from the intestines of Egyptian mummies over 2000 years old which successfully germinated to form vegetative cells.

If conditions become unfavourable, some prokaryotes (e.g. anthrax bacteria) can produce protective resting stages, called **endospores**. Endospores can survive extremes of drought and temperature and, being small and light, are easily dispersed to new environments where conditions may be favourable for their germination into normal vegetative cells once more. Some prokaryotic cells show **conjugation** in which genetic material from one cell passes through a thin strand of cytoplasm that is temporarily formed between two cells. The genetic material from the donor cell may become incorporated into that of the recipient cell, but this is not a form of sexual reproduction. Table 25.1 and Fig 25.3 summarise the major differences between prokaryotic and eukaryotic cells.

Table 25.1 Major differences between prokaryotic and eukaryotic cells

Feature	Prokaryotic cells	Eukaryotic cells
size	usually small, 1–10 µm	usually large, 10–100 µm
genetic material	uncertain whether DNA is coated with protein	DNA coated with protein, called histone
	forms a circular genophore	forms thread-like chromosomes
	DNA in nucleoid, which is not bounded by membrane	DNA in nucleus which is bounded by a membranous nuclear envelope
cell structure	ribosomes are small 70S* structures	cytoplasmic ribosomes are larger 80S* structures
	few internal membranes or membrane-bound organelles	internal membranes form endoplasmic reticulum, and membrane-bound organelles, such as mitochondria, Golgi apparatus and chloroplasts, may be present
	flagella made from the protein flagellin, extend through the cell surface membrane and do not have a 9 + 2 arrangement	undulipodia are enclosed within the cell surface membrane and have an internal 9 + 2 arrangement of fibres of tubulin and many other proteins
	cell wall nearly always present and made of polymers of sugars and amino acids (peptidoglycans)	cell wall may be present and made of cellulose or chitin
cell metabolism	include obligate anaerobes, facultative anaerobes and aerobes	almost all are obligate aerobes
	some aerobic and some anaerobic photosynthesisers with end products which include sulphur, sulphates and oxygen	all those which photosynthesise excrete oxygen
	great variation in metabolic pathways	marked similarity in metabolic pathways
cell division	binary fission without mitotic spindle or microtubules	asexual division by mitosis with mitotic spindle or system of microtubules

S* stands for Svedberg unit – a measure of sedimentation rate. 70S has a slower sedimentation rate than 80S.

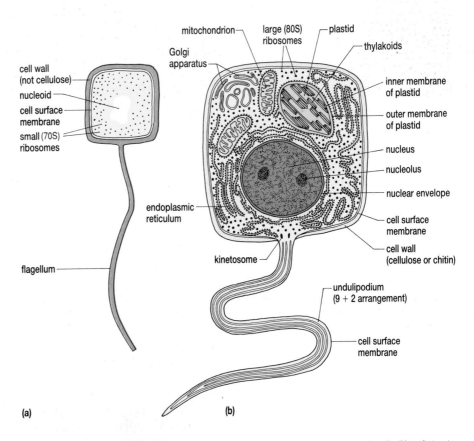

Fig 25.3 Diagrammatic summary of the ultrastructure of typical **(a)** prokaryotic **(b)** eukaryotic cells. (Note that not every cell has the structures shown.)

25.1 What is the difference between a prokaryotic flagellum and a eukaryotic undulipodium?

25.2 Copy Table 25.2. Write a tick (✔) if the feature is found in the group; write a cross (✘) if it is not found.

Table 25.2   A comparison of prokaryotes and eukaryotes

Feature	Prokaryotes	Eukaryotes
1. A cell wall is present, with cytoplasmic membranes visible under the electron microscope		
2. Photosynthetic pigment is found in lamellae within organelles called plastids		
3. Photosynthetic pigment is on lamellae that are not separated from the cytoplasm by membranes		
4. Mitochondria are present		
5. Nuclei present and can divide by mitosis		
6. Respiratory enzymes are concentrated on a cytoplasmic membrane but not in separate organelles		
7. Nuclear material is separated from the cytoplasm by a nuclear envelope		

25.3 Sterilisation involves the destruction of bacteria on working surfaces, equipment and media. Suggest why bacterial spores may make sterilisation difficult.

25.2 KINGDOM PROKARYOTAE (bacteria)

The kingdom Prokaryotae consists solely of bacteria and it is so named because bacteria, and only bacteria, have prokaryotic cells. Bacteria were probably the first living organisms (fossil bacteria have been found in rocks dated to be 3400 million years old) and for about half of the Earth's history were the only living organisms on Earth. Although their size means that they are invisible to the naked eye, bacteria are still the most abundant organisms on Earth.

Different species of bacteria are found in such diverse habitats as the Dead Sea (where the high salinity precludes other organisms), distilled water, the snow at the top of mountains, the depths of the oceans, hot water springs and the skin, nasal passages and gut of humans. They are even found in the jet fuel tanks of aeroplanes. Of course, no single species of bacterium could colonise all these habitats: each bacterium has its own specific ecological niche.

A single gram of fertile soil may contain 10^8 bacteria and a 1 cm^2 film of mucus scraped from your own gums is likely to contain 10^9 bacteria.

SPOTLIGHT

FABULOUS CHEMISTS

Bacteria are of great natural economic importance. Different species of bacteria are responsible for the decomposition processes involved in all the inorganic ion cycles described in Chapter 30, for food spoilage and for many diseases of humans, their crops and livestock. In addition, humans commercially exploit bacteria during such activities as food manufacture, the synthesis of a number of useful chemicals (including such antibiotics as streptomycin and Chloromycetin) and the transfer of artificial DNA constructs into target cells during recombinant DNA technology (more popularly called **genetic engineering**).

Bacteria can live on a bewildering range of food materials. For example, there are bacteria in the genus *Pseudomonas* which can use, among other substances, camphor and octane as energy sources. Often, the genes coding for the enzymes required to catabolise these strange carbon compounds are carried not on the main chromosome, the genophore, but on smaller loops of DNA called **plasmids**. In addition to encoding catabolic enzymes these plasmids may also carry genes which encode antibiotic resistance. The widespread use of antibiotics represents a potent selective force in favour of bacteria which carry such plasmids. Consequently, antibiotic resistance is now becoming a major medical problem with some strains of bacteria, e.g. *Salmonella*, becoming resistant to a wide range of antibiotics. Thus, one strain of *S. typhimurium* isolated from poultry is now resistant to the six antibiotic drugs doctors may use to fight it.

Feeding types

The diversity of habitats which bacteria can colonise is aided by their wide range of feeding types, summarised in Table 25.3. Photosynthesis is carried out by members of three phyla:

- Anaerobic Phototrophic Bacteria
- Chloroxybacteria
- Cyanobacteria.

Table 25.3   Feeding categories of bacterial phyla

	Autotrophic (CO$_2$ used as carbon source)	Heterotrophic (organic molecules used as carbon source)
Phototrophic (light is the energy source)	**Photoautotrophs** Chloroxybacteria Cyanobacteria most Anaerobic Phototrophic Bacteria	**Photoheterotrophs** certain Anaerobic Phototrophic Bacteria, i.e. purple non-sulphur bacteria
Chemotrophic (chemical energy is the energy source)	**Chemoautotrophs** Chemoautotrophic Bacteria Methanocreatrices	**Chemoheterotrophs** all other phyla, e.g. Actinobacteria, Micrococci, Myxobacteria, Omnibacteria, Spirochaetae

Bacteriologists have not followed the nomenclature and taxonomic practice of biologists, but they have tried to make the taxonomic level of phylum comparable. The names of some bacterial groups are not in Latin.

 What is the general name given to organisms which make their food by photosynthesis?

In the Chloroxybacteria and Cyanobacteria, photosynthesis is a similar process to that which occurs in plants (see Chapter 7), with water used as a source of electrons and oxygen released as an end product. Anaerobic Phototrophic Bacteria are only able to photosynthesise in anaerobic conditions, using hydrogen gas, hydrogen sulphide, lactate, pyruvate or ethanol as a source of electrons and releasing such end products as elemental sulphur and sulphate ions.

Chemoautotrophic Bacteria derive their energy by oxidising inorganic molecules or ions. Some of these chemoautotrophs are important in releasing sulphates and nitrates in the inorganic ion cycles described in Chapter 30. **Chemoheterotrophic Bacteria** release enzymes which digest organic molecules in their environment. Most molecules, including those synthesised by humans, can be broken down by Chemoheterotrophic Bacteria.

The term **biodegradable** refers largely to the activity of bacteria which are able to obtain energy from the breakdown of substances which humans have made.

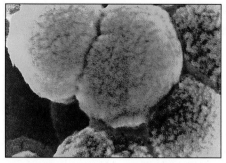

(a)

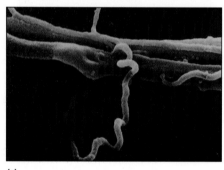

(b)

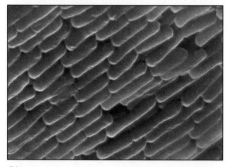

(c)

Fig 25.4 Scanning electronmicrographs of the three common bacterial forms: **(a)** cocci **(b)** bacilli **(c)** spirilla.

 4 Is bacterial digestion intracellular or extracellular?

Respiration

Bacterial species may be aerobic or anaerobic. **Obligate aerobes** must have access to molecular oxygen because it acts as the final electron acceptor of their electron transport chain during respiration (see Chapter 6). **Anaerobic bacteria** may respire in one of two ways

- by using an **alternative terminal electron acceptor** in the electron transport chain, e.g. denitrifying bacteria in the nitrogen cycle reduce nitrate (NO_3^-) to gaseous nitrogen (N_2)
- by **fermentation** which does not involve an electron transport system at all.

 5 What is the ecological significance of denitrification? Why does soil waterlogging increase the rate of denitrification?

Obligate anaerobes, e.g. *Clostridium*, have enzymes that are inhibited by oxygen so they cannot grow at all in oxygen whilst **aerotolerant anaerobes**, e.g. the lactic acid bacteria which sour milk by fermentation, are tolerant of oxygen though they do not use it in respiration. **Facultative aerobes** can respire aerobically in the presence of oxygen but by fermentation in the absence of oxygen.

 6 How does this range of respiratory methods correlate with the success of bacteria as a group?

Morphology

Individual bacterial cells may have one of three classic forms: spheres, called **cocci**; rods, called **bacilli**; spirals, called **spirilla** (Fig 25.4). These cell shapes are maintained with the help of a cell wall. The contents of a cyanobacterial cell as seen with the electron microscope are shown in diagrammatic form in Fig 25.5.

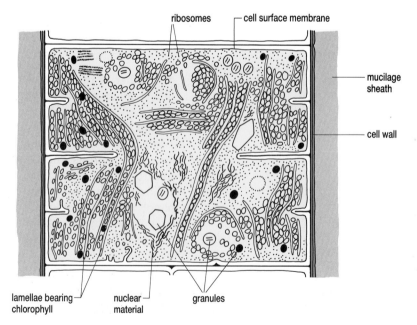

Fig 25.5 Diagram of a cyanobacterial cell as it would be seen with a transmission electron microscope.

PROKARYOTAE AND PROTOCTISTA

IDENTIFYING BACTERIA

The chemical properties of bacterial cell walls are the basis of one of the commonest diagnostic tests used in bacteriology, called the **Gram stain**. The wall of gram-negative bacteria is thinner than that of gram-positive bacteria and contains lipids. Lipids dissolve in polar solvents such as ethanol. During the Gram stain test, heat-fixed smears of bacteria are stained with crystal violet solution and then washed in ethanol. Because they have lipid in their cell walls, gram-negative cells lose the crystal violet stain whereas gram-positive cells do not. The distribution of gram-negative and gram-positive forms among prokaryotes is shown in Fig 25.6.

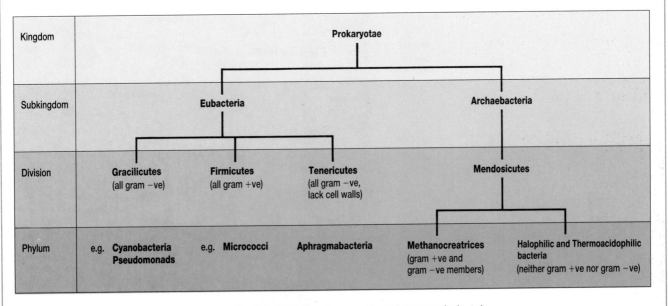

Fig 25.6   Distribution of gram-positive and gram-negative bacteria.

Not all bacteria exist as autonomous, isolated cells. During the 1980s the conventional view that most species of bacteria live as independent, single-celled individuals was challenged by investigators whose studies suggested that an individual bacterial cell is more like a component cell of a multicellular organism than a free-living autonomous organism. Some of the multicellular aggregations formed by bacteria are shown in Fig 25.7.

In some of these aggregations, cell specialisation occurs. Some species of Cyanobacteria are able to fix nitrogen as well as photosynthesise. However, the release of oxygen during photosynthesis prevents nitrogen fixation, because molecular oxygen inactivates the key nitrogenase enzyme of the Cyanobacteria. Only about 5–10% of the cells in a cyanobacterial filament can fix nitrogen: these nitrogen-fixing cells are called **heterocysts**. Unlike normal cells, heterocysts lack chlorophyll-containing thylakoids and are formed only in the absence of nitrogen-containing organic compounds.

 7   On the basis of Figs 25.2, 25.5 and 25.6 what appear to be the principal differences between Cyanobacteria and other types of bacteria?

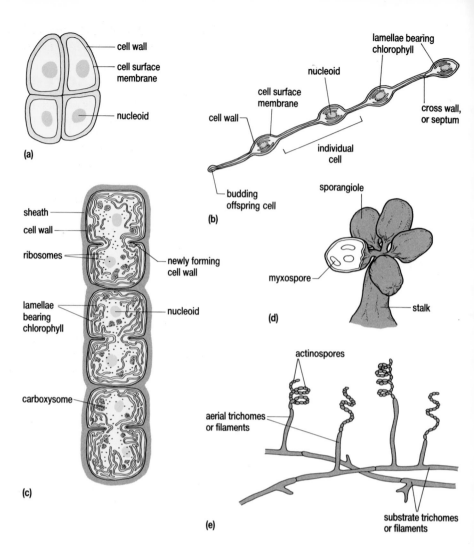

Fig 25.7 Bacteria may form multicellular aggregations, including
(a) tetrads of Micrococci
(b) stalked budding structures of Anaerobic Phototrophic Bacteria
(c) filaments of Cyanobacteria
(d) multicellular fruiting bodies of Myxobacteria
(e) mycelia of horizontal and vertical filaments (trichomes) of Actinobacteria.

Subkingdom Archaebacteria

The major differences between Archaebacteria and other bacteria are chemical, particularly in the composition of their ribosomal RNA. This subkingdom contains just one division (Mendosicutes) with two phyla (Table 25.4).

Table 25.4 Members of the subkingdom Archaebacteria

Phylum name	Cell form	Habitat
Methanocreatrices	bacilli, spirilla and cocci	worldwide in marine and freshwater sediments, including sewage, and in the gut of many animals
Halophilic and Thermoacidophilic	bacilli, cocci and irregularly shaped (pleiomorphic)	worldwide in salt works and brine; hot-water springs

Members of the phylum **Methanocreatrices** are anaerobic chemo-heterotrophs which are able to use only three organic compounds as food:

PROKARYOTAE AND PROTOCTISTA

formate, methanol and acetate. They are termed **methanogenic** because they form methane in a reaction which provides them with energy:

$$CO_2 \; + \; 4H_2 \; \rightarrow \; \underset{\text{methane}}{CH_4} \; + \; 2H_2O$$

Much of the methane in the Earth's atmosphere (in excess of one part per million) is produced by these bacteria. Most comes from methanogenic bacteria in swamps and muddy sediments, but about 30% comes from the methanogenic bacteria living in the rumen of herbivores, such as cows.

The **Halophilic** and **Thermoacidophilic** bacteria do not produce methane. Halophilic bacteria grow well only in high sodium chloride concentrations. They have proteins which only function in, and cell surface membranes which are adapted to, high salt concentrations. Thermoacidophilic bacteria grow in acidic, hot-water springs. One species, *Thermoplasma acidophilum*, is found in the famous hot water springs of Yellowstone National Park.

Subkingdom Eubacteria

These are all bacteria other than the Archaebacteria: most of the general account of bacteria given above refers to members of the subkingdom Eubacteria. Eubacteria are separated into three divisions (Table 25.5) on the basis of their cell walls (see Spotlight : Identifying bacteria).

Table 25.5   The divisions and phyla of the subkingdom Eubacteria

Division	Phylum	Notes
Gracilicutes	Spirochaetae	Tightly coiled cells with many internal flagella (Fig 25.4 (c)). Leptospirosis and syphilis are spirochaete infections.
	Thiopneutes	Obligate anaerobes, using sulphates as electron acceptors in cell respiration.
	Anaerobic Phototrophic Bacteria	Photosynthesisers, described above (Fig 25.7(b)).
	Cyanobacteria	Filaments of photosynthetic cells (Fig 25.7(c)) .
	Chloroxybacteria	Like chloroplasts of plants, though some are filamentous.
	Nitrogen-fixing Aerobic Bacteria	Mostly flagellated cells which fix nitrogen. They include the root-nodule bacterium *Rhizobium*.
	Pseudomonads	Rod-shaped decomposers.
	Omnibacteria	Diverse group including the dysentery-causing Enterobacteria and cholera-causing vibrios.
	Chemoautotrophic Bacteria	Crucial in the carbon, nitrogen and sulphur cycles.
	Myxobacteria	Form motile colonies embedded in slime and erect fruiting bodies (Fig 25.7(d)).
Firmicutes	Fermenting Bacteria	Obligate anaerobes which are chemoheterotrophic. Includes the oxygen-tolerant, milk-souring lactic acid bacteria and oxygen-intolerant *Clostridium* species which cause gas gangrene and botulism.
	Aeroendospora	Aerobic bacteria which form heat-resistant spores. Includes the large genus *Bacillus*
	Micrococci	Aerobic spherical cells (Fig 25.7(a))
	Actinobacteria	Includes those which form fungus-like mycelia of horizontal and vertical filaments, the latter bearing actinospores (Fig 25.7(e)).
Tenericutes	Aphragmabacteria	Lack cell walls, so are resistant to penicillin. Most cause disease in mammals and birds.

QUESTIONS

25.4 (a) Explain the terms chemoautotrophic, chemoheterotrophic, photoautotrophic and photoheterotrophic.
(b) Distinguish between aerobic respiration, anaerobic respiration and fermentation.

25.5 (a) Identify the three classic bacterial cell shapes.
(b) Explain why it is incorrect to state that bacteria are unicellular.

25.6 Explain what is meant by Gram-positive and Gram-negative bacteria.

25.7 Name one criterion which is used to separate Archaebacteria and Eubacteria.

25.8 Which phylum of bacteria has members that
(a) lack cell walls?
(b) are photosynthetic?
(c) produce resistant endospores?
(d) fix nitrogen?

25.3 KINGDOM PROTOCTISTA

This kingdom contains 27 phyla, including

- unicelluar, filamentous, colonial and multicellular forms
- plant-like and animal-like organisms
- slime moulds.

Only a range of the phyla in this kingdom is described below: some of the major features of these phyla are summarised in Table 25.6.

Phylum Dinoflagellata (dinoflagellates)

These have two undulipodia which project through their silica-impregnated cellulose cell walls (Fig 25.8). One undulipodium encircles the cell in a groove around its middle (the **girdle**) whilst the other trails behind the cell. Many dinoflagellates are bioluminescent, producing bright blue-green flashes of light when their water is disturbed by swimming fish or humans. Some dinoflagellates produce toxins which may accumulate in marine food chains (see Chapter 29). Many also have a red pigment which, during algal blooms, results in red tides.

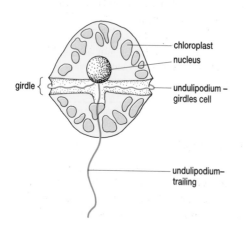

girdle {

chloroplast
nucleus
undulipodium – girdles cell
undulipodium– trailing

Fig 25.8   A photosynthetic dinoflagellate. Notice the two undulipodia (shown in blue).

Table 25.6 Main features of some phyla in the kingdom Protoctista

Phylum name	Habitat	Body form	Cellulose wall	Feeding	Motility	Reproduction
Dinoflagellata	planktonic	unicellular	present	photosynthesis	undulipodia	asexual
Bacillariophyta	planktonic	unicellular, filamentous	absent; silicaceous shells (2)	photosynthesis	usually none	asexual and sexual
Euglenophyta	aquatic	unicellular or colonial	absent	photosynthesis & heterotrophic	undulipodia	asexual
Zoomastigina	aquatic or parasitic	unicellular or colonial	absent	heterotrophic	undulipodia	asexual and sexual
Apicomplexa	parasitic in animals	unicellular	absent	heterotrophic	none in adult	asexual and sexual
Rhizopoda	aquatic or parasitic	unicellular	absent	heterotrophic (phagocytosis)	pseudopodia	asexual
Ciliophora	aquatic	unicellular or colonial	absent	heterotrophic (phagocytosis)	undulipodia (cilia)	asexual and sexual
Chlorophyta	aquatic	unicellular	present	photosynthesis	none in adult	asexual and sexual
Rhodophyta	marine	multicellular	present	photosynthesis	none in adult	asexual and sexual
Phaeophyta	marine	multicellular	present	photosynthesis	none in adult	asexual and sexual
Gamophyta	freshwater	filamentous	present	photosynthesis	none in adult	asexual and sexual

ALGAL BLOOMS

During the dry, hot British summers of 1989 and 1990 there were many reports about algal blooms both in reservoirs, which affect drinking water, and in the sea. In fact, the organisms involved in each ecosystem are different. In the **reservoirs** the organisms causing the problem are not algae at all but **cyanobacteria** such as *Microcystis* sp. which can produce quite powerful toxins.

The sea blooms are the result of the rapid growth of dinoflagellates such as *Gymnodinium breve* which produce the famous red tides. This organism can secrete powerful neurotoxins which can kill fish. Shellfish, e.g. mussels which filter feed, can accumulate high levels of these toxins in their bodies which can prove harmful, indeed potentially fatal, to anybody eating them.

Phylum Bacillariophyta (diatoms)

These organisms have two shells which overlap, rather like the parts of a Petri dish. The pectic materials impregnated with silica, from which the shells are made, produce such intricate patterns that slides of mounted diatoms are often used to test microscope lenses for optical aberrations (Fig 25.9).

Fig 25.9 Diatoms have characteristic ridges. They are used to test optical microscopes – if the ridges are not clear the optics need to be checked.

The dinoflagellates and diatoms are the most important members of the **phytoplankton**, the photosynthetic organisms which float in the upper layers of seas and lakes. Together these two phyla account for about 70% of all photosynthesis which occurs on Earth.

8 Suggest what body shapes would be advantageous to phytoplanktonic organisms.

Phylum Euglenophyta (euglenoids)

This phylum, named after its best-known genus *Euglena*, contains mostly unicellular organisms which move by the whip-like action of their undulipodium (Fig 25.10). Most euglenoids photosynthesise. They have a photoreceptor, called an eyespot, which enables them to move towards illuminated regions. In the absence of light, many euglenoids are capable of feeding on dissolved or particulate food.

9 State two differences between euglenoids and dinoflagellates.

'Protozoa'

Protozoa (meaning first animals) are heterotrophic unicells which were once considered to belong to a single phylum of the animal kingdom. They are now no longer classified as such, but the term protozoa is still often used as a convenient short-hand to refer to all the phyla outlined below.

10 Why can *Euglena* not be grouped with the protozoans?

Phylum Zoomastigina (flagellates)

These are unicells which have at least one, and may have up to several thousand, undulipodia which they use for movement, sensitivity and food capture. Whereas many are free-living in water and soils, some are symbiotic. Zoomastigina of the genus *Trypanosomas* (Fig 25.11) cause a potentially fatal disease called sleeping sickness. These parasites are transmitted to the blood of an uninfected person via the tsetse fly, a blood-feeding insect.

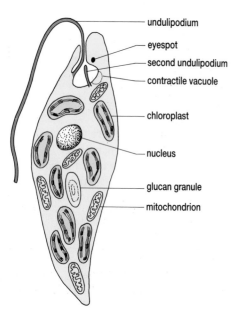

undulipodium
eyespot
second undulipodium
contractile vacuole
chloroplast
nucleus
glucan granule
mitochondrion

Fig 25.10 Generalised structure of an euglenoid.

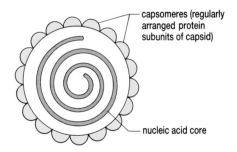

capsomeres (regularly arranged protein subunits of capsid)

nucleic acid core

Fig 25.19 The structure of a generalised virion.

Viruses are utterly dependent on a host cell and represent the ultimate in parasitism. The protein capsid of each virus is specifically adapted to adsorb to a receptor on the surface of its host cell. Following adsorption, the viral nucleic acid (sometimes accompanied by other virion components) enters the cell. Once inside the host's cytoplasm, one of two events may occur.

1. Viral nucleic acid takes control of the activities of the host cell's metabolism. As a result, the host cell manufactures more viral protein coats and viral nucleic acid and assembles them to form new viruses. Eventually, the assembled virions escape from the host cell; they may do so as a consequence of exocytosis (see Section 2.6), by passing harmlessly through the host's cell surface membrane or following lysis of the host cell (the lytic cell). In the latter case, lysis often results when a viral gene coding for lysozyme is translated by the host cell.

2. Viral nucleic acid attaches to the genophore (nuclear material) of the host cell, where it is now termed a **provirus**. It may remain in this state without influence, being harmlessly replicated by the host cell during its own cell cycle (see Fig 4.12). For reasons which are still not clearly understood, a provirus can become detached from its host's DNA and start the lytic cycle described above. Certain types of human cancer are thought to be caused by a provirus: the Herpes virus also occurs as a provirus in human cells.

Many studies of proviruses have been performed on those which infect bacteria. Bacteria-infecting viruses are called **bacteriophages** (**phages** for short). A phage which causes lysis of a bacterium is termed **virulent** (Fig 25.20) whereas one which becomes attached to the bacterial genophore is termed **temperate** (Fig 25.21). Since a provirus (or prophage) can detach from its host's genophore and initiate the lytic cycle described above, bacteria carrying prophages are termed **lysogenic**.

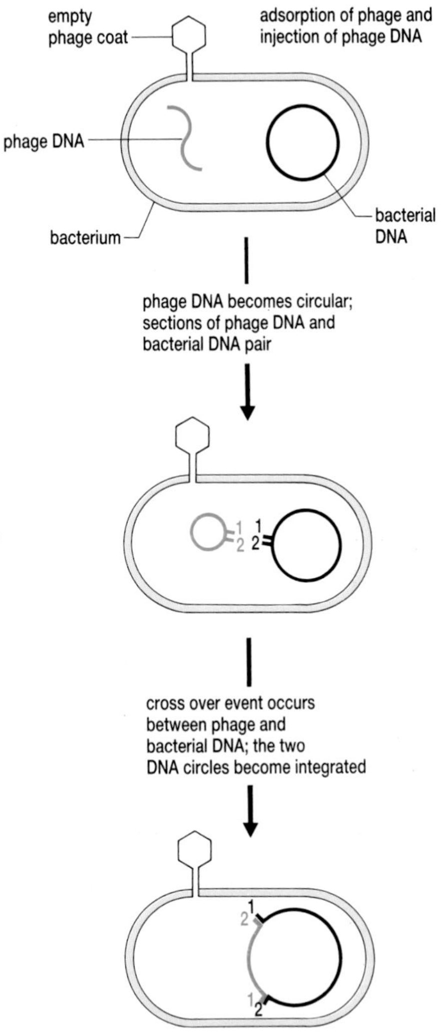

empty phage coat

adsorption of phage and injection of phage DNA

phage DNA

bacterium

bacterial DNA

phage DNA becomes circular; sections of phage DNA and bacterial DNA pair

cross over event occurs between phage and bacterial DNA; the two DNA circles become integrated

Fig 25.21 The formation of a prophage from a temperate bacteriophage. By replacing part of the bacteriophage DNA with, say, a piece of human DNA, temperate bacteriophages can be used by genetic engineers to carry foreign DNA into a bacterial host cell.

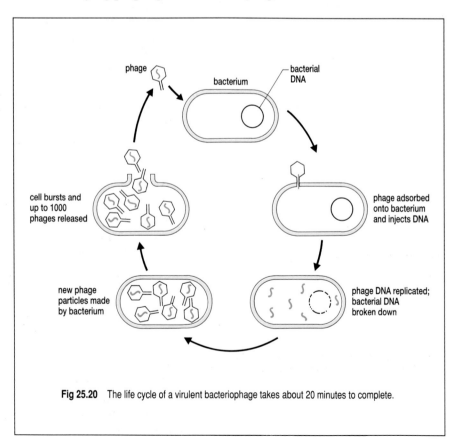

phage

bacterium

bacterial DNA

cell bursts and up to 1000 phages released

phage adsorbed onto bacterium and injects DNA

new phage particles made by bacterium

phage DNA replicated; bacterial DNA broken down

Fig 25.20 The life cycle of a virulent bacteriophage takes about 20 minutes to complete.

AIDS

Acquired immune deficiency syndrome (AIDS) (see Section 13.3) is caused by a retrovirus called Human Immunodeficiency Virus-1 (HIV-1). The life cycle of this organism is shown in Fig 25.22.

HIV-1 has an outer lipid coat, derived from the membrane of the infected cell, which surrounds the protein core containing the genetic material. This lipid coat fuses with the cell surface membrane of the target cell, releasing the viral RNA into the cell along with an enzyme, **RNA reverse transcriptase**. This enzyme, which is unique to the retroviruses, makes the infected cell turn the viral RNA into viral DNA. This DNA copy, when it is termed a provirus, can insert itself into one of the chromosomes in the nucleus of the cell. The provirus can now remain dormant, the latent stage of infection, for months, perhaps even years. Eventually, for some unknown reason, the provirus is 'switched on' and starts to direct the synthesis of new viruses. The new viruses are then budded off from the cell surface membrane and infect new cells.

Fig 25.22 Life cycle of the HIV-1 virus, the causative agent of AIDS.

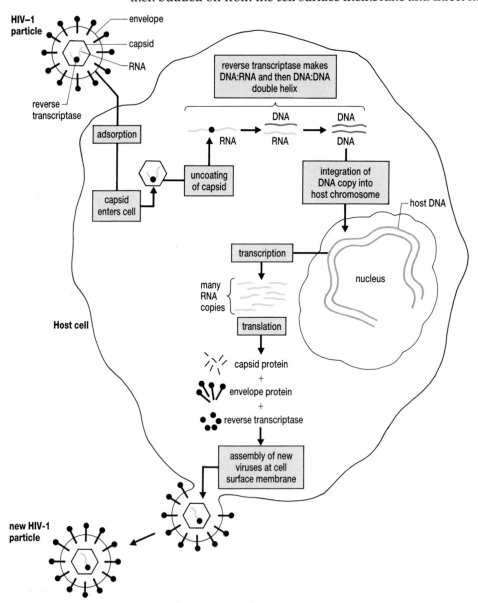

25.15 One type of Herpes virus attacks cells in the mucous membranes of the mouth and lips. During times of stress, this virus may cause infectious cold sores.

(a) Suggest how this Herpes virus causes periodic eruptions of cold sores.

(b) Suggest why a second type of Herpes virus causes sores on the mucous membranes of the genitals but not of the mouth and lips.

25.16 Argue for or against the statement 'Viruses are not alive'.

SUMMARY

Unlike eukaryotic cells, prokaryotic cells lack a membrane-bound nucleus and membranous organelles. Although functionally similar, the flagella of prokaryotes lacks the 9 + 2 arrangement of eukaryotic undulipodia and the cell wall is of different composition. Prokaryotic cells never undergo mitosis.

Only bacteria possess prokaryotic cells. They are classified into 17 phyla in the kingdom Prokaryotae. Although some bacteria are pathogenic, the majority are free-living. They are important in the inorganic ion cycles within ecosystems. The prokaryotes are a physiologically diverse group, including chemoautotrophic, chemoheterotrophic, photoautotrophic and photoheterotrophic nutritional types. Heterotrophic bacteria can utilise a wide range of food types. The Prokaryotae also includes aerobic and anaerobic forms. This physiological diversity aids the colonisation of a wide range of habitats, such that few environments on Earth are not inhabited by bacteria. However, they show little morphological diversity. Three classic cell shapes are rods, spheres and spirals. Many bacteria form filaments or colonies of cells.

The kingdom Protoctista consists of all the eukaryotes which are not classified as fungi, plants or animals. Protoctists show a variety of forms, including unicellular, filamentous, colonial and multicellular organisms. The latter may grow up to 100 metres in length. Protoctista includes the phyla: Zoomastigina, Apicomplexa, Rhizopoda and Ciliophora, which contain non-photosynthetic consumers, as well as Dinoflagellata, Bacillariophyta, Rhodophyta, Phaeophyta, Gamophyta and Chlorophyta, which are photosynthetic. Another group of the Protoctista, Euglenophyta, are photosynthetic but are able to utilise dissolved or particulate food in the absence of light.

Although they appear to be unicellular, viruses show few of the characteristics of life. Each consists of a protein capsid surrounding a nucleic acid core. All viruses are obligate cell parasites; those parasitising bacteria are called bacteriophages. Once inside a specific host cell, the viral nucleic acid may join the host's nuclear material to become a provirus which is harmlessly replicated through generations of host cells. However, the viral nucleic acid at some stage takes control of the host's physiology, so that only new viral particles are formed. Eventually, replicated viruses are released, often accompanied by the lysis of the host cell.

Chapter 25: Answers to Quick Questions

1 They could only be resolved by electron microscopes.
2 Binary fission in prokaryotes does not involve the complex chromosome movements and spindle formation seen in eukaryotic mitosis; DNA is distributed randomly.

3 Photoautotrophs.

4 Extracellular.

5 Reduces soil fertility. Microbes will switch to nitrate as a terminal electron acceptor since less oxygen is available in waterlogged soils.

6 Enabled bacteria to colonise a very wide range of habitats ranging from those which lack any oxygen to those where oxygen is plentiful.

7 Other bacteria are smaller and cyanobacterial cells appear to have a more complex internal structure.

8 Ones which maximised their surface area to volume ratio, i.e. flat discs rather than spheres. Star shapes would be even better; this would increase the efficiency with which materials could diffuse into and out of the cells whilst helping to prevent the cell from sinking away from the sunlit upper waters.

9 Euglenoids have only one functional undulipodium and lack cell walls.

10 *Euglena* is, primarily, an autotroph not a heterotroph.

11 Maximises surface area to volume ratio and so aids diffusion (see 8 above).

12 Sunlight, CO_2, inorganic ions, water. With the exception of sunlight all the others will be absorbed directly by the thallus cells from sea water.

13 Holdfast anchors seaweed to rocks in the face of turbulent wave action. Flexible, but tough, shape allows alga to be moved by waves without damage. Flat lamina provides large surface area for photosynthesis.

Chapter 26

FUNGI AND PLANTAE

LEARNING OBJECTIVES

When you have studied this chapter you should be able to:

1. describe the characteristic features of the kingdom Fungi;

2. describe the structure of a typical fungal mycelium and outline its physiology;

3. outline the main features of the five fungal phyla;

4. describe the characteristic features of the kingdom Plantae;

5. explain the importance of transport vessels, supporting tissues, siphonagamous fertilisation and seeds in adapting plants to life on land;

6. outline the life cycle and main features of the phyla Bryophyta, Filicinophyta, Coniferophyta and Angiospermophyta.

The vast majority of the organisms considered in the last chapter are aquatic: they live in water. In contrast, the majority of the organisms considered in this chapter are terrestrial: they live on land. So, in addition to introducing two further major groups of organisms, fungi and plants, this chapter also sets out to consider the ways in which organisms are adapted to live under terrestrial conditions.

26.1 KINGDOM FUNGI

All members of this kingdom are eukaryotic organisms which lack undulipodia at any stage of their life cycle and which form spores. The simplest fungi are single-celled but the majority consist of thread-like, slender filaments surrounded by an outer wall of chitin, called **hyphae** (Fig 26.1). A large mass of hyphae is called a **mycelium** and is the vegetative form of most fungi. Some groups of fungi have hyphae which are **coenocytic**, i.e. have several nuclei within each cell; other groups of fungi have hyphae which are **septate**, i.e. are divided into cell-like compartments

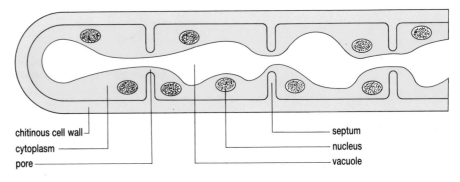

chitinous cell wall

cytoplasm

pore

septum

nucleus

vacuole

Fig 26.1 Diagrammatic section through a septate fungal hypha. Notice that the septa are incomplete and that the single cells are multinucleate.

by cross walls of chitin, called **septa**. Even when septa are present, pores allow the cytoplasm to stream between compartments (see Fig 26.1). Fungal nuclei are usually haploid. The main features of five fungal phyla are shown in Table 26.1.

Table 26.1   Phyla of the kingdom Fungi (fungi)

Phylum	Hyphae	Sexual reproduction	Asexual reproduction
Zygomycota (zygomycetes)	aseptate	conjugation of mating hyphae (gametangia) forms a thick-walled **zygospore**. The many nuclei within the zygospore fuse in pairs; haploid spores are then formed by meiosis	haploid **sporangiospores** are carried within sporangia on vertical hyphae, **sporangiophores**
Ascomycota (ascomycetes)	septate	conjugation of mating hyphae forms a sac-like **ascus** containing four **ascospores**	segmentation of a hypha to form **conidiospores**
Basidiomycota (basidiomycetes)	two types: aseptate primary mycelium and septate secondary mycelium	uninucleate segments conjugate to form a **dikaryotic secondary mycelium**; after fusion of nuclei, this forms haploid **basidiospores** inside club-like **basidia**	
Deuteromycota (imperfect fungi)	septate	none	forms **conidiospores**
Mycophycophyta (lichens)	septate	conjugation to form **ascospores**	forms small fragments (**soredia**) with at least one algal cell surrounded by fungal hyphae

The development of fungal mycelia

A fungal mycelium develops from a single **spore** (Fig 26.2). The spore germinates to produce a single hypha which, providing nutrients are readily available, begins to branch. Providing nothing obstructs its growth the mycelium will continue to grow outwards until it has exhausted all the nutrients available in its environment.

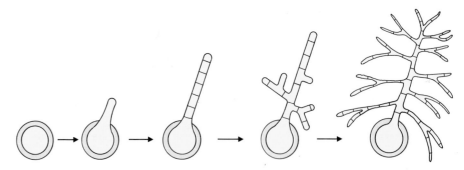

Fig 26.2   The growth of a fungal mycelium from a single spore. Initially, a single hypha extends outwards; by continued growth and branching, an extensive network, or mycelium, is soon formed.

1   What is the effect of branching on the pattern of growth of a mycelium?

Metabolic activity is concentrated at the tips of the hyphae. The hyphal tips are also the areas that absorb nutrients. Most fungi are saprophytic heterotrophs (see Section 8.1). They obtain their food by releasing enzymes onto dead organic matter and then absorbing the digested food across their cell walls and cell surface membranes. Along with saprophytic bacteria, such fungi are important decomposers in natural ecosystems and are, therefore, essential to the cycling of inorganic ions and the maintenance of soil fertility.

 2   **Fungal hyphae are delicate, unprotected structures. Suggest the major environmental hazard that might damage hyphal tips on land.**

Some fungi are parasitic, including species which cause the rusts, smuts and blights of crop plants and human diseases, e.g. thrush, ringworm and athlete's foot. Many more fungi form mutualistic associations with plants. Almost all healthy terrestrial plants have fungi associated with their roots, forming **mycorrhizae** which aid transport of inorganic ions from the soil into the plants' roots.

Reproduction

The majority of fungi reproduce using specialised cells called **spores**. Fungal spores are different from the spores produced by plant sporophytes. Fungal spores can be produced both by mitosis, in which case they are called **asexual spores**, and by meiosis, producing **sexual spores**. An enormous variety of spore types exists (Fig 26.3) but their common feature is that they are agents of dispersal and/or survival. Thus some fungi, e.g. *Mucor* and *Rhizopus*, produce thick-walled resting spores which allow the fungi to survive unfavourable conditions.

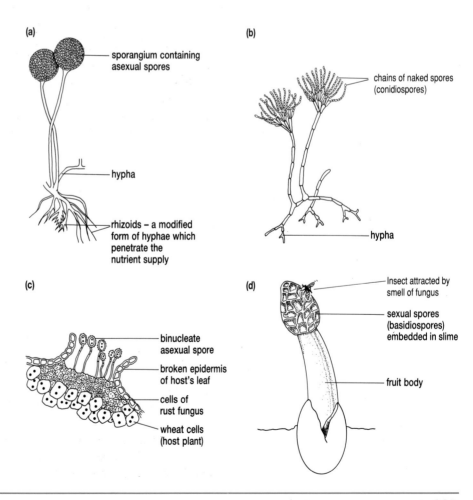

Fig 26.3 Types of fungal spore
    (a) wind-dispersed asexual spores of a terrestrial fungus, e.g. *Rhizopus*
    (b) conidiospores of *Penicillium*
    (c) thick-walled asexual spores of wheat rust *(Puccinia)*
    (d) the fruiting body of a stinkhorn *(Phallus impudicans)* bearing a slimy and unpleasantly odourous mass of basidiospores.

Asexual reproduction

This involves the production of hyphae which grow vertically and carry spores (Fig 26.4). These spores are produced asexually by mitosis and may be enclosed within a structure known as an **ascus**, **basidium** or **sporangium** depending on the fungal phylum, or may be naked **conidiospores** (see Table 26.1). Special adaptations for the **dispersal** of spores and some examples are given below.

- Sporangia of the zygomycetes burst, releasing thousands of spores into the air.
- Conidiospores are released singly.
- Thick-walled spores are produced which can be blown by the wind, e.g. parasitic rusts (see Fig 26.3(c)).
- Sporangia may be produced which can be shot into the air. For example, a genus of dung fungus, *Philobus* (Zygomycota), can shoot its sporangia up to 2 metres in the air. These sporangia stick to plants which may be eaten by herbivores. The spores pass unharmed through the herbivore's gut and are deposited in the faeces, where they form new mycelia.

Sexual reproduction

In fungi this can occur by a variety of methods. However, one consistent difference between fungi and plants is that fungi do not exhibit a regular alternation of generations (see Chapter 25). The diploid phase in the fungal life cycle is usually very brief. Sometimes it consists only of the zygote which immediately undergoes meiosis. This means that the nuclei in fungal hyphae are mainly haploid, a feature which has important implications for sexual reproduction.

Sexual reproduction in most fungi involves conjugation (see Fig 26.4). Here hyphae from mycelia of different mating types meet and swellings develop on the hyphae of the different mating strains. The walls separating these swellings break down and their cytoplasm, which contains haploid

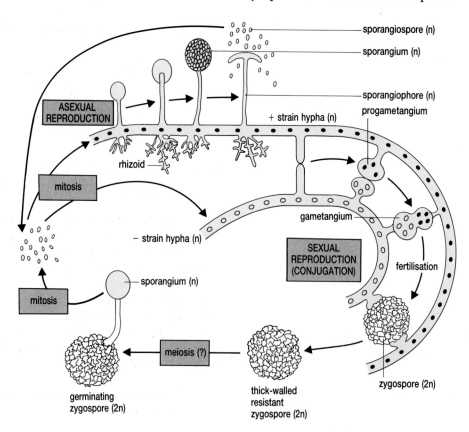

Fig 26.4 Life cycle of the bread mould, *Rhizopus*, demonstrating the production of asexual sporangiospores and, following conjugation, sexual zygospores.

nuclei, mixes. If two or more of these haploid nuclei subsequently fuse, fertilisation has been achieved without the need to produce swimming gametes.

 3 Why will this be advantageous in terrestrial environments?

In ascomycetes and basidiomycetes, fusion of the haploid nuclei does not occur immediately after the mixing of the hyphal contents of two different mating strains. Rather, the parental nuclei grow and divide separately within the hypha giving rise to a **dikaryon** where each cell contains a pair of nuclei, one from each parent. Eventually, the nuclei within each cell fuse, forming a diploid zygote which immediately undergoes meiosis to produce four haploid spores. When they germinate these spores will develop directly into haploid hyphae.

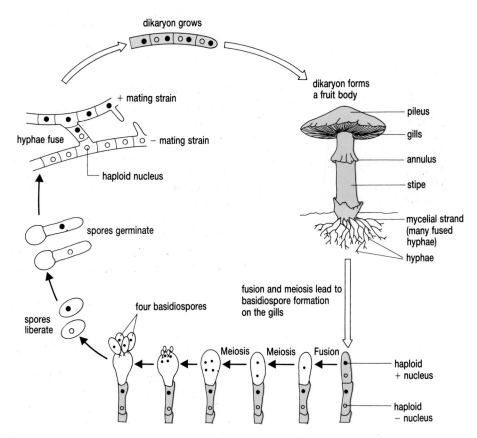

Fig 26.5 Stages in sexual reproduction of a typical basidiomycetes, the field mushroom (Agaricus). From the dikaryon mycelium in the soil, a large fruiting body is formed. Fusion of nuclei occurs followed by meiosis so that millions of haploid basidiospores are produced on the exposed gill surfaces.

The common cultivated mushroom, *Agaricus bisporus*, is a species which typifies the basidiomycetes. As the dikaryon grows, it forms a fruiting body; this is the mushroom we eat. Following fusion of nuclei within the dikaryon, meiosis produces the brown basidiospores on the leaf-like gills below the cap of the mushroom (Fig 26.5). The fruiting body has quite a complex structure and contains a variety of hyphal tissues which provide both support and protection against desiccation. Thus, in the case of mushrooms, hyphal fusion occurs below ground but their sexual spores are released once nuclear fusion and division have occurred in a robust, structure above ground.

 Suggest why this reproductive strategy may be advantageous to mushrooms.

The stinkhorn, *Phallus impudicans* (Basidomycota), produces sexual spores (basidiospores) which are entangled in a smelly slime (see Fig 26.3(d)). This attracts flies which pick up the spores on their bodies and so disperse them when they fly away.

FUNGAL HERBICIDES

Fungal diseases are a major economic threat to crop plants. However, plant scientists are beginning to develop fungi which can attack agricultural weeds such as wild oats and cleavers.

Mycoherbicides already in use include one named Collego which employs the fungus *Colleotrichum gloeosporiodes aeschynomene*. Developed by George Templeton in Fayetteville, Arkansas, USA, it was first introduced in 1982 to control northern jointvetch (*Aeschynomene virginica*), whose black seeds contaminate harvests of rice and soya beans. Marketed as a dry formulation consisting of 15% viable spores and 85% inert ingredients, Collego can be stored for long periods at room temperatures. Each package contains 757×10^{12} spores, which will treat about 2.5 hectares. Farmers mix the formulation with a wetting agent and add about 250 dm^3 of water. The farmer then sprays the suspension, which gives about 18×10^6 m^{-2} spores, from the air at a time when the crop is well watered and relative humidity is likely to be high for the following 12 hours. Within a week or two the vetch plants begin to show lesions that gradually encircle the stem. Most of the vetch plants die within 5 weeks. A single application of Collego in the growing season of the crop is all that is needed; with rice this is when the weed has just emerged above the crop canopy. A few stunted plants may survive treatment but cannot keep up in competition with the crop. Growers have taken enthusiastically to Collego, and the fungus has reduced the input of pesticides by nearly 500 000 dm^3 per annum since its introduction.

Fig 26.6 A novel mycoherbicide. A fungus deadly to a single species of weed could rid this wheat crop of cleavers.

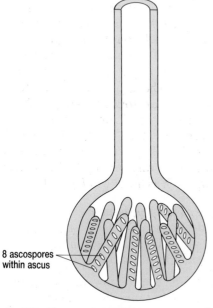

8 ascospores within ascus

Fig 26.7 The ascus, inside which are ascospores, of the Dutch elm disease fungus, a member of the Ascomycota.

Phylum Zygomycota (zygomycetes)

This phylum is so called because its members produce resting sexual **zygospores** as well as asexual **sporangiospores**. It contains the bread moulds *Mucor* and *Rhizopus* (whose life cycle is shown in Fig 26.4). Zygomycetes have coenocytic hyphae. Most are saprophytes but the phylum includes fungi which are parasites of protoctists, animals and other fungi.

Phylum Ascomycota (ascomycetes)

As well as many colourful bread moulds, this phylum contains the yeasts, truffles and the fungus which causes Dutch elm disease (*Ceratocystis ulmi*). The phylum is named after the sac-like **ascus** which its members produce after hyphae of opposite mating types have conjugated. Within the ascus, fertilisation occurs to form a diploid nucleus which immediately undergoes meiosis to produce four haploid nuclei. These nuclei undergo one or two mitotic divisions and form encapsulated **ascospores** which are dispersed (Fig 26.7). Asexual reproduction also occurs when the tips of vertical hyphae segment to form **conidiospores**.

Yeasts are unusual ascomycetes since they are unicellular (Fig 26.8). Although two yeast cells may conjugate to form an ascus containing four ascospores, they usually reproduce by budding (see Fig 19.5). Even if molecular oxygen is present, yeasts ferment sugars to ethanol (see Chapter 6). If molecular oxygen is present, this ethanol is subsequently respired aerobically to form carbon dioxide and water. Yeast fermentation has been used for centuries to raise bread and produce alcoholic drinks.

(a)

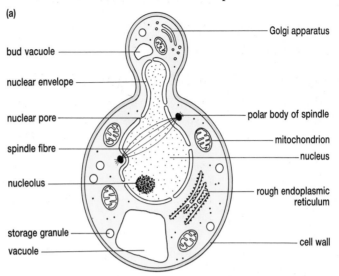

(b)

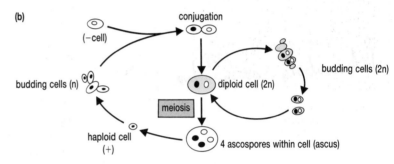

Fig 26.8   (a) Drawing of an electronmicrograph of a yeast cell which is beginning to reproduce asexually (budding).
            (b) Life cycle of yeast showing conjugation, also budding of haploid and diploid cells.

Phylum Basidiomycota (basidiomycetes)

Rusts, smuts, mushrooms and puffballs are basidiomycetes. Following conjugation of opposite mating types, haploid **basidiospores** are formed inside a club-like **basidium**. The phylum is named after these basidia.

The familiar mushrooms and puffballs (Fig 26.9) are dense aggregations of hyphae (called **basidiocarps**) that arise from an underground mycelium. They carry the basidia with their spores. The cap (**pileus**) of many mushrooms has gills on which the basidia are formed (see Fig 26.5); puffballs do not have gills. Liberated spores germinate to form a new mycelium. This mycelium grows outwards from a single spore in a roughly circular pattern. The mycelium becomes ring-like after the older parts in the middle of the mycelium die. The spore-carrying fruiting bodies reflect this pattern of growth, forming fairy rings such as the one in Fig 26.10.

Phylum Deuteromycota (deuteromycetes)

The members of this phylum are fungi that do not show sexual reproduction. For this reason they have been referred to as imperfect fungi. Like the ascomycetes, they form conidiospores (see Fig 26.3(b)) which germinate to form mycelia of septate hyphae. The penicillin-producing genus *Penicillium*, the mould *Aspergillus*, the causative fungus of athlete's foot (*Trichophyton*) and the soil inhabiting *Rhizoctonia* which causes damping off and root rot of cultivated plants are all deuteromycetes.

Fig 26.9   A ripe puffball (a basidiomycetes) may contain 5×10^{12} spores which are released when the puffball is touched, splits or decomposes.

Peanuts provide a suitable substrate on which *Aspergillus* grows. As the mould grows, it releases aflatoxins, potent cancer-causing chemicals. Shipments of peanuts are, therefore, always checked for the presence of aflatoxins before being sold.

Fig 26.10   The fairy ring of fungal fruiting bodies reflects the circular growth of the mycelium from a single spore.

Phylum Mycophycophyta

This phylum contains the lichens (Fig 26.11), a mutualistic relationship between a fungus (**mycobiont**) and a photosynthetic partner (**phycobiont**). The phycobiont manufactures food and may fix nitrogen whilst the hyphae of the mycobiont provide anchorage, products of digested organic remains and protection against desiccation. In most lichens, the mycobiont is an ascomycete and the phycobiont is a green 'alga' or cyanobacterium. Lichens colonise inhospitable places such as bare rocks (their importance in primary succession is described in Section 31.1), the barks of trees or the cold polar regions and mountain tops.

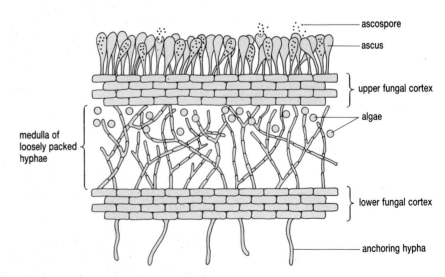

Fig 26.11   Cross-section of a lichen – a fungus and a photosynthetic alga in a mutualistic relationship.

FUNGI AND PLANTAE

Though resistant to drought, lichens must have alternating wet and dry periods. They grow very slowly and are susceptible to even slight air pollution. For this reason, they are often used as monitors of environmental pollution.

QUESTIONS

26.1 Name two features of Fungi which distinguish this kingdom from other eukaryotes.

26.2 Explain each of the following terms: hypha; mycelium; septate; coenocytic; dikaryon.

26.3 Use the account of asexual and sexual reproduction in yeasts to produce a diagram of the life cycle of yeasts in a similar style to Fig 26.4.

26.4 In what ways do fungi affect natural ecosystems?

26.5 Copy the following table and complete the missing terms **A** to **I**.

Phylum	Hyphae	Asexual spores	Sexual spores
A	B	uncommon	basidiospores
C	D	conidiospores	none formed
E	F	sporangiospores	G
Ascomycota	septate	H	I

26.6 Many fungi which are parasites of plants produce **haustoria**. Haustoria are hyphae which push into plant cells, although they do not break either the cellulose cell wall or cell surface membrane.
(a) Suggest two functions of haustoria.
(b) Suggest how haustoria feed.

ANALYSIS

Life on land

This exercise helps you to consolidate your knowledge and understanding of biology and apply it to a new problem. It will provide practice in the skills of recall, understanding, selection and organisation of material and the interpretation of information given in diagrammatic form.

Terrestrial plants live under different conditions from aquatic 'algae'. This exercise draws attention to these different environments and the ways that plants in particular have overcome the problems of life on land.

Fig 26.12 Kelp, *Laminaria*, at low tide. When submerged at high tide the fronds of the kelp plant float upwards, supported by the water.

Fig 26.13 This plant has leaves with a waxy upper cuticle – an adaptation to prevent water loss.

When swimming, you may get a sensation of lightness, the feeling of floating. This illustrates a crucial difference between aquatic and terrestrial environments: water is denser than air. This means that the support available to an organism varies considerably between terrestrial and aquatic environments.

(a) How is this difference in support reflected in Fig 26.12?

Although land organisms can rest on solid ground or burrow into soil, to raise their bodies even a little way above the surface of the ground requires some sort of supporting tissue, a **skeleton**.

(b) What structures act as the skeleton of a plant?

To survive, grow and reproduce, organisms require certain basic resources.

(c) List the basic resources needed by heterotrophs (e.g. fungi, animals) and autotrophs (e.g. plants).

Consider a marine alga. It can obtain the inorganic ions and gases it needs directly from the sea by absorbing them over the whole of its body surface. Now consider a terrestrial plant.

(d) Where will a terrestrial plant obtain the gases, inorganic ions and water it needs to grow?

Terrestrial plants obtain their resources from a two-phase environment: air and soil. To achieve this they need to occupy space both above and below ground. The underground part of a plant (typically a root system) may need to be very large if it is to absorb a sufficient amount of water and inorganic ions.

(e) What other function will such a root system serve in plants with large aerial parts?

Photosynthesis is one function of the aerial part of the plant. However, this separation of function into an aerial photosynthetic part and an absorptive underground part poses an additional problem, i.e. transport. Plants have to transport water and inorganic ions from their roots to their leaves and the products of photosynthesis from their leaves to their roots (see Chapter 12).

(f) Why is transport not a problem for the large marine alga *Laminaria*

Land organisms also face two other problems
* desiccation
* transfer of gametes.

One solution to the problem of desiccation is a waterproof coating over the whole of the aerial part of the plant (Fig 26.13).

(g) Why would total waterproofing not be a satisfactory solution to preventing desiccation if plants are to continue to live?

Look at the life cycle of the sea lettuce *Ulva*, a marine protoctist, shown in Fig 26.14. Here, dispersal of the gametes depends on the availability of water. However, motile, swimming gametes cannot be released into the air and they will dry up on land. Land organisms, both plants, animals and fungi, must therefore restrict sexual reproduction to damp places or possess mechanisms for transferring gametes under dry, terrestrial conditions.

(h) What special mechanism do flowering land plants have for transferring gametes?

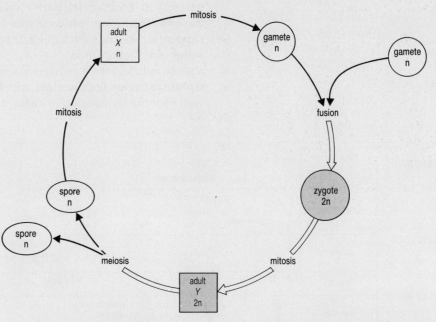

Fig 26.14 Life cycle of sea lettuce *Ulva*, a marine seaweed. The alternation of generations shown here is also found in all land plants.

QUESTIONS

26.7 (a) How will the thermal conditions differ between aquatic and terrestrial environments?

(b) Why might changes in temperature present a particular problem for terrestrial plants?

(c) Name three ways in which terrestrial plants are adapted to overcome extremes of temperature.

26.8 'Stomata represent a trade-off between the conflicting demands of desiccation and gas exchange'. Discuss this statement.

26.2 KINGDOM PLANTAE

Basic features

Plants are multicellular eukaryotes that develop from an **embryo**, i.e. a young, multicellular organism developed from a zygote and supported by maternal non-reproductive tissue. Within their life cycle, plants alternate between

- a sexually reproducing, haploid **gametophyte** generation
- an asexually reproducing, diploid **sporophyte** generation.

Both generations are composed of eukaryotic cells connected by plasmodesmata through their cellulose cell walls. Plants are photosynthetic; many of their cells possess chloroplasts which contain chlorophylls *a* and *b*, xanthophylls and a number of yellow and red carotenoids.

Plants are thought to have evolved from chlorophyte ancestors some 400 million years ago (Fig 26.15) and, although many live in water for part of their life cycle, they are primarily terrestrial organisms. One group of plants, the **bryophytes**, is poorly adapted for life on land. All other plants, which belong to a second group called **tracheophytes**, are better adapted to life on land. Their adaptations may include:

- special transporting vessels which carry water and inorganic ions from the soil to their aerial parts (**xylem**) and photosynthate from their leaves to their roots (**phloem**) (see Chapter 12)
- supporting tissue, which holds them upright in the absence of buoyant water (see Section 18.5)
- a waxy cuticle, which reduces water loss by evaporation
- **siphonagamous fertilisation**, which eliminates the dependence on free water for the fertilisation of swimming gametes.

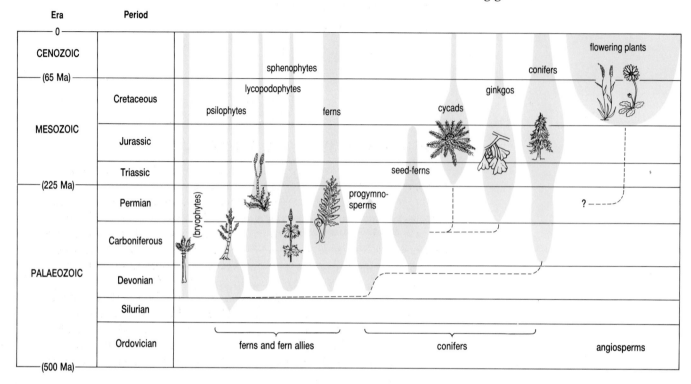

Fig 26.15 The fossil record of land plants. The widths of the coloured areas are proportional to the number of species. Note that today by far the greatest number of species are angiosperms. Ma means millions of years ago, i.e. before the present.

Table 26.2 summarises the main features of a number of plant phyla. Other plant phyla occur but contain few species: by far the majority of plants belong to the phylum Angiospermophyta.

Table 26.2   The main phyla of the kingdom Plantae (plants)

Phylum	Transport tissue	Gametophyte	Male gametes	Sporophyte	Spores
Bryophyta (liverworts and mosses)	absent	dominant; thallus with rhizoids	free-swimming male; non-motile female	capsule, seta and foot within mature gametophyte	homosporous
Filicinophyta (ferns)	present	small prothallus; free living	free-swimming male; non-motile female	dominant; stem, root, leaves	homosporous, some species heterosporous
Coniferophyta (conifers)	present	male in pollen grain, female in embryo sac of cone	non-motile until inside style	trees; cone-bearing; naked seeds	heterosporous; microspores and megaspores
Angiospermophyta (angiosperms)	present	male in pollen grain female in embryo sac of flower	non-motile until inside style	trees, shrubs, herbs; flowers and fruits; enclosed seeds	heterosporous; microspores and megaspores

PLANTS

1. multicellular organisms which develop from an embryo

2. eukaryotic cell features

3. photosynthetic nutrition

4. cellulose cell walls

5. alternation of generations

NATURE'S TREASURE HOUSES

By far the greatest diversity of plants occurs in tropical rain forests. However, the rain forests are being destroyed at a rapid rate. Perhaps the greatest tragedy associated with this destruction is the loss of genes from wild plants which plant breeders could have incorporated into domesticated plants in order to confer useful characteristics, for example pest resistance. In addition, many of the plants in tropical rain forests may be of considerable medical importance. Not only humans benefit from the pharmacological properties of many rain forest plants. Chimpanzees consume a plant called *Aspilia* even though it appears to have no nutritional value to the chimpanzee. Chemical analysis reveals that *Aspilia* leaves contain a potent antibiotic, thiarubine-A, which also has antifungal and worming properties. The destruction of the rain forests means that we may lose plants that could provide us with potent drugs before we even discover them.

Fig 26.16 All these products are produced from plants found in rain forests. Properly managed rain forests can be immensely and sustainably productive.

A generalised plant life cycle

The following terms are essential in understanding the life cycle of plants.

Spore	A haploid cell that can be dispersed; it is the product of meiosis in a diploid cell.
Gamete	A haploid cell, produced by mitosis of haploid cells. It can fuse with another gamete (fertilisation) to produce a diploid zygote.
Sporophyte	A multicellular plant that is diploid and produces spores.
Gametophyte	A multicellular plant that is haploid and produces gametes.

Note that these terms have been defined in terms of the plant kingdom. Thus, a fungal spore may have a quite different function and chromosome content to that of a plant.

(a) How do human gametes differ from those of plants?

Look again at Fig 26.14 which shows the life cycle of the green seaweed *Ulva*.

(b) Which of the two adults, X and Y, is the sporophyte and which the gametophyte?

Alternation of generations	Terrestrial plants also have a life cycle like that shown in Fig 26.14. A multicellular gametophyte (X) alternates in the cycle with a multicellular sporophyte (Y). This is called **alternation of generations.**

All animals have life cycles which are different from that shown in Fig 26.14. The life cycle typical of many animals (and of brown seaweeds, e.g. *Fucus*) is shown in Fig 26.17.

(c) What are the main differences between the life cycles shown in Figs 26.14 and Fig 26.17?

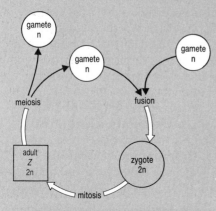

Fig 26.17 Life cycle typical of most animals and also of brown seaweeds, e.g. *Fucus*.

26.9 **(a)** The coal reserves of Britain were produced mainly during the Carboniferous period. What sort of fossil plants would you expect to find in coal from a British mine? (Note: refer to Fig 26.15.)

(b) Adaptive radiation of both flowering plants and insects occurred at the beginning of the Cenozoic era. Suggest how evolution of these two groups might be related.

26.10 Identify the major evolutionary trends which have enabled plants to colonise land.

26.3 BRYOPHYTA and FILICINOPHYTA

Phylum Bryophyta (bryophytes)

Within this phylum there are two commonly seen classes

- mosses
- liverworts.

Unlike all other plant phyla, the gametophyte plant is the main vegetative stage in the life cycle of bryophytes. The gametophyte of a liverwort is a flat thallus (Fig 26.18) whereas that of a moss is erect with small 'leaves' (Fig 26.19). Since the presence of xylem and phloem tissue is a characteristic of roots, stems and leaves, bryophytes technically lack these structures. However, bryophytes do have structures that serve similar functions; instead of roots, a bryophyte is anchored to the ground by unicellular or filamentous **rhizoids**. Without xylem and phloem, bryophytes rely on diffusion for the transport of substances through their body.

Fig 26.18   Like all liverworts, the gametophyte of *Marchantia* is a flat thallus. This photograph shows the female plant which has umbrella-shaped archegonia. Sperm from a male plant swim through a film of water into these archegonia and fertilise their egg cells.

Fig 26.19   Mosses have erect gametophytes. This photograph shows gametophytes of the moss *Polytrichum* with sporophytes growing from their tips.

 5   **Suggest why most bryophytes are no more than 2–3 cm tall.**

The life cycle of a moss

Look at Fig 26.20 which shows the life cycle of a typical moss. Gametophytes produce multicellular **gametangia**: egg-cell-producing **archegonia** and sperm-producing **antheridia**. The life cycle develops through the following stages.

- Attracted by the release of specific proteins from the neck of a mature archegonium, sperm use their two flagella to swim through a film of water into the neck of the archegonium where one fertilises the ovum.
- The resulting zygote is retained within the archegonium, where it develops into a diploid embryo.
- The embryo develops into a sporophyte which grows upwards from the gametophyte and may be dependent on it for food.
- At maturity, the sporophyte produces haploid spores by **meiosis** within a spore capsule. These spores are dispersed and, on suitable ground, germinate to form a new haploid gametophyte.

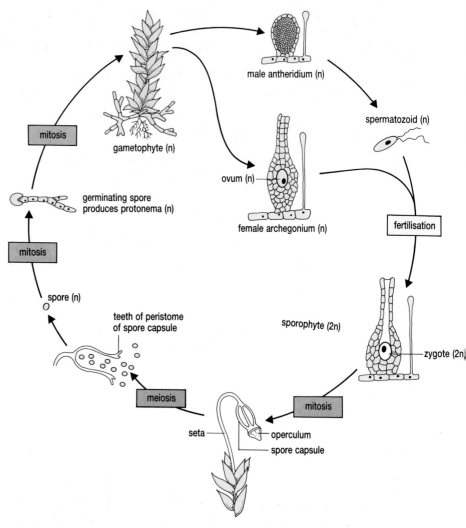

Fig 26.20   Life cycle of a moss.

KEY FEATURES

MOSSES AND LIVERWORTS

1. dominant gametophyte generation
2. no true roots but rhizoids for anchorage
3. no true leaves
4. no vascular tissue
5. sperm have two undulipodia

Fig 26.21   This giant fern in New Zealand is a living reminder of prehistoric flora.

 6   Why is the method of sexual reproduction in the moss life cycle poorly adapted for life on land?

Phylum Filicinophyta (ferns)

Ferns are **tracheophytes**. This means that their sporophytes (but not their gametophytes) possess xylem and phloem tissue: it is probably for this reason that the sporophyte generation is the dominant vegetative stage in the life cycle of ferns and all other vascular plants. Since they possess xylem and phloem, fern sporophytes have true roots, stems and leaves. Their leaves (**fronds**) are large and subdivided into a number of leaflets (**pinnae**) which may be subdivided into smaller **pinnules**: they usually have a thin waterproof cuticle. Leaves grow from thick underground stems (called **rhizomes**), from which **adventitious roots** develop.

The life cycle of a fern

The life cycle of a typical fern is shown in Fig 26.22. Notice that the sporophyte is a large independent plant. The plants seen in the woods or sold as house plants represent the sporophyte generation.

 7   How does the status of the sporophyte differ in ferns and bryophytes?

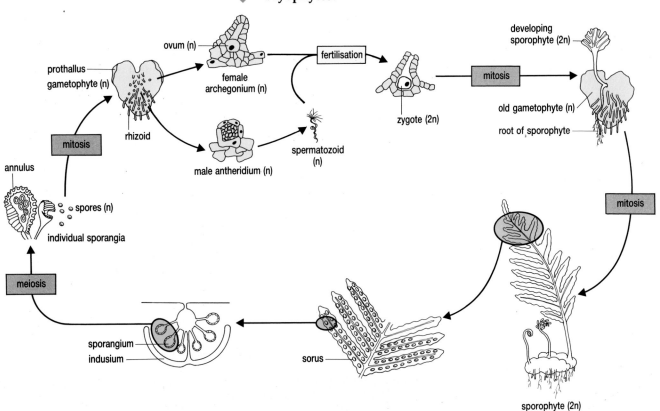

Fig 26.22   Life cycle of a fern.

Despite the different status of the sporophyte and gametophyte generations, the fern life cycle is basically the same as that of a bryophyte. Study Fig 26.22, noting the following information.

- At the appropriate time of the year the diploid fern sporophytes develop spore-forming bodies, **sporangia**, in clusters called **sori** (singular sorus). Sori may form on the ventral surface of ordinary fronds or on special reproductive fronds and are often covered by protective tissue called **indusia** (singular indusium).

Ferns in tropical forests may reach a height of 25 metres; such tree ferns are relatives of those which lived 300 million years ago in the Carboniferous forests that made our coal deposits.

- Meiosis within the sporangia produces haploid spores, which are dispersed by the wind when each sporangium breaks open along its annulus.
- In a suitable habitat, spores germinate to form a gametophyte.
- A fern gametophyte, called a **prothallus**, is a small, heart-shaped film of cells. It is anchored to the ground by rhizoids which form species-specific associations with fungi.
- Archegonia and antheridia develop on the ventral surface of the gametophyte and produce egg cells and sperm respectively. Using their undulipodia, the sperm swim through a film of water into the archegonia, where fertilisation occurs.
- The young embryo (developing sporophyte) absorbs nutrients from the parent gametophyte until its own roots and leaves are sufficiently developed to support it. After this, the gametophyte withers and dies.

So, the fern life cycle still involves motile, male gametes swimming through a film of water to female gametes, a situation reminiscent of bryophytes.

 Is the mode of sexual reproduction found in ferns any better adapted for life on land than that found in bryophytes?

Ferns are considered to be more advanced land plants than bryophytes because their sporophytes have

- a well-developed vascular system (xylem and phloem), providing both support and transport
- a waterproof cuticle and stomata which control water loss
- roots, stems and leaves.

 Suggest why most ferns are only found growing in damp places.

Ferns, e.g. bracken (*Pteridium aquilinium*) which cover large areas of open moorland (Fig 26.23), spread by vegetative reproduction of the sporophyte using an underground system of creeping rhizomes.

Fig 26.23 If this moorland is to be conserved it has to be managed. Left untouched, this area would quickly become carpeted with bracken which would destroy the habitat for existing moorland animals as well as for plants.

FUNGI AND PLANTAE

The evolution of heterospory

The sporophytes of most ferns produce only a single type of spore (Fig 26.24(a)), a condition known as **homospory**. These spores develop to form gametophytes which produce both antheridia and archegonia. A few ferns, such as the north Australian water fern (*Masilea*), show **heterospory** (Fig 26.24(b)). Heterosporous sporophytes produce two types of spore: the smaller **microspores** develop into male gametophytes and the larger **megaspores** develop into female gametophytes. Notice that the gametophytes begin to develop inside the protective coat of the sporophyte and that the female gametophyte tends to be retained by the sporophyte. Heterospory is common in seed plants which are considered in the next section.

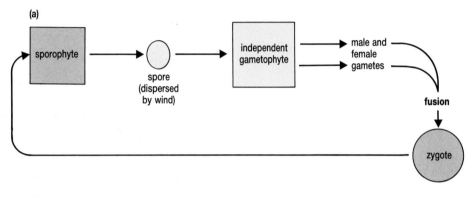

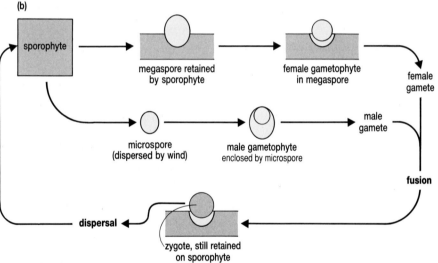

Fig 26.24 The evolution of heterospory within ferns.
(a) A 'primitive' fern which produces only one type of spore.
(b) A more 'advanced' fern which produces two types of spores.

 How does the life cycle of *Masilea* differ from that of ferns such as *Pteridium*?

KEY FEATURES

FERNS

1. dominant sporophyte generation
2. true leaves, stems and roots with vascular tissue
3. complex leaves with waterproof cuticle
4. gametophyte generation is photosynthetic and free-living
5. sperm have one undulipodium
6. some types show heterospory

26.11 Identify the similarities and differences in the life cycle of a bryophyte (see Fig 26.20) compared with that of the green seaweed *Ulva* (see Fig 26.14).

26.12 How are the bodies of erect mosses supported?

26.13 Some bryophytes are found in quite dry habitats, e.g. chalk downlands and stone walls, but their growth is restricted to the spring and autumn. How can you reconcile this observation with the statement that bryophytes are poorly adapted to terrestrial conditions?

26.14 Distinguish between
(a) gametophyte and sporophyte
(b) archegonium and antheridium.

26.15 Explain why the following instruction in a practical examination is technically incorrect. 'Remove one leaf from a moss plant and mount it in a drop of water on a microscope slide.'

26.16 Which plant phylum has members with a dominant gametophyte? Why does this limit the ability of plants in this phylum to grow in dry environments?

26.17 Redraw the stages in the life cycle of a fern (omit the drawings of the plants). Draw a second diagram to represent the stages in the life cycle of a heterosporous fern.

26.18 List the three major trends shown in the evolution of heterospory.

26.4 SEEDS AND SEED PLANTS

A seed consists of an embryo plant, a store of food (the endosperm) for the embryo plant's growth and a protective seed coat (testa) which keeps the seed in a state of dormancy until conditions are suitable for germination. Only a few plant phyla produce seeds; they are called 'seed plants'. The two major groups are

- conifers (phylum Coniferophyta) which produce seeds not enclosed in an ovary (naked seeds)
- angiosperms (phylum Angiospermophyta), the only phylum to produce seeds protected within the tissue of an ovary, which forms a fruit.

A plant is able to form seeds because the female gametophyte plant is always retained within the ovules of the sporophyte. It is thus protected from desiccation. Fertilisation of egg cells also occurs within the sporophyte. Male gametes are carried inside **pollen grains** from their site of production on one sporophyte to a sporophyte which contains the female gametophyte. At no stage in the transfer of pollen or subsequent fertilisation of the egg cell is free water needed. Hence, seed plants are better adapted for life on land than are non-seed plants.

Other seed-bearing groups of plants are the cycads, the ginkgos and the gnetophytes. Ginko trees do grow in Britain and have fan-shaped leaves.

 What advantages do seeds confer on a land plant?

Phylum Coniferophyta (conifers)

Most conifers are evergreen trees. Although many species grow in the tropics, conifers dominate the cold and dry conditions found at the tops of mountains and in the far northern hemisphere (see Figs 31.8 and 31.9). They are able to grow in these environments because

- they can photosynthesise in cold conditions
- they produce a resin which prevents freezing of their cytoplasm

- they have small needle-shaped leaves with a thick waxy cuticle that lose little water by evaporation.

The life cycle of a typical conifer, the pine, is shown in Fig 26.25. The tree itself is the diploid sporophyte. It produces sporangia on the surface of modified leaves, called **sporophylls**. These sporophylls develop in clusters, the pine **cones** with which you may be familiar (Fig 26.26). Conifers are **heterosporous**: microspores (pollen grains) develop in the two microsporangia (pollen sacs) on each microsporophyll of small male cones; megaspores (embryo sacs) develop in the two megasporangia (ovules) on each megasporophyll of large female cones.

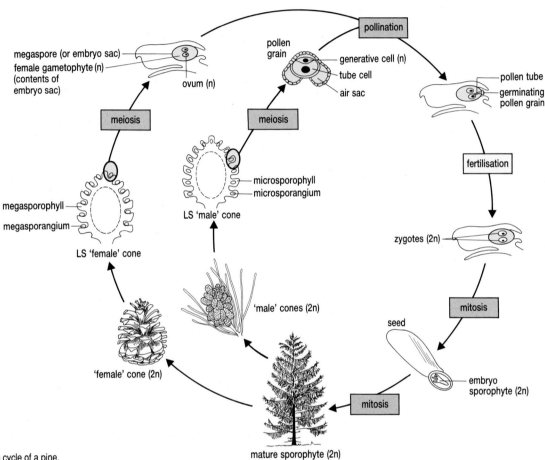

Fig 26.25 Life cycle of a pine.

Fig 26.26 Most conifers bear both male **(a)** *(left)* and female **(b)** *(right)* cones. The individual scales of each cone are the sporophylls which carry either two microsporangia or two megasporangia. The needle-like leaves also visible in this photograph have such a small surface area to volume ratio that little water evaporates from them.

FUNGI AND PLANTAE

Male cones release their pollen grains during the reproductive season and then disintegrate. Each pollen grain carries a male gametophyte and has two air sacs which enable it to be carried by the wind. Female cones do not release their megaspores. A female gametophyte develops within each embryo sac of the female cone. If a pollen grain lands on the megasporophyll of a suitable female cone, it grows a pollen tube, through which the male gametes pass to the egg cell. Following fertilisation within the embryo sac, the ovule of a fertilised egg cell becomes a seed which is dispersed from the female cone by the wind. The seed contains tissue from three generations: the sporophyte tree provides the seed coat, the female gametophyte contributes the stored food and the embryo is a new sporophyte generation. On suitable soil the seed germinates to form a new sporophyte.

 In what ways does the life cycle of conifer make these plants better adapted to terrestrial conditions than either mosses or ferns?

KEY FEATURES

CONIFERS

1. dominant sporophyte generation with small needle-like leaves

2. heterosporous – megasporophylls in 'female' and 'male' cones

3. pollen and siphonagamous fertilisation

4. seeds without enclosing fruit (naked seeds)

Phylum Angiospermophyta (angiosperms)

With over a quarter of a million species, angiosperms are the dominant plants on Earth today: nearly every tree, shrub and garden plant with which you are familiar is likely to be an angiosperm. Of the features which contribute to this success, the most conspicuous is their **flowers**. The process of reproduction in angiosperms is described in some detail in Section 20.4; the angiosperm life cycle is summarised in Fig 26.27. Each flower is formed by the dominant sporophyte plant and contains whorls of modified leaves: sepals, petals, stamens and carpels. Stamens are microsporophylls and form microspores (pollen grains). Carpels are megasporophylls and produce megaspores (embryo sacs). Since the male gametophyte is represented in this life cycle by a germinated pollen grain, both male and female gametophyte occur within the tissue of the carpel. After fertilisation a seed is formed which is enclosed within a fruit. The importance of fruits is described in Section 20.5.

 What structure is produced by conifers and angiosperms which makes the transfer of gametes completely independent of water?

Angiosperms are divided into two important classes, depending on the number of leaves in their seeds.

- **Monocotyledons** (class **Monocotyledoneae**) have only one seed leaf; their mature leaves are usually elongated with parallel veins, e.g. grasses, cereals, lilies and palms.
- **Dicotyledons** (class **Dicotyledoneae**) have two seed leaves; their mature leaves have a leaf stalk (petiole) and a leaf blade (lamina) with a net-like (reticulate) pattern of veins, e.g. most trees, shrubs and herbs.

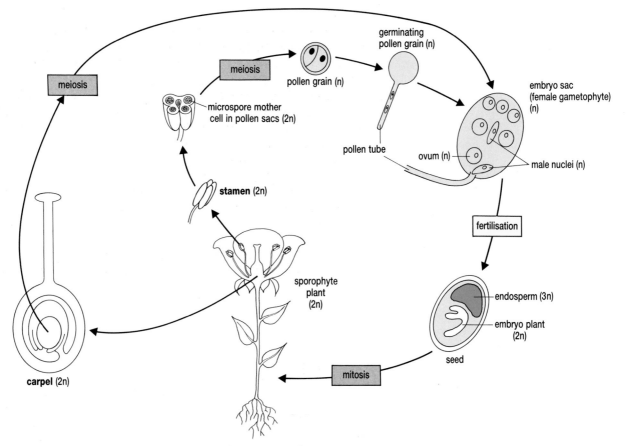

Fig 26.27 Life cycle of an angiosperm.

KEY FEATURES

ANGIOSPERMS

1. dominant sporophyte generation

2. heterosporous – megasporophylls and microsporophylls produced in flowers

3. pollen and siphonagamous fertilisation

4. seeds enclosed in fruit (swollen ovary wall)

26.19 Conifers and angiosperms produce seeds.
 (a) What is a seed?
 (b) What is a naked seed and in which plant phylum are naked seeds produced?
 (c) In what ways is the life cycle of a seed plant different from that of a non-seed plant?

26.20 Name one structure which is formed from a sporophyll in
 (a) a conifer
 (b) an angiosperm.

26.21 Woody pine cones are often used as indicators of weather conditions since their woody scales open outwards in dry weather but close again in wet weather.
 (a) Which part of the plant is a woody cone?
 (b) Suggest the advantage to the conifer plant of this behaviour of pine cones.

26.22 A flower is often described as the organ of sexual reproduction of an angiosperm. Is this true? Explain your answer.

26.23 Some people regard fungi as pale plants. Explain, as if to a non-biologist, why fungi are classified in a different kingdom from plants.

SUMMARY

Fungi are eukaryotic organisms which lack undulipodia at any stage of their life cycle and which form spores. Their bodies usually consist of a mass (mycelium) of thread-like hyphae, each surrounded by chitinous cell walls. Fungi are heterotrophic and must obtain their food by absorbing soluble nutrients through their cell walls and cell surface membranes. Some fungi do this by absorbing fluids from other living organisms but the majority are saprophytes, secreting enzymes onto their food and absorbing the soluble products of digestion. In this way, fungi are important decomposers in biological communities. Some fungi are the causative agents of diseases of plants and humans. Many fungi form mutualistic relationships either with the roots of tracheophytes (to form mycorrhizae) or with photosynthetic unicells (in lichens).

All fungi except the deuteromycetes are capable of sexual reproduction. Although they lack any recognisable sex, nuclei from the haploid hyphae of different mating types fuse to form zygotes. These undergo meiosis to form haploid spores which are dispersed. All fungi except the basidiomycetes form asexual spores, either within sporangia or as naked conidiospores. Sexual and asexual spores develop into haploid mycelia in suitable environments.

Plants are photosynthetic eukaryotes which develop from embryos. Their life cycle is complex and includes two adults: a haploid, sexually repro-ducing gametophyte and a diploid, asexually reproducing sporophyte.

Most plants live on land. Among their adaptations for life on land are: supporting tissues to hold the body erect; transport tissue; a waterproof covering to reduce water loss by evaporation; independence of water for the transfer of male gametes.

Bryophytes include liverworts and mosses. The dominant stage of their life cycle is the gametophyte plant which lacks the terrestrial adaptations previously listed. For this reason, bryophytes are mostly confined to places which are alternately wet, for the swimming of the male gametes, and dry, for the dispersal of their asexual spores.

Other plants have dominant sporophytes which possess supporting and transporting tissues, i.e. they are tracheophytes. Ferns are simple vascular plants which, like bryophytes, need a film of water for their swimming male gametes. Conifers and angiosperms do not have swimming male gametes: instead their male gametes are carried inside pollen grains and achieve fertilisation by migrating to the female gamete within the tissue of a recipient plant.

Following fertilisation within the body of a sporophyte plant, conifers and angiosperms form seeds. A seed contains an embryo plant, a food store from the gametophyte plant and a protective covering from the sporophyte plant. The seeds of angiosperms are enclosed within fruits. Monocotyledons (monocots) and dicotyledons (dicots) are types of angiosperm which are characterised by their number of seed leaves (cotyledons) and pattern of leaf veination.

Chapter 26: Answers to Quick Questions

1 It grows in a circular formation, exploiting all the available substratum.
2 Friction against particles of soil or desiccation.
3 Water is lacking or absent in terrestrial environments.
4 Facilitate opposite mating types actually meeting and fusing without the need for an aquatic environment. Dispersal of subsequent spores takes place above ground, aided by air currents.
5 Bryophytes lack vascular tissues and so rely on diffusion to meet their transport needs.
6 A moss depends on free water for the gametes to move from antheridia (male gametangia) to archegonia (female gametangia).
7 Sporophyte is the dominant adult in ferns; gametophyte is the dominant adult in bryophytes.
8 No; water is still needed for the movement of sperm.
9 Because fern gametophytes rely on free water for sexual reproduction.
10 *Masilea* produces two types of spore, leading to two types of gametophyte; *Pteridium* produces a single type of spore.
11 Seeds provide
   • a means of remaining dormant and so avoiding adverse environmental conditions, e.g. drought or cold
   • a means of dispersal
   • a food source for the developing embryo.
12 Conifers are independent of water for transfer of male gametes.
13 Pollen grain and pollen tube which develops from germinating pollen grain.

Chapter 27

ANIMALIA

LEARNING OBJECTIVES

When you have studied this chapter you should be able to:

1. describe the major features of the kingdom Animalia;

2. explain the meaning and biological significance of the terms diploblastic and triploblastic; bilateral and radial symmetry; acoelomate and coelomate; metameric segmentation;

3. compare and contrast structural and functional features of the major animal phyla.

Animals are multicellular eukaryotes which lack cell walls. Animals obtain their nutrition by breaking down complex molecules from the bodies of other organisms, i.e. they are heterotrophs (see Chapter 8). This breakdown is called **digestion** and partly occurs inside a **digestive cavity** (enteron) following **ingestion** of the food. Unlike plants, animals are diploid throughout their life cycle: only their gametes are haploid. Fertilisation is always anisogamous, i.e. it occurs between two different types of haploid gamete (see Chapter 19). The zygote that results from fertilisation always divides by mitosis to form a hollow ball of cells, called a **blastula**. Development from a blastula is the best criterion for classifying an organism as an animal.

27.1 BASIC PATTERNS

The terms outlined in this section are used repeatedly in the account of animal phyla which comes later in the chapter. It is advisable to understand these concepts before reading further.

Symmetry

Radially symmetrical

This term describes animals which have body parts that are arranged in a circle around an imaginary line through their mouth and digestive cavity (Fig 27.1(a)). This pattern of symmetry is found in

*   sponges (phylum **Porifera**)
*   jelly fish, corals and sea anemones (phylum **Cnidaria**)
*   starfish and sea urchins (phylum **Echinodermata**).

Most animals in these groups are non-moving (**sessile**) or are carried by water currents. **Radial symmetry** is advantageous to such animals since the stimuli to which they must respond (e.g. food, chemicals, light) are equally likely to come from any direction.

Bilaterally symmetrical

This term describes animals which have body parts that are arranged on either side of an imaginary line through their mouths and digestive cavity

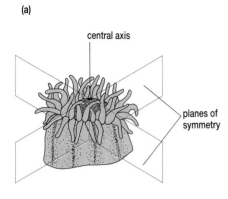

central axis

planes of symmetry

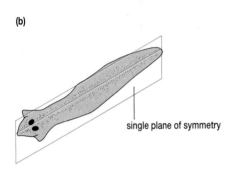

single plane of symmetry

Fig 27.1 Animals may be **(a)** radially symmetrical **(b)** bilaterally symmetrical.

(Fig 27.1(b)). **Bilateral symmetry** is an adaptation to active movement in which muscles on either side of the body act antagonistically. The front end of active animals is usually the first to enter a new environment. Active animals tend to have most of their chemoreceptors, light receptors and mechanoreceptors at their front end, forming a definite head. The tendency to have complex sensory structures at the front end is known as **cephalisation**.

1 Are humans radially or bilaterally symmetrical?

Body layers

Every animal develops from an embryonic structure called a **blastula**. During its development, a blastula forms groups of cells that give rise to definite structures in the adult animal.

The blastulae of poriferans and cnidarians form only two groups of cells that give rise to two layers of cells in the adult body: the outer **ectoderm** and inner **endoderm**. This body pattern is called **diploblastic** (Fig 27.2(a)).

The blastulae of animals other than poriferans and cnidarians form three groups of cells that give rise to three layers of cells in the adult body: **ectoderm, endoderm** and **mesoderm**. This body pattern is called **triploblastic** (Fig 27.2(b)). Ectoderm gives rise to the animal's epidermis on the outside and its nervous system; endoderm develops into the animal's gut and digestive glands; mesoderm gives rise to the muscular system and other body tissues (see Section 21.2).

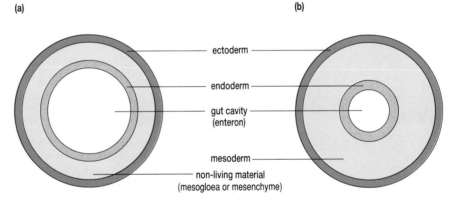

Fig 27.2 The body forms of animals **(a)** diploblastic with two layers of cells **(b)** triploblastic with three layers of cells.

2 Suggest why triploblastic animals tend to be more complex than diploblastic ones.

Body cavities

Platyhelminths have mesoderm which completely fills the space between their endoderm and ectoderm. The only cavity in their bodies is their gut (Fig 27.3(a)). These animals are described as **acoelomate**. In some triploblastic animals, a split develops within the embryonic mesoderm. This split separates one part of the mesoderm (**somatic**) which retains contact with the ectoderm from another part of the mesoderm (**splanchnic**) which retains contact with the endoderm. This split is called the **coelom** and forms a fluid-filled cavity (Fig 27.3(b)). The fluid, known as **coelomic fluid**, is secreted by a layer of mesodermal cells (**peritoneum**) that lines the coelom. Animals with a coelom are termed **coelomate** animals.

Since the body wall muscles that bring about locomotion are part of the somatic mesoderm and those which cause movement of the gut are part of the splanchnic mesoderm, a coelom enables locomotion and gut movements to be independent from each other. This is thought to have been highly advantageous in the evolution of animal movement. The coelomic fluid also acts as a hydrostatic skeleton in soft-bodied animals (see Section 18.3) and may circulate or store metabolites and waste materials. In many coelomate animals, internal organs expand into the coelom.

A nematode has a body cavity that is formed from a space between its embryonic endoderm and ectoderm. Since this space is not mesodermal, it is not a true coelom: it is called a **pseudocoelom**.

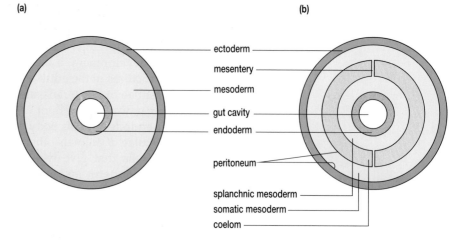

(a)   (b)

ectoderm
mesentery
mesoderm
gut cavity
endoderm
peritoneum
splanchnic mesoderm
somatic mesoderm
coelom

Fig 27.3   **(a)** The acoelomate body form has one body cavity – the gut. **(b)** The coelomate body form has two body cavities – the gut and the coelom.

ANALYSIS

The development of body cavities

This exercise involves interpreting diagrammatic information. Fig 27.4 shows the stages in the development of the cell layers and body cavities from the gastrula to the later stages of a nematode, an annelid, a crustacean and a fish. The gastrula typically has three cavities within it: **archenteron** (archetypal enteron or gut) surrounded by endoderm; **coelom** surrounded by mesoderm (this is not always present); **blastocoel** (the original cavity of the blastula).

Copy Table 27.1, study Fig 27.4 and then answer the questions about the four types of animals listed.

Table 27.1

| | Group of animals | | | |
	Nematode	Annelid	Crustacean	Fish
Embryonic body cavities				
archenteron				
gut				
coelom				
blastocoel				
pseudocoel				
haemocoel				
Primary body layers				
ectoderm				
endoderm				
mesoderm				

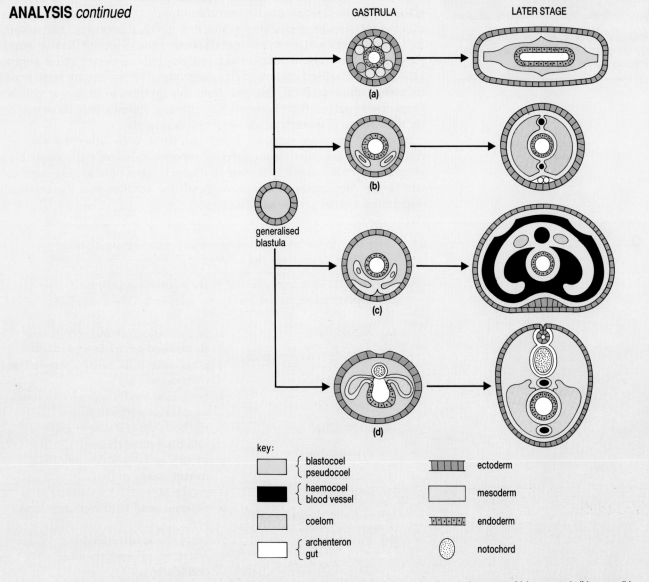

Fig 27.4 Development from the blastula to the gastrula and a later stage of **(a)** a nematode **(b)** an annelid **(c)** a crustacean **(d)** a fish.

(a) Which of the embryonic cavities (i) blastocoel (ii) archenteron (iii) coelom, forms the adult body cavity of each group of animals?

(b) Which of the three primary body layers (i) ectoderm (ii) endoderm (iii) mesoderm borders or encloses each adult body cavity?

(c) What does the blastocoel become in each of the four groups of animals listed?

(d) Fill in Table 27.1 by placing a tick (✔) if each structure is present and a cross (✗) if it is absent.

Segmentation of body

A tapeworm has a body made up of a large, irregular number of segments or **proglottids**. These proglottids are of different ages, those at the head being younger than those further way from the head. As older, gravid (full of fertilised egg cells) proglottids are shed from the end furthest from the head, new proglottids are formed just behind the head. Each proglottid is a reproductive unit, with its own reproductive system, and has an identical part of the body musculature and nervous system.

KEY CONCEPT

Metameric segments are fixed in number and are of the same age. Proglottids are irregular in number and of different ages.

In many groups of coelomate animals, the mesoderm (and often the ectoderm associated with it) becomes divided to form a small, fixed number of similar segments or **somites** along the body. This is called **metameric segmentation** or **metamerism**. Each somite has a similar portion of the musculature, blood vessels and nerves of other somites. However, specialised organs, such as reproductive organs, are present only in some somites. It is thought that this ability to separate different specialised parts of their body within different somites may be one reason for the success of metamerically segmented animals.

Internally, somites may be separated from each other by sheets of connective tissue called **septa**. Internal separation is rarely complete: blood vessels and many organ systems run through septa of adjoining segments. Unlike the proglottids of a tapeworm, all the somites in a metamerically segmented animal are the same age.

QUESTIONS

27.1 Only one characteristic distinguishes animals from all other organisms. Name this characteristic.

27.2 Match each term in the list (**a**) to (**h**) with the appropriate description from the list 1 to 8.

(**a**) acoelomate	1. body made of two layers of cells
(**b**) bilaterally symmetrical	2. fluid-filled cavity in mesoderm
(**c**) cephalisation	3. absence of fluid-filled cavity in the mesoderm
(**d**) coelomate	4. body divided by septa into a fixed number of segments
(**d**) coelomate	5. the body on one side of an imaginary line through the mouth and digestive cavity is a mirror image of that on the other side of that line
(**e**) diploblastic	
(**f**) metamerically segmented	6. definite head with well-developed sense organs
(**g**) radially symmetrical	7. body tissues arranged in ectoderm, mesoderm and endoderm
(**h**) triplobastic	8. body can be divided into mirror-image parts about three or more vertical planes

27.3 Look carefully at the three animals A, B and C shown in Fig 27.5. For each, state whether it is (i) radially symmetrical, bilaterally symmetrical or asymmetrical and (ii) segmented or non-segmented.

A
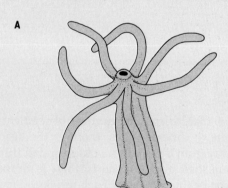

B

C
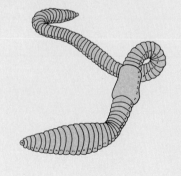

Fig 27.5 Three animals: see Question 27.3.

Parazoans are different from other animals because their cells are not arranged into tissues and organs. If squeezed through a fine sieve, the cells of a parazoan can be separated; if left in suitable conditions, the cells will reaggregate to form a functional animal again.

Phylum Porifera (sponges)

This phylum contains all the sponges. About 10 000 species of sponge are known to occur; all live in sea water except 150 or so freshwater species. Like plants, adult sponges are always sessile. Some sponges have a definite shape, producing tubes, cups and fans (Fig 27.6), whereas other sponges have no definite shape and grow as a crust on a rock surface. In spite of this difference in shape, sponges have a common body plan (Fig 27.7). The

Fig 27.6　Sponges grow into a wide variety of shapes, sizes and colours.

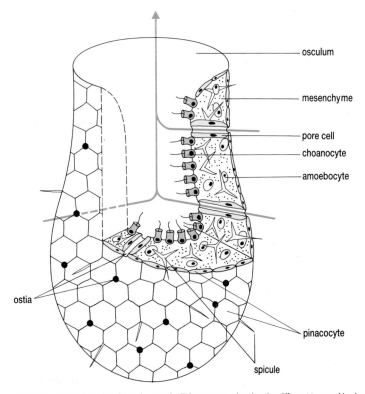

Fig 27.7　Vertical section through part of a living sponge showing the different types of body cell.

name of this phylum arises from the numerous pores (**ostia**) in the body of all sponges. Water is drawn into the sponge through these pores and forced out again through much larger pores, called **oscula** (singular osculum). Between the ostia and the oscula, water travels through canals in which oxygen and excretory products are exchanged by diffusion and micro-organisms are filtered out for food.

 What type of symmetry do sponges exhibit?

As you can see in Fig 27.7, sponges have only three types of cells in their bodies: pinacocytes (epithelial cells), choanocytes (collar cells) and amoebocytes (amoeboid cells). The **pinacocytes** form an epithelium that covers the outer surface of a sponge and lines the ostia. The pinacocytes around the ostia are contractile and can regulate the flow of water through them. The **choanocytes** form an inner layer of the sponge. Choanocytes have a 'collar' of microvilli surrounding a single undulipodium. It is the beating of these undulipodia that creates the current of water through the sponge: the 'collar' of microvilli acts as a sieve, trapping food. Separating these two layers of cells is a gelatinous layer, called **mesenchyme**, which contains amoebocytes and skeletal needles, called **spicules**. **Amoebocytes** move freely around the body of a sponge, storing food, secreting a skeleton of spicules and producing reproductive cells.

 Is the body of a sponge divided into endoderm, ectoderm and mesoderm?

Reproduction

Sponges are usually hermaphrodite, producing gametes of both sexes. Cross fertilisation occurs when clouds of egg cells and sperm are released into the water. Following fertilisation, a free-swimming larva is formed which develops into an adult sponge after it has attached to a suitable surface.

KEY FEATURES

PORIFERA (SPONGES)

1. body composed of relatively independent cells with little coordination between them, i.e. lacks tissues

2. body has ostia, oscula and canals through which water is circulated

3. body includes three types of cell: pinacocytes, choanocytes and amoebocytes

4. adult is sessile, larva is free-swimming

5. digestion is intracellular; gas exchange and excretion are by simple diffusion

6. sexual and asexual reproduction

The rest of the animals discussed in this chapter belong to the subkingdom **Eumetazoa**. All members of this subkingdom have specialised cells that are organised into tissues and organs.

27.4 For centuries, dead natural sponges have been used by humans for washing themselves. How does a sponge's structure make it suitable for this purpose?

27.5 Are sponges multicellular animals or aggregations of individual animal cells? Explain your answer.

27.3 PHYLUM CNIDARIA (jellyfish, corals and sea anemones)

Fig 27.8 Cnidarians include **(a)** sea anemones, **(b)** corals **(c)** jellyfish and **(d)** the Portuguese man-of-war, which is really a colony of cnidarians.

Basic features

A variety of cnidarians are shown in Fig 27.8. In spite of their apparent diversity, these cnidarians have just two body forms, the **polyp** and the **medusa** (Fig 27.9). Whilst some cnidarians have both body forms in their life cycles (Fig 27.10), others have only one. Notice in Fig 27.9 the external and internal appearance of a polyp and a medusa: both forms are radially symmetrical. Cnidarians are diploblastic: their outer layer of epithelial cells is separated from their inner layer of digestive cells by jelly-like **mesogloea**.

(a)

(b)

(c)

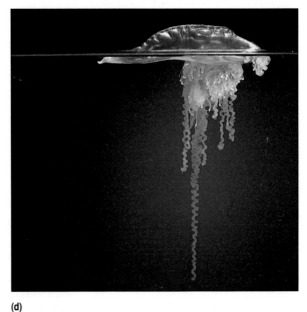

(d)

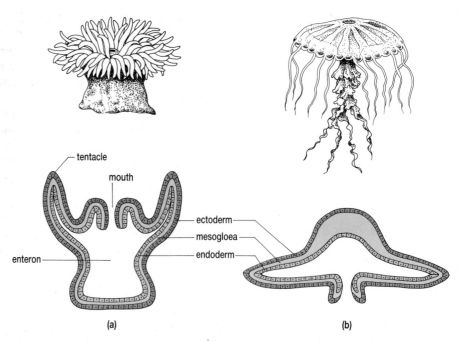

Fig 27.9 The body of a cnidarian is either a polyp **(a)** or a medusa **(b)**. The vertical sections show the underlying similarity of these body forms.

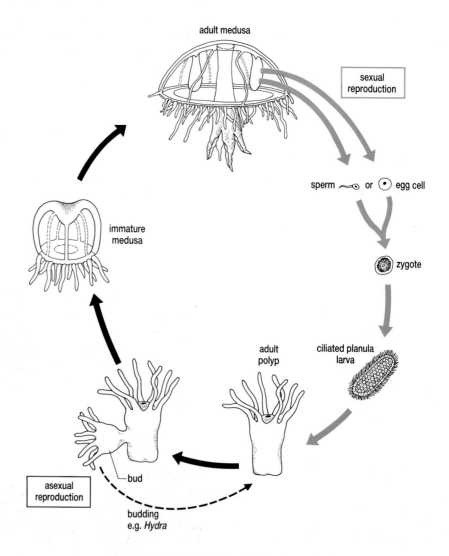

Fig 27.10 Life cycle of cnidarian with both polyps and medusae.

Although they lack a centralised nervous system, a cnidarian has unmyelinated nerve cells within the mesogloea, forming a **nerve net**. Since most of its cells are in contact with the water in which it lives, a cnidarian can successfully obtain oxygen and excrete waste products by diffusion over its entire body surface. Cnidarians do not have specialised gas exchange or excretory systems.

5 List three differences between polyps and medusae.

Feeding and digestion

The diploblastic body wall of a cnidarian surrounds its digestive chamber, called the **enteron**. Cnidarians are all carnivores. However, they do not chase their prey. Instead, when a suitable prey organism touches its tentacles, a cnidarian paralyses it using stinging bladders (**nematocysts**) which occur within specialised epidermal cells, called **nematoblasts** (Fig 27.11). When a suitable prey animal touches the **cnidocil** (trigger) of an undischarged nematocyst, its poison tube and barbs are turned inside out so that they shoot from the nematocyst, impaling the prey animal and injecting it with poisonous venom. Once discharged, nematocysts are replaced using differentiated interstitial cells from the ectoderm. A few cnidarians are able to harm humans in this way: the Portuguese man-of-war shown in Fig 27.8(d) is one of them. Having paralysed its prey, a cnidarian uses its tentacles to push the paralysed prey through its mouth and into its enteron. Endodermal cells secrete protease enzymes onto this food: the same cells ingest small, partly digested fragments of food and continue their digestion intracellularly. Eventually, undigested remains of the food are **egested** from the enteron through the mouth.

One of the most dangerous cniderians is the box jellyfish found off the coast of Queensland, Australia. Its sting is so painful that it can kill people by causing severe shock and heart attacks.

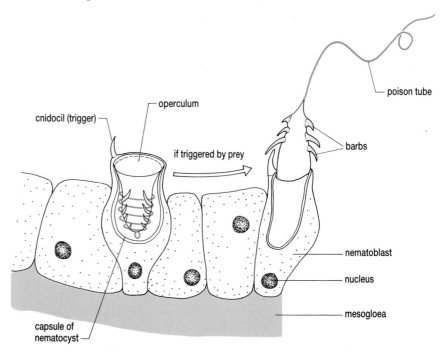

Locomotion

Movement of the tentacles, as with movement of the rest of the body, is caused by contraction of the cells in the two body layers of a cnidarian. Cells in each layer have long extensions, called **muscle tails**, the activity of which is coordinated by nerve cells in the mesogloea. The muscle tails of cells in the ectoderm are arranged longitudinally, running from the foot to the mouth; those of cells in the endoderm are arranged laterally, running around the body. These muscle tails act antagonistically. Contraction of muscle tails in the ectoderm is accompanied by relaxation of muscle tails in the endoderm and results in shortening of the body.

 6 **How could a cnidaria cause its body to elongate?**

Reproduction

A typical cnidarian life cycle is shown in Fig 27.10. Sexual reproduction occurs in the medusa. Gametes are released into the water and external fertilisation gives rise to motile larvae. Eventually, a motile larva settles on a suitable surface and develops into a sessile polyp. The polyp reproduces asexually by forming miniature medusae as outgrowths, called buds. The buds eventually break free from the polyp and develop into sexually mature adults. However, this life cycle is modified in many groups of cnidarians. For example, *Hydra viridis* has no medusal stage in its life cycle: the polyp develops gonads as outgrowths of its wall and reproduces sexually, as well as asexually by budding. Many cnidarians exist as colonies of individuals, e.g. Portuguese man-of-war (Fig 27.8(d)) in which individuals specialise for one particular function. Only some of the individuals (**gonozooids**) in such a colony reproduce, giving rise to an entire new colony.

KEY FEATURES

CNIDARIA (JELLYFISH, CORALS, SEA ANEMONES)

1. diploblastic

2. possess true tissues but lack organs

3. body is radially symmetrical

4. possess two body forms: sedentary polyp and free-floating medusa

5. capture prey using tentacles armed with nematocysts

6. digestive cavity has single opening

7. simple nerve net coordinates muscle tails

8. sexual and asexual reproduction

QUESTIONS

27.6 Suggest an advantage of the alternating life cycle of the cnidarian polyp and medusa.

27.7 In a medusa the muscle tail cells are all ectodermal and form a ring just inside the edge of the bell (see Fig 27.9).
   (a) What will happen to the medusa when this ring of muscles contracts?
   (b) How might this help to propel the medusa forwards?
   (c) How will this ring of muscles be stretched once it is fully contracted?

27.4 PLATYHEL-MINTHES, NEMATODA AND ANNELIDA

'Worms' is not a taxon name. A 'worm' means a soft-bodied elongated creature.

Animals that are long and thin are often called worms, e.g. tapeworm, earthworm, roundworm. However, animals with this kind of shape may belong to one of several different phyla. In this section three animal phyla are discussed.

Phylum Platyhelminthes (flatworms)

Basic features

Platyhelminthes (Greek, meaning flatworm) are so named because they are dorso-ventrally flattened. As a result of this flattening, flatworms have a large surface area to volume ratio and no cell in their bodies is very far from either the gut or the permeable ectoderm.

> **7** Since a flatworm lacks a circulatory system, why may dorso-ventral flattening be of great advantage?

Like cnidarians, a flatworm has a gut cavity with only a single opening. The larvae of certain flatworms are also very similar to those of cnidarians. These similarities might reflect a common evolutionary ancestor of cnidarians and flatworms. However, there are many important biological differences between these two phyla. Unlike cnidarians flatworms are triploblastic, bilaterally symmetrical and show a degree of cephalisation. They do not, however, have coeloms, i.e. they are acoelomate animals.

Feeding and digestion

Free-living flatworms, the turbellarians (class **Turbellaria**), are carnivores. Once attracted to chemicals released by the damaged body of a prey animal, they evert a pharynx (Fig 27.12(a)) and through this pump proteolytic enzymes onto the prey. A flatworm also uses its muscular pharynx to pump semi-fluid food into its branched intestine, where intracellular digestion finally occurs. Undigested remains of its food are egested through the pharynx. The branched intestine reaches most parts of the body of a flatworm and, in the absence of a circulatory system, branching is probably an important adaptation enabling the rapid distribution of digested food.

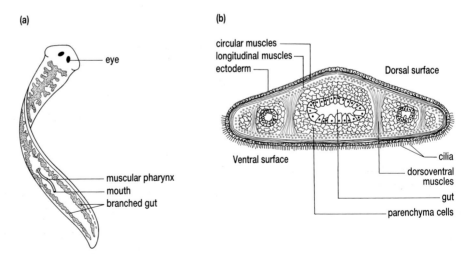

(a)

- eye
- muscular pharynx
- mouth
- branched gut

(b)

- circular muscles
- longitudinal muscles
- ectoderm
- Dorsal surface
- Ventral surface
- cilia
- dorsoventral muscles
- gut
- parenchyma cells

Fig 27.12 **(a)** The branched intestine of a flatworm (*Dendrocoelom lacteum*) is clearly visible through its cream-coloured body wall.
(b) A transverse section through a flatworm. The principal muscles are shown in blue.

Many flatworms are parasitic. Feeding in the flukes (class **Trematoda**) is similar to that in free-living turbellarians: these parasites use muscular pharynxes to pump body fluids from their hosts into their own digestive systems. Tapeworms (class **Cestoda**) are so highly adapted to a parasitic way of life in the intestines of their hosts that they lack digestive systems of their own. A tapeworm absorbs digested food over its entire body surface.

SPOTLIGHT

A TAPEWORM'S CAMOUFLAGE

Tapeworms are very common parasites in mammals. In humans they can reach several metres in length. Although they rarely kill their hosts, a heavy infestation can seriously weaken an animal, making it more prone to disease or predation.

Given that a tapeworm is, in effect, a large piece of foreign protein, it is surprising that the mammalian immune system (see Chapter 13) does not attack and destroy it. One way the tapeworm avoids this is to camouflage itself by regularly changing the proteins in its outer layers, so preventing the immune system from producing appropriate antibodies in time to destroy it.

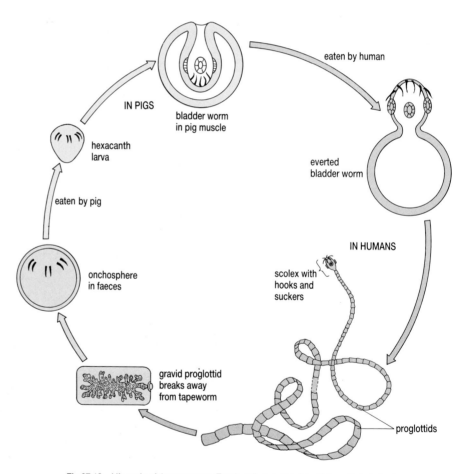

Fig 27.13 Life cycle of the tapeworm, *Taenia solium*, a parasite with two alternate hosts.

Locomotion

The epidermis of free-living turbellarians secretes thick mucus in which the worms beat the undulipodia, which cover their ventral surface. In this way, turbellarians seem to glide over a surface as they move without changing their shape: they can also swim using eel-like movements of their body wall muscles (Fig 27.12(b)).

Adult parasitic flatworms lack external undulipodia. Whilst they often use suckers and hooks to attach to their hosts, parasitic flatworms are not sessile animals. Tapeworms use similar muscular movements to turbellarians as they make regular movements up and down the small intestines of their hosts.

 Suggest how a flatworm could use its muscles to produce eel-like movements of its body.

Adaptations to parasitism

Most Platyhelminthes are internal parasites. Among their many adaptations to this way of life are organs to penetrate the host's body organs to maintain a hold on the host's body surface, a resistant cuticle over the epidermis and the secretion of chemicals to neutralise the host's defence reactions. Flatworm parasites also have complex life cycles in which large numbers of offspring are produced and in which larval parasites spend part of the life cycles in different hosts from the adult parasites. The life cycle of a tapeworm is summarised in Fig 27.13 and shows many of these parasitic adaptations.

KEY FEATURES

PLATYHELMINTHES (FLATWORMS)

1. triploblastic
2. acoelomate
3. dorso-ventrally flattened; bilaterally symmetrical
4. digestive system (where present) is branched and has a single opening
5. simple sense organs and nervous system
6. sexual and asexual reproduction
7. many are internal parasites (flukes and tapeworms)

Phylum Nematoda (nematodes or roundworms)

Basic features

Nematodes have long, thin, thread-like bodies (Greek *nema* meaning thread) with no segmentation (Fig 27.14(a)). Their bodies are covered by a tough, protective cuticle of protein. Though females of the parasitic nematode *Dioctophyme renale* may be one metre long, nematodes are usually microscopic. As they lack circulatory systems, gas exchange and excretion are by diffusion.

A forkful of fertile soil might contain half a million nematodes.

 How are the bodies of nematodes adapted to facilitate their method of gas exchange and secretion?

Feeding and digestion

A nematode has a simple tube-like gut with separate mouth and anus (see Fig 27.14(a)). Whilst a tube-like gut allows food to be ingested as soon as it

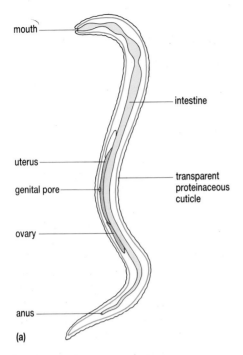

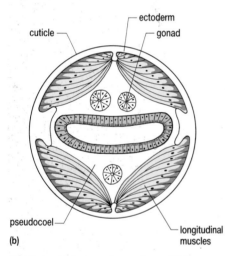

Fig 27.14 (a) Features of a roundworm. (b) A transverse section through the body. The principal muscles are shown in pink.

is detected and digested in a one-way process, the gut of a roundworm is not specialised to perform different functions along its length. Many parasitic roundworms, such as the hookworm, have specialised mouth-parts that enable them to attach to the intestinal walls of their hosts. Free-living roundworms are important decomposers in terrestrial and aquatic environments.

10 **How is the gut of a roundworm fundamentally different to that of a cnidarian and a flatworm?**

Many roundworms are parasites: different species of roundworm infect humans and domestic plants and livestock. Hookworm larvae enter humans by boring through the skin on their feet, enter the bloodstream and travel to the intestine where their feeding causes continuous bleeding. *Trichinella spirosis* is a roundworm which causes trichinosis in humans. Its cysts may be ingested in undercooked, infected pork. Once in a human intestine, its larvae hatch and burrow through the gut wall into the bloodstream and finally encyst in the muscle tissue, causing bleeding and muscle damage.

Locomotion

Nematodes are pseudocoelomate, with fluid-filled spaces in the mesodermal cells, the **pseudocoel**, lying between their gut walls and their muscle cells (Fig 27.14(b)). The cuticle, which is only slightly elastic, exerts a high pressure on the fluid in the pseudocoel which, therefore, acts as a hydrostatic skeleton. The body walls of a nematode have longitudinal muscle but no circular muscle. Contraction of the longitudinal muscle on each side of the body results in thrashing and coiling movements. These movements are an inefficient way of moving unless the environment exerts pressure on the worm: nematodes can move quite rapidly through sandy mud.

Reproduction

Nematodes reproduce only sexually. The sexes are always separate, the male usually being smaller than the female. Males and females copulate, resulting in the internal fertilisation of egg cells. The fertilised egg cells are deposited by females and may hatch immediately or may remain dormant until conditions are suitable for their growth. Females of some species deposit over 100 000 egg cells per day.

KEY FEATURES

NEMATODA (NEMATODES OR ROUNDWORMS)

1. triploblastic
2. pseudocoelomate
3. thread-like body with protective cuticle
4. simple tube-like gut with mouth and anus
5. no circulatory system; gas exchange and excretion by diffusion
6. reproduction always sexual; sexes separate

Phylum Annelida (annelids or segmented worms)

Basic features

A representative group of annelids is shown in Fig 27.15.

- Ragworms and fanworms belong to the class **Polychaeta**
- Earthworms form the class **Oligochaeta**
- Leeches form the class **Hirudinea**.

The bodies of annelids are ringed by external segments that usually correspond with the pattern of internal metameric segmentation. The internal segments are separated by septa and contain identical copies of nerve ganglia, excretory structures and muscles (Fig 27.16). Annelids also have external bristles of chitin (**chaetae**) that are used in locomotion.

 What is the pattern of repeated segments of identical age called?

Annelids are coelomate animals. Within the coelom in each segment, they have ciliated, funnel-shaped **nephridia** that remove excretory waste from the coelomic fluid and discharge it through external pores (see Fig 27.16). Annelids have circulatory systems of blood vessels with valves. Blood is moved along by the peristalsis-like contractions of five 'hearts' which are actually short, expanded segments of blood vessels.

Fig 27.15 (a) Fanworms, (b) ragworms (c) earthworms and (d) leaches are all annelids.

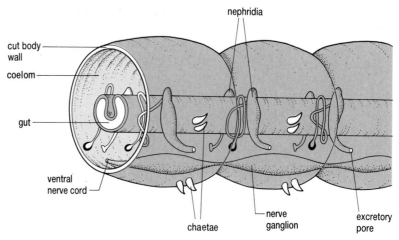

Fig 27.16 Stereogram representing repetition of structures in the body segments of an earthworm (an annelid); the body is metamerically segmented. (Note that muscle blocks and blood vessels are not shown.)

ANALYSIS

Comparing worms

This exercise involves the use of observational and interpretative skills. Look at Fig 27.17 which shows the ragworm *Nereis* (a polychaete).

Compare the structure of this animal with that of a roundworm (Fig 27.14) and answer the following questions.

(a) Are the positions of the mouth and anus similar?
(b) Do the two animals have the same pattern of symmetry?
(c) Do they have the same external appearance?
(d) How do the body cavities of the two animals differ? Do these cavities serve a similar function?
(e) How many sets of muscles does the ragworm have? Are they arranged in the same way as those of a roundworm?
(f) How does a transverse section of the body of the ragworm differ from that of a roundworm?
(g) Does the ragworm rely on diffusion to meet its gas exchange and excretory needs? Justify your answer.

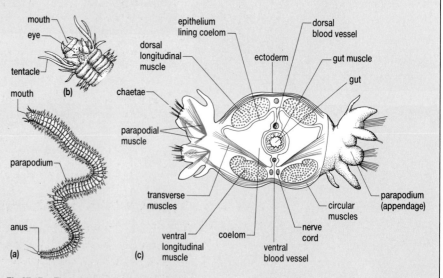

Fig 27.17 The ragworm, *Nereis*, an annelid worm.
(a) The whole worm seen from above (b) The mouth and associated parts (c) A transverse section through the body. The principal muscles are shown in pink.

Feeding and digestion

The gut of an annelid is specialised along its length to process food in different ways. In an earthworm, food is drawn into a muscular pharynx; passed along an oesophagus to be stored in the crop; slowly released into the muscular gizzard, where it is crushed; then digested and absorbed in the intestine. Other annelids have similar guts but differ in the way in which they take in their food. Fanworms filter food particles from the water in which they live using water currents produced by cilia on their tentacles; ragworms and leeches have well-developed teeth with which they trap their prey (ragworms, see Fig 27.17 (b)) or bite their hosts as they suck blood (leeches).

Locomotion

All annelids have well-developed circular and longitudinal muscles in their body walls. Antagonistic contractions of these muscles against the incompressible hydrostatic skeleton of the coelomic fluid enable an earthworm to change the shape of each segment individually. By protruding its chaetae from shortened segments, an earthworm can anchor part of its body whilst it moves other parts forward (see Figs 18.6 and 18.7). Ragworms are free-swimming polychaetes. They perform eel-like movements of their bodies as they swim, their movement being aided by paddle-like parapodia which protrude one from either side of each segment (see Fig 27.17). Leeches may swim using eel-like undulations of their bodies or may crawl along in a series of looping movements. During the looping movement, leeches attach to the substratum alternately using the suckers at the front and back of their bodies.

 Which sets of muscles may be involved in producing the sideways undulation of the body used by a ragworm for locomotion? Look for antagonistic muscle sets and a hydroskeleton.

Reproduction

Except for the leeches, annelids can regenerate lost parts and reproduce asexually by budding. Annelids reproduce sexually. The sexes are separate in polychaetes and external fertilisation occurs when millions of swarming polychaetes release their gametes into the sea. Although oligochaetes and leeches are usually hermaphrodite, copulation is always between different individuals and leads to internal fertilisation. Whereas the fertilised egg cells of polychaetes produce ciliated, free-swimming larvae (called **trochophores**), those of oligochaetes and leeches are incubated in **cocoons** and hatch as young adults. This is thought to be an adaptation to life on land.

 Suggest the main function of the trochophore larvae of polychaetes.

Indeed, hirudin is the most potent anti-thrombin known and it offers considerable promise as a pharmaceutical drug for treating certain blood disorders. Hirudin acts by specifically inhibiting the enzyme thrombin which converts soluble fibrinogen into insoluble fibrin, the protein which makes the fibrous meshwork of the clot (Chapter 10).

The usefulness of hirudin is hindered by the lack of leeches; it is certainly undesirable to see the slaughter of the remaining leeches to extract the compound. This problem may be overcome by inserting the gene encoding hirudin into a bacterium using genetic engineering techniques. Unfortunately, the genetically engineered hirudin produced by the bacteria is not identical to natural hirudin, the former having an activity six times lower compared to natural hirudin.

An alternative approach is being undertaken by a Welsh biotechnology company, which has set up a captive breeding programme for *H. medicinalis*. Hopefully, these twin approaches of genetic engineering and leech farming will reduce the demands placed on natural populations of this animal.

KEY **F**EATURES

ANNELIDA (ANNELIDS OR SEGMENTED WORMS)

1. triploblastic
2. coelomate
3. metamerically segmented
4. possess chitinous chaetae which are used in locomotion
5. possess simple blood circulatory system
6. tubular gut with regions specialised for different functions
7. gas exchange by diffusion through skin or through gills; simple excretory system
8. sexual reproduction

QUESTIONS

27.8 Flatworms are triploblastic; cnidarians are diploblastic.
(a) Explain what this means.
(b) What advantages do flatworms have over cnidarians as a result of being triploblastic?

27.9 Coelomate worms have a blood circulatory system but acoelomate worms do not. Suggest why.

27.10 Explain how an annelid benefits from (a) its coelom (b) having a body that is metamerically segmented.

27.11 Formerly, flatworms, nematodes and annelids were grouped in a single phylum called Vermes (worms). Make a table to show the differences between each, now these are classified as three different phyla. Include at least six features in your table.

27.5 PHYLUM ARTHROPODA (arthropods)

Of the million or so species of arthropod that are known, about half are insects and it is estimated there may be up to nine million more species of insects yet to be identified!

Both in number of individuals and in number of species, arthropods are the most common animals on Earth. Because of the vast number of species, arthropod classification is complex. Table 27.2 shows names of the major taxa which you are expected to know and their relationship with other taxa which you will not be expected to know. The basic differences between the three large groups (superclasses) concerns the arrangement and mechanics of their appendages.

- crustaceans (superclass **Crustacea**)
- insects, centipedes and millipedes (superclass **Uniramia**)
- spiders (superclass **Chelicerata**).

Table 27.2 Classification of the arthropods is complex because there are so many of them. The emboldened terms are ones with which you should be familiar

Phylum Arthropoda (crustaceans, insects and arachnids)

superclass **Crustacea** (crustaceans)			Mainly aquatic ; head and thorax fused to form cephalothorax; part of body enclosed in carapace; two pairs of antennae; one pair of compound eyes; variable number of thoracic legs; gas exchange through gills.
superclass Uniramia	class	**Insecta** (insects)	Mainly terrestrial; head, thorax and abdomen obvious; one pair of antennae; one pair of compound eyes; thorax has three pairs of legs and two pairs of thoracic wings; gas exchange through tracheae.
	class	**Chilopoda** (centipedes)	Terrestrial; distinct head with one pair of poison jaws; simple eyes only; one pair of antennae; one pair of legs on each segment of body; numerous body segments.
	class	**Diplopoda** (millipedes)	Terrestrial; simple eyes only; one pair of antennae; two pairs of legs on each segment of body; numerous body segments.
superclass Chelicerata	class	**Arachnida** (spiders)	Terrestrial; body composed of prosoma and opisthosoma; simple eyes only; lack sensory antennae; gas exchange through lung books; four pairs of walking legs.

 14 Compare the beetle, spider and crayfish, shown in Fig 27.18. How many walking legs are present in each animal? Are the legs similar in general structure?

Basic features

Arthropoda are so called because they have jointed limbs (Greek *arthron* meaning joint, *pons* meaning foot). Although each joint moves only in one plane (see Fig 18.8), the whole limb is very flexible because successive joints along each limb move in different planes. Jointed limbs are made necessary by another diagnostic arthropod feature: the external skeleton. An arthropod's exoskeleton is made chiefly of protein and the nitrogen-rich polysaccharide **chitin**. It is secreted by cells in the epidermis and hardens on contact with air or water. The hardening of the outer exoskeleton is helped by the further secretion of tanning agents (including calcium carbonate) onto its surface. The penetration of these tanning agents through part of the exoskeleton gives rise to the different parts of the exoskeleton shown in Fig 27.19. As the animal grows, its exoskeleton is shed and a new one is produced, resulting in the pattern of discontinuous growth shown in Fig 21.2(b).

 15 One of the problems of living on land is desiccation, i.e. drying out. Suggest how this problem might have been solved by insects and spiders.

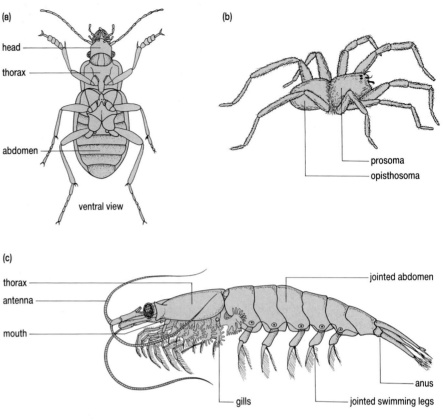

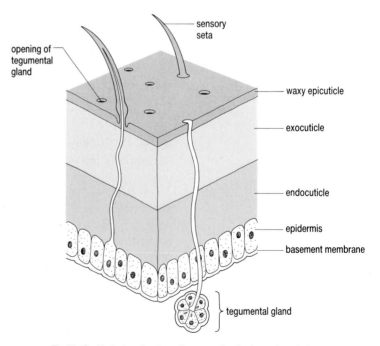

Fig 27.18 **(a)** A beetle – an insect **(b)** A spider – an arachnid **(c)** A crayfish – a crustacean.

Fig 27.19 Vertical section through a generalised arthropod exoskeleton.

Like the annelids, the body of arthropods is metamerically segmented. However, the number of segments is small and segments are often fused and specialised for distinct functions such as feeding, reproduction and locomotion. The body of most arthropods is made up of two or three regions: the **head** (cephalum), chest (**thorax**) and **abdomen**. The head usually has jointed mouthparts and well-developed sense organs, including eyes and tactile antennae. The thorax carries the pairs of jointed legs that are used for walking or swimming. Adult insects usually have two pairs of wings on their thoraxes. The abdomen carries the reproductive organs.

Arthropods are coelomate and have blood circulatory systems. However, during development the blood system becomes the dominant body cavity, called the **haemocoel** (see Fig 10.3(b)). Gas exchange cannot occur through the exoskeleton; permeable gas exchange surfaces are restricted to gills on the legs of aquatic crustaceans, tracheae running throughout the bodies of insects, centipedes and millipedes, and lung books in spiders. Most aquatic crustacea also excrete ammonia through their gills; insects, centipedes and millipedes use Malpighian tubules to excrete uric acid (see Fig 11.25).

 How does the arthropod circulatory system differ from that of an annelid?

Feeding and digestion

Some crustacea use bristle-like setae on their legs and mouthparts to filter small particles of food from the water in which they live, e.g. barnacles. Most arthropods use modified legs and jointed mouthparts to process their food so that it can be swallowed. Insects show a wide range of mouthparts adapted for biting, chewing, piercing, sucking and even for releasing enzymes onto their food. The gut of an arthropod is specialised along its length to process food in different ways, in a similar way to that of an annelid.

Locomotion

Arthropods use their jointed legs to swim, burrow, jump and walk. The movement of their limbs is described in Section 18.3. Alone among non-chordate animals, insects can fly. In a few insects, such as locusts, muscles attach directly to the base of the wings: their contraction causes a relatively slow wing beat of about 20 s^{-1}. The majority of insects move their wings using muscles that change the shape of each thoracic segment. By making the thoracic segment become an oscillating system, an insect's wings move very quickly: the wing muscles are used only to change the angle of the wings.

 How might the haemocoel assist in the locomotion of insects and spiders?

Reproduction

Reproduction is sexual and the majority of arthropods have separate sexes. Both external and internal fertilisation occurs among the groups of crustaceans: most crustaceans lay eggs that pass through a series of larval stages before becoming adults. Fertilisation is internal in other groups of arthropods. Copulatation does not always occur. Male arachnids and centipedes, for example, deposit sperm in packages, called **spermatophores**. These are picked up by the female and inserted into her own genital opening.

The life cycle of most insects shows metamorphosis, during which various larval forms (**instars** or **nymphs**) are formed. Metamorphosis allows young and adult to exploit different environments and foods, thereby avoiding direct competition. Groups which show **hemimetabolous metamorphosis** (Fig 27.20(a)) lay eggs that hatch into nymphs that are miniature, sexually immature versions of the adult (**imago**). Because their wings develop externally, these insects are termed **exopterygota** (note that this is not a taxonomic group since insects from many unrelated taxa are exopterygotous). They include locusts, cockroaches and mayflies. Insects which show **holometabolous metamorphosis** (Fig 27.20(b)) lay eggs that hatch into larvae quite unlike the adult form. These larvae develop through a series of moults before finally forming **pupae**. Within a pupa, the larval body is broken down and reorganised into the adult body. Because their wings develop internally, this group of insects is classed as **Endopterygota** (because all the insects in this group share other important taxonomic features, this is a taxonomic group). Endopterygota includes flies, moths and butterflies. The hormonal control of metamorphosis is outlined in Chapter 14.

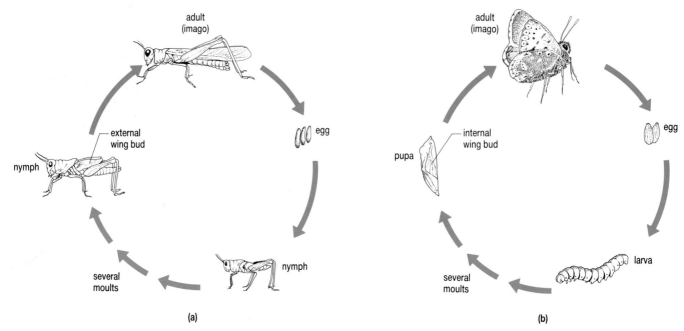

Fig 27.20 Insect life cycles **(a)** hemimetabolous metamorphosis – nymphs increase in size until they become sexually mature adults. **(b)** holometabolous metamorphosis – the insect has different life forms of egg, larva, pupa and imago.

Fig 27.21 Mites feeding on skin scales. A double bed can contain two million of these mites; unfortunately, mites may cause asthma and dermatitis – allergic reactions.

KEY **F**EATURES

ARTHROPODA (ARTHROPODS)

1. triploblastic

2. coelomate, but major body cavity is haemocoel

3. metamerically segmented, though segments not aways clear externally

4. possess chitinous exoskeleton, leading to discontinuous growth pattern

5. possess jointed limbs that are used for feeding as well as for locomotion

6. gas exchange through gills, tracheae or lung books

7. nitrogenous excretion through gills or Malpighian tubules

8. sexual reproduction, metamorphosis common

QUESTIONS

27.12 It is thought that arthropods evolved from annelids (see Fig 27.27).
   (a) Identify five features of annelids and arthropods that suggest they may have a common evolutionary origin.
   (b) Name three ways in which arthropods differ from annelids.

27.13 The most successful terrestrial animals are insects. Outline the problems of living on land and explain how their arthropod features enable insects to be successful in a terrestrial environment.

27.14 (a) Obtain specimens or photographs of as many of the following insects as you can
butterfly or moth (order **Lepidoptera**)
fly (order **Diptera**)
bee (order **Hymenoptera**)
locust (order **Orthoptera**)
dragonfly (order **Odonata**)
cockroach (order **Dictyoptera**)

(b) Construct a simple dichotomous key to distinguish between these orders of insect.

27.6 PHYLUM MOLLUSCA (molluscs)

The phylum of molluscs contains the largest animal without a backbone known to exist: *Architeuthis* (the giant squid) may grow over 20 metres long.

The foot of a mollusc is a muscular region of its body. It is used for different functions in the different classes of mollusc. Only a gastropod 'walks' on its foot.

With over 100 000 living species, this phylum is second only to the arthropods in terms of its numbers and variety.

Basic features

The variety of animals in this phylum is shown in Fig 27.22.

- slow moving slugs and snails (class **Gastropoda**)
- sedentary bivalves (class **Pelycopoda**)
- fast-swimming squid (class **Cephalopoda**).

In spite of this diversity, the bodies of all molluscs share a common pattern. The molluscan body is unsegmented but is divided into a **head**, a muscular **foot** and a **visceral mass** that contains the intestines and branches of the coelom which individually surround the heart, excretory and reproductive organs. The visceral mass is covered by a fold of the body wall, called the **mantle**, which secretes a calcium carbonate shell. The shell of many molluscs surrounds and protects the animal but some gastropods and most cephalopods have internal shells.

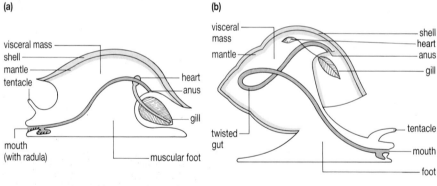

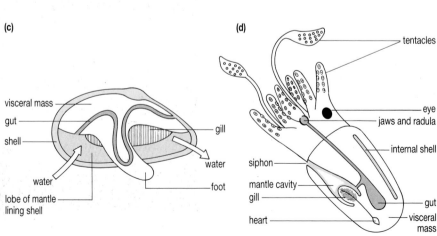

Fig 27.22 A variety of molluscs
(a) *Littorina* – a gastropod
(b) A transverse section through *Littorina* to show the longitudinal muscles of the foot
(c) A bivalve mollusc, with one shell valve removed
(d) A longitudinal section of a squid to show the arrangement of the body organs.

18 Which conspicuous part of the bodies of both gastropods and cephalopods is less obvious in a bivalve?

Typically, gills or 'lungs' lie in a cavity formed between the foot and the mantle. Since most molluscs are aquatic, gas exchange occurs freely over the surface of their soft bodies as well as through their respiratory surfaces. As with fish, the flow of blood through the gills of a pelycopod is in the opposite direction to the flow of water over the gills (**counter-current**).

 19 Are the gills of fish and molluscs analogous or homologous structures?

Like annelids and arthropods, a mollusc has a central nervous system with ganglia around its oesophagus. These ganglia are so large in cephalopods that they form a genuine brain which contains special centres for the coordination of vital activities and simple reflexes. The brain of a cephalopod is enclosed within a protective, skull-like case of cartilage.

 20 Which is the dominant body cavity of molluscs?

Feeding and nutrition

Molluscs have jaws and unique ribbon-like 'tongues' called **radulae**. The radula is covered by chitinous teeth (Fig 27.23) and is used by gastropods to rasp tiny fragments of food which are then ingested. The radula is continuously replaced as it wears away but is so strong that some gastropods, e.g. dogwhelk, can use their radulae to bore through the shells of crustacea and the other molluscs on which they prey.

Pelycopods are filter-feeders. Cilia on their gills draw currents of water through their shells (see Fig 27.22(c)). As water passes over their gills, gas exchange occurs and small particles of food are trapped. These particles are wrapped in mucus and passed to the mouth using the small muscular foot. Cephalopods are active predators. The foot of a cephalopod forms tentacles with sucker-like discs. Prey grasped in the tentacles is immobilised by poisonous saliva and then torn apart by beak-like jaws before being ingested. Molluscs move food through their guts using beating cilia as well as muscle contractions. The molluscan gut is adapted along its length to process food in different ways, including separating food from debris, storing food, digesting and absorbing food.

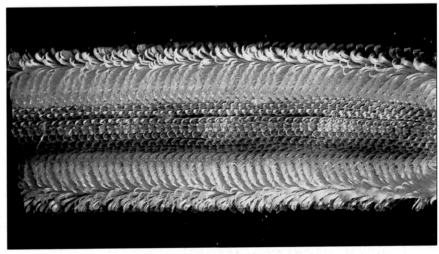

Fig 27.23 Light micrograph of a radula of a mollusc.

What is peculiar about the gut of gastropods compared to that of bivalves and cephalopods (see Fig 27.22)?

Locomotion

Most adult pelycopods are sessile, though some can use a small foot to burrow into sand, mud, wood or rock.

A gastropod uses complex, wave-like contractions of its foot to crawl along. The foot secretes copious quantities of mucus which is essential for this type of movement. Whilst octopuses can use their tentacles to crawl, cephalopods usually 'jet-propel' their bodies by forcing water from their mantle cavities.

 Look at the section of a gastropod's foot shown in Fig 27.22(b). Suggest the probable mechanism by which the longitudinal muscles of the foot are stretched.

Reproduction

Reproduction is sexual. In most molluscs, the sexes are separate; some gastropods and pelycopods are hermaphrodite. Whilst pelycopods discharge their gametes into the water in which they live, gastropods and cephalopods usually show courtship rituals following which packets of sperm (**spermatophores**) are transferred from one individual to another. The eggs of most aquatic molluscs develop via free-swimming larvae. The eggs of land snails and cephalopods develop directly into adults without a free-living larval stage.

Scallops are unusual pelycopods because they lack a foot and when startled or attacked they can 'swim' by flapping their shells. The large muscles needed to achieve this peculiar form of locomotion make scallops highly prized as food, both for humans and other predatory animals, e.g. starfish.

KEY FEATURES

MOLLUSCA (MOLLUSCS)

1. triploblastic

2. coelomate, but coelom reduced and divided into compartments around heart, renal and reproductive systems

3. no metameric segmentation

4. body divided into head, foot and visceral mass

5. mantle may secrete a shell and forms a cavity for organs of gas exchange

6. radula, a unique scraping tongue-like organ, leads into a ciliated gut

7. sexual reproduction

27.7 DEUTEROSTOMES: ECHINODERMS AND CHORDATES

All the animal phyla described so far share two features in the way their embryos develop.

1. The third division (cleavage) of the embryo is a spiral division.
2. The pore of their blastula eventually forms a mouth.

The next two phyla are different from all previous phyla. In these phyla the third cleavage is a radial division and the pore of the blastula eventually forms an anus. Echinoderms and chordates are termed deuterostomes (see the Analysis: Evolutionary pathways).

Phylum Echinodermata (echinoderms)

Basic features

Representatives of three major classes of echinoderms are shown in Fig 27.24.

- brittle stars and starfish (class **Stelleroidea**)
- sea urchins (class **Echinoidea**)
- sea cucumbers (class **Holothuroidea**)

Fig 27.24 **(a)** Sea stars, **(b)** sea urchins **(c)** sea cucumbers **(d)** starfish are all echinoderms.

Although their free-swimming larvae are bilaterally symmetrical, adult echinoderms show a five-way (pentamerous) radial symmetry. As described in Section 27.1, radial symmetry is an adaptation to a sessile habit and most adult echinoderms crawl only very slowly. An echinoderm has an unsegmented body covered by a thin epidermis beneath which is a firm endoskeleton of plates of calcium carbonate (Fig 27.25(b)). This endoskeleton may have projecting spikes, as with sea urchins (see Fig 27.24).

23 Suggest an advantage in adult echinoderms showing no cephalisation.

The body of an echinoderm has a unique water-vascular system which functions in respiration, excretion, locomotion and food handling. Sea water enters through an opening (**madreporite**) on the animal's surface and passes through canals along each arm of its body to tiny, tube-like feet (Fig 27.25(a)). Gas exchange and excretion occurs by diffusion over the surface of these feet.

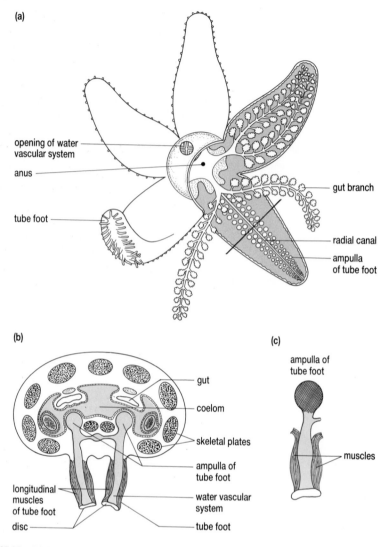

(a)

opening of water vascular system

anus

tube foot

gut branch

radial canal

ampulla of tube foot

(b)

gut

coelom

skeletal plates

ampulla of tube foot

longitudinal muscles of tube foot

water vascular system

disc

tube foot

(c)

ampulla of tube foot

muscles

Fig 27.25 (a) A starfish seen from above. Two arms have been dissected to reveal parts of the gut, ampullae of tube feet and radial canal.
(b) A transverse section (taken at the point indicated by the rule) through the arm of a starfish to show gut cavities and coelom.
(c) A tube foot and its associated muscles.

Feeding and nutrition

Echinoderms may be filter-feeders, grazers, predators, or sand or mud ingestors. Starfish feed on pelycopods, using a relentless pull of their tube feet to force the shells apart before everting their stomachs into the gaping pelycopods and releasing digestive enzymes. Plates around the mouth of a grazing sea urchin form an intricate jaw system, called 'Aristotle's lantern', which is used to graze algae off rocks. Echinoderms have relatively simple intestines with little differentiation along their length and few glands.

Locomotion

Many echinoderms are sessile. Motile echinoderms use their tube-feet (Fig 27.25(c)), which may have sucker-like endings, to slowly crawl along.

 How could a starfish (a) extend and retract a tube foot (b) move a tube foot from side to side?

Reproduction

Most echinoderms can regenerate lost parts: asexual reproduction occurs by fission in some starfish and sea cucumbers. Sexual reproduction also occurs. The sexes are usually separate: males and females release their gametes into the sea water and external fertilisation occurs. A ciliated, bilaterally symmetrical larva forms from each fertilised egg cell and eventually develops into an adult.

SPOTLIGHT

SEA OTTERS AND SEA URCHINS

The western sea board of North America is characterised by the development of large underwater forests of giant kelp. These seaweed forests provide nursery grounds where young fish develop into adults in relative safety. In addition, the kelp forests provide a home for a unique mammal, the sea otter (Fig 27.26). The thick fur of this animal made it a target for fur traders and its habit of feeding on commercially exploited shellfish, e.g. abalone, also made it a natural enemy of fishermen. The result was that large numbers of sea otters were killed.

However, the disappearance of the sea otter also coincided with the disappearance of large areas of kelp forest and an increase in the numbers of sea urchins, a favourite otter food. The sea urchins were now grazing so many of the young kelp plants that there were insufficient to replace the adults – hence the disappearance of the kelp forest. Fortunately, sea otters are now protected and the kelp forest is reappearing in areas where sea otters are becoming re-established.

Fig 27.26   A sea otter lying on its back in a kelp bed eating a sea urchin.

ECHINODERMATA

1. triploblastic

2. coelomate, but coelom reduced to individual compartments around various body systems

3. adult shows pentamerous radial symmetry

4. lack metameric segmentation or head

5. have water vascular systems with tube feet that are used in gas exchange, excretion, locomotion and food handling

6. have sub-epidermal calcareous plates and spines

7. simple gut

8. asexual reproduction by fission and sexual reproduction

Phylum Chordata (chordates)

Basic features

Since this phylum contains fish, amphibians, reptiles, birds and mammals (including humans) it is likely to be the animal phylum with which you are most familiar. Chordates are bilaterally symmetrical, triploblastic coelomates. They have metamerically segmented bodies, best shown by the repeated subunits of the human backbone (vertebrae) and the regular muscle blocks (**myotomes**) on either side of the body of a fish.

 25 What would be the effect if the myotomes on one side of a fish contracted whilst those on the other side relaxed?

Unlike all other animals, chordates have hollow nerve cords which are dorsal to their digestive tracts and hearts which are ventral to their digestive tracts. However, most of the other diagnostic features are absent from the human body, but they were present during embryological development. At some stage of their life, all chordates have a **notochord**, a stiff but flexible rod of vacuolated cells that passes the length of the body and provides attachment sites for muscles. A chordate has gill clefts in its pharynx: these may form functional gill openings or may appear only as grooves during embryonic development. Finally, all chordates develop tails, supported initially by their notochords, which extend beyond the anus. It is during the embryological development of a human that the notochord, gill grooves and (lastly) the post-anal tail are developed then lost.

ANALYSIS

Evolutionary pathways

This is a comprehension exercise.

Taxonomists attempt to determine evolutionary relationships between organisms and to reflect this in their classifications. Although adult echinoderms show little resemblance to adult chordates such as humans, taxonomists believe they share with humans a common evolutionary pathway which is different from that of flatworms, annelids, arthropods and molluscs (Fig 27.27). During its early development, an animal embryo undergoes division (**cleavage**). In all

ANIMALIA

animals, the first two cleavages are at right angles to each other. However, the third cleavage is by **radial cleavage** in the embryos of echinoderms and chordates but by **spiral cleavage** in flatworms, annelids, arthropods and molluscs (Fig 27.28). Further into development, the embryo forms a hollow ball of cells (the **blastula**). The pore that opens into this hollow ball (the **blastopore**) becomes the mouth in flatworms, annelids, arthropods and molluscs but the anus in echinoderms and chordates. The former group are termed **protostomes** ('first mouths') whereas echinoderms and chordates are termed **deuterostomes** ('second mouths'), since they develop mouths after their anuses.

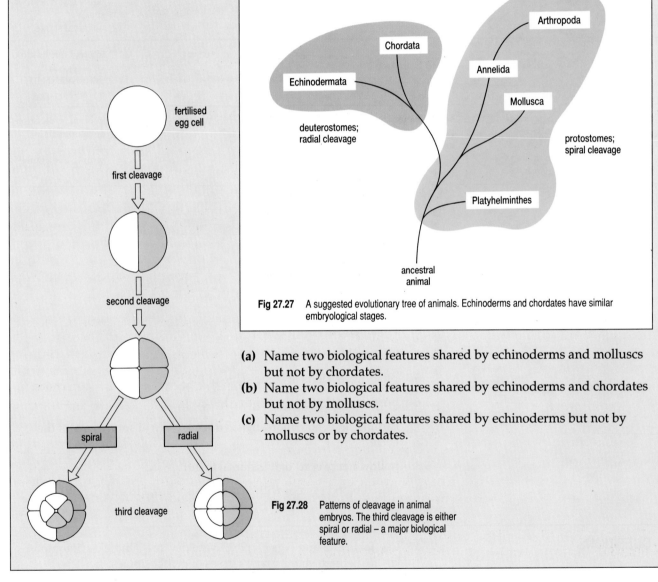

Fig 27.27 A suggested evolutionary tree of animals. Echinoderms and chordates have similar embryological stages.

(a) Name two biological features shared by echinoderms and molluscs but not by chordates.
(b) Name two biological features shared by echinoderms and chordates but not by molluscs.
(c) Name two biological features shared by echinoderms but not by molluscs or by chordates.

Fig 27.28 Patterns of cleavage in animal embryos. The third cleavage is either spiral or radial – a major biological feature.

Chordates are grouped into four subphyla, two without brains and skulls (**acraniate**) and two with brains and skulls (**craniate**). The largest of these subphyla, **Gnathostomata** ('jaw-mouths'), contains chordates with brains, jaws and paired limbs. Since biological features of many gnathostomes are described throughout this book, they will not be repeated here. Characteristics of the major classes of the Gnathostomata are summarised in Table 27.3.

Table 27.3 Classification of the Gnathostomata (commonly called vertebrates)

Class	Skin	Skeleton and limbs	Gas exchange	External ear	Reproduction
Chondrichthyes (cartilaginous fish)	tooth-like scales	cartilaginous skeleton, fleshy fins	gills with separate gill openings	absent	external fertilisation, larval stages
Osteichthyes (bony fish)	bony scales	bony skeleton, fins supported by rays of bone	gills covered by a single bony operculum	absent	external fertilisation, larval stages
Amphibia (amphibians)	soft	bony skeleton, paired pentadactyl limbs	gills in larva, skin and lung in adult	absent	external fertilisation, larval stages (tadpoles)
Reptilia (reptiles)	scaley	bony skeleton, paired pentadactyl limbs	lungs	absent	internal fertilisation, yolky egg in leathery shell, no larval stages
Aves (birds)	scales and feathers	bones with air sacs, paired pentadactyl limbs, front limbs form wings	lungs	absent	internal fertilisation, yolky egg in calcareous shell, no larval stages
Mammalia (mammals)	hairy with sebaceous, sweat and mammary glands	bony skeleton, paired pentadactyl limbs	lungs	present	internal fertilisation, development in uterus, birth to live young

KEY FEATURES

CHORDATA

1. triploblastic
2. coelomate
3. metamerically segmented
4. bilaterally symmetrical
5. pharyngeal clefts at some stage of life cycle
6. post-anal tail at some stage of life cycle
7. cartilaginous notochord that is dorsal to gut at some stage of life cycle
8. hollow nerve cord that is dorsal to gut
9. closed blood system with heart that is ventral to gut

QUESTIONS

27.15 Use Table 27.3 and Figs 27.29–27.34 to devise a simple dichotomous key to distinguish the major classes of gnathostomes.

27.16 For each animal phylum in this chapter, identify at least one way in which members of the phylum are economically important to humans.

27.17 Copy and complete Table 27.4 to serve as a summary of some of the work you have done in this chapter. For each feature in the table, put a tick (✔) if the feature is present or a cross (✘) if it is absent. The first row has been completed as an example.

Fig 27.29 A cartilaginous fish (class Chondrichthyes).

Fig 27.30 A bony fish (class Osteichthyes).

Fig 27.31 An amphibian (class Amphibia).

Fig 27.32 A reptile (class Reptilia).

Fig 27.34 A mammal (class Mammalia).

Fig 27.33 A bird (class Aves).

ANIMALIA

Table 27.4

Phylum	Locomotory Muscles	Nature of skeleton	Gut	Pseudocoelom	Coelom	Blood system or haemocoel	Metameric segmentation
Cnidaria	musculo-epithelial cells	enteron (hydrostatic)	✗	✗	✗	✗	✗
Platy-helminthes	longitudinal	parenchyma (hydrostatic)					
Nematoda	longitudinal	pseudocoelom (hydrostatic)					
Annelida	longitudinal, circular and parapodial	coelom (hydrostatic)					
Arthropoda	longitudinal and antagonistic appendage muscles	exoskeleton					
Mollusca	longitudinal	haemocoel (hydrostatic)					
Echinodermata	tube feet muscles	coelom and water vascular system (hydrostatic)					
Chordata	longitudinal and antagonistic appendage muscles	endoskeleton					

SUMMARY

Animals are multicellular organisms that feed on the bodies of other organisms. During its embryological development, an animal forms a hollow ball of cells, called a blastula. In protostomes, which includes the flatworms, annelids, arthropods and molluscs, the pore leading into this ball of cells develops into a mouth. In deuterostomes, which includes the echinoderms and chordates, this pore develops into an anus.

Sessile and slow-moving animals tend to be radially symmetrical, enabling stimuli from all directions to be detected simultaneously. Thus, adult poriferans, cnidarians and echinoderms are radially symmetrical. Faster-moving animals tend to be bilaterally symmetrical. Sense organs and their associated nerves tend to be concentrated at the end of the animal which reaches a new environment first, resulting in a definite head (cephalisation).

The major animal phyla can be arranged to show an increase in the size and complexity of their bodies. Sponges have relatively few cell types with little coordination between them, cnidarians have more cell types arranged in two layers (diploblastic) whilst other animal phyla have many more cell types arranged in three body layers (triploblastic). Coordination between these cell types becomes increasingly complex in the higher animal phyla, enabling more complex behaviour.

Animals in all phyla other than sponges have digestive cavities. A triploblastic animal may have a further cavity in its body formed by a split between its embryonic ectoderm and endoderm (pseudocoelom) or a split in its embryonic mesoderm (coelom). Animals without such a secondary cavity are termed acoelomate. Flatworms are acoelomate; nematodes are pseudocoelomate; annelids, arthropods, molluscs, echinoderms and chordates are all coelomate. By separating the muscles of the gut from those of the body wall, a coelom prevents interference between digestion and locomotion. In arthropods and echinoderms, the fluid-filled coelom is small compared with a blood-filled haemocoel.

The bodies of annelids, arthropods and chordates are metamerically segmented. This means that the mesoderm is divided into a fixed number of similar segments, or somites, each containing copies of nerve ganglia, muscles and excretory organs. Metameric segmentation is most complete in the annelids, where it aids movement. In arthropods and chordates, only some organ systems show any sign of segmentation.

Chapter 27: Answers to Quick Questions

1   Bilaterally symmetrical.
2   They have more cell layers therefore there is the possibility of a greater degree of cellular specialisation.
3   Radial, like most sessile organisms.
4   No, there is no division into cell layers (tissues).
5   (i)   In the medusa the mouth points down (is ventral); in a polyp it points up (is dorsal).
     (ii) Medusae have relatively more mesoglea than polyps.
     (iii) The enteron cavity occupies a greater proportion of body volume in the polyp than in the medusa.
     (iv) The medusa is free-swimming; the polyp is sessile.
6   By contracting the circular muscle tails in the endoderm whilst relaxing the longitudinal muscle tails in the ectoderm. (Note: the water in the enteron forms a hydrostatic skeleton.)
7   Increases the surface area to volume ratio and reduces the distance over which gases and excretory products have to diffuse.
8   By alternate contractions of the longitudinal muscles on each side of its body.
9   They are long and thin. Hence they have a high surface area to volume ratio.
10 The gut of a roundworm has both a mouth and an anus, the others have only one opening.
11 Metameric segmentation.
12 The dorsal and ventral longitudinal muscles. The fluid-filled coelom will act as a hydroskeleton.
13 Dispersal.
14 Beetle – 6, spider – 8 and crayfish – 10. Yes.
15 By waterproofing the cuticle.
16 It is an open system, that of the annelid is closed.
17 By acting as a hydrostatic skeleton.
18 The head.
19 Analagous, i.e. they have the same function.
20 Haemocoel.
21 It is twisted, an effect called torsion, so that the anus opens above the head.
22 The fluid in the haemocoel acts as a hydrostatic skeleton. Local

contractions of the longitudinal muscles distort the body and increase the pressure of the fluid in the haemocoel. This pressure will stretch the muscles when they relax.

23 Echinoderms are adapted to perceive stimuli from all directions rather than just one direction. Hence the sense organs need to be distributed over the whole body rather than concentrated in one region – a head.

24 (a) Extend the foot by contracting the muscles around the ampulla and relaxing the longitudinal muscles. This would force fluid from the ampulla into the tube foot, so extending it. Retraction is the opposite process.

   (b) By alternate contraction of the longitudinal muscles on each side of the foot. The fluid in the foot would act as a hydrostatic skeleton.

25 The fish's body would be bent into a curve.

Theme 6
EXAMINATION QUESTIONS

6.1 Read the following passage:

Two American research workers have spent the last nine years comparing genetic material, DNA, from a thousand bird species. The work has revealed many evolutionary relationships that were confused or concealed by traditional taxonomy based on physical structure.

The intention of the work is to measure the true "genetic distance" between species and then to draw an evolutionary tree showing when they diverged from common ancestors. The anatomical characters used by traditional taxonomists are often misleading because of convergent evolution.

The biologists use a tool called "DNA-DNA hybridisation". They heat DNA, extracted from birds' red blood cells, to separate its strands. Single DNA strands from two different species are then combined to give a hybrid double strand. Differences between the nucleotide sequences of the different strands weaken the bond between them. When heated, therefore, the hybrid dissociates at a lower temperature than pure DNA from either species. This difference in dissociation temperature can be used as a measure of the genetic distance between the two species. This in turn can be translated into the approximate date when the species diverged from a common ancestor.

(adapted from Science Report, *The Times*, 6 September 1983)

Using information in the passage and your own knowledge, answer the following questions.
- **(a)** What is meant by:
  - (i) *species* (line 3)? (2)
  - (ii) *"genetic distance"* (line 8)? (1)
  - (iii) *convergent evolution* (line 12/13)? (2)
  - (iv) *nucleotide* (line 19)? (1)
- **(b)** DNA is extracted from birds' red blood cells (line 16). Why would research workers not use this method to obtain mammals' DNA? (1)
- **(c)** Explain how
  - (i) the two strands of a DNA molecule are normally linked: (3)
  - (ii) the structure of DNA ensures exact replication during interphase of mitosis: (2)
  - (iii) nucleotides can code for up to twenty different amino acids. (3)
- **(d)** (i) Suggest why hybrid DNA dissociates at a lower temperature than the pure form (lines 21 and 23) (3)

- (ii) How may this knowledge be used to determine the relationship between two species? (2)
- **(e)** Similar relationship studies have involved egg-white.
  - (i) What category of chemical substance, found in egg white, would you expect to be used in these studies? (1)
  - (ii) How would you expect this chemical to differ between different species of birds? (2)
  - (iii) Name **one** technique that could be used to analyse this chemical so that relationships of species could be inferred. (1)

(AEB 1987)

6.2 **(a)** Explain the meaning of the following terms.
  - (i) Allele
  - (ii) Crossing over
  - (iii) Polyploidy (9)
- **(b)** When seeds from a cross between two tomato plants were germinated they produced the following plants:

    32 tall plants with red fruits
    29 dwarf plants with yellow fruits

  It was known that both tallness and red fruit were dominant characters but the genotypes of the parents of the cross were not known. Explain in detail why this result was obtained. (11)

(ULSEB 1988)

6.3 **(a)** Explain the differences between the members of **each** of the following pairs of genetical terms and give **one** example of **each** term to illustrate your answer.
  - (i) complete and incomplete dominance
  - (ii) continuous and discontinuous variation
  - (iii) chromosomal mutation and crossing-over
  - (iv) polyploidy and haploidy (12)
- **(b)** Crosses between ginger female cats and black male cats produce only tortoiseshell females and ginger-coloured males. A single gene controls expression of colour in cats.
  - (i) Give a reasoned explanation of these results and show the genotypes of the parents, their gametes and the offspring produced in these crosses.
  - (ii) Is it possible to have tortoiseshell male cats? Explain your answer. (8)

(JMB 1989)

6.4 **(a)** Explain what is meant by the following genetic terms:
  - (i) allele; (1)
  - (ii) linked genes; (2)
  - (iii) test cross (1)

In maize, the allele for coloured grain *C* is dominant to *c*, the allele for colourless grain. The allele *S* for smooth grain is dominant to *s*, the allele for wrinkled grain.

(b) A maize plant homozygous for coloured, smooth grain was crossed with another plant that was homozygous for colourless, wrinkled grain and all the resulting F_1 generation had coloured, smooth grain. The F_1 plants were self-fertilised to produce the F_2 generation.

    (i) Show this cross in the form of a diagram, giving the genotypes of the parents, their gametes and the F_1 plants. (3)

    (ii) Give the genotype of each type of gamete produced by the F_1 plants. (1)

    (iii) What proportion of the F_2 generation would be expected to have each of the following phenotypes:
        coloured, wrinkled grain,
        coloured, smooth grain? (2)

(c) (i) The drawing shows a maize plant with male and female flowers.

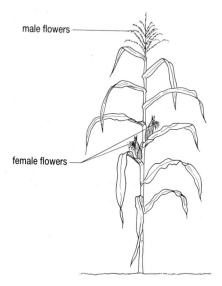

male flowers

female flowers

Write a set of instructions describing how you would make a cross between two maize plants to investigate the inheritance of grain colour. (3)

    (ii) Suggest **two** reasons why maize is a particularly suitable organism for use in genetic experiments. (2)

(AEB 1990)

6.5 In a genetics investigation using the fruit-fly *Drosophila melanogaster*, the following experiments were performed;

Experiment 1:
Male flies showing the two recessive characteristics claret eyes (cl) and vestigial wings (vg) were mated with female flies which were true breeding for wild type eyes (cl⁺) and wild type wings (vg⁺).

The female offspring of this first cross were then mated to their male parents. The results of this second cross were as shown:

Experiment 2:
Male flies showing the two recessive characteristics purple eyes and vestigial wings were mated to female flies which were true breeding for wild type eyes and wild type wings. The female offspring of this first cross were then mated to their male parents and the following results were observed:

6.5 **Experiment 1**

Characteristics	Number
claret eyes, vestigial wings	72
claret eyes, wild type wings	80
wild type eyes, vestigial wings	76
wild type eyes, wild type wings	84

Experiment 2

Characteristics	Number
purple eyes, vestigial wings	128
purple eyes, wild type wings	21
wild type eyes, vestigial wings	17

(a) (i) Using suitable genetic cross diagrams (and the symbols given) explain fully the reasons for the results of the two crosses obtained in Experiment 1. Give the genotypes and phenotypes of the offspring of the first and second crosses.

    (ii) What deductions can be made about the positions of the claret eye allele and the vestigial wing allele on the *Drosophila* chromosomes? (8, 4)

(b) (i) Explain why the results observed in the second cross of Experiment 2 are different from those observed in Experiment 1.

    (ii) What can be deduced about the relative positions of the purple eye allele and the vestigial wing allele on the *Drosophila* chromosomes? (4, 10)

(c) Comment on the significance of these results in relation to Mendel's Second Law of Inheritance (the law of 'Independent Assortment'). (4)

(UODLES 1987)

6.6 A chemical which affects the natural clotting of blood was used to try to control a population of rats. However, it was found that some rats were resistant to the chemical and that the proportion of resistant rats increased during the time the chemical was used. Assume that this resistance is related to a single gene with two alleles *R* and *r*. Rats resistant to the chemical have the dominant allele R in their genotype, while the non-resistant rats do not.

The graph shows the percentage frequency of the r allele in a rat population over 20 generations. Chemical control treatment was started in generation 5 and continued to generation 20.

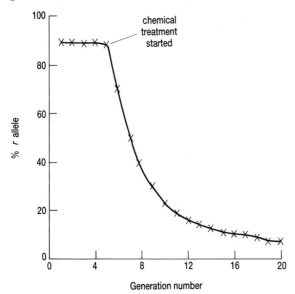

(a) (i) Give **two** factors which may contribute to the maintenance of a constant proportion of the alleles R and r in the population in generations 1–5.

(ii) In generation 4 the r allele has a frequency of 90%. Using the Hardy-Weinberg formula, calculate the expected frequency of the RR, Rr and rr genotypes in this generation.   (5)

Genotype	RR	Rr	rr
Expected frequency			

(b) (i) Explain why the frequency of the r allele in the rat population changes between generations 6 and 20.

(ii) Why is the rate of loss of r alleles greater between generations 6 and 12 than between generations 12 and 20?   (4)

(c) The frequency of r alleles transmitted from one generation of rats to the offspring can be calculated from the equation:

$$\frac{\text{frequency of } Rr \text{ genotypes in parent population}}{2} + \text{frequency of } rr \text{ genotypes in parent population}$$

$$= \text{frequency of } r \text{ alleles in offspring population}$$

(i) Why has the frequency of the Rr genotypes to be halved in this equation?

(ii) The genotypic frequencies in generation 8 are:
RR = 0.294 Rr = 0.612 rr = 0.094
Using the equation above, calculate the

expected frequency of the r allele in generation 9.

(iii) When the expected frequency of the r allele in generation 9 was compared with its actual frequency, the values were found to differ. When the difference between the values was tested statistically, a probability value of p >5% was obtained. Explain what this value of p indicates.   (4)

(JMB 1986)

6.7 The typical form of the European Swallow-tail butterfly has yellow patches on its wings, but in a rare variety called *nigra* these areas are shaded black.

A cross between a typical male and a *nigra* female produces 14 typical offspring and 6 *nigra*.

(a) As a first hypothesis to explain this result it was suggested that the *nigra* variety is caused by a recessive allele of a single gene, which is not sex-linked.

(i) On the basis of this hypothesis, a 1:1 ratio would be expected in the offspring. Construct a genetic diagram to show the genotypes of the parents in this cross, and to show how this ratio could be obtained. Use the letters **N** and **n** to represent the two alleles of the gene.   (3)

(ii) Complete the following table, filling in the expected numbers (E) of each phenotype on the basis of this hypothesis, and the differences between the observed and expected numbers (O–E).   (2)

Phenotype	Observed number	Expected number	Difference
	O	E	O–E
typical	14		
nigra	6		

(iii) Using the following formula, calculate the value of chi-squared (X^2). Show your working.   (3)

$$X^2 = \sum \frac{(O - E)^2}{E}$$

(iv) Probability levels (P) corresponding to some values of X^2 in this case are shown below.

X^2	0.004	2.71	3.85	6.63
P(%)	95	10	5	1

What does your calculated value of X^2 indicate in relation to this particular cross?   (1)

(b) A second hypothesis was suggested, in which the *nigra* variety is produced by interaction between two unlinked genes, **A** and **B**. One of the genes has alleles **A** and **a** and the

other gene has alleles **B** and **b**.
The *nigra* phenotype is seen only in individuals which are homozygous for the **a** allele and *either* homozygous *or* heterozygous for the **B** allele (*either* **aa BB** or **aa Bb**).
This second hypothesis suggested that the genotypes of the original parents in the above cross were as follows.

The typical parent **AaBb**
The *nigra* parent **aaBb**

Construct a genetic diagram to show the genotype and phenotype ratios expected from this cross on the basis of the second hypothesis. (4)

(c) (i) Comparison of the observed results with those expected from the second hypothesis gives a value for X^2 of 0.48. Compare this with the value calculated for the first hypothesis and suggest which hypothesis has the greater probability of being correct. Justify your answer. (2)
(ii) State *two* aspects of the data given that might reduce the reliability of the conclusion drawn in (c) (i). (2)

(d) Using only the *nigra* offspring from the original cross, suggest a further investigation to test the two hypotheses. Explain how the results of the investigation might show one the other of the hypotheses to be incorrect. (3)

(ULSEB 1989)

6.8 The diagram shows part of a family tree of a mammal. Several members have the condition called albinism where there is a lack of normal pigment in the body.

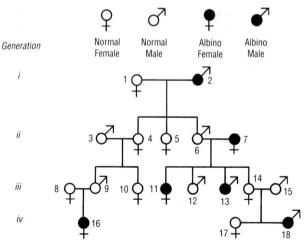

Generation — Normal Female, Normal Male, Albino Female, Albino Male

(a) Using the information in the family tree, explain why it is unlikely that albinism is controlled by any of the following (in each case give one reason):
(i) an autosomal dominant gene;
(ii) a sex-linked (X-linked) recessive gene;
(iii) a gene on the Y chromosome.

(b) Assume that pigment levels in this mammal are controlled by a single autosomal gene which has two alleles. The albino condition is produced by the recessive allele, *a*, while the normal pigment level is controlled by the dominant allele, *A*. What are the possible genotypes of the following members of the family tree?
(i) number 2 (in generation I);
(ii) number 3 (in generation II);
(iii) number 12 (in generation III)

(c) The same family was scored for inability to distinguish red and green colours, a condition called colour-blindness. This is controlled by a sex-linked gene: individuals with normal colour vision carry at least one *B* allele of this gene, while colour-blind individuals do not carry any *B* alleles. Using the symbols *B* and *b*, give the likely genotypes of:
(i) a female homozygous for normal pigmentation and normal colour vision;
(ii) an albino male with normal colour vision;
(iii) a male with normal pigmentation whose mother was colour-blind.

(d) Derive and give the phenotype ratio produced in the F_2 of the following mating:

Female		*Male*
Colour-blind and homozygous for normal pigmentation	×	Normal colour vision and albino

6.9 The **A, B, O** blood groups in humans are controlled by multiple alleles of a single autosomal gene. The gene locus is usually represented by the symbol I. There are three alleles represented by the symbols I^A, I^B and I^o. Alleles I^A and I^B are equally dominant and I^o is recessive to both.

(a) State all the possible genotypes of blood groups **A** and **O**.
(b) If a group **O** man married a Group **AB** woman, state the possible blood groups that their children could have.
(c) (i) Explain, using the above symbols, the possible blood groups of the children whose parents are both heterozygous, the father for blood group **A** and the mother for blood group **B**.
(i) If two of these children are non-identical twins, what is the probability that both will have blood group **AB**? (10)

(UCLES 1988)

6.10 (a) Describe the factors which contribute to evolutionary change in organisms. (13)
(b) Indicate how human activities can influence this change. (7)

(WJEC 1989)

6.11 (a) List any six characteristics of each of the following phyla.
  (i) Chordata,
  (ii) Annelida,
  (iii) Arthropoda. (6)
(b) Outline the ways in which the features of the various skeletal and gas exchange systems influence the range of habitats and modes of locomotion apparent in the following phyla:
  (i) Chordata, (5)
  (ii) Annelida, (4)
  (iii) Arthropoda. (5)

(NISEC 1988)

6.12 The drawings show four different species of Protoctista found in a pond.

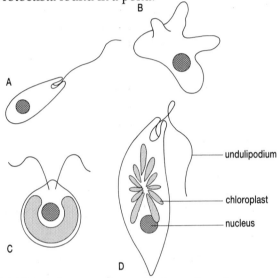

(a) Using features shown in these drawings construct a dichotomous key which would enable you to distinguish these four organisms from each other. (4)
(b) Organism **A** is drawn at a magnification of 400 times. Calculate its actual length in micrometres. Show your working. (3)
(c) Describe **one** method by which you could estimate the diameter of organism **C** using a light microscope. (4)

(AEB 1990)

6.13 (a) Why do biologists classify organisms? (4)
(b) Define a 'natural classification' and discuss the principles used to create one, giving appropriate examples (11)
(c) Do you regard viruses as plants, animals or neither. Give your reasons. (3)

(UCLES 1983)

6.14 In this question, phenotypes may be described by showing the alleles which are expressed, e.g. Phenotypes with the alleles AABBCC and phenotypes with the alleles AaBbCc would **both** be shown as ABC.

Three pairs of alleles, A/a, B/b and C/c are carried on homologous chromosomes as indicated in the diagram.

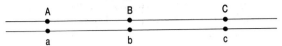

When an individual with this genotype was crossed with another which was recessive for all three genes, the following offspring were produced:

Offspring Phenotype	Number
ABC	284
Abc	50
ABc	76
AbC	2
aBc	3
abC	81
aBC	44
abc	260

(a) Give the phenotypes that would result from a cross-over between;
  (i) the A/a and B/B loci;
  (ii) the B/b and C/c loci. (2)
(b) Explain how the aBc and the AbC phenotypes could arise. (2)
(c) How many of the offspring show crossing over between;
  (i) the A/a and B/b loci;
  (ii) the B/b and C/c loci? (4)
(d) Calculate the cross-over value between:
  (i) the A/a and B/b loci;
  (ii) the B/b and C/c loci. (2)

(UODLES 1988)

6.15 Phenylketonuria (PKU) and albinism occur in humans when genetic abnormalities have caused errors in the normal metabolism of the amino acids phenylalanine and tyrosine respectively. In the diagram below the solid lines show the normal pathway and the broken lines show the abnormal metabolic pathways which lead to phenylketonuria and albinism.

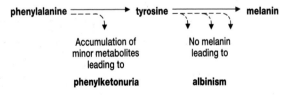

(a) (i) Explain why a genetic error may lead to the inability to metabolise an amino acid. (2)
  (ii) Assume that a single pair of alleles governs the occurrence of PKU. Show by a labelled diagram how normal parents could have a child with PKU. (4)

(b) In the following pedigrees, open symbols indicate people who are normal for skin pigmentation and solid symbols indicate albino people. Albinism is not sex linked.

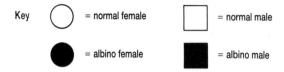

Key   ◯ = normal female   ☐ = normal male

⬤ = albino female   ■ = albino male

Use **A** and **a** to represent alleles in your answers to the following questions.
(i) Give the genotypes of the parents in the following family. (2)

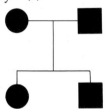

(ii) Give the genotypes of the parents (1 and 2) and the wife (3) of the albino son (4) in the following family. (3)

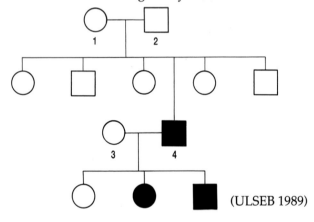

(ULSEB 1989)

6.16 Giving suitable examples in **each** case, describe the differences between
(i) acoelomate and coelomate animals; (6)
(ii) radial and bilateral symmetry; (5)
(iii) genus and species: (4)
(iv) monocotyledonous and dicotyledonous plants. (5)

(WJEC 1988)

6.17 For each of the following write brief notes to explain the observations in terms of evolution. Your notes should clearly indicate the major steps in a logical explanation of each phenomenon.
(a) Existing members of the camel family (Camelidae) have a discontinuous distribution, being represented by the camel (a highly specialised desert herbivore) in Africa and Asia and by the llama (a relatively unspecialised large herbivore) in South America. The earliest fossil remains of camel-

like animals have been found in North America. (6)
(b) The large herbivore niche is filled in Africa by antelope but in the Galapagos Islands by giant tortoise. (5)
(c) In Australia there are many indigenous (native) mammals and all but a few are marsupials. Many of these marsupial forms bear a close resemblance to placental mammals in other areas of the world (e.g. the marsupial banded anteater and the placental anteater; the marsupial Tasmanian wolf and the placental hyena). (5)
(d) The peppered moth (*Biston betularia*) is normally a light colour with a mottled pattern which blends well with the lichen covered bark on which it normally rests. The first melanic (dark) individuals of this moth were reported in 1849. After that time the number of the melanic form steadily increased in industrial areas, until by 1970 it formed over 95% of the population in Manchester. In the north of Scotland and the extreme south west of Britain, on the other hand, the light form accounts for nearly 100% of the population. (4)

(NISEC 1989)

Answers Guidelines

6.1 Many examination boards set comprehension questions to test recall, understanding and the ability to apply knowledge. Like most comprehension exercises, this one involves subject material from several Themes of this book.

(a) (i) A group of organisms that are actually or potentially capable of interbreeding; to produce fertile offspring;
(ii) the degree to which their DNA (and hence their genes) are different;
(iii) the acquisition of similar features by different groups; analogous structures/processes; named example (e.g. eye of chordates and of cephalopod molluscs); (maximum 2 marks)
(iv) unit of DNA structure consisting of pentose, phosphate and organic base;
(b) Unlike birds, mammalian red blood cells lack nuclei;
(c) (i) hydrogen bonds; between organic bases; A–T and C–G;
(ii) organic bases pair together in a specific way; A–T and C–G;
(iii) three organic bases together form the code for one amino acid/base triplet; since there are four bases, this gives 64 (4^3) possible combinations/

codes; code is degenerate/some base triplets code for the same amino acid;

(d) (i) Arrangement of bases on the two strands is not exactly complementary, so fewer hydrogen bonds in hybrid than in normal DNA; less energy needed to weaken fewer hydrogen bonds;

(ii) The greater the difference in sequence of bases the lower the temperature at which the strands fall apart; high dissociation temperature means birds from which DNA came are more closely related than those with low dissociation temperature;

(e) (i) protein (Note the question asks for a category, albumen is a specific protein and not a category of chemical substance);

(ii) Differences in protein structure should mirror DNA differences; since DNA is code for proteins;

(iii) electrophoresis;

Phenotype of parents	tall, red fruits	×	dwarf, yellow fruits;
Genotype of parents	(TR) (tr)		(tr) (tr);
Gamete	(TR) (tr)		(tr);
Genotype of offspring (1)	(TR) (tr)		(tr) (tr);
Phenotype of offspring (1)	tall, red fruits		dwarf, yellow fruits;
Ratio of phenotypes	1	:	1

The actual ratio of phenotypes in the offspring (32:29) is a very close fit to the expected ratio of 1:1; variation from this ratio is by chance. (N.B. If you try all other possible crosses you will find they do not fit the data. In an examination, you would either cross them out so that they will not be marked or clearly explain that they do not fit the data so that the examiner can see that this is not your explanation of the results.)

6.2 This is a more traditional genetics question. Like many genetics questions, this question involves deduction. The best way to find the answer is to try out all possible assumptions until you find one that fits the results given – a skill which requires some confidence in examinations!

(a) (i) alternative form/DNA base sequence of a gene; alleles result in production of different proteins; named example (e.g. Hb^N results in formation of normal haemoglobin, Hb^S results in formation of sickle-cell haemoglobin);

(ii) occurs in prophase I of meiosis; chromatids of homologous chromosomes entangle, break and rejoin onto opposite chromosome; results in linked gene being separated;

(iii) more than two copies of each homologous chromosome/three or more times the haploid chromosome number; in each cell; most common in plants;

(b) Let T represent the allele for tallness and t represent the allele for dwarfness and let R represent the allele for red fruits and r represent the allele for yellow fruits; Since tall stems and red fruits were inherited together, this cross must involve linkage with the linkage groups (TR) and (tr); Since there are no tall plants with yellow fruits or dwarf plants with red fruits, no crossing over can have occurred;

Theme 7

ECOLOGY

The word ecology has become a cliché. Many people think it is synonymous with the activities of environmental pressure groups such as Greenpeace and Friends of the Earth. This is not the case. Ecology is the scientific study of the interactions that determine the distribution and abundance of organisms. It seeks to discover where plants and animals live, how many there are and what they do.

Ultimately ecology seeks to explain these patterns of distribution and abundance in terms of interactions between organisms and their environment. In this theme we shall look at four areas.

- *Populations and factors controlling their size and distribution.*
- *Energy flow through ecosystems* – the way in which energy is captured by producers and passed to consumers.
- *Nutrient cycles* – natural geochemical cycles and the effect of pollutants.
- *Patterns in time and space* – different ecosystems around the world and their changes over a period of time.

(Left) A food chain in action.
(Right) A tropical rainforest: the most diverse community on Earth.

Chapter **28**

POPULATIONS AND FACTORS CONTROLLING THEIR SIZE AND DISTRIBUTION

LEARNING OBJECTIVES

When you have studied this chapter you should be able to:

1. define and explain the terms habitat, population, community, ecosystem, carrying capacity and biotic potential;

2. explain how population size changes and discuss the concepts of environmental resistance and carrying capacity;

3. distinguish between biotic and abiotic factors and discuss their effect on population size;

4. use the terms ecological niche and resource partitioning to interpret ecological data.

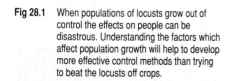

Fig 28.1 When populations of locusts grow out of control the effects on people can be disastrous. Understanding the factors which affect population growth will help to develop more effective control methods than trying to beat the locusts off crops.

For the past hour, between half-a-million and a million starlings – impossible to estimate – had been flying in every direction. From daytime feeding grounds up to 30 kilometres away, they had converged on Abbey Park to roost.

New Scientist 18 June 1987

If you live in a large town you may have seen large flocks of birds collecting together in the evening like the ones in Abbey Park, a 30 hectare green space in the centre of Leicester, England. Not all animal groups are as conspicuous as these starlings, but you will probably find it hard to think of an animal species in which, in its natural habitat, individuals live alone. Even male sperm whales, which seem to live singly in the warm waters off the coast of Ecuador, are in constant contact with groups of females which are about one day's swim away. You only have to look at a field or woodland to realise that members of plant species also live in groups.

Why do organisms live in groups? Why are some species, such as starlings, so common whilst others, for example sperm whales, are rare? Why does the size of these groups change from year to year? Why do groups of the same species differ in size? These are the types of questions you will be considering in this chapter. However, some important ecological terms need to be defined first.

28.1 ECOLOGICAL LEVELS

So far in this book we have looked at individual organisms at a number of different levels, molecular, cellular, tissue and so on. Similarly ecologists investigate and try to explain how the world around us works at a number of different levels. It will help you to become familiar with these ecological levels at the beginning of this theme on ecology so they will be defined in this Section, using a woodland as a source of examples.

Fig 28.2 A typical deciduous woodland in spring. Many different plants and animals live in this habitat. They are affected by biotic factors (e.g. competition for space) and abiotic factors (e.g. amount of sunlight).

Habitat

An organism's habitat is the place where it lives, its address. Our woodland is the habitat for a whole host of organisms. Many organisms will only occupy a small part of the total habitat, for example, the snail in our woodland. This small part of the total habitat is called a **microhabitat**.

1 **What is the microhabitat of a snail?**

Each habitat will have certain distinctive features which affect the organisms living in it. On the one hand, there are **physico-chemical** or **abiotic** factors: climate, soil, type of water (marine, fresh, running, still) and so on. On the other hand, there are **biotic** factors, which are determined by the organisms which share the habitat. For example, organisms may eat each other, compete with each other for food or provide shelter.

Biotic and abiotic factors are not independent of each other. For example, the trees in a woodland affect the humidity, temperature and amount of sunlight there. So trees, a biotic factor, influence the physico-chemical features of the habitat which, in turn, will affect the other organisms living in the woodland. Understanding this complexity of interaction between organisms and their habitat is one of the challenges faced by ecologists.

 How might other organisms affect the microhabitat of bluebells?

Population

A population is a group of individuals of the same species which occupy a particular habitat. For example, you could talk about the population of beetles living in a rotten tree stump or the population of starlings which roost in Abbey Park.

Community

A community is a collection of populations of plants and animals which occur together in both space and time. For example, our woodland is a community because a number of populations live together in it. In a woodland there are likely to be populations of many plant species, notably trees, shrubs and herbs, together with populations of fungi and many animal species, especially insects.

 What is the difference between a habitat and a community?

Ecosystem

An ecosystem is an ecological unit which consists of several habitats and their associated communities, so it includes both biotic and abiotic components. Ecosystems do not have visible boundaries around them which you could draw on a map. For example, we would describe our woodland as an ecosystem but it is part of a larger terrestrial ecosystem which is in turn part of the island ecosystem of Britain. More importantly, ecosystem is a useful concept because it is a convenient way of describing the links between organisms and their environment. Indeed the best way to think about an ecosystem is as a network of habitats and communities

---

QUESTIONS

28.1 Fill in each of the blanks in the following sentences using the appropriate term: community, ecosystem(s), habitats(s) or population(s).
   (a) A moorland .... contains grassland, scrub and stream .....
   (b) Highland streams are a .... for trout and salmon.
   (c) A cat harbours a .... of fleas in its fur.
   (d) Ladybirds are members of the predator .... that feeds upon .... of greenfly, reducing their numbers.
   (e) The grassland .... of the Serengeti in Africa, is grazed by a .... of large herbivores including wildebeest, zebra, topi and buffalo.

28.2 What communities can you distinguish in the pond shown in Fig 28.3? Which organisms belong to each community?

---

linked by flows of energy and nutrients between them. We will return to this important idea in the next chapter.

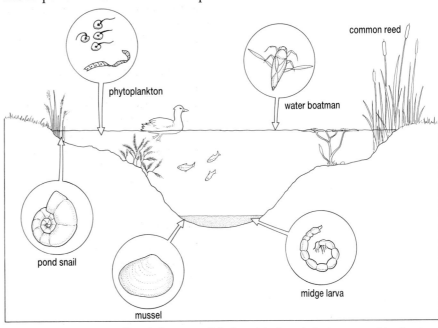

Fig 28.3 A pond ecosystem. Phytoplankton are small, floating autotrophs; water boatmen are predatory insects.

28.2 CHANGES IN POPULATION SIZE

This chapter is mainly concerned with how biotic and abiotic factors affect population size. Such understanding may be of great practical significance. For example, the number of algae in a lake may increase rapidly, producing a choking algal bloom (Fig 28.4(a)). Fig 28.4(b) shows what happened after sheep were introduced into Tasmania in 1814, while Fig 28.4(c) shows fluctuations in the number of great tits in Sweden. All these examples illustrate changes in population size but as ecologists we need to know why the patterns of change differ.

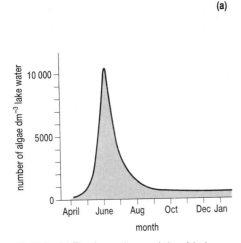

(a)

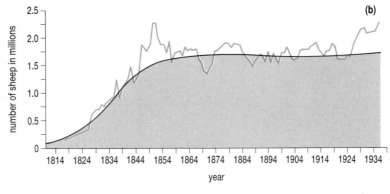

(b)

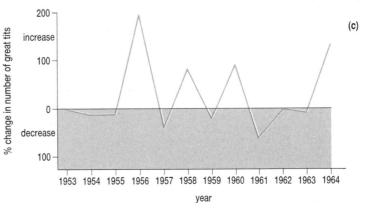

(c)

Fig 28.4 (a) The changes in a population of the brown alga, *Dinobryon divergens*. This sort of growth is typical of an algal bloom.
(b) Changes in the size of a population of sheep following their introduction into Tasmania in 1814. The black line shows the trend in population size; the blue line shows year-to-year variation.
(c) Fluctuations in breeding numbers of great tits, *Parus major*, in Veluwe, Sweden, between 1953 and 1964. The black line shows the trend in population size, the blue line year-to-year variation.

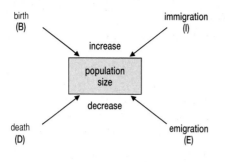

The four factors affecting population size are shown in Fig 28.5. Organisms can join the population by **birth** and **immigration**, they can leave it by **death** and **emigration**.

 Using Fig 28.5, explain why a population (a) grows (b) declines (c) stays the same size.

Exponential growth

Under ideal conditions, plenty of food, the correct temperature and so on, a population can grow at its maximum rate, its **biotic potential**.

Such growth starts off slowly and then increases rapidly, producing an exponential, or J-shaped, growth curve, (Fig 28.6(b)).

 Can you think what reproductive process might produce such a pattern of growth?

Fig 28.5 shows how four factors affect changes in population size. To make the model easier to understand we will ignore the effects of immigration and emigration on population size and concentrate just on births and deaths.

Rate of change in population size = (births – deaths) × number in population

If there are consistently more births than there are deaths, the number in the population will increase. As this value is itself used in the equation, the population will grow at an ever increasing rate, i.e. **exponentially**.

Exponential growth (also called logarithmic growth) is different from arithmetic growth (in which the rate of increase remains constant) and leads to much faster population growth (Fig 28.6).

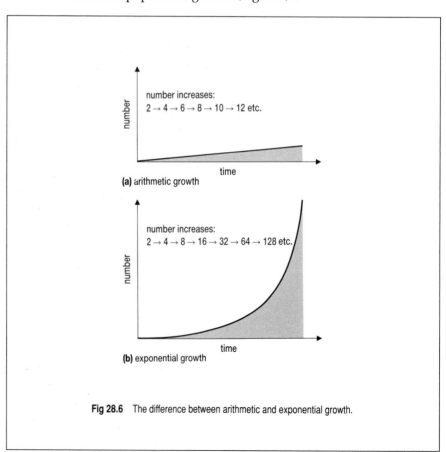

Fig 28.6 The difference between arithmetic and exponential growth.

When does exponential growth occur?

This exercise will help you develop skills in data analysis and interpretation.

Staphylococcus aureus is a species of bacterium which lives on human skin and is harmless to normal healthy people. In a suitably warm and nutritious environment, such as a bowl of warm custard, each bacterial cell can divide to form two new cells every twenty minutes.

(a) Starting with one cell, draw a graph to show how the bacterial population would increase in size.
(b) How big would the population be after six hours?
(c) Some strains of *S. aureus* produce poisonous proteins (exotoxins). Why do you think that so many cases of food poisoning have been traced to food which has been left standing in warm kitchens for many hours before it was served?

In the early years of the nineteenth century a plant called the swamp stonecrop (*Crassula helmsii*) was introduced from Australia to be grown in ornamental ponds in Britain. In the 1950s it was found in a number of natural waterways in Essex and is now a serious pest, forming a dense mat of vegetation which smothers other plant life. The number of sites at which it grows is doubling every three to five years and British water authorities spend hundreds of millions of pounds each year trying to eliminate it from their waterways. Following their initial colonisation of British waterways, the populations of swamp stonecrop must have grown exponentially.

(d) Except where a species colonises a new habitat, exponential growth is not often seen in natural populations. Why do you think this is?

Limits to growth

A growth curve typical of what happens if a sample of bacteria is transferred (inoculated) onto a growth medium and then kept under suitable conditions is shown in Fig 28.7. The reason the bacterial population stopped increasing was that its environment could not support the vast number of new cells being produced. Before long there would be a food shortage in the growth vessel, for example. In addition, many bacteria produce waste products which are toxic to themselves, so the number of births decreases and the number of deaths increases, causing a population decline. Such an accumulation of toxins and population decline occurs when yeast cells ferment sugar solutions, explaining why wines normally have a concentration of only 10–13% alcohol.

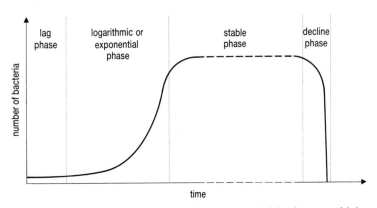

Fig 28.7   A typical bacterial growth curve. Maximum rate of growth is during the exponential phase.

These examples clearly show that it is changes in the organism's environment which prevent the population from realising its biotic potential: the J-shaped growth curve has become an S-shaped **logistic growth curve**.

Carrying capacity and environmental resistance

The maximum population size a particular environment can support, under a particular set of conditions, is called that environment's **carrying capacity**. The environmental factors that reduce the growth rate of a population are called the **environmental resistance**. Such factors include competition, disease, predation and an unfavourable climate. Now look carefully at Fig 28.8. As a population begins to encounter environmental resistance, notice how the J-shaped growth curve, which the population would follow if it was realising its biotic potential, bends away from the steep slope to give the more gentle S-shaped **logistic growth curve**. Eventually the curve levels off and the population then fluctuates about some average carrying capacity for the environment.

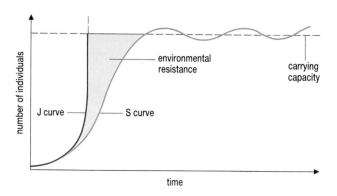

Fig 28.8   The J-shaped population growth curve becomes S-shaped when the population encounters environmental resistance due to one or more limiting factors.

The causes of population fluctuations

The number of individuals in a population changes according to fluctuations in the number of births and deaths (again we will ignore migration). To understand what causes a plague of mice or locusts, we have to find out what causes the changes in the numbers of births and deaths. For example, the huge increases in the number of great tits shown in Fig 28.4(c) followed bumper crops of beech seeds, the staple winter food of this bird. The extra food availability following bumper crops meant that fewer juvenile birds starved to death in the winter. Consequently, more birds survived until the following spring to breed and the population increased in size. In boom years, then, the number of births exceeds the number of deaths.

 Suggest why the populations of some species, e.g. seed-eating birds, are characterised by 'boom and bust' whilst others, e.g. the Tasmanian sheep, remain relatively constant from year to year.

Populations do not keep increasing for ever and our great tits are no exception, but why do populations decrease in size? The answer involves a quite simple principle. Population size is limited by increases in deaths or decreases in births. To really understand the implications of this idea you need to understand two new terms.

- Birth rate (natality) = $\dfrac{\text{number of births}}{\text{number of adults in the population}}$

- Death rate (mortality) = $\dfrac{\text{number of deaths}}{\text{number of adults in the population}}$

These important concepts are applied in the study on owls. Read through this study and use the principles it demonstrates to think about why the owl population of Wytham Wood decreased in size.

Case Study: The tawny owls of Wytham Wood

The adult tawny owls inhabiting Wytham Wood near Oxford establish territories in the autumn. If an owl does not occupy a territory its chances of surviving through the winter are very small. Adult birds live for many years occupying the same territory.

The death rate of newly fledged owls (those that have just left the nest and are ready to set up a territory) is much greater in years when the wood already has a large adult population and competition for territories is high. When the adult population is reduced, perhaps by a severe winter, the chances of a newly fledged bird establishing a territory are much higher and the death rate is correspondingly lower. So in Fig 28.9 the death rate of the newly fledged owls would be represented by the black line. A higher percentage of the young owls dies as the population size increases, i.e. this death rate is **density-dependent**.

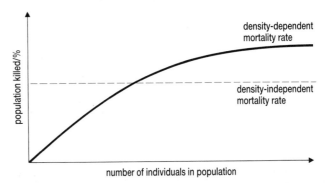

Fig 28.9 Density-dependent and density-independent mortality rates.

Now consider the adult owls which are killed by winter cold. The percentage of adults which die is unaffected by the size of the population, i.e. this death rate is **density-independent**. This time a fixed percentage of the population is killed irrespective of the population density, represented by the blue line in Fig 28.9. This example illustrates two important points

- density-dependent factors are always biotic (in this case competition between the owls for territories)
- density-independent factors may be abiotic (e.g. winter cold), or biotic.

 What would happen to the death rate of the young owls if there were a plague (i.e. huge increase in numbers) of mice?

To summarise, populations of plants and animals respond to changes in environmental conditions such as an excess or shortage of food.

If conditions are favourable then:

$$\underset{\text{(B)}}{\substack{\text{number}\\\text{of births}}} + \underset{\text{(I)}}{\substack{\text{number}\\\text{of immigrants}}} \text{ will exceed } \underset{\text{(D)}}{\substack{\text{number}\\\text{of deaths}}} + \underset{\text{(E)}}{\substack{\text{number}\\\text{of emigrants}}} \text{ and the population will increase}$$

If conditions are unfavourable then:

$$\underset{(B)}{\underset{\text{of births}}{\text{number}}} + \underset{(I)}{\underset{\text{of immigrants}}{\text{number}}} \quad \text{will be less than} \quad \underset{(D)}{\underset{\text{of deaths}}{\text{number}}} + \underset{(E)}{\underset{\text{of emigrants}}{\text{number}}} \quad \text{and the population will decrease}$$

The balance between population growth and decline is determined by abiotic and biotic factors which are summarised in Fig 28.10.

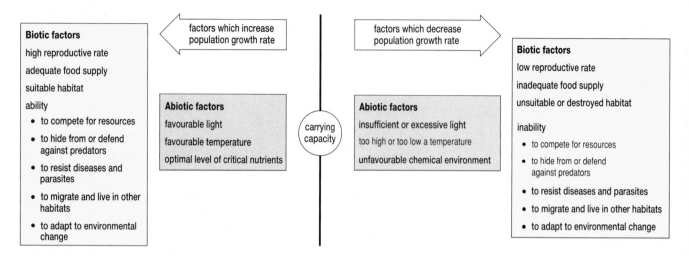

Fig 28.10   Population size is a balance between factors that increase numbers and factors that decrease numbers.

28.3 ABIOTIC FACTORS

These are the non-living components of an organism's habitat. They include

- physical factors (sunlight, rainfall, temperature, wind, water currents)
- chemical factors (all the elements and compounds in the habitat including those put there by people, e.g. poisons and pesticides).

We shall examine these factors in more detail but first we should develop a general idea about how abiotic factors affect individuals and populations.

A general model

Look at Fig 28.11. This shows a generalised model of how organisms might react to an abiotic factor. The abiotic factor could be light intensity, the concentration of a nutrient in the soil, water current speed or temperature, but it would still have a similarly shaped curve. The model shows that each species, and each individual in a species, has a particular **range of tolerance** to variations in chemical and physical factors in its environment. This tolerance range includes some optimum range of values in which the organisms thrive and where their populations will be biggest. At values slightly above or below the optimum the population will be smaller, i.e. the carrying capacity of the environment is lower.

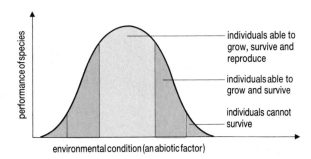

Fig 28.11 A generalised graphical representation of the way in which the performance of a species is related to the intensity of an environmental condition. Above or below some threshold the species cannot survive. The species only does well over a narrow range where it can reproduce.

 8 **What will happen to a population once values exceed the upper or lower limits of the species range of tolerance?**

Different species have different ranges of tolerance for a particular abiotic factor such as temperature. The range of tolerance will also be affected by the physical condition and age of the individual. For example, an animal weakened by disease or shortage of food will be less tolerant of severe cold than a healthy individual. Similarly, young animals are usually less tolerant than healthy adults.

On a global scale, abiotic factors determine the distribution of different types of ecosystem: this pattern is discussed in Chapter 31.

 9 **Which will be more affected by extremes of temperature, an ectotherm (e.g. a snake) or an endotherm (e.g. a mammal)?**

Physical factors

Temperature

This influences the rate of enzyme-controlled reactions, each enzyme having an optimum temperature at which it works best (Section 5.2). Environmental temperatures influence the ability of organisms in a population to survive and reproduce, especially if the organisms are ectotherms. The result is that population growth is highest in a temperature range which is around the optimum for the enzymes of that species.

There can be a marked difference in temperature within a single habitat. For example, the temperature on the south side of a tree growing in the northern hemisphere is usually higher than that on its north side. Thus, a gradient of temperature occurs around the tree which may affect sessile organisms living on the tree trunk. At some point around the tree, the temperature will cause these organisms to suffer physiological stress. As a

Sex determination in alligators and crocodiles depends upon the temperature at which the eggs develop. At temperatures below 30 °C all embryos develop into females, above 33 °C all embryos develop into males. At intermediate temperatures different sex ratios develop.

result, their growth rate will be lower than that of other organisms growing at a different position around the tree.

Fig 28.12   Some organisms, such as the prokaryotes which live in these hot water springs in Yellowstone National Park, are adapted to living at very high temperatures.

Light

This affects the survival of all green plants since it is their source of energy for photosynthesis. The intensity and wavelength of light changes as it passes vertically through a community, since plants in the upper parts of the habitat absorb light for their own photosynthesis. Consequently, only populations of plants which can tolerate shade can survive on the floor of a woodland. The ability of plants to flower, and for their seeds to germinate, is also affected by light (Section 17.2). Whilst light intensity and wavelength have a lesser effect on animals, light duration (daylength) is important to animals, affecting the synchronisation of reproduction in some species and the time available for food collection in others.

 Why might it be to an individual's advantage to synchronise its reproduction with the rest of the population?

Currents

Both water and air currents influence populations. In air, wind increases the loss of water and heat from the bodies of organisms. In water, the currents may dislodge organisms which lack either a strong attachment to the substratum or are weak swimmers.

Substratum

The material on which organisms live affects their ability to survive. Sand is easily blown away by currents of air, offering an unstable surface; streams with boulders have different inhabitants from those with gravel beds.

 How might stream organisms overcome the problems posed by the current?

ANALYSIS

Navigation using trees

This exercise develops the skill of devising investigations to test a hypothesis.

If you are lost in a wood you are supposed to be able to tell which way is North by looking at the trees. Moss on the tree trunks supposedly grows better on the north side of the trunk.

(a) Devise and, preferably, carry out an investigation to test this idea.
(b) Why do you think moss might grow better on the north side of a tree?
(c) Would moss grow on the north side of a tree in New Zealand?

Chemical factors

A large number of chemical factors affect organisms, including the availability of inorganic ions and presence of poisons. However, we will consider just two – water and oxygen.

Water availability

This is an important environmental factor. For example, water is an essential metabolite and is needed for gamete transfer. In terrestrial environments, water availability depends on such factors as rainfall, the rate of evaporation and the rate of loss through the soil.

 How might farmers affect the rate of water supply to their crops?

Oxygen availability

Whilst superabundant in the atmosphere, the supply of oxygen can become limited in two environments.

- **Water** – Oxygen dissolves very poorly in water. It is more soluble in cold water than in warm and in fresh rather than salt water. Fast flowing streams, because the water is turbulent, dissolve more oxygen than lakes (Fig 28.13).
- **Soil** – Well-aerated soil has air spaces containing oxygen which is needed for the respiration of plant roots. If soil becomes waterlogged, these air spaces are filled by water and plant roots are starved of oxygen.

 Why do plant roots need oxygen?

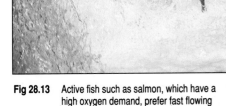

Fig 28.13 Active fish such as salmon, which have a high oxygen demand, prefer fast flowing cold water in which to spawn.

Edaphic factors

These are the physico-chemical factors which affect soil. Many of the factors mentioned above, such as temperature and the availability of water and oxygen, are edaphic factors when they apply to the soil. Assessment of the impact of edaphic factors is usually made in relation to plant growth, since it is traditionally farmers who have been most interested in this part of the environment.

Plants need to absorb many **inorganic ions** from the soil water around their roots. Nitrates and phosphates are needed for the production of protein and nucleic acid during growth, magnesium is needed for the production of more chlorophyll and so on. The effect that variations in the concentration of one inorganic ion have on the growth of plant seedlings is shown in Fig 28.14. The seedlings which had been in abnormally low and high concentrations did not grow as well as those in the middle of the concentration range.

Fig 28.14 The effect of nutrient availability on plant growth.

Soil pH

A keen gardener knows that soil pH affects crop growth. For example, beans grow best in soil of pH 6. Section 5.2 describes how pH affects the rate of enzyme-controlled reactions within cells. Soil pH does not directly affect the pH within the roots of a plant, since plant cells are selectively permeable (see Fig 2.1). Instead it affects the availability of inorganic ions which are needed as metabolites, e.g. Fe^{3+}. The concentration of many inorganic ions in the soil solution is directly related to the pH of the soil (Fig 28.15). One harmful effect on plants is thought to result from the release of toxic Al^{3+} ions from soils which have been acidified by acid rain (Chapter 30). This free aluminium has two effects

- it is highly toxic
- it precipitates phosphates out of the soil solution so making an essential nutrient, phosphorus, unavailable to the plants and soil micro-organisms.

Relationships between abiotic factors

Although abiotic factors are often presented as if they were discrete entities, this is not the case. For example, two edaphic factors, soil pH and inorganic ion concentration, interact with each other (Fig 28.15). Furthermore, the concentration of inorganic ions in the soil also depends

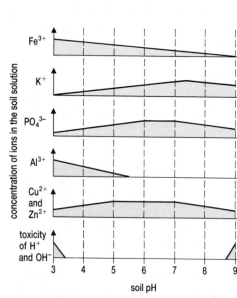

Fig 28.15 pH influences the concentration of inorganic ions in the soil solution.

on the activity of decomposers, mainly bacteria, a biotic factor. The activity of these bacteria also depends on soil pH. The interrelationships between these factors are shown in Fig 28.16. You should always look for such interactions when you think about the effects of abiotic and biotic factors on organisms.

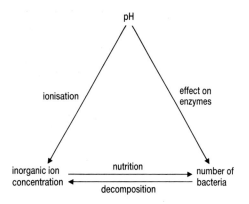

Fig 28.16 One way in which abiotic and biotic factors interact in the soil.

QUESTIONS

28.6 **(a)** Construct a diagram based on Fig 28.16 to summarise the interrelationships between some of the environmental factors in a woodland.

28.7 **(a)** Interpret the pattern of growth shown by the seedlings in Fig 28.14 .
(b) What does Fig 28.14 tell you about the effects of essential nutrients on plants when they are present in high concentrations?

28.8 In 1990 Britain agreed to stop dumping sewage sludge, which contains high concentrations of heavy metals (e.g. copper, zinc and lead) into the sea. This means that much more of this sludge will have to be disposed of by spraying it onto agricultural land.

(a) What possible problems can you foresee arising from this practice?
(b) Will these problems be greater on acid or alkaline soils?

28.4 BIOTIC FACTORS

The size of all populations is affected in some way by the presence of other organisms in the habitat. This section examines some of these biotic factors.

Food availability

All organisms need organic chemicals such as carbohydrates, lipids and proteins to stay alive (see Chapter 1). Thus, food availability is a biotic factor which limits the size of consumer populations. For organisms which cannot synthesise their own, there is only one natural source of these compounds; the body or products of another organism. Many animals eat a number of different species of food organism, limiting the effect which any one food source might have on their own population growth. Some animals eat very few species of food organism, for example, the Australian koala only eats leaves of the eucalyptus tree and pandas are thought to eat only one species of bamboo.

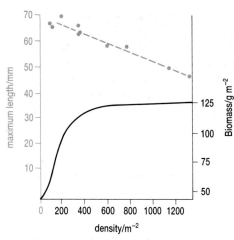

Fig 28.17 Maximum length (blue line) and biomass (black line) of the limpet, *Patella cochlear*, in relation to density. As population density rises intraspecific competition reduces the length of individual limpets, maintaining a constant population biomass.

 Animals with restricted diets often show large fluctuations in the size of their populations compared with animals which have a range of foods in their diet. Suggest why.

Competition

In most natural habitats there is a limit to the availability of essential resources such as food, space in which to grow or water. If there is not enough of one resource for all the members of a population, competition for that resource will occur. The effect of this competition is to reduce the growth rate of individuals and, ultimately their reproductive capacity, to a lower level than they could achieve if there were no competition. Consequently, competition reduces the rate of population growth, i.e. it is a form of environmental resistance.

Intraspecific competition occurs between members of the same species. Since they are likely to have very similar resource requirements, they will compete for any that are in short supply. Such intraspecific competition results in changes in the frequency of favourable and unfavourable genes in the process of natural selection: this is described in Chapter 23. The effect of intraspecific competition on growth in the limpet, *Patella cochlear*, is shown in Fig 28.17.

Interspecific competition occurs between members of different species.

ANALYSIS

Competition between barnacles

This exercise gives practice in interpretive skills.

Many rocky shores support populations of small crustaceans, called barnacles. As adults, these animals are fixed to a rock surface and feed on floating microscopic organisms during the hours when they are covered by the sea. One common Atlantic species, *Chthamalus stellatus*, is able to grow over all parts of rocky shores. A second species, *Balanus balanoides*, cannot tolerate the dry conditions at the top of the shore but can live lower down the shore.

(a) What do you think is the key resource for which barnacles, like all sessile organisms, will compete?

Balanus is the larger of the two and has a faster growth rate: as a result, its shell prizes *Chthamalus* off the rocks as it grows or *Balanus* simply grows over *Chthamalus* (Fig 28.18).

(b) Why, if an area of rock lower down the shore is kept clear of *Balanus*, will *Chthamalus* be able to thrive?

(c) The dogwhelk, *Nucella lapillus*, is a predatory gastropod which feeds on barnacles and is common on the lower shore. How is this reflected in the distribution of *Balanus*?

Chthamalus stellatus — mean high tide

Balanus balanoides — mean low tide

Fig 28.18 The distribution of two species of barnacle, *Chthamalus stellatus* and *Balanus balanoides*, in the intertidal zone. Desiccation is the primary controlling factor in the upper part of the zone for *Balanus*; competition with *Balanus* limits *Chthamalus* to the upper part of its zone of larval settlement. Predation by dogwhelks and competition with plants are the main factors determining the lower limit for *Balanus*.

Scramble and contest competition

Sometimes competition is an obvious event. For example during the autumn rut, male red deer compete for access to females with which to mate. After their ritualised combat, one animal will retreat and only the victor will win the right to mate. Competition of this type, where one competitor gains the whole resource, is called **contest competition**.

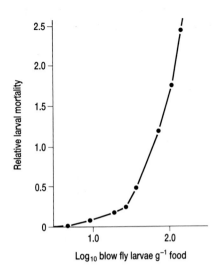

Fig 28.19 Populations of blowfly larvae kept in a laboratory with fixed amounts of food.

Competition does not always involve such combats, however. Increasing the population density of blowfly larvae, without increasing the amount of food which they are given, affects their growth (Fig 28.19). At first, despite an increase in density, larval mortality remains fairly stable. Instead of starving to death, most larvae obtain enough food to survive, but their mean mass decreases. Competition in which all organisms obtain less of the scarce resource is called **scramble competition**. In an extreme case, some organisms might obtain so little of a particular resource that they die.

Predation and grazing

Predators and grazers consume all or parts of other organisms. **Predators** eat all or parts of animals, called **prey**, whereas **grazers** consume plants. Predators are usually animals but a few plants, such as the Venus flytrap, can also be regarded as predators.

15 Is there any fundamental ecological difference between a predator and a grazer? Justify your answer.

Predators and grazers are usually less abundant than the organisms on which they feed, for reasons which are discussed in Chapter 29. The organisms which they eat are either killed or their reproductive rate is slowed, so predators and grazers can reduce the population size and growth rate of the organisms they eat. Since the population size of the predators and grazers is affected by the abundance of their food, the populations of consumers and consumed will affect each other. Because of this, it has been predicted that predator and prey populations might be expected to show regular cycles (Fig 28.20(a)). Whether the cycle shown in Fig 28.20(b) results from the effects of predator and prey populations is still a matter of much scientific debate.

16 Suggest how such a cycle might be produced. Think in terms of Fig 28.5, and ignore migration.

Interestingly the Snowshoe hare population tends to fluctuate when there are no lynxes at all, so there are clearly other factors also involved in this particular predator-prey cycle.

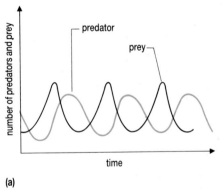

(a)

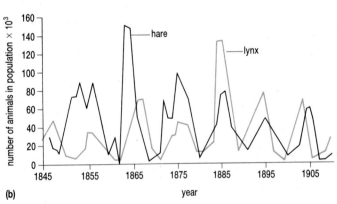

(b)

Fig 28.20 (a) Hypothetical predator – prey cycles generated by a computer running a mathematical model.
(b) Snowshoe hare and lynx cycles in Canada between 1845 and 1905.

Parasitism and symbiosis

Like predators and grazers, parasites feed on the bodies of other living organisms. The parasite is usually smaller than the organism on which it feeds and often lives permanently on the outside (ectoparasite) or inside (endoparasite) of its body (hence the name of **host** for the infected organism). Not all organisms which live on or in the body of a host are parasites, however. If you look at Table 28.2 you will see the ways in which various host organisms interact with organisms (called **symbionts**) which live in or on their bodies. Although symbionts always benefit from the relationship, parasites harm their host.

Table 28.2 Relationships between hosts and organisms which live on or in their bodies (symbionts) showing the effect of the relationship on each organism

Type of association	Effect of association on each organism		Example
	host	symbiont	
mutualism	benefits	benefits	lichens (an alga and fungus)
commensalism	not harmed	benefits	barnacle on belly of whale
parasitism	harmed	benefits	malarial protozoan in human

 17 **Why is it to a parasite's advantage not to kill its host?**

By weakening its host, however, a parasite increases the host's susceptibility to other forms of environmental resistance. For example, a severe winter is more likely to kill parasitised organisms than healthy ones and a predator is more likely to catch a parasitised animal than a healthy one.

Parasites which are introduced into a new habitat can cause devastating effects on their host population. Chestnut blight is a disease caused by a fungus. The fungus occurred naturally in China where it had little effect on the Chinese chestnut trees. In 1895 a number of young trees were imported from China and planted in the New York Zoological Park. At least one of the trees must have been infected, since the fungus spread to the American chestnut. Within 50 years the fungus had wiped out nearly the entire chestnut population of the United States of America. A similar thing happened in Britain with Dutch elm disease which is also caused by a fungus.

There is evidence to suggest that, given time, the relationship between a parasite and its host evolves to minimise the harm to the host. For example, when European settlers introduced the myxomatosis virus into Australia, all rabbits died within a short time of infection. However, a number of years after introduction, infected rabbits were found to survive longer.

 18 **Can you suggest why this happened?**

SPOTLIGHT

Fig 28.21 Biological control is most successful in controlled environments like green-houses. For example, whitefly can be successfully controlled using a parasitic wasp. The black scales are parasitised whitefly larvae.

BIOLOGICAL CONTROL

If you have a garden you know the constant effort needed to maintain the garden the way you planned it. Plants which you do not want to grow (**weeds**) do grow and a variety of parasites and grazers attack your crops. In a small garden these pests can be eliminated by hand but in large agricultural systems this is impractical. A large number of agrochemicals exist which can be sprayed onto crops to deter the growth of competitors, grazers and parasites. Since these chemicals may harm the environment in a number of ways, there is growing world-wide concern about their continued use. An alternative way to control pest populations is to exploit their natural biological enemies, such as predators or parasites. The use of natural enemies to eradicate populations of pests is called **biological control**.

Pesticides: friend or foe?

It is estimated that 45% of annual human food supply is lost to pests such as weeds, insects, slugs, snails, fungal and viral diseases. To feed people adequately such losses must be reduced. One way of doing this is to use pesticides. Some different types of pesticide are listed in Tables 28.3 and 28.4.

Pesticides are extremely effective at killing pests. In addition to protecting crops, DDT and other insecticides have probably saved more human lives, by killing disease-carrying insects like mosquitos (malaria) and lice (typhus), than any other synthetic chemical. Unfortunately pesticides can

- kill the natural enemies of pests; this may reduce the environmental resistance on a minor pest so that its population erupts to pest proportions
- persist in ecosystems and damage wildlife
- damage people: pesticides are poisons
- result in pest populations developing genetic resistance to pesticides, so you have to use more and more to kill them.

Table 28.3   Major insecticides

Chemical group	Examples	Persistence
Chlorinated hydrocarbons	DDT, DDE, DDD, aldrin, dieldrin endrin, heptachlor, toxaphene, lindane, chlordane, kepone, mirex	2–15 years
Organophosphates	malathion, parathion, Azodrin, Phosdrin, methyl parathion, Diazinon, TEPP, DDVP	normally 1–12 weeks, but some can last several years
Carbamates	carbaryl (Sevin), zineb, maneb, Baygon, Zectran, Temik, Matacil	days to weeks
Pyrethroids	pyrethrums extracted from flowers, used directly or modified chemically	days to weeks

Table 28.4   Major herbicides

Mode of action	Examples	Effects
contact	triazine (atrazine)	kill foliage by interferring with photosynthesis
systemic	phenoxy compounds (2,4–D, 2,4,5–T, Silvex), substituted ureas (diuron, norea, and fenuron)	absorption results in excess growth, plants die because they cannot obtain enough nutrients to sustain their greatly accelerated growth
soil sterilant	Treflan, Dymid, Dowpon, Sutan	kill soil micro-organisms essential for plant growth; most also act as systemic herbicides

ANALYSIS

Controlling the alfalfa aphid

This exercise will help you to develop comprehension skills and the ability to apply what you have learnt so far in this chapter to a real pest problem.

Lucerne, or alfalfa, is a crop grown in many parts of the world as fodder for cattle. In 1977 the spotted alfalfa aphid was found in Australian lucerne crops, after which it became a serious pest. The Australian Agricultural Council established a three point programme to eradicate the pest.

1. Short-term chemical control using pesticides to kill the spotted alfalfa aphid.
2. Medium-term introduction of exotic natural enemies of the aphid to achieve biological control of the aphid.
3. Selection of aphid-resistant strains of lucerne plants from which cultivars suited to Australian conditions could be bred.

A cultivar is a variety of plant which, although produced from natural species, is maintained by cultivation.

Within six years the spotted alfalfa aphid had ceased to be a problem on irrigated lucerne crops in many areas of Australia, thanks to an introduced parasite, *Trioxys*.

(a) Why were insecticides considered a short-term but not a long-term solution to this problem?

(b) *Trioxys* did not eliminate the alfalfa aphid: it reduced it to a very low population density. Why would it not be a good idea to eliminate the aphid completely?

(c) Even though it was still present, why did the aphid cease to be a problem?

(d) Why is option 3 considered the best long-term solution?

QUESTIONS

28.9 Table 28.5 shows the results of replicate experiments in which equal numbers of two species of flour beetle were kept in a fixed mass of flour until one species had died out. The replicates were kept at different, fixed temperatures and humidities. Identify the biotic and abiotic factors affecting each species of flour beetle.

Table 28.5

Temperature /°C	RH/%	Number of times each species died out / %	
		Tribolium castaneum	*Tribolium confusum*
34	70	0	100
34	30	90	10
29	70	14	86
29	30	87	13
24	70	69	31
24	30	100	0

28.10 Distinguish between

(a) intraspecific competition and interspecific competition
(b) contest competition and scramble competition.

28.11 Humans have bacteria in the lower part of their intestine (colon) which break down the undigested remains of our food, producing vitamin K. Their excretory products may damage the lining of the colon. Discuss whether this relationship is an example of mutualism, commensalism or parasitism and justify your answer.

28.12 Grain-eating birds are a serious agricultural pest. These birds normally reproduce during the summer but their population numbers fall during the winter months.

(a) Suggest three biotic factors and one abiotic factor which might cause these bird populations to fall during the winter.

(b) Farmers often hunt and kill these birds after their breeding season but before the winter. Assuming that these birds are not migratory, suggest why this pattern of hunting might result in there being a larger breeding population of these birds the following spring than if they had not been hunted.

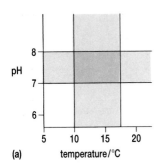

(a)

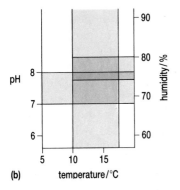

(b)

Fig 28.22 Ecological niches
(a) If two environmental variables are considered (temperature and pH).
(b) If three environmental variables are considered (pH, temperature and humidity).

The ecological niche

Information about habitat requirements can be obtained from a knowledge of the distribution of populations. Much more accurate information about habitat requirements is obtained from laboratory experiments, however. Imagine that a series of experiments has been performed on a plant species in which it was found that the plant could survive only in a temperature range of 10–15 °C and in soils of pH 7.0–8.0. These data could be shown graphically (Fig 28.22).

Since the plant cannot live in any environment where the temperature or pH is outside its tolerance range, the blue area of Fig 28.22(a) shows the type of environment in which populations of this plant can survive. It is possible to add data about another environmental tolerance range to produce a graph with three axes (Fig 28.22(b)). The blue, shaded area in which the plant can grow is defined even further by the addition of these new data. Imagine it were possible to add more data to this diagram with a fourth axis, fifth axis and so on. A graph would be produced which would be multi-dimensional. The space in the middle of this graph represents precisely the environmental conditions needed by this plant species and is a very accurate description of the plant's required habitat. The name given to this imaginary, multi-dimensional space is the **ecological niche** of that plant species.

For most purposes, ecologists restrict their use of the term niche to include only those aspects of the environment which they are investigating. Thus, the use of the terms trophic (feeding) niche, spatial niche and so on. In this way 'niche' describes the role of an organism within an ecosystem, e.g. browser or grazer. A particular trophic niche can be filled by different organisms in different habitats (Table 28.6).

Table 28.6   Different organisms fill roughly the same trophic niche in different ecosystems

	North America	Europe	Africa	Australia
grazer	bison	deer	antelope	kangaroo

 Can you think of four animals which occupy a seed (nut)-eating trophic niche?

Resource partitioning

This chapter has previously described how organisms in a single population compete with each other for any limited environmental resource. Some members of one population have inherited characteristics which make them better able to compete for limited resources (Chapter 23). These organisms tend to have more offspring and a change in allele frequency occurs in that population (the process of natural selection). Since members of one population differ in their ability to compete for limited resources, it is not hard to imagine that members of different species must do so to an even greater extent. Just as with natural selection, it is likely that one species of organism will compete more efficiently and the weaker species will die out. If this is so, species with the same ecological niche will not co-exist. This theory is known as the **competitive exclusion principle**.

Experiments to test the value of this theory have produced conflicting results. If you look back to question 28.9 concerning flour beetles, you will see how even two environmental variables can greatly alter the results of competition experiments. Two sea birds, the cormorant and the shag, are very similar in appearance, nesting sites and breeding times. They would seem to occupy the same ecological niche and yet they are found living together around the rocky coasts of Britain. Whilst they share some items of their diet, Fig 28.23 shows that they avoid competition for food by eating different prey. Subtle differences such as this may explain why many similar species seem to live together in single ecosystems. The division of environmental resources between them is known as **resource partitioning**. The potential niche (**fundamental niche**) of a population is restricted to a **realised niche** by the presence of a competitor.

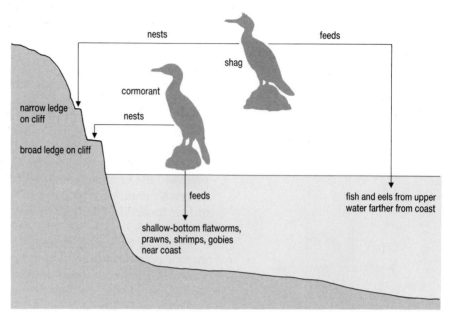

nests

feeds

shag

cormorant

narrow ledge
on cliff

nests

broad ledge on cliff

feeds

fish and eels from upper
water farther from coast

shallow-bottom flatworms,
prawns, shrimps, gobies
near coast

Fig 28.23 The niches of two morphologically similar birds, the shag and cormorant. Notice how differences in diet and nesting habits means that the niches do not overlap.

Similar resource partitioning has been found among herbivores in the New Forest, a mixed woodland of about 37 500 hectares in Hampshire, England. The forest is inhabited throughout the year by ponies. During the spring and summer, farmers also graze their cattle in parts of the forest. The ponies and cattle show preferences for certain areas of the forest so that only about 7% of its area is grazed by them. This restriction might be expected to increase competition between the two species. However, the cattle usually graze areas of over 10 hectares which have been cultivated. Ponies are seldom seen grazing these sites, being more common on roadside verges and streamside lawns. The ponies will also feed on browsing material, such as holly and gorse, which the cattle avoid. All these behaviour patterns will reduce the competition between the species and allow them to co-exist.

QUESTIONS

28.13 Define the term ecological niche.

28.14 Re-read the information about competition between barnacles in Section 28.4. From the limited information given, describe the fundamental niche of *Chthamalus stellatus* and explain how its realised niche is determined.

28.15 What would you expect to happen if the ecological niches of two populations overlapped
(a) at their boundary
(b) so that one niche was entirely within the second, larger one?
Explain your answers.

28.16 A large proportion of chickens sold for human consumption are infected with *Salmonella*, a genus of bacterium that lives harmlessly in the bird's intestine but causes food-poisoning in humans. A new technique has been developed to rid chickens of this bacterium. It involves coating the food of young chicks with a culture of bacteria from the gut of *Salmonella*-free chickens. After this treatment the young chicks cannot be infected with *Salmonella*. Use your knowledge from this chapter to suggest how this technique works.

28.17 It is probable that the human population reached five thousand million during 1987. Look at Fig 28.24 which shows how the human population of the world has grown over the past 10 000 years. It is clearly exponential growth. In this chapter you have learned of a number of ways in which exponential growth of populations is prevented.

(a) Suggest the factors which have allowed the human population to grow exponentially over the past 10 000 years.

(b) Use your library to find out if the rate of population growth is the same in all countries. If not, find out what effect this has on the age structure of the population in these countries.

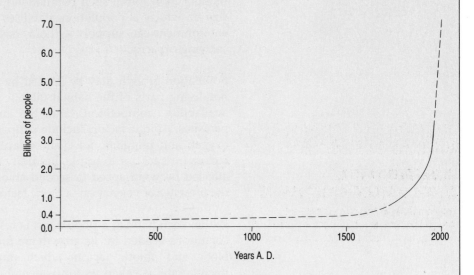

Fig 28.24 The growth of the human population.

28.18 Find out how the following farming activities affect the abiotic and/or biotic factors which affect crop growth:
(a) ploughing, discing and harrowing **(b)** improving soil drainage **(c)** liming **(d)** seed drilling – this ensures that seeds are planted the correct distance apart **(e)** the addition of organic (e.g. manure) and inorganic fertiliser **(f)** the use of pesticide sprays. (Another book in this series, G. Hayward *Applied Ecology*, is a good starting point to find out more.)

SUMMARY

An environment in which organisms live is called a habitat. Organisms of one species live together within a habitat in a potentially interbreeding group are called a population. Populations of different species which share one habitat form a community. A woodland may be regarded as a community because it contains populations of different species living together, or as a habitat, because it forms an environment in which an individual population may live.

Populations within a community interact with each other and with the non-living parts of the habitat. The populations within a habitat, the non-living parts of the habitat and the interactions between them are collectively called an ecosystem.

The number of individuals in a population is termed population size. Changes in population size = (births – deaths) + (immigrants – emigrants). Ignoring immigration and emigration, the fastest rate at which a population could increase in size is called its biotic potential. In an ideal environment, populations grow exponentially. In natural habitats, environmental limits, such as lack of food or space, reduce the rate of growth. These environmental limits to population growth are called environmental resistance. If the limits to population growth vary depending on population size, they are said to be density-dependent; if they are unaffected by population size they are said to be density-independent.

Following a time lag after their introduction into a new environment, population numbers usually increase exponentially before reaching a stable number. A graph of such population growth is called a logistic curve. The size at which a population stabilises is near the maximum which the environment can support without itself being destroyed and is called its carrying capacity.

Population growth may be limited by living organisms (biotic factors) or non-living parts of the habitat (abiotic factors). Biotic factors include food availability, competition, and the presence of predators, grazers or parasites. Abiotic factors include temperature, pH, the availability of water, oxygen and inorganic ions, and edaphic (soil) factors. Biotic and abiotic factors interact, for example, humidity in a woodland (an abiotic factor) is affected by wind speed (a second abiotic factor) and both are affected by the presence of vegetation (a biotic factor).

Within each habitat, a population is restricted to a microhabitat where the conditions needed for its growth are found. An exact description of all the biotic and abiotic factors which must be present in a population's microhabitat is called its fundamental ecological niche. For simplicity, the term ecological niche is often used to describe one aspect of a population's environmental requirements, e.g. feeding or spatial requirements.

Populations with identical ecological niches cannot co-exist because one population always exploits the environment more successfully than the other (this is the competitive exclusion principle). Populations with apparently similar ecological niches do co-exist because subtle resource partitioning reduces competition for one or more environmental requirements. Each population restricts its competitor's fundamental ecological niche to a realised ecological niche.

Chapter 28: Answers to Quick Questions
1 The ground layer.
2 Trees shade the forest floor, reduce the wind speed, increase the relative humidity. Other plants compete for water, nutrients and light. Some animals eat them.
3 A habitat is an organism's address – where it lives. A community is the organisms which share the habitat.

4 **(a)** grow when B + I > D + E
   (b) decline when B + I < D + E
   (c) stay same B + I = D + E
5 A doubling in population size, $2 \rightarrow 4 \rightarrow 8 \rightarrow 16$, for example.
6 The biotic and abiotic factors affecting the sheep remain reasonably constant, while those affecting the great tit vary greatly from year to year.
7 No effect, unless plague of mice reduced territory size of the adult owls. It is competition for space that is important here.
8 Decrease in size and eventually become extinct.
9 Cold affects an ectotherm more. High temperatures affect both.
10 So it has someone to mate with. So all the young are born together, which in a herd or a gull colony would reduce the risk from predators.
11 Avoid it by hiding under boulders or in plants, cling to boulder plants using hooks and suckers, be streamlined.
12 Increase by irrigation; remove by drainage.
13 Plant roots take up nutrients by active transport. This requires ATP. More ATP is made by aerobic respiration than by anaerobic respiration.
14 If they cannot switch to an alternative food when their current supply runs out, their populations crash. If there is plenty of food the population will grow rapidly.
15 No, both consume all or parts of other organisms.
16 At low densities of lynx, mortality of snowshoe hares is less than natality, so the population of hares increases. Lynx now have plenty to eat so their natality increases and the lynx population grows. At high lynx densities, predation of snowshoe hares is high and the hare population declines. Lynx now have less food, their natality decreases and their mortality increases leading to a decline in the lynx population.
17 The longer the host stays alive the more eggs/offspring the parasite can produce.
18 The rabbit population consists of individuals which are resistant to the virus. The myxomatosis virus has become less virulent.
19 Squirrels, mice, birds (e.g. finches), humans, but there are, of course many more.

Chapter **29**

ENERGY FLOW THROUGH ECOSYSTEMS

LEARNING OBJECTIVES

When you have studied this chapter you should be able to:

1. define the terms autotroph, heterotroph, standing crop, biomass, gross primary productivity, net primary productivity, secondary productivity, trophic level;

2. construct food chains and food webs from data about ecosystems;

3. describe how energy is transferred through the trophic levels of a food chain;

4. solve energy budget problems;

5. construct and compare pyramids of numbers, biomass and productivity.

In this and the next chapter, we are going to look at how the living and non-living components of ecosystems interact. Life on Earth depends on two processes

- the flow of energy from the sun through ecosystems.
- the cycling of nutrients in ecosystems

As we will see, all the various components of the living world, the biosphere, are linked by these two fundamental processes.

29.1 ENERGY AND NUTRIENT SUPPLY

Earth receives energy in the form of electromagnetic radiation from the sun. Most of the sun's energy is reflected, absorbed and radiated by the atmosphere (including the ozone layer), by clouds and by the Earth's surface, but about 1% of it is available to the Earth's organisms. Humans cannot make use of this energy directly. If you lie on a beach sunbathing you are burned by the sun, not energised by it. Only organisms which have pigments that trap the sun's energy and couple it to biochemical processes can make direct use of the sun's energy. These organisms are called **autotrophs** (self-feeders) and their photosynthetic activity is summarised in Chapter 7. All other organisms must rely on being able to release the energy from the high-energy molecules made by these autotrophs. They are called **heterotrophs** (other feeders). Thus, energy from the sun is trapped by the photosynthesis of autotrophs and converted into the chemical energy of carbohydrates.

 Eventually, this chemical energy is released as heat. How?

Fig 29.1 Unlike plants, heterotrophs cannot use sunlight energy directly to make food. The main effect is to stimulate the production of melanin against the dangerous UV radiation contained in sunlight.

Photosynthesis only produces carbohydrate but plants need to make other organic molecules, for example proteins and nucleic acids, in order to grow. Such **biosynthesis** requires a supply of nutrients, especially phosphorus, nitrogen and sulphur. These inorganic materials can be taken up by plants from the soil or water in the form of ions, such as nitrate, sulphate and phosphate. However the supply of these materials on Earth is fixed; it is the same now as it was 4.54 billion years ago when the Earth was formed.

 What does this suggest must happen to the nutrients in an ecosystem if the organisms are not to run out of them?

This information is summarised in the context of a simple terrestrial ecosystem in Fig 29.2. Energy from the sun is trapped by autotrophs, in this case plants, and, together with inorganic nutrients, is used to make complex biochemical compounds. The blue arrows show that, whether through death and decomposition or through being eaten, the nutrients are eventually re-used by other plants. The yellow arrows show that the sun's energy is progressively lost as heat.

Two important points emerge from this summary.

1. Energy flows only once through an ecosystem and is eventually lost as heat: nutrients are continually recycled.

2. Since energy can neither be created nor destroyed (the Law of Conservation of Energy), if we know that x kilojoules (kJ) of sunlight energy fall on a square metre of grass in a given time we should be able to work out where it all goes. We can therefore act like energy accountants, tracking the fate of the energy as it flows through an ecosystem. You will practise doing this later.

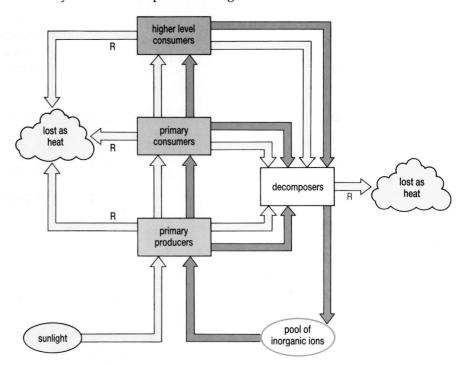

Fig 29.2 Summary of energy flow (yellow arrows) and nutrient cycling (blue arrows) in an ecosystem; R = respiration. Energy from the sun is progressively lost as heat. Nutrients are eventually recycled thanks to the decomposers.

QUESTION

29.1 Look at the pond ecosystem shown in Fig 28.3. Which organisms are **(a)** autotrophs, **(b)** heterotrophs?

29.2 TROPHIC LEVELS

Energy and nutrients pass through ecosystems from autotrophs (**primary producers**) to heterotrophs. Some heterotrophs are animals which feed exclusively on plants; they are called **primary consumers**.

 What other name could you give to a primary consumer?

Other heterotrophs are carnivores which eat animals. Most, termed **secondary consumers**, or first carnivores, eat herbivores: a smaller number, termed **tertiary consumers** (higher carnivores), eat other carnivores. Each of these feeding roles, primary producer, primary consumer, secondary consumer and tertiary consumer, is called a **trophic level**. Two important groups of organisms do not fit neatly into a single trophic level. One of these is the group of animals, including earthworms, which feed on the partly broken down remains (**detritus**) of both animals and plants. These animals, called **detritivores**, are simultaneously primary and secondary consumers. The second group of organisms is the **decomposers**. These organisms, mainly bacteria and fungi, normally digest the dead remains and waste products of every type of organism.

 Which trophic level(s) do you occupy?

ENERGY FLOW THROUGH ECOSYSTEMS

Food chains and food webs

The trophic levels within an ecosystem are linked together. A primary producer may be eaten by a primary consumer, which in turn may be eaten by a secondary consumer, and so on. This feeding sequence is called a **food chain**. Some examples of food chains are given in Fig 29.3.

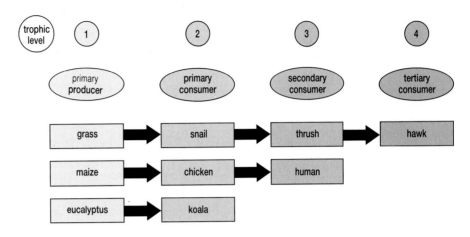

Fig 29.3 Some examples of food chains. The trophic levels are joined by the feeding sequence of animals within the ecosystem.

The first two trophic levels of an Australian food chain might be: eucalyptus – koala. This accurately represents the diet of the koala; it only eats eucalyptus leaves. However, most herbivores eat more than one type of food, as do most carnivores. Thus, in any ecosystem, each population of consumers will be a member of a number of different food chains. The feeding interrelationships in an ecosystem are shown in Fig 29.4. You can find food chains within the diagram but they intersect to produce a complicated network. This more accurate representation of feeding interrelationships is called a **food web**.

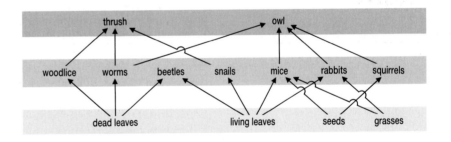

Fig 29.4 A simple woodland food web. Herbivores and carnivores eat more than one food species; thus the food chain becomes a complicated network. The stability of an ecosystem is related to the complexity of its food web.

 5 **If eucalyptus plants were harmed by parasites or fire, what would happen to the koalas?**

The complexity of food webs tends to make them resistant to change. Consider the woodland food web. If the number of rabbits was drastically decreased by a disease such as myxomatosis, the owl population would probably not starve; owls could eat more worms, squirrels and mice. Although at first sight this might suggest that the populations of squirrels and mice would decrease, the absence of competition from rabbits might result in them getting more food and reproducing more. In spite of a change in the rabbit population, the community has not suffered.

ENERGY FLOW THROUGH ECOSYSTEMS

Constructing food chains and webs

In this exercise you have to convert written information into a diagrammatic form.

The following relationships may be observed on a British rocky shore, some when the tide is in, some when the tide is out:

(i)   limpets grazing diatoms (small algae) attached to rocks
(ii)   dogwhelks eating barnacles and mussels
(iii)   crabs consuming dead mussels in crannies in rocks
(iv)   barnacles feeding on zooplankton
(v)   mussels feeding on phytoplankton
(vi)   periwinkles feeding on diatoms attached to seaweeds
(vii)   gulls feeding on dead crabs
(viii) turnstones (a wading bird) feeding on dogwhelks, limpets and periwinkles.

(a) Which of the organisms in (i) to (viii) are herbivores; first carnivores; autotrophs; secondary consumers; tertiary consumers?
(b) Construct two complete food chains from the description given in (i) to (viii) above. Note that you do not need to include all the listed species.
(c) Construct a food web of this rocky shore community.

Bioaccumulation

A consumer can make use of most of the digested food which it absorbs, either to make new cells or as a source of energy. Some of the absorbed food products cannot be used, however, and are usually eliminated fairly quickly. This ability to get rid of unusable food products is vital to consumers, particularly if they are harmful.

Some compounds are not eliminated rapidly and accumulate within an organism's body. The accumulation of harmful compounds within food chains is called **bioaccumulation**. Many of the harmful compounds which accumulate in animal tissues are introduced into the environment by humans. Chlorinated hydrocarbons, such as DDT and dieldrin, were first used to control insect populations during the 1940s and were found to be much more effective than existing insecticides. The use of DDT was considered to be so important to world health that its discoverer was awarded a Nobel prize. Chlorinated hydrocarbons take a long time to break down. As a result a single application of pesticide would continue to kill insects for a long time, which is advantageous. However, it also means that they would remain in ecosystems for a long time. Animals which could not excrete them stored these compounds in their fat. It became clear during the late 1960s and early 1970s that compounds such as DDT and dieldrin were being accumulated in food webs and were causing clinical poisoning in tertiary consumers. Tertiary consumers in most food webs were affected (Fig 29.5).

 Why do you think it is mainly tertiary consumers which are affected?

In Britain, concern was expressed about the populations of predatory birds, already rare, whose numbers declined from the time of introduction of these insecticides. High concentrations of DDT, DDE (an intermediate in the breakdown of DDT) and dieldrin were found in the fat, muscle and liver of these birds. Whilst these concentrations were not lethal, they did affect the bird's ability to produce a thick shell around its eggs.

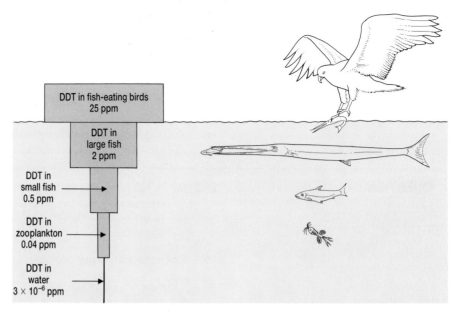

DDT in fish-eating birds
25 ppm

DDT in
large fish
2 ppm

DDT in
small fish
0.5 ppm

DDT in
zooplankton
0.04 ppm

DDT in
water
3×10^{-6} ppm

Fig 29.5 The bioaccumulation of DDT in the fatty tissue of organisms in an estuary. The harmful pesticide has become concentrated as it passed through the food chain. The tertiary consumers, the fish-eating birds, are most affected.

7 **Why would this inability to produce thick egg shells reduce the population size of these birds?**

In the USA, the use of DDT and dieldrin was banned in 1973. Similar bans were imposed by most European countries shortly afterwards. Any new pesticide or herbicide must undergo strict tests to demonstrate that it will not be accumulated in food chains before it is granted a government licence in these countries. However, in many developing countries the use of chemicals like DDT and dieldrin continues. It is argued that in these areas the cost in loss of human lives through insect-transmitted diseases, such as malaria and African sleeping sickness, or starvation caused by loss of crops devastated by insect pests, outweighs the environmental objections to their use. This argument may be valid but concern about the use of these chemicals continues. This concern has been heightened by the discovery that food webs in Antarctica, an ice-bound area which is neither cultivated nor permanently populated, accumulate toxic chemicals.

ANALYSIS

Agricultural ecosystems

This exercise is designed to encourage you to apply knowledge you already have to a new situation.

In order to feed our own population, humans have destroyed diverse ecosystems which contain many species and have complex food webs. Having cleared the ground we plant our crops. The use of large harvesting machines in many countries is only efficient if there are large areas with a single species of crop plant, i.e. a **monoculture**.
(a) Monocultures are more susceptible to outbreaks of pests, e.g. weeds and aphids, which cause economic damage, than diverse ecosystems. Can you suggest why?
(b) Why must so much energy be used in the manufacture of pesticides and herbicides and in farm machinery to ensure the success of crops grown in monoculture?
(c) Suggest some long-term disadvantages of using pesticides.

29.3 PRIMARY PRODUCTION

Think about an ecosystem, say the pond shown in Fig 28.3. The organisms in it all need a supply of energy to survive, grow and reproduce. In our pond the plants, fish, insects and birds are all linked by their common need for energy. In the pond, energy enters the ecosystem via the compounds made by autotrophs. The autotrophs are the primary producers in our ecosystem. There are some important definitions you need to know.

- The **primary productivity** of an ecosystem is the rate at which biomass is produced per unit area per unit time by plants, the primary producers.
- **Biomass** is the mass of organisms per unit area of ground (or water) and is usually expressed either in units of dry mass (e.g. tonnes ha^{-1}) or in units of energy (e.g. J m^{-2}).
- **Standing crop** is made up of the bodies of the living organisms within a unit area. So in our pond ecosystem we could talk about the standing crop of primary producers, of water boatmen, or of   biomass.
- **Secondary productivity** is the rate of production of biomass by heterotrophs.

Gross and net primary production

At the equator, sunlight energy reaches the Earth's upper atmosphere at an intensity of 1.4 kJ m^{-2} s^{-1} (Fig 29.6). This practically constant energy input, called the **solar flux**, is roughly equivalent to the energy required to run about one and a half bars of a two-bar electric fire. Over the course of an entire year the solar flux amounts to 5.434×10^{24} J, enough sunlight energy to produce millions and millions of tonnes of carbohydrate. Surprisingly though only about 0.5–1% of this sunlight energy is actually used in primary production.

ENERGY FLOW THROUGH ECOSYSTEMS

Look at Fig 29.6. Why are you more likely to become sun burnt at the equator than in Britain?

Much of the solar flux never actually reaches the Earth's surface (Fig 29.7).

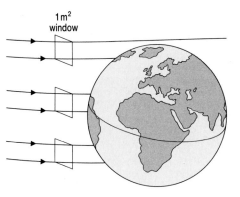

Fig 29.6 Solar energy or solar flux. Every second 1.4 kJ of energy pass through each 1 m² window in the upper atmosphere above the equator. Note: because of the curvature of the Earth the radiation is more intense at the equator than at the poles.

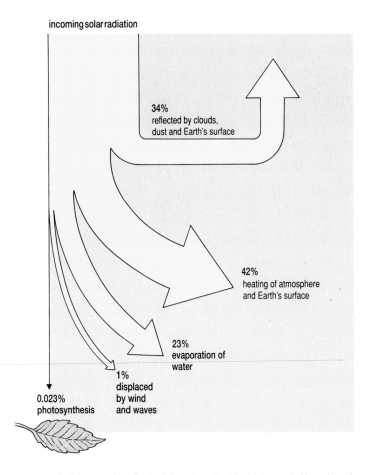

Fig 29.7 The fate of solar flux reaching the Earth. Less than 1% of the total amount of incoming solar radiation is used in photosynthesis.

 What happens to the sunlight which does reach the ground?

The fate of the solar flux which hits a leaf is shown in Fig 29.8. Notice how only a tiny proportion of the solar flux is actually converted into the chemical energy of newly synthesised carbohydrate. This captured energy is called the **gross primary production (GPP)**, which is defined as the total amount of energy fixed by photosynthesis. It can be measured either in units of energy (e.g. $J\ m^{-2}\ day^{-1}$) or of dry organic mass (e.g. $kg\ ha^{-1}\ yr^{-1}$).

As Fig 29.8 shows, a proportion of the GPP will be respired by the plant itself and will be lost as **respiratory heat (R)**. The difference between GPP and R is the **net primary production (NPP)**. This represents the rate of production of new biomass available for consumption by heterotrophic organisms (bacteria, fungi and animals).

An ecologically important equation
GPP = NPP + R.

Measuring primary productivity

The amount of solar energy trapped by primary producers, and then made available to heterotrophs as NPP, is an important factor in determining the amount of biomass that a particular ecosystem can support. So it is important that ecologists measure GPP, NPP and R if they are to follow the fate of solar energy as it flows through the ecosystem.

An early attempt to do this was the work of Edgar Transeau on maize in 1926. His work is summarised in the Analysis below which you should work through before continuing.

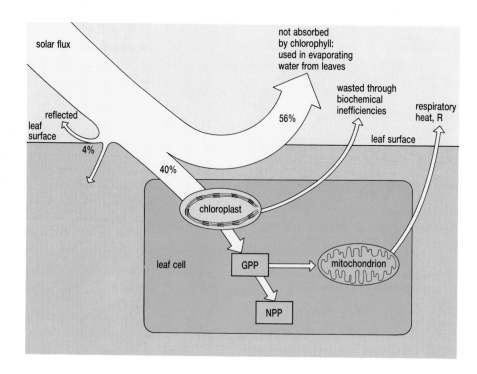

Fig 29.8 The fate of solar flux striking a leaf. GPP = Gross primary production, NPP = Net primary production. Of the solar radiation striking the leaf surface, only 40% reaches the chloroplast and contributes towards primary production.

ANALYSIS

Estimating primary production

This is an exercise in numerical analysis.

Transeau estimated that one hectare (10 000m²) of his maize field contained 25 000 plants. He dried and weighed several plants at the end of the growing season and so estimated the biomass of his plants as 14 800 kg ha^{-1}. This standing crop of plants was produced from the seed in one growing season so it represents the net primary production. Chemical analysis of the maize plants revealed that 14 800 kg of the plant tissue was equivalent to 6610 kg carbon.

(a) If all this carbon were fixed by photosynthesis, why is it equivalent to 16 524 kg of glucose?

Transeau then measured the respiration rate of maize plants during the night and so estimated the amount of glucose which had been used in respiration and which he had, therefore, not been able to weigh at the end of the growing season. This gave him a value for R of 5053 kg glucose ha^{-1} yr^{-1}.

(b) Why did Transeau have to estimate the respiration rate of maize plants during the night?

ENERGY FLOW THROUGH ECOSYSTEMS

Since 1 kg glucose releases 15 790 kJ of heat energy when it is burnt you are now in a position to balance the energy budget for this field:

$$NPP = 16\ 524 \text{ kg glucose ha}^{-1}\text{yr}^{-1} = 2.6 \times 10^8 \text{ kJ ha}^{-1}\text{yr}^{-1}$$
$$= 26\ 000 \text{ kJ m}^{-2}\text{yr}^{-1}$$

$$R \quad = 5053 \text{ kg glucose ha}^{-1}\text{yr}^{-1} = 8.0 \times 10^7 \text{ kJ ha}^{-1}\text{yr}^{-1}$$
$$= 8000 \text{ kJ m}^{-2}\text{yr}^{-1}$$

(c) Calculate the GPP for this maize field.

(d) Transeau estimated that 2×10^6 kJ sunlight energy fell on each square metre of his maize field during the growing season. Calculate the percentage of this sunlight energy which is represented by GPP, NPP and R of the maize crop.

(e) Why is so little of the sunlight energy converted into NPP?

Modern methods for estimating primary productivity are still basically the same as those used by Transeau. For example, to estimate the productivity of a field of grass you would do the following.

1. Collect all the plant material, including the roots, in a given area, say 1 m².
2. Dry the plant material in an oven at 105 °C until it reaches constant mass.
3. Weigh the dried plant material. This is the standing crop in kg m⁻².
4. Repeat this procedure say one month later. The difference between the two values represents the NPP though you would need to make allowances for the plant material which was eaten or that had died and dropped off and which you could not therefore collect.

(f) Why do you need to dry the plant material before you weigh it?

(g) What would you need to find out to express the standing crop in kJ m⁻²?

(h) What losses of plant material could occur which would cause you to underestimate NPP?

(i) What do you need to measure before you can estimate GPP? Suggest how you might do this.

The fate of NPP

Since only net primary production is available for consumption by heterotrophs, NPP values are usually used to compare the productivity of different ecosystems. Mean values for the NPP of some ecosystems are shown in Table 29.1.

Table 29.1   Mean NPP in different ecosystems

Ecosystem	NPP/g m⁻² yr⁻¹
desert	70
open ocean	125
upwelling zones of seas	500
temperate grassland	600
temperate deciduous forest	1200
estuaries	1500
tropical rain forest	2200
algal beds and reefs	2500

There are two routes for the fate of NPP. It can remain in the plant itself and will then be measurable as an increase in plant biomass. Alternatively it can enter the world of heterotrophs either by being eaten directly, like grass being grazed by cows, or by dying, falling off the plant and so becoming food for detritivores and decomposers. We will follow these heterotrophic routes in more detail in the next section

PUTTING PLANTS TO WORK

Humans use NPP in many ways in addition to eating it!

- As fuel – coal and oil represent the NPP of plants which lived millions of years ago. In Brazil almost two million cars are powered not by petrol but by ethanol (alcohol) made by fermenting the juice of sugar cane.
- As clothing – cotton shirts are the processed NPP of cotton plants whilst wool and leather represent part of the secondary productivity of sheep and cows made after consuming NPP.
- For writing and packaging – paper and cardboard are the NPP of trees.
- To make a range of products – plastics manufacture uses oil as a starting point; car tyres are made out of rubber, part of the NPP of rubber trees; ropes are made of fibres from hemp plants.

The list is endless. Clearly the debt we owe plants is enormous. We simply could not survive without this renewable resource.

Fig 29.9 Coppicing is an ancient method of managing woodlands. The central trunk of the tree is removed encouraging the growth of many side shoots which can be used for fuel, to make charcoal and to construct fencing, i.e. NPP is being converted into useful products.

QUESTIONS

29.4 **(a)** In what units could you express primary and secondary productivity?
(b) Classify the following organisms as primary or secondary producers
(i) cabbages (ii) chickens (iii) mushrooms (iv) earthworms
(v) cyanobacteria (blue-green bacteria).

29.5 Suggest four reasons why the NPP of evergreen tropical rain forest is greater than that of temperate deciduous forest.

29.6 Provide an ecological explanation of the old saying 'All flesh is grass'.

29.4 ENERGY BUDGETS

In this section we need to consider what happens to the energy bound up in NPP after it has been eaten by heterotrophs. Look at Fig 29.10. This shows the fate of the energy bound up in grass after it has been eaten by a cow. Some of the grass will pass through the cow and be egested as faeces, so this is the fate of at least some of the energy. The grass which is digested will be assimilated by the cow. Three things can happen to the energy contained in the assimilated grass.

1. It can be used for respiration, in which case it will be lost as heat.

2. It can be used to produce more cow, i.e. it becomes secondary production.
3. Some of the assimilated energy will be lost in the cow's urine.

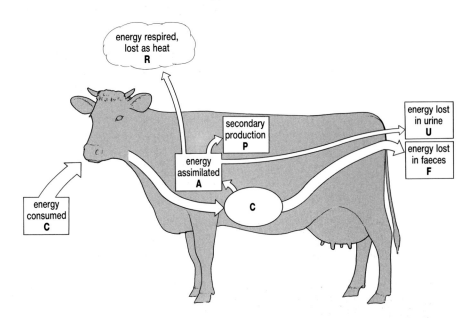

Fig 29.10 The fate of energy in the grass consumed by a cow. Some of the energy consumed is lost in urine, faeces and respiratory heat. The remaining energy is termed the production of the trophic level.

We can write an equation which accounts for the fate of all the energy consumed by the cow

$$\text{Energy consumed} = \text{energy used in respiration} + \text{secondary production} + \text{energy lost in urine} + \text{energy lost in faeces}$$

$$C = R + P + U + F$$

10 **Why must this equation be correct?**

Now imagine that, rather than just looking at the fate of the energy entering one consumer, we were to look at the fate of all the energy entering a trophic level. As Fig 29.11 shows the above equation would still apply. All the energy entering the trophic level (C) must be accounted for in terms of R, P, F and U.

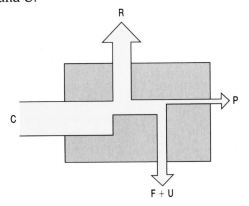

Fig 29.11 The pattern of energy flow through a trophic level.
C = energy entering the trophic level(consumed)
R = respiratory heat loss
F = energy lost in faeces
U = energy lost in urine (usually very small)
P = production of the trophic level

We could expand Fig 29.11 to show the fate of the energy as it flows through an entire food chain. Faeces and urine do not lie about: they are decomposed and their energy enters the decomposer trophic level. Similarly, not all herbivores are eaten by predators. Some die and their remains again become food for decomposers. Fig 29.12, which is still not a full representation since it does not, for example, show consumption of detritivores by carnivores, does emphasise that all the energy entering an ecosystem can be accounted for: it will either be lost during respiration or stored. This key concept rests on three crucial assumptions.

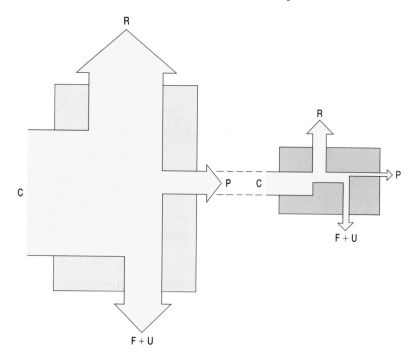

Fig 29.12 Energy flow through the trophic levels of a food chain. As food is eaten, energy flows through the food chain from primary producers to primary consumers to secondary consumers to tertiary consumers and from all trophic levels to the decomposers.

Assumption 1

The biomass of heterotrophs remains the same, on average, from year to year.

 Is this true for the human population?

Assumption 2

The decomposers (bacteria and fungi) which are the ultimate consumers of all dead organic matter break down all the material they receive each year, i.e. there is no accumulation of dead organic material in the soil or the sediment at the bottom of lakes. This is simply not the case. In peat bogs, for example, the cold, wet, acidic conditions mean that the rate of decomposition is slowed down to such an extent that only about 10% of the material entering the decomposer food chain is broken down each year. The rest accumulates as peat.

 What geological evidence is there that organic matter must also have accumulated in the past?

Assumption 3

The ecosystem is a closed one. This means that all the energy entering the food web results from the activities of the primary producers in the ecosystem. Again, if you think about a real ecosystem like a lake you will soon realise this is not the case. In addition to the primary production of its own plants the lake will receive additional inputs of organic matter from, for example

- leaves from the trees surrounding it
- bait from anglers
- faeces from animals coming to drink its water.

Such additional inputs to aquatic ecosystems are described as being **allochthonous** whilst inputs from its own primary producers are described as being **autochthonous**.

 Suggest why in fast flowing streams allochthonous material is a far more important energy source than autochthonous material.

Another consideration for assumption 3 is that ecosystems can lose energy. For example, primary producers could be swept out of the lake or birds could consume the fish as they passed through on their migration.

ANALYSIS

Balancing the books

This is an exercise requiring the skills of data analysis.

The Law of Conservation of Energy enables us to act like energy accountants. If we know how much energy goes into an ecosystem, we should be able to work out where it all goes. If at the end of the day the books don't balance, then

- either we have mislaid some energy
- or the ecosystem is an open one, i.e. it is either gaining energy from somewhere else or exporting it.

Some students obtained the data given in Table 29.2 for a large, freshwater stream in Britain.

Table 29.2 Ecology data measured for a large freshwater stream, all values are in units $kJ\ m^{-2}\ yr^{-1}$

	GPP	R	NPP
plants	90 000	55 000	(x)
	Assimilation	Respiration	Production
herbivores	12 000	10 300	1 700
first carnivores	1 500	1 300	200
higher carnivores	100	60	40
decomposers	20 000	18 000	2 000

(a) Calculate the missing value (x).
    (i) Compare the efficiency of secondary production by herbivores and first carnivores. Account for any difference that you calculate.
    (ii) Assume that total biomass is constant and that the total amount of detritus remains the same. Is this ecosystem self-contained as far as energy is concerned?
(b) Look at Fig 29.13 summarising the energy flow through this ecosystem, in which boxes V to Z represent the trophic levels involved. By reference to Table 29.2, identify each box.

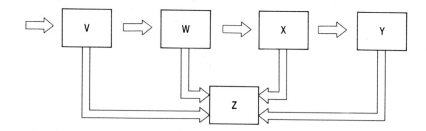

Fig 29.13 A freshwater stream food chain.

29.7 Imagine a small, isolated, uninhabited tropical island, covered with forest. Assume that the total amount of plant, animal and dead organic material remains the same from year to year. What happens to the annual NPP?

29.8 For which aquatic heterotrophs, in particular, will allochthonous material represent a major food source?

29.9 **(a)** Write an equation to show the relationship between assimilation (A), respiration (R) and production (P) for a heterotroph.
(b) From the productivity data in Table 29.3 calculate the missing values (v), (w) and (x).

Table 29.3 **Productivity data**

Heterotroph population	Assimilation / kJ m^{-2} yr^{-1}	Respiration / kJ m^{-2} yr^{-1}	Production /kJ m^{-2} yr^{-1}
1	(v)	65	5
2	132	(w)	12
3	24	16	(x)

29.10 (a) Using the data given in Fig 29.14 calculate the percentage of the energy which would be available for use by humans.
(b) (i) Why would the efficiency of secondary production be greater in an ectotherm, such as a fish, than in an endotherm, such as a bullock? (ii) What are the implications of this for farming animals for meat?

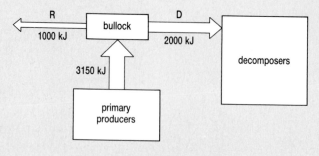

Fig 29.14 Energy flow through a bullock.

29.5 ECOLOGICAL OR TROPHIC PYRAMIDS

If you compare two cars you might consider the size of their engines, their colour, the number of passengers they carry. But how do you compare two food webs? In what sense can trees and phytoplankton, or herrings and owls, be comparable? Clearly, some simplifications need to be considered to compare ecosystems.

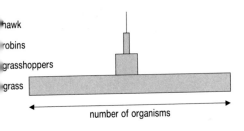

Fig 29.15 A pyramid of numbers represents the numbers of individual plants and animals present in the food web. It does not take into account the relative size of any organisms.

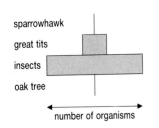

Fig 29.16 An inverted pyramid of numbers. As in a pyramid of numbers, the numbers of individuals have been counted. The primary producer is one oak tree – hence the inverted pyramid.

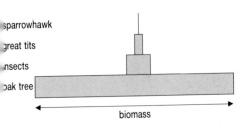

Fig 29.17 A pyramid of biomass for a woodland food web. Biomass is a measurement of body size (dry mass/g m⁻²). One oak tree is very large.

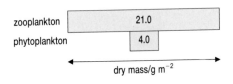

zooplankton | 21.0
phytoplankton | 4.0

dry mass/g m⁻²

Fig 29.18 An inverted pyramid of biomass for the English Channel. A small biomass of phytoplankton when measured at a given time supports a large biomass of zooplankton. In reality, the phytoplankton is growing very fast and its productivity, or amount of energy it can pass on to the next trophic level, is very high.

Pyramids of numbers

One way to simplify food webs might be to consider not the organisms themselves, but what they do in the ecosystem. One way to do this would be to assign organisms to trophic levels and then to count the number of organisms in each trophic level. If you do this for a grassland food web then you arrive at a picture like the one in Fig 29.15.

It is tempting to think that Fig 29.15 represents a general rule for all ecosystems: you might have a greater number of primary producers than herbivores, a greater number of herbivores than carnivores and so on. But this view of the world does not take into account the relative size of producers and consumers.

Pyramids of biomass

Look at Fig 29.16. Here we have an **inverted pyramid of numbers**: one oak tree supports many herbivores. A pyramid of biomass takes body size into account. Such a pyramid for our oak tree is shown in Fig 29.17. This now seems to make sense but it requires much more data to construct, you have to find the dry mass of the organisms in your sample and then their energy content. Nonetheless it does seem to be logical: big organisms support lots of little ones.

 14 Identify one other inverted pyramid of numbers which would give you an upright pyramid of biomass?

Pyramids of energy

The apparent rule of nature, that a large biomass supports a small biomass, is broken by the pyramid shown in Fig 29.18. Here a huge biomass of zooplankton seems to be supported by a much smaller biomass of phytoplankton. How can we resolve this apparent paradox?

To produce a pyramid like Fig 29.18 we would need to collect a sample of sea water and establish the standing crop of phyto- and zooplankton in it. But we would be just taking a snapshot, not taking into account how fast the standing crop of phytoplankton is producing more phytoplankton. This is a difficult, but vital, ecological concept to grasp. Here is an analogy.

Rather than considering phytoplankton, consider a lawn. Throughout the summer we mow the lawn twice a week and collect all the grass clippings. If we observe the lawn each week, just after it has been mown, it appears that the standing crop is not very large. But if we look at the pile of clippings we appreciate how productive the lawn is. The same is true of the phytoplankton. Its standing crop may not be high but its productivity, equivalent to the pile of grass clippings, is enormous. The zooplankton are not being supported by the standing crop of phytoplankton we measure at the moment when we take our single sample but by their productivity over time. So what we need is a **pyramid of energy** or **productivity** (Fig 29.19).

15 What would happen to the pyramid of energy if the rate of reproduction of the phytoplankton was slowed down? Suggest three factors which could cause such a reduction in phytoplankton reproduction.

Pyramids of energy require much more data to construct since many samples, separated by intervals of time, are needed to estimate productivity, but they provide a coherent view of how an ecosystem works. Production by primary producers must be greater than secondary producers and so on. Here energy is being used in a new and interesting way: as a common ecological currency. We do not need to consider numbers or differences in size because all organisms can be converted to their energy equivalent.

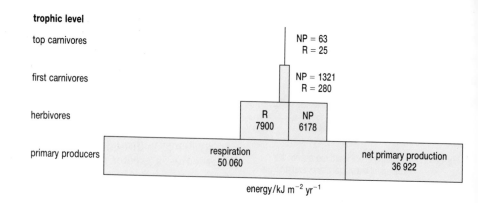

top carnivores   NP = 63
R = 25

first carnivores   NP = 1321
R = 280

herbivores   R 7900   NP 6178

primary producers   respiration 50 060   net primary production 36 922

energy/kJ m^{-2} yr^{-1}

Fig 29.19   A pyramid of energy for a pond ecosystem, Silver Springs, Florida, in kJ m^{-2}yr^{-1}. The energy of each trophic level of any ecosystem has two parts: NP (net production) and R (respiration).

SPOTLIGHT

SHOULD WE ALL BECOME VEGETARIANS?

The loss of energy in food webs, shown by ecological pyramids, has important implications for human food production. Since more energy is available to primary consumers than to secondary consumers, more energy will be available to humans if we eat plants than if we eat animals. A vegetarian diet will support many more people than a diet of meat (Fig 29.20), and in many of the poorer developing countries meat is seldom eaten. This does not mean that all animal husbandry is wasteful of resources, however. For example, in the upland areas of northern England, Scotland and Wales, the soil and climate will not support a crop plant which humans can eat. In these areas it makes ecological sense to raise animals, such as deer and sheep, which can tolerate the poor quality vegetation growing there, and then to eat them or their products.

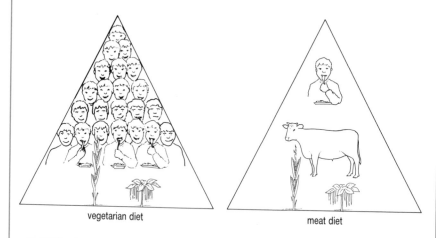

vegetarian diet   meat diet

Fig 29.20   As a result of the loss of energy at each trophic level, over 20 times the number of people can be adequately fed on a vegetarian diet than on a diet of meat.

The length of food chains

Trophic pyramids also pose one more fundamental question – what determines the length of food chains? There is an enormous range of food chains in the world but it is rare to find food chains with more than about five trophic levels – sharks are not eaten by super sharks. Why should this be so? The answer is that no-one really knows. One idea is based on the so-called '10% rule'. This theory suggests that, since only about one tenth of the energy consumed by one trophic level is available for consumption by the next, the food chain simply runs out of energy. By the fifth trophic level, there is not enough energy left to support another trophic level.

 Where does the missing 90% of the energy go?

This idea can be put into an experimental context. If it is true then we would expect to find the longest food chains in the most productive environments and the shortest food chains in the least productive. For example, if you look at the data in Table 29.1 you might expect the food chains in upwellings, areas of the sea which have very high NPP, to be longer than those in the open ocean. This is not the case, oceanic food chains are longer than those in upwellings! So the inefficiency with which energy is transferred from one trophic level to another is, at the most, only part of the answer. This aspect of the ecology of ecosystems still awaits resolution.

ANALYSIS

Energy flow in a freshwater ecosystem
This exercise involves the interpretation of diagrammatic information.

The data in Fig 29.21 were obtained from ecological studies on a 4 km stretch of the River Thames just outside Reading.

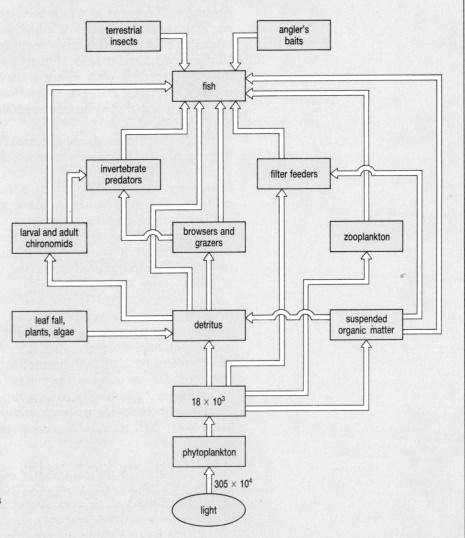

Fig 29.21 A simplified energy flow diagram for organisms living in the River Thames near Reading. The figures represent kJ m^{-2} yr^{-1}.

(a) How many trophic levels are present in the diagram?

(b) To what extent do the energy transfers in the diagram support the 10% rule? (Calculate the energy entering the final trophic level as % of light energy and compare with value expected if 10% rule applied.)

(c) Identify two food chains from the diagram.

(d) Can productivity pyramids be drawn for these food chains?

QUESTION

29.11 Construct a table which compares the advantages and disadvantages of using pyramids of numbers, biomass and energy to compare ecosystems.

SUMMARY

Ecosystems are fuelled by the continuous flow of energy from sunlight and the constant recycling of inorganic nutrients.

Energy from sunlight enters a biological community when it is trapped in organic compounds by autotrophs during photosynthesis. These organisms, also known as primary producers, may be eaten by primary consumers which, in turn may be eaten by secondary consumers and so on. Each of these feeding roles is called a trophic level.

Feeding relationships in which each trophic level is represented by only one group of organisms are called food chains. In most natural ecosystems, feeding relationships are much more complex and are called food webs. Human agricultural practice in many parts of the world attempts to replace diverse ecosystems with simple food chains. Since a food web is more resistant to change than is a simple food chain, the maintenance of these agricultural systems requires the input of a great deal of energy.

Persistent toxic chemicals are accumulated by organisms in food webs such that secondary and tertiary consumers may be poisoned.

The energy from sunlight which is trapped in the organic compounds made by primary producers during photosynthesis is called the gross primary productivity of an ecosystem. Some of these compounds are used by producers during respiration so that only a proportion of the trapped energy, the net primary productivity, is held in their tissues. The energy content of a sample of producers is the standing crop of those producers at that time.

In general, only a small proportion of the net primary productivity is transferred to primary consumers and then from primary consumers to subsequent trophic levels when they feed. As a result, producers are usually more abundant than primary consumers, which are more abundant than secondary consumers. The storage of energy in each trophic level can be represented graphically as a pyramid of energy and the number of organisms in each trophic level as a pyramid of numbers. Pyramids of biomass can be constructed to show the dry mass of organisms in each trophic level.

Decomposers digest the dead bodies and wastes of organisms. In doing so they capture some of the energy lost during each transfer between trophic levels and free inorganic nutrients for recycling.

Chapter 29: Answers to Quick Questions

1 Respiration
2 Nutrients, unlike energy, must be continually recycled in an ecosystem.
3 Herbivore
4 Primary consumer when you eat plant food. Secondary consumer when you eat animal flesh (meat), milk and eggs. Tertiary consumer when you eat some fish (e.g. pike, shark), and squid.
5 The koalas would die out.
6 They consume animals which have already concentrated the pesticides in their own bodies.
7 Eggs would break so reducing the birth rate.
8 Sunlight is more intense at the equator.
9 Absorbed by the soil or reflected back to space by rocks, water and ice. It also evaporates water.
10 Because of the Law of Conservation of Energy.
11 It depends on the country, but worldwide the answer is no.
12 Coal, oil
13 Fast flowing streams have very few primary producers. It is a difficult habitat for plants to root and grow in.
14 A single host with many symbionts.
15 Biomass of primary producers would decline as zooplankton grazed standing crop. Then the biomass of the zooplankton would decline. Reduction in the availability of light, nutrients, fall in temperature.
16 Into faeces, urine and respiratory heat losses.

Chapter 30

ECOSYSTEMS AND NUTRIENT CYCLES

LEARNING OBJECTIVES

When you have studied this chapter you should be able to:

1. describe a general outline of a biogeochemical (nutrient) cycle;

2. outline the water, carbon, nitrogen and phosphorus cycles;

3. explain how humans interfere with natural biogeochemical cycles;

4. define the term pollution and discuss how human activities pollute natural ecosystems.

Photosynthesis produces carbohydrate but plants cannot live on carbohydrate alone. They have to synthesise a whole host of other compounds: proteins, nucleic acids, chlorophyll and so on. To manufacture these compounds from carbohydrate, plants need access to additional macronutrients, for example nitrogen, phosphorus and sulphur, in a form which they can absorb and use. In addition, they also need micronutrients such as zinc, molybdenum and copper which play important roles as activators of enzymes (see Chapter 5). Unlike sunlight which is available in limitless supplies from outside the Earth, there is no external source of nutrients, which means that nutrients must be recycled and used again and again. This idea is encapsulated in the familiar song:

> 'Where 'as t'a been since I saw thee
> On Ilkley moor baht 'at?'
> 'I've been a courtin' Mary Jane'
> 'Tha's goin' to catch thee death of cold.
> Then we shall have to bury thee.
> Then t'worms will come and eat thee up.
> Then t'ducks will come and eat up t'worms.
> Then we shall come and eat up t'ducks.
> Then we shall all have eaten thee.'

30.1 A GENERAL MODEL OF A NUTRIENT CYCLE

Elements which are important to organisms cycle between living (biotic) and non-living (abiotic) phases (Fig 30.1). Some, like the carbon, water (hydrological) and nitrogen cycles, operate on a global scale. Other cycles, like the phosphorus cycle which does not involve the atmosphere, are more localised: see Fig 30.2 which represents a generalised version of such a cycle. Note the following three points.

1. There are two linked cycles: one in the sea and one on land.
2. The cycles are leaky. Nutrients from the land leak into the sea (that is why it is salty) as dissolved ions are carried down through the soil (a

process called **leaching**), washed into rivers and so transported out to sea. The marine cycle loses nutrients through the processes of **sedimentation**. Materials carried into the sea by rivers can settle out directly, forming marine sediments which eventually turn into rock. In addition, nutrients can fall to the bottom of the sea locked up in the skeletons and remains of plants and animals. Again these can be turned into rock so removing the nutrient from the cycle.

3. The loss of nutrients from the land is made good by the weathering of rocks to form new soil. Note that both rock formation and weathering are slow processes. This will be important later on when we look at the impact of humans on ecosystems.

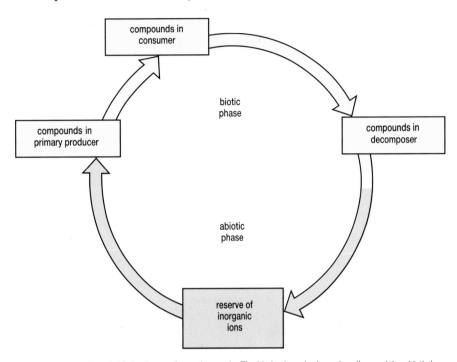

Fig 30.1 The biotic and abiotic phases of a nutrient cycle. The biotic phase is shown in yellow and the abiotic in blue. Note the same colour convention is used in other figures.

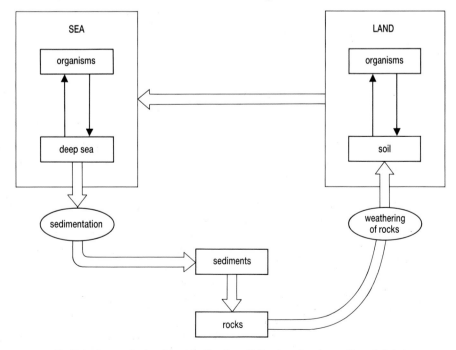

Fig 30.2 A generalised nutrient cycle. Transfers with the atmosphere have not been included.

This chapter summarises four cycles – the carbon, nitrogen, phosphorus and water (hydrological) cycle. In each case you will be presented with a diagram which contains a large amount of information.

ANALYSIS

The carbon cycle

This exercise requires you to interpret diagrammatic information.

Look at Fig 30.3 carefully then answer the following questions.

(a) How does carbon enter the biotic phase of the cycle?
(b) In what form does carbon move between the compartments in the biotic phase of the cycle?
(c) How, and in what forms, can carbon leave the biotic phase of the cycle?

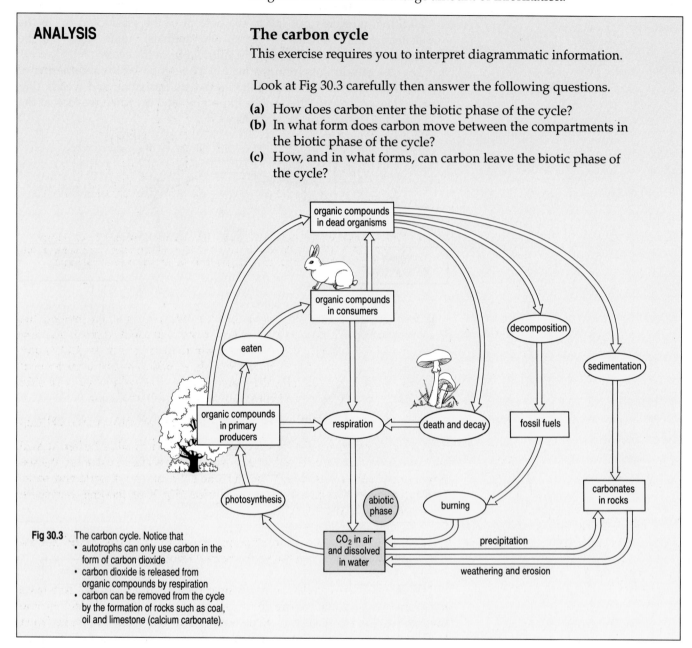

Fig 30.3 The carbon cycle. Notice that
- autotrophs can only use carbon in the form of carbon dioxide
- carbon dioxide is released from organic compounds by respiration
- carbon can be removed from the cycle by the formation of rocks such as coal, oil and limestone (calcium carbonate).

The greenhouse effect

Two human activities affect the carbon cycle:

- burning of fossil fuels, which annually releases about as much carbon dioxide as 5% of the respiration of all animals
- burning and ploughing of tropical rainforests.

At present some six million hectares of forest are cleared each year. Burning the trees obviously releases carbon dioxide. However, more important is the ploughing of the soil after the trees have been cleared. This stimulates the activity of soil micro-organisms which rapidly break down the organic material in the soil releasing CO_2 in the process. The resulting increase in the concentration of atmospheric CO_2 may lead to the so-called greenhouse effect (Fig 30.4).

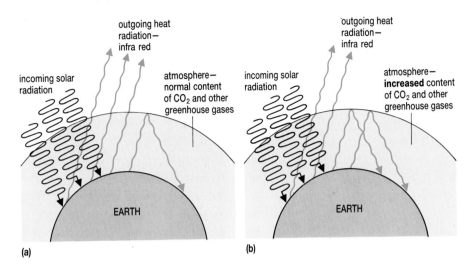

Fig 30.4 The greenhouse effect **(a)** carbon dioxide and other gases trap heat so warming the Earth, a process essential to life on this planet. **(b)** Increasing the level of CO_2 in the atmosphere alters the heat balance by reducing the loss of heat to space, resulting in an increase in the temperature of the Earth's atmosphere.

It may be that ecological balances come into play and so prevent the Earth heating up. There are three possible ways this could happen. The first is that carbon dioxide dissolves in the ocean to form carbonate (CO_3^{2-}) and hydrogencarbonate (HCO_3^-) ions. Look at the equation below. If the level of carbon dioxide in the air falls, the reaction in the equation below will shift to the left, maintaining equilibrium and releasing CO_2 to the air.

$$CO_2\,(g) + H_2O(l) \rightleftharpoons HCO_3^-\,(aq) + H^+(aq) \rightleftharpoons CO_3^{2-}\,(aq) + 2H^+(aq)$$

This means that the ocean is an enormous carbon dioxide buffer. It will soak up extra CO_2 or release it if atmospheric levels fall. However, the sea now appears to be less able to mop up the extra carbon dioxide and so its concentration in the atmosphere is increasing (Fig 30.5), leading to concern about the greenhouse effect.

 Suggest why the concentration of carbon dioxide in the atmosphere (Fig 30.5) fluctuates so much within a year?

The second theory is that some of the carbonate and hydrogencarbonate ions formed when carbon dioxide dissolves in sea water will also become incorporated into animal and plant skeletons which, after they have fallen to the bottom of the ocean, may become compressed into new limestone, calcium carbonate. Coral reefs, for example, are largely limestone. A third theory suggests that the extra carbon dioxide in the atmosphere might increase the growth of phytoplankton in the sea, so removing CO_2 from the atmosphere.

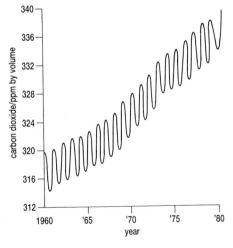

Fig 30.5 Atmospheric concentrations of carbon dioxide; measured at the Mauna Loa Observatory in Hawaii.

QUESTIONS

30.1 Look at Fig 30.6 which shows the complete carbon cycle.
 (a) Which compartment contains the most carbon and in what chemical form is it?
 (b) How much carbon moves into and out of this compartment?

30.2 Explain how phytoplankton could remove carbon from the cycle on a long-term basis.

ECOSYSTEMS AND NUTRIENT CYCLES

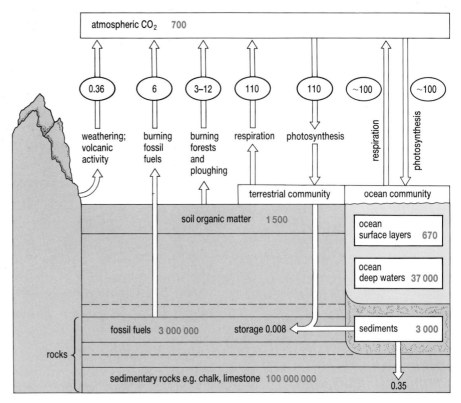

Fig 30.6 The complete carbon cycle including carbon dioxide inputs to the atmosphere from human activities. Transfers of carbon (in circles) are in 10^9 tonnes yr^{-1}. Quantities of carbon in blue are in 10^9 tonnes.

30.2 THE NITROGEN CYCLE

This cycle, shown in Fig 30.7, contains a number of important microbial processes which need some explanation.

Putrefaction

Putrefying bacteria and fungi break down organic nitrogen compounds in dead tissues, faeces and urine to ammonia. The ammonia is converted to ammonium ions (NH_4^+). Let us start with organic nitrogen compounds present in living organisms.

 What form of nitrogen is absorbed by plants from the soil?

Nitrification

If there is no oxygen available or if the soil is very cold the process stops here and only ammonium ions will be available to plant roots. However, in the presence of oxygen, two further reactions involving bacteria occur, the process called **nitrification**:

$$NH_4^+ \xrightarrow{\textit{Nitrosomonas}} \underset{\text{(nitrite)}}{NO_2^-} \xrightarrow{\textit{Nitrobacter}} \underset{\text{(nitrate)}}{NO_3^-}$$

Denitrification

Another group of soil bacteria, for example *Pseudomonas denitrificans*, convert nitrates to gaseous nitrogen. This occurs most rapidly under anaerobic conditions, for example when the soil is waterlogged.

Nitrogen fixation

Some bacteria possess the enzyme nitrogenase which enables them to convert gaseous nitrogen into ammonia. Examples of such bacteria are *Rhizobium*, which form symbiotic relationships with leguminous plants, free-living species, e.g. *Azotobacter* and cyanobacteria, e.g. *Nostoc*.

$$N_2 + 3H_2 \xrightarrow{\text{nitrogenase}} 2NH_3$$

The free-living species, which are responsible for 90% of the nitrogen fixation in the soil, use the ammonia to make their own proteins. The root nodule bacteria, e.g. *Rhizobium*, derive carbohydrate from their plant host and pass ammonium ions directly to the plant, a sort of in-built fertiliser factory!

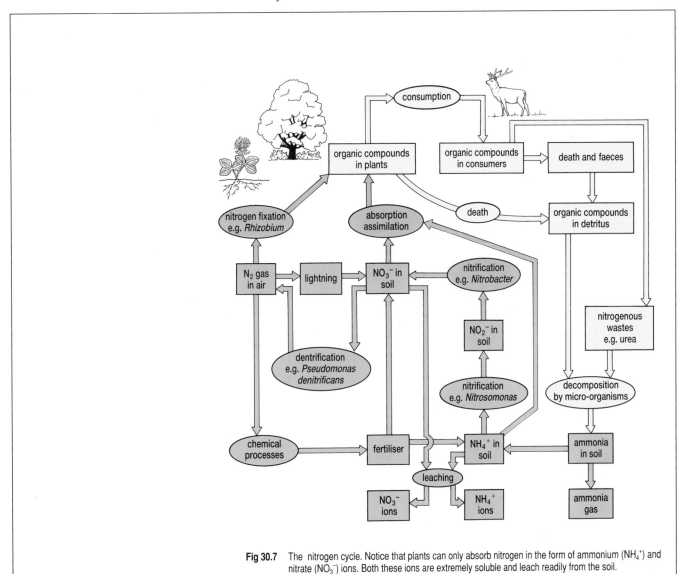

Fig 30.7 The nitrogen cycle. Notice that plants can only absorb nitrogen in the form of ammonium (NH_4^+) and nitrate (NO_3^-) ions. Both these ions are extremely soluble and leach readily from the soil.

Each year about 175 million tonnes of nitrogen are biologically fixed, exceeding by a factor of four the amount applied as fertiliser. This massive amount of nitrogen fixation is probably catalysed by no more than a few kilograms of nitrogenase.

Humans have invented their own form of nitrogen fixation called the Haber process. This involves an enormous chemical factory, high temperatures and pressures and the use of expensive catalysts to produce ammonia which is then converted into inorganic fertiliser by treating it with either nitric or sulphuric acid.

 List four ways in which soils can lose nitrogen.

Fig 30.8   Synthetic, inorganic fertilisers are widely used to increase crop yields.

ANALYSIS

Other cycles

This exercise involves the interpretation of diagrammatic information and the application of that information to a new situation.

Look at Figures 30.9 and 30.10 representing part of the phosphorus and water (hydrological) cycles.

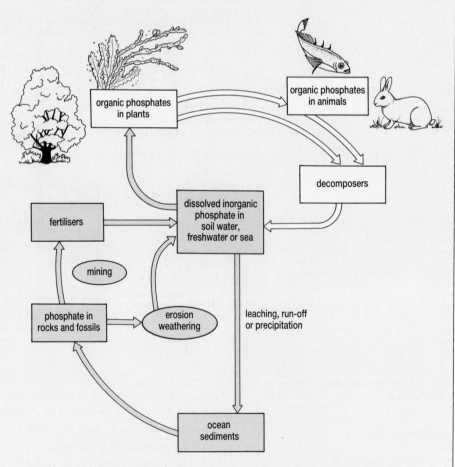

Fig 30.9   The phosphorus cycle. Since the phosphate ion reacts with, and becomes fixed to, soil particles this ion is not particularly soluble in the soil and is therefore not easily lost from the soil by leaching.

ECOSYSTEMS AND NUTRIENT CYCLES

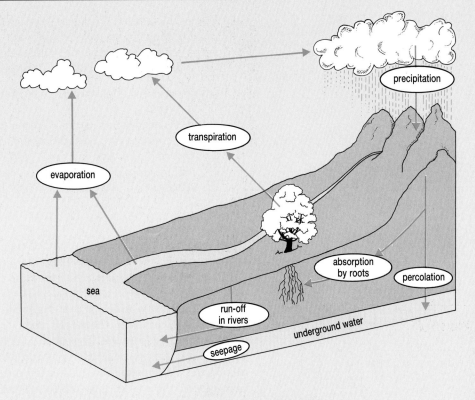

Fig 30.10 The hydrological cycle

(a) In what major way does the phosphorus cycle differ from the nitrogen cycle?

(b) How are phosphates added to, and lost from, the cycle?

(c) In what ways does the water cycle differ from the other cycles we have looked at?

(d) How are the hydrogen and oxygen locked into organic compounds by photosynthesis released?

(e) Use the following description and the knowledge you have gained in this chapter to produce an outline diagram of the sulphur cycle.

Sulphur cycles between organic sulphur compounds in living organisms and inorganic sulphate ions in soil or water. Sulphate is easily washed out of the soil. The products of decomposition of organic sulphur compounds are gases which escape to the atmosphere where they are converted to sulphur dioxide. Additional sources of sulphur dioxide include volcanic eruptions which release sulphur locked up in rocks. Sulphur dioxide reacts in the atmosphere to produce sulphuric acid. This substance dissolves in water to produce hydrogen and sulphate ions.

QUESTIONS

30.3 Why will growing and then ploughing a leguminous crop such as clover into the soil increase soil fertility?

30.4 The conversion of ammonium ions to nitrate ions is an oxidation process involving bacteria. Explain how this process will be affected by (i) waterlogging of the soil (ii) an acid pH (iii) soil temperature (iv) a lack of organic matter in the soil.

ECOSYSTEMS AND NUTRIENT CYCLES

30.5 State which of the following types of organism are essential for the cycling of nitrogen and explain why.
(i) plants (ii) herbivores (iii) nitrogen-fixing bacteria
(iv) denitrifying bacteria (v) decomposers.

30.6 (a) When soils are ploughed which stores of carbon are being depleted?
(b) Why, in the short-term, will ploughing increase the supply of phosphorus and nitrogen in the inorganic soil pools?
(c) (i) What do you think the long-term effects of ploughing on soil fertility might be?
(ii) How could a farmer remedy this situation?

30.3 ROBBING NUTRIENT CYCLES: HARVESTING

Vegetables grown in your own garden use nutrients from the garden soil in order to grow. These nutrients become a part of the plant's tissues. When you remove these plants and eat them, the nutrients temporarily become a part of your body. It is highly unlikely that you will either use your vegetable plot as a toilet or that you will be buried and decompose in it. The soil nutrients will, therefore, never return to the soil of your plot. In one growing season the loss of nutrients in this way may not have a significant effect on the fertility of your garden. Over many years, however, there will be a considerable loss of the nutrients essential for plant growth. Throughout the world, large areas of land are used to grow crops. Once harvested, these crops are transported and sold for human consumption. The loss of valuable soil nutrients in areas like these is immense and, unchecked, would make the soil totally unfit for plant growth.

Fig 30.11 Giant onions. These vegetables contain a vast amount of nutrients taken from the soil. These need to be replaced to maintain soil fertility and so grow another mammoth crop next year.

Fertilisers

One way to make up the nutrients which are lost during harvesting is to add more in the form of a fertiliser (Fig 30.12). Fertilisers may be of two types: decomposing organic material, which is often farmyard manure or treated sewage, and manufactured inorganic compounds containing the

ECOSYSTEMS AND NUTRIENT CYCLES

essential ions which plants need. Use of the former mimics the way in which nutrient cycles would normally occur. The decomposing organic matter becomes a part of the soil, helping to improve the soil's structure and water-retaining capacity, and releases nutrients slowly as it decomposes. Manufactured inorganic fertilisers do not improve a soil's structure and are relatively short-lived in the soil.

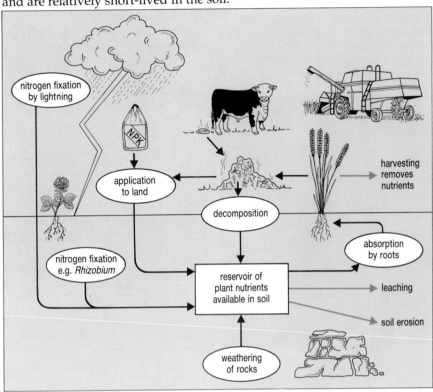

nitrogen fixation by lightning

NPK

application to land

decomposition

harvesting removes nutrients

nitrogen fixation e.g. *Rhizobium*

reservoir of plant nutrients available in soil

absorption by roots

leaching

soil erosion

weathering of rocks

Fig 30.12   The loss and addition of plant nutrients in soils. NPK is an artificial fertiliser containing nitrogen (N), phosphorus (P) and potassium (K).

QUESTION	30.7 Farmers often only need to apply phosphate fertilisers to the soil once a year. By contrast, they have to apply nitrate fertilisers several times in the year. Explain this practice.

30.4 FLOODING NUTRIENT CYCLES: POLLUTION

Pollution occurs whenever substances contaminate air, soil or water, interfering with natural cycles in ecosystems. These pollutants may be synthetic, as in the case of the chlorinated hydrocarbons described in Chapter 29. Pollution also occurs when natural substances are produced in unnatural amounts, e.g. carbon dioxide and sulphur dioxide, or when substances which occur naturally on Earth are released into the environment in a form in which they seldom occur, e.g. lead and asbestos.

Eutrophication

To be taken up by plants, nutrients must be soluble in water. This means that they will be easily washed from the soil (**leaching**) and, in solution, follow the water cycle into streams, rivers and, eventually, into lakes or the sea. As a result, the concentration of dissolved inorganic ions in waterways increases and the populations of algae and photosynthetic bacteria in the water undergo exponential growth. The resulting algal blooms cloud the water and form a scum on its surface.

 What effect will this have on plants growing on the bottom of rivers and lakes?

KEY CONCEPT

EUTROPHICATION

Eutrophication is a natural process in which a water course becomes enriched in nutrients and organic matter. It often leads to an increase in the biomass of primary producers (Fig 30.13). The process is speeded up by sewage effluent and leachate from fertilised fields, leading to serious pollution problems.

The algae and bacteria die faster than they are eaten and the decomposer populations increase exponentially. Most of the oxygen dissolved in the water is used up by decomposers so that the organisms which need a plentiful oxygen supply, i.e. have a high biochemical oxygen demand or BOD, such as many fish and crustaceans, die. Their dead remains provide even more fuel for growth of the decomposer populations. Eventually, when the oxygen is gone, anaerobic bacteria thrive. Many of these produce poisonous gases, such as hydrogen sulphide which smells of rotten eggs, which further foul the water.

 Why is oxygen depletion in a eutrophicated water course faster at night than during the day?

Fig 30.15 Water quality can be monitored by examining the number and types of invertebrates. Here a kick sample is being taken in a stream to collect invertebrates on the bottom of the stream. The sample is incubated in the dark at 20 °C for five days. The amount of oxygen removed from the water during this time is the Biochemical Oxygen Demand.

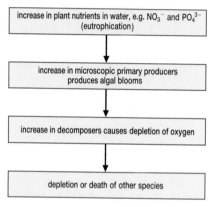

Fig 30.13 The consequences of eutrophication.

increase in plant nutrients in water, e.g. NO_3^- and PO_4^{3-} (eutrophication)

↓

increase in microscopic primary producers produces algal blooms

↓

increase in decomposers causes depletion of oxygen

↓

depletion or death of other species

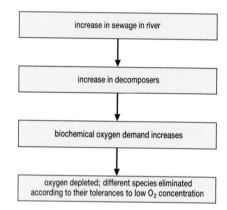

Fig 30.14 The effect of sewage on aquatic ecosystems.

increase in sewage in river

↓

increase in decomposers

↓

biochemical oxygen demand increases

↓

oxygen depleted; different species eliminated according to their tolerances to low O_2 concentration

Sewage and its treatment

In a balanced ecosystem, the valuable inorganic nutrients in urine and faeces would return to the same area from which they were absorbed by plants through the appropriate nutrient cycles. Your urine and faeces is disposed of where you live, which is not necessarily the same area of the world from which your food, containing the nutrients, originated. The same is true of the accumulated urine and faeces of animals which are farmed in 'factory' conditions. Raw sewage is harmful to human health because it contains disease-causing organisms (**pathogens**) like cholera. Raw sewage also has a high biochemical oxygen demand (BOD). This means that if we pump it directly into a river or lake, without first treating it, eutrophication will occur (Fig 30.14).

 Why are the water samples used in BOD analysis incubated in the dark?

A sewage purification plant is a managed, artificial ecosystem. Biological communities within different parts of the sewage works contain populations of aerobic and anaerobic bacteria, algae, protozoa and annelid worms. Within this purification plant, all the nutrient cycles occur and

pathogens are destroyed so that the effluent is less noxious than the material entering. It does, of course, have a very high concentration of inorganic nutrients so that, when deposited into local waterways, it may contribute to eutrophication. In particular, synthetic detergents, which contain substances called tripolyphosphates, are a major source of phosphorus – the limiting nutrient in most freshwater ecosystems.

Who killed the Norfolk Broads?

This is a comprehension exercise.

The broads are a series of lakes set in the flat, farmland of Norfolk. They were formed some 700 years ago when peat workings were flooded. The water was then crystal clear, about 4 metres deep and the bottom of the lakes were covered in plants. The concentration of phosphorus in the form of phosphate ions in the water would have been very low, perhaps 20 μg dm^{-3}. Up to the year 1800, the surrounding grass marshland was grazed by sheep. After 1800 more and more of the land was ploughed up to grow cereals, barley and wheat. Ploughing the grassland released large amounts of nitrogen into the soil which then leached into the broads. At roughly the same time Norwich suffered an outbreak of cholera and responded by pumping its sewage into the River Yare which feeds the broads.

Initially the increased levels of phosphorus and nitrogen proved to be beneficial. By 1900 the phosphate ion concentration in the River Yare was 100 μg dm^{-3} promoting the growth of a lush vegetation of reeds and water plants which, in turn, supported insects and birds.

After World War II the Norfolk Broads habitat declined.

The Haber process made inorganic nitrogen fertilisers extremely cheap: they had previously been very expensive. Farmers abandoned their reliance on organic manure fertilisers, turned to inorganic sources and brought more grassland into cereal cultivation.

Norwich grew and so did the amount of sewage it pumped into the River Yare which, by now, contained large quantities of synthetic detergents. Tourism, in particular pleasure boating, increased. The propellors of the boats stirred up the sediment on the bottom of the broads while the faeces, urine and washing-up water of the people they carried went straight into the broads.

By the late 1960s almost all the 41 broads were smothered each year with thick carpets of green phytoplankton. Originally the planktonic algae were probably kept in check by water mites. It seems that the water mite numbers were severely reduced by fish predation and then by pesticides and now the algae make the water so turbid that the water plants rooted to the bottom are deprived of light and die.

The Norfolk Broads are now one of the most severely contaminated freshwater ecosystems in the world. Phosphate levels commonly exceed 500 μg dm^{-3}, compared with 50 μg dm^{-3} in uncontaminated East Anglian water. Nitrate concentrations in the upper reaches of the River Bure, which feeds many of the broads, are 8 μg dm^{-3}, ten times the level in uncontaminated water.

Simply stopping the inputs of nutrients will not help. The bottom of the broads is now covered with a layer of black, anaerobic mud one metre thick. The mud is full of phosphates. If you clean up the water the sludge will release its phosphates. The mud contains enough phosphates to last several hundred years.

(a) Why did ploughing up the grassland increase nitrate but not phosphate levels in the broads?

(b) Why did the increase in primary productivity of the broads not start until the levels of phosphates rose? Where did the phosphates come from?

(c) Why did a small amount of eutrophication actually increase the beauty and amenity value of the broads?

(d) Explain why the Haber process and the introduction of synthetic detergents increased the rate of eutrophication.

(e) Explain the origin of the thick layer of mud which now lies at the bottom of the broads.

(f) The following strategies have been suggested to clean up the broads.
  (i) Suck the mud up from the bottom of the broads and dispose of it elsewhere.
  (ii) Exclude fish from the broads.
  (iii) Take surrounding farmland out of cultivation, return it to grassland and graze animals on it
  (iv) Install better sewage treatment plants
  Explain how each of these might help to clean up the broads?

(g) 'Who, is to blame for the disaster in the broads? Everyone who demands cheap, intensively grown food and who wants their wastes disposed of as cheaply as possible.' (Dr. Rosalind Boar). Do you agree with this statement? How true is it of other pollution problems you know about?

Acid rain

Carbon dioxide is not the only gas produced when fuels burn. Sulphur dioxide and nitrogen oxides are also waste products from the burning of coal and oil. Whilst plants need both sulphur and nitrogen to make amino acids, they cannot use either of these gases. Both gases react with water in the atmosphere to produce acids, sulphur dioxide producing sulphuric acid and nitrogen oxides producing nitric acid (Fig 30.16). The result is acid rain; precipitated water with dissolved acids.

 What are the major sources of oxides of nitrogen (NO_x) and sulphur dioxide?

Acid rain which falls on land can have drastic effects, as you may be aware if you live in a heavily industrialised area and look at the effect that years of pollution have had on the carved stonework of old buildings. In Baltimore, USA, for example, Rodin's famous statue 'The Thinker' has been moved indoors to protect it from further damage by corrosive rain. In the soil, acidic water dissolves the nutrients needed for plant growth, causing leaching. It also dissolves normally inert toxic metals, such as aluminium, which may poison plants directly or become concentrated in food chains. In Sweden for example, the kidneys and liver of moose grazed on plants grown in acidified water sometimes contained enough cadmium to be lethal to anyone eating them. Plants which are poisoned and weakened by lack of nutrients may be more susceptible to infections by pathogens and parasites and begin to die. An environmental group in 1985 recorded the state of health of two species of trees in Britain, see Table 30.1. This poor health is thought to be the result of acid rain (Fig 30.17).

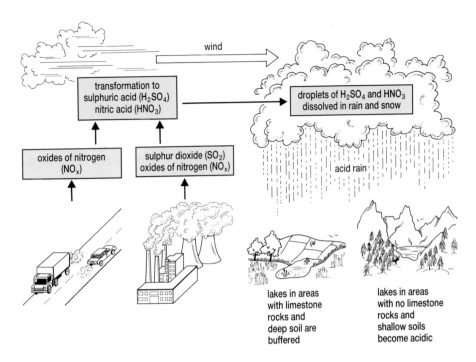

wind

transformation to
sulphuric acid (H₂SO₄)
nitric acid (HNO₃)

droplets of H₂SO₄ and HNO₃
dissolved in rain and snow

oxides of nitrogen
(NOₓ)

sulphur dioxide (SO₂)
oxides of nitrogen (NOₓ)

acid rain

lakes in areas
with limestone
rocks and
deep soil are
buffered

lakes in areas
with no limestone
rocks and
shallow soils
become acidic

Fig 30.16 The formation of acid rain. Notice that it is only lakes in areas poor in limestone (calcium carbonate) that suffer from the effects of acid rain. Can you suggest why?

 List four ways in which acid rain may affect the health of trees.

Table 30.1   The results of a survey on the state of Britain's trees carried out in 1985 by an environmental group

Tree species	Trees in survey/%			
	healthy	partial dieback	advanced dieback	complete dieback
beech	31	53	14	2
yew	22	52	22	4

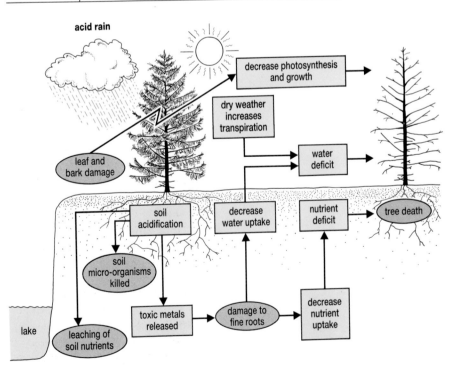

acid rain

decrease photosynthesis
and growth

dry weather
increases
transpiration

water
deficit

leaf and
bark damage

soil
acidification

decrease
water uptake

nutrient
deficit

tree death

soil
micro-organisms
killed

lake

leaching of
soil nutrients

toxic metals
released

damage to
fine roots

decrease
nutrient
uptake

Fig 30.17 The effects of acid rain on the soil and on trees.

Country	Mass of sulphur dioxide emitted/ kg per head of population per year
Eastern Germany	240
Czechoslovakia	201
Hungary	153
Yugoslavia	133
Finland	119
Bulgaria	112
Spain	99
Soviet Union	91
Italy	90
Denmark	89
Britain	83
Belgium	82
Poland	76
France	60
Sweden	60
Western Germany	58
Netherlands	31
Switzerland	18

Because the prevailing winds over Britain blow air to north-eastern Europe, much of Britain's sulphur dioxide falls as acid rain elsewhere. You can see from Table 30.2 that other countries in Europe will also be inadvertent exporters of acid rain. The coniferous forests of Scandinavia, Poland, Bohemia and western Germany have recently been devastated and acid rain is the major suspected cause, a factor contributing to a number of European nations signing an agreement to reduce their sulphur dioxide emissions by 30% of their existing levels by 1993 (the so-called 30 Club). Similar effects attributable to acid rain have been observed in Canada and the eastern states of the USA, prompting the US government, in 1980, to set up a ten-year National Acid Precipitation Assessment Program to study the causes of acid rain.

The acidic water percolating through soil runs into streams, rivers and lakes. Unless local rocks contain calcium carbonate which reacts with the acid, acting as a buffer, the pH of these waterways falls. At pH 5 most aquatic communities are destroyed. The result is lakes in which the water is crystal clear because it contains no life. In the Adirondack mountains of New York, USA, 200 lakes were said to be in this condition in 1986 and many waterways in Europe are similarly affected. The total cost of dealing with acid rain problems throughout the world is incalculable.

FURTHER RESEARCH

Start a file for each of the environmental issues discussed in this Chapter. Scan the library copies of one 'reliable' newspaper and a weekly science publication from the past month and make a note of each time an article about one of these environmental issues occurs. Would you say that these issues are being dealt with adequately by world governments, or do the articles raise cause for concern?

QUESTIONS

30.8 Before World War II, it was common for farmers in Britain to rotate their crops. Four crops were grown and the land was divided into quarters, one quarter for each crop. Each year, over a period of four years, each quarter was sown with a different crop. This followed a sequence such that after four years each quarter would have had all four crops grown on it. The crops were root crops (e.g. turnips, mangolds), barley, clover mixed with grass and wheat. The root crops were fed to livestock (cattle, pigs and sheep) during the winter, often in the field. The clover-grass mixture was either made into hay and fed to livestock over the winter or ploughed into the soil. Barley was harvested mainly for feeding livestock, wheat for human food. The manure from livestock kept indoors over the winter and bedded on the straw from the barley and wheat crops was spread on the fields in the spring.

(a) Explain how this rotation conserved the farm's resources of inorganic ions.

(b) Suggest why modern farming methods no longer include a substantial amount of crop rotation.

30.9 Explain what is meant by eutrophication? Name three human activities which will lead to eutrophication.

30.10 Define the term pollution and list the ways in which humans may pollute their environment.

Nutrient cycles describe the pathways traced by inorganic nutrients as they flow between living (biotic) and non-living (abiotic) phases of an ecosystem.

Water evaporates from its reservoir in the oceans and is precipitated back to the oceans or onto land. During this cycle some water is taken up by organisms. A small amount of this is chemically altered by organisms, but water is eventually released during their respiration or decomposition.

Carbon enters communities during photosynthesis and becomes incorporated into the bodies of producers; it is then passed through food webs. Carbon is released back into the atmosphere as carbon dioxide produced during respiration and decomposition.

From its atmospheric reservoir, nitrogen gas is converted to ammonium and nitrate ions by lightning and by cyanobacteria and nitrogen-fixing bacteria. Once absorbed in this form by producers, the nitrogen becomes incorporated into molecules, such as proteins and nucleic acids, and is passed through food webs. Nitrogen is returned to the environment through excretion and the activity of decomposers. Some decomposers release nitrate ions, some ammonium ions and some convert nitrate ions into nitrogen gas, which rejoins the atmospheric reservoir.

Humans affect local nutrient cycles when they harvest food which is transported for consumption elsewhere. They also harm local nutrient cycles through pollution. Pollution occurs when humans release unnatural substances into the environment or release natural substances in greater amounts than can be absorbed by local nutrient cycles.

Chapter 30: Answers to Quick Questions

1   These probably correspond to seasonal fluctuations in the rate of photosynthesis.
2   As nitrate and ammonium ions.
3   Leaching; denitrification; absorption by plants; as ammonia gas.
4   Algae will absorb light and carbon dioxide (including hydrogen-carbonate ions) thus reducing the growth rate of the bottom-dwelling plants.
5   Because photosynthesis, which oxygenates the water, ceases at night.
6   If the sample were incubated in light the effects of decomposer respiration would be masked by photosynthetic oxygenation of the water sample.
7   NO_x comes from both cars and power stations; SO_2 from power stations.
8   (i)   Direct damage to leaves.
    (ii) Release of toxic metals bound to soil particles, which kills tree roots.
    (iii) Disturbance of nutrient cycles by killing essential soil micro-organism, so depriving the trees of essential nutrients.
    (iv) Increase in the susceptibility of trees to damage by pests, diseases and frost.

Chapter 31

ECOSYSTEM PATTERNS

LEARNING OBJECTIVES

When you have studied this chapter you should be able to:

1. describe and explain primary and secondary succession in terrestrial and aquatic ecosystems;

2. account for patterns of zonation seen on mountain sides, rocky shores and in woodlands;

3. discuss the major factors influencing the distribution of major terrestrial biomes;

4. explain the relative stability of the sea and large lakes and describe the abiotic factors affecting communities in them.

A central idea in biology is that organisms have evolved so that they are adapted to where they live, to their ecological niche. Each organism has its own particular suite of characteristics which enables it to live under certain environmental conditions. In a wood, some species must have full sunlight to survive, others thrive only in the shade. This chapter examines how these individual requirements result in ecological patterns on a global scale; patterns that do not remain static but change with time as organisms alter their own environment.

31.1 PATTERNS IN TIME: SUCCESSION

Human activities constantly reform our environment: land has been reclaimed from the sea in Holland by building large dikes and in Japan by dumping refuse into the sea to make a platform on which to build; large valleys are filled with water to make reservoirs; tropical rainforests in South America and Indonesia are cleared so that beef can be ranched or citrus trees grown. Natural processes also reform our environment: eutrophication of lakes, resulting from the leaching of inorganic ions from the surrounding land, causes gradual changes in the lake communities; the explosion of Mount St. Helens in Washington State, USA, in 1980 obliterated the coniferous forest on its north side; volcanic islands, such as those in the Galapagos and Hawaiian archipelagos, emerge from the oceans.

The communities originally present are often destroyed when the environment changes. In time, however, new communities develop. Initially these new communities may be dominated by a small number of colonising species (**pioneers**) which are usually autotrophic. These simple communities interact with the abiotic environment around them, causing it to change. In turn, these changes in the abiotic environment result in changes in the community and so on. This process of **succession** continues through a number of stages, called **seres**, until a final stage, the **climax community**, is formed. Some of the important changes which occur during succession are summarised in Table 31.1. Succession may take one of two forms. **Primary succession** occurs on bare rock or in sand or a clear glacial

KEY CONCEPT

SUCCESSION

Succession is a process in which communities of plant and animal species are replaced in a particular area over time by a series of different and usually more complex communities.

pool where there is no existing community. **Secondary succession** occurs after an existing ecosystem has been disturbed, e.g. by fire or flooding.

 Will primary or secondary succession be more common?

Table 31.1   Summary of the changes that occur during succession

Sere	Species diversity	Number of interactions between populations	Resistance of community to environmental change
pioneer community ↓	few species	few	susceptible ↓
subclimax community ↓			resistant ↓
climax community	many species	many	resistant

Succession on bare rock

Bare rock breaks down physically and chemically during the process of **weathering**. Cracks form in the rock as it expands and contracts during the different seasons. Water in the cracks expands as it freezes during the winter, further crumbling the rocks. Rain, which is normally slightly acidic because of dissolved carbon dioxide from the air, dissolves some minerals. Whilst the resulting rock face is inhospitable to most plants, lichens (a mutualistic relationship between a filamentous fungus and an alga) are hardy enough to survive there. They are able to do so because they are almost self-sufficient: the hyphae of the fungus penetrate tiny cracks in the rock, providing a firm attachment and absorbing inorganic nutrients from the rock, the alga photosynthesises.

 What benefit does (a) the fungus and (b) the alga get from this relationship?

Fig 31.1   Reclamation of spoil heaps depends upon a sound knowledge of the processes of succession, which allows manipulation of fertiliser levels and planting programmes to achieve the end product you see here.

After the activity of the lichen has enlarged the cracks in the rock and filled them with decomposing dead lichen, the rock is hospitable enough for the growth of drought-resistant, sun-tolerant mosses. Moss plants form a dense mat which traps tiny particles of rock, bits of organic debris and water. The death of moss plants adds to the nutrient content of the mat so that, eventually, the mat will support the germination of the seeds of large colonising angiosperms, whose bodies will also contribute to the growing layer of soil. As larger woody shrubs begin to grow in the newly formed soil, the mosses and lichens may be shaded or covered by decaying leaves and other vegetation. This illustrates an important aspect of succession: at each stage until the climax is reached, populations alter the environment in ways that encourage their direct competitors. Eventually trees are able to take root in the deeper rock crevices and the sun-loving shrubs disappear. Within centuries a mature forest ecosystem may thrive on what was once bare rock. A succession on bare rock in the upper Michigan region of the USA is summarised in Fig 31.2.

time →

lichen,
bare rock

bluebell,
yarrow

blueberry,
juniper

jack pine,
black spruce,
aspen

balsam fir,
paper birch,
white spruce,
i.e. climax forest

Fig 31.2 Summary of primary succession on bare rock in Michigan, USA. At each stage until its climax is reached, populations alter their environment enabling larger organisms to thrive.

Fig 31.3 Catalytic convertors fitted to car exhausts should reduce this environmental problem in the future. However, catalytic convertors increase fuel consumption, so adding more CO_2 to the atmosphere, exacerbating the greenhouse effect!

ECOSYSTEM PATTERNS

MONITORING AIR POLLUTION

The growth of cities during the industrial revolution meant that air pollution from the burning of wood and later of coal became an increasingly serious problem. In 1911 at least 1150 Londoners died from the effects of smog, a mixture of coal smoke and fog. Smog consists mostly of a mixture of sulphur dioxide, sulphuric acid and suspended solid particles. Most of the deaths were due to chronic bronchitis. In 1952, 4000 people died from the effects of smog and the smog disasters of 1956, 1957 and 1962 killed a further 2500 people. As a result of successive Clean Air Acts, London now has much cleaner air. Lichens are well-adapted to life on bare rocks, being able to absorb rapidly inorganic nutrients dissolved in water that runs off the rock surface. This ability results in their death in highly polluted areas because they quickly absorb lethal concentrations of soluble toxins. Thus the distribution of lichens can be used as a way of monitoring air pollution.

ANALYSIS

Sand dune succession

This exercise involves the interpretation of information in diagrams and its translation to written prose.

Investigating succession is a problem because this biological process requires a long period of time. However, if you go to certain coasts around Britain you will find dunes formed by sand blowing off the beach. Sand dunes tend to occur in 'fields'. The dunes furthest from the sea were formed first, i.e. they are the oldest: those nearest the shore formed last, i.e. they are the youngest. If you walk inland from the sea you are effectively walking back in history.

(a) Using the information in Fig 31.4, outline the succession of plant species which occurs on sand dunes.

(b) How do the abiotic conditions within a sand dune change as the succession proceeds?

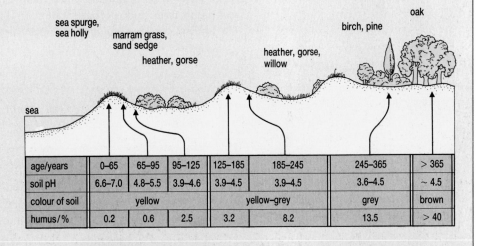

age/years	0–65	65–95	95–125	125–185	185–245	245–365	> 365
soil pH	6.6–7.0	4.8–5.5	3.9–4.6	3.9–4.5	3.9–4.5	3.6–4.5	~ 4.5
colour of soil	yellow			yellow–grey		grey	brown
humus/%	0.2	0.6	2.5	3.2	8.2	13.5	> 40

Fig 31.4 Sand dune succession. The environment is altered by each stage making it more suitable for different types of plant.

Succession in ponds and lakes

Rivers and streams carry sediments to the ponds and lakes into which they empty. These settle to the bottom making the pond or lake shallower. As the level of sediment rises, submerged plants are replaced by those with floating leaves and then by marsh plants which have their roots in the water but their stems and leaves in the air. The stems and roots of these plants trap more sediment and their decomposition helps to enrich the sediment as it develops into a soil. Depending on the drainage and pH of these developing soils, a number of different communities will eventually develop, as summarised in Table 31.2.

Table 31.2   The climax communities which develop as a result of succession from static freshwater to dry land

pH	Poor drainage (wetlands)	Good drainage (drylands)
acidic	bog	heath
basic	fen	grassland

 How do the pond plants early in the succession 'prepare the way' for their competitors?

Secondary succession

Secondary succession tends to happen more quickly than primary succession because the abiotic environment is already suitable for plant growth and because there are some vegetative remains from the previous community.

Secondary succession in an abandoned field

The early plant colonisers, referred to as the **open pioneer community**, are fast-growing annuals. The only animals will be those belonging to the soil ecosystem and their predators such as birds. After a few years, perennials begin to grow. Herbaceous plants are the first to grow but woody plants follow, forming a **closed herb community**. This community is replaced about fifteen years later by **scrub vegetation**, consisting mainly of shrubs and small trees and, perhaps 150 years later, a climax forest will be found. The animal community changes with the plant community (Fig 31.5).

QUESTIONS

31.1 Define the terms primary succession, secondary succession, sere and climax ecosystem.

31.2 Why can lichens colonise bare rocks which are inhospitable to other plants? Suggest why lichens which live on rocks are seldom found in later stages of succession.

31.3 Suggest, with reasons, whether each of the following will (i) increase (ii) decrease or (iii) stay the same as succession proceeds:

(a) nutrient content of the soil or water
(b) productivity
(c) number of species (species diversity)
(d) biomass
(e) rate at which populations replace each other.

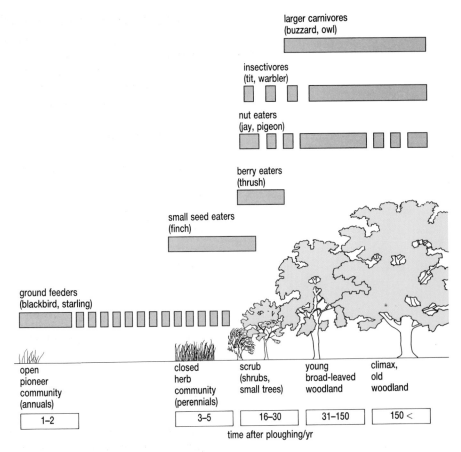

larger carnivores
(buzzard, owl)

insectivores
(tit, warbler)

nut eaters
(jay, pigeon)

berry eaters
(thrush)

small seed eaters
(finch)

ground feeders
(blackbird, starling)

open pioneer community (annuals)	closed herb community (perennials)	scrub (shrubs, small trees)	young broad-leaved woodland	climax, old woodland
1–2	3–5	16–30	31–150	150 <

time after ploughing/yr

Fig 31.5 Secondary succession in an abandoned ploughed field. The climax community is woodland. The animal community changes with the plant community as illustrated by the birds in the diagram.

31.2 ZONATION

A pattern is clearly visible in the photograph of a rocky shore (Fig 31.6). The bands of different colour across the photograph are present because each level on the shore is dominated by a different type of organism. This pattern, called **zonation**, is the result of interactions between the animals and plants living on the shore and their physical environment. In particular

KEY CONCEPT

ZONATION
A change in species composition through space.

Fig 31.6 Different organisms live under different conditions leading to the pattern of zonation seen on this rocky shore.

- organisms living at high levels on the shore are adapted to tolerate exposure to air for long periods when the tide is out, i.e. they can withstand **desiccation**;
- organisms which live lower down the shore, where they are exposed to air for shorter periods, cannot withstand prolonged desiccation;
- organisms compete for space on the rock on which to attach themselves;
- organisms which live lower down the shore compete more effectively for space than the upper shore organisms which are, therefore, excluded from the lower shore. This phenomenon, competitive exclusion, is discussed in Section 28.4.

The fine balance between tolerance to desiccation and competitive ability is a major cause of the zonation pattern seen on rocky shores.

 Consider a plant living half way down the shore shown in Fig 31.6. Predict (i) how tolerant of desiccation and (ii) how good a competitor for space it might be compared to plants (a) higher (b) lower on the shore.

Zonation also occurs in forests. The light intensity, wind speed, temperature and humidity gradually change from the floor to the canopy of a woodland or forest, creating a whole range of microhabitats within it. This results in the **stratification**, a sort of vertical zonation, shown in Fig 31.7. Most of the photosynthesis within this woodland occurs in the canopy layer and most decomposition occurs at ground level. Some consumers occupy microhabitats at a particular point within this zonation pattern, others may move freely from one layer to another.

Fig 31.7 Section through a woodland to show stratification. A large range of micro-habitats is created within the woodland.

canopy layer (over 5 m)

shrub layer (2–5 m)

field layer (0–2 m)

ground layer (< 3 cm)

QUESTION

31.4 Rooted plants which are totally submerged grow in the middle of a freshwater pond. Taking samples of plants along a line fixed from the middle of the pond to its dry edge (a line transect), areas dominated by floating plants would be followed by areas dominated by rooted plants with floating leaves, rooted swamp plants with most of their stems out of the water and then marsh plants. Does this pattern represent zonation or succession? Explain your answer.

ECOSYSTEM PATTERNS

The parts of the Earth which are able to support life, collectively termed the **biosphere**, contain four major habitats: terrestrial, marine, freshwater and estuarine. You may have been lucky enough to visit other parts of the world which are quite different from your own, but even travelling around the country in which you live, you are likely to appreciate that different areas have different biological communities living there.

 Suggest why ecologists tend to classify terrestrial ecosystems according to their plant life.

Large areas of land with similar conditions and characteristic types of plants adapted to those conditions are called **biomes**. The distribution of these biomes is the result of an interaction between the communities inhabiting them and their physical environment. Temperature and rainfall are important in determining the distribution of biomes (Fig 31.8). In spite of the vast areas involved, and the complicating effects of mountain ranges, there is a global pattern to these biomes (Fig 31.9).

Fig 31.8 The relationship between major terrestrial biomes and climate.

Temperature and light intensity

The intensity of the sun's energy falling on the Earth is fairly predictable. A beam of radiation from the sun falling on the equator of the Earth during the spring or autumn solstice is almost perpendicular (Fig 29.6). The intensity of solar radiation, and hence of light, is greater here than towards the poles, where a beam of the same width falls over a greater area of the Earth. Both the land and the sea are, therefore, much cooler at the poles than at the equator.

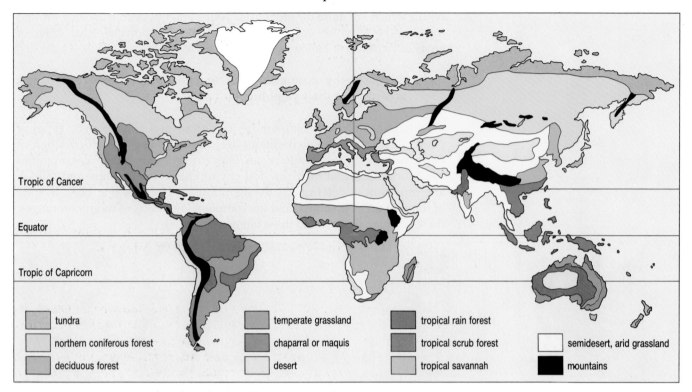

Fig 31.9   The global distribution of major terrestrial biomes. Can you locate the chaparral (or maquis) type?

Legend:
- tundra
- northern coniferous forest
- deciduous forest
- temperate grassland
- chaparral or maquis
- desert
- tropical rain forest
- tropical scrub forest
- tropical savannah
- semidesert, arid grassland
- mountains

6 Starting at the equator and walking due north on the blue line shown in Fig 31.9 which biomes would you pass through?

Precipitation

Whilst less predictable than temperature, precipitation also has a fairly regular pattern. Look at the photograph of the Earth taken by the crew of Apollo 11 (Fig 31.10). There are bands of heavy cloud over the equator, covering the rain forests of Zaire, and over the temperate regions.

7 What evidence is there from this photograph that the distribution of tropical rain forests and deserts is determined by rainfall?

Fig 31.10   The distribution of cloud cover as seen from space.

There is little or no cloud over the poles, which receive little precipitation. The Arctic and Antarctic are covered in ice because the temperature is too low for the little snow that does fall there to melt, not because it snows a lot. The intense cold means that the soil water is frozen in the winter and so is unavailable to plants. Thus northern forest is dominated by conifers; evergreens which, through their possession of waxy, needle-like leaves, are adapted to withstand the long periods of drought when all water is frozen.

8 In which other biome might conifers do well?

Further north, below the tundra, is a permanently frozen layer of soil, the **permafrost**, often no more than 50 cm below the soil surface, which limits the depth to which roots can grow. During the summer the melted snow and ice cannot drain through this layer so a huge marsh is produced, providing an ideal breeding ground for insects such as mosquitoes. The tundra's vegetation consists of plants which can tolerate shallow soil at, or near to, freezing point. Small perennial plants, dwarf willows and reindeer moss (which is actually a lichen, not a moss) dominate this tundra.

 Suggest why, despite its hostile climate, the tundra is a major breeding ground for migratory birds and mammals, e.g. caribou, in the summer.

The effects of mountain ranges

These global patterns are affected by changes in the air currents caused by large land masses, e.g. the annual rainfall in the Snowdon mountain range of North Wales is about five times heavier than that on the island of Anglesey which lies in its rain shadow 30 km to the north. If the rain shadow is particularly intense then a desert may form. This accounts for the northerly extension of the desert in western USA (Fig 31.9).

 Suggest one reason to explain the northerly extent of the deserts in Asia, e.g. the Gobi. Will these be hot or cold deserts?

The effects of seasons

Away from the hot, wet areas covered in tropical rain forest, distinct dry and wet seasons occur. These areas are characterised by **tropical deciduous forests**, the trees losing their leaves in times of drought.

Savannah areas and **temperate grasslands** characteristically receive all their rainfall during a short, rainy season. For the rest of the year they are characterised by drought and fires are frequent.

 Why might these conditions explain the small number of trees in these areas?

The effect of the sea

In areas where the rainfall is less than 25 cm yr^{-1} but which are close to the sea, the rainy season is longer and spring and autumn mists reduce evaporation. Here small trees or large bushes with thick, waxy or hairy leaves can grow in addition to grasses. In North America, the term chaparral is used to describe this type of vegetation. Similar biomes are common all around the Mediterranean Sea, where they are called maquis (after the French word for brushwood).

 What advantage do the trees of the chaparral get from their 'thick, waxy or hairy leaves'?

The effect of people

Humans have had a dramatic effect on the world's biomes. For example the **long-grass prairies (grasslands)** of Iowa and Missouri in USA lack trees, not because there is too little rainfall, but because of burning by humans to maintain grazing land. In fact, the North American Indians used to burn to maintain the grazing land for herds of bison. Now even this managed ecosystem has practically disappeared because of the intensive cultivation of crops, in particular maize.

The western **short-grass prairies** in states such as Colorado and Wyoming, home of the American western legends, are now chiefly used for cattle and sheep grazing. Along their boundaries, overgrazing has destroyed the grass and allowed bushes, such as sagebrush which cattle will not eat, to dominate. These areas are now becoming **cool desert**.

MAINTAINING DIVERSITY

Conserving desirable environments, such as these hayfields in the Yorkshire Dales with their rich diversity of plants and animals, requires careful management. The plants are annuals so the fields must not be cut or grazed until the plants have flowered and set seed.

In the spring, moderate amounts of animal manure must be applied: too much fertiliser would encourage the growth of a few vigorous grass species which would then exclude other plants.

Developing such management strategies requires a detailed understanding of the ecological processes operating in the environment and the cooperation of local farmers, not merely letting nature have her own way.

Fig 31.11   Traditional hayfields in the Yorkshire Dales. The diversity of plants and animals requires careful management over many years.

Deforestation

The trees which dominate forest ecosystems have a profound effect on all the other populations within it. They provide food and shelter to many animal populations. Being taller than all the other plants, they absorb most of the incident light, so that only shade-tolerant plants can grow beneath them. Their roots spread over a large area, not only depriving other plants of inorganic nutrients, but also absorbing most of the water. This water evaporates from their leaves during transpiration, creating a humid environment within the forest. The roots also help to hold the soil together.

Humans use trees to provide
- fodder for livestock
- a source of fuel
- building materials
- raw materials for processes such as paper making.

ECOSYSTEM PATTERNS

Fig 31.12 The devastating soil erosion you can see here is the direct result of removing the trees whose roots had previously bound the soil together.

As a result of overgrazing, over-collection of wood or clearance to make new roads or farmlands, many of the world's forests are being destroyed at an alarming rate. The effects of such deforestation include

- erosion of soil, which is no longer held in place by tree roots
- loss of soil fertility, since organic compounds in the plants are removed and inorganic nutrients are no longer recycled by the decay of leaves
- loss of soil water which evaporates from a bare soil surface and is lost to the atmosphere.

Tropical rain forests are particularly badly affected. These forests, because of the favourable climatic conditions, have the highest primary productivity of any terrestrial ecosystem (Table 29.1). Competition for light is intense and tropical rainforests are dominated by tall broad-leaved evergreens. The rapid growth is sustained because inorganic nutrients are absorbed by roots almost as soon as they are released by decomposers. As a result, virtually all of the nutrients are tied up in the vegetation: the soil is very infertile.

The number of species of animals and plants (**species diversity**) is greater in tropical rain forests than in any other ecosystem. About two-thirds of the world's species live in rain forests occupying about 6% of the Earth's land area. The plants in tropical rain forests are a major source of medicinal chemicals (one quarter of the chemicals and medicines sold by your local pharmacist will contain compounds derived from plants in tropical rain forests). In spite of this, they are being cleared at a rate of over 2% of their total area each year, representing the annual destruction of rain forest with an area equivalent to that of England, Scotland and Wales. At this rate, these rain forests could be totally eradicated within the twenty-first century.

QUESTIONS	**31.5** **(a)** What is a biome?
	(b) Are all the biomes discussed in Section 31.3 examples of climax communities?
	31.6 For each biome
	(a) list the climatic conditions under which it develops (e.g. hot, dry, seasonal)

ECOSYSTEM PATTERNS

(b) describe the vegetation present (e.g. grass with a few trees)

(c) suggest the adaptations you would expect the plants and animals to have, to meet the demands of the climate.

31.7 Explain why tropical rain forests have infertile soil.

31.8 The Earth is supposed to be warming up because of the greenhouse effect. What might this do to the pattern of biomes shown in Fig 31.9?

31.9 If you climbed a high mountain in the tropics you would pass from tropical rain forest at the bottom, through deciduous woodland and grassland to a tundra-like alpine vegetation near the top. Account for these changes.

31.4 LIFE IN THE SEA

Abiotic conditions in the seas are more stable than on land so there is not so strong a pattern of biomes as on land. Since water can absorb a lot of energy before warming up and lose a lot of energy before cooling down, the temperature of large bodies of water is more stable than the air temperature. The maximum density of water occurs at 4 °C so that, as water freezes, there is usually a layer of liquid water beneath the ice. With favourable temperatures and ready access to water, the major factors affecting life in aquatic ecosystems are the availability of energy and of inorganic nutrients. The major marine ecosystems are shown in Fig 31.13.

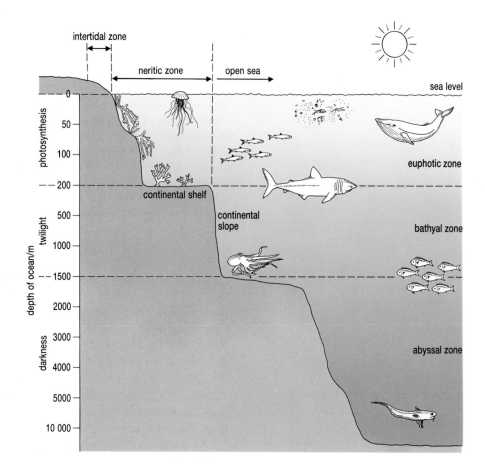

Fig 31.13 Major marine ecosystems. Many organisms are found in the euphotic zone where the amount of light is enough to support photosynthesis. (Note the change in scale on the axis.)

Energy

Water absorbs light: the presence of turbulence, silt or micro-organisms increases this absorption. The result is that only the uppermost layer of water (the **euphotic zone**) has enough light to support photosynthesis (Fig 31.13). Below this layer the usual source of energy is chemical energy in the excrement or bodies of organisms which sink or swim there. Most organisms are found in the euphotic zone where photosynthesis by the **phytoplankton** (free-floating photosynthesisers) supports the food webs of **zooplankton** (free-floating animals) and **pelagic** (free-swimming) animals.

 13 Explain why large seaweeds are only found in shallow coastal waters.

Nutrient supply

Phytoplankton take up inorganic nutrients from the environment during photosynthesis. These nutrients may then pass to the higher trophic levels of food chains if the phytoplankton are eaten. When marine algae and animals die their remains sink. As the dead remains, aptly called **marine snow**, fall to the bottom of the sea they decompose, releasing nutrients. Decomposition is completed by the activities of benthic organisms, particularly bacteria. However, the nutrients are in the darkness of the deep sea and the organisms which need them, the phytoplankton, are at the top of the sea: the euphotic layer rapidly becomes devoid of inorganic nutrients.

 14 Why can the demands of the phytoplankton not be met by upward diffusion of the nutrients released by decomposition on the ocean floor?

Nutrients are returned to the euphotic zone by three mechanisms.

1. Storms and their wave action help to stir up the water column.
2. Upwelling occurs, for example, when winds blowing parallel to a shore line cause deep, cold ocean water rich in nutrients to rise to the surface. Upwelling also occurs where oceanic currents diverge, for example along the equator.
3. Rivers continually remove inorganic ions from the land and carry them to the sea.

 15 Why are the most productive fishing areas of the world found in areas of upwelling?

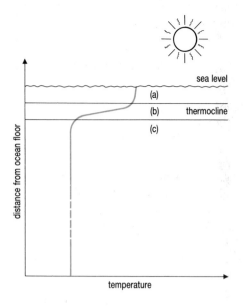

Fig 31.14 Vertical section through a sea to show the thermocline. (The thermocline has been drawn disproportionally wide.)

Thermocline

The mixing of sea water, needed to return nutrients to the euphotic zone, is hindered by the warming effects of the sun. In tropical waters, the intense heating effect of the sun produces a surface layer of warm, less dense water which floats without mixing on the cool, denser water beneath it. The boundary between the two, the **thermocline** (Fig 31.14), effectively reduces any upward diffusion of nutrients. The same phenomenon occurs in temperate waters during the summer.

16 How does temperature change with increased depth in zones (a), (b) and (c) in Fig 31.14?

Coastal waters

Near to shore, marine communities are affected by local geographical features such as the presence of estuaries and calm water bays, resulting in salt marshes, mud flats and the like. In the area of coast which is covered and uncovered by daily tidal movements (the **intertidal zone**) the substratum has a great effect on communities.

17 What will be the major problem facing organisms which live in estuaries?

In some tropical coastal waters with a particular combination of nutrient availability, depth and shelter from wave action, **coral reefs** are found. Fig 31.15 shows the calcareous skeletons of various species of algae and corals which make up the reef and provide an extremely sheltered environment for the anchorage of many algae and the animals they support. Coral reefs are sensitive ecosystems, however. Many are being destroyed, either through the action of natural forces, such as the destruction of the coral-building cnidarians by the predatory crown of thorns starfish (*Acanthaster planci*) in the Great Barrier Reef of Australia, or by human action, such as the silting of many coral reefs around Hawaii as house building erodes the land.

Fig 31.15 A coral reef: the most diverse marine community in the world.

ECOSYSTEM PATTERNS

31.10 Explain the terms euphotic layer, thermocline, plankton, pelagic and benthos.

31.11 Explain why the oceans do not produce such clear biomes as occur on land.

31.12 Photographs in holiday brochures show tropical seas which are clear and blue. The seas around Britain are green and relatively murky. Suggest why, even without pollution, these differences occur.

31.13 The primary productivity of Lake Windermere, in the English Lake District, is higher than that of Wast Water, a nearby lake of similar size. Assuming that large lakes are subject to similar constraints as are oceans, suggest what abiotic factor(s) might cause these differences in productivity.

FURTHER RESEARCH

Research and prepare a report which discusses both the pros and cons of the following suggestion.

> The deep ocean represents an enormous, self-sustaining ecosystem well away from human settlements so why should we not use it to dispose of our radiation and toxic waste?

SUMMARY

Ecosystems change over a period of time as a result of interactions between their biotic and abiotic phases. Ecological succession is an orderly change in an ecosystem through a series of stages, or seres, culminated in a climax community. During succession, an increase in species diversity occurs and the community becomes more resistant to environmental changes. The nature of climax communities can be predicted from a knowledge of the abiotic factors affecting the ecosystem.

Primary succession occurs when an environment with no previous community is colonised. Secondary succession occurs in an environment where a previous community did exist but has been disrupted or destroyed and is much faster than primary succession, since the habitat already contains a rich supply of inorganic nutrients. Both primary and secondary succession are started by specific colonising populations, called pioneers.

The type of community within a habitat is largely determined by abiotic factors, principally intensity of sunlight, temperature and availability of water and inorganic ions. Terrestrial communities are dominated by plants. Large regions of the continents that have similar climates have similar vegetation. These regions are called biomes.

Biomes are mainly influenced by patterns of rainfall and temperature, which vary with latitude. The consistently warm, moist conditions of equatorial regions favour the growth of tropical rain forests whereas at latitudes of 20° to 30° north and south, where rainfall is minimal, major deserts occur. At increasing latitudes, chaparral, grassland and temperate deciduous forest biomes occur. In extreme northern latitudes coniferous forest and tundra biomes are found. Ice is found around both poles.

The global pattern of biomes is affected by the topography of the continents, such that desert biomes are often found in the rain shadow of mountain ranges, and coniferous forest biomes are found at high altitudes.

The change in climate at different altitudes results in a series of biomes on a mountain side. A similar zonation of microhabitats occurs within an individual ecosystem, such as a rocky shore or woodland.

Life in water is affected most by the availability of sunlight and of inorganic nutrients. Light does not travel far through water; consequently, most life in the sea is found near its surface. Inorganic nutrients are either washed off the land through leaching or are brought from the sea bed to the surface by upwellings. In open oceans, the availability of inorganic nutrients is further limited by a boundary, the thermocline, which forms between the warm surface layer and a cool deeper layer, where decomposition and nutrient recycling occur. Thus, marine life is most common in the shallow water around continents. Coral reefs are found in some tropical coastal waters and provide a sheltered environment which supports a characteristic ecosystem.

Chapter 31: Answers to Quick Questions

1 Secondary succession which occurs when gardens or forest clearings are uncultivated, after fires or when valleys flood.
2 The fungus will receive organic nutrients and oxygen from the alga whilst providing the alga with attachment, inorganic nutrients and protection against desiccation.
3 Trap and enrich sediments into which other plants can then root.
4 (a) More tolerant of desiccation compared to plants lower on the shore but a poorer competitor for space.
(b) Less tolerant of dessication compared to plants higher on the shore but a better competitor for space.
5 Plants do not move and are therefore easier to study than animals.
6 Tropical rain forests, tropical savanna, semi-desert, desert, semi-desert, maquis, coniferous forest, deciduous forest, coniferous forest.
7 Clouds cover the areas of tropical rain forest whilst desert areas are clear.
8 Semi-desert conditions.
9 During the summer the continuous sunlight in these areas ensures high productivity and so plenty of food for migratory birds and mammals.
10 Effects of rain shadows, in the case of the Gobi, behind the Himalayas. Cold deserts.
11 Trees require more rain throughout the year to grow and cannot stand being continually burnt.
12 Reduction in water loss.
13 Large seaweeds need to be fixed to rocks otherwise they will drift rapidly in ocean currents. However, a seaweed which was fixed to the bottom of the ocean would be in perpetual darkness.
14 The distances involved are too great and the thermocline acts as a barrier.
15 Nutrient-rich water found in upwellings will support rapid primary production which will, in turn, support a large number of secondary and tertiary consumers, e.g. fish.
16 (a) Little change in temperature with increased depth.
(b) A rapid change in temperature with increased depth.
(c) Temperature practically constant with respect to increased depth.
17 Changes in salinity as the tide ebbs and flows.

Theme 7
EXAMINATION QUESTIONS

7.1 The figure shows a profile of the four growth phases in the life cycle of the heather, *Calluna vulgaris*. The phases are identified as Pioneer (P), Building (B), Mature (M), and Degenerate (D).

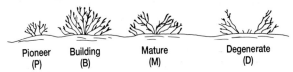

Pioneer (P)   Building (B)   Mature (M)   Degenerate (D)

(a) The bar charts show the percentage cover of *Calluna*, other dwarf shrubs, and grasses in areas typical of each growth phase.

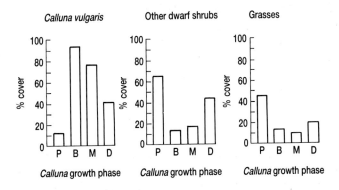

Calluna vulgaris   Other dwarf shrubs   Grasses
Calluna growth phase

   (i) Describe how the estimates of percentage cover could be carried out.

   (ii) Account for the differences in the percentage cover of *Calluna* between
1. the pioneer and the building phases
2. the mature and the degenerate phases.   (5)

(b) Explain the percentage cover values obtained for other dwarf shrubs and grasses at the pioneer and building phases.   (4)

(c) Heather moors are subject to management so as to provide ideal conditions for game birds such as the red grouse. Adult grouse feed mainly from young, succulent heather shoots but there is evidence that the chicks eat many insects, especially crane flies. The main form of management is by burning. This is carried out after about 12 years, usually in the autumn or early winter. Estate managers prefer to burn after two or three days of heavy rain followed by two or three days of good drying conditions.

   (i) State, giving a reason for your answer, which growth phase of *Calluna* would provide the best food supply for adult grouse.

   (ii) Which phase would provide the best shelter for nesting?

   (iii) Suggest why the young broods of grouse often move away from the heather into more exposed, but damper areas, especially after recent burning?

   (iv) Why do you think estate managers prefer to burn the heather after a period of rain followed by good drying conditions?   (6)
(WJEC 1989)

7.2 The table summarises the results of an ecological investigation on some animals living on nettle plants.

Species	Population density/ number m^{-2}	Mean individual mass/g	Notes on food
Plants Nettle	23	38.0	Autrotrophic
Insects Large nettle aphid	5420	0.002	Feeds from phloem of nettle
7-spot ladybird	21	0.047	Aphids
Flower bug	–	–	Aphids
Bush cricket	–	–	Nettle, aphids, flower bugs
Birds Whitethroat	–	–	Feeds generally on insects

– data not available

(a) **(i)** Draw a food web linking all the species listed in this table.   (2)

   (ii) From your food web, name **one** organism that feeds at two trophic levels.   (1)

(b) **(i)** Calculate the biomass of the nettle plants, the aphids and the ladybirds. Write your answers in a suitable table.   (2)

   (ii) For these three organisms, sketch a pyramid of numbers and a pyramid of biomass.   (2)

   (iii) Explain why a pyramid of energy would differ from the corresponding pyramid of biomass.   (3)

(c) Give **two** reasons why not all the light energy reaching the nettle plant can be used for photosynthesis.   (2)

(d) Explain how nitrogen contained in an animal such as a bush cricket is made available to nettle plants.   (4)

The passage of nutrients through a food chain was investigated by labelling the dominant plants in an ecosystem with radioactive phosphorus, ^{32}P. The graph overleaf shows the amount of radioactivity in different organisms in the six weeks following the end of the period of labelling.

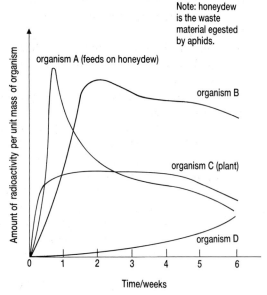

Note: honeydew is the waste material egested by aphids.

(e) Suggest a *biological* explanation why
  (i) radioactive phosphorus was used in this investigation rather than a radioactive isotope of carbon. (2)
  (ii) the amount of radioactivity in the plant falls after 4 weeks. (2)
(f) Draw a food chain linking organisms **B, C** and **D**. Explain how you arrived at your answer. (2)
(g) What is the evidence from the graph and the table to support the view that radioactive phosphorus may be transported in the phloem? (2)

(AEB 1990)

7.3 (a) Explain what is meant by the term 'ecosystem'. (4)
  (b) With reference to examples from *named* ecosystems explain what you understand by each of the following.
  (i) Pyramid of biomass
  (ii) Edaphic factors
  (iii) Decomposers
  (iv) Succession (16)

(ULSEB 1991)

7.4 The European spruce sawfly (*Gilpinia hercynia*) has recently become a pest of spruce trees (*Picea spp.*) in coniferous plantations in Britain. The adult and larval sawfly are found on spruce from July until October, large infestations causing considerable damage to needles.
Many species of bird prey on the sawfly larvae. Insectivorous birds eat mainly live larvae, but if they cannot find live food they will take dead larvae. Seed eaters feed larvae to their young. Planting deciduous trees increases the numbers of small birds.

The sawfly larvae are attacked by the Nuclear Polyhedrosis Virus (NPV) which kills them. There are two methods of detecting the virus:
Method A–Using the light microscope, at the limit of its resolution, groups of viruses can be seen in the gut of infected larvae, and in the droppings of birds which have eaten them.
Method B–Larvae are fed with leaves which have been sprayed with a solution of bird droppings. If these larvae become infected with the virus, this indicates that the bird droppings were also infected. This method of assessment can be carried out at any time of year.
Large quantities of the virus can be produced in the laboratory, using culture techniques.
The data below are from a study of the relationships between spruce sawfly, NPV and 6 bird species. Birds were caught and their droppings collected. The droppings were then assessed for the presence of the virus using both the methods describe above.

Life cycle and habitats of the spruce sawfly.

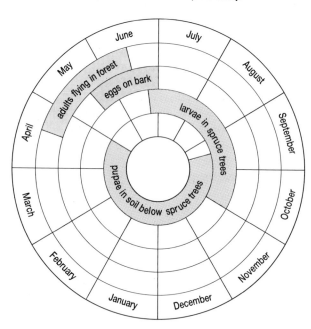

Table: Incidence of NPV in dropping of different bird populations

Species		Month	Number caught	Number of birds with virus in their droppings	
				Method A	Method B
Coal Tit	(I)	January	2	1	1
		September	20	13	20
Blue Tit	(I)	January	2	0	1
		September	3	1	3
Goldcrest	(I)	January	5	0	5
		September	14	7	14
Robin	(I)	January	1	0	1
		September	7	1	7
Chaffinch*	(S)	September	23	2	13
Bullfinch*	(S)	September	1	0	1

* – no data available in January
(I) – Insectivorous
(S) – Seed Eater

(a) Summarise the information provided in the Figure. (4)

(b) (i) Account for the differences in viral infection between the chaffinch and the bullfinch. (2)

    (ii) Comment on these differences. (2)

(c) Account for the different incidence of the virus in the droppings of insectivorous birds in both January and in September. (2)

(d) Laboratory experiments show that, three hours after feeding on an infected larva, no virus is retained within the bird's body. Despite this, bird droppings contain the virus in January. Explain this, giving your reasons. (5)

(e) (i) State which of the two viral estimation methods is less efficient. (1)

    (ii) Suggest a reason for the apparent lack of reliability. (2)

(f) How might a forester exploit the bird/sawfly/virus system as a means of biological control of the spruce sawfly? (2)

(SEB 1989)

7.5 **(a)** (i) What is meant by 'net primary production' and what is its ecological significance?

    (ii) For a **named** ecosystem or habitat, outline how net primary production could be estimated. (8)

(b) (i) What factors can limit the size of populations? Explain how they can do so.

    (ii) For a given population, explain briefly how its size and geographical distribution may have evolutionary significance. (12)

(JMB 1989)

7.6 Identical plots of land were cleared of weeds prior to sowing pea seeds. The plots were kept weed-free for different periods immediately after sowing. After a period of nine weeks, all the plants were harvested and the pea plants and weeds in each plot were weighed. The results are shown in the graph below.

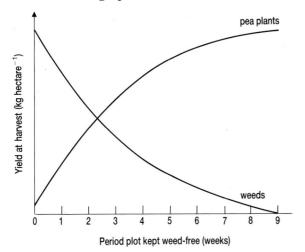

(a) (i) What conclusions can be drawn about the competition between the pea plants and the weeds?

    (ii) Describe how you might estimate the total mass of weeds growing in a large field. (5)

(b) Competition can also occur between plants of the same species. Design an experiment to test the hypothesis that maize plants grown at a high density produce fewer grains per plant than maize plants grown at a low density. (6)

(c) Assuming the hypothesis in **(b)** was supported by experimental evidence and that maize plants respond to competition with weeds in a similar way to pea plants, state the advice you would give to a farmer who wishes to obtain the maximum yield of maize from a field. (2)

(WJEC 1988)

7.7 It may be possible to increase the use of plant biomass as a source of fuel. Scientists have therefore investigated the production and re-cycling of plant biomass in three forest ecosystems by determining the dry mass of certain components. The results are given in the table below.

Types of forest	Total dry mass in kg m^{-2}			
	Living plant biomass	New plant material per year	Plant litter production per year	Humus content of soil
Coniferous	26.6	0.7	0.5	4.5
Deciduous	40.7	0.9	0.7	1.5
Tropical rain	52.5	3.3	2.5	0.2

(a) (i) Comment on the differences in total living plant biomass between the three ecosystems (3)

    (ii) Compare and comment on the production of new plant materials per year in relation to total living plant biomass in the three ecosystems. (4)

(b) In each ecosystem the mass of plant litter produced per year is less than the mass of new plant material produced per year. Suggest *two* reasons for this. (2)

(c) From the data for tropical rain forest, what predictions might be made concerning the annual production of new herbivore biomass in this ecosystem? Explain your answer. (3)

(d) The conversion of plant litter into humus is an important stage in the re-cycling of minerals. What do the ratios of litter production to humus content suggest about the relative rates of mineral re-cycling in the three ecosystems? (3)

(e) Using the information in the table, compare these three forest ecosystems in terms of their suitability as sources of biomass for fuel. (5)

(ULSEB 1990)

7.8 The diagram shows some aspects of nitrogen cycling in a tropical rain forest. Figures are in kilograms of nitrogen per hectare per year (kg nitrogen ha^{-1} year^{-1}).

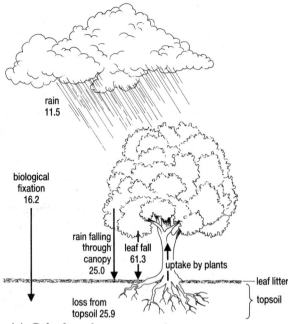

(a) Calculate the amount of nitrogen available for uptake by the plants. Show your working. (2)

(b) **(i)** In what form is nitrogen present in leaf litter?

(ii) In what form is nitrogen taken up by most plants?

(iii) Briefly describe the role of microorganisms in converting the nitrogen contained in leaf litter to a form able to be taken up by plants. (5)

(c) Account for the increase in nitrogen in water draining through the canopy. (1)

(AEB 1989)

7.9 The figure below shows the feeding relationships in a marine tidal community. There are two predators shown, a starfish and a snail. On each arrow a pair of numbers is given. The figure on the left in each bracket is the percentage of the total number of prey consumed and the figure on the right is the percentage of the total energy intake. For example, 63% of the prey taken by the starfish are acorn barnacles and they represent 12% of the total energy intake of the starfish. In an investigation, all the starfish were removed from an experimental area on the shore. Soon afterwards, the number of acorn barnacles increased and they occupied most of the space in the experimental area. Later, the bivalves and goose barnacles replaced the acorn barnacles, with the bivalves occupying most of the space. The original community was thus virtually reduced to two species.

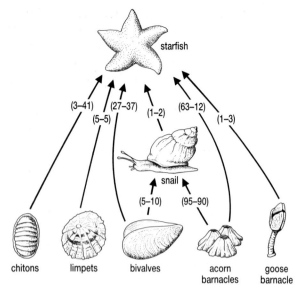

By reference to the information given

(a) Suggest a reason why the acorn barnacles became so abundant soon after removal of the starfish.

(b) Suggest reasons why the bivalves did not take over most of the space in the presence of the starfish.

(c) Indicate the major significance of predation by the starfish to the community as a whole. (10)

(UCLES 1988)

7.10 Here are three pyramids of numbers from different food chains.

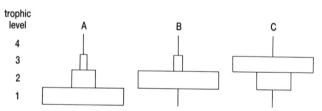

(a) **(i)** Name the four trophic levels in pyramid A and give a typical example of an organism found at each level. (The organisms you suggest should all come from the same food chain.)

(ii) Account for the shapes of pyramids B and C. (4)

(b) The following pyramids were estimated for a single food chain in a terrestrial ecosystem.

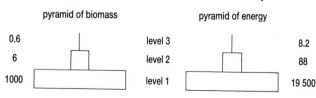

(i) In what units might the figures for each pyramid be expressed?

(ii) Calculate the energy content per unit biomass of organisms in levels 1 and 3.

(iii) Calculate the percentage energy flow from level 2 to level 3 in this ecosystem. Give one reason why this value is not 100%? (6)

(c) The table below relates the number of fish per unit area to the annual rate of population growth. An annual rate of population growth of 1.0 signifies no net change in population.

Number of fish per unit in year 1	10	15	20	25	30	35	40	45
Annual rate of population growth	0.5	0.75	1.0	4.3	5.0	3.9	1.5	0.1
Number of fish per unit in year 2								

(i) Calculate the missing values in the table.
(ii) Consider the column in the table with the highest number of fish per unit area in Year 2. How many fish should be harvested in Year 2 in order to maintain the same annual rate of population growth in Year 3? Explain your answer.
(iii) Suppose that there are 40 fish per unit area, of which 25 are harvested prior to reproduction. What would be the annual rate of population growth as a result of this cropping activity? What would be the likely effect of continuing this level of cropping activity on the future of the population? (5)

(JMB1987)

7.11 (a) Define the term *population*. (2)
(b) With reference to examples of **named** organisms, discuss the factors which limit the sizes of populations in nature. (11)
(c) Describe, with examples, how parasites and predators are used in attempts to control pests. (5)

(UCLES 1988)

7.12 The graph below shows how net primary production of the marine alga *Hormosira banksii* varies with depth and sea temperature.

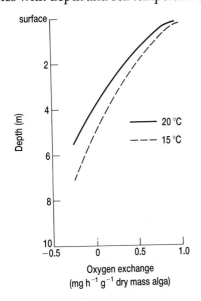

(a) What is meant by the term net primary productivity? (1)
(b) (i) Explain why oxygen exchange can be taken as an indicator of net primary productivity.
(ii) On the graph, oxygen exchange is expressed as mg h^{-1} g^{-1} dry mass of alga. Describe how you would find the dry mass of a sample of the fresh material in the laboratory.
(iii) Using the graph, state the relationship between net primary productivity and depth. Suggest an explanation for this relationship. (5)
(c) (i) Show clearly on each curve the position of the compensation point. Give the reason for your choice of position.
(ii) Account for the difference in the positions of the compensation points at each temperature.
(iii) Would *H. banksii* be able to survive at greater depths in warmer or colder seas? Explain. (4)

(JMB 1987)

7.13 (a) Explain the following ecological terms:
(i) ecosystem
(ii) producer
(iii) consumer. (3)
(b) Describe the way in which energy flows through a typical ecosystem. (4)
(c) Quantitative studies of ecosystems often involve the construction of pyramids such as:
(i) pyramids of numbers,
(ii) pyramids of mass and
(iii) pyramids of energy (productivity).
Outline the advantage and disadvantage of each of the above methods of depicting relationships between different trophic levels. (4)
(d) Describe the way in which nitrogen is cycled in a typical ecosystem. (5)
(e) The high levels of productivity in modern intensive crop production demand the use of large quantities of artificial fertilisers which are produced industrially. Describe how the manufacture and use of fertilisers may result in:
(i) atmospheric pollution
(ii) eutrophication. (4)

(NISEC 1989)

Answer Guidelines to Question 7.1
Factual recall of moorland ecology could not be expected from this syllabus. This question therefore involves the ability to apply your knowledge and understanding of ecology to interpret and explain the information given in the illustrations. People with

specialist knowledge might object to ling (*Calluna vulgaris*) being referred to as heather (*Erica cinerea*).

(a) (i) place a quadrat frame over the ground; either at random or along a line transect; estimate how much of the area of ground within the frame is covered by each type of plant;

(ii) 1: in the pioneer phase the stems and leaves of the heather plants are small and cover little area whereas in the building phase they are large and cover almost the entire area within the quadrat frame;

2: in the degenerate phase the stems in the middle of the plant have died and so will not be counted as covering the area within the quadrat frame whereas during the mature phase there are healthy stems in this part of the heather plant;

(b) Pioneer: all the plants are in competition; the heather plants are small/pyramid shaped so they do not block out all the light so that other plants can grow;

Building: the stems of the heather plant are large/hemispherical shape and shade the ground so that few other plants can photosynthesise and grow; the roots of the heather are now extensive and they absorb most of the available water/inorganic ions;

(c) (i) Building phase; more heather available than pioneer stage and growth of young stems still occurring/productivity greatest;

(ii) degenerate phase has enough space for nests

(iii) shortage of food; young shoots emerging in damper areas provide food for grouse;

(iv) rain enables seeds of undesired plants to germinate, burning then kills them;

ANSWERS TO IN-TEXT QUESTIONS, TUTORIALS AND ANALYSES

Theme 1
Chapter 1

1.1 **(a)** a sodium ion has one less electron than a sodium atom;

(b) an oxygen molecule is made of two oxygen atoms joined by a double covalent bond;

(c) each element is made of atoms with a unique number of protons, neutrons and electron;

1.3 it has a hydrogen ion concentration of $10^{-4.3}$ mol dm^{-3};

1.5

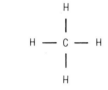

1.7 it is polar and so can becomes part of water's hydrogen-bonded structure;

1.8 sweat only cools if it evaporates. It sweat were pure water (it has dissolved solutes), it would remove 2.26 MJ kg^{-1} K^{-1} of heat from the surface of the skin as it turned into vapour;

1.10 **(a)** lactose + water → glucose + galactose;

(b) $C_{12}H_{22}O_{11} + H_2O \rightarrow C_6H_{12}O_6 + C_6H_{12}O_6$;

(c)

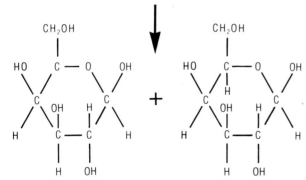

1.11 cows do have enzymes, amylases, which can break α 1,4 glycosidic bonds in starch but which cannot

recognise β 1,4 glycosidic bonds in cellulose; the bacteria in the rumen do produce enzymes, called cellulases, which can recognise and break β 1,4 glycosidic bonds;

1.12 insulation against heat loss; waterproofing; insulation against mechanical shock; insulation against electrical leakage; source of chemical energy; as components of membranes; hormones;

1.14 the phosphate group is polar and mixes with water;

1.15 the oil may protect the fish against heat loss; being lighter than water, the oil will give the fish buoyancy; oil is liquid at sea water temperatures;

1.16 carbohydrates are assembled by specific enzymes – it is the enzymes which are coded by DNA; the saccharide units in carbohydrates are mainly the same;

1.17 **(a)** and **(b)** – see Fig 1.27
(c) peptide bond;

1.18 see Fig 1.28

1.19 see Fig 1.26

1.21 DNA has deoxyribose but RNA has ribose; DNA has thymine instead of uracil; DNA is made of two antiparallel strands (double helix) but RNA is a single strand which bends back on itself;

1.22 **(a)** the hydrogen-bonded base pairs (C–G and A–T);
(b) the sugar-phosphate-sugar backbone;

1.23 **(b)** is true, since adenine pairs with thymine and cytosine with guanine A/T = C/G = 1;

Tutorial: Monosaccharides (single sugars)

(a) glyceraldehyde is a triose; ribose is a pentose; fructose and glucose are hexoses;

(b) aldose; it has a terminal aldehyde (C O) group;

(c) number the carbon atoms 1 to 6, starting at the top;

(d)

(e) the hydrogen and hydroxyl groups are on different sides of cabon atom 1 (^{1}C);

(f) α form; the hydrogen group is uppermost on ^{1}C;

Analysis: Using sugars for respiration

If you use a pencil, you can push a full fermentation tube to the bottom of an inverted test tube. By *quickly* turning the test tube around, you should be able to assemble the apparatus as shown in the Figure.

Prepare an active yeast culture and several sugar solutions of identical concentration. Pipette an equal volume of each sugar solution separately into labelled fermentation tubes and add an equal volume of yeast culture to each tube. Top up the fermentation tubes with water if they are not absolutely full.

Push each tube into an inverted test tube and quickly turn the test tubes upright. Measure and record the height of the air bubble in each fermentation tube. Incubate the tubes at 40 °C for two hours and re-measure the height of the air bubble in each fermentation tube. The difference in heights is related to the rate at which each sugar has been used by the yeast as a repiratory substrate.

Chapter 2

2.1 (a) chemical analysis;

(b) area of phospholipid is twice that of cell surface membrane of cell from which it was extracted;

(c) freeze-fracture technique;

2.2 (a) hydrophilic mixes with water; hydrophobic does not mix with water;

(b) a pore running through an intrinsic protein in the cell surafce membrane which allows the passage of water-soluble molecules and ions through the membrane;

2.4 protein; different specific sites result from different structure of proteins on surface; phospholipids and carbohyrates lack the structural variety of proteins;

2.5 (a) the greater the difference the faster the rate;

(b) the larger the molecule the slower the rate;

(c) the higher the temperature the faster the rate (until such a high temperature is reached that the membrane is destroyed);

(d) the less polar, the faster the rate (they will dissolve in the lipid of the membrane);

2.6 lipids dissolve in and pass through the phospholipid bilayers; ions are surrounded by water and so have to diffuse through hydrophilic channels in the centre of intrinsic protein molecules;

2.7 by providing hydrophilic channels and carrier molecules for facilitated diffusion;

2.8 salt and sugar solutions are hypertonic to the contents of the cells of decomposers; water leaves the cells of decomposers by osmosis; without adequate water, cell activity stops;

2.9 $\Psi = 0$;

2.10 (a) $Z \rightarrow Y \rightarrow X$;

(b) the difference in potential is $X \rightarrow Y = 10$ kPa and $Y \rightarrow Z = 15$ kPa; osmosis is greatest between Y and Z since the difference in water potential is the greatest;

2.12 for system A: $\Psi_{cell} = -50$ kPa $+ 20$ kPa $= -30$ kPa; for system B: $\Psi_{cell} = -40$ kPa $+ 15$ kPa $= -25$ kPa; water passes from B to A (less negative to more negative potential);

2.13 (a) both involve intrinsic proteins in membrane which combine with substance to be transported and undergo a conformational change;

(b) conformational change involves ATP hydrolysis during active transport but not during facilitated diffusion;

2.14 rate of uptake of glucose is greater than that of xylose, i.e. uptake mechanism is selective; rate of uptake of xylose unaffected by cyanide, therefore is by passive transport (e.g. diffusion); uptake of glucose is reduced by cyanide, therefore is by a mechanism involving active transport (i.e. involving ATP);

2.15 A intrinsic protein; B hydrophobic hydrocarbon tail of phospholipid; C polysacharide of glycocalyx; D hydrophilic head of phospholipid; E extrinsic protein; F diffusion of lipid-soluble substance; G diffusion of water-soluble substance; H facilitated diffusion; I active transport;

2.16 (a) they stopped (they use energy from ATP hydrolysis);

(b) it would have increased by diffusion;

(c) influx of Na^+ ions would make cell potential more negative so it would take in water by osmosis, increasing volume and pressure on cell surface membrane;

Analysis: The experiments of Gorter and Grendel

(a) the hydrophilic heads (circles in the diagram) mix with water and the hyrophobic tails avoid the water;

(b) the two would have the same value;

(c) the phospholipids are arranged in a double layer (a bilayer) with the hydrophobic ends in the middle;

Analysis: The factors affecting diffusion

(a) temperature; concentration difference; molecular size;

(c) the higher the temperature the faster the rate of diffusion; the lower the molecular weight the faster the rate of diffusion; the greater the difference in concentration the faster the rate of diffusion;

(d) as the gases diffuse down the tube their concentration decreases in an exponential fashion i.e. $1 \rightarrow \frac{1}{10} \rightarrow \frac{1}{100}$; since rate of diffusion is proportional to concentration difference, this gives rise to a curvilinear rather than a straight line relationship;

Analysis: Haemodialysis

(a) 200 cm^3 $min^{-1} = 12$ dm^3 $hour^{-1} = 36$ dm^3 in three hours;

(b) diffusion;

(c) so that suitable diffusion gradients are maintained (i.e. the dialysis fluid always has a lower concentration of solutes than the blood;

(d) so that the blood does not lose heat as it is purified;

Analysis: The problems of living in sea water

(a) plot percentage of added freshwater (added fresh water/%) on the horizontal axis and use two scales on the vertical axis for the two dependent variables;

(b) as more freshwater is added the water potential of the cell becomes less negative; more water therefore enters the cell by osmosis; the cell does not burst because the contractile vacuole increases its activity;

(c) contractile vacuole activity is not sufficient to remove all the water which enters the organism, hence the relative body volume increases;

(d) large quantities of water are taken in as the animal feeds; this water is then removed by the contractile vacuoles;

Analysis: Water potential, turgor and plasmolysis

(a) it is a reaction of the cell wall against the expanding cell;

(b) (i) incipient plasmolysis occurs where the pressure potential becomes zero (cell no longer pushing on cell wall);

(ii) full turgor occurs when the pressure
potential is at a maximum, (further entry of water
prevented);

Chapter 3

3.2 high resolving power gives greater detail of small
objects; specimens must be dead and subjected to
distorting effects of processing, heavy metal salts and
vacuum;

3.4 processing of specimens for viewing in electron
microscopes is very harsh, possibly leading to the
production of artefacts;

3.5 it contained a nucleus during its development;

3.7 see information in Section 3.3;

3.8 (a) nuclei, mitochondria and lysosomes;
(b) ribosomes;

3.9 amino acids diffuse into cytosol; amino acids taken up
by ribosomes during protein synthesis; protein passed
into sacs of rough ER for modification and transport;

Analysis: The structure of cells as seen by the light microscope

(a) it is chemically different from the cytoplasm (contains
an acid (DNA) which has reacted with stain);

(b) too thin to be resolved by light;

(d) cell wall; cell vacuole; tonoplast; large granules;

Analysis: Identifying and measuring organelles

(a) cell nucleus;

(b) A nuclear envelope; B nuclear pore; C nucleolus; D
starch granule; E granum; F stroma; G crista;

(c) eukaryotic; has nuclear envelope and cell organelles;

(d) plant; has chloroplast and cell wall;

(e) measured from the photograph, its width is 36 mm;
since this figure is magnified 9000 times, its real size is
$(36 \times 10^{-3}/9000)$ m, i.e. 4×10^{-6} m = 4 µm;

(f) mitochondrion;

(g) measured from the photograph, the mean width is
about 8 mm = 8×10^{-3} m; therefore to find the real
width you must divide this value by the
magnification given in the figure;

$$\frac{8 \times 10^{-3}}{3 \times 10^4} \text{ m} = 2.7 \times 10^{-7} \text{ m} = 0.27 \text{ µm} = 270 \text{ nm}$$

Analysis: Following a biochemical pathway

(a) label carbon atoms in glucose using ^{14}C; stop reaction
at fixed intervals; extract sugars; separate sugars
using chromatography; place chromatogram on film
in dark for two days; develop film and identify
radioactively labelled compounds;

(b) (i) neither supports nor refutes because entire
homogenate was oxygenated (i.e. both cytosol
and mitochrondria);

(ii) supports because whole pathway can proceed in
oxygenated homogenate, but does not support
two-stage pathway;

(iii) supports because pyruvate breakdown to carbon
dioxide can only have occurred in mitochondria
(N.B. this is the only experiment in which you can
define where the reaction occurred);

(iv) supports because glucose to pyruvate is an
anaerobic process;

Chapter 4

4.1 (a) proteinaceous enzymes determine cell's activities
and their activity is crucially dependent on their
shape, which is determined by their primary
structure;

(b) sequence of three RNA bases which specify a
single amino acid;

(c) start (methionine), isoleucine, arginine,
phenylalanine, serine, glycine;

4.2 DNA is removed along with nucleus yet protein
synthesis continues (because mRNA in cytoplasm is
destroyed only slowly);

4.3 (a) true; DNA → RNA occurs inside nucleus;

(b) true; see Fig 4.10;

(c) false; amino acids are carried by specific tRNA
molecules;

(d) true: once they reach the terminal end of a mRNA
molecule they can rejoin the 'start' end again;

(e) false: ribosomes attached to mRNA during
translation, transcription does not involve
ribosomes;

(f) false: it occurs in the cytoplasm;

4.4 (a) 66 ((20×3) + 3 for 'start' and 3 for 'stop');

(b) at least 66 (see Section 4.3);

(c) 12

4.5

TAC	CAT	AGC	ACA	AGG	ATG
ATG	GTA	TCG	TGT	TCC	TAC
UAC	CAU	AGC	UCU	AGG	AUG
AUG	GUA	UCG	TGT	UCC	UAC
Tyr	His	Ser	Ser	Arg	Met

4.6 (a) false; the code is degenerative – several codons
code for a single amino acid;

(b) false: 'stop' and 'start' codons are also present;
bases for introns are also present although introns
are removed from functional protein;

4.7 (a) 2 ($8^2 = 64$);

(b) 30 (one per amino acid);

(c) 64 (one per codon);

4.8 not bound by histones; circular, not linear;

4.9 see Fig 4.13

4.10 **comparison:** both have promoter lying near functional
gene to which RNA polymerase attaches during
transcription; both have a part of the DNA that affects
attachment of RNA polymerase to promoter gene;
contrast: prokaryote has operator gene which
eukaryote does not have; prokaryote has regulator
gene which eukaryote lacks; eukaryote has
transcriptional enhancer region which prokaryote
lacks; chemical from outside cell prevents binding of
substance to prokaryotic gene (sugar) whereas it binds
to eukaryotic gene (oestrogen);

4.12 DNA replication has already occurred, so each
structure consists of two copies of the chromosome
(called chromatids while they are still held together);

4.15 (a) condensed when dividing so become large
enough to be resolved;

(b) products of DNA replication;

4.16 interphase – chromosomes replicate
prophase – chromsomes condense and spindle forms
metaphase – chromosomes line up on equator of
spindle;
anaphase – chromatids pulled to opposite ends of
spindle;

telophase – new nuclear envelope forms, spindle disappears;
cytokinesis – cytoplasm divides;

4.17 **(a)** (i) between homologous chromosomes;
(ii) between chromatids from a different chromosome;
(b) only one chromosome from each homologous pair is now present in each cell;

4.18 ✓ ✓ ✓;
✗ ✓ ✗;
✗ ✓ ✗;
✗ ✓ ✗;
✓ ✗ ✓;

4.19 **(a)** mitosis, late telophase;
(b) mitosis, anaphase;
(c) mitosis, metaphase;
(d) about to undergo mitosis, early prophase;

4.20 repeated DNA replication occurs but is never followed by separation of chromatids;

Analysis: How do we know DNA carries the genetic information?
(a) has passed inside bacterial cell in phage DNA;
(b) during replication of DNA, some of phosphate is used to make new DNA molecules; phage A had ^{32}P but phage B has only ^{30}P;
(c) radioactively labelled compound that enters the cell must carry the genetic code; it is ^{32}P – labelled DNA;
(d) results for phage A and phage B would have been reversed;

Analysis: The evidence for semi-conservative replication
(a) organic bases of nucleotides;
(b) each new molecule has one heavy chain from the parental DNA and one light chain made from ^{14}N;
(c) in each Generation 1 DNA molecule there is a light and a heavy chain; these separate during replication and each produces a new light chain; the Generation 2 cell which inherited the heavy chain will have a hybrid DNA, that which inherited the light chain will have only light DNA;
(d) the results are as predicted in (c) assuming that semi-conservative replication had occurred;
(e) (i) 1 heavy, 1 light, 1 heavy, 3 light;
(ii) 2 hybrid; 4 intermediate between light and hybrid;
(f) 4 hybrid, 14 light;

Theme 2

Chapter 5

5.5 catalyse the decomposition of 50 000 molecules of hydrogen peroxide per second at O °C;
5.6 **(a)** rate of reaction at 20 °C = $1/8 \times 5$ = 0.025 cm^3 min^{-1}; rate of reaction at 30 °C = $1/4 \times 5$ = 0.050 cm^3 min^{-1};
(b) Q_{10} = 0.050/0.025 = 2.0;
5.7 **(a)** up to the optimum temperature the increasing kinetic energy of the substrate and enzyme molecules means more collisions and so a faster rate of reaction;

(b) above the optimum temperature the enzyme begins to denature;
5.9 Your answer should relate to the denaturation of the enzyme in the washing powder at temperatures higher than that recommended.
5.10 **(a)** malonate is a competitive inhibitor of succinic dehydrogenase – notice the chemical similarity of succinate and malonate;
(b) competitive inhibitors become less effective when their concentration is low relative to the substrate; design a test to find out if the reaction rate is affected by malonate concentration;
5.11 **(a)** (i) heating denatures enzymes that cause decay;
(ii) sealing stops entry of airborne micro-organisms;
(b) if the insect had been in the can prior to sealing, its enzymes would have been denatured; if its amylase is found to be active, the insect almost certainly fell into the can after it had been opened;

Analysis: An enzyme investigation
(a) distilled water contains no impurities that might affect the reaction; a water bath maintains a constant temperature, 37 °C is likely to be about the optimum temperature for the amylase; during the 30-minute period the contents of all tubes will stabilise at 37 °C; all conformational changes in the enzyme molecules are likely to have occurred within ten minutes;
(b) to maintain a constant pH irrespective of chemical changes in the tubes;
(c) add a drop of test solution to a drop of iodine in potassium, iodide solution; starch is indicated by a change from yellow to blue-black;
(d) about pH 7; to find the optimum pH more precisely, use pH buffer at smaller intervals, e.g. 6.1, 6.2, 6.3 ... 7.8, 7.9, 8.0;
(e) a replicate that is identical to the experiment in all respects save the one under investigation;
(f) no starch hydrolysis would have occurred during the experiment;
(g) they lack enzyme and so confirm that pH affects the activity of enzyme and not the intrinsic reaction of starch;
(h) your account should include: preparation of buffer solution with pH value at optimum using distilled water; preparation of starch suspension and enzyme solution using this buffer solution; set up range of ·tubes containing a fixed volume of enzyme solution, each one in a water bath at a different fixed temperature; incubation of starch suspensions at same temperature as amylase; incubation for 30 minutes; pipette a fixed volume of starch suspension into each enzyme solution; test for starch each minute; use of control, e.g. volume of buffer solution with no enzyme treated exactly as each enzyme solution;

Chapter 6

6.3 pH greater in mitochondrial matrix than in intermembranal space;
6.4 **(a)** 6/7 = 0/86
(b) 36/51 = 0.71;

6.5 **(a)** 1 mol glucose ... 3000 kJ, 1 mol AT... 30 kJ;
efficiency of anaerobic respiration =
$2 \times 30/3000 = 2\%$;

(b) efficiency of aerobic respiration = $38 \times 30/3000 = 38\%$; aerobic respiration is 19 times more efficient;

(c) Over the first 10 hours, the glucose is metabolised to ethanol and the yeast population increases in number. After a time lag, the ethanol is then metabolised to CO_2 and H_2O and a second increase in the number of yeast cells occurs;

6.6 **(a)** the sprinter has enough ATP in her/his muscles to last the race;

(b) to repay the oxygen debt;

(c) the cells of the sprinter's liver;

(d) lactate inhibits muscle contraction; glycogen is stored in the liver;

6.7 a sealed can is an anaerobic environment in which *Clostridium botulinum* can thrive; its toxin accumulates in the food without signs of spoilage and may be eaten;

Tutorial: Oxidation and reduction

(a) $Fe^{2+} \rightarrow Fe^{3+} + e^-$;

(b) NAD^+ oxidised; NADH reduced; FAD oxidised; $FADH_2$ reduced;

(c) Cu^{2+} reduced, NADH oxidised; FAD reduced;

Analysis: The energy yield from glucose

(a) $38 \text{ mol} \times 30 \text{ kJ mol}^{-1} = 1140 \text{ kJ}$;

(b) $(1140 \div 3000) \times 100\% = 38\%$;

(c) 1860 kJ;

Analysis: The ATP tally during aerobic respiration

(a)

Glycolysis	4,	2,	2,	0
Link reaction	0,	0,	2,	0
Krebs cycle	2,	0,	6,	2

(b) 30 (3 per NADH molecule);

(c) 4 (2 per NADH molecule);

(d) 34;

(e) 6 (see table);

(f) 2 (see table above);

(g) 38;

(h) the intermediates in glucose catabolism, e.g. NADH and pyruvate, may be used in other reactions in the cell, e.g. the manufacture of amino acids;

(i) the most likely advantage is that energy is released in 'packets' which can be used without excessive heat loss;

Analysis: Measuring respiratory quotients

(a) decrease as they are used in respiration;

(b) absorbed by the KOH;

(c) decreases;

(d) pressure will fall in left-hand tube;

(e) manometer fluid will be pushed upwards in the left-hand side of the U-tube;

(f) volume of oxygen used = $\pi r^2 h$, where r is internal radius of the tube and h is the distance moved by the manometer fluid;

(g) pressure is affected by temperature changes (the so-called 'General Gas Law');

(h) to act as a thermobar, i.e. to replicate and nullify any volume changes resulting from changes in atmo-

spheric pressure or temperature;

(i) the volume change is now (CO_2 given off – O_2 used), therefore, volume of CO_2 produced = volume change in second experiment minus volume change in first experiment;

(j) RQ = 18/20 = 0.9, suggesting proteins are being respired;

(k) air leaks at any of the joints; inconstant bore diameter of manometer tubing; inaccuracy in making readings of manometer levels; KOH failing to absorb all the CO_2; damage to organisms by contact with KOH;

Analysis: The regulation of glucose catabolism

(a) decrease;

(b) inactivate;

(c) activate ICDH and increase activity of Kerbs cycle;

(d) NAD promotes ICDH activity, NADH inhibits ICDH activity;

Chapter 7

7.1 it synthesises energy-rich organic molecules using the energy from sunlight and organic molecules as substrates;

7.2

NADP	CO_2
H_2O	ATP
ADP + P	NADPH
ATP	GALP
NADPH	ADP + P
O_2	NADP

7.3 **(a)** light and chlorophyll are used to reduce CO_2 to form carbohydrate;

(b) the source of electrons is hydrogen sulphide, not water;

7.4 **(a)** ^{16}O: carbon dioxide is reduced to form carbohydrate;

(b) ^{18}O: water is a source of electrons used to reduce NAD; its breakdown (photolysis) releases oxygen.

7.5 in both experiments the oxygen-sensitive bacteria are located where oxygen is released by photosynthesising chloroplasts; in Experiment 1, photosynthesis (and associated oxygen release) occurs where the spot of red light strikes part of a chloroplast but not where it strikes colourless cytoplasm; in Experiment 2, photosynthesis occurs at those parts of the algal filament which are exposed to either red or blue light; (see Fig 7.7 to remind yourself of the absorption and action spectra of chlorophyll).

7.6 **(a)** true; **(b)** true; **(c)** true; **(d)** false; **(e)** false;

7.7 **(b), (d), (e)**;

7.8 ATP provides the energy to 'power' the reduction reactions; NADPH provides electrons to reduce it; CO_2 is the substrate;

7.10 **(a)** C_4 since the C_3 pathway is slow at low CO_2 concentrations;

(b) C_4 since closed stomata results in high O_2 concentrations and low CO_2 concentrations within leaf air spaces;

(c) C_4 since oxygen is a competitive inhibitor of RuBP carboxylase;

(d) C_4 since plenty of light is available for ATP

production (ATP is needed to form PEP);

7.11 no light-dependent reactions occur in these cells; cells which made grana would waste energy and be at a disadvantage over cells with poorly developed grana;

7.12 devise an experiment to measure the rate of photosynthesis;
  (a) at different intensities of white light;
  (b) at identical intensities of light of different wavelengths (using coloured light bulbs or filters);

7.13 (a) increases humidity as well as the concentration of CO_2;
  (b) reduces water loss while the cutting develops new roots;

7.14 (a) the chloroplasts will be towards the bottom of the cell;
  (b) the chloroplasts will be concentrated at the top of the cell;

7.15 reduce light penetration into leaf; reduce diffusion of CO_2 into leaf through stomata;

Analysis: Plant pigments and light absorption

(a) to do this calculation you must first measure the distance moved by the solvent (from origin line to solvent front); now measure the distance from the origin line to the centre of each spot of pigment; the Rf value is found by dividing the distance moved by each pigment by the distance moved by the solvent front;

(b) all wavelengths between 400 nm and 700 nm are absorbed but maximum absorption occurs around 430 – 460 nm (violet/blue) and 670 – 690 nm (red); once the blue and red parts of the spectrum have been absorbed, the remaining light is mainly green;

(c) they broadly correspond; the rate of photosynthesis is largely determined by the amount of light of different wavelengths absorbed;

(d) traps light of wavelengths which are not absorbed by chlorophyll, so acting as an accessory pigment;

Analysis: Herbicides and photosynthesis

(a) X – in a school or college laboratory it would not be convenient to measure this;
(b) F;
(c) X – it could inhibit the effect of CMV, you can't tell from the data;
(d) F – CMV inhibits photolysis in PSII;
(e) X;
(f) F;

Analysis: Determining the steps in the Calvin cycle

(a) instantly kills them, stopping photosyntheses after a precisely measured time interval;

(b) radiation from ^{14}C affects the photographic film, causing dark spots to appear after the film is developed;

(c) $^{14}CO_2$;

(d) by reduction of $^{14}CO_2$ during the Calvin cycle;

(e) after 5 seconds the five compounds identified in (a) are formed; after 15 seconds glycolate, phosphoglycolate, alanine and aspartate have been formed; after 30 seconds glutamate, glycine, serine, sucrose and glucose monophosphate have been formed. (Use

Chapter 1 to find the nature of these molecules and suggest the way that they might be synthesised, one from another.)

Tutorial: The law of limiting factors

(a) light intensity; light wavelength; CO_2 concentration; temperature; availability of water; availability of inorganic ions;

(b) (i) light and temperature are the most likely limiting factors;
   (ii) water availability and CO_2 concentration (closed stomata);

(c) no; increases in light intensity do not increase the rate of photosynthesis;

(d) CO_2 concentration and temperature are the most likely; (see answer to (a) above);

(e) your experiment should allow you to measure the rate of photosynthesis at different light intensities; you should repeat the experiment at controlled temperatures of, e.g. 10 °C, 15 °C, 20 °C etc;

(f) light intensity;

(g) the rate of photosynthesis increased faster in Q and in P; the rate of photosynthesis stabilised at a higher value in Q than in P;

(h) temperature is higher in Q;

(i) A–B was light intensity and temperature; B–C was temperature;

(j) CO_2 concentration;

(k) A–B was light intensity; B–C was some other factor (cannot say which from available data);

(l) in comparison with Curve R, Curve P was limited by low temperature and low CO_2 concentration; Curve Q was limited by low CO_2 concentration;

(m) Giving concise summaries is an examination skill worth practising. Try this one for yourself and check it with your tutor.

(n) use a gas/paraffin/solid fuel burner;

(o) other factors may be limiting the rate of photosynthesis, so further expenditure on one that is not limiting will not increase the rate of photosynthesis;

(p) no; these methods need a closed environment;

Chapter 8

8.1 (a) food molecules are too large to pass through cell surface membranes;
  (b) food is present in a digested form in the intestine of their host;

8.2 (a) herbivore; (b) parasite; (c) saprophyte;

8.3 digestive enzymes work away from the cytoplasm of the animal's own cells;

8.5 those without which the average mass was significantly less than the control (complete diet), i.e. leucine, phenylalanine, threonine;

8.6 by contraction of circular muscles behind the food; fibre makes the food bulkier, enabling it to be 'squeezed' along more easily;

8.7 dental caries = tooth decay, perodontal disease = gum inflammation; children's teeth are less calcified than those of adults; adults have had longer to damage gums by inefficient brushing;

8.11 (a) see Fig 8.12 and Table 8.8
  (b) see Fig 8.15 and Table 8.16

8.14 Ability to produce amylase may be inherited and may be affected by the presence of amylose in the diet. We do not know the diet of the Tswans in Botswana and so cannot make valid conclusions about the effect of inheritance or diet by comparing Tswans and Europeans. Alternatively amylase may be an inducible enzyme and Tswans may have a diet richer in amylose than Europeans. This view is supported by the data on Bushmen. It is fair to assume that the bushpeople of the Kalahari are genetically homogenous; it can be assumed that the difference between the mean amylase activity of the two groups of bushpeople results from the different dietary regime during captivity.

Analysis: A balanced diet

(a) You should make point-by-point comparisons rather than separate accounts for females and for males. The following represent some of the major comparisons to be made.

   (i) the recommended energy intake for females is less than that for males of the same age; in both sexes, the total energy requirement increases up to the age of 14 (though *rate* of increase gets less); in females this increase is threefold, in males it increases by a factor of 3.4; between the ages of 14 and 17, the energy requirements of females remains constant whereas in males it continues to increase;

   (ii) the recommended intake of protein is less for females than that for males of the same age; in both sexes, the total protein requirement increases up to the age of 14 (though *rate* of increase gets less); in females this increase is almost threefold, in males it increases by a factor of 3.5; between the ages of 14 and 17, the protein requirements of females remains constant at 53 g whereas in males it continues to increase;

   (iii) the recommended calcium intake is identical in females and males of the same age; it remains constant between the ages of 1 to 8, increases from 600 to 700 mg between the ages of 9 and 14 and drops again between the ages of 15 to 17;

(b) (i) as humans become larger they require more energy to maintain their bodies; as males tend to be larger than females of the same age, they use more energy and so need a greater energy intake; the adolescent growth spurt tends to be later in males than females, helping to explain the continued increase in energy requirement after 14 years in males; after puberty, females have a thicker layer of subcutaneous fat; as a result they lose less energy as heat than males so do not need to replace this energy;

   (ii) protein is needed for growth; although growth continues until adulthood, the rate of growth decreases steadily from birth and this is reflected in the rate of increase in recommended protein intake; males tend to be larger throughout life, explaining their higher protein requirement; females reach their adult size earlier than males, explaining the constant recommended protein intake after the age of 14 when that for males

continues to increase;

   (iii) calcium is needed in the blood for blood clotting and by muscles for their contraction as well as for the growth of bones and teeth; throughout childhood, 600 mg of calcium per day is sufficient for the increases in the volume of blood and the size of the skeleton and muscular systems that accompany the growth of a relatively small person; the growth spurt during puberty requires more calcium because the body is now relatively large; after this growth spurt, 600 mg is sufficient to maintain calcium levels against losses, e.g. in urine;

(c) pregnancy increases the requirement for all these nutrients (give data from the table) because growth of the fetus requires energy and protein and bone development requires extra calcium; milk production requires even more energy and protein than fetal growth; the demands of the infant are greater than when it was a fetus because it is larger; calcification of the infant's skeleton continues at a rate similar to that of the fetus; any mothers that did not produce adequate milk would have had weak infants that were less likely to survive; milk production is likely to have been acted upon by natural selection;

(d) basal metabolism and metabolism during sleep are unaffected; energy expenditure is higher during more active work and more energy-containing food must be consumed to stay in energy balance;

(e) all efficiency indicators, such as metabolic rate, heart rate, muscle size, lung volume, decrease with age; as a result, the need for nutrients to fuel these activities decreases;

Analysis: Measuring the energy content of food

(a) $500 \times 5 \times 4.18 = 10.45$ MJ;

(b) the amount of water in bread is variable and contains no biologically useful energy;

(c) to ensure its complete oxidation;

(d) inaccuracies in measuring volume, mass and temperature; heat losses; energy lost as light and sound; food incompletely burned;

Analysis: Molecular adaptations to diet

(a) (i) incubate with starch suspension and test for appearance of maltose (e.g. red-brown colour with iodine solution or brick red colour after boiling with Benedict's reagent);

   (ii) incubate with a protein and test for disappearance of protein (e.g. lack of violet colour/ring with Biuret reagent);

   (iii) incubate with lipid and test for disappearance of lipid (e.g. change in turbidity);

(b) *Glossina* produces only proteases;

(c) A: feeds on leaf cells (full range of enzymes); B: feeds from phloem cells (only sucrose);

(d) larva: it produces protease whereas the adult does not; adult produces disaccharidases whereas the larva does not;

(e) enzyme activation;

(f) the activity of pancreatic amylase rose sharply from 1.8 units mg^{-1} to over 4.0 units mg^{-1};

(g) yes; the solid black lines are always higher than the broken lines;

(h) the amylase of rats fed on a 60% sucrose diet throughout increased after one week; the amylase activity of rats fed on a 20% sucrose diet throughout fell after the first week but rose after a further one week; the amylase activity of rats changed from a 60% to a 20% sucrose diet fell; the amylase activity of rats changed from a 20% to a 60% diet rose faster than that of rats fed entirely on a 60% diet and reached a higher level six weeks after the change in diet;

Theme 3

Chapter 9

9.1 **(a)** ensures O_2 always removed from, and CO_2 always brought to, gas exchange surface, so maintaining steep diffusion gradient;
   (b) the faster the blood flow the steeper the diffusion gradient;
9.2 large SA : vol ratio; thin; permeable; short distance from capillaries;
9.3 aquatic so surface likely to be permeable therefore diffusion of gases across surface; thalloid therefore SA : vol ratio and all cells near gas exchange surface; relatively low metabolic rate therefore no need for further mechanisms to increase diffusion rate;
9.4 when the liver and intestines are in a more forward position the air pressure inside the lungs exceeds external air pressure, leading to exhalation; vice versa for inhalation;
9.5 gills would stick together so their SA : vol ratio would become very small;
9.6 Note that the graph shows three repeated units of a cycle. The answers given below relate to the first of these units only. The key to answering this question is to realise that water flows from a greater to a lower pressure. Since the pressure in the opercular cavity is lower than that in the buccal cavity most of the time, there is an almost continuous flow of water over the gills. In fact, the period in each cycle over which the opercular cavity is higher than the buccal cavity is so short (only a fraction of a second) that the inertia of the water and resistance of the gills probably prevent any reverse flow. Movement of water is brought about by a 'double pump': a buccal 'force pump' that pushes water over the gills and an opercular 'suction pump' that pulls water into the opercular cavity.
   (a) at the point, just before the second intersection of the curves, where the black curve shows the pressure beginning to rise in the buccal cavity as water enters;
   (b) from the beginning of the graph, where the pressure inside the buccal cavity begins to rise above zero, to the first intersection of the two curves, where the pressure in the opercular cavity becomes greater than that in the buccal cavity;
   (c) at the points at which the rising black curve crosses the x-axis;
   (d) water passes over the gills most of the time but it is *pushed* over the gills in the period from the beginning of the graph (mouth closed) to the first intersection of the curves (pressure in buccal cavity less than that in opercular cavity); from the

second intersection of the two lines to the point at which the black curve rises above zero, water is pulled over the gills by expansion of the opercular cavity;
   (e) whenever the pressure in the opercular cavity is above zero;
9.7 **(a)** tracheoles carry gases straight to, and from, cells by diffusion therefore they cannot be very long;
   (b) valve can be closed so that water loss can be controlled; setae trap a still layer of moist air which will reduce water lost by diffusion of water vapour;
9.9 the living cells are in the leaves, which are thin and flat, and in the outer layer of the woody stem, which is well served by lenticels;

Analysis: Surface area to volume relationships
(a)

side/cm	1	2	3	4	5	10
SA/cm^2	6	24	54	96	150	600
vol/cm^3	1	8	27	64	125	1000
SA : vol	6.0	3.0	2.0	1.5	1.2	0.6

(b) as length of cube increases, surface area and volume increase exponentially; volume increases at faster rate; SA : vol ratio decreases exponentially;
(c) rate of diffusion in larger organism will be slower since diffusion distance to cell is greater and SA : vol ratio is less than for smaller organism;
(d) the diffusion distance is small and the SA : vol ratio is large;
(e) probably not since the rate of uptake of metabolites and excretion of wastes must double;
(f) SA = 34 cm^2, volume = 8 cm^3, SA : vol ratio = 4.25; the elongated, thin form of this 'animal' has an increased surface area to volume ratio compared with a cuboidal animal of similar volume;
(g) because it is long and thin, it has a large SA : vol ratio; it is also relatively inactive;

Analysis: measurement of human lung volumes
(a) 500 cm^3;
(b) IRV = 1500 cm^3; ERV = 1500 cm^3;
   vital capacity = 3500 cm^3;
(c)

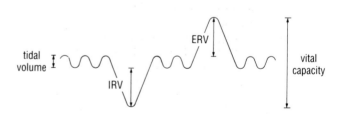

(d) (3500 + 1500) = 5000 cm^3 = 5 dm^3;
(e) vital capacity would decrease; low inspiratory reserve volume would result from oedema;
(f) expiratory volume less than inspiratory volume/expiration slower than inspiration;
(g) subject uses up O_2 and soda lime absorbs CO_2;
(h) (i) 33 breaths in 180 seconds = 11 breaths minute^{-1};
   (ii) mean volume of each breath is 1.5 dm^3 so minute volume = (11 × 1.5) dm^3 = 16.5 dm^3;

Chapter 10

10.1 (a) increased body size increases diffusion distance whilst high metabolic rate increases requirements for oxygen and increases production of waste products like CO_2; a circulatory system is essential to transport these metabolites and wastes to and from exchange surfaces;

(b) ants, like all insects, rely on diffusion to meet their oxygen requirements; this process could not meet the metabolic needs of an ant the size of an elephant since the distance over which diffusion would have to take place would be too great;

10.2 auxiliary 'hearts' in its gills (brachial hearts);

10.4 (a) ability to clot;

(b) your list should include the contents of plasma (Table 10.1) minus the fibrinogen;

(c) see Section 1.6 on amphoteric nature of proteins

(d) it is made more negative;

10.5 see Fig 10.8

10.6 (a) prevent extensive bleeding – see Fig 10.9; to prevent entry of pathogens;

(b) increased numbers of thrombocytes increases the possibility of a thrombosis forming in coronary arteries or an embolus, formed elsewhere in the body, blocking a coronary artery;

10.7 (a) a curve which shows how oxygen is carried by haemoglobin/much of a sample of haemoglobin is in the form of oxyhaemoglobin at different oxygen tensions;

(b) curve for myoglobin is to left of curve for haemoglobin; at any oxygen tension, more oxygen is carried by myoglobin than by haemoglobin, i.e. myoglobin has a higher affinity for oxygen than does haemoglobin;

(c) to left; fetal haemoglobin has higher affinity for oxygen (otherwise diffusion from maternal to fetal blood would be inadequate);

(d) the high affinity of haemoglobin for oxygen ensures that a steep diffusion gradient into the insect's tissues is maintained despite low external oxygen tensions;

10.8 (a) (i) decreasing pH of blood increases oxygen dissociation so oxygen dissociation curves shift to the right (Bohr shift);
(ii) decreased pH of blood increases rate of breathing;

(b) removing CO_2 increases blood pH; when a diver submerges blood CO_2 levels begin to build up, decreasing blood pH; however, it will take longer for the pH to decrease to the point where chemoreceptors linked to the respiratory centre are stimulated;

10.9 (i) increases difference between hydrostatic pressure and colloid osmotic potential in capillaries, so more fluid is forced from capillaries;

(ii) reduces colloid osmotic potential with same effect as (i);

(iii) reduces return of tissue fluid in lymph vessels;

10.10 (a) cells remove oxygen, and also metabolites like glucose and amin acids; release wastes like CO_2 and urea;

(b) similar to plasma but with no proteins and reduced levels of metabolites; increased levels of waste;

Analysis: Single and double circulations

(a) decreases (see Fig 10.23);

(b) (i) pressure will be less;
(ii) becomes slower;
(iii) diffusion gradients becomes less steep;
(iv) exchange of metabolites will become less efficient;

(c) fish have a low metabolic rate;

(d) fall in blood pressure in lungs overcome by return to heart; fast blood flow is therefore maintained; diffusion gradients remain steep;

Analysis: Why erythrocytes?

(a) travels faster in blood vessels, so maintaining steep diffusion gradients across exchange surfaces;

(b) the distance between them is small so that their reactions occur more quickly;

(c) cells would lose water to the blood by osmosis;

Analysis: Blood pressure

(a) the pressure during ventricular systole is 15.79 kPa, that during diastole is 10.53 kPa;

(b) (i) would increase; because contracting muscles press on veins, pushing more blood back to heart; so stroke volume of heart would automatically rise; and because heart rate would increase in response to increased CO_2 concentration resulting from exercise;
(ii) it depends on the position of the person (see (c) and (d) below)

(c) to get to the head and leg, blood must pass through blood vessels; friction against the inner wall of blood vessels causes a reduction in blood pressure;

(d) blood flows against or with gravity;

(e) rabbits are quadrupeds, their brains are not adapted to withstand the reduced blood pressure of an upright posture;

Tutorial: The heart beat

(a) atrial blood pressure increases;

(b) stops blood flowing backwards into veins;

(c) gives time for wave of contraction to pass over both atria;

(d) atrioventricular valves will shut, semi-lunar valves open; ventricles prevent blood flowing back into the atria which is, instead, forced into the aorta and pulmonary artery;

(e) atrial pressure < pressure in veins;
atrial pressure > ventricular pressure, atrioventricular valves open;
arterial walls recoil, arterial pressure > ventricular pressure;

(f) (i) each heart beat takes about 0.7 seconds so heart rate per minute is about 85 beats per minute;
(ii) represented by Q, about 0.04 seconds;

Chapter 11

11.1 (a) traps layer of air and reduces heat loss by convection;

(b) traps layer of air and reduces heat gain during day and heat loss during night;

11.2 (a) they cannot control heat gain and loss through skin (although they have behavioural means, e.g. gaping of alligators);

(b) their blubber is adapted to withstand excessive heat loss when in contact with the cold sea; on land they cannot lose enough heat to avoid death by overheating;

11.3 Note that all these answers relate to the advantages or disadvantages on which natural selection acts.

(a) large animals have a smaller SA : vol ratio than small animals and so lose relatively less heat;

(b) extremities with a large SA : vol ratio lose too much heat in cold climates;

(c) both groups are relatively inactive, so generate little heat; the temperature control mechanism of babies is poorly developed; both groups may have little subcutaneous fat;

(d) sweat only has a cooling effect when it evaporates; dense fur prevents evaporation;

(e) heat passes from the artery to blood in the vein, returning to the animal's body before the arterial blood reaches the foot, so reducing heat loss; a good example of a counter current exchange mechanism;

11.4 on land blood flows to the iguana's skin, 'collecting' heat from the Sun; in the sea blood flow to skin is reduced, so reducing heat loss;

11.5 (a) water is also being reabsorbed;

(b) the filtrate concentration in the descending limb increases as Na^+ and Cl^- diffuse in from the surrounding tissues; filtrate concentration decreases in the ascending limb as Cl^- is pumped out and Na^+ diffuses out, see Fig 11.19;

(c) filtrate concentration increases as water is reabsorbed from the filtrate, particularly from the collecting duct;

11.7 so that they lose more water in their urine and their blood volume decreases;

11.8 freshwater Osteichthyes produce large volumes of dilute urine since their rate of ultrafiltration is high (this gets rid of the water that they take in by osmosis from their hypotonic surroundings); marine Osteichthyes produce small volumes of concentrated urine since their rate of ultrafiltration is low (this helps them conserve as much water as possible);

11.9 (a) Bowman's capsule ensures efficient ultrafiltration; excretion of urea is efficient since urea needs to be diluted less than more toxic waste products such as ammonia; since urea storage gives blood a water potential which is the same as sea water, fish will not lose water by osmosis at their permeable surfaces;

(b) although osmosis will not occur at permeable surface, diffusion of sodium ions will occur from sea water into fish; rectal gland eliminates these sodium ions;

11.10 see Fig 11.26 and associated text;

11.12 (a) liver cells remove alcohol from blood and break it down using an enzyme, alcohol dehydrogenase; this takes time; in the meantime alcohol molecules enter the urine via filtration;

(b) water reabsorption in the nephron will concentrate alcohol in the urine;

Analysis : The process of ultrafiltration

(a) filtrate is identical to plasma except for the presence of plasma proteins which are too large to be filtered;

(b) blood cells;

(c) increases it;

Analysis: Reabsorption in the first convoluted tubule

(a) all substances, other than ammonia, are being reabsorbed from the filtrate;

(b) the concentration of ammonia increases as water is absorbed; and ammonia is secreted into the filtrate;

(c) sodium 3.3 g dm^{-3}; potassium 1.3 g dm^{-3}; calcium 0.13 g dm^{-3}; ammonia 0.5 g dm^{-3}; glucose $\approx$ 0 g dm^{-3}; urea 23.3 g dm^{-3};

(d) urea is toxic and is being excreted; sodium has a major effect on the osmotic potential of the blood;

(e) microvilli give large SA : vol ratio; increases efficiency of reabsorption;

(f) mitochondria produce ATP, suggesting active transport occurs in these cells;

(g) there is more glucose in the plasma than in the tubular fluid;

(h) in the first convoluted tubule;

(i) active; it occurs very rapidly and can be inhibited by another substance;

(j) other substances are reabsorbed from the ultrafiltrate but glucose is not; the same amount of glucose in a smaller volume of ultrafiltrate gives a greater concentration of glucose;

Analysis: Urine production

(a) (i) between 1 and 2, the absorption of water by the blood resulted in a decrease in ADH secretion, so less water was reabsorbed from the second convoluted tubule and collecting duct; between 3 and 5, excretion of surplus water meant that the water content of the blood was back to its optimum and ADH secretion has resumed;

(ii) the same amount of salt in a larger volume of urine represents a decreased salt concentration;

(b) (i) absorption of both water and solutes from the gut would result in no change in the water potential of the blood; however, the blood volume would increase, so that the rate of ultrafiltration and volume of urine would rise; the salt concentration of the urine would not change;

(ii) level of activity; diet; previous intake of fluids; mass of each person;

(iii) second convoluted tubule secretes H^+ into urine;

Chapter 12

12.1 (a) lack waterproof covering so water loss would be too great in dry environments; thalloid body is particularly susceptible to drying (large SA : vol ratio);

(b) small enough for diffusion to be adequate means of transport; thalloid body gives large SA : vol ratio for uptake of water and ions; permeable surface allows uptake of water and ions over whole body;

12.3 **(a)** loss of water by evaporation from leaves of plant:

(b) *Curve a*; evaporation and diffusion of water are so slow at temperatures close to 0 °C that no transpiration occurs; increases in temperature cause faster evaporation of water and diffusion of water vapour, transpiration is therefore faster; at temperatures above 40 °C the stomata are closed so that evaporation slows;

Curve b: high humidities reduce the water vapour gradient out of the leaf and so reduce the rate of transpiration;

Curve c: steady state opening of stomata occurs during periods of illumination;

12.4 **(a)** see Figs 12.17 and 12.19;

(b) symplast–diffusion through cytoplasm, apoplast–diffusion through cell walls;

12.5 as a result of root pressure;

12.6 largely by evaporation of water through stomata; cohesion of water molecules ensures a continuous column of water is pulled up;

12.7 see Fig 12.20 and surrounding text

12.8 spruce and pine have spine-like leaves with few, sunken stomata and thick cuticle, beech and oak have broad leaves with many stomata and thin cuticles; pine has larger leaves and more stomata than spruce;

12.9 **(a)** slowed, uptake is active;

(b) not slowed, transport is passive;

(c) slowed, exchange across surface membranes of cells in outside of root must occur and this is an active process;

(d) slowed, uptake into cytoplasm of endodermis is active;

12.10 **(a)** support it since less resistance to flow;

(b) fail to support it since sieve plate protein is an essential feature of Thaine's hypothesis;

12.11 **(a)** refute theory since mass flow involves all solutes and occurs in one direction only;

(b) refute theory since all components of cytoplasm stream together in one direction only;

(c) support and is part of Thaine's proposal;

12.12 Xylem vessels are dead, have no cross walls and have strengthened walls; vessels do not restrict movement of water, vessels are able to withstand pressures generated by transpiration stream whilst also providing structural support for the plant; sieve tube elements contain little cytoplasm and few organelles to interfere with transport;

Analysis: Evidence that transport occurs through xylem and phloem

(a) (i) results of experiments 1, 2 and 4;

(ii) results of experiments 1 and 3;

(iii) results of experiment 3;

(iv) results of experiment 3;

(b) in vascular tissue to region just behind bud; then in developing phloem tissue; then by diffusion through undifferentiated cells;

(c) use radioactive isotopes of ions; produce autoradiographs of sections through apex of stem;

Analysis: Water loss from leaves

(a) detach leaves from healthy, well-watered plants;

ensure several leaves of each species used; weigh each leaf at regular intervals; for each leaf; calculate loss of mass between readings; find mean loss of mass for the two species;

(b) (i) water was rapidly lost from spaces in mesophyll and from surface of cell walls;

(ii) water is lost more slowly from inside the cells;

(iii) *Pelargonium* has adaptations to reduce water loss, *Phaseolus* does not; since Plant 2 has a slower rate of water loss, it is probably the *Pelargonium*;

Chapter 13

13.1 an organism which causes disease;

13.2 **(a)** see Fig 13.2 and text

(b) leaves empty niches which pathogenic bacteria can invade;

(c) final, specific immune response most important;

13.3 some macrophages are phagocytic, others release chemicals which attract phagocytes, dilate blood vessels (histamines), promote phagocytosis, promote blood clotting (see Fig 13.6 for exact details);

13.4 some granulocytes are phagocytic, others, e.g. eosinophils, release enzymes which attack larger parasites; the granulocytes must come into contact with the pathogen;

13.5 **(a)** (i) an antigen is any molecule which promotes an immune response; it may be a whole organism, e.g. a virus, or part of an organism, e.g. a protein on the cell surface membrane of a bacterium;

(ii) an antigen produced by your own body;

(iii) a protein which binds to a specific antigen;

(b) see bulleted list on page 298;

(c) receptor sites of B and T cells are highly specific;

13.6 (i) (a); (e); (f);

(ii) (c); (d); (g);

(iii) (b);

13.7 both B and Tc cells have specific recognition site on surface membrane; B cell receptor has four poly-peptide chains Tc cell has only two; Tc cell needs specific marker alongside antigen B cell does not need this; Tc cell's main target is infected body cell B cell's target is cell from another organism; B cells release antibodies (humoral immunity) Tc cells do not (cellular immunity) but do release perforin; perforin joins surface membrane of target cell leading to lysis, antibodies cause neutralisation, precipitation or agglutination, depending on B cell;

13.8 combine information from Figs 13.7 and 13.12.

13.9 the first injection results in the formation of a small number of clone cells; these rapidly disappear from the body leaving memory cells; the second injection triggers the specific memory cells to produce vast numbers of offspring cells each of which releases antibody;

13.10 **(a)** antigens still present on surface of treated bacterium or virus; immunised person produces own antibodies against these antigens without risk of infection by active bacterium or virus; memory B cells remain in blood and are active against future infections of active bacterium or virus;

(b) to cause increase in number of appropriate B cells and production of long-term memory cells;

(c) increase titre of antibodies; so that response to antigen is immediate;

13.11 because baby's own immune system is not fully functional;

13.12 the initial exposure to the bacterium was the first, so the person had no immunity; s/he recovered because s/he made appropriate antibodies (clonal selection of appropriate B cells); on the second visit the diner was already immune to that bacterium so no further infection; s/he was not immune to the flu virus and suffered the symptoms before making specific antibodies.

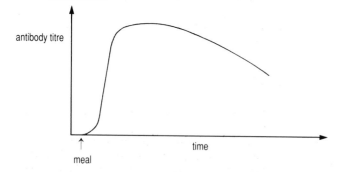

13.13 person with blood group A has antibodies to red blood cells carrying B antigen; antibodies react to B cells, leading to blood clotting; O cells have no antigen on the surface, so no clotting takes place;

13.14 **(a)** proteins on cell surface membranes of transplanted organs recognised as foreign by recipient; reactions by B and T cells cause destruction of cells in transplanted organ;

(b) your discussion should include the concepts that immunosuppressant drugs reduce rejection of transplanted tissue but also reduce ability to destroy antigens involved in infection or cancer;

Analysis: The order of battle
The order is **(b)**, **(d)**, **(a)**, **(c)**, **(e)**;

Analysis: The clonal selection hypothesis
(a) immunisation with an antigen results in the activation of the few cells with surface receptors that are specific to that antigen; as a result, these cells multiply to form vast numbers of clones;

as they pass down their respective columns, cells with these surface receptors attach to the beads with their specific antigen: they do not therefore emerge from the column; this explains the absence of antibodies against Type I antigen in cells emerging from column 1 and of antibodes against Type VIII antigen in cells emerging from column 4 (numbering from left to right);

cells with surface receptors for Type I antigens are not held in the column of beads coated with Type VII antigen, explaining the high level of Type I antibody production by cells emerging from column 2; (Adapt this explanation yourself to account for the high level of Type VIII antibody production by cells emerging from column 3.)

the presence of a few committed cells within the circulatory system is shown by the low level of antibody production by cells emerging from columns 1 and 4;

Theme 4

Chapter 14

14.3

	adrenaline	other hormones	nerves
response time	rapid	slow	rapid
return to normal	rapid	slow	rapid
effect of response	widespread	widespread	localised

14.4 receptor sites (proteins) for insulin in surface membranes of cells in many organs; those for glucagon only in surface membranes of liver cells;

14.5 provides link between nervous and endocrine systems;

14.7 iodine is taken up only by thyroid cells (Note, this makes radioactive iodine a very specific form of treatment for cancer of the thyroid);

14.8 A *anterior* pituitary; B beneath hypothalamus; C thyroid stimulating hormone; D thyroid; E either side of larynx; F increase metabolism; G adrenal medulla; H adrenaline/noradrenaline;

14.9 see Table 14.5

14.10 **(a)** causes larval stage to occur after each moult;

(b) the adults reproduce, so prevention of adult stage reduces population growth; only the adult flea bites its host; it is usually the larvae of pests of stored food that do the damage;

14.11 it might metamorphose into an adult salamander;

Analysis: Calcium and hormones
(a) parathormone acts on the renal tubules to increase the absorption of Ca^{2+} ions (therefore less in urine) and increase the excretion of PO_4^{3-} ions (therefore more in urine and less in blood); as you can see from the bar chart in Fig 14.11, these effects occur about one hour after injection of parathormone; parathormone also causes an increase in the release of Ca^{2+} ions and PO_4^{3-} ions from bone and their absorption from the gut; these increases explain the eventual increase in concentration of Ca^{2+} and PO_4^{3-} ions in the serum; note that these effects occur about $3\frac{1}{2}$ hours after injection of parathormone;

(b) (i) increase;
(ii) decalcification;
(iii) Ca^{2+} ions decrease but PO_4^{3-} ions increase;

Chapter 15

15.2 impulse and current result from flow of charged particles; axon carries impulse in same way as live wire carries current; myelin sheath reduces electrical leakage as does plastic (brown) coating of live wire;

15.3 see Fig 15.5 and associated text

15.4 example of a conditioned reflex; cats associate the sound of the tin opener with food;

15.5 **(a)** a membrane having unequal charges on each side and hence a potential difference across it;

(b) the size of the potential difference across the surface membrane of an axon when it is not

15.6 by active transport of sodium ions out of, and potassium ions into, the axon;

15.7 see Fig 15.8

15.8 wires leak; each action potential always has same strength (all-or-none response);

15.9 if it occurs at all, it always has the same strength;

15.11 by spatial summation; see end of Tutorial: Synaptic transmission for further details

15.12 some neurones are inhibitory; decision-making area of brain communicates using these neurones to prevent release reflex;

15.13 membranes surround brain and spinal cord; meningitis interferes with transmission of nerve impulses in the brain;

15.14 (a) myelinated axons;
(b) cell bodies, unmyelinated axons;

15.15 (a) receptor neurones to spinal cord; relay neurones in ascending tract to receptor area of right cerebral hemisphere; relay cells to speech centre of left cerebral hemisphere; effector neurones to muscles of mouth and larynx;
(b) communication between left and right cerebral hemisphere is via corpus callosum (see Table 15.4); if this were cut, no impulses could pass from sensory area to right hemisphere to speech area in left cerebrum (Note that in tests such as this, the person was able to draw the object with her/his left hand);

15.17 see Figs 15.5 and 15.23

15.18 parasympathetic stimulation during relaxed state; increase levels of sympathetic stimulation result in increased levels of fear; sympathetic stimulation during frightened state (noradrenaline);

Tutorial: The nerve impulse

(a) (i) concentration of chloride and sodium ions is low, whereas concentration of organic anions and potassium ions is high;
(ii) opposite to (i);
(iii) chloride and sodium ions are more concentrated outside, potassium and organic anions more concentrated inside;

(b) active transport;
(c) goes from –70 mV to +40 mV;
(d) inside the membrane will become positive relative to the outside;
(e) as the action potential spike;
(f) inside the membrane will become increasingly negative relative to the outside; see Fig 15.8

Analysis: The speed of propagation of nerve impulses

(a) mammalian neurones conduct impulses faster than amphibian ones of same diameter; unmyelinated neurones carry impulses slower than myelinated ones of same diameter; unmyelinated neurones with a large diameter carry impulses faster than unmyelinated neurones with a smaller diameter;

(b) mammals have a faster metabolic rate, so more energy for active transport of ions; impulse conduction is saltatory in myelinated neurones; resistance to flow is greater in axons of small diameter (just as is the case with electrical currents in wires);

(c) to ensure all the water is ejected from the mantle cavity in order to provide the greatest thrust;
(d) large diameter axons to distant muscle but small diameter axons to those near stellate ganglion; myelinated axons to distant muscles but unmyelinated ones to those near stellate ganglion;
(e) dissect out neurones; see if myelinated; measure diameter;

Tutorial: Synaptic transmission

(a) Ca^{2+} is more concentrated outside the synaptic knob;
(b) pumped out by an active transport mechanism;
(c) the inside becomes more positive relative to the outside, i.e. an action potential is produced, see Fig 15.8;
(d) so that the post-synaptic membrane repolarises rapidly;

Chapter 16

16.1 receptor → receptor neurone → central processes, e.g. brain;

16.2 see Section 16.1 Electrical charges in receptors;

16.3 light falling on groups of rods is transduced into an impulse in only one receptor neurone so the power of resolution is low;

16.4 A – movement of head; B – crista; C – cupula; D – sound; E – organ of Corti; F – tectorial membrane; G – gravity; H – macula; I – otoliths;

16.5 (a) vibration of ossicles → vibration of oval window → vibration of perilymph → vibration of Reissner's membrane → vibration of endolymph → vibration of basilar membrane;
(b) only those which cause the basilar and tectorial membranes to move apart, i.e. vibration of basilar membrane;

16.6 (a) by detecting vibrations in different parts of the cochlea;
(b) destroying sensitive hair cells;

16.7 (a) the pressure of the water presses the ear drum inwards causing pain; blowing the nose in this way forces air up the eustachian tube thus increasing the pressure in the middle ear;
(b) just as the inertia of the endolymph causes a deflection of the moving cupula when we start to spin, its inertia causes it to continue moving when we stop spinning; this deflects the cupula making us think we are still moving; it deflects it in the same direction as we were spinning which would normally happen when we start to move in the opposite direction and the cupula is deflected by stationary endolymph;

16.8 (a) what one eye misses, the other eye sees;
(b) the brain 'fills in' with whatever occupies the general background;

16.9 several rod cells eventually synapse with a single ganglion cell so little resolution; cone cells synapse individually to single ganglion cells so greater resolution (greater visual acuity);

16.10 (a) most mammals do not have colour vision; they are sensitive to dim light but not to colour; their visual acuity is low;
(b) 'seeing' objects using rod cells leads to lower

visual acuity; rod cells are insensitive to colour;

16.12 initially pupil dilated to allow maximum light to enter; ciliary muscle contracted to enable him to focus on the watch; vision using rods; when he glances out of windowpupil will constrict, ciliary muscles relax, vision using cones; when he glances back at his watch eye reverts to initial condition;

16.13 A – heart (myocardium); B – gut, tubes of urogenital system, blood vessels; C – attached to skeleton; D – cylindrical with side arms; E – spindle-shaped; F – cylindrical; G – with striations; H – without striations; I – with striations; J – no need since myogenic; K – needed; L – needed;

16.14 see Fig 16.21

16.15 see Tutorial: The sliding filament theory for main points

Analysis: Sight and age

(a) $1/9.0 = 0.11$ m (= 110 cm);

(b) (i) humans become progressively long-sighted as they age;

(ii) reading glasses must converge the light rays;

(c) (i) the lens will lose its ability to recoil to its former relaxed state so we become long-sighted;

(ii) since the suspensory ligaments become stretched they are always in tension and so unable to allow the lens to recoil to its former relaxed state;

Analysis: The structure of striated muscle

(a) I band = region of thin actin filaments; dark band is where actin and myosin overlap. H zone is myosin only;

(b) Z line is anchorage of actin filaments; M line is where adjacent myosin filaments join together halfway along their length;

(c) A band stays same width as it represents length of myosin filaments; H zone shortens because actin and myosin overlap to a greater extent during contraction; I band shortens because more of the actin overlaps with the myosin;

Tutorial: The sliding filament theory

(a)

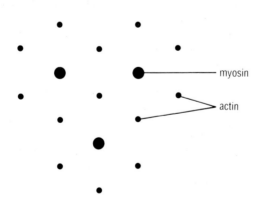

(b) without ATP, no actin-myosin cross bridges can break and reform;

(c) creatine phosphate is used as an energy store by muscle tissue; it has a high energy yield on hydrolysis and breaks down faster than, say, glucose;

Chapter 17

17.1 (a) rapid, involves depolarisation of a membrane;

(b) involves pumping of K^+ or H^+ ions out of cell; involves cell expansion/contraction resulting from osmosis;

17.2 prevents damage in *M. pudica*; enables *D. muscipula* to catch and digest prey (source of nitrates since it lives in nitrate-deficient soils);

17.3 both are slow working and produced in one specific place; plant growth substances can have multiple effects depending on their concentration and site of action;

17.4 (a) encourages growth of roots;

(b) taken up more by broad leaves than by narrow leaves of grass; stimulates excessive growth of broad-leaved plants; broad-leaved plants exhaust energy supplies and die;

(c) stimulate development and retention of fruit;

(d) promote abscission of fruit;

(e) inhibit senescence of leaves so vegetables remain fresh;

17.5 (a) in darkness P_{660} accumulates; this probably stimulates germination;

(b) in darkness P_{660} still accumulates; stimulates growth;

(c) in shade caused by taller plants, P_{660} still accumulates; stimulates growth of internodes so that plant becomes taller than its competitors;

(d) P_{730} now accumulates; inhibits growth of internodes and stimulates chloroplast production;

17.6 (a) short day plant;

(b) in regime 3, long night interrupted by light which converts P_{660} to P_{730} – exposure to P_{660} now not long enough; in regime 4, exposure to P_{660} after light flash now above critical value;

(c) far-red light does not interfere with P_{660} so plant would flower;

Analysis: Phototropism

A: your diagram should show elongation of the coleoptile but no bending;

B: mica prevents movement of auxin to shaded side of coleoptile;

C: because of position of cut tip of coleoptile, auxin passes down one side of coleoptile only – elongation is greater on this side and results in bending;

D: the thin glass plate prevents lateral movement of auxin so both sides of the coleoptile receive the same amount of auxin and elongation without bending occurs;

Analysis: Abscission

(a) (i) lateral bud smaller on treated side than on untreated side; vascular tissue developing from lateral bud on untreated side but not on treated side; abscission layer forming on untreated side but not on treated side;

(ii) your crosses should show apex of bud on untreated side and in abscission layer;

(b) (i) best way would be to apply agar block containing hormone against cut end of petiole so that there is no aerosol effect to other petiole, or paint the hormone onto the plant in the form of an oily paste;

(ii) leave both petioles untreated/leave both leaves intact on a control plant;

(c) (i) B; since it inhibits abscission;
(ii) abscisic acid;

Analysis: The control of flower induction

(a) cocklebur;
(b) P_{730};
(c) it would slowly be converted to P_{660};
(d) it would rapidly be converted to P_{730};
(e) P_{660} stimulates flowering in cocklebur;
(f) if a critical length of uninterrupted exposure to P_{660} inhibits flowering in spinach then the flash of light would interrupt this by producing P_{730};
(g) the length of darkness;
(h) only the last flash of light counts since the flash of far-red light will rapidly convert P_{730} back to P_{660};

Chapter 18

18.2 see text for definitions: (a) nasty; (b) taxis; (c) kinesis; (d) tropism;
18.3 see Fig 18.3
18.4 flagellum is longer than cilium; flagellum produces a series of waves along its length whereas cilium has distinct effective and recovery strokes; cilia work in groups and their basal bodies are connected by neuronemes whereas flagella occur singly and have independent basal bodies;
18.5 muscles cannot expand of their own accord; therefore they must be pulled by an antagonist;
18.6 the thrust of the foot against the ground causes a reaction in the opposite direction (Newton's third law); this reaction has two components which push the body upwards and forwards;
18.7 for support and to act as an incompressible structure against which muscles can act;
18.8 your comparison might include such features as the chemical composition; the presence or absence of cells; presence of joints; position inside or outside the body; degree of support and protection; attachment of muscles; effect on growth;
18.9 hard bond around shaft resists tension and compression; trabeculae of spongy bone resist stress; large surface area enables muscle attachment; hyaline cartilage at epiphysis resists friction; head of femur and cavity of pelvic girdle (acetabulum) form a ball and socket joint;
18.10 see Figs 18.15, 18.16 and associated text;
18.11 stems of herbaceous plants rely on hydroskeletons for support not lignified cells;
18.12 central stele of root resists pulling as aerial parts move in wind; peripheral bundles of stem resist tension and compression as aerial parts bend in wind; network of vascular bundles prevent collapse of thalloid leaf;

Theme 5

Chapter 19

19.1 sexual reproduction involves fusion of gametes, asexual does not;
19.2 this is not sexual reproduction since no division

process is involved;
19.3 both are forms of asexual reproduction but multiple fission results from the mitotic division of an individual cell whereas sporulation involves the production of spores by specialised structures, e.g. sporangia; in plants sporulation involves meiosis not mitosis; spores often dormant;
19.4 if more than one bud grew, two aerial stems would form in the second year; however, since these would not produce new carrots (these plants are biennials) no separate plants would develop;
19.6 (a) independent of free water for swimming gametes; gametes brought together so more likely to fuse; enables protection of delicate reproductive parts during their development;
(b) involves greater energy expenditure by one parent; may make one parent (pregnant mammal) more vulnerable; complex mechanisms needed to enable gamete transfer; recipient's body may be hostile environment for motile gamete;
19.7 advantage = reproduce without another member of the same species being present. However, disadvantage is that such continual inbreeding could lead to inbreeding depression in a species which is normally an outbreeder;
19.8

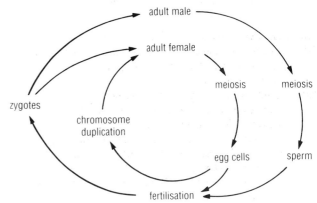

Analysis: Reproductive myths
(a) otherwise the chromosome number would double at each fertilisation;
(b) during gamete production in mammals; during spore production in flowering plants;
(c) mitosis;
(d) no;
(e) mitosis produces cells with the same number of chromosomes as the parent cell whereas meiosis produces cells with half the chromosome number of the parent cell;
(f) mitosis; this is asexual reproduction of cells following sexual reproduction by the gametophytes;
(g) meiosis; asexual;
(h) no;
(i) no, they are the products of meiosis;
(j) yes;
(k) 2048;
(l) oyster;
(m) seed production;
(n) that more than two of them survive and germinate;

Chapter 20

20.1 primary sex organs are where gametes are produced, e.g. testes and ovaries; accessory sex organs store and transport gametes, e.g. vas deferens, fallopian tubes; secondary sexual characteristics include deep voice and facial hair in men, development of breasts in women;

20.3 male reproductive system is adapted to deliver a large quantity of sperm in one ejaculation at any time whereas the female reproductive system is adapted to deliver one gamete, at regular intervals and provide an environment in which that gamete can be fertilised and develop;

20.4 **(a)** *females* – FSH stimulates development of follicles, LH stimulates continued follicular development and development of corpus luteum;
males – FSH stimulates sperm production, LH stimulates intestial cells to produce testosterone;

(b) would suppress sperm production but not testosterone production, so male would be infertile but would still develop male secondary sexual characteristics;

20.5 1 – (g); 2 – (d); 3 – (h); 4 – (c); 5 – (f); 6 – (b); 7 – (a); 8 – (e);

20.6 see Figs 20.8 (a) to (d) and associated text

20.7 see Fig 20.16

20.8 see Section 20.5, Pollen production

20.9 see Fig 20.23

20.10 see Table 20.2

20.11 pollination involves the arrival of a pollen grain on the stigma of a plant; fertilisation involves the fusion of gametes which occurs inside the embryo sac;

20.12 fusion of one of the male nuclei with the two polar nuclei to form the primary endosperm cell; fertilisation produces a zygote, this triple fusion does not;

20.13 embryo is next sporophyte generation; fruit and seed coat are tissue of parent sporophyte; nucellus/endosperm is derived from tissue of gametophyte (embryo sac);

Analysis: Male sex hormones

(a) testosterone; its androgenic potency is 100% and it is most concentrated in blood plasma;

(b) 5 α-dihydroxytestosterone and testosterone;

(c) oestrogen; it is a hormone more important in female development;

(d) apart from human characteristics such as pubic hair, facial hair and voice breaking, your answer might include male sexual behaviour (including aggressiveness in mammalian species which form herds); increase in size; increase in musculature; sexual features such as antlers or coloured fur; growth of testes;

(e) outside the seminiferous tubules/interstitial cells;

Analysis: The oestrous cycles of ewes and rats

(a) yes, since the same hormones are involved in both ewes and humans and they have a similar effect; however, the human oestrus cycle is longer than that of the ewe and occurs throughout the year; humans do not show some of the behaviour associated with oestrus in ewes but there is an increase in basal body temperature at the time of ovulation; the order of release of hormones is also different;

(b) the LH peak occurs at the beginning of the oestrus cycle in ewes not in the middle as in humans;

(c) FSH promotes release of oestrogen by the developing follicles; oestrogen then inhibits further FSH production;

(d) increasing levels of one or all of the three other hormones, LH, FSH or oestrogen, inhibit the function of the corpora lutea, inhibiting progesterone production at (1); low levels of one or all of these three hormones remove this inhibiting effect at (6) leading to an increase in progesterone production;

(e) one cycle is very rapid, only 3–4 days long; the basic pattern of oestrogen and LH production is similar, but there is a double not a single peak of progesterone production in the rat;

Analysis: Male fertility

(a) 10% of 3.4 cm^3 = 0.34 cm^3 = 340 mm^3;
340 × 100 000 = 34 000 000 (or 34 × 10^6) sperm;

(b) total length of sperm in Fig 20.3 = (50 + 4 + 5) μm = 59 μm; distance travelled = 19 cm = 19 × 10^4 μm; this distance is (19 × 10^4)/59 = 3220.3 sperm lengths;

(c) 3220 sperm length in 10 800 seconds represents 0.3 sperm lengths per second;

(d) not enough survive to reach the egg cell;

(e) this man is possibly fertile;

Analysis: Drawing floral diagrams

First you need to work out the number of parts in each whorl. Use this information to construct diagrams like **(a)** Fig 20.16; **(b)** Fig 20.17;

Analysis: Ensuring cross-pollination

(a) half flower;

(b) pin-eyed have long style, thrum-eyed have short style; pin-eyed have anthers half-way down the flower, thrum-eyed have anthers at the top of the flower;

(c) pollen from pin-eyed is low down on the bee's proboscis, corresponding to the level of the stigma in the next thrum-eyed flower visited; pollen from thrum-eyed is high up on the bee's proboscis, corresponding to level of the stigma in the next pin-eyed flower visited; pollen is always in wrong place to pollinate another flower of the same type;

(d) moth transfers pollinia to stigma of next flower visited;

(e) long floral spur means that moth has to insert its head further into flower to obtain nectar; this makes it more likely that moth's head will come into contact with anthers and stigma, so increasing the probability of pollination;

Chapter 21

21.2 **(a)** continuous growth occurs uninterrupted throughout life, although its rate may vary; discontinuous growth is interrupted;

(b) (i) suitable temperature, O$_2$ supply, source of calcium (for growth of shell), abundant food;
(ii) the body of a mollusc is not completely

enclosed by its shell – there is an aperture; material can be added to the shell continuously and the animal can grow continuously within it;

21.4 growth is an increase in cell number/dry mass; development is an increase in cell/tissue complexity;

21.5 the grey crescent contains substances that are needed for normal growth and development of the embryo;

21.6 although the gametophyte is quite unlike the sporophyte, it is not a non-reproductive dispersal stage; alternation of generations is best considered as an unusual form of direct development;

21.7 indirect development requires less energy expenditure per egg cell by the female parent than does direct development; death rate is higher in indirect than direct development; larva of indirect development life cycle may be a better dispersal stage; larva and adult may avoid competition for food in indirect cycle but not in direct cycle;

21.8 **(a)** the blastocyst has differentiated into inner mass and trophoblast;
(b) cells are now too specialised;

21.9 the central nervous system, essential for co-ordination of the activity of all other fetal systems, e.g. circulation;

21.10 the placenta is an exchange surface for oxygen, carbon dioxide, nutrients and waste products; it provides anchorage; it is an endocrine organ; it is also a filter which prevents bacteria from reaching the fetus, though some viruses, e.g. HIV and the rubella (German measles) virus, can get through;

21.11 water uptake, mobilisation of food stores (Fig 21.17); growth of radicle and plumule (Fig 21.19);

21.13 **(a)** compare with labels in Fig 21.24(b)
(b) both result from division of meristematic tissue; in root, meristematic tissue is in pericycle, therefore lateral root emerges from inner part of main root; in stem, meristematic tissue is in lateral/axillary bud, therefore lateral stem emerges from outside of main stem;

Analysis: Differentiation
(a) the nucleus of a tadpole's intestinal cell contains the genetic material needed for the growth and differentiation of a fertilised egg cell;
(b) by UV radiation; so that no nuclear DNA remained in the egg cell – the only nuclear DNA was from the intestinal cell;
(c) cytoplasm contains enzymes and mitochondrial DNA which might have affected development of egg cell;
(d) the nucleus of an intestinal cell has all the DNA needed for normal growth and development of a *Xenopus* egg cell;
(e) it is likely, but not certain, that the conclusion would be valid for closely related species;

Analysis: Germination
(a) dry mass is best indicator of amount of matter in seed;
(b) plot time on the *x*-axis and dry mass of embryo and endosperm on the *y*-axis;
(c) endosperm dry mass decreased by only 6% in the first 2 days of the experiment, but the rate of decrease of mass increased dramatically in the next 3 days, averaging 17.5% of the original endosperm dry mass

per day; embryo growth is exponential over the five days of the experiment, the total embyo mass increasing 11 fold in 5 days;
(d) gets less (202 g on day 0 to 107 g on day 5);
(e) endosperm is used up as embryo grows; some of endosperm forms structures in embryo (e.g. proteins and cellulose); some of endosperm is lost as CO_2 and H_2O during respiration;
(f) well containing the unboiled seed would have a clear zone around it, the rest of the agar would be blue-black;
(g) amylase from unboiled seed diffuses into agar and digests starch; amylase in boiled seed has been denatured;

Analysis: Writing continuous prose
Marks would be awarded for *selecting* relevant facts, principles or concepts, *organising* them into a logical and coherent order and *communicating* them in a correct and unambiguous way. Some examination boards (e.g. AEB) allocate marks separately for these skills but many do not. Mark schemes for essays always have more marks available than the maximum for the question.
(a) storage in cotyledons/endosperm/nucellus; starch; (starch) granules in cells; lipids; (lipid) droplets in cytoplasm of cells; proteins; (proteins) in colloidal suspension in cytoplasm; (maximum 6 marks)
(b) water taken up through micropyle; water needed as solvent/as metabolite in hydrolysis; abscisic acid levels fall; gibberellin levels rise; genes for enzymes transcribed; hydrolytic enzymes produced/become soluble; starch hydrolysed to maltose/glucose; maltose diffuses to embryo and is hydrolysed to glucose; fats hydrolysed to fatty acids (and glycerol); proteins hydrolysed to amino acids; (maximum 6 marks)
(c) metabolic rate increases; new cells produced at root and shoot tip; new cells elongate causing increase in length; radicle breaks through testa; root tip protected by root cap; radicle grows downwards/positively geotropic; and grows towards water/positively hydrotropic; growth of lateral roots; plumule grows upwards/negatively geotropic at first; once above ground, plumule grows towards light/positively phototropic; plumule protected by coleoptile in monocots; plumule protected by bending in dicots/no bending in monocots; elongation of hypocotyl in epigeal germination; and cotyledons emerge above soil; elongation of epicotyl in hypogeal germination; and cotyledons remain below ground; once in light, chloroplasts/chlorophyll develop(s); chlorophyll in cotyledons/first true leaves; tropisms controlled by auxins; (maximum 8 marks)

Analysis: Ageing trees
(a) compare with labels in Fig 21.22(c)
(b) you should have found two annual rings making this a 2-year-old tree (note that the large xylem vessels represent the spring wood);

Theme 6

Chapter 22

22.1 **(a)** (ii); **(b)** (i); **(c)** (iii);

22.2 the nucleic acid code controlling seed coat colour is called the gene for seed coat colour, the nucleic acid codes for green and yellow seed coat colour are alleles of that gene;

22.5 See Fig 22.5;

Phenotype of parents	long-winged	vestigial-winged

Genotype of parents LL ll

Gametes L l

Genotype of offspring (1) Ll X Ll

Gametes L l L l

Genotype of offspring (2) LL Ll Ll ll

Phenotype of offspring (2) long long long vestigial

Ratio 3 long : 1 vestigial

22.6 **(a)** e.g. *B* represents the allele for black and white coat colour; *b* represents the allele for red and white coat colour;

(b) this is a traditional test-cross (see Fig 22.15), mating the black and white bull with homozygous recessive cows (red and white); the appearance of any calves with red and white coats indicates the bull is heterozygous (*Bb*). Your genetics diagram will begin as shown in Fig 22.6;

Phenotype of parents red and white cow black and white bull

Genotype of parents bb BB or Bb

Gametes b B or B and b

22.7 **(a)** See Fig 22.7;

Phenotype of parents group MN group MN

Genotype of parents $G^M G^N$ $G^M G^N$

Gametes G^M G^N G^M G^N

Genotype of offspring (1) $G^M G^M$ $G^M G^N$ $G^M G^N$ $G^N G^N$

Phenotype of offspring (1) group M group MN group MN group N

(b) no; since fertilisation occurs at random, the chance of any of their children being of a

particular blood group, is 0.25 (M), 0.5 (MN) and 0.25 (N);

22.8 **(a)** e.g. G^R represents the allele for red coat colour, G^W represents the allele for white coat colour;

(b) (i) See Fig 22.8;

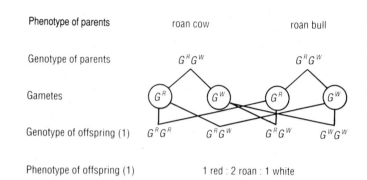

Phenotype of parents roan cow roan bull

Genotype of parents $G^R G^W$ $G^R G^W$

Gametes G^R G^W G^R G^W

Genotype of offspring (1) $G^R G^R$ $G^R G^W$ $G^R G^W$ $G^W G^W$

Phenotype of offspring (1) 1 red : 2 roan : 1 white

(ii) you should find a ratio of 1 roan : 1 white in the offspring (1);

22.9 **(a)** $Hb^A Hb^S$;

(b) 1 $Hb^A Hb^A$: 2 $Hb^A Hb^S$: 1 $Hb^S Hb^S$;

(c) the children must get an Hb^S allele from both parents but homozygotes ($Hb^S Hb^S$) are unlikely to survive to reproduce;

22.10 **(a)** you should find a ratio of 1 pink : 1 white;

(b) you should find a ratio of 1 red : 1 pink;

22.11 **(a)** e.g. Hb^T and Hb^t;

(b) (i) you should find a ratio of 1 : 2 : 1 :

(ii) half the offspring will be $Hb^T Hb^T$ (no anaemia) and half the offspring will be $Hb^T Hb^t$ (thalassemia minor);

22.12 **(a)** the alleles of the gene controlling production of this enzyme are codominant;

(b) e.g. E^N and E^T;

(c) valid suggestions include: medical counselling to advise couples of the risk, which may result in some couples deciding against having children together; test for enzyme concentration in fetus followed by termination of pregnancy may be possible;

22.13 **(a)** you should have found that any child of the father must contain the I^A or the I^B allele, since a baby of blood group O has the gentoype $I^o I^o$, the baby cannot belong to this couple;

(b) the father could be genotype $I^B I^o$ so, without any further evidence the baby could be theirs;

22.14 **(a)** you should have found a ratio of long wings and grey bodies: long wings and ebony bodies: vestigial wings and grey bodies: vestigial wings and ebony bodies = 1:1:1:1;

(b) a test cross;

22.15 Using, for example, the symbols D and d to represent the alleles for dark hair and albino respectively, and S and s to represent the alleles for long hair and short hair, you should have found:

(a) both parents are *DdSs*, giving an approximate 9:3:3:1 ratio;

(b) the parents are *DDSs* and *DDss*;

(c) the parents are *DdSs* and *ddSs*;

(d) both parents are *ddSS*;

(e) both parents are *DdSS*;

(f) both parents are *DDSs*;

(g) the parents are *DdSs* and *Ddss*;

22.16 (a) *HT, Ht, hT* and *ht* in a ratio of 1:1:1:1;

(b) *hT* and *Ht* in a ratio of 1:1;

(c) most of the gametes will be *hT* and *Ht* in a ratio of 1:1 but there will be a few recombinants of genotype *HT* and *ht*, again in equal numbers;

22.18 (a) they are codominant;

(b) see Fig 22.13 for similar pattern of inheritance of flower colour in *Antirrhinum*;

22.19 C^O represents an X chromosome with the orange allele, C^B represents an X chromosome with the black allele and Y represents a Y chromosome which lacks a coat colour locus; calico males are not possible because the Y chromosome has no coat colour locus; see Fig 22.19;

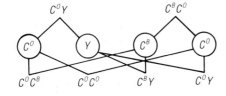

Phenotype of parents	ginger male		calico female	
Genotype of parents	$C^O Y$		$C^B C^O$	
Gametes	C^O	Y	C^B	C^O
Genotype of offspring (1)	$C^O C^B$	$C^O C^O$	$C^B Y$	$C^O Y$
Phenotype of offspring (1)	calico female	orange female	black male	orange male

22.20 they are pleotropic;

22.21 (a) they affect three phenotypic characteristics; pleitropy;

(b) no; the *W* allele is dominant over the *w* allele in its effect on seed appearance but the two alleles are codominant in their effect on starch content and sugar content of the seed;

Analysis: The inheritance of leaf spotting in Blue-eyed Mary

(a) 34 : 14 = approx. 3 : 1;

(b) adapt Fig 22.12 using symbols you have devised yourself

(c) heterozygous, e.g. *Bb*;

(d) use a test cross; adapt Fig 22.15 using your own symbols;

Analysis: Genetics maps

(a)

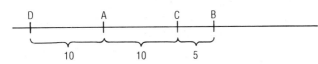

(b)

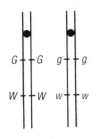

(c) grey bodies and straight wings: ebony bodies and curled wings = 1:1;

(d) COV = (number of recombinants/total number of offspring) × 100% = 22/100 × 100% = 22%. This value is close to, but not identical to, the stated value of 20.7%; the difference is due to random sampling error in a fairly small sample.

Chapter 23

23.1 (a) 0.36;

(b) 0.6;

(c) 0.4;

(d) 0.16;

(e) 25% of them are homozygous and 75% of them are heterozygous;

23.2 (a) 1 − 0.07 = 0.93;

(b) $0.07^2 = 0.0049$; $2pq = 2 \times 0.93 \times 0.07 = 0.13$;

23.4 it is likely to maintain the local dialect;

23.6 (a) X; Y; XY and a sperm with no sex chromosome (O);

(b) XX – normal female; XY – normal male; XXY – a male with Klinefelter's syndrome (see Table 23.3); XO – a female with Turner's syndrome (see Table 23.3);

23.9 neo-Darwinism incorporates modern knowledge about inheritance; Darwin thought 'unfit' individuals died without reproducing whereas we now believe that differential fertility is sufficient to give individuals different fitnesses;

23.10 apart from transgenic genetic engineering, you could use artificial selection, e.g. grow generations of British beans in a cool climate planting only those beans in each generation which lack speckles, or grow the cream bean plants in a cool climate planting only those beans in each generation which are large;

23.11 (a) humans mistake it for a rice plant and so do not remove it;

(b) by natural selection; any resemblance to a rice plant would increase its chances of survival and reproduction; it would produce more offspring than non-mimics; the offspring would also be mimics.

Analysis: Karyotypes

(a) (i) heparin prevents blood clotting; spindle inhibitor prevents separation of contracted chromosomes;

(ii) it is composed of two chromatids;

(b) (i) from homologue 2 at middle right;

(ii) one of the number 5 homologues; from the fragment of 4 at top left;

(c) AF; BD; CE;

(d) male (XY) with Down's syndrome (three copies of chromosome 21);

Analysis: The strength of selection

(a) you will probably find that there are more of the green caterpillars;

(b) you may find that the more common the green the more they are eaten (birds form a 'mental image' of the food for which they are searching;

(c) it is likely that the birds will have avoided the non-green caterpillars;

(d) if the ratio of distasteful caterpillars to mimics is high, it is likely that birds will encounter a distasteful model rather than an edible mimic and so avoid all similarly coloured caterpillars;

Chapter 24

24.1 **(a)** see Section 24.1;
(b) (i) see if they would breed to form fertile offspring;
(ii) look for homologous similarities in anatomy, protein structure, etc;
24.4 X✓; X✓; ✓X; ✓X; X✓;
24.5 each has a chromosome number which is a multiple of 20; the species probably arose by polyploidy;
24.6 **(a)** see Section 24.3;
(b) go to your library and look at the way in which books are grouped and see if you can answer this
(c) natural; it shows the origins of these rocks;
24.8 binomial name; name of genus and species in italics; name of genus has upper case initial; name of person who described and named it is not in italics; L. shows that *Nucella* was named by Linnaeus; Darwin named *Elminius*;

Analysis: Clines
(a) the frequency of cyanogenic plants increases from north to south;
(b) there is a direct correlation, i.e. the higher the mean January temperature the higher the frequency of cyanogenic plants;
(c) (i) cyanogenesis is a protection against predation by molluscan herbivores, so cyanogenic plants are at a selective advantage;
(ii) in cool northern areas molluscan herbivores are few and clover plants are usually damaged by winds and frost; in these conditions it is a waste of energy to produce cyanide and cyanogenic plants are at a selective disadvantage;
(d) suitable tests you could perform include: population estimates of molluscan herbivores in northern and southern areas; controlled laboratory experiments to test the survival of cyanogenic and acyanogenic clover plants when exposed to herbivores, to cold and to wind damage;

Analysis: The evolution of copper tolerance in *Agrostis tenuis*
(a) by random mutation;
(b) before soils were contaminated by copper ions the copper tolerance gene probably put plants at a disadvantage compared with those with its normal allele, the copper tolerance gene was therefore selected against; when human activities produced copper-contaminated mine tips, plants with the gene for copper tolerance would be at an advantage over those with the normal allele and so would be selected for;
(c) these individuals were at a selective disadvantage in competition with the non-copper-tolerant plants (e.g. the latter might grow faster);
(d) copper-tolerant plants will complete their life cycle before non-copper tolerant plants flower; as a result,

the two populations become reproductively isolated; genetic changes in one population will not be mirrored in the other;

Analysis: Why should anyone care about taxonomy?
(a) to allow easy communication about organisms; to reconstruct evolutionary relationships; to allow generalisations about groups of organisms; to enable predictions to be made about newly discovered organisms;
(b) they are not anthropocentric (based on human interests) but attempt to show evolutionary relationships;
(c) most of their genes will be the same, or very similar;
(d) organisms are likely to behave in a similar way to others in their taxon;
(e) this question asks why *you* think it is important, but see (a) above for clues:

Chapter 25

25.1 see Table 25.1
25.2 ✓ ✓; 2 X ✓; 3 ✓ X; 4 X ✓; 5 X ✓; 6 ✓ X; 7 X ✓;
25.3 spores are resistant to most methods of killing bacteria;
25.4 **(a)** see Table 25.3
(b) aerobic respiration uses molecular oxygen as the terminal electron acceptor in an electron-transport chain; anaerobic does not use oxygen in the electron-transport chain; fermentation is respiration without an electron-transport chain;
25.5 **(a)** see Fig 25.4
(b) many form multicellular aggregations – see Fig 25.7
25.6 see Spotlight: Identifying bacteria
25.7 e.g. composition of ribosomal RNA;
25.8 see Table 25.5
25.9 see Fig 25.6
25.10 unicellular; lack cell wall; heterotrophic;
25.11 **(a)** chain of cells instead of thallus; no projections from *Ulva* or *Laminaria*;
(b) chain of cells less susceptible to damage by surface waves and projections give large surface area to volume ratio for buoyancy; benthic algae must have organ for attachment to substratum;
25.12 light intensity and wavelength are different in deep water from the surface; accessory pigments overcome lack of light intensity and trap more blue light that penetrates deep water;
25.13 see bulleted list in text
25.14 see Figs 25.17 and 26.14
25.15 **(a)** DNA of the Herpes virus attaches to the DNA of the host cells, cold sores develop when the viral DNA is transcribed in a manner similar to that for the HIV shown in Fig 25.22;
(b) surface receptors on cells lining the mouth are different from those on cells lining the genitals and attach specifically to outer capsule of only one type of virus;
25.16 living organisms are usually defined according to their ability to grow, reproduce, feed, move, respire, excrete and detect stimuli; since viruses do not show these abilities, you could argue either that viruses are

not living or that their minimalist activities are the result of natural selection on obligate intracellular parasites;

Analysis: The evolution of multicellular bodies

(a) e.g. specialisation of different cells (division of labour) leads to greater efficiency; greater size leads to better food capture (larger prey caught or stronger water current created) and more efficient locomotion;

(b) e.g. coordination between cells needed; cells become reliant on other cells for their survival; SA : volume ratio may become disadvantageous for exchange; size becomes too large for movement by undulipodia;

(c) (i) flat thallus;
(ii) spherical or cylindrical body pattern;

Chapter 26

26.1 lack undulipodia at any stage of life cycle; chitinous cell wall; (plants also produce spores)

26.2 see Section 26.1

26.3 see Fig 26.8(b)

26.4 decomposers are important in mineral cycles; mutualistic mycorrhizae aid uptake of ions by plant roots; mutualistic lichens important in colonisation of bare rock (see Section 31.1); parasitic fungi depress the reproductive potential of their host populations;

26.5 see Table 26.1; **A** Basidiomycota; **B** aseptate primary and septate secondary; **C** Deuteromycota; **D** septate; **E** Zygomycota; **F** aseptate; **G** zygospores; **H** conidiospores; **I** ascospores;

26.6 (a) anchorage; absorption of food;
(b) from the information given, you could validly suggest either that they absorb soluble material from cell sap of host or that they secrete digestive enzymes into their host and absorb the products of digestion;

26.9 (a) bryophytes and ferns;
(b) many insects feed on nectar and pollen and also carry pollen from plant to plant;

26.10 waterproof cuticle, development of vascular tissue, development of supportive tissue, development of true roots and reduction in dependence on water for fertilisation are among the most important trends;

26.12 mainly by turgid cells;

26.13 spring and autumn are wet seasons when the problem of desiccation is less than in summer; these mosses miss the advantage of bright sunlight for rapid photosynthesis during summer;

26.15 without vascular tissue, bryophytes lack true leaves;

26.18 retention of megaspore within sporophyte; development of gametophytes within spores; reduced dependence on water;

26.21 (a) (female cone) megasporophyll;
(b) before pollination, they open when pollen grains are likely to be blown by the wind instead of washed to the ground by rain; after pollination and fertilisation, seeds are released in dry weather when they are likely to be carried further from the parent plant;

26.22 the flower contains the megasporagia and microsporangia of the sporophyte plant, i.e. it is an organ of asexual reproduction; however, since the female gametophytes are retained within the megasporangia, you could also argue that sexual reproduction occurs in the flower as well;

Analysis: Life on land

(a) *Laminaria* falls over when exposed to the air at low tide;

(b) turgid cells; cell walls, especially those that are thickened with extra cellulose or with lignin (see Section 18.5);

(c) heterotrophs need water, inorganic ions, vitamins, carbohydrates, proteins, lipids; autotrophs need water, inorganic ions, a carbon source (e.g. CO_2 in plants) and an energy source (e.g. light in plants);

(d) oxygen from the air and soil air; carbon dioxide from the air; inorganic ions dissolved in soil water; water from soil;

(e) anchorage;

(f) all parts of the *Laminaria* are permeable and photosynthesise;

(g) they could not absorb and excrete gases during respiration and photosynthesis;

Tutorial: A generalised plant life cycle

(a) they are formed by meiosis, not by mitosis;

(b) X is the gametophyte; Y is the sporophyte;

(c) only one generation of adult in Fig 26.17 whereas two in Fig 26.14; gametes formed by meiosis in Fig 26.17 but by mitosis in Fig 26.14;

Chapter 27

27.1 develop from blastula;

27.2 (a) 3; (b) 5; (c) 6; (d) 2; (e) 1; (f) 4; (g) 8; (h) 7;

27.3 **A** radially symmetrical; **B** asymmetrical; **C** bilaterally symmetrical;

27.5 they lack tissues, show little coordination between cells and can be separated into individual cells which reaggregate; however, there is division of labour between cells and no cell can survive independently; using these criteria you could argue that they are multicellular or that they are a colony of mutualistic individuals;

27.6 medusa disperses to new environment, so reducing competition in polyp population; polyp enables rapid asexual growth of population in optimum environment;

27.7 (a) force water from bell;
(b) jet propulsion (equal and opposite force against medusa);
(c) by water as it re-enters empty bell;

27.9 since the coelom increases the diffusion distance between gut and outside of the body, an advantage of a blood circulatory system is that it enables rapid transport between the two regions;

27.10 (a) locomotory movements and gut movements do not interfere with each other; provides a hydrostatic skeleton;
(b) enables specialisation of different parts of the body (somites);

27.11 use features in Section 27.4 to construct your table;

27.12 (a) both groups are triploblastic, coelomate,

ANSWERS TO IN-TEXT QUESTIONS, TUTORIALS AND ANALYSES

metamerically segmented, possess chitinous
bristles (chaetae in annelids, setae in
arthropods), have simple tube-like hearts; (see
also Analysis: Eevolutionary pathways);
  (b) e.g. jointed limbs, exoskeleton, division of body
into regions;
27.13 problems include support (water is much denser than
air), prevention of desiccation, gas exchange and
exchange of gametes in the absence of water;
arthopod features include support of exoskeleton,
waxy cuticle which reduces water loss, excretion of
uric acid with very little water in urine, trachaeal
system which penetrates cells in the tissues, internal
fertilisation etc;
27.15 there are lots of keys you could construct, e.g.

1	scales present	2
	scales absent	5
2	feathers present	Aves
	feathers absent	3
3	pentadactyl limbs present	Reptilia
	pentadactyl limbs absent	4
4	scales tooth-like	Chondrichthyes
	scales bony	Osteichthyes
5	skin has hair and sebaceous glands	Mammalia
	skin lacks hair and sebaceous glands	Amphibia

27.17 ✓ ✗ ✗ ✗ ;
✓ ✓ ✗ ✗ ;
✓ ✗ ✓ ✗ ✓ ;
✓ ✗ ✓ ✓ ✓ ;
✓ ✗ ✓ ✓ ✗ ;
✓ ✗ ✓ ✓ ✗ ;
✓ ✗ ✓ ✓ ✓ ;

(a) nematode – blastocoel; annelid – coelom;
crustacean – haemocoel; fish – coelom;
(b) nematode – mesoderm and endoderm;
annelid – mésoderm; crustacean – mesoderm;
fish – mesoderm;
(c) nematode – pseudocoelom; lost during development
in the others;
(d)

Analysis: The development of body cavities

	Group of animals			
	Nematode	Annelid	Crustacea	Fish
Embryonic body cavities				
archenteron	✓	✓	✓	✓
gut	✓	✓	✓	✓
coelom	✗	✓	✓	✓
blastocoel	✓	✓	✓	✓
pseudocoel	✓	✗	✗	✗
haemocoel	✗	✗	✓	✗
Primary body layers				
ectoderm	✓	✓	✓	✓
endoderm	✓	✓	✓	✓
mesoderm	✓	✓	✓	✓

Analysis: Comparing worms
(a) mouth terminal in both but anus only terminal in
Nereis;
(b) both are bilaterally symmetrical;
(c) no; the ragworm is metamerically segmented and has
appendages (parapodia) on each segment, the
roundworm lacks both features;
(d) coelom in ragworm but pseudocoelom in nematode;
like the pseudocoelom, the coelom separates body
wall muscles from gut muscles; the coelom also
allows growth of organs, including excretory
nephridia that filter the coelomic fluid;
(e) 2 pairs of dorsal and 2 pairs of ventral longitudinal
muscles, 1 pair of transverse muscles plus circular
muscles and parapodial muscles; roundworm has
similar pairs of longitudinal but no transverse,
circular or parapodial muscles;
(f) compare Figs 27.14 and 27.17(c)
(g) no; it has a circulatory system which transports gases
and excretory products (see Fig 27.17(c)

Analysis: Evolutionary pathways
(a) reduced coelom; no metameric segmentation;
(b) radial cleavage; anus derived from blastopore;
(c) any two from: (pentamerous) radial symmetry;
water-vascular system; sub-epidermal calcareous
plates and spines; asexual reproduction by fission;

Theme 7

Chapter 28

28.2 e.g. bottom community contains mussels and midge
larvae; bank community contains common reeds and
pond snails; swimming community contains fish,
water boatmen and phytoplankton;
28.3 **(a)** time/years should be plotted on the *x*-axis;
(b) the eagles breed in their second year;
(c) exponential; colonisation of a new habitat
(island);
28.7 Your answer should relate to the effect of each ion
deficiency on the growth of the roots and of the stem
and on the extent to which chlorophyll is produced.
28.8 **(a)** the heavy metal ions will enter human food
chains and may be accumulated to clinically
harmful levels in the tissues of humans and their
livestock;
(b) the harmful cations are more likely to be released
in acid soils;
28.9 **biotic:** interspecific competition occurs between these
beetles; **abiotic:** at relatively high temperatures and
relative humidity levels, *T. confusum* competes better;
it has poor tolerance of low temperatures and RH
levels; *T. castaneum* survives better at low
temperatures and low RH levels;
28.11 the bacteria benefit from the relationship since they
get food from the host; since the host gains vitamin K,
the relationship may be considered to be mutualistic;
however, since the host may be harmed it may also be
considered a parasitic relationship; No classification
system is perfect!
28.12 **(a)** biotic: intraspecific competition; lack of food;

predators; abiotic: low temperature;

(b) without shooting, there would be intense competition for food and many birds would die; as a result of shooting, the competition for food would be less severe and fewer birds would starve to death;

28.15 (a) one species might compete better in the area of overlap at the boundary and eliminate the second species; if neither 'won' in the overlap, both might retreat to the remainder of their niche as a 'refuge';

(b) the species with the smaller niche may competitively exclude the species with the larger niche, forming a 'microniche'; alternatively, the species in the smaller niche may become extinct;

28.16 the new bacteria establish a flora in the gut of chickens that competitively excludes subsequent *Salmonella* invasions;

Analysis: When does exponential growth occur?

(a) if the cell divides every twenty minutes the following data should be plotted:

Time/min	0	20	40	60	80	100	120	140	160	180	200	etc.
Cell no		1	2	4	8	16	32	64	128	256	512	1024

Time should be plotted on the x-axis and cell number on the y-axis.

(b) 262 144 ($= 2^{18}$);

(c) the warmth and food source provide ideal conditions for the growth of some bacteria; as they increase in number exponentially, sufficiently high concentrations of exotoxin accumulate which cause food poisoning when eaten;

(d) natural populations coexist in communities that are in ecological balance; scarcity of space, food, light etc., or presence of herbivores, predators and parasites prevent exponential growth of any one population in the community;

Analysis: Navigation using trees

(a) devise an objective way of estimating the extent to which an area of tree trunk is covered by alga, e.g. a plastic sheet with a grid of one hundred squares can be used as a quadrat frame to estimate percentage cover; estimate the percentage cover of alga on the north, east, south and west sides of several tree trunks in the same locality; calculate the mean percentage cover in each compass position;

(b) facing directly away from the equator, the intensity of sunlight is very low; as a result, north-facing areas are damp; algae are poorly adapted to withstand dessication and survive better in damp areas;

(c) probably; but the equator is to the north in the southern hemisphere; here the south facing side of the tree receives least sunlight and is dampest;

Analysis: Competition between barnacles

(a) space;

(b) *Chthamalus* can tolerate the physical conditions lower down the shore; it is interspecific competition with *Balanus* that prevents *Chthamalus* growing there;

(c) fewer *Balanus* where *Nucella* occurs;

Analysis: Controlling the alfalfa aphid

(a) the long-term use of pesticides may cause great harm to local ecosystems, may accumulate in human food chains and aphid becomes resistant;

(b) the alfalfa aphid probably competes with other insects; elimination of the former may enable population explosions of the latter so that they become serious pests;

(c) its population was so small that the damage it caused to the alfalfa crop was not of economic significance;

(d) once established, no further interference with the environment is needed;

Chapter 29

29.3 the caterpillar population was controlled by the predatory wasp population; when the wasp population was eradicated by DDT, the caterpillar population underwent exponential growth and out-ate its food supply (the thatched roofs); DDT in the insects accumulated within food chains, leading to toxic levels in populations of the first (gecko) and second (cat) carnivores; the rat population had previously been controlled by the cat population;

29.5 e.g. greater light intensity in tropics results in higher rate of photosynthesis; higher temperatures (effect on enzymes) in tropics increases rate of photosynthesis; leaves always present, so photosynthesis not seasonal in tropical evergreens; less energy used to replace leaves in evergreens, so R is lower;

29.7 some is eaten; some is recycled following decomposition of dead plant material and animal faeces;

29.8 detritivores and animals which feed on decomposers;

29.9 (a) $A = R + P$;

(b) $v = 70$ kJ M^{-2} year^{-1}; $w = 120$ kJ M^{-2} year^{-1}; $x = 8$ kJ M^{-2} year^{-1}

29.10 (a) only 150 kJ is available for human consumption; this is $(150/3150) \times 100\% = 4.8\%$ of the energy in the primary producers;

(b) (i) endotherms maintain a high body temperature by producing heat, consequently they have a much higher rate of respiration than ectotherms;

(ii) farming ectotherms, e.g. fish, molluscs, is more energy-efficient than farming endotherms, e.g. mammals and birds; more food is needed to farm endotherms and more fuel energy is needed to keep them warm;

29.11 Your table might look like the one below.

Criterion	number	biomass	energy
Sample time	Samples needed only once	Samples needed only once	Samples needed throughout the year
Ease of use	Relatively easy, count numbers in samples	Difficult, need to find dry mass of samples	Very difficult, need to burn samples in bomb calorimeter
Takes sizes of organisms into account	No, so may be misleading	Yes, so provides information for better comparison of tropic levels	Not relevant, organisms reduced to energy equivalence
Validity of data	Snapshots of when samples taken only	Snapshots of when samples taken only	Year-round information, so can be used to generalise

Analysis: Constructing food chains and webs

(a) Herbivores: limpets; mussels; periwinkles.
First carnivores: barnacles; dogwhelks and crabs when they eat mussels; turnstones when they eat limpets and periwinkles.
Secondary carnivores: turnstones when they eat dogwhelks.

(b) e.g. diatom → limpet → turnstone:
phytoplankton → mussel → crab → gull.

(c)

Analysis: Agricultural ecosystems

(a) many populations result in many and complex interrelationships; this gives resistance to change; monocultures lack complex patterns of interrelationships and so are susceptible to change;

(b) to resist change in the ecosystem;

(c) accumulate in human food chains; kill natural enemies of pests; kill beneficial organisms, etc;

(d) like a natural ecosystem, there is a complex pattern of interrelationships between populations;

(e) it is cheap; there is less pollution; the events in (c) above are less likely to happen;

Analysis: Estimating primary production

(a) since molar mass of glucose is 180 g mol^{-1} and molar mass of carbon is 12 g mol^{-1}, mass of carbon constitutes $(6 \times 12)/180 = 0.4$ of mass of glucose; 6610 kg of carbon was found in the maize so, if all the carbon was present in glucose, there would have been $6610/0.4 = 16\,525$ kg of glucose;

(b) oxygen is taken up for respiration at the same rate during the day and night; however, during the day oxygen is also given off by photosynthesis and masks respiration; at night no photosynthesis occurs;

(c) GPP = NPP + R therefore GPP = $(26\,000 + 8000)$ kJ m^{-2} $year^{-1}$ = 34 000 kJ m^{-2} $year^{-1}$;

(d) GPP = $(34\,000/2 \times 10^{6}) \times 100\% = 1.7\%$ of energy in sunlight;
NPP = $(26\,000/2 \times 10^{6}) \times 100\% = 1.3\%$ of energy in sunlight;
R = $(8000/2 \times 10^{6}) \times 100\% = 0.4\%$ of energy in sunlight;

(e) most of the sun's energy is of the wrong wavelength and so is not absorbed by plants; some is reflected; some warms up the plant or evaporates water from the plant; some falls on the ground and warms the soil;

(f) the water content of plants is variable and unrelated to their energy content;

(g) the energy content of the plant material; this could be done by burning a known mass of material in a calorimeter;

(h) plant parts may fall off or be eaten between collections;

(i) energy lost in respiration; measure rate of CO_2 evolution at night;

Analysis: Balancing the books

(a) $x = (90\,000 - 55\,000)$ kJ m^{-2} $year^{-1}$ = 35 000 kJ m^{-2} $year^{-1}$;
(i) $1700/12\,000 = 14.2\%$ for herbivores; $200/1500 = 13.3\%$ for first carnivores; lower efficiency for carnivores may result from extra energy spent in search for, and catching, prey;
(ii) no; input = 90 000 kJ m^{-2} yr^{-1}, output = 55 000 + 10 300 + 1300 + 60 + 18 000 = 84 660 kJ m^{-2} yr^{-1}; the ecosystem is exporting energy;

(b) V = plants; W = herbivores; X = first carnivores; Y = higher carnivores; Z = decomposers;

Analysis: Energy flow in a freshwater ecosystem

(a) 5; producers, herbivores, first carnivores, second carnvores (fish act as both first and second carnivores) and detritivores (organisms, including decomposers, which feed on the detritus);

(b) generally the energy transferred is much less than 10% of the energy received at each trophic level;

(c) e.g. phytoplankton → zooplankton → fish; leaves → detritus → browsers and grazers → invertebrate predators → fish;

(d) in spite of the detail in the diagram, there is not enough to construct productivity pyramids;

Chapter 30

30.1 **(a)** rocks; carbonates (mainly calcium carbonates);
(b) 0.34×10^{9} tonnes $year^{-1}$;

30.2 fixed during photosynthesis: falls to benthos;

30.4 (i) if water fills the soil air spaces, less oxygen is available for bacterial oxidations;
(ii) low pH will reduce the effectiveness of bacterial enzymes involved in oxidation processes;
(iii) soil temperature has a direct effect on the rate of enzyme-controlled reactions;
(iv) organic matter is needed by bacteria as a source of energy and of ammonium ions;

30.5 *not essential* herbivores (ii): they do not change nitrogen from one form to another;
essential plants (i): convert nitrogen from inorganic to organic form; decomposers (v): release ammonium from nitrogen-containing organic compounds; nitrogen-fixing bacteria (iii): convert gaseous nitrogen to inorganic ions; denitrifying bacteria (iv): produce gaseous nitrogen from inorganic ions; you might consider that the cycle would continue perfectly well without both (iii) and (i);

30.6 **(a)** carbon compounds in the form of humus;
(b) increase air supply in soil, encouraging aerobic bacteria; which breakdown organic material releasing nitrates and phosphates;
(c) (i) the greater rate of decay caused by ploughing would lead to increased soil fertility only in the medium-term; in the long-term, too rapid a rate of decay results in most ions being in biotic phase of cycle (e.g. tropical rain forests) or they are lost by leaching;

(ii) add fertiliser to replace ions lost from soil;

30.7 nitrate fertilisers are rapidly lost from soils by leaching, phosphates remain bound to the soil;

30.8 (a) different plants absorb ions from different parts of the soil profile and at different rates; clover, a legume, fixes nitrogen in its root nodules; animal faeces replace some of the ions removed from the soil during harvesting;

(b) human population is too large to be efficiently fed by small-scale farming methods; economically large crop yields are obtained by energy-intensive, large-scale monoculture techniques;

Analysis: The carbon cycle

(a) carbon dioxide in air; carbon dioxide and hydrogencarbonate ions in water;

(b) organic molecules such as carbohydrates, proteins, lipids; hydrogencarbonate ions in body fluids and cytosol;

(c) as carbon dioxide during respiration or following burning; as hydrogencarbonate ions in urine; as organic molecules if fossilisation occurs (though you might still regard this as being part of the biotic cycle since decay may resume);

Analysis: Other cycles

(a) phosphorous cycle simpler because it contains no natural gaseous compounds;

(b) added to cycle as phosphate ions are released during erosion of rocks; lost from cycle as rocks form from freshwater and ocean deposits;

(c) the water remains chemically unchanged as it passes through the cycle;

(d) by respiration (including that by decomposers);

(e)

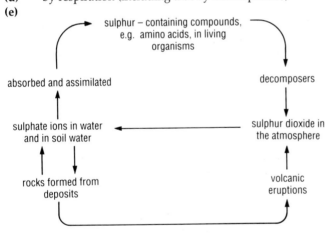

Analysis: Who killed the Norfolk Broads?

(a) decomposition of plant remains releases both nitrates and phosphates but only nitrates are leached from soil into rivers;

(b) in freshwater ecosystems phosphate is a limiting nutrient; they came from sewage;

(c) because it promoted plant growth which in turn supported increased insect, bird and probably fish populations; more vegetation may be considered pretty; more insects may allow increases in the insectivorous fish populations, aiding angling;

(d) Haber process meant nitrate fertilisers could be made

cheaply, leading to their over application and loss of nitrate ions from soil by leaching; detergents contain sodium polyphosphates a rich source of phosphate ions;

(e) dead, undecomposed algae and rooted plants;

(f) (i) remove phosphate bound to mud;
(ii) fish eat water mites which control algal population;
(iii) animal manure can be used as fertiliser that is not quickly leached; dependence on inorganic fertilisers less with grass than with cereals; (iv) ions resulting from sewage treatment would still get into the river but the biochemical oxidation demand of the river would decrease if raw sewage is broken down before entry to river;

(g) you may agree or disagree with this statement but you need to research other pollution problems in your library to justify your answer;

Chapter 31

31.2 fungal filaments provide anchorage and protection against desiccation which plants would lack; competition with plants eliminates lichens in later stages of succession;

31.3 (a) increase; death and decay of more organisms;
(b) increase; more nutrients in soil;
(c) increase; more complex environment provides more niches;
(d) increase; larger individuals and more individuals;
(e) decrease; community becomes more stable as it becomes more complex;

31.7 decomposition is very fast and inorganic ions are absorbed almost as soon as they are released; as a result most inorganic ions are present in the tissues of plants and the soil contains very few; (would you expect removal of the trees to grow other crops to be successful in the long-term?)

31.8 if the Earth becomes warmer, communities which are found in cooler regions may disappear; (this may lead to a disappearance of common, and popular, organisms, e.g. wild daffodils from southern Britain).

31.9 the temperature and humidity drops in the mountain ascent – compare these changes with the axes of Fig 31.8 and relate the environmental conditions to the communities shown in the figure;

31.11 conditions in the sea are more stable than those on land;

31.12 because of the thermocline, tropical seas are relatively unproductive and so are clear of phytoplankton; seas around Britain are more productive and rich in plankton;

31.13 Lake Windermere is eutrophic (rich in inorganic ions which support plant growth, such as nitrates) whereas Wast Water is oligotrophic (few inorganic ions);

Analysis: Sand dune succession

(a) succession is shown by listing the communities from left (youngest) to right (oldest);

(b) the table shows changes in pH (becomes more acidic) and humus content (increases); from the diagram you can infer changes in humidity, wind speed etc. in the oak woodland;

Index

INDEX